Stern's Introductory Plant Biology

Fifteenth Edition

James E. Bidlack
UNIVERSITY OF CENTRAL OKLAHOMA

Shelley H. Jansky
UNIVERSITY OF WISCONSIN – MADISON

McGraw Hill

STERN'S INTRODUCTORY PLANT BIOLOGY, FIFTEENTH EDITION

Published by McGraw-Hill Education, 2 Penn Plaza, New York, NY 10121. Copyright © 2021 by McGraw-Hill Education. All rights reserved. Printed in the United States of America. Previous editions © 2018, 2014, and 2011. No part of this publication may be reproduced or distributed in any form or by any means, or stored in a database or retrieval system, without the prior written consent of McGraw-Hill Education, including, but not limited to, in any network or other electronic storage or transmission, or broadcast for distance learning.

Some ancillaries, including electronic and print components, may not be available to customers outside the United States.

This book is printed on acid-free paper.

2 3 4 5 6 7 8 9 LWI 24 23 22 21 20

ISBN 978-1-260-24083-2 (bound edition)
MHID 1-260-24083-5 (bound edition)
ISBN 978-1-260-48862-3 (loose-leaf edition)
MHID 1-260-48862-4 (loose-leaf edition)

Portfolio Manager: *Michael Ivanov*
Product Developer: *Erin DeHeck*
Marketing Manager: *Kelly Brown*
Content Project Managers: *Jeni McAtee/Rachael Hillebrand*
Buyer: *Sandy Ludovissy*
Design: *David Hash*
Content Licensing Specialists: *Beth Cray*
Cover Image: © *Bruce Fritz, USDA-ARS*
Compositor: *MPS Limited*

All credits appearing on page or at the end of the book are considered to be an extension of the copyright page.

Library of Congress Cataloging-in-Publication Data

Cataloging-in-Publication Data has been requested from the Library of Congress

The Internet addresses listed in the text were accurate at the time of publication. The inclusion of a website does not indicate an endorsement by the authors or McGraw-Hill, and McGraw-Hill does not guarantee the accuracy of the information presented at these sites.

mheducation.com/highered

Contents in Brief

About the Authors   ix
Preface   x

1   What Is Plant Biology?   1
2   The Nature of Life   11
3   Cells   27
4   Tissues   51
5   Roots and Soils   63
6   Stems   82
7   Leaves   101
8   Flowers, Fruits, and Seeds   120
9   Water in Plants   143
10   Plant Metabolism   160
11   Growth and Development   187
12   Meiosis and Alternation of Generations   211
13   Genetics and Molecular Biology   220
14   Plant Breeding, Propagation, and Biotechnology   243
15   Evolution   261
16   Plant Names and Classification   275
17   Domain (Kingdom) Bacteria, Domain (Kingdom) Archaea, and Viruses   291
18   Kingdom Protista   316
19   Kingdom Fungi   345
20   Introduction to the Plant Kingdom: Bryophytes   370
21   The Seedless Vascular Plants: Ferns and Their Relatives   385
22   Introduction to Seed Plants: Gymnosperms   408
23   Seed Plants: Angiosperms   427
24   Flowering Plants and Civilization   447
25   Ecology   473
26   Biomes   498

Appendix 1   Scientific Names of Organisms Mentioned in the Text   A1
Appendix 2   Biological Controls   A20
Appendix 3   Useful and Poisonous Plants, Fungi, and Algae   A26
Appendix 4   Houseplants and Home Gardening   A51
Appendix 5   Metric Equivalents and Conversion Tables   A75
Appendix 6   Periodic Table of the Elements   A77

Glossary   G1
Index   I1

Contents

About the Authors ix
Preface x

1 What Is Plant Biology? 1

1.1 Introduction to Plant Biology 2
Key Theme: ecology Who Needs Plants? 3
1.2 The Relationship of Humans to Their Environment 4
1.3 Botany as a Science 6
1.4 Diversification of Plant Study 7
Key Theme: general Plant Biology on the Internet 9
SUMMARY 10
REVIEW QUESTIONS 10
DISCUSSION QUESTIONS 10

2 The Nature of Life 11

2.1 Attributes of Living Organisms 12
2.2 Chemical and Physical Bases of Life 13
Key Theme: molecular The Skinny on Fats 21
SUMMARY 25
REVIEW QUESTIONS 26
DISCUSSION QUESTIONS 26

3 Cells 27

3.1 Introduction to Cells 28
3.2 Cells 28
3.3 Eukaryotic versus Prokaryotic Cells 31
3.4 Cell Structure and Communication 31
3.5 Cellular Components 34
3.6 Cellular Reproduction 42
Key Theme: cellular Microscapes 46
3.7 Higher Plant Cells versus Animal Cells 48
SUMMARY 49
REVIEW QUESTIONS 50
DISCUSSION QUESTIONS 50

4 Tissues 51

4.1 Introduction to Tissues 52
4.2 Meristematic Tissues 52
Key Theme: molecular Chimeras and Variegated Leaves 53
4.3 Tissues Produced by Meristems 53
Key Theme: ecology The THC in Marijuana 60
SUMMARY 61
REVIEW QUESTIONS 62
DISCUSSION QUESTIONS 62

5 Roots and Soils 63

5.1 Introduction to Roots and Soils 64
5.2 How Roots Develop 64
5.3 Root Structure 65
5.4 Specialized Roots 69
Key Theme: ecology Plants Need Roots 72
5.5 Mycorrhizae 73
5.6 Root Nodules 74
5.7 Human Relevance of Roots 75
5.8 Soils 75
Key Theme: ecology Metal-Munching Plants 79
SUMMARY 80
REVIEW QUESTIONS 81
DISCUSSION QUESTIONS 81

6 Stems 82

6.1 Introduction to Stems 83
6.2 External Form of a Woody Twig 83
6.3 Origin and Development of Stems 84
Key Theme: ecology Standing in Fields of Stone 86
6.4 Tissue Patterns in Stems 87
Key Theme: ecology Dendroclimatology 90
6.5 Specialized Stems 93
6.6 Wood and Its Uses 96
SUMMARY 99
REVIEW QUESTIONS 100
DISCUSSION QUESTIONS 100

7 Leaves 101

7.1 Introduction to Leaves 102
7.2 Leaf Arrangements and Types 103
7.3 Internal Structure of Leaves 103
7.4 Stomata 105
7.5 Mesophyll and Veins 106
7.6 Specialized Leaves 108
Key Theme: ecology More on Leaf Structure 109
7.7 Autumnal Changes in Leaf Color 115
7.8 Abscission 115
7.9 Human and Ecological Relevance of Leaves 116
Key Theme: anatomy Glass Cuts from Grass? 117
SUMMARY 118
REVIEW QUESTIONS 118
DISCUSSION QUESTIONS 119

8 Flowers, Fruits, and Seeds 120

8.1 Introduction to Flowers, Fruits, and Seeds 121
8.2 Differences between Dicots and Monocots 123
8.3 Structure of Flowers 124
8.4 Fruits 126
Key Theme: ecology Goober Peas 127
8.5 Fruit and Seed Dispersal 133
8.6 Seeds 137
Key Theme: physiology The Seed That Slept for 1,200 Years 140
SUMMARY 141
REVIEW QUESTIONS 142
DISCUSSION QUESTIONS 142

9 Water in Plants 143

9.1 Introduction to Water in Plants 144
9.2 Molecular Movement 144
Key Theme: physiology Measuring Water Potential and Psychrometry 147
9.3 Water and Its Movement through the Plant 149
9.4 Regulation of Transpiration 151
9.5 Transport of Food Substances (Organic Solutes) in Solution 154
Key Theme: physiology Xylem and Phloem Synopsis 156
9.6 Mineral Requirements for Growth 156
SUMMARY 159
REVIEW QUESTIONS 159
DISCUSSION QUESTIONS 159

10 Plant Metabolism 160

10.1 Introduction to Plant Metabolism 161
10.2 Enzymes and Energy Transfer 161
10.3 Photosynthesis 162
Key Theme: physiology Photosynthesis and Pizza 172
10.4 Respiration 176
10.5 Additional Metabolic Pathways 181
10.6 Assimilation and Digestion 182
Key Theme: ecology Photosynthesis, Global Warming, and Tropical Rain Forests 184
SUMMARY 185
REVIEW QUESTIONS 186
DISCUSSION QUESTIONS 186

11 Growth and Development 187

11.1 Introduction to Growth and Development 188
11.2 Nutrients, Vitamins, and Hormones 188
Key Theme: physiology Plant Hormones Beyond "The Classic Five" 195
11.3 Hormonal Interactions 196
11.4 Plant Movements 197
11.5 Photoperiodism 205
11.6 Temperature and Growth 207
11.7 Dormancy and Quiescence 208
SUMMARY 209
REVIEW QUESTIONS 210
DISCUSSION QUESTIONS 210

12 Meiosis and Alternation of Generations 211

12.1 Introduction to Meiosis and Alternation of Generations 212
12.2 The Phases of Meiosis 213
Key Theme: evolution Why Plants Have Sex Lives 214
12.3 Alternation of Generations 216
Key Theme: evolution Ordering Complexity 218
SUMMARY 219
REVIEW QUESTIONS 219
DISCUSSION QUESTIONS 219

13 Genetics and Molecular Biology 220

13.1 Introduction to Genetics and Molecular Biology 221
13.2 Molecular Genetics 222

Key Theme: molecular Massive DNA Sequencing 224
Key Theme: molecular The Polymerase Chain Reaction (PCR) 226
13.3 Cytogenetics 231
13.4 Mendelian Genetics 232
13.5 Quantitative Traits 238
13.6 Extranuclear DNA 239
13.7 Linkage and Mapping 239
13.8 The Hardy-Weinberg Law 241
SUMMARY 241
REVIEW QUESTIONS 242
DISCUSSION QUESTIONS 242

14 Plant Breeding, Propagation, and Biotechnology 243

14.1 Crop Plant Evolution 244
14.2 Plant Breeding 246
Key Theme: molecular Genome Editing 253
14.3 Plant Propagation 254
SUMMARY 260
REVIEW QUESTIONS 260
DISCUSSION QUESTIONS 260

15 Evolution 261

15.1 Introduction to Evolution 262
15.2 A Brief Overview of the Early Development of Evolutionary Concepts 264
15.3 Charles Darwin 265
15.4 Evidence for Evolution 267
15.5 Microevolution—Evolution within Species 267
15.6 Macroevolution—How Species Evolve 269
15.7 The Role of Polyploidy in Evolution 271
Key Theme: evolution Our Daily Bread 273
15.8 Discussion 273
SUMMARY 273
REVIEW QUESTIONS 274
DISCUSSION QUESTIONS 274

16 Plant Names and Classification 275

16.1 Introduction to Plant Classification 276
16.2 Development of the Binomial System of Nomenclature 276
16.3 Development of the Kingdom Concept 279
16.4 Classification of Major Groups 280
16.5 Species Concepts 281
16.6 The Future of Plant Classification 289
SUMMARY 289
REVIEW QUESTIONS 290
DISCUSSION QUESTIONS 290

17 Domain (Kingdom) Bacteria, Domain (Kingdom) Archaea, and Viruses 291

17.1 Introduction to Domain (Kingdom) Bacteria, Domain (Kingdom) Archaea, and Viruses 292
17.2 Features of Domains (Kingdoms) Bacteria and Archaea 292
17.3 Domain (Kingdom) Bacteria—the True Bacteria 296
17.4 Human Relevance of the Unpigmented, Purple, and Green Sulfur Bacteria 297
Key Theme: ecology The Social Life of Prokaryotes 298
17.5 Class Cyanobacteriae—the Cyanobacteria (Blue-Green Bacteria) 303
Key Theme: ecology Cyanobacteria and Algae Blooms 306
17.6 Class Prochlorobacteriae—the Prochlorobacteria 307
17.7 Domain (Kingdom) Archaea—the Archaebacteria 307
17.8 Viruses 309
Key Theme: molecular Plant Viruses 310
17.9 Viroids and Prions 313
SUMMARY 314
REVIEW QUESTIONS 315
DISCUSSION QUESTIONS 315

18 Kingdom Protista 316

18.1 Introduction to Kingdom Protista 317
18.2 Features of Kingdom Protista 317
Key Theme: evolution Green Plant Phylogeny 318
18.3 Algae 318
18.4 Phylum Chlorophyta—the Green Algae 319

18.5 Phylum Chromophyta—the Yellow-Green Algae, Golden-Brown Algae, Diatoms, and Brown Algae 325
18.6 Phylum Rhodophyta—the Red Algae 330
18.7 Phylum Euglenophyta—the Euglenoids 331
18.8 Phylum Dinophyta—the Dinoflagellates 332
18.9 Phylum Cryptophyta—the Cryptomonads 334
18.10 Phylum Prymnesiophyta (Haptophyta)—the Haptophytes 334
18.11 Phylum Charophyta—the Stoneworts 334
18.12 Human and Ecological Relevance of the Algae 335
Key Theme: ecology Biofuels from Algae 336
18.13 Other Members of Kingdom Protista 339
18.14 Phylum Myxomycota—the Plasmodial Slime Molds 339
18.15 Phylum Dictyosteliomycota—the Cellular Slime Molds 341
18.16 Phylum Oomycota—the Water Molds 341
SUMMARY 343
REVIEW QUESTIONS 344
DISCUSSION QUESTIONS 344

19 Kingdom Fungi 345

19.1 Introduction to the Kingdom Fungi 346
19.2 Distinctions between Kingdoms Protista and Fungi 346
19.3 Kingdom Fungi 347
Key Theme: ecology Bat White Nose Syndrome 354
19.4 Lichens 365
SUMMARY 368
REVIEW QUESTIONS 369
DISCUSSION QUESTIONS 369

20 Introduction to the Plant Kingdom: Bryophytes 370

20.1 Introduction to the Plant Kingdom 371
20.2 General Characteristics of Bryophytes 371
20.3 Phylum Hepaticophyta—Liverworts 373
20.4 Phylum Anthocerophyta—Hornworts 377
20.5 Phylum Bryophyta—Mosses 377
Key Theme: ecology Hibernating Mosses 381
20.6 Human and Ecological Relevance of Bryophytes 382
SUMMARY 382
REVIEW QUESTIONS 383
DISCUSSION QUESTIONS 384

21 The Seedless Vascular Plants: Ferns and Their Relatives 385

21.1 Introduction to the Seedless Vascular Plants 386
21.2 Phylum Psilotophyta—the Whisk Ferns 386
21.3 Phylum Lycophyta—the Ground Pines, Spike Mosses, and Quillworts 388
21.4 Phylum Equisetophyta—the Horsetails and Scouring Rushes 394
21.5 Phylum Polypodiophyta—the Ferns 398
21.6 Fossils 404
Key Theme: ecology Ferns and Fossil Fuels 405
SUMMARY 406
REVIEW QUESTIONS 407
DISCUSSION QUESTIONS 407

22 Introduction to Seed Plants: Gymnosperms 408

22.1 Introduction to Gymnosperms 409
22.2 Phylum Pinophyta—the Conifers 410
Key Theme: ecology Resilient and Useful Gymnosperms 411
22.3 Other Gymnosperms 416
22.4 Human Relevance of Gymnosperms 419
Key Theme: evolution A Living Fossil? 423
SUMMARY 425
REVIEW QUESTIONS 426
DISCUSSION QUESTIONS 426

23 Seed Plants: Angiosperms 427

23.1 Introduction to Angiosperms 428
23.2 Phylum Magnoliophyta—the Flowering Plants 428

Key Theme: molecular The Difference between "*n*" and "*x*" in Plant Life Cycles 435
23.3 Pollination Ecology 437
23.4 Herbaria and Plant Preservation 442
SUMMARY 445
REVIEW QUESTIONS 446
DISCUSSION QUESTIONS 446

24 Flowering Plants and Civilization 447

24.1 Introduction to Flowering Plants and Civilization 448
24.2 Origin of Cultivated Plants 448
24.3 Selected Families of Flowering Plants 449
24.4 Dicots 450
24.5 Monocots 466
Key Theme: ecology Wild Rice—More Than Just Food 469
Key Theme: ecology Coffee and Caffeine 470
SUMMARY 470
REVIEW QUESTIONS 472
DISCUSSION QUESTIONS 472

25 Ecology 473

25.1 Introduction to Ecology 474
25.2 Plants and the Environment 474
25.3 Life Histories 479
25.4 Natural Cycles 479
25.5 Succession 483
Key Theme: ecology Plant Population Ecology 483
25.6 Impact of Humans on Plant Communities 488
25.7 Loss of Biodiversity 491
Key Theme: ecology John Muir, Father of America's National Park System 494
25.8 Restoration of the Land 495
SUMMARY 495
REVIEW QUESTIONS 496
DISCUSSION QUESTIONS 497

26 Biomes 498

26.1 Introduction to Biomes 499
26.2 Major Biomes of the World 499
Key Theme: ecology Alpine Flora as an Indication of Climate Change: The GLORIA Project 505
SUMMARY 508
REVIEW QUESTIONS 509
DISCUSSION QUESTIONS 509

Appendix 1 Scientific Names of Organisms Mentioned in the Text A1
Appendix 2 Biological Controls A20
General Controls A20
Specific Controls A22
Companion Planting A22
Additional Reading A22
Appendix 3 Useful and Poisonous Plants, Fungi, and Algae A26
Wild Edible Plants, Fungi, and Algae A26
Poisonous Plants and Fungi A26
Medicinal Plants, Fungi, and Algae A26
Hallucinogenic Plants A44
Spice Plants A44
Dye Plants A44
Additional Reading A50
Appendix 4 Houseplants and Home Gardening A51
Growing Houseplants A51
Common Houseplants A52
Growing Vegetables A61
Common Vegetables and their Nutritional Values A62
Pruning A67
Major Types of Grafting A68
Additional Reading A74
Appendix 5 Metric Equivalents and Conversion Tables A75
Appendix 6 Periodic Table of the Elements A77

Glossary G1
Index I1

About the Authors

Jim Bidlack, Kingsley Stern, and Shelley Jansky at Kingsley's office residence in Paradise, California. © Jan Monelo

Introductory Plant Biology was originally written by Kingsley R. Stern (1927–2006), who spent more than 40 years as a devoted botanist and teacher. It is estimated that Dr. Stern educated 15,000 students through classroom/laboratory teaching and inspired thousands of botanists as the author of this book, which has sold more than 285,000 copies since the first edition was published in 1979. Kingsley's enthusiasm for the botanical world captivated those around him for many decades. He will long be remembered for his attention to detail and dedication to high standards, along with a refreshing sense of humor. It was always Kingsley's aspiration that those who read *Stern's Introductory Plant Biology* will share his lifelong love of botany.

In late 1999 and early 2000, Drs. Jim Bidlack and Shelley Jansky joined Kingsley Stern in editing and updating this textbook. They worked with him on several editions and have since carried Dr. Stern's legacy forward to educate and inspire young botanists. This fifteenth edition reflects the same accuracy, content, and enthusiasm of the Stern writing style, along with revisions and updates to make it an up-to-date and enjoyable resource and textbook for plant biology.

James E. Bidlack

© James E. Bidlack

Jim Bidlack received a bachelor of science degree in agronomy, with a soil and crop option, from Purdue University in 1984 and continued his education with a master's degree in crop physiology at the University of Arkansas in 1986. Upon completing a Ph.D. in plant physiology at Iowa State University in 1990, Jim joined the teaching faculty at the University of Central Oklahoma (UCO) where he is a professor of biology. His first paper was published from undergraduate research at Purdue University on the use of synthetic growth regulators to stimulate seed germination. Subsequent work in Arkansas, Iowa, and Oklahoma focused on soybean physiology, cell wall chemistry, alternative crops, and photovoltaic cells, as well as teaching responsibilities in botany. Equipment and student salaries for Jim's research projects have been funded by grants from the National Science Foundation and the United States Department of Agriculture. About 25 refereed publications, as well as hundreds of abstracts and popular articles, have resulted from this work. Jim has been recognized with UCO's Presidential Partner's Excellence in Teaching Award; University Merit Awards in Service, Research, and Teaching; the Biology Club Teaching Award; and the Pre-Med Teaching Award. Some of Jim's additional responsibilities have included participation on NSF and USDA review panels, editor of the Biology Discipline for the Multimedia Educational Resource for Learning and Online Teaching (MERLOT), vice president of The Genome Registry, and president of Metabolism Foundation.

Shelley H. Jansky

© Shelley H. Jansky

Shelley Jansky received a bachelor's degree in biology from the University of Wisconsin–Stevens Point in 1982, and a master of science degree and Ph.D. in plant breeding and plant genetics from the University of Wisconsin–Madison in 1984 and 1986, respectively. Her graduate work focused on developing methods to incorporate genes from wild relatives of potato into the cultivated potato. Then, she spent four years as an assistant professor at North Dakota State University, teaching courses in plant breeding and plant propagation, and performing research in the potato breeding program. She taught courses in botany, genetics, and horticulture, and continued to perform potato genetics research at the University of Wisconsin–Stevens Point from 1990 until 2004. She was the chair of the Department of Biology and was promoted to associate professor in 1992 and professor in 1995. In 2004, she moved to Madison, where she is a research geneticist with the U.S. Department of Agriculture and a professor in the Department of Horticulture at the University of Wisconsin–Madison. Her research program focuses on using disease resistance and nutritional quality genes from wild relatives for the improvement of cultivated potato varieties. She received the University of Wisconsin–Stevens Point Excellence in Teaching Award in 1992 and the University Scholar Award in 2000. She has published 112 refereed research articles and 11 book chapters.

Preface

Plants and algae are essential for life on earth as it exists today. They provide our world with oxygen and food, make an essential contribution to water and nutrient cycling in ecosystems, provide clothing and shelter, and add beauty to our environment. Some scientists believe that if photosynthetic organisms exist on planets beyond our solar system, it would be possible to sustain other forms of life that depend upon them to survive.

Botany today plays a special role in many interests of both major and nonmajor students. For example, in this text, topics such as global warming, ozone layer depletion, acid rain, genetic engineering, organic gardening, Native American and pioneer uses of plants, pollution and recycling, houseplants, backyard vegetable gardening, natural dye plants, poisonous and hallucinogenic plants, nutritional values of edible plants, and many other topics are discussed. To intelligently pursue such topics, one needs to understand how plants grow and function. To this end, the text assumes little prior knowledge of the sciences on the part of the student, but covers basic botany, without excessively resorting to technical terms. The coverage, however, includes sufficient depth to prepare students to go further in the field, should they choose to do so.

The text is arranged so that certain sections can be omitted in shorter courses. Such sections may include topics such as soils, molecular genetics, and phylum Bryophyta. Because botany instructors vary greatly in their opinions about the depth of coverage needed for photosynthesis and respiration in an introductory botany course open to both majors and nonmajors, these topics are presented at three different levels. Some instructors will find one or two levels sufficient, whereas others will want to include all three.

Both majors in botany and nonmajors who may initially be disinterested in the subject matter of a required course frequently become engrossed if the material is related repeatedly to their popular interests. This is reflected, as intimated above, in the considerable amount of ecology and ethnobotany included with traditional botany throughout the book.

Organization of the Text

A relatively conventional sequence of botanical subjects is followed. Chapters 1 and 2 cover introductory and background information; Chapters 3 through 11 deal with structure and function; Chapters 12 and 13 introduce meiosis, genetics, and molecular biology. Chapter 14 discusses plant propagation and biotechnology; Chapter 15 introduces evolution; Chapter 16 deals with classification; Chapters 17 through 23 stress, in phylogenetic sequence, the diversity of organisms traditionally regarded as plants; and Chapter 24 deals with ethnobotanical aspects and other information of general interest pertaining to 16 major plant families or groups of families. Chapters 25 and 26 present an overview of the vast topic of ecology, although ecological topics and applied botany are included in the preceding chapters as well. Some of these topics are broached in anecdotes that introduce the chapters, while others are mentioned in text boxes as well as the appendices.

Learning Aids

A chapter outline is provided at the beginning of each chapter and learning outcomes are shown for major sections within the text. The end of each chapter includes a summary, review questions, and discussion questions to help with the learning experience. New terms are defined as they are introduced, and those that are boldfaced are included, with their pronunciation, in a glossary. A list of the scientific names of all organisms mentioned throughout the text is given in Appendix 1. Appendix 2 deals with biological controls and companion planting. Appendix 3 includes wild edible plants, poisonous plants, medicinal plants, hallucinogenic plants, spices, tropical fruits, and natural dye plants. Appendix 4 gives horticultural information on houseplants, along with brief discussions on how to cultivate vegetables. Nutritional values of the vegetables are included. Appendix 5 covers metric equivalents and conversion tables and Appendix 6 includes a periodic table of the elements.

New to This Edition

A major new feature in the fifteenth edition is the presentation of learning outcomes at the beginning of every major heading. This will guide students regarding important concepts in each section. Most of the major changes in the book have been made as a result of new discoveries and technologies in plant biology. This edition still retains the hallmark style and pedagogy that make it one of the most enduring and popular introductory plant biology books on the market, and it now has a more contemporary look, writing style, and appearance. For instructors using Internet resources such as Connect and the textbook website offered by McGraw-Hill, all learning outcomes are directly tied to assessment within the question and test banks. With encouragement from reviewers, the new material has been eloquently incorporated directly into the textbook narrative and as text boxes, to provide a blend of historically important principles as well as modern developments in plant biology. Some of the more interesting components that make this fifteenth edition more accurate and up-to-date with our current understanding of plant biology include:

Chapter 1 (What Is Plant Biology?): The introduction has been extensively rewritten to address contemporary topics, such as climate change, the opioid crisis, and plant genetic resources. A section has been added on gene editing and bioinformatics. The chapter has been updated with eight new images.

Chapter 2 (The Nature of Life): The section on features of living organisms has been combined and condensed from eight to five topics. The opening image has been updated, as have four additional images.

Chapter 3 (Cells): A new image of a compound microscope has been added. The discussion of the cell wall has been updated.

Chapter 4 (Tissues): A text box has been added to discuss THC in marijuana. The distinction between vessels and vessel elements has been clarified.

Chapter 5 (Roots and Soils): Two new images have been added.

Chapter 7 (Leaves): The introduction, relating leaves to solar power plants, has been updated. A discussion of compounds produced by marijuana has been added.

Chapter 8 (Flowers, Fruits, and Seeds): Pedicle and peduncle have been defined, and their definitions have been added to the glossary. A higher-quality image of carrot seeds has been added.

Chapter 9 (Water in Plants): A text box has been added to compare and contrast xylem and phloem. The section on water flow into roots has been updated.

Chapter 11 (Growth and Development): Several sections have been updated, including senescence, plant movements, gravitropism, and photoperiodism. A discussion of growing degree days was added.

Chapter 12 (Meiosis and Alternation of Generations): The GISH/FISH text box has been replaced with a box that describes classification schemes based on life cycles. Six new images have been added.

Chapter 13 (Genetics and Molecular Biology): A new figure has been added to illustrate chromosomal rearrangements and aneuploidy/polyploidy.

Chapter 14 (Plant Breeding, Propagation, and Biotechnology): The model for the domestication of the first cereal grains has been updated. The opening image has been replaced with one that shows a production field for salad greens.

Chapter 15 (Evolution): Seven new images have been added.

Chapter 16 (Plant Names and Classification): The opening image has been replaced, and three new images have been added.

Chapter 17 (Domain (Kingdom) Bacteria, Domain (Kingdom) Archaea, and Viruses): A pie chart has been added to show the estimated distribution of major groups of organisms on Earth. Ten new images have been added.

Chapter 18 (Kingdom Protista): This chapter has a new key theme box discussing kleptoplasty in dinoflagellates that cause red tides. Ten new images have been added.

Chapter 19 (Kingdom Fungi): A new key theme box has been added to introduce the reader to white nose syndrome, which has killed millions of bats. Six new images have been added.

Chapter 20 (Introduction to the Plant Kingdom: Bryophytes): The introduction has been extensively rewritten to introduce the bryophytes in the context of the establishment of land plants. Throughout the chapter, the concept of alternation of generations is highlighted. A section has been added on primitive humans preserved in bogs. The image of moss sporophytes has been enlarged.

Chapter 25 (Ecology): The introduction to the carbon cycle has been revised and updated. New climate change data and examples have been added. Graphs showing atmospheric carbon dioxide and temperature across time have been added.

Appendix 2 (Biological Controls): The introduction has been rewritten to discuss the consequences of improper use of pesticides. The section on the use of resistant plant varieties has been updated. The section on the control of weeds has also been updated.

Appendix 4 (Houseplants and Home Gardening): This appendix has been extensively updated. New information is presented on LED lights, transplanting of seedlings, direct seeding, temperature and seed germination, the effect of overwatering on root growth, fertilizer application, pest control, and inoculation of soil with *Rhizobia*.

FOR INSTRUCTORS

You're in the driver's seat.

Want to build your own course? No problem. Prefer to use our turnkey, prebuilt course? Easy. Want to make changes throughout the semester? Sure. And you'll save time with Connect's auto-grading too.

65%
Less Time Grading

Laptop: McGraw-Hill; Woman/dog: George Doyle/Getty Images

They'll thank you for it.

Adaptive study resources like SmartBook® 2.0 help your students be better prepared in less time. You can transform your class time from dull definitions to dynamic debates. Find out more about the powerful personalized learning experience available in SmartBook 2.0 at **www.mheducation.com/highered/connect/smartbook**

Make it simple, make it affordable.

Connect makes it easy with seamless integration using any of the major Learning Management Systems—Blackboard®, Canvas, and D2L, among others—to let you organize your course in one convenient location. Give your students access to digital materials at a discount with our inclusive access program. Ask your McGraw-Hill representative for more information.

Padlock: Jobalou/Getty Images

Solutions for your challenges.

A product isn't a solution. Real solutions are affordable, reliable, and come with training and ongoing support when you need it and how you want it. Our Customer Experience Group can also help you troubleshoot tech problems—although Connect's 99% uptime means you might not need to call them. See for yourself at **status.mheducation.com**

Checkmark: Jobalou/Getty Images

FOR STUDENTS

Effective, efficient studying.

Connect helps you be more productive with your study time and get better grades using tools like SmartBook 2.0, which highlights key concepts and creates a personalized study plan. Connect sets you up for success, so you walk into class with confidence and walk out with better grades.

Study anytime, anywhere.

Download the free ReadAnywhere app and access your online eBook or SmartBook 2.0 assignments when it's convenient, even if you're offline. And since the app automatically syncs with your eBook and SmartBook 2.0 assignments in Connect, all of your work is available every time you open it. Find out more at **www.mheducation.com/readanywhere**

> *"I really liked this app—it made it easy to study when you don't have your textbook in front of you."*
>
> - Jordan Cunningham, Eastern Washington University

Calendar: owattaphotos/Getty Images

No surprises.

The Connect Calendar and Reports tools keep you on track with the work you need to get done and your assignment scores. Life gets busy; Connect tools help you keep learning through it all.

Learning for everyone.

McGraw-Hill works directly with Accessibility Services Departments and faculty to meet the learning needs of all students. Please contact your Accessibility Services office and ask them to email accessibility@mheducation.com, or visit **www.mheducation.com/about/accessibility** for more information.

Top: Jenner Images/Getty Images, Left: Hero Images/Getty Images, Right: Hero Images/Getty Images

Acknowledgements

Over 285 reviewers for the past few editions, along with reviewers for this fifteenth edition, have helped to revise and update *Stern's Introductory Plant Biology*. Additional contributions and encouragement were provided by colleagues at the University of Central Oklahoma and University of Wisconsin; and the design, editorial, and production staffs of McGraw-Hill Education. The authors extend thanks to the following reviewers who provided recent feedback on the text and the illustrations. Their help has been very useful in shaping the fifteenth edition of *Stern's Introductory Plant Biology*. These reviewers include the following:

Robin Buckallew, *Central Community College, Hastings, NE*
David H. Demezas, *University of Wisconsin-Fond du Lac*
Paul Hankamp, *College of San Mateo*
Hon H. Ho, *State University of New York at New Paltz*
Mort Javadi, *Columbus State Community College*
Manjushri Kishore, *Heartland Community College*
Parul Khurana, *Indiana University East*
Shawn Krosnick, *Tennessee Tech University*
Jeffrey K. Lake, *Adrian College*
Rizana M. Mahroof, *South Carolina State University*
Henri Roger Maurice, *University of Southern Indiana*
Teena S. Michael, *University of Hawaii WCC*
James E. Mickle, *North Carolina State University*
Michael Plotkin, *Mt. San Jacinto College*
Flona Redway, *Barry University*
Brian W. Teasdale, *Kean University*

Upon reaching this fifteenth edition, we would also like to once again extend gratitude to the reviewers of earlier editions, who have provided considerable comments and suggestions. Although too numerous to include here, their contributions have been much appreciated. The following is a list of reviewers for recent editions:

Donovan Bailey, *New Mexico State University*
Terese Barta, *University of Wisconsin-Stevens Point*
Lisa Boucher, *University of Nebraska-Omaha*
Charles Cannon, *Texas Tech University*
William Cook, *Midwestern State University*
Roger del Moral, *University of Washington*
Melinda Findlater, *Texas Tech University*
Dr. Gabrielle Forbes, *Southwest Texas Junior College*
Stephen W. Fuller, *University of Mary Washington*
Ed Gasque, *University of Wisconsin-Stevens Point*
Rick Hammer, *Hardin-Simmons University*
Paul Hankamp, *College of San Mateo*
Allan Scott Holaday, *Texas Tech University*
Harry Janes, *Rutgers University*
Terry Jordan, *College of Southern Maryland*
Ari Jumpponen, *Kansas State University*
John Kiss, *Miami University*
Jeffrey K. Lake, *Adrian College*
Fengshan Ma, *Wilfrid Laurier University*
Silvia Maciá, *Barry University*
Rizana M. Mahroof, *South Carolina State University*
Karen McGinnis, *Florida State University*
James Mickle, *North Carolina State University*
Carrie Monlux, *California State University-Chico*
Clark L. Ovrebo, *University of Central Oklahoma*
Bob Peregoy, *Spokane Community College*
Dennis T. Ray, *University of Arizona*
Flona Redway, *Barry University*
Traesha R. Robertson, *Texas Tech University*
Connie E. Rye, *East Mississippi Community College*
Devinder Sandhu, *University of Wisconsin-Stevens Point*
Mikel Stevens, *Brigham Young University*
Qiang Sun, *University of Wisconsin-Stevens Point*
Leslie Towill, *Arizona State University*
Carol Wake, *South Dakota State University*
Ligia Arango, *Stone Child College*
Joseph Arditti, *University of California–Irvine*
Mark H. Armitage, *Azusa Pacific University*
Janice Asel, *Mitchell Community College*
Tasneem K. Ashraf, *Cochise College–Sierra Vista*
Ralph A. Backhaus, *Arizona State University*
Nina L. Baghai-Riding, *Delta State University*
Randy G. Balice, *New Mexico Highlands University*
Susan C. Barber, *Oklahoma City University*
Paul W. Barnes, *Southwest Texas State University*
Sharon Bartholomew-Began, *West Chester University*
Robert W. Bauman, Jr., *Amarillo College*
Dorothea Bedigian, *Washington University*
Patricia Bedinger, *Colorado State University*
Maria Begonia, *Jackson State University*
Robert A. Bell, *University of Wisconsin–Stevens Point*
Cynthia A. Bottrell, *Scott Community College*
Richard R. Bounds, *Mount Olive College*
Richard G. Bowmer, *Idaho State University*
Rebecca D. Bray, *Old Dominion University*
James A. Brenneman, *University of Evansville*
George M. Briggs, *State University of New York*
Michelle Briggs, *Lycoming College*
George M. Brooks, *Ohio University*
Suzanne Butler, *Miami-Dade College*
William J. Campbell, *Louisiana Technical University*
Ajoy G. Chakrabarti, *South Carolina State University*
Brad S. Chandler, *Palo Alto College*
Gregory Chandler, *University of North Carolina–Wilmington*
James A. Christian, *Louisiana Technical University*
Richard Churchill, *Southern Maine Technical College*
Jerry A. Clonts, *Anderson College*
John Cruzan, *Geneva College*
Kenneth J. Curry, *University of Southern Mississippi*
David B. Czarnecki, *Loras College*
Stephen S. Daggett, *Avila College*
Raviprakash G. Dani, *Texas Tech University*
Roy Darville, *East Texas Baptist University*
Cynthia Dassler, *Ohio State University*
Bill D. Davis, *Rutgers University*

Jerry D. Davis, *University of Wisconsin–LaCrosse*
John W. Davis, *Benedictine College*
Semma Dhir, *Fort Valley State University*
Rebecca M. DiLiddo, *Mount Ida College*
Susan C. Dixon, *Walla Walla College*
Ben L. Dolbeare, *Lincoln Land Community College*
Patricia M. Dooris, *Saint Leo College*
Donald Drake, *University of Hawaii*
Tom Dudley, *Angelina College*
Jan Federic Dudt, *Bartlesville Wesleyan College*
Diane Dudzinski, *Washington State Community College*
Kerry B. Dunbar, *Dalton State College*
Robert Ediger, *California State University–Chico*
H. Herbert Edwards, *Western Illinois University*
William Eisinger, *Santa Clara University*
Inge Eley, *Hudson Valley Community College*
Thomas E. Elthon, *University of Nebraska–Lincoln*
Frederick B. Essig, *University of South Florida*
G. F. Estabrook, *The University of Michigan*
James Ethridge, *Joliet Junior College*
Paul G. Fader, *Freed-Hardeman University*
Bruce Felgenhauer, *University of Louisiana–Lafayette*
Jorge F. S. Ferreira, *Southern Illinois University–Carbondale*
David G. Fisher, *Maharishi University of Management*
Rosemary H. Ford, *Washington College*
James Garner, *Horry-Georgetown Technical College*
Sibdas Ghosh, *University of Wisconsin–Whitewater*
Mike Gipson, *Oklahoma Christian University*
Katherine Glew, *University of Washington*
Richard Glick, *Winston-Salem State University*
Charles Good, *Ohio State University*
David L. Gorchov, *Miami University of Ohio*
Scott A. Gordon, *University of Southern Illinois*
Govindjee, *University of Illinois*
Steve Greenwald, *Gordon College*
Timothy C. Hall, *Texas A & M University*
Mark Hammer, *Wayne State College*
Susan Han, *University of Massachusetts*
Laszlo Hanzely, *Northern Illinois University*
Joyce Phillips Hardy, *Chadron State College*
Nancy E. Harris, *Elon College*
David Hartsell, *Phillips Community College*
Jill F. Haukos, *South Plains College*
David L. Herrin, *University of Texas–Austin*
Peter Heywood, *Brown University*
Jeffrey P. Hill, *Idaho State University*
L. Michael Hill, *Bridgewater College*
H. H. Ho, *State University of New York–New Paltz*
Elisabeth A. Hooper, *Truman State University*
Susan Houseman, *Southeastern Community College*
Lauren D. Howard, *Norwich University*
Vernon R. Huebschwerlen, *Reedley Community College*
Patricia L. Ireland, *San Jacinto College, South*
William A. Jensen, *Ohio State University*
Cindy Johnson-Groh, *Gustavus Adolphus College*
Chad Jordan, *North Carolina State University*

Toney Keeney, *Southwest Texas State*
Sekender A. Khan, *Elizabeth City State University*
Joanne M. Kilpatrick, *Auburn University–Montgomery*
Helen G. Kiss, *Miami University*
John Z. Kiss, *Miami University of Ohio*
Kaoru Kitajima, *University of Florida*
Sharon Klavins, *University of Wisconsin–Platteville*
Roger C. Klockziem, *Martin Luther College*
Robert L. Koenig, *Southwest Texas Junior College*
David W. Kramer, *Ohio State University–Mansfield*
Robert N. Kruger, *Mayville State University*
Martin LaBar, *Southern Wesleyan University*
Vic Landrum, *Washburn University*
James M. Lang, *Greenville College*
Brenda Price Latham, *Merced College*
Cheryl R. Laursen, *Eastern Illinois University*
Peter J. Lemay, *College of the Holy Cross*
Donald C. Leynaud, *Wabash Valley College*
Barbara E. Liedl, *Central College*
John F. Logue, *University of South Carolina–Sumter*
Elizabeth L. Lucyszyn, *Medaille College*
Karen Lustig, *Harper College*
Erin D. MacKenzie, *Weatherford College*
Paul Mangum, *Midland College*
Steve Manning, *Arkansas State University–Beebe*
Michael H. Marcovitz, *Midland Lutheran College*
Bernard A. Marcus, *Genesee Community College*
David Martin, *Centralia College*
Margaret Massey, *Mississippi University for Women*
William J. Mathena, *Kaskaskia College*
Alicia Mazari-Andersen, *Kwantlen University College*
Andrew McCubbin, *Washington State University*
Joseph H. McCulloch, *Normandale Community College*
Julie A. Medlin, *Northwestern Michigan College*
Richard G. Merritt, *Houston Community College*
Andrew S. Methven, *Eastern Illinois University*
Timothy Metz, *Campbell University*
David H. Miller, *Oberlin College*
David W. Miller, *Clark State Community College*
Subhash C. Minocha, *University of New Hampshire*
Beth Morgan, *University of Illinois, Urbana-Champaign*
Dale M. J. Mueller, *Texas A & M University*
Lytton John Musselman, *Old Dominion University*
Nusrat H. Naqvi, *Southern University*
Joanna H. Norris, *University of Rhode Island*
Chuks A. Ogbonnaya, *Mountain Empire College*
Jeanette C. Oliver, *Flathead Valley Community College*
Sebastine O. Onwuka, *Lesley College*
A. D. Owings, *Southeastern Louisiana University*
Julie M. Palmer, *University of Texas–Austin*
Richard A. Palmer, *Fresno City College*
Carolyn Peters, *Spoon River College*
Martha M. Phillips, *The College of St. Catherine*
Jerry L. Pickering, *Indiana University of Pennsylvania*
Wayne S. Pierce, *California State University–Stanislaus*
Indiren Pillay, *Southwestern Tennessee Community College*

Mary Ann Polasek, *Cardinal Stritch University*
Kumkum Prabhakar, *Nassau Community College*
Tyre J. Proffer, *Kent State University*
Francis Putz, *University of Florida*
V. Raghaven, *The Ohio State University*
Mohammad A. Rana, *St. Joseph College*
Margene M. Ranieri, *Bob Jones University*
W. T. Rankin, *University of Montevallo*
Linda Mary Reeves, *San Juan College*
Maralyn A. Renner, *College of the Redwoods*
Tom Reynolds, *University of North Carolina–Charlotte*
Stanley A. Rice, *Southeastern Oklahoma State University*
Dennis F. Ringling, *Pennsylvania College of Technology*
Daryl Ritter, *Okaloosa-Walton Community College*
Suzanne M. D. Rogers, *Salem International University*
Wayne C. Rosing, *Middle Tennessee State University*
Robert G. Ross, *University of Puerto Rico*
Jimmy Rozell, *Tyler Junior College*
Manfred Ruddat, *University of Chicago*
Patricia Rugaber, *Coastal Georgia Community College*
Frances Rundlett, *Georgia State University*
Thomas H. Russ, *Charles County Community College*
Dennis J. Russell, *University of Alaska Southeast*
C. L. Sagers, *University of Arkansas*
A. Edwards Salgado, *Christian Brothers University*
Thomas Sasek, *Northeast Louisiana University*
Michael A. Savka, *University of West Florida*
Neil W. Sawyer, *University of Wisconsin–Whitewater*
Neil Schanker, *College of the Siskiyous*
Renee M. Schloupt, *Delaware Valley College*
Bruce S. Serlin, *DePauw University*
Barbara Greene Shipes, *Hampton University*
Richard H. Shippee, *Vincennes University*
Brian R. Shmaefsky, *Kingwood College*
Shaukat M. Siddiqi, *Virginia State University*
Dilbagh Singh, *Blackburn College*
Del William Smith, *Modesto Junior College*
James Smith, *Boise State University*
Joanna M. K. Smith
Steven Smith, *University of Arizona*
Nancy Smith-Huerta, *Miami University*
F. Lee St. John, *Ohio State University–Newark*
Spencer S. Stober, *Alvernia College*
Marshall D. Sundberg, *Emporia State University*
Eric Sundell, *University of Arkansas–Monticello*
Stan R. Szarek, *Arizona State University*
Mesfin Tadesse, *Ohio State University*
Max R. Terman, *Tabor College*

R. Dale Thomas, *Northeast Louisiana University*
Stephen L. Timme, *Pittsburgh State University*
Richard E. Trout, *Oklahoma City Community College*
Jun Tsuji, *Sienna Heights College*
Gordon E. Uno, *University of Oklahoma*
Rani Vajravelu, *University of Central Florida*
John Vanderploeg, *Ferris State University*
Staria Vanderpool, *Arkansas State University*
Delmar Vander Zee, *Dordt College*
C. Gerald Van Dyke, *North Carolina State University*
Leon Walker, *University of Findlay*
Ami Lea Wangeline, *Colorado State University*
Betty J. Washington, *Albany State University*
Edgar E. Webber, *Keuka College*
Christopher R. Wenzel, *Eastern Wyoming College*
Cherie Wetzel, *City College of San Francisco*
Ingelia White, *Windward Community College*
Donald L. Williams, *Sterling College*
Justin K. Williams, *Sam Houston State University*
Marvin Williams, *University of Nebraska–Kearney*
Dwina W. Willis, *Freed-Hardeman University*
James A. Winsor, *The Pennsylvania State University*
Chris Wolverton, *Ohio Wesleyan University*
Kathleen Wood, *University of Mary Hardin-Baylor*
Richard J. Wright, *Valencia Community College*
Todd Christian Yetter, *Cumberland College*
Brenda Young, *Daemen College*
Rebecca Zamora, *South Plains College*

Teaching and Learning Supplements

Stern's Introductory Plant Biology Lab Manual, 15th Edition by James Bidlack

(ISBN: 9781260488630/MHID: 1260488632)

The laboratory manual that accompanies *Stern's Introductory Plant Biology* has been revised and updated. It is written for the student who is entering the study of botany. The exercises utilize plants to introduce biological principles and the scientific method. They are written to allow for maximum flexibility in the order in which the labs are implemented during the semester.

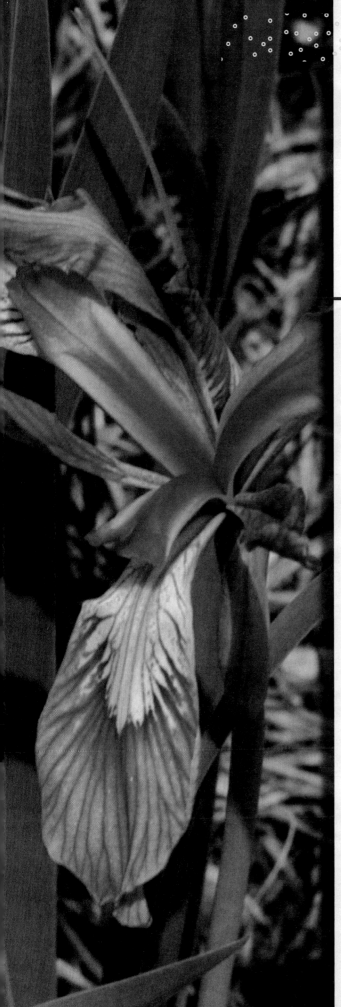

CHAPTER

1

What Is Plant Biology?

1.1 Introduction to Plant Biology
1.2 The Relationship of Humans to Their Environment
   Key Theme 1.1: Who Needs Plants?
      Human and Animal Dependence on Plants
1.3 Botany as a Science
   Hypotheses
1.4 Diversification of Plant Study
   Key Theme 1.2: Plant Biology on the Internet

OVERVIEW

This chapter introduces you to botany: what it is, how it developed, how it relates to our everyday lives, and its potential for the future. The discussion includes a brief introduction to some common questions about plants and their functions, an examination of the scientific method, and a brief look at botany after the invention of the microscope. It concludes with a brief survey of the major disciplines within the field of botany.

A mountain iris (*Iris missouriensis*) growing along a slope near the roadside in the Carson National Forest, New Mexico. *Courtesy of Cliff Pelchat*

1.1 Introduction to Plant Biology

Learning Outcomes
a. List some questions addressed by studies of plants.
b. Describe potential effects of global climate change on living organisms.

Plant life constitutes more than 98% of the total biomass of the earth. Plants and other green organisms are exclusively equipped to produce oxygen while packing the sun's energy into compounds needed for life. Conveniently, the backbone of these compounds is carbon, captured by plants from carbon dioxide given off by all living organisms. All life on the earth depends on green organisms. If some major disease killed all green organisms, then all animals would soon starve. Even if some alternative source of energy were available, animal life would suffocate within 11 years—the time estimated for all the earth's oxygen to be completely used up if it were not replaced. How can plants build leaves and wood and roots from an invisible gas? Why do plants produce oxygen as a by-product?

Our tropical rain forests, which once covered 14% of the earth's land surface, now occupy only 6% of land area. They are disappearing at the rate of several acres a minute as the plant life is cleared for agriculture, wood supplies (primarily for fuel), cattle ranching, and other human activities such as mining for gold. Experts estimate that all rain forests may be destroyed within 40 years. Rain forests are home to 50% of all the species of living organisms; it is estimated that 137 species are destroyed every day due to rain-forest deforestation. What will be the long-term impacts of these activities, and can they be reversed? There is no doubt that our climate is changing. There is much debate, though, about the effects that global warming will have on life as we know it. Are those who proclaim that global warming will eventually have disastrous effects on modern civilization and living organisms simply exaggerating, or is there a scientific basis for the claims?

In addition to providing 90% of the world's food, plants provide important narcotic and medicinal drugs. Of course, these drugs save lives and improve the quality of life for millions of people. However, some are addictive and can lead to serious challenges for societies and individuals. The opioid epidemic in the United States took over 42,000 lives and cost $504 billion in 2016. In 2017, the Department of Health and Human Services declared the epidemic to be a public health emergency. Some opioids are derived from capsules of the opium poppy (Fig. 1.1). Why do plants produce both beneficial and harmful compounds? What are scientists doing to discover and harness the incredible diversity of compounds produced by plants?

We think of animals as consumers of plants, but not the other way around. Do plants ever eat animals? Yes, carnivorous plants consume many types of small animals,

Figure 1.1 Opium poppy capsules that were gashed with a razor blade have released droplets of opium-containing latex. © *mafoto/Getty Images*

including insects, frogs, rodents, and birds (Fig. 1.2). These plants are typically found in bogs, where minerals are difficult to obtain. The animals are important sources of nutrients but not calories. How have plants adapted to grow in extreme environments? What are the limits and costs of these adaptations? Will we be able to tap into plant genetic resources that will allow us to grow crops in marginal environments?

California redwood (*Sequoia sempervirens*) and giant sequoia (*Sequoiadendron giganteum*) trees can grow to heights of 90 meters (300 feet) or more (Fig. 1.3). Water is heavy and must be carried from the ground to a height equivalent to the length of a football field. This is a very expensive activity, yet these trees lift gallons of water high into the canopy every day and still have enough energy for growth and maintenance. The trees actually expend little energy for water transport. Instead, they take advantage of energy from the sun to pull water from the roots to the leaves. How do plants harness energy from the sun to both build biomass and transport water?

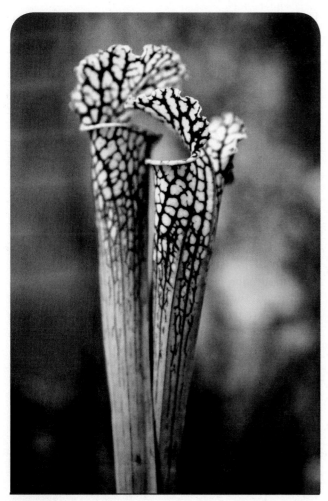

Figure 1.2 These pitcher plants are examples of carnivorous plants that grow in nutrient-poor environments. Small insects trapped in their pitchers are digested and used as a source of minerals.
© 2009 Pia Liikala/Getty Images

Figure 1.3 Sequoia trees in California may grow for thousands of years, and some reach heights of nearly 100 meters (330 feet).
© mark52/123RF.com

This book addresses these questions and many others about living organisms. In addition to plants, we will investigate the lives and impacts of a diverse array of lifeforms, including bacteria, fungi, and protists, including algae. The journey is constantly changing as our understanding of the natural world expands. Use this book as a solid resource, but continue to explore, observe, and learn beyond its pages.

KEY THEME 1.1 ecology

Who Needs Plants?

Human existence depends on plants. In fact, we need plants more than we need animals. In addition to providing oxygen essential to all aerobic organisms, plants act as the foundations of our civilizations, providing food, animal feed, shelter, clothing, industrial products, and medicines. Despite our dependence on plants, though, human activities threaten their survival. Given the fundamental need for plants by humans, it is surprising that conservation efforts in past decades have instead focused on so-called charismatic megafauna (large mammals such as tigers and whales). In recent years, though, there has been a refocusing of conservation strategies to be more inclusive of all members of the biotic world, including plants, fungi, bacteria, and algae. This new direction reflects the growing recognition that preservation of all biological diversity is important for the survival of the human race.

1.2 The Relationship of Humans to Their Environment

Learning Outcomes

a. Explain how plants and animals depend on each other.
b. Describe the effects of human population growth on the global environment.

It has been estimated that the total human population of the world was fewer than 20 million in 6000 B.C. During the next 7750 years, it rose to 500 million; by 1850, it had doubled to 1 billion; and 70 years later, it had doubled again to 2 billion. In 2011, the human population size reached a milestone, exceeding 7 billion people. The earth remains constant in size, but the human population continues to grow. In May 2018, 7.6 billion people inhabited the earth.

In feeding, clothing, and housing ourselves, we have had a major impact on our environment. We have drained wetlands and cleared natural vegetation from vast areas of land. California, for example, now has less than 5% of the wetland it had 100 years ago. We have dumped wastes and other pollutants into our waters and added pollutants to the atmosphere.

If we are to survive on this planet beyond the 21st century, there is little question that humans have to stop increasing in numbers, and the many unwise agricultural and industrial practices that have accompanied the mushrooming of human populations must be replaced with practices more in tune with restoring some ecological balance. Agricultural practices of the future will have to focus on maintaining soil health. Harvesting of timber and other crops will have to be done in a manner that prevents topsoil erosion. Industrial pollutants will have to be rendered harmless and recycled whenever possible.

Many products that now are still largely discarded (e.g., garbage, paper products, glass, metal cans) will also have to be recycled on a much larger scale. Biological pest controls (discussed in Appendix 2) will have to be used whenever possible. Water and energy conservation will have to be universally practiced, and rare plant species, with their largely unknown gene potential for future crop plants, will need to be saved from extinction by preservation of their habitats and by other means. The general public will have to be made even more aware of the urgency for wise land management and conservation—which will be especially needed when pressures are exerted by influential forces promoting unwise measures in the name of "progress"—before additional large segments of our natural resources are irreparably damaged or lost forever. Alternatives appear to be nothing less than death from starvation, respiratory diseases, poisoning of our food and drink, and other catastrophic events that could ensure the premature demise of large segments of the world's population.

Scientists and, increasingly, the general public are alarmed about the effects of human carelessness on our environment. Damage to forests and lakes caused by acid rain, contamination of ground water by nitrates and pesticides, reduction of the ozone shield, major global climatic changes, depletion of aquifers, and loss of biodiversity have gained widespread publicity.

Human and Animal Dependence on Plants

Our dependence on green organisms to produce the oxygen in the air we breathe and to remove the carbon dioxide we give off doesn't stop there. However, plants are also the sources of products that are so much a part of human society that we largely take them for granted. We know, of course, fruits, vegetables, and grains are plants (Fig. 1.4); but all foods, including meat, fish, poultry, eggs, cheese, and milk, to mention just a few, owe their existence to plants. Condiments, such as spices (Fig. 1.5), and luxuries, such as perfumes, are produced by plants, as are some dyes, adhesives, digestible surgical stitching fiber, food stabilizers, beverages (Fig. 1.6), and emulsifiers.

Our houses are constructed with lumber from trees, which also furnish the cellulose for paper, cardboard, and synthetic fibers. Some of our clothing, camping equipment, bedding, curtains, and other textile goods are made from fibers of many different plant families (Fig. 1.7). Coal is fossilized plant material, and oil comes from microscopic green organisms or animals that either directly or indirectly were plant consumers. All medicines and drugs at one time came from plants, fungi, or bacteria, and many important ones, including most antibiotics, still do (Fig. 1.8). Microscopic organisms play a vital role in

Figure 1.4 Vegetables and herbs are major sources of nutrients in the human diet. © *foodandmore/123RF.com*

Figure 1.5 Some spices are derived from plants. © *profotokris/123RF.com*

Figure 1.7 The white fibers in cotton bolls are the source of textiles and fabrics. The bolls also contain seeds, which produce oils used for the production of margarine and shortening. After the oils have been extracted, the remaining "cotton cake" is used for cattle feed. *Courtesy of Derrick Oosterhuis*

Figure 1.6 Coffee berries are picked by hand when they are red. The seeds are extracted for roasting after the berries are fermented. © *Alexander Mazurkevich/Shutterstock*

Figure 1.8 The tiny beads of fluid on the surface of this *Penicillium* colony contain penicillin, which is widely used as an antibiotic. © *Kingsley Stern*

recycling both plant and animal wastes and aid in the building of healthy soils. Others are responsible for human diseases and allergies.

Although shortages of oil and other fossil fuels may sometimes be politically or economically manipulated, there is no question that these fuels are finite and eventually will disappear. Accordingly, the development of alternative energy sources is receiving increased attention. In 2015, 12.5% of world energy consumption was renewable, and it is predicted that this will increase to 13.7% by 2020.

The United States is the largest producer of ethanol in the world. Corn, switchgrass, and other sources of carbohydrates are currently used in the manufacture of ethanol, which is blended with gasoline. Most cars in the United States can run on fuel containing up to 10% ethanol. Currently, ethanol fuel in the United States is mainly produced from corn, but there are concerns about losing food cropland to produce fuel. In addition, the energy and pollution balance of ethanol production is under debate. Cellulosic ethanol, which is derived from inedible plant fiber such as wood chips or switchgrass, may overcome some of these concerns.

What about plants and the future? As you read this, the population of the earth already has exceeded 7 billion persons, every one of whom needs food, clothing, and shelter in order to survive. To ensure survival, we may need to learn not only how to cultivate food plants but also how to use plants to remove pollutants from the water, air, and soil (Fig. 1.9), to make land productive again, and to renew

Figure 1.9 Pollution of land, water, and soil threatens plant and animal life. © papulov74/123RF.com

urban areas. In addition, we need to minimize the destruction of plant habitats caused primarily by the huge increase in the number of the earth's inhabitants. This subject and related matters are further discussed in Chapter 25.

At present the idea that humanity may not be able to save itself may seem radical, but there are a few who have suggested that it might become necessary in the future to emigrate to other planets. Experiments with portable oxygen generators have been in progress for many years. Tanks of water teeming with tiny green algae are taken aboard a spacecraft and installed so that they are exposed to light. The algae not only produce oxygen, which the spacecraft inhabitants can breathe, but also consume the waste carbon dioxide produced by respiration. As the algae multiply, they can be fed to a special kind of shrimp, which in turn multiply and become food for the space travelers. Other wastes are recycled by different microscopic organisms. When this self-supporting arrangement, called a *closed system,* is perfected, the range of spacecraft should greatly increase because heavy oxygen tanks will not be necessary, and the amount of food reserves will be reduced.

Today, teams of botanists, anthropologists, and medical doctors are interviewing medical practitioners and herbal healers in remote tropical regions and taking notes on various uses of plants by the local inhabitants. These scientists are doing so in the hope of preserving at least some plants with potential for contributions to modern civilization before disruption of their habitats results in their extinction.

1.3 Botany as a Science

Learning Outcomes

a. Explain how the study of plants became a science.
b. Describe the steps in a typical scientific experiment.

The study of plants, called **botany**—from three Greek words, *botanikos* (botanical), *botane* (plant or herb), and *boskein* (to feed), and the French word *botanique* (botanical)—appears to have had its origins with Stone Age peoples who tried to modify their surroundings and feed themselves. At first, their interest in plants was mostly practical and centered around how plants might provide food, fibers, fuel, and medicine. Eventually, however, an intellectual interest arose. People became curious about how plants reproduced and how they were put together. This inquisitiveness led to plant study becoming a **science,** which, broadly defined, is simply "a search for knowledge of the natural world." **Botanists** are scientists who study plants.

A science is distinguished from other fields of study by its basis in measurable observations. The scientific method is a series of steps that involves first asking a *question,* then formulating a *hypothesis,* followed by conducting experiments, and finally developing a *theory.*

Hypotheses

A **hypothesis** is simply a tentative, unproven explanation for something that has been observed. It may not be the correct explanation—testing will determine whether it is correct or incorrect. To be accepted by scientists, the results of any experiments designed to test the hypothesis must be repeatable and capable of being duplicated by others.

A scientific experiment is typically carried out with a test group of plants and a control group. The test group receives the experimental treatment, while the control group is treated the same in all ways except that it does not receive the treatment. For example, if you would like to test the effect of light on seed germination, you would divide your seed lot and place some in the light and some in the dark. All other environmental conditions, such as temperature and humidity, would be the same for both sets of seeds. If you observe a germination difference between the two sets of seeds, then it may be due to light.

When a hypothesis is tested, *data* (bits of information) are accumulated and may lead to the formulation of a useful generalization called a *principle.* Several related principles may lend themselves to grouping into a *theory,* which is not simply a guess. A theory is a group of generalizations (principles) that help us understand something. We reject or modify theories only when new principles increase our understanding of a phenomenon. An example of a theory is the concept of natural selection and evolution. This theory describes how the diversity of life on earth came about and provides the foundation for many principles you will encounter in this textbook.

While the testing of hypotheses is a rigorous, well-defined process, the development of hypotheses is not. This aspect of science is creative because it involves finding new ways to look at our natural world. Often, the most successful scientists are those who can think outside the box, integrating observations across fields or organisms to create a hypothesis that explains a complex phenomenon. While scientific advances are often attributed to luck, there is a growing body of evidence that the methods used by scientists actually harness the unexpected to the benefit of science. When the results of an experiment do not follow predictions, that experiment will be repeated with extra care and from a new perspective. If the unexpected observation is observed again, then the scientist must develop a new hypothesis.

1.4 Diversification of Plant Study

Learning Outcomes

a. List the major disciplines of plant science.
b. Design a research question that would require studies in both plant physiology and plant genetics.

Plant anatomy, which is concerned chiefly with the internal structure of plants, was established through the efforts of several scientific pioneers. Early plant anatomists of note included Marcello Malpighi (1628–1694) of Italy, who discovered various tissues in stems and roots, and Nehemiah Grew (1628–1711) of England, who described the structure of wood more precisely than any of his predecessors (Fig. 1.10).

Today, knowledge of plant anatomy is used to help us find clues to the past, as well as for many practical purposes. For example, the related discipline of *dendrochronology* deals with determining past climates by examining the width and other features of tree rings. We can also learn much from archaeological sites by matching tree rings found in the wood of ancient buildings to the rings of wood of known age. Plant anatomy is also used to solve crimes. Forensic laboratories may use fragments of plant tissues found on clothing or under fingernails to determine where a crime took place or if certain people could have been present where the crime was committed. The anatomy of leaves, stems, and other plant parts is used to unravel and sort out relationships among plants. A form of plant anatomy, known as *paleobotany,* focuses on plant fossils to help us understand how plants evolved.

Figure 1.10 This highly magnified view shows bark (top) and wood (bottom) structure. © STEVE GSCHMEISSNER/SPL/Getty Images

Plant physiology, which is concerned with plant function, was established by J. B. van Helmont (1577–1644), a Flemish physician and chemist, who was the first to demonstrate that plants do not have the same nutritional needs as animals. In a classic experiment, van Helmont planted a willow branch weighing 5 pounds in an earthenware tub filled with 90.7 kilograms (200 pounds) of dry soil. He covered the soil to prevent dust from settling on it from the air. The willow produced roots and grew, and after 5 years, he reweighed the willow and the soil. He found that the soil weighed only 56.7 grams (2 ounces), less than it had at the beginning of the experiment, but that the willow had gained 76.7 kilograms (169 pounds). He concluded that the tree had added to its bulk and size from the water it had absorbed. We know now that most of the weight came as a result of photosynthetic activity (discussed in Chapter 10), but van Helmont deserves credit for landmark experimentation in plant physiology.

Modern plant physiologists use cloned *genes* (units of heredity that are found mostly within the nuclei of cells) to learn more about plant functions, including how plants conduct materials internally; how temperature, light, and water are involved in growth; why plants flower; and how plant growth regulatory substances are produced, to mention just a few.

During past centuries, Europeans who explored other continents took large numbers of plants back home with them, and it soon became clear to those working with the plants that some sort of formalized system was necessary just to keep the collections straight. Several *plant taxonomists* (botanists who specialize in the identifying, naming, and classifying of plants) proposed ways of accomplishing this, but we owe much of our present system of naming and classifying plants to the Swedish botanist Carolus Linnaeus (1707–1778) (see Fig. 16.2).

Plant taxonomy involves describing, naming, and classifying organisms. **Plant systematics** is a related field but is broader than taxonomy. It is the science of developing methods for grouping organisms. Plant taxonomy is the oldest branch of plant study, but Linnaeus did more for the field than any other person in history. Thousands of plant names in use today are those originally recorded in Linnaeus's book *Species Plantarum,* published in 1753. An expanded account of Linnaeus and his system of classification is given in Chapter 16.

There are still thousands of plants, fungi, and other organisms that have not yet been described or even discovered. Although it is already too late to identify species that were not described before they became extinct, plant taxonomists around the world have united to try to identify and describe as many new organisms as possible—many with food, medicinal, and other useful potential—before much more of their natural habitat disappears. Other plant taxonomists, through the use of *cladistics* (analysis of shared features) and molecular techniques, are refining our knowledge of plant relationships. The molecular knowledge and techniques are also contributing to the improvement of many of our food crops.

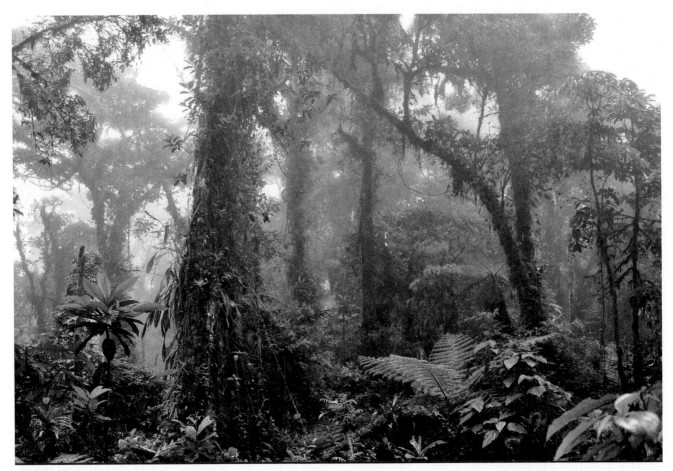

Figure 1.11 Plant ecologists, plant geographers, and other biologists recognize large communities of plants and animals that occur in areas with distinctive combinations of environmental features. These areas, called *biomes,* are represented here by the tropical rain forest, which, although occupying about 6% of the earth's surface, is home to more than half of the world's species of organisms. © *OMeidl/Getty Images*

The discipline of **plant geography,** the study of how and why plants are distributed where they are, did not develop until the 19th century (Fig. 1.11). The allied field of **plant ecology,** which is the study of the interaction of plants with one another and with their environment, also developed in the 19th century.

The study of the form and structure of plants, **plant morphology,** was developed during the 19th century. During the 20th century, much of our basic knowledge about the forms of plants and their life cycles was incorporated into the plant sciences as we know them today. During this time, the number of scientists engaged in investigating plants also greatly increased.

Genetics, the science of heredity, was founded by the Austrian monk Gregor Mendel (1822–1884), who performed classic experiments with pea plants. Today, branches of genetics include *plant breeding,* which has greatly improved the yields and quality of crop plants, and *genetic engineering* (Fig. 1.10). Genetic engineering may involve the transfer of genes from one organism to another. This technique has improved disease resistance, quality, and herbicide tolerance in some crop plants. Recently, gene editing techniques have emerged, providing geneticists with the power to alter specific genes. In recent years, the field of **genomics,** which focuses on genes and their function, has burst onto the genetics scene and now impacts nearly all genetics research. In parallel, **bioinformatics** combines biology, statistics, and computer science to analyze the huge data sets being generated by DNA and RNA sequencing efforts.

Cell biology, the science of cell structure and function, received a boost from the discovery of how cells multiply and how their components perform and integrate a variety of functions, including that of sexual reproduction. The mid-20th-century development of *electron microscopes* (see Chapter 3) further spurred cell research and led to vast new insights into cells and new forms of cell research that continues to the present.

Economic botany and **ethnobotany,** which focus on practical uses of plants and plant products, had their origin in antiquity as humans discovered, used, and eventually cultivated plants for food, fiber, medicines, and other purposes. Today there is increased interest in herbal medicines (see Appendix 3) and many other uses of plants by the general public. Research is being conducted in collaboration with indigenous peoples with an eye to discovering new medicines and other useful plant products previously unknown in developed countries.

KEY THEME 1.2 general — Plant Biology on the Internet

People use the Internet for direct communication, education, and recreation, among other things, and do so through the use of various devices. Wireless connections allow access to information virtually anywhere and by any device that can pick up a signal. Computers have revolutionized the study of every field of plant biology. Laptop computers, tablets, and cell phones with Internet access (Box Figure 1.2A) can be taken to the field to help navigate through the wilderness, locate and identify specimens, collect data, and interact directly with scientists at other locations to exchange information. Even more popular are the numerous handheld devices with Internet access and applications that provide excellent learning experiences.

What botanical information is available on the Internet? Just about anything a user can imagine about plant biology is available on the Internet and, if it isn't, it can be added. Sites that include materials on botany education, botanical gardens, plant databases, and videos about plants are a few of the many resources available on the Web. The following are some interesting websites that can be explored. Try them out sometime! Maybe you'll find a good idea for a research paper.

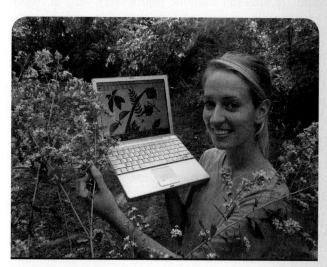

Box Figure 1.2A A botany student uses the Internet to identify and learn about a botanical specimen in the field. © *James E. Bidlack*

1. The Multimedia Educational Resource for Learning and Online Teaching (MERLOT) includes many learning objects for plant biology and other disciplines. This website includes a search engine as well as peer-reviewed learning materials for many topics in higher education. http://www.merlot.org/
2. The Missouri Botanical Garden is one of the oldest botanical institutions in the United States. It is a center for botanical research and science education. http://www.missouribotanicalgarden.org/
3. The New York Botanical Garden is situated on 250 acres in the Bronx and includes 27 outdoor gardens and plant collections, a beautiful Victorian conservatory, and a 40-acre presettlement forest. http://www.nybg.org/
4. The Arnold Arboretum of Harvard University is the nation's oldest arboretum. The site includes a catalog of over 5,000 kinds of woody plants cultivated in the arboretum as well as educational and visitor information. http://www.arboretum.harvard.edu/
5. Australian National Botanic Gardens provides a wealth of botanical and biological information about Australia. http://www.anbg.gov.au/gardens/
6. The United States Department of Agriculture contains news and information about the nation's agricultural economy. There is an excellent section on the history of American agriculture from 1776 to 2000. http://www.usda.gov/
7. GardenWeb is an information center for gardening enthusiasts. http://www.gardenweb.com/
8. California Flora Database contains geographic and ecological distribution information for 7,975 California vascular plant taxa, as well as additional habitat information for rare taxa and species of the Sierra Nevada. http://www.calflora.org/
9. Common Conifers of the Pacific Northwest provides information about the conifers of Oregon, including a dichotomous key for their identification. http://www.oregonstate.edu/trees/
10. Carnivorous Plants Database includes over 3,000 entries, giving an exhaustive nomenclatural synopsis of all carnivorous plants. http://www.omnisterra.com/bot/cp_home.cgi
11. Plants in Motion provides short video clips of different types of plant movement, including germination, flowering, and tropic responses. http://plantsinmotion.bio.indiana.edu/

There is still a vast amount of botanical information to be discovered. Dozens of research journals publish thousands of botanical scientific papers every year. In recent years, open source journals have allowed free online access to research findings for scientists all over the world. While scientists have identified about 2 million species of organisms, the earth likely contains about 9 million species.

SUMMARY

1.1 Introduction to Plant Biology
- Why do some plants produce poisons, while others are edible and useful? Why are some plants carnivorous? How and why do plants respond to their environment? What is the future of tropical rain forests? What can be done about global warming and other environmental problems? This book addresses these questions and more about plants.

1.2 The Relationship of Humans to Their Environment
- Human populations have increased dramatically in the past few centuries, and the disruption of the balance of nature by the activities directly or indirectly associated with the feeding, clothing, and housing of billions of people threatens the survival of not only humans but other organisms as well.
- We are totally dependent on green organisms because they alone can convert the sun's energy into forms that are usable by, and vital to the very existence of, animal life.
- We largely take plants and plant products for granted. Animals, animal products, many luxuries and condiments, and other useful substances such as fibers, lumber, coal, medicines, and drugs either depend on plants or are produced by them.
- To ensure human survival, everyone soon may need to acquire some knowledge of plants and how to use them. Plants will undoubtedly play a vital role in space exploration as portable oxygen generators.
- Teams of scientists are interviewing medical practitioners and herbal healers in the tropics to locate little-known plants used by local inhabitants before the plants become extinct.

1.3 Botany as a Science
- Botany, the study of plants, apparently began with Stone Age peoples' practical uses of plants. Eventually, botany became a science as intellectual curiosity about plants arose.
- Science involves the observation, recording, organization, and classification of facts. The verifying or discarding of facts is done chiefly from known samples. The scientific method involves following a routine series of steps and generally assuming and testing hypotheses.

1.4 Diversification of Plant Study
- The microscope has had a profound effect on studies in the biological sciences and led to the discovery of cells.
- Plant anatomy and plant physiology developed during the 17th century. J. B. van Helmont was the first to demonstrate that plants have nutritional needs different from those of animals. During the 17th century, Europeans engaged in botanical exploration on other continents and took plants back to Europe.
- During the 18th century, Linnaeus produced the elements of a system of naming and classifying plants. In recent years, molecular and cladistical investigations have resulted in modifications of Linnaeus's system.
- During the 19th century, plant ecology, plant geography, and plant morphology developed, and by the beginning of the 20th century, genetics and cell biology became established. Much remains yet to be discovered and investigated.

REVIEW QUESTIONS

1. How and to what extent have humans affected their natural environment?
2. What is meant by the *scientific method*?
3. To what extent is animal life dependent on green organisms for its existence?
4. In terms of biological experiments, what are *hypotheses* and *controls*?
5. What is the oldest branch of botany, and why did it precede other branches?
6. What are the basic features of each of the other branches of botany?

DISCUSSION QUESTIONS

1. Since humans survived on wild plants for thousands of years, might it be desirable to return to that practice?
2. What factors are involved in determining if and when humans might not be able to sustain themselves on this planet?
3. How do you suppose that Stone Age peoples discovered medicinal uses for plants?
4. Why do you suppose that many of the early botanists were also medical doctors?
5. Consider the following hypothesis: "The majority of mushrooms that grow in grassy areas are not poisonous." How could you go about testing this hypothesis scientifically?

CHAPTER 2

The Nature of Life

2.1 Attributes of Living Organisms
    Organization and Structure
    Growth
    Reproduction
    Response to Stimuli
    Metabolism
    Adaptation to the Environment

2.2 Chemical and Physical Bases of Life
    The Elements: Units of Matter
    Molecules: Combinations of Elements
    Valence
    Bonds and Ions
    Acids, Bases, and Salts
    The pH Scale
    Energy
    Chemical Components of Cells
    Monomers and Polymers
  Key Theme 2.1: The Skinny on Fats

OVERVIEW

This chapter begins with a discussion of the attributes of living organisms. These include organization and structure, reproduction, response to stimuli, metabolism and growth, and adaptation to the environment. Then it examines the chemical and physical bases of life. A brief look at the elements and their atoms is followed by a discussion of compounds, molecules, valence, bonds, ions, acids, bases, and salts. Forms of energy and the chemical components of cells are examined next. The chapter concludes with an introduction to macromolecules: carbohydrates, lipids, proteins, and nucleic acids.

The instructions for all life on earth are encoded in DNA. The double helix is a simple molecule, as shown in this artistic rendition, that directs the incredibly complex array of activities of all living things. © *MedicalRF.com*

2.1 Attributes of Living Organisms

Learning Outcomes

a. List the attributes of living organisms.
b. Identify the basic components of a cell.
c. Describe the two factors that control plant growth.
d. List the attribute of living organisms that prevents extinctions.
e. Distinguish between plant and animal responses to stimuli.
f. Distinguish between respiration and photosynthesis; digestion and assimilation.
g. Explain how plants can move although they are rooted in the ground.
h. Define the word *molecule*.
i. List the components of the environment that contribute to adaptation by plants.

Organization and Structure

The activities of living organisms originate in tiny structural units called *cells,* which consist of *cytoplasm* (a thick fluid with a consistency similar to gelatin) bounded by a very thin cell membrane. All living cells contain genetic material that controls their development and activities. In the cells of most organisms, this genetic material, known as *DNA* (deoxyribonucleic acid), is housed in the *nucleus,* which is suspended in the cytoplasm. In bacteria, however, the DNA is not contained in a nucleus. Instead, it is suspended in the cytoplasm. The cells of plants, algae, fungi, and bacteria have a *cell wall* outside the cell membrane. The cell wall provides support and rigidity. Cells are discussed in more detail in Chapter 3.

Cells are composed of large numbers of **molecules** (the smallest units of elements or compounds). Even the most complex nonliving object has only a tiny fraction of the types of molecules of the simplest living organism. Typically there are more than 1 trillion molecules in a single cell. The molecules are not simply mixed, like the ingredients of a cake or the concrete in a sidewalk, but are organized into compartments, membranes, and other structures within cells. Furthermore, the arrangements of these molecules in living organisms are highly structured and complex (Fig. 2.1). Bacteria, for example, are considered to have the simplest cells known, yet each cell contains a minimum of 600 different kinds of protein as well as hundreds of other substances, with each component playing an important role in the function of the cell. When flowering plants and other larger living objects are examined, the complexity of organization is overwhelming, and the number of molecule types can run into the millions.

Growth

Some have described **growth** as simply an increase in *mass* (*matter*—the basic "stuff" of the universe), usually accompanied by an increase in *volume.* Most plant growth results from

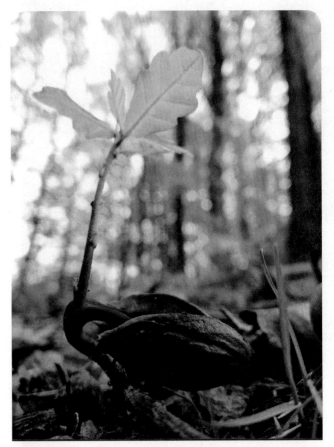

Figure 2.1 An oak tree begins as a single cell that grows in a highly organized set of steps into a complex organism. © *bogdan ionescu/Shutterstock*

the production of new cells and leads to variations in *form* due to genetics and the environment. If you plant two varieties of tulips near each other and grow them under identical conditions, they are likely to differ in size, color, and other characteristics due to differences in genetic makeup. On the other hand, if you plant bulbs of the same variety, they may also look different from each other, especially if you treat them differently. That is, they are exposed to different environments. If you water one just enough to allow it to grow, while you water and fertilize the other one freely, you might expect the second one to grow larger and produce more flowers than the first. A plant's growth, therefore, is controlled by both its genetic makeup and the environment in which it is grown. Various aspects of growth are discussed in Chapter 11.

Reproduction

Dinosaurs were abundant 160 million years ago, but none exist today. Hundreds of mammals, birds, reptiles, plants, and other organisms are now listed as endangered or threatened species, and many of them will become extinct within your lifetime. All of these once-living or currently threatened organisms have one feature in common: it became impossible or difficult for them to reproduce. **Reproduction** is such an obvious feature of living organisms that it is taken for granted—until it no longer takes place.

When organisms reproduce, the offspring always resemble the parents: guppies never have puppies—just more guppies—and a petunia seed, when planted, will not develop into a pineapple plant. Also, offspring tend to resemble their parents more than they do other individuals of the same kind. The laws governing these aspects of inheritance are discussed in Chapter 13.

Response to Stimuli

If you stick a pin into a pillow, you certainly don't expect any reaction from the pillow, but if you stick the same pin into a friend, you know your friend will react immediately, because responding to stimuli is a major characteristic of all living things. You might argue, however, that when you stuck a pin into your houseplant nothing happened, even though you were fairly certain the plant was alive. You might not have been aware that the houseplant did indeed respond, but in a manner very different from that of a human. Plant responses to stimuli are generally of a different nature than those of animals. If the houseplant's food-conducting tissue was pierced, it probably responded by producing a plugging substance called **callose** in the affected cells. Some studies have shown that callose may form within as little as 5 seconds after wounding. Also, a mass of cells called **callus,** which form much more slowly, may be produced at the site of the wound. Responses of plants to injury and to other stimuli, such as light, temperature, and gravity, are discussed in Chapters 9 through 11.

Metabolism

Metabolism is *the collective product of all the biochemical reactions taking place within an organism.* All living organisms undergo metabolic activities, which include the production of new cytoplasm, the repair of damage, and normal cell maintenance. The most important activities include **respiration,** an energy-releasing process that takes place in all living things; **photosynthesis,** an energy-harnessing process in green cells that is, in turn, associated with energy storage; **digestion,** the breakdown of food molecules; and **assimilation,** the conversion of raw materials into cytoplasm and other cell substances. These topics are discussed in Chapters 9 through 11.

Adaptation to the Environment

Organisms respond to the air, light, water, and soil of their environment, as will be explained in later chapters. They are also, after countless generations of natural selection (as discussed in Chapter 15), genetically adapted to their environment in many subtle ways. Genetic changes that occur across generations in response to the environment are called adaptations. Some weeds (e.g., dandelions) can thrive in a wide variety of soils and climates (Fig. 2.2), whereas many species now threatened with extinction have adapted to their environment so specifically that they cannot tolerate even relatively minor changes.

Figure 2.2 Weedy plants such as dandelions can often grow in a variety of environments. © *hydromet/123RF.com*

2.2 Chemical and Physical Bases of Life

Learning Outcomes

a. List the three characteristics of matter.
b. Explain how atoms are bound together in a molecule.
c. Describe the relationship between electron number and valence.
d. Distinguish among covalent, ionic, and hydrogen bonds.
e. Contrast acids with bases.
f. Define pH.
g. Describe the first and second laws of thermodynamics.
h. List the four main elements found in living things.
i. Distinguish between monomers and polymers.

The Elements: Units of Matter

The basic "stuff of the universe," called matter, occurs in three states—*solid, liquid,* and *gas.* In simple terms, matter's characteristics are as follows:

1. It occupies space.
2. It has *mass,* which is commonly associated with weight.
3. It is composed of **elements.** There are 98 elements that occur naturally on our planet. At least 20 more elements have been produced artificially. Only a few of the natural elements (e.g., nitrogen, oxygen, gold, silver, copper) occur in pure form; the others are found combined together chemically in various ways. Each element has a designated symbol, often derived from its Latin name. The symbol for copper, for example, is **Cu** (from the Latin *cuprum*); and for sodium, **Na** (from the Latin *natrium*). The symbols for carbon, hydrogen, and oxygen are **C, H,** and **O,** respectively.

The smallest stable subdivision of an element that can exist is called an **atom.** Atoms are so minute that until the mid-1980s, individual atoms were not directly visible to us with even the most powerful electron microscopes. Atoms consist of several kinds of subatomic particles. Each atom has a tiny **nucleus** consisting of **protons,** which are particles with positive electrical charges, and other particles called **neutrons,** which have no electrical charges. Both protons and neutrons have a small amount of mass and are composed of quarks. If the nucleus, which contains nearly all of the atom's mass, were enlarged so that it was as big as a beach ball, the atom, which is mostly space, would be larger than a professional football stadium (Fig. 2.3). Because each atom is mostly space, solid objects are not as "solid" as they appear. Objects that hit each other are not actually contacting solid surfaces. Instead, negative charges on the objects repel each other. Without these charges, the objects would pass through each other.

Atoms are extremely long-lived. It is estimated that they survive for about 10^{35} years. Accordingly, the atoms in every living thing were once found in stars. Each tree you see outside your window probably contains a billion atoms, many of which may well have been in the bodies of your ancestors.

Each atom of an element has a specific number of protons in its nucleus, ranging from one in hydrogen, the lightest element, to 92 in uranium, the heaviest natural element. This number is referred to as the *atomic number.* The atomic number is often shown as a subscript to the left of the chemical symbol. For example, nitrogen, which has seven protons in its nucleus, has its atomic number of seven shown as $_7N$. The combined number of protons and neutrons in a single atom is referred to as its *atomic mass* (Table 2.1). The atomic mass number is commonly shown as a superscript to the left of the chemical symbol. For example, the atomic mass of nitrogen, which has seven protons and seven neutrons in its nucleus, is shown as ^{14}N, and when both the atomic number and the atomic mass are shown, the chemical symbol appears as $^{14}_{7}N$.

Electrons, which are little more than negative electric charges, whirl around an atom's nucleus. Electron masses are about 1,840 times lighter than those of both protons and neutrons and are so minute that they are generally disregarded. Because opposite electric charges attract each other, the positive electric charges of protons attract the negative electric charges of electrons and determine the paths of the electrons whirling around the nucleus.

The region occupied by electrons around the nucleus is called an **orbital.** Each orbital has an imaginary axis and is somewhat cloudlike, but it doesn't have a precise boundary, so an electron's position within an orbital at any time is uncertain. This has led to an orbital's being defined as *a volume of space in which a given electron occurs 90% of the time.* Electrons actually occupy all space in an orbital simultaneously, so they do not circle around the nucleus like planets. In addition, according to the quantum leap theory of physics, an electron can move instantaneously from one orbital to another without visiting the space between them!

Electrons may be located in one or more energy levels of an atom, and their distance from the nucleus depends on their energy level. Each energy level is usually referred to as an *electron shell.* The outermost electron shell determines how or if an atom reacts with another atom. Only two electrons can occupy the first and lowest energy level associated with the innermost orbital; this orbital is more or less spherical and is so close to the nucleus that it is often not shown on diagrams of atoms. One to several additional orbitals, which are mostly spindle shaped (like the tips of cotton swabs), generally occupy much more space. Up to eight electrons can be held by the second energy level, and although the third and fourth energy levels can hold more than eight electrons each, they can become unstable if more than eight electrons are present. If an electron in one orbital is provided with more energy, it can jump to an orbital farther away from the nucleus. Conversely, if an electron releases energy, it drops to an energy level closer to the nucleus. The electrons of each orbital tend to repel those of other orbitals, so the axes of all the orbitals of an atom are oriented as far apart from each other as possible; the outer parts of the orbitals, however, actually overlap more than shown in diagrams of them. Orbitals usually have diameters thousands of times more extensive than that of an atomic nucleus (Fig. 2.4).

Because each atom usually has as many electrons as it does protons, the negative electric charges of the electrons balance the positive charges of the protons, making the atom electrically neutral. The number of neutrons in the atoms of an element can vary slightly, so the element may occur in forms having different weights but with all forms behaving alike chemically. Such variations of an element are called **isotopes.** The element oxygen has seven known isotopes. For instance, the nucleus of one of these isotopes contains eight protons and eight neutrons; the nucleus of another isotope holds eight protons and 10 neutrons (Fig. 2.5). If the number of neutrons in an isotope of a particular element

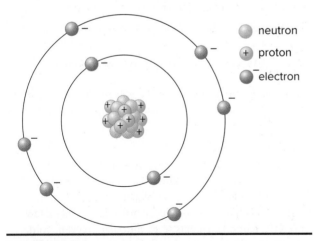

Figure 2.3 In this model of an oxygen atom, the nucleus in the center consists of eight electrically neutral neutrons and eight positively charged protons. Eight negatively charged electrons whirl around the nucleus. In a real atom the electrons would not be spaced or confined as shown in this simple diagram. The nucleus is one-millionth of one-billionth the diameter of the atom.

Table 2.1 Atomic Numbers, Masses, and Functions of Some Elements Found in Plants

Element	Atomic Number	Usual Atomic Mass	Some Functions
Hydrogen (H)	1	1	Part of nearly all organic molecules
Carbon (C)	6	12	Forms skeleton of organic molecules
Nitrogen (N)	7	14	Part of amino acids, nucleic acids, and chlorophyll
Oxygen (O)	8	16	Essential for most respiration; part of most organic molecules
Magnesium (Mg)	12	24	Basic element of chlorophyll
Phosphorus (P)	15	31	Part of ATP (a molecule involved in energy exchange); part of nucleic acid molecules
Sulfur (S)	16	32	Stabilizes a protein's three-dimensional structure
Potassium (K)	19	39	Helps stabilize ion balance in cells
Calcium (Ca)	20	40	Important in the structure of cell walls
Iron (Fe)	26	56	Involved in electron transport during respiration

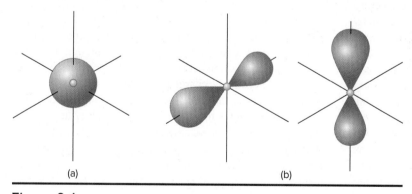

Figure 2.4 In these models of orbitals, the two electrons closest to the atom's nucleus occupy a single spherical orbit in image *a*. In image *b*, additional orbitals are dumbbell-shaped, with axes that are perpendicular to one another. The atom's nucleus is at the intersection of the axes.

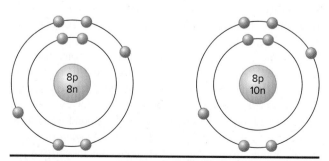

Figure 2.5 In these two-dimensional illustrations of isotopes of oxygen, the one on the left has 8 protons and 8 neutrons, while the one on the right has 8 protons and 10 neutrons. As mentioned in Fig. 2.3, the nucleus is proportionally much smaller in an atom.

varies too greatly from the average number of neutrons for its atoms, the isotope may be unstable and split into smaller parts, with the release of a great deal of energy. Such an isotope is said to be *radioactive*.

Molecules: Combinations of Elements

The atoms of most elements can combine with other atoms of the same or different elements; in fact, most elements do not exist independently as single atoms. When two or more elements are united in a definite ratio by chemical bonds, the substance is called a **compound**. Table salt (sodium chloride, NaCl), for example, is a compound consisting of sodium and chlorine atoms combined in a 1:1 ratio.

A molecule consists of two or more atoms bound together and is the smallest independently existing particle of a compound or an element. The molecules of the gases oxygen and hydrogen, for example, exist in nature as combinations of two atoms of oxygen (O_2) or two atoms of hydrogen (H_2), respectively. Water molecules (H_2O) consist of two atoms of hydrogen and one atom of oxygen (Fig. 2.6). Molecules are in constant motion, with an increase or decrease in temperature speeding up or slowing down the motion. The more molecular movement there is, the greater

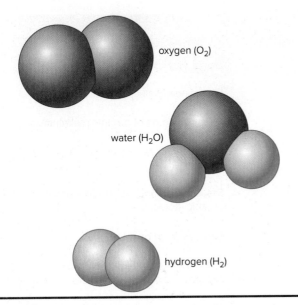

Figure 2.6 In these models of oxygen, water, and hydrogen molecules, each sphere represents the electron cloud of the outer orbital.

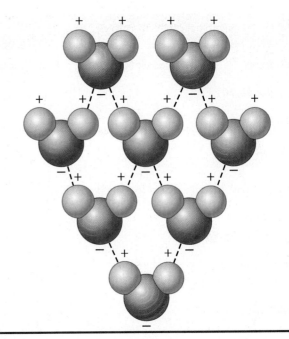

Figure 2.7 The asymmetrical shape of water molecules and the resulting unequal sharing of electrons in the bond between the oxygen and hydrogen atoms cause one side of a water molecule to have a slight positive charge and the other side a slight negative charge. Such molecules are said to be polar. In this image, the partial positive charge is at the top and the partial negative charge is at the bottom.

the chances are that some molecules will collide with each other. Also, the chances of random collisions increase in proportion to the density of the molecules (i.e., the number of molecules present in a given space).

Random collisions between molecules capable of sharing electrons are the basis for all chemical reactions. The reactions often result in the formation of new molecules. Each chemical reaction in a cell usually takes place in a watery fluid and is controlled by a specific *enzyme*. Enzymes are organic *catalysts* (a catalyst speeds up a chemical reaction without being used up in the reaction; enzymes are discussed later in this chapter).

When a water molecule is formed, two hydrogen atoms bond to an oxygen atom. The electrons of the three atoms are shared and form an electron cloud around the core, giving the molecule an asymmetrical shape. Although the electron and proton charges balance each other, the asymmetrical shape and unequal sharing of the electrons in the bond between oxygen and hydrogen cause one side of the water molecule to have a slight positive charge and the other to have a slight negative charge. Such molecules are said to be *polar*. Because negative charges attract positive charges, polarity affects the way in which molecules become aligned toward each. The oxygen portion of a water molecule is negatively charged, while the hydrogen portion is positively charged. Consequently, compounds that can be dissociated into charged components will dissolve in water. For example, table salt (NaCl) dissolves in water because it breaks down into Na^+, which is attracted to the O^- part of the water molecule, and Cl^-, which is attracted to H^+.

Water molecules form a cohesive network as their slightly positive hydrogen atoms are attracted to the slightly negative oxygen atoms of other water molecules (Fig. 2.7). The cohesion between water molecules is partly responsible for their movement through fine (capillary) tubes, such as those that carry water from roots to leaves through the wood of trees. The attraction between the hydrogen atoms of water and other, negatively charged molecules, such as those of fibers, also causes *adhesion* (attraction of charged molecules to each other) and is the basis for water's ability to wet substances. When there is no attraction between water and other substances (e.g., between water and the waxy surface of a cabbage leaf), the cohesion between the water molecules results in droplets beading in the same way that raindrops bead on a freshly waxed car.

Valence

The ability of an element to combine with another one is based on electron number and is called **valence.** For example, atoms of the element calcium have a valence of two, while those of the element chlorine have a valence of one. In order for the atoms of these two elements to combine, there must be a balance between electrons lost or gained (i.e., the valences must balance); it takes *two* chlorine atoms, for example, to combine with *one* calcium atom. The compound formed by the union of calcium and chlorine is called *calcium chloride*. It is customary to use standard abbreviations taken from the Latin names of the elements when giving chemical formulas or equations. Calcium chloride is shown as $CaCl_2$, indicating that one atom of calcium (Ca^{++}) required two atoms of chlorine (Cl^-) to form a calcium chloride molecule.

Bonds and Ions

Bonds hold atoms together. Bonds can form in several ways. The number of electrons in an atom's outermost energy level determines how many chemical bonds it can form. If the number of electrons in the outermost energy level is less than eight, then the atom may lose, gain, or share electrons, resulting in an outermost energy level that contains the maximum number of electrons. Three types of chemical bonds are especially significant for living organisms:

1. **Covalent bonds** form when two atoms complete their outermost energy level by sharing a pair of electrons in the outermost orbital; the electrons hold two or more atomic nuclei together and travel between them, keeping them at a stable distance from each other. For example, the single orbital of a hydrogen atom, which has just one electron, is usually filled by attracting an electron from another hydrogen atom. As a result, two hydrogen atoms share their single electrons, making a combined orbital with two electrons. The combined orbital, with its two hydrogen atoms, forms a molecule of hydrogen gas. The covalent bond is shown as a single line, so hydrogen gas (H_2) is depicted as H—H.

   Except for hydrogen and helium, which have only one orbital, elements can have up to four more orbitals in each energy level. Carbon atoms, for example, have six electrons—two in the innermost orbital and one in each of the four outer orbitals of the second shell; carbon can share four electrons through covalent bonding. When four hydrogen atoms bond to one carbon atom, a molecule of methane gas (CH_4) is formed. To illustrate the bonds, the structural formula for CH_4 is shown as follows:

   When one pair of electrons is shared, the bond is said to be *single*. When two pairs of electrons are shared, the bond is referred to as *double*, and *triple* bonds are formed when three pairs of electrons are shared. Double bonds are shown in structural formulas with double lines (e.g., C=C), and triple bonds are shown with three lines (e.g., C≡N). In covalent bonds involving molecules such as those of hydrogen (H_2), where electrons are shared equally, the bonds are said to be *nonpolar*. However, *polar covalent bonds* (e.g., those of a water molecule) are formed when electrons are closer to one atom than to another and therefore are shared unequally. Because the electrons are shared unequally, parts of the molecule are not electrically neutral and are slightly charged. Covalent bonds are the strongest of the three types of bonds discussed here and are the principal force binding together atoms that make up some important biological molecules discussed later in this chapter (Fig. 2.8).

2. **Ionic bonds.** In nature, some electrons in the outermost orbital are not really shared but instead are transferred from one atom to another, particularly between elements that can strongly attract or easily give up an electron. Molecules that lose or gain electrons become positively or negatively charged particles called *ions*. Ionic bonds form whenever one or more electrons are donated to another atom and result whenever two oppositely charged ions come in contact. Ions are shown with their charges as superscripts. For example, table salt (sodium chloride) is formed by ionic bonding between an ion of sodium (Na^+) and an ion of chlorine (Cl^-). The sodium becomes a positively charged ion when it loses one of its electrons, which is gained by an atom of chlorine. This extra electron makes the chlorine ion negatively charged, and the sodium ion and chlorine ion become bonded together by the force of the opposite charge (Fig. 2.9).

   Some ions, such as those of magnesium (Mg^{++}), give up two electrons and therefore have two positive charges. Such ions can form ionic bonds with two single negatively charged ions such as those of chlorine (Cl^-), forming magnesium chloride ($MgCl_2$). Many biologically important molecules exist as ions in living matter.

3. **Hydrogen bonds** form as a result of attraction between positively charged hydrogen atoms in polar molecules and negatively charged atoms in other polar molecules. Negatively charged oxygen and/or nitrogen atoms of one molecule may attract positively but weakly charged hydrogen atoms of other molecules, forming a weak bond. Hydrogen bonds are very important in nature because of their abundance in many biologically significant molecules. They have, however, only about 7% to 10% of the strength of covalent bonds. Hydrogen bonds help cellular processes by maintaining the shapes of proteins such as enzymes, which make different compounds fit together precisely to complete a chemical reaction.

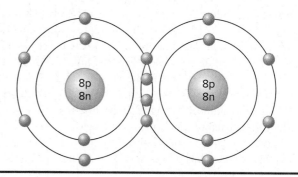

Figure 2.8 In this covalent bond between two oxygen atoms, electrons are shared as outer shells of atoms overlap. In this instance, two pairs of electrons are shared between the two atoms, and the shared electrons are counted as belonging to each atom.

Figure 2.9 In this example of ionic bonding between a sodium atom and a chlorine atom, the sodium becomes positively charged when it loses one of its electrons, which is gained by an atom of chlorine. The gained electron makes the chlorine ion negatively charged, and the two ions become bonded together by the attraction of opposite charges.

Acids, Bases, and Salts

Water molecules are held together by weak hydrogen bonds. In pure water, however, some molecules dissociate into equal numbers of hydrogen (H^+) and hydroxyl (OH^-) ions.

Acids, which include things that taste sour, like cranberry or lemon juice, are chemicals that release hydrogen ions (H^+) when dissolved in water, resulting in more hydrogen than hydroxyl ions. Some acids, such as the acetic acid of vinegar, release relatively few hydrogen ions and are said to be weak. Strong acids such as sulfuric acid dissociate almost completely into hydrogen and sulfate ions.

Bases (also referred to as *alkaline compounds*) usually feel slippery or soapy. They are defined as *compounds that release negatively charged hydroxyl ions (OH^-) when dissolved in water*. Sodium hydroxide (NaOH) is a base that dissociates in water to positively charged sodium ions (Na^+) and negatively charged hydroxyl ions (OH^-). Bases can also be defined as *compounds that accept H^+ ions*.

The acidity or alkalinity of soils affects many aspects of plant growth, and this can have a great impact on root and shoot dynamics in any environment. Similarly, the acidity or alkalinity of the fluids inside cells has to be stable, or chemical reactions vital to life can't take place.

The pH Scale

The concentration of H^+ ions present is used to define degrees of acidity or alkalinity on the **pH** scale. The scale ranges from 0 to 14, with each unit representing a tenfold change in H^+ concentration (Fig. 2.10). Pure water has a pH of 7—the point on the scale where the numbers of H^+ and OH^- ions are equal. This is the neutral point.[1] The lower a number is below 7, the higher the acidity; conversely, the higher a number is above 7, the higher the alkalinity. Vinegar, for example, has a pH of 3, tomato juice has a pH of 4.3, and egg white has a pH of 8. Precipitation with a pH of less than 4.5 is referred to as *acid rain*. Acid rain (discussed in Chapter 25) is associated with industrial emissions and has caused damage to vegetation, soil organisms, and buildings in many parts of the world, including North America.

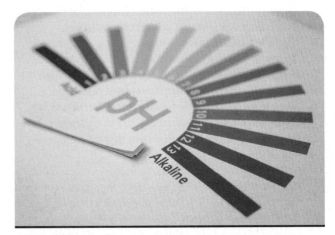

Figure 2.10 The pH scale ranges from 0 (acidic) to 14 (basic).
© phototake/Getty Images

When an acid and a base are mixed, the H^+ ions of the acid bond with the OH^- ions of the base, forming water (H_2O). The remaining ions bond together, forming a salt. If hydrochloric acid (HCl) is mixed with a base—for example, sodium hydroxide (NaOH)—water (H_2O) and sodium chloride (NaCl), a salt, are formed. That is, an acid plus a base produces a salt plus water. The reaction is represented by symbols in this equation:

$$HCl + NaOH \rightarrow H_2O + NaCl$$

Energy

Energy is the capacity to do work or to produce a change. Energy exists in several forms and is required for growth, reproduction, movement, repair, and other activities of living things. On earth, the sun is the ultimate source of energy to fuel life.

Thermodynamics is the study of energy and its conversions from one form to another. Scientists apply two laws of thermodynamics to energy. The *first law of thermodynamics* states that energy is constant—it cannot be gained or lost—but it can be converted from one form to another.

---

1. Note that although distilled water is theoretically "pure," its pH is always less than 7 because carbon dioxide from the air with which it is in contact dissolves in it, forming carbonic acid (H_2CO_3); the actual pH of distilled water is usually approximately 5.7.

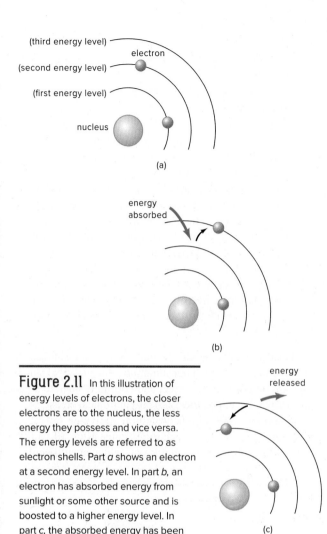

Figure 2.11 In this illustration of energy levels of electrons, the closer electrons are to the nucleus, the less energy they possess and vice versa. The energy levels are referred to as electron shells. Part *a* shows an electron at a second energy level. In part *b*, an electron has absorbed energy from sunlight or some other source and is boosted to a higher energy level. In part *c*, the absorbed energy has been released, with the electron dropping back to its original level.

Among its forms are *chemical, electrical, heat,* and *light* energy to fuel life.

The *second law of thermodynamics* states that when energy doesn't enter or leave a given system and is converted from one form to another, it (energy) flows from a high to a low state. For example, heat will always flow from a hot coffee cup (high energy) to your hand (lower energy), but never from your hand to the hot coffee cup. Furthermore, energy will be released. The total amount of energy in the universe, however, remains constant. Reactions that release energy are necessary for cells to function and grow.

Forms of energy include *kinetic* (motion) and *potential* energy. Potential energy is defined as the "capacity to do work owing to the position or state of a particle."

Although all electrons have the same weight and electrical charge, their amount of potential energy varies. Electrons with the least potential energy are located within the orbital closest to the atom's nucleus, and electrons with the most potential energy are in the outermost orbital (Fig. 2.11). Some of the numerous energy exchanges and carriers that occur in living cells are discussed in later chapters.

The Nature of Life

Chemical Components of Cells

The living substance of cells consists of cytoplasm and the structures within it. The internal structures, which vary in size, are discussed in Chapter 3. About 96% of cytoplasm and its included structures is composed of the elements carbon, hydrogen, oxygen, and nitrogen; 3% consists of phosphorus, potassium, and sulfur. The remaining 1% includes calcium, iron, magnesium, sodium, chlorine, copper, manganese, cobalt, zinc, and minute quantities of other elements. Plants are able to use simple molecules to build very large, complex molecules through the process of metabolism.

The large molecules typically have "backbones" of carbon atoms and are said to be *organic*. Most chemical reactions in living organisms are based on organic compounds. Molecules that contain no carbon atoms are called *inorganic*. Exceptions to this classification include carbon dioxide (CO_2) and sodium bicarbonate ($NaHCO_3$). The name "organic" was given to most of the chemicals of living things when it was believed that only living organisms could produce molecules containing carbon. Today, many organic compounds can be produced artificially in the laboratory, and scientists sometimes hesitate to classify as organic or inorganic some of the 9 million carbon-containing compounds thus far identified. Most scientists, nevertheless, agree that inorganic compounds usually do not contain carbon.

Monomers and Polymers

Most cell components are composed of large molecules called macromolecules, or **polymers** ("many units"). Polymers are formed when two or more small units called **monomers** ("single units") bond together. The bonding between monomers occurs when a hydrogen (H^+) is removed from one monomer and a hydroxyl (OH^-) is removed from another, creating an electrical attraction between them. Because the components of water (H^+ and OH^-) are removed (*dehydration*) in the formation (*synthesis*) of a bond, the process is referred to as *dehydration synthesis*. Dehydration synthesis reactions are controlled by enzymes. In summary, dehydration synthesis extracts water from compounds and uses energy.

Hydrolysis, which is essentially the opposite of dehydration synthesis, occurs when a hydrogen from water becomes attached to one monomer and a hydroxyl group to the other to form a water molecule. Energy is released when a bond is broken by hydrolysis. This energy may be stored temporarily or used in the manufacture or renewal of cell components. In summary, hydrolysis adds water to compounds and releases energy.

Four of the most important classes of polymers found in cells are *carbohydrates, lipids, proteins,* and *nucleic acids.*

Carbohydrates

Carbohydrates are the most abundant organic compounds in nature. They include sugars and starches and contain C, H, and O in a ratio of approximately 1C:2H:1O (CH$_2$O). The number of CH$_2$O units in a carbohydrate can vary from as few as three to as many as several thousand. There are three basic kinds of carbohydrates:

1. *Monosaccharides* are simple sugars with backbones consisting of three to seven carbon atoms. Among the most common monosaccharides are *glucose* (C$_6$H$_{12}$O$_6$) and *fructose*, which is an *isomer* of glucose. Isomers are molecules with identical numbers and kinds of atoms, but with different structures and shapes. Accordingly, fructose, which is found in fruits, has the same C$_6$H$_{12}$O$_6$ formula as glucose, but the different arrangement of its atoms gives it different properties, such as a slightly sweeter taste. Glucose, which is produced by photosynthesis in green plant cells, is a primary source of energy in the cells of all living organisms (Fig. 2.12).

2. *Disaccharides* are formed when two monosaccharides become bonded together by dehydration synthesis. The common table sugar **sucrose** (C$_{12}$H$_{22}$O$_{11}$) is a disaccharide formed from a molecule of glucose and a molecule of fructose; a molecule of water is removed during synthesis. The removal of a molecule of water during the formation of a larger molecule from smaller molecules is referred to as a *condensation reaction*. Sucrose is the form in which sugar is usually transported throughout plants and is the form of sugar stored in the roots of sugar beets and the stems of sugarcane.

3. *Polysaccharides* are formed when more than two monosaccharides bond together. Polysaccharide polymers sometimes consist of thousands of simple sugars attached to one another in long, branched or unbranched chains, or in coils. For example, *starches*, which are the main carbohydrate reserve of plants, are polysaccharides that usually consist of several hundred to several thousand glucose units. When many glucose molecules bond together to become a starch molecule, each glucose gives up a molecule of water. The formula for starch is (C$_6$H$_{10}$O$_5$)$_n$; n indicates the number of units of glucose in the starch molecule. In order for a starch molecule to become available as an energy source in cells, it has to be hydrolyzed; that is, water is added as starch is broken into glucose molecules and energy is released.

Throughout the world, starches are major sources of carbohydrates for human consumption—the principal starch crops are potatoes, wheat, rice, and corn in temperate areas and cassava, sweet potatoes, and taro in tropical areas.

Cellulose, the chief structural polymer in plant cell walls, is a polysaccharide consisting of 3,000 to 10,000 unbranched chains of glucose molecules.

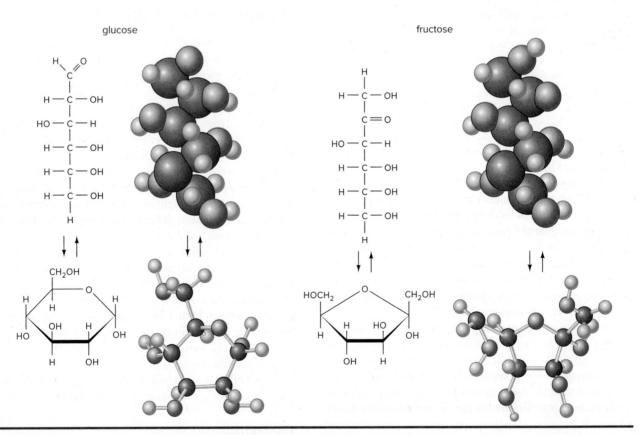

Figure 2.12 In these structural formulas and space-filling models of glucose (left) and fructose (right) molecules. H = hydrogen, C = carbon, O = oxygen, the numbers of atoms and locations of bonds are easy to see in the upper linear diagrams, but when these molecules are in solution, they are in the form of rings, as shown in the lower diagrams. Unless indicated otherwise, each junction in a ring contains a carbon atom.

Although cellulose is very widespread in nature, its glucose units are bonded together differently from those of starch, and most animals digest it much less readily than they do starch. Organisms that do digest cellulose, such as the protozoans living in termite guts, caterpillars, and some fungi, produce special enzymes capable of facilitating the breakdown of bonds between the carbons and the glucose units of the cellulose; the organisms then can digest the released glucose.

Lipids

Lipids are fatty or oily substances that are mostly insoluble in water because they do not have polar regions. They typically store about twice as much energy as similar amounts of carbohydrate and play an important role in the longer-term energy reserves and structural components of cells. Like carbohydrates, lipid molecules contain carbon, hydrogen, and oxygen, but there is proportionately much less oxygen present. Examples of lipids include **fats,** which are solid at room temperature (Fig. 2.13), and **oils,** which are liquid. An oil molecule is produced when a unit of *glycerol*—a three-carbon compound that has three hydroxyl (—OH) groups—combines with three *fatty acids*. A fatty acid has a carboxyl (—COOH) group at one end and typically has an even number of carbon atoms to which hydrogen atoms can become attached (Fig. 2.14).

Most fatty acid molecules consist of a chain with 16 to 18 carbon atoms. If hydrogen atoms are attached to every

Figure 2.13 Sources of lipids in the diet include cooking oil, cheese, butter, and ice cream. © *McGraw-Hill Education/John Thoeming*

available bonding site of these fatty acid carbon atoms, as in most animal fats such as butter and those found in meats, the fat is said to be *saturated*. If there is at least one double bond between two carbons and, consequently, there are fewer hydrogen atoms attached, the fat is said to

---

KEY THEME 2.1 molecular — The Skinny on Fats

Doctors encourage us to limit our intake of saturated fats and *trans* fats in order to reduce our risk of heart disease caused by high cholesterol. Lipids from different natural sources vary in the proportion of their saturated fatty acids. Typically, it is healthier to consume lipids derived from plants than from animals, because plant oils, such as canola, sunflower, corn, olive, and soybean, have a low amount of saturated fatty acids (Box Table 2.1A). However, oils from some tropical plants, such as coconut and palm, are high in saturated fatty acids. Moreover, the food-processing industry sometimes hydrogenates fatty acids to make plant oils solid at room temperature. This results in high levels of *trans* fatty acids, which are also linked to heart disease. In contrast to saturated fats and *trans* fats, unsaturated fats (found in many plant oils) do not raise "bad" cholesterol (low-density lipoprotein) levels. These lipids are healthy to consume in moderation.

Box Table 2.1A Source and Percentage of Saturated Fatty Acids

Source	Saturated Fatty Acids
Canola oil	7%
Sunflower oil	10%
Corn oil	13%
Olive oil	13%
Soybean oil	15%
Palm oil	50%
Coconut oil	86%
Margarine (from plants)	15%
Butter (from dairy)	63%

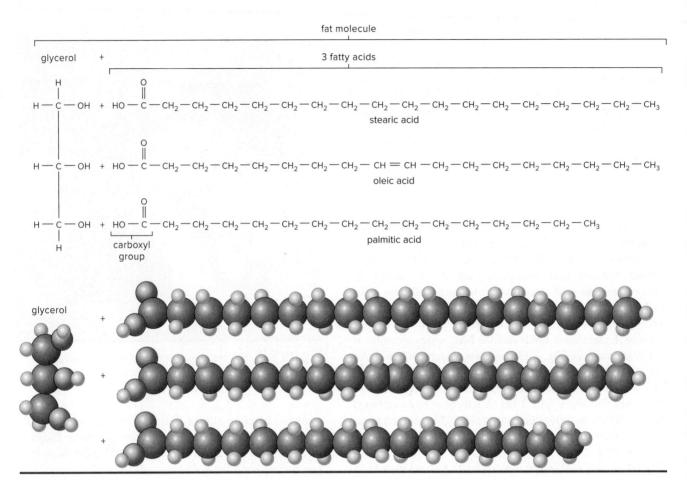

Figure 2.14 Structural formula and space-filling model of the components of a fat molecule. H = hydrogen, C = carbon, O = oxygen. A typical fatty acid is 4 nanometers long.

be *unsaturated*. If there are three or more double bonds between the carbons of a fatty acid, as in some vegetable oils such as those of canola, olive, or safflower, the fat is said to be *polyunsaturated*. Unsaturated vegetable oils can become saturated by bubbling hydrogen gas through them, as is done to produce margarine. Like polysaccharides and proteins (discussed in the next section), lipids are broken down by hydrolysis.

Waxes are lipids consisting of very long-chain fatty acids bonded to a very long-chain alcohol other than glycerol. Waxes, which are solid at room temperature, are found on the surfaces of plant leaves and stems. They are usually embedded in a matrix of *cutin* or *suberin,* which are also lipid polymers that are insoluble in water. The combinations of wax and cutin or wax and suberin function in waterproofing, reduction of water loss, and protection against microorganisms and small insects.

Phospholipids are similar to fats, but one of the three fatty acids is usually replaced by a phosphate group; this can cause the molecule to become a polarized ion. When phospholipids are placed in water, they form a double-layered sheet resembling a membrane. In fact, phospholipids are important components of all membranes found in living organisms.

Proteins, Polypeptides, and Amino Acids

The cells of living organisms contain from several hundred to many thousands of different kinds of **proteins,** which are second only to cellulose in making up the dry weight of plant cells. Each kind of organism has a unique combination of proteins that gives it distinctive characteristics. There are, for example, hundreds of kinds of grasses, all of which have certain proteins in common and other proteins that make one grass different from another. The hundreds of kinds of daisies are distinguished from each other and from grasses by their particular combinations of proteins.

Proteins consist of carbon, hydrogen, oxygen, and nitrogen atoms, and sometimes also sulfur atoms. Proteins regulate chemical reactions in cells and comprise the bulk of protoplasm apart from water. Protein molecules are usually very large and consist of one or more *polypeptide* chains with, in some instances, simple sugars or other smaller molecules attached.

Polypeptides are chains of amino acids. There are 20 different kinds of **amino acids,** and from 50 to 50,000 or more of them are present in various combinations in each protein molecule. Each amino acid has two special groups of atoms plus an *R group*. One group is called the *amino*

Figure 2.15 Structural formula and space-filling model of the amino acid glycine.

group (—NH₂); the other, which is acidic, is called the *carboxyl group* (—COOH). The structure of an R group can vary from a single hydrogen atom to a complex ring. Some R groups are polar, while others are not, and each is distinctive for one of the 20 amino acids. Glycine (Fig. 2.15) is representative of general amino acid structure. Amino acids are linked together by **peptide bonds,** which are covalent bonds formed between the carboxyl carbon of one amino acid and the nitrogen of the amino group of another in a dehydration reaction.

Plants can synthesize amino acids they need from raw materials in their cells, but animals have to supplement from plant sources some amino acids they need, because they can manufacture only a few amino acids themselves.

Each polypeptide usually coils, bends, and folds in a specific fashion within a protein, which characteristically has three levels of structure and sometimes four (Fig. 2.16):

1. A linear sequence of amino acids fastened together by peptide bonds forms the *primary structure* of a protein.
2. As hydrogen bonds form between oxygen atoms of carboxyl groups and hydrogen atoms of amino groups in different molecules, the polypeptide chain can coil to form a staircase-like spiral, called an *alpha helix*. The helix is one type of *secondary structure* that may form. Other secondary structures include polypeptide chains that double back to form hydrogen bonds between two lengths in what is referred to as a *beta sheet*, or *pleated sheet*.
3. *Tertiary structure* develops as the polypeptide further coils and folds. The tertiary structure is maintained by interactions and bonds among R groups.
4. If a protein is composed of more than one kind of polypeptide, a fourth, *quaternary structure* forms when the polypeptides associate.

The three-dimensional structure of a protein may be somewhat flexible in solution, but anything that disturbs the normal pattern of bonds between parts of the protein molecule can *denature* the protein. Denaturing alters the characteristic coiling and folding and adversely affects the protein's function and properties. Denaturing may be reversible, but if it is brought about by high temperatures or harsh chemicals, it may kill the cell of which the protein is a part. For example, boiling an egg, which is mostly protein, brings about an irreversible denaturing: the solid egg proteins can't be restored to their original semiliquid condition.

Figure 2.16 The four levels of protein structure using the plant protein ribulose bisphosphate carboxylase/oxygenase as an example. (a) The primary structure consists of a chain of amino acids bonded together. (b) As the amino acid chain grows, rotation of the chain occurs to form an alpha helix, which is stabilized by hydrogen bonds. (c) The coil or helix folds further and interacts with other amino acids in the chain to form a somewhat globular structure. (d) Many chains combine into a single functional protein molecule.

Storage Proteins

Some plant food-storage organs, such as potato tubers and onion bulbs, store small amounts of proteins in addition to large amounts of carbohydrates. Seeds, in particular, however, usually contain proportionately larger amounts of proteins in addition to their complement of carbohydrates and are very important sources of nutrition for humans and animals. One example of an important protein source in human and animal diets is wheat gluten (to which, incidentally, some humans become allergic). The gluten consists of a complex of more than a dozen different proteins.

A seed's proteins get used during germination and its subsequent development into a seedling. Some legume seeds

Figure 2.17 Facilitation of a reaction by an enzyme using the enzyme (EcoRV) as an example. EcoRV recognizes a specific sequence in its substrate (DNA) and forces it to break. EcoRV is a restriction enzyme commonly used for gene cloning (Chapter 14). (a) The enzyme (EcoRV, on the bottom, in grey) recognizes the sequence GATATC in the DNA molecule. (b) The enzyme binds with the substrate at that region. (c) The enzyme twists, causing a kink in the substrate. (d) The substrate breaks. The enzyme itself is unaffected by the interaction.

may contain more than 40% protein, but legumes are deficient in certain amino acids (e.g., methionine), and a human diet based on beans needs to be balanced with other storage proteins (e.g., those found in unpolished rice) to furnish a complete complement of essential amino acids. Some seed proteins, such as those of jequirity beans (*Abrus precatorius*—used in India to induce abortions and as a contraceptive), are highly poisonous.

Enzymes

Enzymes are mostly large, complex proteins that function as organic catalysts under specific conditions of pH and temperature. By breaking down bonds and allowing new bonds to form, they facilitate cellular chemical reactions, even at very low concentrations, and are absolutely essential to life. None of the 2,000 or more chemical reactions in cells can take place unless the enzyme specific for each one is present and functional in the cell in which it is produced. Enzymes increase the reaction rate by as much as a billion times, and without them, the chemical reactions in cells would take place much too slowly for living organisms to exist. Enzymes are often used repeatedly and usually do not break down during the reactions they accelerate.

Enzyme names normally end in *-ase* (e.g., maltase, sucrase, amylase). The material whose breakdown is catalyzed by an enzyme is known as the *substrate*. Maltose is a very common disaccharide composed of two glucose monomers. The enzyme maltase catalyzes the hydrolysis of maltose (its substrate) to glucose. Enzymes work by lowering the *energy of activation,* which is the minimal amount of energy needed to cause molecules to react with one another. An enzyme brings about its effect by temporarily bonding with potentially reactive molecules at a surface site. The reactive molecules temporarily fit into the active site, where a short-lived complex is formed. The reaction occurs rapidly, often at rates of more than 500,000 times per second. The complex then breaks down as the products of the reaction are released, with the enzyme remaining unchanged and capable of once more catalyzing the reaction (Fig. 2.17).

Many enzymes, derived mostly from bacteria and fungi, have very important industrial uses. For example, waste treatment plants, the dairy industry, and manufacturers of detergents all use enzymes that have been mass-produced by microorganisms in large vats. One such commercially marketed enzyme, known as *Beano*®, is produced by the activities of *Aspergillis,* a mold. Beano breaks down complex sugars found in beans, broccoli, and many other vegetables consumed by humans. A few drops of the enzyme placed on these foods while they are being consumed effectively reduces the gas produced when enzymes in human digestive tracts are otherwise unable to accomplish the breakdown.

Nucleic Acids

Nucleic acids are exceptionally large, complex polymers originally thought to be confined to the nuclei of cells but now known also to be associated with other cell parts. They are vital to the normal internal communication and functioning of all living cells. The two types of nucleic acids—deoxyribonucleic acid (DNA) and ribonucleic acid (RNA)—are briefly introduced here and discussed in more detail in Chapter 13.

Deoxyribonucleic acid (DNA) molecules consist of double helical (spiral) coils of repeating subunits called **nucleotides.** Each nucleotide is composed of three parts: (1) a base containing nitrogen, (2) a five-carbon sugar, and (3) a phosphate (phosphoric acid) molecule. The phosphate of one nucleotide is attached to the sugar of the next nucleotide (Fig. 2.18). Four kinds of nucleotides, each with

Figure 2.18 A nucleotide of DNA is comprised of double strands of subunits which consist of a nitrogenous base, a five-carbon sugar, and a phosphate. This nucleotide contains cytosine as its nitrogenous base. The phosphate of one nucleotide is attached to the sugar of the next nucleotide.

a unique nitrogenous base, occur in DNA. DNA molecules contain, in units known as **genes,** the coded information that precisely determines the nature and proportions of the myriad substances found in cells and the ultimate form and structure of the organism itself. If this coded information were written out, it would fill over 1,000 books of 300 pages each—at least for the more complex organisms. DNA molecules can replicate (duplicate themselves) in precise fashion. When a cell divides, the hereditary information contained in the DNA of the new cells is an exact copy of the original and can be passed on from generation to generation without change, except in the event of a *mutation* (discussed in Chapter 13).

Ribonucleic acid (RNA) is similar to DNA but differs in its sugar and one of its nucleotide components. It usually occurs as a single strand. Different forms of RNA are involved in protein synthesis. DNA and RNA are discussed in more detail in Chapter 13.

SUMMARY

2.1 Attributes of Living Organisms

- Activities of living organisms are organized in cells. Structure and growth are among the attributes of living organisms. Growth is an increase in volume; it results primarily from the production of new cells. Variations in form may be inherited or may result from a response to the environment.
- Reproduction produces offspring that are always similar in form to their parents; if reproduction ceases, the organism becomes extinct.
- Plants generally respond to stimuli more slowly than animals.
- All living organisms carry out metabolic activities, including respiration, digestion, assimilation, production of new cytoplasm, and in green organisms, photosynthesis. Metabolic activities support growth, which is an increase in volume mainly through the production of new cells.

2.2 Chemical and Physical Bases of Life

- The basic "stuff of the universe" is called matter, which occurs in solid, liquid, or gaseous form. It is composed of elements, of which the smallest stable subdivision is an atom. Atomic nuclei contain positively charged protons and uncharged neutrons; the nuclei are surrounded by much larger orbitals of negatively charged electrons. Isotopes are forms of elements that have slight variations in the number of neutrons in their atoms.
- Valance determines the capability of an element to combine with another one. Atoms can bond to other atoms, and those of most elements do not exist independently. Compounds are substances composed of two or more elements combined in a definite ratio by chemical bonds; molecules are the smallest independently existing particles. In a covalent bond, pairs of electrons link two or more atomic nuclei. If a molecule loses or gains electrons, it becomes an ion, which may form an ionic bond with another ion.
- Acids release positively charged hydrogen ions when dissolved in water. Bases release negatively charged hydroxyl ions when dissolved in water. The pH scale is used to measure degrees of acidity or alkalinity. Salts and water are formed when acids and bases are mixed.
- Energy is defined as the ability to produce a change or as the ability to do work. Its forms include chemical, electrical, heat, light, kinetic, and potential energy. The farther away from the nucleus an electron is, the greater the amount of energy required to keep it there.
- Cells are composed of carbon, hydrogen, oxygen, and nitrogen, with a little phosphorus, potassium, and other elements. A plant may convert simple molecules to very large, complex molecules. Organic molecules are usually large polymers that have carbon backbones.
- Carbohydrates contain carbon, hydrogen, and oxygen in a ratio of 1C:2H:1O. Carbohydrates occur as monosaccharides (simple sugars) and disaccharides (two simple sugars joined together). Polysaccharides may consist of many simple sugars condensed together; others are more complex. When simple sugars bind together, they form starch and release water. Starch is broken down by hydrolysis during digestion. This requires the addition of water.
- Lipids (e.g., fats, oils, and waxes), which are insoluble in water, consist of a unit of glycerol or other alcohol with three fatty acids attached. They contain carbon, hydrogen, and oxygen, with proportionately much less oxygen than is found in carbohydrates. Saturated fats have hydrogen atoms attached to every available bond of their carbon atoms; if double bonds prevent the maximum number of hydrogen atoms from attaching, the fat is said to be polyunsaturated. In phospholipids, a phosphate group replaces one fatty acid.
- Proteins are usually large molecules composed of subunits of amino acids. Each amino acid has an amino group (—NH2) and a carboxyl group (—COOH). These groups bond amino acids together, forming polypeptide chains; the bonds are called peptide bonds. Enzymes are large protein molecules that function as organic catalysts. Their names end in *-ase*. Some have important industrial uses.
- Two nucleic acids (DNA and RNA) are associated primarily with the activities of cell nuclei. DNA and RNA molecules consist of chains of nucleotides.

Four kinds of nucleotides, each with a unique nitrogenous base, occur in DNA. Helical coils of DNA contain coded information that determines the nature and proportions of substances in cells and the ultimate form and structure of the organism. The sugar and nucleotide in RNA differ from those of DNA.

REVIEW QUESTIONS

1. What distinguishes a living organism from a nonliving object, such as a rock or a tin can?
2. What is meant by the term *organic?*
3. How are acids, bases, and salts distinguished from one another?
4. Distinguish among carbohydrates, lipids, and proteins.
5. What is energy, and what forms does it take?
6. How are polymers formed?
7. How is a protein molecule different from a nucleic acid molecule?

DISCUSSION QUESTIONS

1. Can part of an organism be alive while another part is dead? Explain.
2. What is the difference between inherited form and form resulting from response to the environment?
3. What might happen if all enzymes were to work at half their usual speed?

CHAPTER 3

Cells

3.1 Introduction to Cells
3.2 Cells
    History
    Modern Microscopes
3.3 Eukaryotic versus Prokaryotic Cells
3.4 Cell Structure and Communication
    Cell Size
    The Cell Wall
    Communication between Cells
3.5 Cellular Components
    The Plasma Membrane
    The Nucleus
    The Endoplasmic Reticulum
    Ribosomes
    Dictyosomes
    Plastids
    Mitochondria
    Microbodies
    Vacuoles
    The Cytoskeleton
3.6 Cellular Reproduction
    The Cell Cycle
    Interphase
    Mitosis
    Key Theme 3.1: Microscapes
3.7 Higher Plant Cells versus Animal Cells

OVERVIEW

This chapter provides a brief review of the history of cell discovery and the development of the cell theory. Differences between prokaryotic and eukaryotic cells are discussed, and observations on cell structure and communication follow. Descriptions are provided for the plasma membrane, nucleus, endoplasmic reticulum, ribosomes, dictyosomes, plastids, mitochondria, microbodies, vacuoles, and cytoskeleton. The chapter next discusses the cell cycle, including interphase as well as mitosis and cytokinesis, and concludes with a brief comparison of plant and animal cells.

A root tip longitudinal section showing various stages of mitosis. © *Steven P. Lynch*

3.1 Introduction to Cells

Learning Outcome

a. List three functions that may result from modification of a cell after it divides.

All living organisms, from aardvarks to zinnias, are composed of cells, and all living organisms, including each of us, also begin life as a single cell. This single cell divides repeatedly until it develops into an organism often consisting of billions of cells. During the first few hours of an organism's development, the cells all look alike, but changes soon take place, not only in the appearance of the cells but also in their function. Some modifications, for example, equip cells to transport food and water, while other cells become modified for secretion of various fluids such as resin or nectar, and still others give strength to tissues such as wood. Some cells may live and function for many years; others mature and degenerate in just a few days. Even as you read this, millions of new cells are being produced in your body. Some cells add to your total body mass (if you have not yet stopped growing), but most replace the millions of older cells that are destroyed every second of your life. The variety and form of cells seem almost infinite, but certain features are shared by most of them. This chapter focuses on those features.

3.2 Cells

Learning Outcomes

a. List four scientists who worked with early microscopes, and explain how each of these scientists led us to our current understanding of cells.
b. Explain the differences between compound microscopes and dissecting microscopes.
c. Describe what features are revealed with transmission electron microscopy and with scanning electron microscopy.

History

Imagine the excitement of the first scientist who observed cells! This discovery was made in 1665 by the English physicist Robert Hooke, who used a primitive microscope (Fig. 3.1) to examine thin slices of cork found in stoppered wine bottles:

> *I took a good clear piece of Cork, and with a Penknife sharpen'd as keen as Razor, I cut a piece of it off, and thereby left the surface of it exceedingly smooth, then examining it very diligently with a Microscope, methought I could perceive it to appear a little porous . . . these pores, or cells . . . were indeed the first microscopical pores I ever saw, and perhaps that were ever seen, for I had not met with any Writer or Person, that had made mention of them before this . . . I had with the discovery of them, presently hinted to me the true and intelligible reason of all the Phaenomena of Cork.*

Figure 3.1 Robert Hooke's microscope, reconstructed from his book *Micrographia*, which was published in 1665. © *Science & Society Picture Library/Contributor/Getty Images*

Hooke compared the boxlike compartments he saw to the surface of a honeycomb and is credited with applying the term *cell* to those compartments. He also estimated that a cubic inch of cork would contain approximately 1,259 million such cells. What Hooke saw in the cork were really only the walls of dead cells, but he also saw "juices" in living cells of elderberry plants and thought he had found something similar to the veins and arteries of animals.

Two physicians, Marcello Malpighi in Italy and Hooke's compatriot, Nehemiah Grew, in England, along with Anton van Leeuwenhoek, reported for 50 years on the organization of cells in a variety of plant tissues. In the 1670s, they also reported on the form and structure of single-celled organisms, which they referred to as "animalcules."

After this period, little more was reported on cells until the early 1800s. This lack of progress was mainly due to imperfections in the primitive microscopes and the crude way in which tissues were prepared for microscopic examination. Microscopes and tissue preparations both slowly improved, however, and by 1809, the French biologist Jean Baptiste de Lamarck had seen a wide enough variety of cells and tissues to conclude that "no body can have life if its constituent parts are not cellular tissue or are not formed by cellular tissue." In 1824, René J. H. Dutrochet, also of France, reinforced Lamarck's conclusions that all animal and plant tissues are composed of cells of various kinds. Neither of them, however, realized that each cell could, in many cases, reproduce itself and exist independently.

In 1831, the English botanist Robert Brown discovered that all cells contain a relatively large body that he called the *nucleus*. Soon after the discovery of the nucleus, the German botanist Matthias Schleiden observed a smaller body within the nucleus that he called the *nucleolus*. Schleiden and German zoologist Theodor Schwann were not the first to understand the significance of cells, but they explained them more clearly and with greater insight than others before them had done. They are generally credited with developing the **cell theory,** beginning with their publications of 1838 to 1839. This theory, in essence, states that all living organisms are composed of cells and that cells form a unifying structural basis of organization.

In 1858, another German scientist, Rudolf Virchow, argued persuasively in a classic textbook that every cell comes from a preexisting cell (*"omnis cellula e cellula"*) and that there is no spontaneous generation of cells. Virchow's publication stirred up much controversy because there had been a widespread belief among scientists and nonscientists alike that animals could originate spontaneously from dust. Many who had microscopes were thoroughly convinced they could see "animalcules" appearing in decomposing substances.

The controversy became so heated that in 1860 the Paris Academy of Sciences offered a prize to anyone who could experimentally prove or disprove spontaneous generation. Just 2 years later, the brilliant scientist Louis Pasteur of France was awarded the prize. Pasteur, using swan-necked flasks, demonstrated convincingly that boiled media remained sterile indefinitely if microorganisms from the air were excluded from the media.

In 1871, Pasteur proved that natural alcoholic fermentation always involves the activity of yeast cells. In 1897, the German scientist Eduard Buchner accidentally discovered that the yeast cells did not need to be alive for fermentation to occur. He found that extracts from the yeast cells would convert sugar to alcohol. This discovery was a big surprise to the biologists of the time and quickly led to the identification and description of enzymes (discussed in Chapter 2), the organic *catalysts* (substances that aid chemical reactions without themselves being changed) found in all living cells. This also led to the belief that cells were little more than miniature packets of enzymes. During the first half of the 20th century, however, further advances were made in the refinement of microscopes and in tissue preparation techniques. Many structures and bodies, besides the nucleus, were observed in cells, and the relationship between structure and function came to be realized and understood on a much broader scale than previously had been possible.

Modern Microscopes

Our investigation of life is enhanced by microscopes that can magnify cells and tissues up to hundreds or even thousands of times their actual size. A better understanding of cell structure and function is also provided by preparing tissues in different ways. While **light microscopes,** similar to the one used by Hooke in 1665, provide basic information about cell structure and some of the bodies within cells, the development of **electron microscopes** has revealed detailed images of tiny structures within cells.

Light microscopes increase magnification as light passes through a series of transparent lenses, made of various types of glass or calcium fluoride crystals. The curvatures of the lens materials and their composition are designed to minimize distortion of image shapes and colors.

Light microscopes are of two basic types: *compound microscopes,* which require most material being examined to be sliced thinly enough for light to pass through, and *dissecting microscopes* (*stereomicroscopes*), which allow three-dimensional viewing of opaque objects. The best compound microscopes in use today can produce magnifications of up to 1,500 times under ideal conditions. Many dissecting microscopes used in teaching laboratories magnify up to 30 times, but higher magnifications are possible with both types of microscopes. Light microscope magnifications of more than 1,500 times, however, are considered "empty" because *resolution* (the capacity of lenses to separate closely adjacent tiny objects) does not improve with magnification beyond a certain point. In general, when using a compound light microscope, one can distinguish *organelles* (bodies within cells) only if they are 2 micrometers or larger in diameter. Techniques such as *phase-contrast* and *fluorescence microscopy* take advantage of how certain types of light interact with a specimen to reveal surface structure and localized key constituents. In the mid-1980s, further refinement of instrumentation enabled the use of lasers and microelectronics in *confocal scanning microscopy*, which enhances target cell features by increasing the contrast and optical resolution of the image. In this chapter, the structures discussed that are most commonly observed with a light microscope include cell walls, nuclei, nucleoli, cytoplasm, chloroplasts, and vacuoles. Light microscopes (Fig. 3.2) will continue in the foreseeable future to be useful, particularly for observing living cells.

Since the 1950s, biological techniques with high-resolution electron microscopes have revealed greater details about cell structure that cannot be studied with conventional light microscopy. Instead of light, electron microscopes use a beam of electrons produced when high-voltage electricity is passed through a wire. This electron beam is directed through a vacuum in a large tube or column. When the beam passes through a specimen, an image is formed on a plate toward the base of the column. Magnification is controlled by powerful electromagnetic lenses located on the column.

Like light microscopes, electron microscopes are of two basic types. *Transmission electron microscopes* (Fig. 3.3a) can produce magnifications of 200,000 or more times, but the material to be viewed must be sliced extremely thin and introduced into the column's vacuum, so living objects can't be observed.

Scanning electron microscopes (Fig. 3.3b) usually do not attain such high magnifications (30 to 10,000 times is the usual range), but surface detail of thick objects can be observed when a scanner makes the object visible on a

Figure 3.2 (a) A stereomicroscope (dissecting microscope). (b) A compound light microscope. *(a) © Richard Hutchings; (b) © Comstock/Alamy*

Figure 3.3 (a) A transmission electron microscope. (b) A scanning electron microscope. *(a–b) Courtesy of JEOL-USA, Inc., Peabody, MA*

cathode tube like a television screen. The techniques for observation with electron microscopes have become so refined that even preserved material can appear exceptionally lifelike, and high-resolution, three-dimensional images can be obtained.

In 1986, the Nobel Prize in physics was awarded to two IBM scientists, Gerd Binnig and Heinrich Rohrer, for their invention in 1982 of a *scanning tunneling microscope*. This microscope uses a minute probe that "tunnels" electrons upon a sample. As the probe is moved across the surface, its height is continually adjusted to keep the flow of electrons constant, and the fluctuations in height are recorded to produce a map of the sample surface. Without doing any damage to the probed area, this microscope reproduces an image of such high magnification that even atoms can become discernible. The probe can scan areas barely twice the width of an atom and theoretically could be used to print on the head of an ordinary pin the words contained in more than 50,000 single-spaced pages of books. Early in 1989, the first picture of a segment of DNA showing its helical structure was taken with a scanning tunneling microscope by an undergraduate student associated with the Lawrence Laboratories in northern California. Several variations of this microscope, each using a slightly different type of probe, have now been produced. Significant new discoveries by cell biologists using one or more of all types of microscopes in their research have become frequent events.

3.3 Eukaryotic versus Prokaryotic Cells

Learning Outcomes

a. State the major difference between prokaryotic and eukaryotic cells.

b. Describe the event(s) that led to large eukaryotic cells with multiple organelles.

Nearly all higher plant and animal cells share most of the various features discussed in this chapter. Some of these features (e.g., nuclei, plastids) are, however, lacking in the cells of some very simple organisms such as bacteria. Cells without nuclei are called **prokaryotic** (*pro* = before; *karyon* = nucleus) to distinguish them from typical **eukaryotic** (*eu* = well or good; *karyon* = nucleus) cells, discussed here. Prokaryotic cells are covered in more detail in Chapter 17. According to **endosymbiotic theory,** some cell components (chloroplasts and mitochondria) evolved when a large eukaryotic cell engulfed independent prokaryotes. Both prokaryotic and eukaryotic organisms can have **cell walls** (rigid boundaries of cells), but only eukaryotic cells contain **organelles** (membrane-bound structures with specialized functions). These features, and other cellular components, are discussed in the sections that follow.

3.4 Cell Structure and Communication

Learning Outcomes

a. Explain why organisms with many small cells might have an advantage over those with fewer, larger cells.

b. List the three major components of secondary plant cell walls.

c. Name and describe the structures that allow plant cells to communicate.

Plant cells typically have a *cell wall* surrounding the **protoplasm,** which consists of all the living components of a cell. These living components are bounded by a membrane called the *plasma membrane.* All cellular components between the plasma membrane and a relatively large body called the *nucleus* are known as **cytoplasm.** Within the cytoplasm is a souplike fluid called *cytosol,* in which various bodies called *organelles* are dispersed. Organelles are persistent structures of various shapes and sizes with specialized functions in the cell; most, but not all, are bounded by membranes (Figs. 3.4 and 3.5).

Cell Size

Most plant and animal cells are so tiny they are invisible to the unaided eye. Cells of higher plants generally vary in length between 10 and 100 micrometers.[1] Remember that the resolution of a light microscope is 2 micrometers, making it useful for the study of eukaryotic cells. Because there are roughly 25,000 micrometers to the inch, it would take about 500 average-sized cells to extend across 2.54 centimeters (1 inch) of space; 30 of them could easily be placed across the head of a pin. Some prokaryotic (bacterial) cells are less than 0.5 micrometer wide, while cells of the green alga called mermaid's wineglass (*Acetabularia*) are mostly between 2 and 5 centimeters long, and fiber cells of some nettles are about 20 centimeters long.

Why are cells so small? Consider that as a cell increases in size, its volume grows much more than its surface area. The increase in surface area of a spherical cell, for example, is equal to the square of its increase in diameter, but its increase in volume is equal to the cube of its increase in diameter. This means that a cell whose diameter increased 10 times would increase in surface area 100 times (10 squared) but in volume 1,000 times (10 cubed). Because all substances enter or leave cells through their surfaces, which are the only contact areas with their surroundings, larger cells are at a disadvantage. Furthermore, the nucleus regulates all aspects of a cell's activities, and the greater the volume of the cell, the longer it takes for instructions from the nucleus to reach the surface. On the other hand, smaller cells have a clear advantage because they have relatively larger surface area to volume ratios, thereby enabling faster and more efficient communication between the nucleus and other parts of the cell.

Full-grown organisms have astronomical numbers of cells. For example, it has been calculated that a single mature leaf of a pear tree contains 50 million cells and that the total number of cells in the roots, stems, branches, leaves, and fruit of a full-grown pear tree exceeds 15 trillion. Can you imagine how many cells there are in a 3,000-year-old redwood tree of California, which may reach heights of 90 meters (300 feet) and measure up to 4.5 meters (15 feet) in diameter near the base?

Some cells are boxlike with six walls, but others assume a wide variety of shapes, depending on their location and function. The most abundant cells in the younger parts of

---

1. See Appendix 5 for conversion tables.

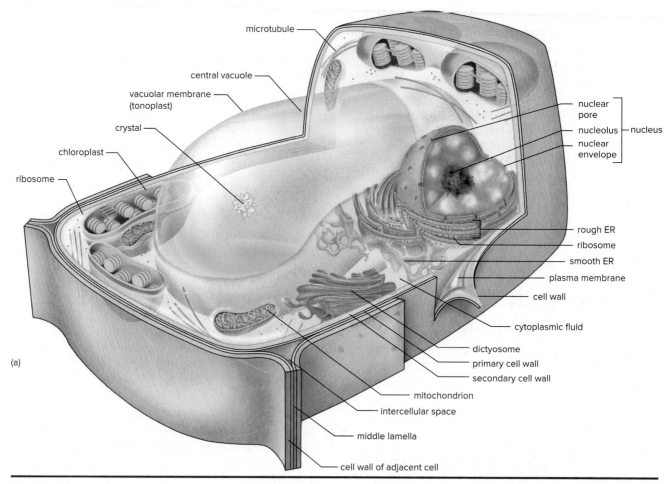

Figure 3.4 Anatomy of a young plant cell. (a) Generalized drawing to show cellular components within a young plant cell. (b) Colored transmission electron micrograph section of a young cell from the root-tip meristem of *Arabidopsis thaliana*: the nucleus (pink) contains a nucleolus (dark); the cytoplasm contains mitochondria (blue), proplastids (juvenile chloroplasts), without developed thylakoid systems (green), dictyosomes (red), and both large and small vacuoles; the cell membrane is also shown with a distinct cell wall around its perimeter. *(b)* © *Biophoto Associates/Science Source*

plants and fruits may be more or less spherical, like bubbles, when they are first formed, but as they press against each other, they commonly end up with an average of 14 sides by the time they are mature. These cell types are discussed in Chapter 4.

The Cell Wall

Cell walls are among the most abundant and diverse of biological structures on earth. In essence, the cell wall provides the basis of food, fuel, and shelter that sustains a majority of life, and accentuates our lives with clothing, furniture, paper, tools, toys, and almost anything you can imagine that is

Figure 3.5 Anatomy of a plant cell. Colored scanning electron micrograph of a section through a plant cell, showing internal structures. The cell is encased in by a cell wall composed of cellulose, hemicellulose, and pectin. Inside the cell are chloroplasts (dark green), which carry out photosynthesis, and the nucleus (orange) which contains the cell's genetic information. At the center of the cell is a large vacuole, which maintains the cell's shape, stores useful materials, and digests waste products. © DR DAVID FURNESS, KEELE UNIVERSITY/Getty Images

made of fiber or wood. In addition to providing plants with mechanical strength, the cell wall maintains cell shape, controls cell expansion, provides protection, regulates transport, and stores food reserves. From the cell's perspective, walls protect the delicate cell contents within. When millions of these cells function together as a tissue, their collective strength is enormous. The redwoods and Tasmanian *Eucalyptus* trees, which are the largest trees alive today, exceed the mass and volume of the largest land animals, the elephants, by more than a hundred times. The wood of one giant redwood tree could support the combined weight of a thousand elephants.

The first cell structure discovered by Robert Hooke in 1665 was the *cell wall,* and among plant cell structures observed with a microscope, the cell wall is the most obvious because it defines the shape of the cell. Many of the prepared specimens observed with a microscope in plant biology are merely stained remnants of once-living cells. But the vast diversity of cell walls within and among species tells a story about the structure and function of each cell. For instance, *epidermal* cells, which form a thin layer on the surfaces of all plant organs, often have unusual shapes and sizes. Some such cells form hairs that may secrete substances that discourage animals from grazing on the plants producing them. Thin-walled cells found beneath the epidermis of leaves are specialized for their function of photosynthesis (discussed in Chapter 10); and thick-walled cells of wood help transport water without collapsing. The structure and function of different plant cells, and the tissues they form, are addressed in Chapter 4.

The main structural component of cell walls is *cellulose,* which is composed of 100 to 15,000 glucose monomers in long *chains* and is the most abundant polymer on earth. As a primary food source for grazing animals and at least indirectly for nearly all other living organisms, it could be said that most life on earth relies directly or indirectly on the cell wall.

In addition to cellulose, cell walls typically contain a matrix of *hemicellulose* (a gluelike substance that holds cellulose fibrils together), *pectin* (the organic material that gives stiffness to fruit jellies), and *glycoproteins* (proteins that have sugars associated with their molecules).

A **middle lamella,** which consists of a layer of pectin, is first produced when new cell walls are formed. This middle lamella is normally shared by two adjacent cells and is so thin that it may not be visible with an ordinary light microscope unless it is specially stained. A flexible primary wall, consisting of a fine network of cellulose, hemicellulose, pectin, and glycoproteins, is laid down on either side of the middle lamella (Fig. 3.6a). Reorganization, synthesis of new molecules, and insertion of new wall polymers lead to rearrangement of the cell wall during growth. *Secondary walls,* which are produced toward the inside of the primary walls, are derived from primary walls by thickening and inclusion of *lignin,* a complex polymer.

Secondary cell walls of plants generally contain more cellulose (40% to 80%) than primary walls. As the cell ages, wall thickness can vary, occupying from as little as 5% to more than 95% of the volume of the cells. During secondary wall formation, cellulose microfibrils become embedded in

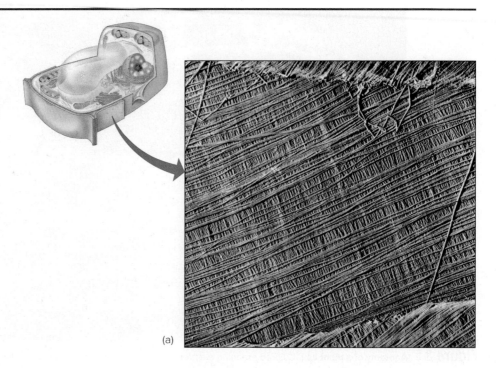

Figure 3.6a A small portion of a cell wall of the green alga *Chaetomorpha melagonium*, showing how cellulose microfibrils are laid down. Each microfibril is composed of numerous molecules of cellulose. ×24,000. *(a) © BioPhoto Associates/Photo Researchers*

lignin, much as steel rods are embedded in concrete to form prestressed concrete (Fig. 3.6b).

Communication between Cells

Cells that manufacture, process, or store food have thin walls, while those involved in support usually have relatively thick walls. Although each living cell is capable of independently carrying on complex activities, it is essential that these activities be coordinated through some means of communication among all the living cells of an organism. Fluids and dissolved substances can pass through primary walls of adjacent cells via **plasmodesmata** (singular: **plasmodesma**), which are tiny strands of cytoplasm that extend between the cells through minute openings (see Fig. 3.20). The translocation of sugars, amino acids, ions, and other substances occurs through the plasmodesmata. The middle lamellae and most cell walls are, however, permeable and permit slower movement of water and dissolved substances between cells.

3.5 Cellular Components

Learning Outcome

a. Draw the following cell structures and organelles, and describe the function of each: plasma membrane, nucleus, endoplasmic reticulum, ribosomes, dictyosomes, plastids, mitochondria, microbodies, vacuoles, and cytoskeleton.

Most chemical reactions that take place in cells occur in the protoplasm, as part of a dynamic series of events that make the plant a living entity. Each organelle within the protoplast has a primary function, and the flow of *metabolites* (products of chemical synthesis or breakdown) from one organelle to another is necessary for a balance of events that take place.

Envision a journey through the plant cell as an exciting voyage in which information is stored primarily in the nucleus, processed in the cytoplasm, and sent on to different parts of the cell. This information can bring about the synthesis of proteins in the cytoplasm, where they become involved in *metabolic reactions* (see Chapter 10), or they may be destined for use in other cellular locations. The packaged proteins may be incorporated in membranes or organelles, and other compounds may be manufactured in specific organelles or may enter from an adjacent cell.

The Plasma Membrane

The outer boundary of the living part of the cell, the **plasma membrane,** is roughly eight-millionths of a millimeter thick. To get an idea of how incredibly thin that is, consider that it would take 12,500 such membranes neatly stacked in a pile to achieve the thickness of an ordinary piece of writing paper. Yet this delicate, semipermeable structure is of vital importance in regulating the movement of substances into and out of the cell. While the plasma membrane may inhibit movement of some substances, it can otherwise allow free movement and can even control movement of other substances into and out of the cell. As a result, the proportions and makeup of chemicals within a cell become quite different from those outside the cell. The plasma membrane is also involved in the production and assembly of cellulose for cell walls.

Evidence obtained since the early 1970s indicates that the plasma membrane and other cell membranes are

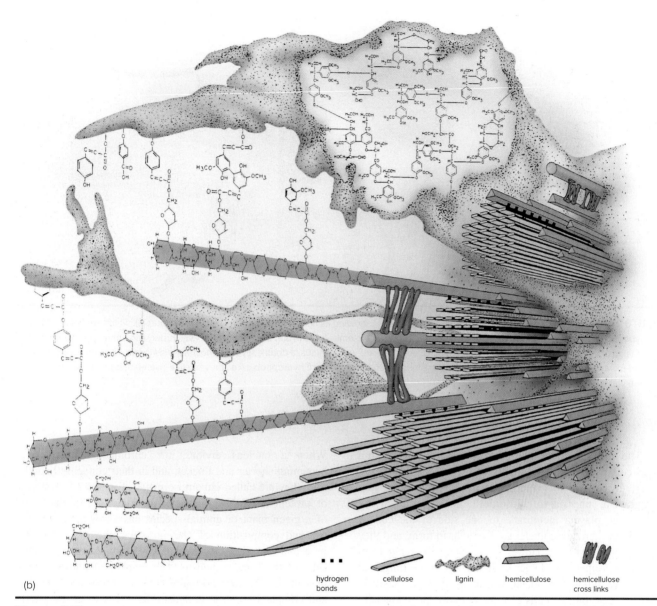

Figure 3.6b Secondary cell wall structure. Components are arranged so that the cellulose microfibrils and hemicellulose chains are embedded in lignin. *(Source: Reproduced by permission of the Oklahoma Academy of Science.)*

composed of *phospholipids* arranged in two layers, with proteins interspersed throughout (Fig. 3.7). This fluid mosaic model for the plasma membrane implies a dynamic structure with numerous components, some of which can migrate and interact directly with each other. Covalent bonds link carbohydrates to both lipids and proteins on the outer surfaces of membranes. Some proteins extend across the entire width of the membrane, while others are embedded or apparently are loosely bound to the outer surface.

The remaining cell contents usually push the plasma membrane up against the cell wall because of pressures developed by *osmosis* (see Chapter 9), but the membrane is quite flexible and often forms folds, which may, in turn, become little, hollow spheres or vesicles that float off into the cell. In fact, experiments have shown that by adding detergents to a continuous membrane, it can be broken up and dispersed, yet it can partially reform when the detergents are removed. The membrane may even shrink away from the wall temporarily, but if it ever ruptures, the cell soon dies.

The Nucleus

The *nucleus* is the control center of the cell. In some ways, it functions like a combination of a computer program and a dispatcher that sends coded messages or "blueprints" originating from DNA in the nucleus with information that will ultimately be used in other parts of the cell. In other words, the DNA in the nucleus provides the original information needed to fulfill the cell's needs. This nuclear information contributes to growth, differentiation, and the myriad activities of the complex cell "factory." The nucleus also stores hereditary information, which is passed from cell to cell as new cells are formed.

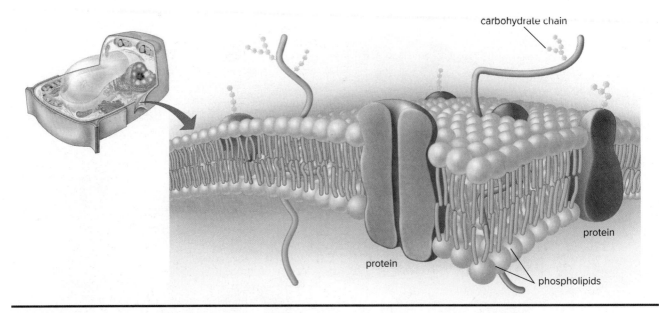

Figure 3.7 A model of a small portion of a plasma membrane, showing its fluid mosaic nature. The proteins (purple), which are coiled chains of polypeptides, are either embedded or on the surfaces. Some of the embedded proteins extend all the way through and may serve as conduits for diffusion of certain ions. In cells, and other places where there are watery fluids, a double layer of *phospholipids* forms. The heads (gold) point outward toward the water. The tails (orange), which are long-chain fatty acids, are hydrophobic (i.e., they "dislike" water) and point inward away from the water. The membrane is about 8 nanometers thick.

The nucleus often is the most conspicuous object in a living cell, although in green cells, chloroplasts may obscure it. In living cells without chloroplasts, the nucleus may appear as a grayish, spherical to ellipsoidal lump, sometimes lying against the plasma membrane to one side of the cell or toward a corner. Some nuclei are irregular in form, and they can vary greatly in size. However, they generally range from 2 to 15 micrometers or larger in diameter. Certain fungi and algae have numerous nuclei within a single extensively branched cell, but the cells of more complex plants usually have a single nucleus.

Each nucleus is bounded by two membranes, which together constitute the **nuclear envelope.** Structurally complex pores, about 50 to 75 nanometers apart, occupy up to one-third of the total surface area of the nuclear envelope (Fig. 3.8). Proteins that act as channels for molecules are embedded within the pores. The pores apparently permit only certain kinds of molecules (for example, proteins being carried into the nucleus and RNA being carried out) to pass between the nucleus and the cytoplasm.

The nucleus contains a granular-appearing fluid called *nucleoplasm,* which is packed with short fibers that are about 10 nanometers in diameter; several different larger bodies are suspended within it. Of the larger nuclear bodies, the most noticeable are one or more **nucleoli** (singular: **nucleolus**), which are composed primarily of RNA and associated proteins.

Other important nuclear structures, which are not apparent with light microscopy unless the cell is stained or is in the process of dividing, include thin strands of **chromatin.**

When a nucleus divides, the chromatin strands coil, becoming shorter and thicker, and in their condensed condition, they are called **chromosomes.** Chromatin is composed of protein and DNA (discussed in Chapter 13). Each cell of a given plant or animal species has its own fixed number and composition of chromosomes; the cells involved in sexual reproduction have half the number found in other cells of the same organism. The number of chromosomes present in a nucleus normally bears no relation to the size and complexity of the organism. Each body cell of a radish, for example, has 18 chromosomes in its nucleus, while a cell of one species of goldenweed has four, and a cell of a tropical adder's tongue fern has over 1,000. Humans have 46 chromosomes in each body cell.

The Endoplasmic Reticulum

The outer membrane of the nucleus is connected to and continuous with the **endoplasmic reticulum.** The endoplasmic reticulum facilitates cellular communication and channeling of materials. Many important activities, such as the synthesis of membranes for other organelles and modification of proteins from components assembled from elsewhere within the cell, occur either on the surface of the endoplasmic reticulum or within its compartments.

The endoplasmic reticulum (often referred to simply as **ER**) is an enclosed space consisting of a network of flattened sacs and tubes that form channels throughout the cytoplasm, the amount and form varying considerably from cell to cell. Transmission electron micrographs of sectioned ER

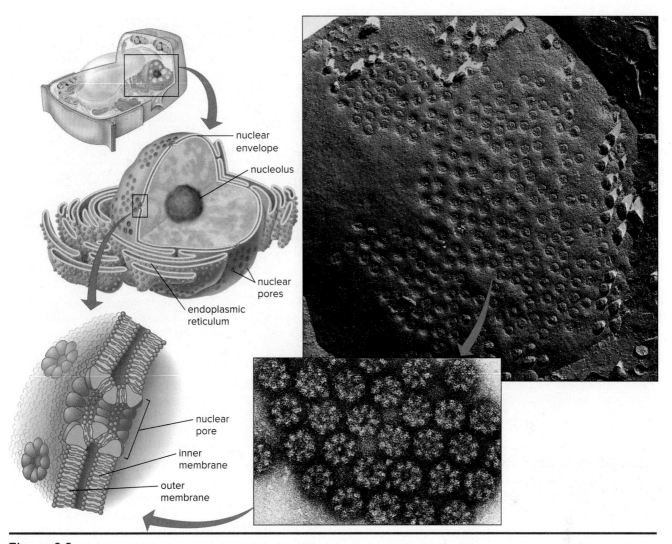

Figure 3.8 Drawing of the nucleus showing the inner and outer membranes, nuclear pores, and nucleolus. The electron micrographs show detail of the nuclear pores, which are about 60 nanometers in diameter. *Courtesy of Ron Milligan*

give it the appearance of a series of parallel membranes that resemble long, narrow tubes or sacs, creating subcompartments within the cell.

Ribosomes (discussed in the section that follows) may be distributed on the outer surface (i.e., the surface in contact with the cytoplasm) of the endoplasmic reticulum. Such endoplasmic reticulum is said to be *rough* and is primarily associated with the synthesis, secretion, or storage of proteins (Fig. 3.9; see also Chapter 13). This contrasts with *smooth* endoplasmic reticulum, which has few, if any, ribosomes lining the surface and is associated with lipid secretion. Both types of endoplasmic reticulum can occur in the same cell and can be interconverted, depending on the demands of the cell. Many enzymes involved in the process of cellular respiration are synthesized on the surface of the endoplasmic reticulum. The enzymes, however, enter other organelles (primarily mitochondria, which are discussed later in this chapter) without passing through the endoplasmic reticulum. The endoplasmic reticulum also appears to be the primary site of membrane synthesis within the cell.

Ribosomes

Ribosomes are tiny bodies that are visible with the aid of an electron microscope. They are typically roughly ellipsoidal in shape with apparently varied and complex surfaces. Each ribosome is composed of two subunits that are composed of RNA and proteins; the subunits, upon close inspection, can be differentiated by a line or cleft toward the center. Ribosomes average only about 20 nanometers in diameter in most plant cells. Unattached ribosomes often occur in clusters of 5 to 100, particularly when they are involved in linking amino acids together in the construction of the large, complex protein molecules that are a basic part of all living organisms.

Ribosomal subunits are assembled within the nucleolus, are released, and in association with special RNA molecules initiate protein synthesis. Once assembled, complete ribosomes may line the outside of the endoplasmic reticulum, but they can also occur unattached in the cytoplasm, chloroplasts, or other organelles. About 55 kinds of protein are found in each ribosome of prokaryotic cells and a

Figure 3.9 Transmission electron micrograph of the cytoplasm in a root tip cell of corn (*Zea mays*). The endoplasmic reticulum, with attached ribosomes, can be seen slightly toward the right portion of the photograph. ×60,000. © Dr Jeremy Burgess/Science Source

Figure 3.10 A dictyosome from *Euglena*, a waterweed. ×40,000. *Courtesy of John Z. Kiss*

slightly higher number in those of eukaryotic cells (see the discussion of various types of RNA in Chapter 13). Unlike other organelles, ribosomes have no bounding membranes, and because of this, some scientists prefer not to call them organelles.

Dictyosomes

Stacks of flattened discs or vesicles known as **dictyosomes** may be scattered throughout the cytoplasm of a cell. Dictyosomes are often bounded by branching tubules that originate from the endoplasmic reticulum but are not directly connected to it (Fig. 3.10). Five to eight dictyosomes per cell are typical, but up to 30 or more may be found in cells of simpler organisms. Aggregations of dictyosomes, constituting the *Golgi apparatus,* occur in protein-secreting animal cells and a few plant cells with similar function. In animal cells, the term *Golgi body* is used to describe dictyosomes, which are named after Camillo Golgi, who discovered the Golgi apparatus in 1898.

Dictyosomes are involved in the modification of carbohydrates attached to proteins that are synthesized and packaged in the endoplasmic reticulum. Complex polysaccharides are also assembled within the dictyosomes and collect in small *vesicles* (tiny, blisterlike bodies) that are pinched off from the margins. These vesicles migrate to the plasma membrane, fuse with it, and secrete their contents outside the cell. Substances secreted by vesicles may include cell-wall polysaccharides, floral nectars, and essential oils found in herbs.

The enzymes needed for the packaging of proteins are produced by the endoplasmic reticulum and further modified within the dictyosomes. One might describe dictyosomes as collecting, packaging, and delivery centers or, perhaps, as "post offices" of the cell.

Plastids

Most living plant cells have several kinds of **plastids,** with the **chloroplasts** (Fig. 3.11a) being the most conspicuous. They occur in a variety of shapes and sizes, such as the beautiful corkscrewlike ribbons found in cells of the green alga *Spirogyra* and the bracelet-shaped chloroplasts of other green algae, such as *Ulothrix* (see Chapter 18). The chloroplasts of higher plants, however, tend to be shaped somewhat like two Frisbees glued together along their edges, and when they are sliced in median section, they resemble the outline of a rugby football.

Figure 3.11 Drawing of leaf mesophyll cell chloroplast and transmission electron micrographs showing variation in chloroplast structure. (a) A chloroplast. ×20,000. (b) Cutaway of a chloroplast. (c) Grana. ×40,000. (d) A few thylakoids. *(a)* © *DR KARI LOUNATMAA/Getty Images; (c)* © *Biology Pics/Science Source*

Although several algae and a few other plants have only one or two chloroplasts per cell, the number of chloroplasts is usually much greater in a green cell of a higher plant. Seventy-five to 125 is quite common, with the green cells of a few plants having up to several hundred. The chloroplasts may be from 2 to 10 micrometers in diameter, and each is bounded by an envelope consisting of two delicate membranes. The outer membrane is apparently derived from endoplasmic reticulum, while the inner membrane is believed to have originated from the cell membrane of a cyanobacterium (discussed in Chapter 17).

Within the chloroplast are numerous **grana** (singular: **granum**), which are formed from membranes and have the appearance of stacks of coins with double membranes. There are usually about 40 to 60 grana linked together by arms in each chloroplast, and each granum may contain from two or three to more than 100 stacked **thylakoids.** In reality, thylakoids are part of an overlapping and continuous membrane system suspended in the liquid portion of the chloroplast (Fig. 3.11b). The thylakoid membranes contain green **chlorophyll** and other pigments. These "coin stacks" of grana are vital to life as we know it, for it is within the thylakoids that the first steps of the important process of *photosynthesis* (see Chapter 10) occur. In photosynthesis, green plants convert water and carbon dioxide (from the air) to simple food substances, harnessing energy from the sun in the process. The existence of human and all other animal life depends on the activities of the chloroplasts.

The liquid portion of the chloroplast is a colorless fluid matrix called **stroma,** which contains enzymes involved in photosynthesis. Genes in the nucleus dictate most of the activities of chloroplasts, but each chloroplast contains a small, circular DNA molecule that encodes for production of certain proteins related to photosynthesis and other activities in the chloroplast and cell. The chloroplast also contains RNA and ribosomes, which facilitate some protein synthesis. Some plastids (e.g., those of tobacco) store proteins. There are usually four or five *starch grains* in the stroma, as well as oil droplets and enzymes.

Figure 3.12 Chromoplasts in the flesh of a red pepper. ×400.
© James E. Bidlack

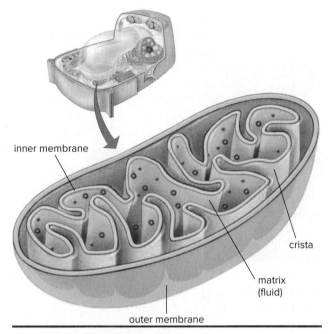

Figure 3.13 A mitochondrion greatly enlarged and cut away to show the cristae (folds of the inner membrane). A mitochondrion is about 2 micrometers long.

Chromoplasts are another type of plastid found in some cells of more complex plants. Although chromoplasts are similar to chloroplasts in size, they vary considerably in shape, often being somewhat angular. They sometimes develop from chloroplasts through internal changes that include the disappearance of chlorophyll. Chromoplasts are yellow, orange, or red in color due to the presence of carotenoid pigments, which they synthesize and accumulate. They are most abundant in the yellow, orange, or some red parts of plants, such as ripe tomatoes, carrots, or red peppers (Fig. 3.12). These carotenoid pigments, which are lipid soluble, are not, however, the predominant pigments in most red flower petals. The anthocyanin pigments of most red flower petals are water soluble.

Leucoplasts are yet another type of plastid common to cells of higher plants. They are essentially colorless and include *amyloplasts,* which synthesize starches, and *elaioplasts,* which synthesize oils. If exposed to light, some leucoplasts will develop into chloroplasts, and vice versa.

Plastids of all types develop from **proplastids,** which are small, pale green or colorless organelles having roughly the size and form of mitochondria (discussed in the next section). They are simpler in internal structure than plastids and have fewer thylakoids, the thylakoids not being arranged in grana stacks. Proplastids frequently divide and become distributed throughout the cell. After a cell itself divides, each daughter cell has a proportionate share. Plastids also arise through the division of existing mature plastids.

Mitochondria

Mitochondria (singular: **mitochondrion**) are often referred to as the powerhouses of the cell, for it is within them that energy is released from organic molecules by the process of *cellular respiration* (the role of mitochondria in respiration is further discussed in Chapter 10). This energy is needed to keep the individual cells, and the plant as a whole, functioning.

Carbon skeletons and fatty acid chains are also rearranged within mitochondria, allowing for the building of a wide variety of organic molecules. Mitochondria are numerous and tiny, typically measuring from 1 to 3 or more micrometers in length and having a width of roughly 0.5 micrometer; they are barely visible with light microscopes. They appear to be in constant motion in living cells and tend to accumulate in groups where energy is needed. They often divide in two; in fact, they all originate from the division of existing mitochondria.

Mitochondria typically are shaped like cucumbers, rods, or balls. A sectioned mitochondrion resembles a scooped-out watermelon, with inward extensions of the rind forming mostly incomplete partitions perpendicular to the surface (Fig. 3.13). The appearance of incomplete partitions results from the fact that each mitochondrion is bounded by two membranes, with the inner membrane forming numerous platelike folds called *cristae*. The cristae greatly increase the surface area available to the enzymes contained in a matrix fluid. The number of cristae, as well as the number of mitochondria themselves, can change over time, depending on the activities taking place within the cell. The matrix fluid also contains DNA, RNA, ribosomes, proteins, and dissolved substances.

Microbodies

Various small bodies distributed throughout the cytoplasm tend to give it a granular appearance. Examples of such components include types of small, spherical organelles called **microbodies,** which contain specialized enzymes and are bounded by a single membrane. **Peroxisomes,** for instance, contain enzymes needed by some plants to survive during hot conditions in a process called *photorespiration* (discussed in Chapter 10), whereas **glyoxisomes** contain enzymes that aid in the conversion of fats to carbohydrates during, for example, the germination of seeds containing fats. If present, peroxisomes are generally found associated with chloroplasts, and glyoxisomes usually are located near mitochondria. During a plant's life cycle, peroxisomes and glyoxisomes may increase in number at stages when the need for them is greatest.

At one time, lipid, fat, or wax droplets commonly found in cytoplasm were believed to be bounded by a membrane; recent evidence, however, suggests no membrane is present, and some, therefore, do not consider them true organelles. Another organelle, called a *lysosome,* stores enzymes that digest proteins and certain other large molecules but is apparently confined to animal cells. The digestive activities of lysosomes are similar to those of the vacuoles of plant cells (discussed next).

Vacuoles

In a mature living plant cell, 90% or more of the volume may be taken up by one or two large central **vacuoles** that are bounded by **vacuolar membranes (tonoplasts)** (Fig. 3.14). The vacuolar membranes, which constitute the inner boundaries of the living part of the cell, are similar in structure and function to plasma membranes.

The vacuole evidently received its name because of a belief that it was just an empty space; hence, its name has the same Latin root as the word *vacuum* (from *vacuus*—meaning "empty"). Vacuoles, however, have a variety of functions, including maintenance of cell pressure and pH, as well as storage of numerous cell metabolites and waste products. Inside the vacuole is a watery fluid called **cell sap,** which is slightly to moderately acidic. Cell sap, which helps maintain pressures within the cell (see the discussion of osmosis in Chapter 9), contains dissolved substances, such as salts, sugars, organic acids, and small quantities of soluble proteins. It also frequently contains water-soluble pigments. These pigments, called *anthocyanins,* are responsible for many of the red, blue, or purple colors of flowers and some reddish leaves. In some instances, anthocyanins accumulate to a greater extent in response to cold temperatures in the fall. They should not be confused, however, with the red and orange *carotenoid* pigments confined to the chromoplasts. Yellow carotenoid pigments (*carotenes*) also play a role in fall leaf coloration (discussed in Chapter 7).

Sometimes, large crystals of waste products form within the cell sap after certain ions have become concentrated there. Vacuoles in newly formed cells are usually tiny and numerous. They increase in size and unite as the cell matures.

Figure 3.14 A small portion of a root cap cell of tobacco. ×100,000. V = vacuole; T = vacuolar membrane (tonoplast); G = dictyosome with vesicles (arrows); M = mitochondrion; ER = endoplasmic reticulum; PM = plasma membrane; CW = cell wall. *Courtesy of John Z. Kiss*

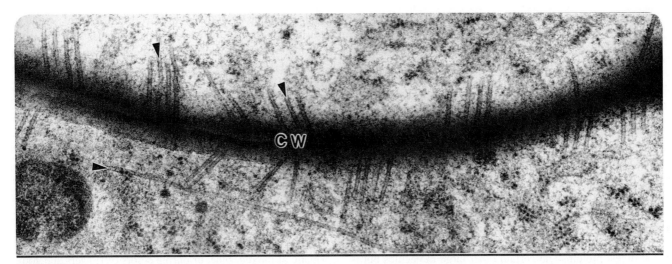

Figure 3.15 A small portion of a plant cell wall (CW) with microtubules more or less perpendicular to it. ×100,000. *Courtesy of John Z. Kiss*

In addition to accumulating the various substances and ions mentioned previously, vacuoles are apparently involved in the recycling of certain materials within the cell and even aid in the breakdown and digestion of organelles, such as plastids and mitochondria.

The Cytoskeleton

The **cytoskeleton** is involved in movement within a cell and in a cell's architecture. It is an intricate network constructed mainly of two kinds of fibers—*microtubules* and *microfilaments*.

Microtubules control the addition of cellulose to the cell wall (Fig. 3.15). They are also involved in cell division, movement of cytoplasmic organelles, control of the movement of vesicles containing cell-wall components assembled by dictyosomes, and movement of the tiny, whiplike *flagella* and *cilia* possessed by some cells (see the section on plant movements in Chapter 11).

Microtubules are unbranched, thin, hollow, tubelike structures that resemble tiny straws. They are composed of proteins called *tubulins* and are of varying lengths, most being between 15 and 25 nanometers in diameter. They are most commonly found just inside the plasma membrane. Microtubules are also found in the special fibers that form the *spindles* and *phragmoplasts* of dividing cells, discussed later in this chapter.

Microfilaments, which play a major role in the contraction and movement of cells in multicellular animals, are present in nearly all cells. They are three or four times thinner than microtubules and consist of long, fine threads of protein with an average diameter of 6 nanometers. They are often in bundles and appear to play a role in the *cytoplasmic streaming* (sometimes referred to as *cyclosis*) that occurs in all living cells. When cytoplasmic streaming is occurring, a microscope reveals the apparent movement of organelles as a current within the cytoplasm carries them around within the walls. This streaming probably facilitates exchanges of materials within the cell and plays a role in the movement of substances from cell to cell. The precise nature and origin of cytoplasmic streaming are still not known, but there is evidence that bundles of microfilaments may be responsible for it. Other evidence suggests that it may be related to the transport of cellular substances by microtubules.

3.6 Cellular Reproduction

Learning Outcome

a. Describe what happens during interphase of the cell cycle as well as the events that take place during the four phases of mitosis.

The Cell Cycle

When cells divide, they go through an orderly series of events known as the **cell cycle.** This cycle is usually divided into **interphase** and **mitosis,** mitosis itself being subdivided into four phases (Fig. 3.16). In a typical onion cell, the duration of the complete cell cycle is about 16 hours. However, it is important to keep in mind that the length of the cycle varies with the kind of organism involved, with the type of cell within an organism, and with temperature and other environmental factors. In most instances, interphase occupies up to 90% or more of the time it takes to complete the cycle. The relatively small amount of time involved in actual division explains why most cells viewed with a microscope are in interphase, and cells in stages of mitosis can be hard to find.

Interphase

Living cells that are not dividing are said to be in *interphase,* a period during which chromosomes are not visible with light microscopes. It is such cells that have been discussed

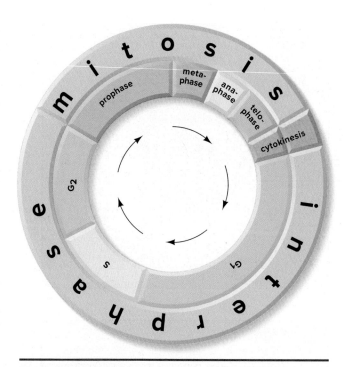

Figure 3.16 A diagram of a cell cycle showing relative amounts of time involved in interphase, mitosis, and cytokinesis.

up to this point. Interphase is the longest phase of the cell cycle and the time in which the physiological function of most cells is undertaken.

For many years, immature cells were considered to be "resting" when they were not actually dividing, but we know now that three consecutive periods of intense activity take place during interphase. These intervals are designated as *gap* (or *growth*) *1, synthesis,* and *gap* (or *growth*) *2* periods, usually referred to as G_1, S, and G_2, respectively.

The G_1 period is relatively lengthy and begins immediately after a nucleus has divided. During this period, the cell increases in size. Also, ribosomes, RNA, and substances that either inhibit or stimulate the S period that follows are produced. During the S period, the unique process of DNA replication (duplication) takes place. Details of this process and of DNA structure are discussed in Chapter 13. In the G_2 period, mitochondria and other organelles divide, and microtubules and other substances directly involved in mitosis are produced. Coiling and condensation of chromosomes also begin during G_2.

Mitosis

All organisms begin life as a single cell. This cell usually divides almost immediately, producing two new cells. These two cells, in turn, divide, with each of them producing two more cells. This process, called *mitosis* (Fig. 3.17), occurs in an organism until it dies. It ensures that the two new cells (*daughter cells*) resulting from a cell undergoing mitosis each have precisely equal amounts of DNA and certain other substances duplicated during interphase.

Strictly speaking, "mitosis" refers to the division of the nucleus alone, but with a few exceptions seen in algae and fungi (discussed in Chapters 18 and 19), the division of the remainder of the cell, called **cytokinesis,** normally accompanies or follows mitosis. Both processes will be considered together here.

In flowering plants, conifers, and other higher plants, mitosis occurs in specific regions, or tissues, called *meristems* (see Chapter 4). Meristems are found in the root and stem tips and in a thin, perforated, and branching cylinder of tissue called the *vascular cambium* (often referred to simply as the *cambium*), located in the interior of some stems and roots a short distance from the surface. In some herbaceous and most woody plants, a second meristem similar in form to the cambium lies between the cambium and the outer bark. This second meristem is called the *cork cambium.* These specific tissues are discussed in Chapters 4, 5, and 6.

When mitosis occurs, the number of chromosomes in the nucleus, whether small or large, makes no difference in the way the process takes place. The daughter cells that result from mitosis each have exactly the same number of chromosomes and distribution of DNA as the parent cell. Mitosis is a continuous process, which may take as little as 5 minutes or as long as several hours from start to finish. Typically, however, it takes from 30 minutes to 2 or 3 hours.

Mitosis is initiated with the appearance of a ringlike *preprophase band* of microtubules just beneath the plasma membrane and is usually divided into four arbitrary phases, primarily for convenience. Descriptions of the phases follow.

Prophase

The main features of **prophase** (Fig. 3.17a) are these: (1) the chromosomes become shorter and thicker, and their two-stranded nature becomes apparent; and (2) the nuclear envelope dissociates, and the nucleolus disintegrates.

Prophase utilizes about as much time as the remaining three phases combined. Before prophase begins, a *preprophase band,* formed from microtubules and microfilaments inside the plasma membrane, develops in a narrow bundle around the nucleus. The beginning of prophase is marked by the appearance of the chromosomes as faint threads in the nucleus. These chromosomes gradually coil or fold into thicker and shorter structures, and soon, two strands, or **chromatids,** can be distinguished for each chromosome. The chromatids are themselves independently coiled and are identical to each other. The coils appear to tighten and condense until the chromosomes have become relatively short, thick, and rodlike, with areas called **centromeres** holding each pair of chromatids together.

The centromere is located at a constriction on the chromosome (Fig. 3.18). A **kinetochore,** which is a dense region composed of a protein complex, is located on the outer surface of each centromere; spindle fibers become attached to the kinetochore. When examined with a light microscope, the centromeres appear to be single structures, but they actually

Figure 3.17 The phases of mitosis as seen in onion root-tip cells. These chromosomes are about 4 micrometers long. (a) Cell (*center*) in prophase. (b) Cell in metaphase. (c) Cell in anaphase. (d) Cell in telophase. *(a–d)* © *Kingsley Stern*

have become double by the G_2 stage of interphase and simply function as a single unit at this point. They may be located almost anywhere on a chromosome but tend to be toward the middle. Sometimes other constrictions appear on individual chromosomes, usually toward one end, giving them the appearance of having extra knobs; these knobs are referred to as *satellites*. The constrictions at the base of the satellites have no known function, but the satellites themselves are useful in helping distinguish certain chromosomes from others in a nucleus.

As prophase progresses, the nucleolus *gradually* becomes less distinct and eventually disintegrates. By the end of prophase, **spindle fibers** consisting of microtubules have developed; these spindle fibers extend in arcs between two invisible *poles* located toward the ends of the cell. The tips of the spindle fibers become anchored at the poles. Other spindle fibers grow from each pole to the center of the cell, where they become attached to a centromere. At the conclusion of prophase, the nuclear envelope has been reabsorbed into the endoplasmic reticulum and has totally fragmented.

In certain simpler organisms, such as fungi and algae, and in animal cells, the cytoplasm just outside the nucleus contains pairs of tiny, keg-shaped structures called *centrioles*. The centrioles are surrounded by microtubules that radiate out from them and arrange cytoplasmic particles in the vicinity into starlike rays, each group of rays collectively called an *aster*. At the beginning of prophase, the aster divides into two parts; one part remains in its original location, while the other part migrates around the nuclear envelope to the opposite side. Centrioles have not been

Figure 3.18 **Diagram of a chromosome at metaphase.** Spindle fibers from opposite ends of the cell become attached at the centromere.

detected in the cells of most of the more complex members of the Plant Kingdom.

Metaphase

The main feature of **metaphase** (see Fig. 3.17b) is the alignment of the chromosomes in a circle midway between the two poles around the circumference of the *spindle* and in the same plane as that previously occupied by the preprophase band. This invisible circular plate is perpendicular to the axis of the spindle and is something like the equator of a globe.

As indicated in our discussion of prophase, spindle fibers can be seen in the area previously occupied by the nucleus after the nuclear envelope has disassociated. They form a structure that looks like an old-fashioned spinning top made of fine threads. Collectively, the spindle fibers are referred to as the **spindle.** The chromosomes become aligned so that

KEY THEME 3.1 cellular Microscapes

Scientists use the scanning electron microscope (SEM) to study the details of many different types of surfaces. Unlike the light microscope or even the transmission electron microscope, which form images by passing either a beam of light or electrons through a thin slice of fixed tissue, the SEM's great advantage is its ability to allow us to look at surfaces of specimens and observe topographical detail not possible with other types of microscopy.

The basic concept of a scanning electron microscope is that a finely focused beam of electrons is scanned across the surface of the specimen. The high-velocity electrons from the beam create an energetic interaction with the surface layers. These electron-specimen interactions generate particles that are emitted from the specimen and can be collected with a detector and sent to a TV screen (cathode ray tube). Particles that form the typical scanning electron image are called *secondary electrons* because they come from the electrons in the specimen itself. The more electrons a particular region emits, the brighter the image will be on the TV screen. The end result, therefore, is brightness associated with surface characteristics and an image that looks very much like a normally illuminated subject. SEM images typically contain a good deal of topographical detail because the electrons that are emitted and produced on the TV screen represent a one-for-one correspondence with the contours of the specimen.

All scanning electron images have one very distinctive characteristic because of this feature of electron emission and display: the images are three-dimensional rather than the flat, two-dimensional images obtained from other types of microscopes. The images can be understood even by the layperson because the eye is accustomed to interpreting objects that are in three dimensions.

Take, for instance, a leaf surface, which looks smooth with an ordinary light microscope. But with a scanning electron microscope, the leaf surface is a rich composition of undulating cell walls, cells joined together like pieces of a jigsaw puzzle,

Box Figures 3.1A and 3.1B (A) A scanning electron micrograph of the surface of a sepal (modified leaf) from a flower of the mouse-ear cress, *Arabidopsis thaliana*. ×2,000. (B) A scanning electron micrograph of the surface of the stigma from a flower of the mouse-ear cress, *Arabidopsis thaliana*. ×200. *(A–B) Courtesy of Daniel Scheirer*

---

their centromeres are in a plane roughly in the center of the cell. This invisible circular plate, called the *equator,* is analogous to the equator of the earth. At the end of metaphase, the centromeres holding the two strands (*sister chromatids*) of each chromosome together separate lengthwise.

Anaphase

Anaphase—the briefest of the phases—involves the sister chromatids of each chromosome separating and moving to opposite poles (see Fig. 3.17c).

Until the end of metaphase, the sister chromatids of each chromosome have been united at their centromeres. Anaphase begins with all the sister chromatids separating in unison and moving toward the poles. The chromatids, which after separation at their centromeres are called **daughter chromosomes,** appear to be pulled toward the poles as their spindle fibers gradually shorten. The shortening occurs as a result of material continuously being removed from the polar ends of the spindle fibers. The centromeres of the daughter chromosomes lead the way, with the chromosomes assuming V shapes as they are pulled through the cytoplasm.

All of the chromosomes separate and move at the same time. Although experiments have shown that a chromosome will not migrate to a pole if the fiber attached to its centromere is severed, other experiments have shown that the chromosomes will separate from one another but not move to the poles, even if no spindle is present. The force or forces involved in this initial separation phenomenon have not yet been identified. It appears, however, that the main movement of the chromosomes to the poles results from a shortening of the spindle fibers.

Telophase

The five main features of **telophase** (see Fig. 3.17d) are these: (1) each group of daughter chromosomes becomes surrounded by a reformed nuclear envelope; (2) daughter chromosomes become longer and thinner and finally squiggly ridges of waxes that look like frosting decorations on a cake, and lens-shaped stomata (Box Figure 3.1A). The stomata even provide a window into the interior of the leaf where deeper cellular layers are visible. Or look at the tentacles seen in Box Figure 3.1B that remind us of some sinister sea creature. No stinging tentacles here, but rather the surface of a small flower of a common weed called mouse-ear cress (*Arabidopsis thaliana*). The "tentacles" are actually stigmatic papillae that trap pollen grains, which are released from the pollen sacs of the flower (see Chapter 8 for details of flower structure).

While biologists utilize the SEM extensively, other types of scientists put it to work in diverse ways as well, whether looking at "moon rocks" brought to earth by the Apollo astronauts or studying the impact craters created by micrometeorite projectiles striking the space shuttle's heat-resistant tiles. In 2011, an astrobiologist with the National Aeronautics and Space Administration (NASA) used SEM to reveal fossilized structures of bacterial remains in a rare meteorite, suggesting evidence of life from someplace other than earth. Some of the bacterial features studied by SEM are similar to those found here on earth, whereas others are not. The SEM images of what may be indications of extraterrestrial organisms are being met with excitement and scrutiny from scientists around the world.

Whether used by biologists or NASA scientists, the scanning electron microscope provides a stunning view of the previously unseen, but nevertheless real, world. As the beauty of nature becomes seen for the first time in startling detail, micrographs do indeed become "microscapes."

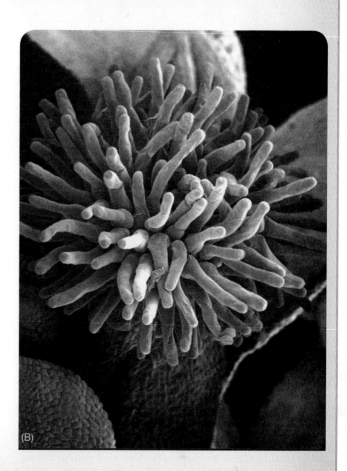
(B)

become indistinguishable; (3) nucleoli reappear; (4) many of the spindle fibers disintegrate; and (5) a *cell plate* forms.

The transition from anaphase to telophase is not distinct, but telophase is definitely in progress when elements of new nuclear envelopes appear around each group of daughter chromosomes at the poles. These elements gradually form intact envelopes as the daughter chromosomes return to the diffuse, indistinct threads seen at the onset of prophase. The new nucleoli appear on specific regions of certain chromosomes.

During telophase, the spindle microtubules gradually break down, and a set of shorter fibers (fibrils), composed of microtubules, develops in the region of the equator between the daughter nuclei. This set of fibrils, which appears somewhat keg-shaped, is called a *phragmoplast*. Dictyosomes produce small vesicles containing raw materials for the cell wall and membranes. Some of these vesicles, which resemble tiny droplets of fluid when viewed with a light microscope, are directed toward the center of the spindle (equator) by the remaining spindle fibers.

The microtubules apparently trap the dictyosome-derived vesicles, which then fuse together into one large, flattened but hollow structure called a **cell plate** (Fig. 3.19). Carbohydrates in the vesicles are synthesized into two new primary cell walls and a *middle lamella*. The middle lamella is shared by what now have become two new **daughter cells.** The cell plate grows outward until it contacts and unites with the plasma membrane of the mother cell. *Plasmodesmata* (minute strands of protoplasm that extend via tiny *desmotubules* through the walls between cells—Fig. 3.20) are formed apparently as portions of the endoplasmic reticulum are trapped between fusing vesicles of the cell plate.

New plasma membranes develop on either side of the cell plate as it forms, and new cell-wall materials are deposited between the middle lamella and the plasma membranes. These new walls are relatively flexible and remain so until the cells increase to their mature size. At that time, additional cellulose and other substances may be added, forming a secondary cell wall interior to the primary wall. In some instances, cell-plate formation does not accompany division of the nucleus.

3.7 Higher Plant Cells versus Animal Cells

Learning Outcomes

a. List three features of cell structure that make plant cells different from animal cells.
b. Explain how differences in plant and animal cells cause them to communicate differently.

All animals have either internal or external skeletons or skeletonlike systems to support their tissues. Animal cells do not have cell walls; instead, the plasma membrane, called the *cell membrane* by most zoologists (animal scientists), forms the outer boundary of animal cells. Higher plant cells have walls that are thickened and rigid to varying degrees, with a framework of cellulose fibrils. Higher plant cells also have plasmodesmata connecting the protoplasts with each other through microscopic holes in the walls. Animal cells lack plasmodesmata, because they have no walls. When higher plant cells divide, a cell plate is formed during the telophase of mitosis, but cell plates do not form in animal cells, which divide by pinching in two.

Other differences are due to the presence or absence of certain organelles. Centrioles, for example, the tiny, paired, keg-shaped structures found just outside the nucleus, occur in all animal cells but are generally absent from higher plant cells. Plastids, common in plant cells, are not found in animal cells. Vacuoles, which are often large in plant cells, are either small or absent in animal cells.

Figure 3.19 How a cell plate is formed. (a) During telophase a phragmoplast (a complex of microtubules and endoplasmic reticulum) develops between the two daughter cell nuclei. The microtubules trap dictyosome vesicles along a central plane. (b) The vesicles fuse into a flattened, hollow structure that becomes a *cell plate*. (c) The cell plate becomes more extensive; two primary cell walls and two plasma membranes form. When the cell plate reaches the mother cell walls, the plasma membranes unite with the existing plasma membrane, and the production of two daughter cells is complete.

Figure 3.20 (a) Transmission electron micrograph of two adjacent cells. ×17,000. (b) and (c) Diagrammatic representations showing the relative locations of the primary cell wall, cell membrane, middle lamella, plasmodesma, and desmotubule. *(a) © Biophoto Associates/Science Source*

SUMMARY

3.1 Introduction to Cells
- All living organisms are composed of cells. Cells are modified according to the functions they perform; some live for a few days, while others live for many years.

3.2 Cells
- The discovery of cells is associated with the development of the microscope. In 1665, Robert Hooke coined the word *cells* for boxlike compartments he saw in cork. Leeuwenhoek and Grew reported frequently during the next 50 years on the existence of cells in a variety of tissues.
- In 1809, Lamarck concluded that all living tissue is composed of cells, and in 1824, Dutrochet reinforced Lamarck's conclusions. In 1833, Brown discovered that all cells contain a nucleus, and shortly thereafter, Schleiden saw a nucleolus within a nucleus. Schleiden and Schwann are credited with developing the cell theory in 1838 and 1839. The theory holds that all living organisms are composed of cells and that cells form a unifying, structural basis of organization.
- In 1858, Virchow contended that every cell comes from a preexisting cell and that there is no spontaneous generation of cells from dust. In 1862, Pasteur experimentally confirmed Virchow's contentions and later proved that fermentation involves the activity of yeast cells. In 1897, Buchner found that yeast cells do not need to be alive for fermentation to occur. This led to the discovery of enzymes.
- Light microscopes can magnify up to 1,500 times. Thinly sliced materials can be viewed with compound microscopes. Opaque objects can be viewed with stereomicroscopes; most magnify up to 30 times.
- Electron microscopes have electromagnetic lenses and a beam of electrons within a vacuum that achieve magnification. Transmission electron microscopes magnify up to 200,000 or more times. Scanning electron microscopes, which can be used with opaque objects, usually magnify up to 10,000 times.
- Scanning tunneling microscopes use a minute probe to scan surfaces at a width as narrow as that of two atoms.

3.3 Eukaryotic versus Prokaryotic Cells
- Eukaryotic cells are the subject of this chapter. Prokaryotic cells, which lack some of the features of eukaryotic cells, are discussed in Chapter 17.

3.4 Cell Structure and Communication
- Cells are minute, varying in diameter between 10 and 100 micrometers. They number into the billions in larger organisms, such as trees. Plant cells are bounded by walls that surround the living protoplast. The cytoplasm contains a souplike fluid, called the cytosol, and all cellular components between the plasma membrane and nucleus.
- A pectic middle lamella is sandwiched between the primary cell walls of adjacent cells. The primary wall and the secondary cell wall, often added inside the primary wall, are composed of cellulose polymers, with hemicelluloses and glycoproteins. Secondary cell walls contain lignin, which strengthens the wall.
- Living cells are in contact with one another via fine strands of cytoplasm called plasmodesmata, which often extend through minute holes in the walls.

3.5 Cellular Components
- A flexible plasma membrane, which is sandwichlike and often forms folds, constitutes the outer boundary of the cytoplasm. It regulates the substances that enter and leave the cell.
- The nucleus is bounded by a nuclear envelope consisting of two membranes that are perforated by numerous pores. Within the nucleus are a fluid called nucleoplasm, one or more spherical nucleoli, and thin strands of chromatin, which condense and become chromosomes when nuclei divide. Each species of organism has a specific number of chromosomes in each cell.
- The endoplasmic reticulum is a system of flattened sacs and tubes associated with the storing and transporting of protein and other cell products. Granular particles called ribosomes, which function in protein synthesis, may line the outer surfaces of the endoplasmic reticulum. Ribosomes also occur independently in the cytoplasm.
- Dictyosomes are structures that appear as stacks of sacs and function as collecting and packaging centers for the cell.
- Plastids are larger green, orange, red, or colorless organelles. Green plastids, known as chloroplasts, contain enzymes that catalyze reactions of photosynthesis. These reactions take place in the membranes of structures that resemble stacks of coins, called thylakoids, as well as in the surrounding matrix, called the stroma. Plastids develop from proplastids, which divide frequently and arise from the division of mature plastids.
- Mitochondria are tiny, numerous organelles that are bounded by two membranes with inner platelike folds called cristae; they are associated with cellular respiration.
- One or more vacuoles may occupy 90% or more of the volume of a mature cell. Vacuoles are bounded by a vacuolar membrane (tonoplast) and contain a watery fluid called cell sap. Cell sap contains dissolved substances and sometimes water-soluble red or blue anthocyanin pigments.

- The cytoskeleton, which is involved in the architecture of cells and internal movement, is composed of microtubules and microfilaments. Microfilaments may be responsible for cytoplasmic streaming.

3.6 Cellular Reproduction

- Cells that are not dividing are in interphase, which is subdivided into three periods of intense activity that precede mitosis, or division of the nucleus. Mitosis is usually accompanied by division of the rest of the cell and takes place in meristems.

- Mitosis is arbitrarily divided into four phases: (1) prophase, in which the chromosomes and their two-stranded nature become apparent and the nuclear envelope breaks down; (2) metaphase, in which the chromosomes become aligned at the equator of the cell; a spindle composed of spindle fibers is fully developed, with some spindle fibers being attached to the chromosomes at their centromeres; (3) anaphase, in which the sister chromatids of each chromosome (now called daughter chromosomes) separate lengthwise, with each group of daughter chromosomes migrating to opposite poles of the cell; and (4) telophase, in which each group of daughter chromosomes becomes surrounded by a nuclear envelope, thus becoming new nuclei, and a wall dividing the daughter nuclei forms, creating two daughter cells.

3.7 Higher Plant Cells versus Animal Cells

- Animal cells differ from those of higher plants in not having a wall, plastids, or large vacuoles. Also, they have keg-shaped centrioles in pairs just outside the nucleus, and they pinch in two instead of forming a cell plate when they divide.

REVIEW QUESTIONS

1. What cellular structures can be observed with the aid of light microscopy and electron microscopy?
2. Why are cells so small, and how is this small size beneficial for transport of substances within and between cells?
3. What is the function of the plant cell wall?
4. What is the difference between protoplasm and cytoplasm?
5. What is the function of a cell nucleus? How does it perform its function?
6. What are plasmodesmata? What is their importance to living plant cells?
7. Describe the major parts and functions of a chloroplast.
8. In a typical complete cell cycle, how long, proportionately, does mitosis take?
9. What cellular structures are responsible for division of cytoplasm, and how does this occur?
10. What are the differences and similarities between plant and animal cells?

DISCUSSION QUESTIONS

1. Would you consider any one type of cell more useful than another? Why?
2. After you have completed your introductory plant science course, do you believe you will be able to determine the function of each of a cell's organelles in a laboratory? Explain.

CHAPTER 4

Tissues

4.1 Introduction to Tissues
4.2 Meristematic Tissues
   Apical Meristems
   Lateral Meristems
   Key Theme 4.1: Chimeras and Variegated Leaves
   Intercalary Meristems
4.3 Tissues Produced by Meristems
   Simple Tissues
   Complex Tissues
   Key Theme 4.2: The THC in Marijuana

OVERVIEW

A discussion of meristems (apical meristems, vascular cambium, cork cambium, intercalary meristems) and non-meristematic tissues (parenchyma, collenchyma, sclerenchyma, secretory tissues, xylem, phloem, epidermis, periderm) forms the body of this chapter.

Macerated section showing tracheary elements from white oak (*Quercus alba*).
© Steven P. Lynch

4.1 Introduction to Tissues

Learning Outcome

a. State the major organs of plants.

Imagine the construction of a new home on a vacant lot. In the beginning, a foundation must be laid and building materials are obtained to build the house. A frame is built according to instructions in the blueprint, and the house begins to take shape. The completion of the frame is followed by the installation of plumbing, electrical wiring, windows, heating and air-conditioning units, vents, and other components. Waterproof walls and a roof are added, and, upon occupation, food and other materials are stored in their appropriate niches.

In a sense, the growth of a plant from a seed is something like the construction of a house. Using raw materials from the soil and a superb manufacturing process, each plant develops a framework, "plumbing," a waterproof covering that includes "windows," vents, means of waste disposal, and food-storage areas. Even a form of air-conditioning, which enables plants to survive and thrive in the hottest summer sun, is included in each mature plant package. The building components of the framework, plumbing, and related features of plants form the body of this chapter.

There are many interesting modifications of higher plants, discussed in the three chapters that follow this one, but, regardless of the outer form, most plants have three or four major groups of organs: **roots, stems, leaves,** and often, **flowers.** Each of these organs is composed of tissues, which are defined as "groups of cells performing a similar function." Any plant organ may be composed of several different tissues; each tissue is classified according to its structure, origin, or function.

Three basic tissue patterns occur in roots and stems (see *woody dicots, herbaceous dicots,* and *monocots,* discussed in Chapters 5 and 6). The next section describes the major kinds of tissues found in higher plants. Then, the major features of the cells within those tissues are outlined.

4.2 Meristematic Tissues

Learning Outcome

a. List and describe plant meristems and where they are found.

Unlike animals, plants have permanent regions of growth called **meristems,** or *meristematic tissues,* where cells actively divide (Fig. 4.1). As new cells are produced, they typically are small, six-sided, boxlike structures, each with a proportionately large nucleus, usually near the center, and with tiny vacuoles or no vacuoles at all. As the cells mature, however, they assume many different shapes and sizes, each related to the cell's ultimate function; the vacuoles increase in size, often occupying more than 90% of the volume of a mature cell.

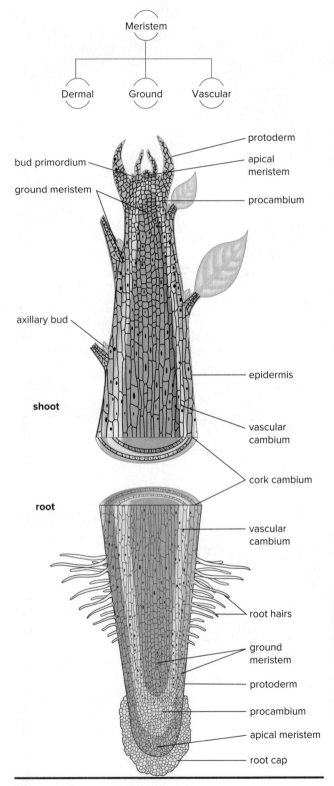

Figure 4.1 Flow chart showing how meristems give rise to tissues, along with a diagram of the longitudinal axis of a plant, depicting the location of meristems.

Apical Meristems

Apical meristems are meristematic tissues found at, or near, the tips of roots and shoots, which increase in length as the apical meristems produce new cells. This type of growth is

KEY THEME 4.1 molecular

Chimeras and Variegated Leaves

In Greek mythology, a chimera is a fire-breathing monster, with a lion's head, a goat's body, and a serpent's tail. The plant world contains real chimeras. Although they are not as sensational as those in mythology, they can be just as amazing. A plant chimera contains cells of more than one genotype in adjacent tissues. The most common example of a chimera is a plant with variegated leaves (Box Figure 4.1A). This chimera arose when a cell near the apical meristem underwent a mutation in a gene essential for chlorophyll production. All leaf cells produced by mitosis of that mutant cell in the apical meristem will be the mutated type, unable to produce chlorophyll. If all cells in that plant were incapable of synthesizing chlorophyll, then the plant would not be able to grow. However, other cells in the apical meristem did not undergo the mutation, so they are capable of producing sugars by photosynthesis. The leaves of this chimeral plant have patches of white and green, reflecting genetic differences in their progenitor cells in the apical meristem. Another example of a chimera is thornless blackberries. In this case the epidermal layer of cells has been mutated, so it produces no "thorns" (which are technically called "prickles"). Only the outer layer of these plants contains the mutation. All cells inside the epidermis contain the genetic information for the thorny genotype. If root cuttings are made from these plants, the adventitious shoots that develop on the root cuttings are not chimeral, so they are thorny.

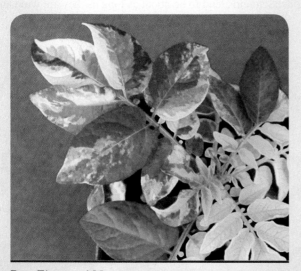

Box Figure 4.1A Variegated leaves in a chimeral potato.
© Shelley H. Jansky

known as *primary growth*. Three *primary meristems*, as well as embryo leaves and buds, develop from apical meristems. These primary meristems are called **protoderm, ground meristem,** and **procambium.** The tissues they produce are called **primary tissues.** Note their locations in Figure 4.1; they are discussed in Chapters 5 and 6.

Lateral Meristems

The *vascular cambium* and *cork cambium,* discussed next, are **lateral meristems,** which produce tissues that increase the girth of roots and stems. Such growth is termed *secondary growth.*

Vascular Cambium

The **vascular cambium,** often referred to simply as the **cambium,** produces *secondary tissues* that function primarily in support and conduction. The cambium, which extends throughout the length of roots and stems in perennial and many annual plants, is a thin cylinder of mostly brick-shaped cells. The cambial cylinder often branches, except at the tips, and the tissues it produces are responsible for most of the increase in a plant's girth as it grows. The individual remaining cells of the cambium are referred to as *initials,* while their sister cells are called *derivatives.* The cambium and its cells and tissues are discussed in Chapters 5 and 6.

Cork Cambium

The **cork cambium,** like the vascular cambium, is a thin cylinder that runs the length of roots and stems of woody plants. It lies outside of the vascular cambium, just inside the outer bark, which it produces. The cork cambium is discussed in Chapters 5 and 6. The tissues laid down by the vascular cambium and the cork cambium are called *secondary tissues,* because they are produced *after* the primary tissues have matured.

Intercalary Meristems

Grasses and related plants have neither a vascular cambium nor a cork cambium. They do, however, have apical meristems, and near **nodes** (leaf attachment areas), they have other meristematic tissues called *intercalary meristems.* The intercalary meristems develop at intervals along stems, where, like the tissues produced by apical meristems, their tissues add to stem length.

4.3 Tissues Produced by Meristems

Learning Outcomes

a. State the difference between simple tissues and complex tissues, and provide examples of each.

b. Describe three types of simple tissues and where they may be found in the plant.

c. Explain structural and functional differences between xylem and phloem.
d. Describe the structure and function of the epidermis and periderm, as well as secretory tissue.

After cells are produced by meristems, the cells assume various shapes and sizes related to their functions as they develop and mature. Some tissues consist of only one kind of cell and are called simple tissues. Complex tissues are composed of two or more cell types.

Simple Tissues

Parenchyma

Parenchyma tissue is composed of *parenchyma cells* (Fig. 4.2), which are the most abundant of the cell types and are found in almost all major parts of higher plants. They are generally spherical in shape when they are first produced, but when they grow and they push against one another, their thin, pliable walls are flattened at the points of contact. As a result, parenchyma cells may be found in various shapes and sizes, with the majority having 14 sides. They tend to have large vacuoles and may contain starch grains, oils, tannins, crystals, and other secretions.

More often than not, parenchyma cells have spaces between them; in fact, in water lilies and other aquatic plants, the intercellular spaces are quite extensive and form a network throughout the entire plant. This type of parenchyma tissue with extensive connected air spaces is referred to as *aerenchyma*.

Parenchyma cells containing numerous chloroplasts (as found in leaves) are collectively referred to as **chlorenchyma** tissue. Chlorenchyma tissue functions mainly in photosynthesis, while parenchyma tissues without chloroplasts function mostly in food or water storage. For example, the soft, edible parts of most fruits and vegetables consist largely of parenchyma.

Some parenchyma cells develop irregular extensions of the inner wall that greatly increase the surface area of the plasma membrane. Such cells, called *transfer cells,* are found in nectaries of flowers and in carnivorous plants, where they apparently play a role in transferring dissolved substances between adjacent cells. Many parenchyma cells live a long time; in some cacti, for example, they may live to be over 100 years old.

Mature parenchyma cells can divide long after they were produced by a meristem. In fact, when a *cutting* (segment of stem) is induced to grow, it is parenchyma cells that start dividing and give rise to new roots. When a plant is damaged or wounded, the capacity of parenchyma cells to multiply is especially important in tissue repair.

Collenchyma

Collenchyma cells (Fig. 4.3), like parenchyma cells, have living cytoplasm and may remain alive a long time. Their walls are generally thicker and more uneven in thickness than those of parenchyma cells. This unevenness is generally due to extra primary wall in the corners. Collenchyma cells often occur just beneath the epidermis; they are typically longer than they are wide, and their walls are pliable as well as strong. They provide flexible support for both growing organs and mature organs, such as leaves and floral parts. The "strings" of celery that get stuck in our teeth, for example, are composed of collenchyma cells.

Sclerenchyma

Sclerenchyma tissue consists of cells that have thick, tough, secondary walls, normally impregnated with **lignin.** Most sclerenchyma cells are dead at maturity and function in support. Two forms of sclerenchyma occur: **sclereids** and **fibers.** Sclereids (Fig. 4.4) may also be randomly distributed in other tissues. For example, the slightly gritty texture of pears is due to the presence of groups of sclereids called *stone cells.* Nut shells and peach pits are composed of sclereids. Sclereids tend to be about as long as they are wide and

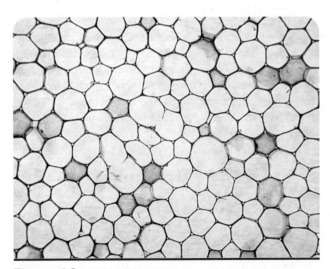

Figure 4.2 Parenchyma cells. They are more or less spherical when first formed, but as their walls touch other parenchyma cell walls, the cells end up with an average of 14 sides at maturity. ×100.
© James E. Bidlack

Figure 4.3 Collenchyma cells. Note the walls are unevenly thickened at the corners. ×100. © James E. Bidlack

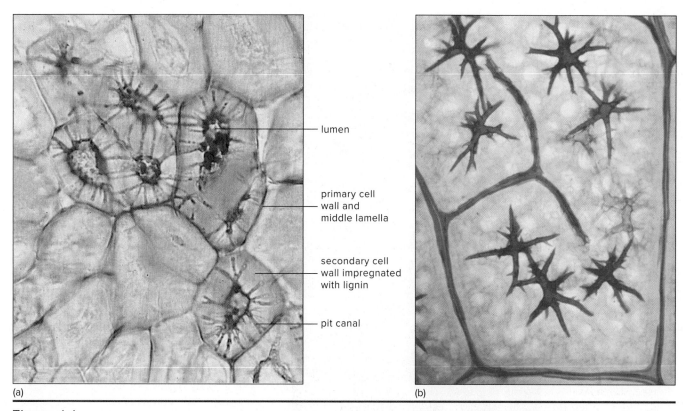

Figure 4.4 (a) Sclereids (stone cells) of a pear in cross section. ×1,000. (b) Sclereids in the leaf of a wheel tree (*Trochodendron aralioides*). ×1,000. *(a) © Kingsley Stern; (b) © G. S. Ellmore*

sometimes occur in specific zones rather than being scattered within other tissues.

Fibers (Fig. 4.5) may be found in a number of tissues in roots, stems, leaves, and fruits. They are usually much longer than they are wide and have a proportionately tiny cavity, or *lumen,* in the center of the cell. At present, fibers from more than 40 different families of plants are used commercially to manufacture textile goods, ropes, string, canvas, and similar products. Archaeological evidence indicates that humans have been using plant fibers for at least 10,000 years.

Complex Tissues

The simple tissues we have discussed thus far consist of one kind of cell, but a few important tissues are always composed of two or more kinds of cells and are referred to as *complex tissues*. Two of the most important complex tissues in plants, *xylem* and *phloem,* transport water, ions, and soluble food (sugars) throughout the plant. Some complex tissues are produced by apical meristems, but in woody plants most are produced by the vascular cambium and are often referred to as *vascular tissues*.

The *epidermis,* which forms a protective layer covering all plant organs, consists primarily of parenchyma or parenchyma-like cells, but it also often includes specialized cells involved in the movement of water and gases in and out of plants, secretory glands, hairs, cells in which crystals are isolated, and others that greatly increase absorptive parts of roots. Therefore, the epidermis and tissues with secretory cells are discussed in this section.

Periderm, which comprises the outer bark of woody plants, consists mostly of cork cells, but it is considered a complex tissue because it contains pockets of parenchyma-like cells.

Xylem

Xylem tissue is an important component of the "plumbing" and storage systems of a plant and is the chief conducting tissue in all organs for water and minerals absorbed by the roots. Xylem consists of a combination of parenchyma cells, fibers, *vessels, tracheids,* and *ray cells* (Fig. 4.6). **Vessel elements** have thick secondary cell walls and are open at their ends. They are stacked end-on-end to form long tubes, called **vessels.** A **perforation plate** with bars or holes may be found between the end walls of each pair of vessel elements. Perforation plates with bars can be seen in the first and third vessel elements illustrated in Figure 4.6b. The flow of fluid through the vessels is not blocked by perforation plates. However, perforation plates may block the movement of objects, such as fungal spores, that may invade the xylem.

Tracheids, which, like vessel elements, are dead at maturity and have relatively thick secondary cell walls, are tapered at each end, with the ends overlapping those of other tracheids. Tracheids do not have openings like those in vessels, but there are usually pairs of lateral *pits* wherever two

Figure 4.5 Fibers. (a) A cross section of a portion of stem tissue from a linden tree (*Tilia* sp.). ×1,000. Note the thickness of the walls of the darker fibers. (b) A longitudinal section through fibers in a *Welwitschia* leaf. ×1,000. *(a)* © Kingsley Stern; *(b)* © G. S. Ellmore

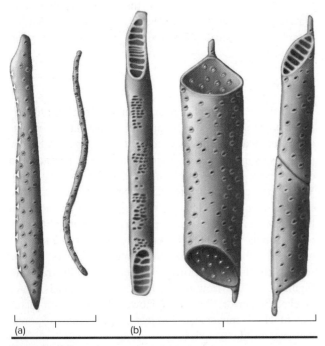

Figure 4.6 (a) Tracheids. ×200. (b) Vessel elements. ×200.

tracheids are in contact with one another (Fig. 4.7). Pits are areas with no secondary cell wall and, as discussed in Chapter 3, they allow water to pass from cell to cell. In some plants, bordered pits may be present and resemble doughnuts in surface view. These bordered pits are composed of a pit membrane and a thickened region called the **torus.** Figure 4.8 illustrates how, in some plants, pit pairs regulate the movement of water between adjacent cells.

In cone-bearing trees and certain other non-flowering plants, the xylem is composed almost entirely of tracheids. The walls of many tracheids, as well as vessel elements, have spiral thickenings that are easily seen with the light microscope (Fig. 4.9). Most conduction through xylem is upward, but some is lateral (sideways) through **rays.** Ray cells, which also function in food storage, are actually long-lived parenchyma cells that are produced in horizontal rows by special *ray initials* of the vascular cambium. In woody plants, the rays radiate out from the center of stems and roots like the spokes of a wheel (see Chapter 6).

Phloem

Phloem tissue (Fig. 4.10), which conducts dissolved food materials (primarily sugars) produced by photosynthesis throughout the plant, is composed mostly of two types of cells without secondary walls. The relatively large cylindrical **sieve tube members** (sometimes referred to as sieve tube elements) have narrower, more tapered **companion cells** closely associated with them. Phloem is derived from

Tissues 57

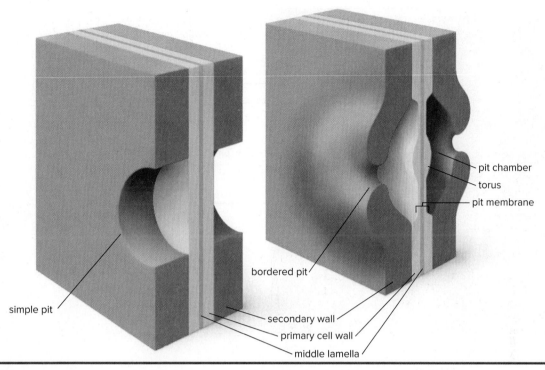

Figure 4.7 Pits are depressions in cell walls where the secondary wall does not form. There may be from one or two to several thousand in a cell. They often occur in pairs, with one on each side of the middle lamella. Some, called *bordered pits* (right), bulge out from the wall and resemble doughnuts in surface view, while others, called *simple pits* (left), do not bulge.

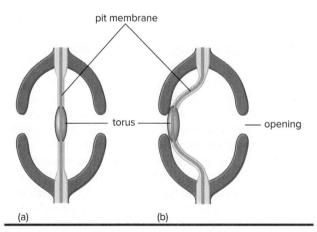

Figure 4.8 How water flow is controlled in adjacent pairs of bordered pits. The pits are separated by a *pit membrane* consisting of the middle lamella and two thin layers of primary walls. (a) Water moves relatively freely through the pit openings and pit membrane when the torus (a thickened region of the pit membrane) is in the center. (b) If the flexible pit membrane swings to one side so that the torus blocks an opening, water movement through the pit pair is restricted.

Figure 4.9 Spiral thickenings on the inside walls of vessel elements. ×400. © *James E. Bidlack*

the parent cells of the cambium, which also produce xylem cells; it often also includes fibers, parenchyma, and ray cells. Sieve tube members, like vessel elements, are laid end to end, forming **sieve tubes.** Unlike vessel elements, however, the end walls have no large openings; instead, these walls are full of small pores and are called **sieve plates.** Cytoplasm can flow between sieve tubes through sieve plates.

Sieve tube members have no nuclei at maturity, even though their cytoplasm is very active in transporting food throughout the plant. The adjacent companion cells contain nuclei, which help to direct the activities of sieve tube members.

Living sieve tube members contain a polymer called *callose* that stays in solution as long as the cell contents are under pressure. If a cell is damaged, the pressure drops,

Figure 4.10 Longitudinal view of part of the phloem of a black locust tree *(Robinia pseudo-acacia)*. ×1,000. © Kingsley Stern

and the callose precipitates. The callose and a phloem protein are then carried to the nearest sieve plate, where they form a *callus* plug that prevents leaking of the sieve tube contents.

Sieve cells, which are found in ferns and cone-bearing trees, are similar to sieve tube members but tend to overlap at their ends rather than form continuous tubes. Like sieve tube members, they have no nuclei at maturity, but they have no adjacent companion cells. Instead, they have adjacent *albuminous cells,* which are equivalent to companion cells and apparently function in the same manner.

Epidermis

The outermost layer of cells of all young plant organs is called the **epidermis.** Because it is in direct contact with the environment, it is subject to modification by the environment and often includes several different kinds of cells. The epidermis is usually one cell thick, but a few plants produce aerial roots called **velamen roots** (e.g., orchids) in which the epidermis may be several cells thick, with the outer cells functioning like a sponge. Such a multiple-layered epidermis also occurs in the leaves of some tropical figs and members of the Pepper Family (Piperaceae), where it protects the plant from desiccation.

Most epidermal cells secrete a fatty substance called **cutin** within and on the surface of the outer walls.

Cutin forms a protective layer called the **cuticle** (Fig. 4.11). The thickness of the cuticle (or, more importantly, wax secreted on top of the cuticle by the epidermis) to a large extent determines how much water is lost through the cell walls by evaporation. The cuticle is also exceptionally resistant to bacteria and other disease organisms and has been recovered from fossil plants millions of years old. The waxes deposited on the cuticle in a number of plants (see Chapter 7) can reach the surface by diffusion, migrate between cells, or travel through microscopic channels in the cell walls. The susceptibility of plants to herbicides may depend on the thickness of these wax layers. Some wax deposits are extensive enough to have commercial value. Carnauba wax, for example, is deposited on the leaves of the wax palm. It and other waxes are harvested for use in polishes.

In leaves, the epidermal cell walls perpendicular to the surface often assume bizarre shapes that, under the microscope, look like pieces of a jigsaw puzzle. Epidermal cells of roots produce tubular extensions called *root hairs* (see Chapter 5) a short distance behind the growing tips. The root hairs greatly increase the absorptive area of the surface.

A different type of hair occurs on the epidermis of aboveground parts of plants. These hairs, called **trichomes,** are outgrowths consisting of one to several cells (Fig. 4.12). Leaves also have numerous small pores, the **stomata,** bordered by pairs of specialized epidermal

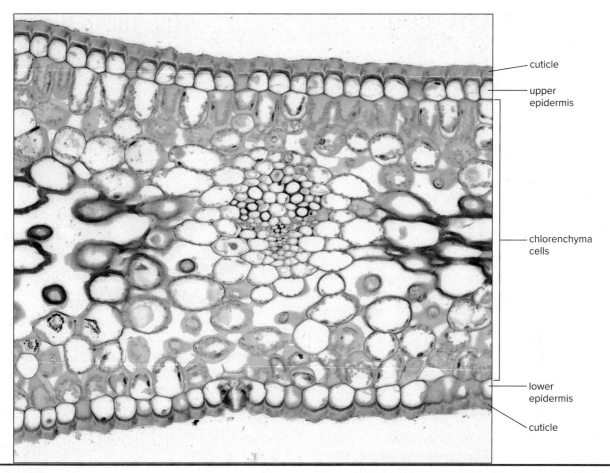

Figure 4.11 A portion of a cross section of a kaffir lily *(Clivia)* leaf, showing the thick cuticle secreted by the epidermis. ×1,000.
© Kingsley Stern

Figure 4.12 Hairs on the surface of an ornamental mint plant. ×50.
© Kingsley Stern

cells called **guard cells** (see Chapters 7 and 9). Guard cells differ in shape from other epidermal cells; they also differ in that they contain chloroplasts. The stomatal apparatus is discussed in Chapters 7 and 9. Some epidermal cells may be modified as **glands** that secrete protective or other substances, or modified as hairs that either reduce water loss or repel insects and animals that might otherwise consume them (Fig. 4.13).

Periderm

In woody plants, the epidermis is sloughed off and replaced by a **periderm** after the cork cambium begins producing new tissues that increase the girth of the stem or root. The periderm constitutes the outer bark and is primarily composed of rectangular and boxlike **cork** cells, which are dead at maturity (Fig. 4.14). Before the cork cells die, they secrete a fatty substance, **suberin,** into their walls. This makes cork cells waterproof and helps them protect the phloem and other tissues beneath the bark from desiccation, mechanical injury, and freezing temperatures. Some cork tissues, such as those produced by the cork oak, are harvested commercially and are used for bottle corks and to manufacture linoleum and gaskets.

KEY THEME 4.2 ecology

The THC in Marijuana

Many of our modern-day medicines are based upon, or derived directly from, plant roots, stems, leaves, and flowers. One such medicine that has attracted much attention in the media, and has been an issue of political debate, is tetrahydrocannabinol (THC), which is primarily derived from the leaves (Box Figure 4.2A) and buds of marijuana (*Cannabis sativa*). The THC is most heavily concentrated in the trichomes, found on the surface of plant tissues, presumably produced as a deterrent to discourage foraging animals from preying upon the leaves and buds of the plant. However, humans have discovered the relaxing effect that can be induced by ingesting or smoking marijuana. Therefore it has been, and continues to be, a very popular prescription and/or recreational drug that is illegal in many states. Knowledge of which plant tissues provide the heaviest concentrations of THC, and how THC can vary with age and environmental conditions, has become a very important aspect of cultivating marijuana, even as the political debate on its use continues.

Box Figure 4.2A Leaf of a marijuana plant. © Emilio100/Shutterstock

In parts of the cork cambium, pockets of loosely arranged parenchyma cells are not impregnated with suberin. These pockets of tissue, called **lenticels** (Fig. 4.14), provide openings in the periderm, allowing gas exchange between the air and the interior of the stem. Periderm is discussed in more detail in Chapter 6.

Secretory Cells and Tissues

All cells secrete substances that can damage the cytoplasm if allowed to accumulate internally. Such materials must be either isolated from the cytoplasm of the cells in which they originate or moved outside the plant body. Often, they consist of waste products that are of no further use to the plant,

(a)

(b)

Figure 4.13 (a) Colored scanning electron micrograph (SEM) of a lavender (*Lavandula* sp.) leaf, showing hair-like trichomes and oil glands that cover its surface. When the leaf is touched, the glands rupture and release the oil that gives lavender its unique fragrance. ca. ×200. (b) Colored SEM of the stem of a tomato plant (*Lycopersicon lycopersicum*), showing glandular trichomes that secrete oils that are irritating to predatory insects. ca. ×150. *(a) Thierry Berrod, Mona Lisa Production/Science Source; (b) © Andrew Syred/Science Source*

Figure 4.14 (a) Lenticels on the bark of a Himalayan cherry tree (*Prunus rufa*). (b) Cross section showing periderm and a lenticel of elderberry (*Sambuscus*). ×100. *(a) © JohnatAPW/Getty Images; (b) © James E. Bidlack*

but some substances, such as nectar, perfumes, and plant hormones (discussed in Chapter 11), are vital to normal plant functions.

Secretory cells may function individually or as part of a **secretory tissue.** Secretory cells or tissues, which are often derived from parenchyma, can occur in a wide variety of places in a plant. Among the most common secretory tissues are those that secrete nectar in flowers; oils in citrus, mint, and many other leaves; mucilage in the glandular hairs of sundews and other insect-trapping plants; latex in members of several plant families, such as the Spurge Family; and resins in coniferous plants, such as pine trees. Latex and resins are usually secreted by cells lining tubelike ducts that form networks throughout certain plant species (see Chapter 6). Some plant secretions, such as pine resin, rubber, mint oil, and opium, have considerable commercial value.

SUMMARY

4.1 Introduction to Tissues
- A group of cells performing a common function is called a tissue.

4.2 Meristematic Tissues
- Apical meristems are found in the vicinity of the tips of roots and stems; the vascular cambium and the cork cambium occur as lengthwise cylinders within roots and stems; intercalary meristems occur in the vicinity of nodes of grasses and related plants.

4.3 Tissues Produced by Meristems
- Tissues produced by meristems consist of one to several kinds of cells. They include parenchyma, collenchyma, sclerenchyma, epidermis, xylem, phloem, periderm, and secretory tissues.
- Parenchyma cells are thin-walled, while collenchyma cells have unevenly thickened walls that provide flexible support for various plant organs.
- Two types of sclerenchyma occur: fibers (which are long and tapering) and sclereids (which are short in length); both types have thick walls and are usually dead at maturity.
- Complex tissues have more than one kind of cell. The principal types are xylem, phloem, epidermis, and periderm.
- Xylem conducts water and minerals throughout the plant. It consists of a combination of parenchyma, fibers, vessels (tubular channels), tracheids (cells with tapering end walls that overlap), and ray cells (involved in lateral conduction).
- Phloem conducts primarily dissolved sugars throughout the plant. It is composed of sieve tubes (made up of cells called sieve tube members), companion cells (that apparently regulate adjacent sieve tube members), parenchyma, ray cells, and fibers. Callose aids in plugging injured sieve tubes. Sieve cells, which have overlapping end walls, and adjacent albuminous cells take the place of sieve tube members and companion cells in ferns and cone-bearing trees.

- The epidermis is usually one cell thick, with fatty cutin (forming the cuticle) within and on the surfaces of the outer walls. The epidermis may include guard cells that border pores called stomata; root hairs, which are tubular extensions of single cells; other hairs that consist of one to several cells; and glands that secrete protective substances.
- Periderm, which consists of cork cells and loosely arranged groups of cells comprising lenticels involved in gas exchange, constitutes the outer bark of woody plants.
- Secretory tissues occur in various places in plants; they secrete substances such as nectar, oils, mucilage, latex, and resins.

REVIEW QUESTIONS

1. What is the function of meristems? Where are they located?
2. How are parenchyma, collenchyma, and sclerenchyma distinguished from one another?
3. Distinguish between epidermis and periderm.
4. What are the functions of xylem and phloem? What cells are involved in their normal activities?
5. What types of substances do secretory cells secrete?

DISCUSSION QUESTIONS

1. Most plant meristems are located at the tips of shoots and roots and in cylindrical layers within stems and roots. What could happen if they were present in leaves?
2. The cambium produces xylem toward the center of a tree and phloem toward the outside. Do you think it would make any difference if the positions of the xylem and phloem were reversed? Why?

CHAPTER 5

Roots and Soils

5.1 Introduction to Roots and Soils
5.2 How Roots Develop
5.3 Root Structure
    The Root Cap
    The Region of Cell Division
    The Region of Elongation
    The Region of Maturation
5.4 Specialized Roots
    Food-Storage Roots
    Water-Storage Roots
    Propagative Roots
    Pneumatophores
    Aerial Roots
    Contractile Roots
    Buttress Roots
    Parasitic Roots
    Key Theme 5.1: Plants Need Roots
5.5 Mycorrhizae
5.6 Root Nodules
5.7 Human Relevance of Roots
5.8 Soils
    Parent Material
    Climate
    Living Organisms and Organic Composition
    Topography
    Soil Texture and Mineral Composition
    Soil Structure
    Key Theme 5.2: Metal-Munching Plants
    Water in the Soil
    Soil pH

OVERVIEW

This chapter discusses roots, beginning with their functions and continuing with the development of roots from a seed. It covers the function and structure of the root cap, region of cell division, region of elongation, and region of maturation (with its tissues). The endodermis and pericycle are also discussed.

Specialized roots (food-storage roots, water-storage roots, propagative roots, pneumatophores, aerial roots, contractile roots, buttress roots, parasitic roots) and mycorrhizae are discussed. This is followed by some observations on the economic importance of roots. After a brief examination of soil horizons, the chapter concludes with a discussion of the development of soil texture, composition, and structure; and water.

Aerial roots of a banyan tree (*Ficus* sp.) near the door of Ta Prohm Temple in Cambodia. © Glowimages

5.1 Introduction to Roots and Soils

Learning Outcome

a. List three major functions of roots.

You have likely seen pictures of the destruction caused by a tornado as it passed through a city, but have you seen what a twister can do to a forest? Large trees may be snapped off above the ground or knocked down, and branches may be stripped bare of leaves. Unless the soil in the area happens to be thin, sandy, or loose, however, you will probably see relatively few trees uprooted and carried away. In the tropics, it is rare to find healthy palm trees uprooted even after a hurricane.

Roots *anchor* trees firmly in the soil, usually through an extensive branching network that constitutes about one-third of the total dry weight of the plant. The roots of most plants do not usually extend more than 3 to 5 meters (10 to 16 feet) down into the earth; those of many herbaceous species are confined to the upper 0.6 to 0.9 meter (2 to 3 feet). The roots of a few plants, such as alfalfa, however, often grow more than 6 meters (20 feet) into the earth. When the Suez Canal was being built, workers encountered roots of tamarisk at depths of nearly 30 meters (100 feet), and mesquite roots have been seen 53.4 meters (175 feet) deep in a pit mine in the southwestern United States. Some plants, such as cacti, form very shallow root systems, but these systems still effectively anchor the plants with a densely branching mass of roots radiating out in all directions as far as 15 meters (50 feet) from the stem.

Besides anchoring plants, roots *absorb water and minerals in solution*, mostly through "feeder" roots found in the upper meter (3.3 feet) of soil. Some plants have roots that, as well as anchoring and absorbing, store water or food, or perform other specialized functions.

Some aquatic plants (e.g., duckweeds and water hyacinths) produce roots in water, and many *epiphytes* (non-parasitic plants that grow suspended without direct contact with the ground—e.g., orchids) produce aerial roots. The great majority of vascular plants, however, develop their root systems in soils. The soils, which vary considerably in composition, texture, and other characteristics, are discussed toward the end of this chapter.

5.2 How Roots Develop

Learning Outcome

a. Describe how dicot root systems differ from those of monocots.

The **embryo** (immature plant) within a germinating seed contains a tiny, rootlike **radicle.** As the seedling grows, the radicle may develop into a long *taproot,* from which thinner *branch roots* arise. Alternatively, many *adventitious roots* develop from the stem above the radicle. Adventitious roots are those that do not develop from another root but develop instead from a stem or leaf. A *fibrous root system,* which may have large numbers of fine roots of similar diameter, then develops from the adventitious roots (Fig. 5.1). Many mature plants have a combination of taproot and fibrous root systems.

One plant can produce an exceptionally large number of roots. For example, a single, mature ryegrass plant may have as many as 15 million individual roots and branch roots,

(a)   (b)

Figure 5.1 Root systems. (a) A fibrous root system of a grass. (b) A taproot system of a California poppy.

with a combined length of 644 kilometers (400 miles) and a total surface area larger than a volleyball court, all contained within 0.57 cubic meter (20 cubic feet) of soil. *Root hairs* (discussed in the section "The Region of Maturation") greatly increase the total surface area of the root.

Many plants, such as peas and carrots, whose seeds have two "seed leaves"—commonly referred to as *dicots*—have taproot systems with one, or occasionally more, *primary roots* from which *secondary roots* develop (see the discussion of primary and secondary tissues in Chapter 4). *Monocotyledonous plants* (e.g., corn and rice, whose seeds have one "seed leaf"—commonly referred to as monocots), on the other hand, have fibrous root systems. Adventitious and other types of roots may develop in both dicots and monocots. In English and other ivies, lateral adventitious roots that aid in climbing are found on the aerial stems, and in certain plants with specialized stems (e.g., rhizomes, corms, and bulbs; see Chapter 6), only adventitious roots are produced.

5.3 Root Structure

Learning Outcomes

a. Describe root regions, including the root cap, region of cell division, region of cell elongation, and region of maturation (including root hairs and all tissues) and list the function of each.

b. Describe the location and function of the endodermis and the pericycle.

Young roots can be divided into four regions, or zones. These regions are called (1) the *root cap,* (2) the *region of cell division,* (3) the *region of elongation,* and (4) the *region of maturation* (Fig. 5.2). The root cap is a distinct structure, but the other three regions do not have sharply defined boundaries. Each region gradually merges into the next.

The Root Cap

The **root cap** is composed of a cluster of parenchyma cells covering the tip of the root. It is quite large and obvious in some plants, while in others, it is nearly invisible. One of its functions is to *protect* the delicate tissues behind it as the young root tip pushes through abrasive soil particles. The root cap has no equivalent in stems. The dictyosomes of the root cap's outer cells secrete and release a slimy substance that lodges in the walls and eventually passes to the outside. The cells, which are replaced from the inside, constantly slough off, forming a slimy lubricant that facilitates the root tip's movement through the soil. This mucilaginous lubricant also supports the growth of beneficial bacteria that add to the nitrogen available to the plant (see the discussion of the nitrogen cycle in Chapter 25).

Root cap cells have an average life of less than a week. If they are removed, a new root cap will be produced. Until the new root cap forms, however, the root seems to grow randomly instead of downward, suggesting that the root cap also functions in the *perception of gravity* (see *gravitropism* in Chapter 11). **Amyloplasts** (plastids

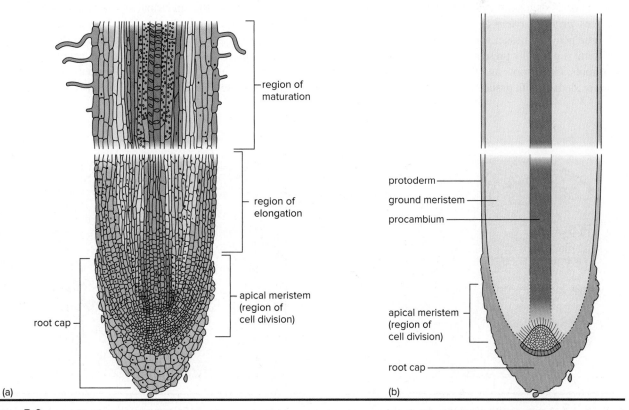

Figure 5.2 A longitudinal section through a dicot root tip. (a) Regions of the root. (b) Locations of the primary meristems of the root.

containing starch grains) act as gravity sensors, collecting on the bottom of root-cap cells. When a root that has been growing vertically is tipped horizontally, the amyloplasts tumble or float down to the new bottom of the cells. The root begins growing downward again within 30 minutes to a few hours. The exact nature of this gravitational response is not known, but there is some evidence that calcium ions in the amyloplasts influence the distribution of growth hormones in the cells.

The Region of Cell Division

Cells in the **region of cell division,** which is composed of an *apical meristem* (a tissue of actively dividing cells) in the center of the root tip, produce the surrounding root cap. Most of the cell divisions take place next to the root cap at the edges of this inverted cup-shaped zone, located a short distance behind the actual base of the meristem. Here the cells divide every 12 to 36 hours, while at the base of the meristem, they may divide only once in every 200 to 500 hours. The divisions are often rhythmic, reaching a peak once or twice each day, usually toward noon and midnight, with relatively quiescent intermediate periods. Cells in this region are mostly cubical, with relatively large, centrally located nuclei and a few very small vacuoles.

In both roots and stems, the apical meristem soon subdivides into three meristematic areas: (1) The **protoderm** gives rise to an outer layer of cells, the *epidermis.* (2) The **ground meristem,** to the inside of the protoderm, produces parenchyma cells of the *cortex.* (3) The **procambium,** which appears as a solid cylinder in the center of the root, produces *primary xylem* and *primary phloem* (Fig. 5.3). *Pith* (parenchyma) tissue, which originates from the ground meristem, is generally present in stems but is absent in most dicot roots. Grass roots and those of most other monocots, however, do have pith tissue.

The Region of Elongation

The **region of elongation,** which merges with the apical meristem, usually extends about 1 centimeter (0.4 inch) or less from the tip of the root. As the name implies, in this region, the cells become several times their original length and somewhat wider. At the same time, the tiny vacuoles merge and grow until one or two large vacuoles, occupying up to 90% or more of the volume of each cell, have been formed. Only the root cap and apical meristem are actually pushing through the soil, because no further increase in cell size takes place above the region of elongation. The usually extensive remainder of each root remains stationary for the life of the plant. If a cambium is present, however, there is normally a gradual increase in girth through the addition of **secondary tissues** produced by the cambium.

The Region of Maturation

Most of the cells mature, or *differentiate,* into the various distinctive cell types of the primary tissues in this region, which is sometimes called the **region of differentiation, or root-hair zone.** The large numbers of delicate hairs that develop from many of the epidermal cells give the root-hair zone its name. The **root hairs,** which absorb water and minerals, adhere tightly to soil particles (Fig. 5.4) with the aid of microscopic fibers they produce and greatly increase the total absorptive surface of the root. Differentiation is discussed further in Chapter 11.

The root hairs are not separate cells; rather, they are tubular extensions of specialized epidermal cells. In fact, the nucleus of the epidermal cell to which each is attached often moves out into the root hair. They are so numerous that they appear as a fluffy mass to the naked eye, typically numbering more than 38,000 per square centimeter (250,000 per square inch) of surface area in roots of plants such as corn; they seldom exceed 1 centimeter (0.4 inch) in length.

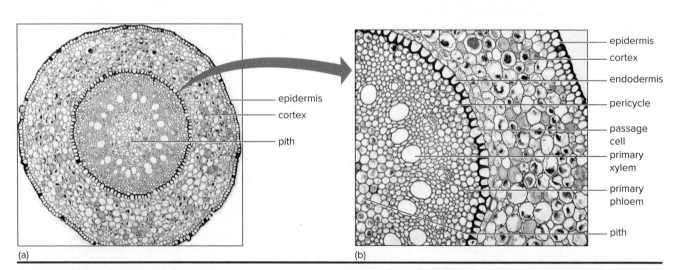

Figure 5.3 A cross section of a monocot (greenbrier—*Smilax*) root. (a) Complete view. ×40. (b) Enlargement showing partial section of the root interior. ×100. *(a-b)* © *James E. Bidlack*

Figure 5.4 (a) A radish (*Raphanus*) seedling shortly after germination, showing the root-hair zone. (b) A diagram of an enlargement of a longitudinal section of a small portion of a root-hair zone, showing root hairs in contact with soil particles. Note that in most soils there is normally considerably less air space between soil particles than is depicted in this image. *(a) © Kingsley Stern*

A single ryegrass plant occupying less than 0.6 cubic meter (20 cubic feet) of soil was found to have more than 14 billion root hairs, with a total surface area almost the size of a football field.

When a seedling or plant is moved, many of the delicate root hairs are torn off or die within seconds if exposed, greatly reducing the plant's capacity to absorb water and minerals in solution. This is why plants should be watered, shaded, and pruned after transplanting until new root hairs have formed. In any growing root, the size of the root-hair zone remains fairly constant, with new root hairs being formed toward the root cap and older ones dying in the more mature regions. The life of the average root hair is usually not more than a few days, although a few may live up to 3 weeks.

The *cuticle* (see Chapter 4), is thin enough on the root hairs and epidermal cells in the region of maturation (Fig. 5.5) to allow water to be absorbed but still sufficient to protect against invasion by bacteria and fungi.

The cells of the **cortex,** a tissue composed of parenchyma cells lying between the epidermis and inner tissues, mostly store food. This tissue, which may be many cells thick, is similar to the cortex of stems except for the presence of an **endodermis** at its inner boundary (Fig. 5.6). The endodermis consists of a single-layered cylinder of compactly arranged cells whose walls are impregnated with lignin and suberin. These substances form bands, called **Casparian strips,** which can be found on the radial and transverse cell walls. The plasma membranes of the endodermal cells are fused to the Casparian strips, which

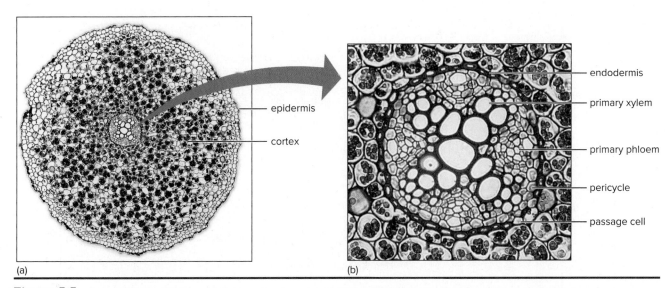

Figure 5.5 A cross section of a dicot (buttercup—*Ranunculus*) root. (a) Complete view. ×40. (b) Enlargement of the root center (vascular cylinder). ×400. *(a-b) © James E. Bidlack*

Figure 5.6 A portion of the endodermis of a buttercup (*Ranunculus*) root. ×1,000. © *James E. Bidlack*

are perpendicular to the root's surface; they prevent water from passing through the otherwise permeable (porous) cell walls. The Casparian strip barrier forces water and dissolved substances to pass through the plasma membranes of the endodermal cells or their plasmodesmata. This regulates the types of minerals absorbed and transported by the root to the stems and leaves. The plasma membranes tend to exclude harmful minerals while generally retaining useful ones.

An endodermis is rare in stems but so universal in roots that only three species of plants are known to lack a root endodermis. In some roots, the epidermis, cortex, and endodermis are sloughed off as their girth increases, but in those roots where the endodermis is retained, the inner walls of the endodermal cells eventually become thickened by the addition of alternating layers of suberin and wax. Later, cellulose and sometimes lignin are also deposited. Some endodermal cells, called **passage cells,** may remain thin-walled and retain their Casparian strips for a while, but they, too, eventually tend to become suberized.

A core of tissues, referred to collectively as the **vascular cylinder,** lies inside the endodermis. Most of the cells of the vascular cylinder conduct water or food in solution, but lying directly against the inner boundary of the endodermis is an important layer of parenchyma tissue known as the **pericycle.** This tissue, which is usually one cell wide, may be wider in some plants. The cells of the pericycle may continue to divide even after they have matured. *Lateral (branch) roots* and part of the *vascular cambium* of dicots arise within the pericycle (Fig. 5.7).

In most dicot and conifer roots, the primary xylem consists of a solid central core of water-conducting cells (e.g., tracheids and vessels). In cross section, this first root xylem usually loosely resembles the rear view of a rocket with fins. The fins, generally referred to as *arms,* tend to taper toward their tips and terminate just inside of the thin, cylindrical pericycle layer. There are usually four of these arms, with some plants having two, three, or several. Lateral roots

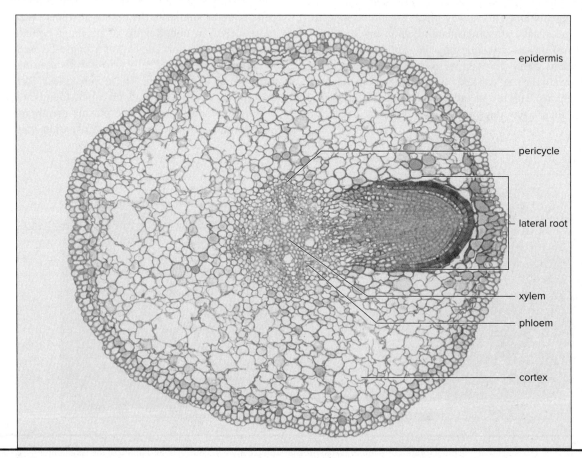

Figure 5.7 A cross section through a dicot (willow—*Salix*) root showing the origin of a lateral (branch) root. © *G. S. Ellmore*

arise toward the outside of these xylem arms, from within the pericycle. Dividing cells cause the pericycle to bulge out into the cortex, and the endodermis usually ruptures before the lateral root emerges. As the new root emerges, it pushes aside or crushes the cortex cells and eventually breaks through the epidermis. Parenchyma cells between the vascular tissues of the originating and lateral roots differentiate into vascular elements and establish connections between the old and new tissues.

The primary xylem surrounds pith parenchyma cells in monocot roots and those of a few dicots; in such plants, the arms may not be well defined. Primary phloem, which conducts food, forms in discrete patches between the xylem arms of both dicot and monocot roots.

A vascular cambium develops from parts of the pericycle and other parenchyma cells between the xylem arms and phloem patches in most dicots and conifers. This cambium at first follows the starlike outline of the primary xylem as it starts producing secondary phloem to the outside and secondary xylem to the inside. Eventually, however, the position of the cambium gradually shifts so that instead of appearing as patches and arms, the secondary conducting tissues appear as concentric cylinders. The primary phloem may be sloughed off and lost as secondary tissues are added.

In woody plants, a second cambium, the *cork cambium,* normally arises in the pericycle outside the vascular cambium and gives rise to *cork* tissue. Cork cells, which are dead at maturity, are impregnated with suberin and are impervious to moisture; similar tissues are produced in stems. Although there are exceptions, monocot roots generally have no secondary meristems and therefore no secondary growth.

In both roots and stems (see Chapter 6), growth may be *determinate* or *indeterminate.* Determinate growth is growth that stops after an organ such as a flower or a leaf is fully expanded or after a plant has reached a certain size. Indeterminate growth occurs in trees and other perennials where new tissues are added indefinitely, season after season.

Natural grafting can take place between roots of different trees of the same species, especially in the tropics. The roots unite through secondary growth when they come in contact with one another, but the details of the uniting process are not yet known. One unfortunate aspect of this grafting is that if one tree becomes diseased, the disease can be transmitted through the grafts to all the other trees connected to it.

5.4 Specialized Roots

Learning Outcome

a. Describe and provide the function(s) of at least five types of specialized roots.

As mentioned earlier, most plants produce a fibrous root system, a taproot system, or, more commonly, a combination

Figure 5.8 Sweet potato (*Ipomoea*) plants. Note the food-storage roots. © KPG_Payless/Shutterstock

of the two types. Some plants, however, have roots with modifications that adapt them for performing specific functions in addition to absorbing water and minerals in solution.

Food-Storage Roots

Most roots and stems store some food, but in certain plants, the roots are enlarged and store large quantities of starch and other carbohydrates (Fig. 5.8), which may later be used for extensive growth. In sweet potatoes and yams, for example, extra cambial cells develop in parts of the xylem of branch roots and produce large numbers of parenchyma cells. As a result, the organs swell and provide storage areas for large amounts of starch and other carbohydrates. Similar food-storage roots are found in the deadly poisonous water hemlocks, in dandelions, and in salsify. In carrots, beets, turnips, and radishes, the food-storage tissues are actually a combination of root and stem. Although the external differences are not obvious, approximately 2 centimeters (0.8 inch) at the top of an average carrot is derived from stem tissue that merges with the root tissue below.

Water-Storage Roots

Some members of the Pumpkin Family (Cucurbitaceae) produce huge water-storage roots. This is particularly characteristic of those that grow in arid regions or in those areas where there may be no precipitation for several months of the year. In certain manroots (*Marah*), for example, roots weighing 30 kilograms (66 pounds) or more are frequently produced (Fig. 5.9), and a root of one calabazilla plant (*Cucurbita perennis*) was found to weigh 72.12 kilograms (159 pounds). The water in the roots is used by the plants when the supply in the soil is inadequate.

Propagative Roots

Many plants produce **adventitious** buds (buds appearing in places other than stems) along the roots that grow near the surface of the ground. The buds develop into aerial stems called *suckers,* which have additional rootlets

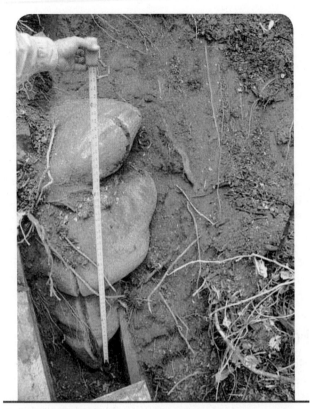

Figure 5.9 A manroot (*Marah*) water-storage root that weighs over 25.3 kilograms (60 pounds). *Courtesy of Paul Furman*

at their bases. The rooted suckers can be separated from the original root and grown individually. Cherries, apples, pears, and other fruit trees often produce suckers. The adventitious roots of rice-paper plants (*Tetrapanax papyrifera*) and tree-of-heaven (*Ailanthus altissima*) can become a nuisance in gardens, often producing propagative roots 10 meters (33 feet) or more from the parent plant. Horseradish (*Rorippa armoracia*), Canada thistle (*Cirsium arvense*), and some other weeds have a remarkable ability to reproduce by both propagative roots and seeds. In the past, this made them difficult to control, but some biological controls being investigated (see Appendix 2) may be an answer to the problem in the future.

Pneumatophores

Water, even after air has been bubbled through it, contains less than one-thirtieth the amount of free oxygen found in the air. Accordingly, plants growing with their roots in water may not have enough oxygen available for normal respiration in their root cells. Some swamp plants, such as the black mangrove (*Avicennia nitida*) and the yellow water weed (*Ludwigia repens*), develop special spongy roots, called **pneumatophores,** which extend above the water's surface and enhance gas exchange between the atmosphere and the subsurface roots to which they are connected (Fig. 5.10). The woody "knees" of the bald

Figure 5.10 (a) Pneumatophores (foreground) of tropical mangroves rising above the sand at low tide. The pneumatophores are spongy outgrowths from the roots beneath the surface. Pneumatophores facilitate the exchange of oxygen and carbon dioxide for the roots, which grow in areas where little oxygen is otherwise available to them. (b) Close-up photograph of pneumatophores. *(a) Majority World/Getty Images; (b) © Majority World/Contributor/Getty Images*

cypress (*Taxodium distichum*), which occurs in southern swamps (see Chapter 22), were in the past believed to be pneumatophores, but there is no conclusive evidence for this theory.

Aerial Roots

Velamen roots of orchids (Fig. 5.11), *prop roots* of corn and banyan trees, *adventitious roots* of ivies, and *photosynthetic roots* of certain orchids are among various kinds of **aerial roots** produced by plants. It was formerly assumed that the epidermis of velamen roots, which is several cells thick, aided in the absorption of rain water. It appears, however, it may function more in preventing loss of moisture from the root. Corn prop roots, produced toward the base of the stems, support the plants in a high wind. Some tropical plants, including the screw pines and various mangroves, produce sizable prop roots extending for several feet above the surface of the ground or water. Debris collects between them and helps create additional soil.

Many of the tropical figs or banyan trees produce roots that grow down from the branches until they contact the soil. Once they are established, they continue secondary growth and look just like additional trunks (Fig. 5.12). Banyan trees may live for hundreds of years and can become very large. In India and southeast Asia, there are several banyan trees that have almost 1,000 root-trunks and have circumferences approaching 450 meters (1,476 feet). The oldest is estimated to be about 2,000 years old.

The vanilla orchid, from which we obtain vanilla flavoring, produces chlorophyll in its aerial roots and, through photosynthesis, can manufacture food with them. The adventitious roots of English ivy, Boston ivy, and Virginia creeper appear along the stem and aid the plants in climbing.

Contractile Roots

Some herbaceous dicots and monocots have contractile roots that pull the plant deeper into the soil. Many lily bulbs are pulled a little deeper into the soil each year as new sets of contractile roots are developed (Fig. 5.13). The bulbs

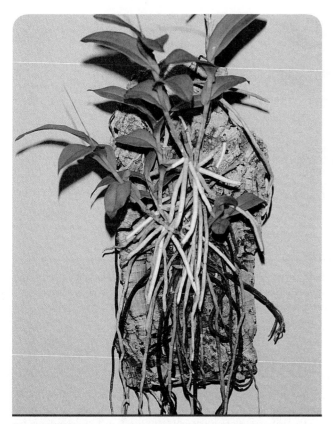

Figure 5.11 The aerial (velamen) roots of orchids have a thick epidermis that reduces water loss from internal tissues. © *Kingsley Stern*

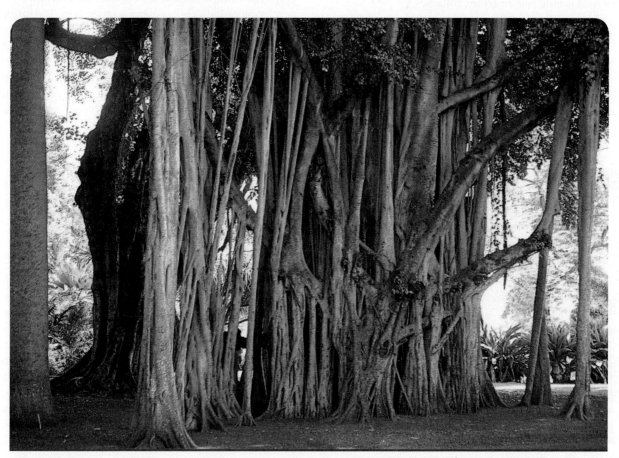

Figure 5.12 A banyan (*Ficus*) tree with many large prop roots that have developed from the branches. © *Kingsley Stern*

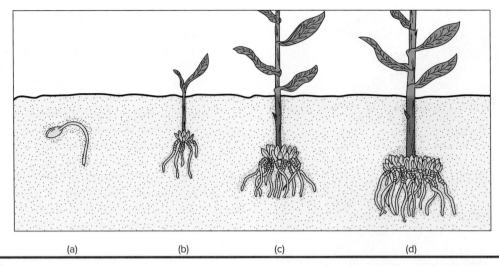

Figure 5.13 A lily bulb over three seasons is pulled deeper into the soil by the action of contractile roots. (a) A small bulb produced during the first growing season. (b) Contractile roots pull the newly formed bulb down several millimeters during the first season. (c) The bulb is pulled down farther the second season. (d) The bulb is pulled down even farther the third season. The bulb will continue to be pulled down in succeeding seasons until it reaches an area of relatively stable soil temperatures.

continue to be pulled down until an area of relatively stable temperature is reached. Plants such as dandelions always seem to have the leaves coming out of the ground as the top of the stem is pulled down a small amount each year when the root contracts. The contractile part of the root may lose as much as two-thirds of its length within a few weeks as stored food is used and the cortex collapses.

Buttress Roots

Some tropical trees growing in shallow soils produce huge, buttresslike roots toward the base of the trunk, giving them great stability (Fig. 5.14). Except for their angular appearance, these roots look like a part of the trunk.

Parasitic Roots

Some plants, including dodders, broomrapes, and pinedrops, have no chlorophyll (which is necessary for photosynthesis) and have become dependent on chlorophyll-bearing plants for their nutrition. They parasitize their host plants via peglike projections called **haustoria** (singular: **haustorium**), which develop along the stem in contact with the host. The haustoria penetrate the outer tissues and establish connections with the xylem and phloem (Fig. 5.15). Some green plants, including Indian warrior and the mistletoes, also form haustoria. These haustoria, however, apparently aid primarily in obtaining water and dissolved minerals from the host plants, because the partially parasitic plants are capable of manufacturing

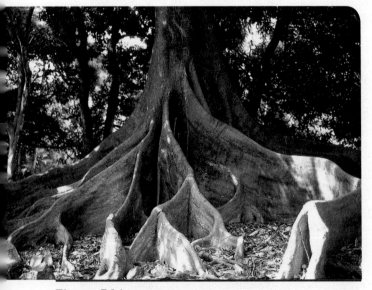

Figure 5.14 Buttress roots of a tropical fig tree. © *Kingsley Stern*

KEY THEME 5.1 ecology

Plants Need Roots

Plants take up water and nutrients from soils through their roots, one of the major avenues of ecological exchange between plants and the environment. Roots also act as organs for storage of water and energy, for support, and for uptake of oxygen in some environments. Plant roots are also sites of mutualistic relationships with fungi (*mycorrhizae*) and with nitrogen-fixing bacteria. The roots of parasitic plants attack host plants. Soils, the primary medium with which most roots interact, are structured through complex relationships between parent mineral material, climate, organisms, and topography over the course of time.

It has been used in the description of some plants, but this is not correct, and more accurate terminology can be applied, depending on where plants obtain their nutrients. Orchids, for instance, are often considered **epiphytes** because they grow on other plants and obtain moisture and nutrients from the air and rain. Indian pipe is referred to as an **epiparasite** because it is a parasite of fungi in the soil; some scientists also use the term **myco-heterotroph** to describe Indian pipe because this plant obtains some or all of its carbon from the fungus rather than from photosynthesis.

5.5 Mycorrhizae

Learning Outcome

a. Explain the function of mycorrhizae, and describe two ways mycorrhizae associate with roots.

More than 95% of land plants have fungi associated with their roots, and these associations are termed **mycorrhizae** (Fig. 5.16). The association is *mutualistic;* that is, both the fungus and the root benefit from it and are dependent upon the association for survival. Mutualism is a form of *symbiosis.* For some mycorrhizae the fungal hyphae penetrate the root cortex but the hyphae do not enter the cells. On the root surface, the hyphae form a thin, encrusting layer called the mantle. This type of mycorrhiza is termed *ectotrophic* and is found in trees like pines and oaks. In another type of mycorrhiza, the hyphae invade the cells of the cortex and a mantle is not present. This type of mycorrhiza is called *endotrophic* and is found in the majority of woody and herbaceous land plants.

Mycorrhizae are essential for the normal growth and development of forest trees, as well as most herbaceous plants, and they facilitate the uptake of water and inorganic ions. The fungus is able to absorb and concentrate phosphorus much better than it can be absorbed by the root hairs. In fact, if mycorrhizal fungi have been killed by fumigation or are otherwise absent, many plants appear to have considerable difficulty absorbing phosphorus, even when the element is abundant in the soil. Hence, making phosphorus readily available for absorption is a very important function of these fungi. In exchange for the phosphorus made available to the plant by fungi, sugars and amino acids are provided by plants, which are essential for the fungi to survive.

Some plants do not seem to need mycorrhizae unless the soil lacks enough essential elements for healthy growth. Plants with mycorrhizae develop few root hairs compared to those growing without an associated fungus. Mycorrhizae have been shown to be particularly susceptible to acid rain (discussed in Chapter 25); this may signal major problems for our coniferous forests in the future if the problem of acid rain is not solved. Methyl bromide, used in the past to sterilize seed beds, kills all soil organisms, including mycorrhizae; its continued use in the United States has been banned.

(a)

(b)

Figure 5.15 (a) Pale stems of a parasitic plant (dodder - *Cuscuta*) twining about other vegetation. The inset shows peglike haustoria that penetrate the tissues of the host plant. (b) Indian pipe (*Monotropa uniflora*), which some refer to as an an epiparasite because it is a parasite of fungi in the soil. *(a)* © *Kingsley Stern (b) Courtesy of Justin Wyatt*

at least some of their own food through photosynthesis. The term **saprophyte,** which refers to organisms that lack chlorophyll and obtain nutrients from organic matter, is most appropriately used to describe the way many fungi obtain nutrition.

Figure 5.16 Mycorrhizae. (a) A longitudinal drawing of a root with ectomycorrhizae (visible on the right outside of the root). (b) A diagram of a cross section of a root with ectomycorrhizae. See Figs. 5.5b and 5.6 for definitive photomicrographs of most of the tissues in this diagram. (c) Photomicrograph of a cross section of a root around which ectomycorrhizae have formed a mantle ×40. The fungal cells have not penetrated deeper than the outermost layers of root cells. (d) A cross section of a few root parenchyma cells with endomycorrhizae ×100. The endomycorrhizae develop and flourish within the parenchyma cells. *(c-d)* © *Kingsley Stern*

5.6 Root Nodules

Learning Outcome

a. State what types of plants may have root nodules, and name the bacteria responsible for enabling nitrogen fixation.

Although almost 80% of our atmosphere consists of nitrogen gas, plants cannot convert the nitrogen gas to usable forms. A few species of bacteria, however, produce enzymes that allow them to convert nitrogen into nitrates and other nitrogenous substances readily absorbed by roots. Members of the Legume Family (Fabaceae), which includes peas, beans,

5.7 Human Relevance of Roots

Learning Outcome

a. List at least five practical human uses of roots.

Roots have been important sources of food for humans since prehistoric times, and some, such as the carrot, have been in cultivation in Europe for at least 2,000 years. Many cultivated root crops are biennials (i.e., plants that complete their life cycles from seed to flowering and back to seed in two seasons). Such plants store food in a swollen taproot during the first year of growth, and then the stored food is used for the production of flowers in the second season. Among the best-known biennial root crops are sugar beets, beets, turnips, rutabagas, parsnips, horseradishes, and carrots. Other important root crops include sweet potatoes, yams, and cassava. Cassava (Fig. 5.18), from which tapioca is made, forms a major part of the basic diet for millions of inhabitants of the tropics. With a minimum of human labor, it yields more starch per hectare (about 45 metric tons, the equivalent of 20 tons per acre) than any other cultivated crop. Minor root crops, including relatives of wild mustards, nasturtiums, and sorrel, are cultivated in South America and other parts of the world.

Several well-known spices, including sassafras, sarsaparilla, licorice, and angelica, are obtained from roots. Sweet potatoes are used to produce alcohol in Japan. Some important red to brownish dyes are obtained from roots of members of the Madder Family (Rubiaceae), to which coffee plants belong. Drugs obtained from roots include aconite, ipecac, gentian, and reserpine, a tranquilizer. A valuable insecticide, *rotenone,* obtained from the barbasco plant, has been cultivated for centuries as a fish poison by primitive South American tribes. When thrown into a dammed stream, the roots containing rotenone cause the fish to float but do not poison them for human consumption. In tobacco plants, nicotine produced in the roots is transported to the leaves. Other uses of roots are discussed in Chapter 24.

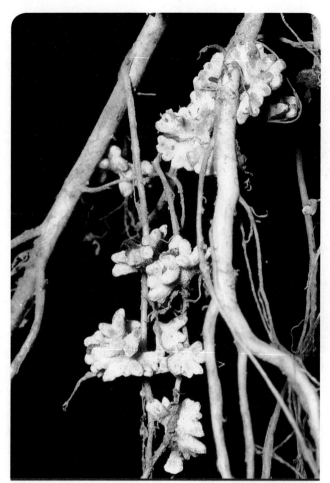

Figure 5.17 Root nodules on the roots of bur clover (*Medicago polymorpha*). The popcorn-like nodules contain bacteria that convert nitrogen from the air into forms that can be used by the plant. ×15.
© Kingsley Stern

alfalfa, and alders, form associations with certain soil bacteria that result in the production of numerous small swellings called **root nodules** that are clearly visible when such plants are uprooted (Fig. 5.17). The nodules contain large numbers of nitrogen-fixing bacteria.

Root exudates stimulate the growth of *Rhizobium* bacteria, which cause root hairs to bend sharply. A bacterium may attach to the concave side of a bend and then invade the cell with a tubular infection thread that does not actually break the host cell wall and plasma membrane. The infection thread grows through to the cortex, which is stimulated to produce new cells that become a part of the root nodule; here the bacteria multiply and engage in nitrogen conversion. (See also the discussion of the nitrogen cycle in Chapter 25.)

Root nodules should not be confused with *root knots,* which are also swellings that may be seen in the roots of tomatoes and many other plants. Root knots develop in response to the invasion of tissue by small, parasitic roundworms (nematodes). Unlike bacterial nodules, root knots are not beneficial, and the activities of the parasites within them can eventually lead to the premature death of the plant.

5.8 Soils

Learning Outcomes

a. Describe the three typical soil horizons and how they can be can be distinguished from each other.
b. Explain how soil constituents affect the availability of water to plants, and discuss the role of soil pH in nutrient availability.

The soil is a dynamic, complex, constantly changing part of the earth's crust, which extends from a few centimeters deep in some places to hundreds of meters deep in others. It is essential not only to our existence but also to the existence of most living organisms. It has a pronounced effect on the plants that grow in it, and they have an effect on it.

If you dig up a shovelful of soil from your yard and examine it, you will probably find a mixture of ingredients,

Figure 5.18 Cassava (*Manihot esculenta*) plants. Note the food-storage roots on the plants that have been dug up. © *Shelley H. Jansky*

including several grades of sand; rocks and pebbles; powdery silt; clay; humus; dead leaves and twigs; clods consisting of soil particles held together by clay and organic matter; plant roots; and small animals, such as ants, pill bugs, millipedes, and earthworms. Also present, but not visible, would be millions of microorganisms, particularly bacteria, fungi, and, of course, air and water.

The soil became what it is today through the interaction of a number of factors: climate, parent material, topography of the area, vegetation, living organisms, and time. Because there are thousands of ways in which these factors may interact, there are many thousands of different soils. The solid portion of a soil consists of mineral matter and organic matter. Pore spaces, shared by variable amounts of water and air, occur between the solid particles. The smaller pores often contain water, and the larger ones usually contain air. The sizes of the pores and the connections between them largely determine how well the soil is aerated.

If you were to dig down 1 or 2 meters (3 to 6 feet) in an undisturbed area, a soil profile of three intergrading regions called *horizons* would probably be exposed (Fig. 5.19). The horizons show the soil in different stages of development, and the composition varies accordingly. The upper layer, usually extending down 10 to 20 centimeters (4 to 8 inches), is called the *topsoil*. It is usually subdivided into a darker upper portion called the *A horizon* and a lighter lower portion below it called the *E horizon*. The *E horizon* is not always present; it usually occurs in older, well-developed soils. The *A* portion contains more organic matter than the layers below. It can be divided into different parts such as the *A1* and *A2* horizons.

*A*1 horizon (crushed fine sandy loam, dark brown)

*A*2 horizon (crushed fine sandy loam, brown)

*Bt*1 horizon (crushed sandy loam, brown)

2*Bt*2 horizon (broken face sandy clay loam, reddish brown)

2*Bt*3 horizon (broken face sandy clay loam, yellowish red)

2*C* horizon (crushed stratified coarse sand, strong brown)

Figure 5.19 A soil profile showing horizons that depict different stages of development. *Provided by Steven McGowen, USDA-NRCS*

The next 0.3 to 0.9 meter (1 to 3 feet) section is called the *B horizon,* or *subsoil.* It usually contains more clay and is lighter in color than the topsoil. As with other soil horizons, lowercase letters after the uppercase letter may be used to describe characteristics within the horizon, such as *Bw* (weakly colored or structured) and *Bt* (translocation of silicate clay), as well as numbers after these labels to designate layers within the horizon. A number preceding the horizon letter may also be used if the soil consists of different parent materials. The *C horizon* at the bottom may vary from about 10 centimeters (4 inches) to several meters (6 to 10 feet or more) in depth; it may even be absent. It is commonly referred to as the *soil parent material* and extends down to *bedrock.*

Parent Material

The first step in the development of soil is the formation of *parent material* from existing rocks that have not yet been broken down into smaller fragments. Parent material accumulates through the weathering of three types of rock, which originate from various sources. These sources and types include volcanic activity (*igneous rocks*); depositing by glaciers, water, or wind (*sedimentary rocks*); and changes in igneous or sedimentary rocks brought about by great pressures, heat, or both (*metamorphic rocks*).

Climate

Climate varies greatly throughout the globe, and its role in the weathering of rocks varies correspondingly. In desert areas, for example, there is little weathering by rain, and soils are poorly developed. In areas of moderate rainfall, however, well-developed soils are common. In some areas of high rainfall, the excessive flow of water through the soil may leach out important minerals. Similar leaching out of important minerals may occur when garden sprinklers are left on all night. Many gardeners and house-plant enthusiasts have stunted or killed the very plants they were trying to foster by "drowning" them with too much water or too frequent watering. As a general rule, plants should not be watered unless the soil surface feels dry.

In areas where there are great temperature ranges, rocks may split or crack as their outer surfaces expand or contract at different rates from the material beneath the surface. When water in rock crevices freezes, it expands and causes further cracks and splits. The breaking up of rocks contributes to the development of soil.

Living Organisms and Organic Composition

There are many kinds of organisms in the soil, as well as roots and other parts of plants. In the upper 30 centimeters (1 foot) of a good agricultural soil, living organisms constitute about one-thousandth of the total weight of the soil. This may not sound significant, but it amounts to approximately 6.73 metric tons per hectare (3 tons per acre). Bacteria and fungi present in the soil decompose organic matter, which accumulates when leaves fall and plants and animals die. (This process is further discussed under the section on composting in Chapter 17.) Roots and all other living organisms produce carbon dioxide, which combines with water in the soil and forms an acid, thereby increasing the rate at which minerals dissolve. Ants and other insects, earthworms, burrowing animals, and birds all alter the soil through their activities and add to its organic content, either through wastes that they deposit or through the decomposition of their bodies when they die. *Humus,* which consists of partially decomposed organic matter, gives some soils a dark color.

The total organic composition of a soil varies greatly. An average topsoil might consist of about 25% air, 25% water, 48% minerals, and 2% organic matter. Soils in low, wet areas, where a lack of oxygen keeps microorganisms from their normal activities, may contain as much as 90% organic matter. Except in legumes and a few other plants, almost all of the nitrogen utilized by growing plants, as well as much of the phosphorus and sulfur, comes from decomposing organic matter. In addition, as organic matter breaks down, it produces acids, which, in turn, decompose minerals. Other roles of organic matter in the soil are discussed in the "Soil Structure" section.

Topography

If the land is steep, soil may wash away or erode through the action of wind, water, and ice as soon as it is weathered from the parent material. It has been estimated that more than 20 metric tons of topsoil per hectare (8.2 tons per acre) are washed away annually from some prime croplands in the central United States.

If an area is flat and poorly drained, pools and ponds may appear in slight depressions when it rains. If these bodies of water cannot drain quickly, the activities of organisms in the soil are interrupted, and the development of the soil is arrested. The ideal topography for the development of soil is one that permits drainage without erosion.

Soil Texture and Mineral Composition

Soil texture refers to the relative proportions of sand, silt, and clay in a given soil (Table 5.1).

Sands are usually composed of many small particles bound together chemically or by a cementing matrix material. Silt consists of particles that are mostly too small to be seen without a lens or a microscope.

Clay particles are so tiny that they can't be seen with even a powerful light microscope, although they can be seen with an electron microscope. Individual clay particles are called *micelles.* Micelles are somewhat sheetlike, negatively charged, and held together by chemical bonds. The negative charges attract, exchange, or retain positively charged ions.

Table 5.1 — Soil Mineral Components as Classified by the U.S. Department of Agriculture

Mineral	Diameter (Range in mm)	Comments
Stones	>76 mm	Do not support plant growth but affect permeability and erosion of the soil
Gravel	76 mm–2.0 mm	
Very coarse sand	2.0 mm–1.0 mm	
Coarse sand	1.0 mm–0.5 mm	
Medium sand	0.5 mm–0.25 mm	Makes soil feel gritty
Fine sand	0.25 mm–0.10 mm	
Very fine sand	0.10 mm–0.05 mm	
Silt	0.05 mm–0.002 mm	A lens or microscope needed to see any but the coarsest silt particles
Clay	<0.002 mm	May absorb water, swell, and later shrink, causing the soil to crack as it dries

Many of the positively charged ions, such as magnesium (Mg^{++}) and potassium (K^+), which are needed for normal plant growth, are absorbed with water by the roots. Clay is plastic in nature because the water that adheres tightly to the surface of the particles acts as both a binding agent and a lubricant. Physically, clay is a **colloid**—that is, a suspension of particles that are larger than molecules but that do not settle out of a fluid medium.

The best agricultural soils are usually loams, which are a mixture of sand, clay, and organic matter. The better loams have about 40% silt, 40% sand, and 20% clay. Light soils have a high sand and low clay content. Heavy soils have high clay content. Coarse soils, which have larger particles, are porous and don't retain much water, while clay soils have high water content and allow little water to pass through.

Over half the composition by weight of mineral matter is oxygen. Other elements commonly present are hydrogen, silicon, aluminum, iron, potassium, calcium, magnesium, and sodium. However, soil obtains hundreds of different mineral combinations from its parent material.

Soil Structure

Soil structure refers to the arrangement of the soil particles into groups called *aggregates*. Aggregates in sands and gravels show little cohesion, but most agricultural soils have aggregates that stick together. Structure develops when colloidal particles clump together, mostly as a result of the activities of soil organisms, freezing, and thawing. If the individual granules do not become coated with organic matter, they may continue to clump until they become clods.

Productive agricultural soils are granular soils with pore spaces that occupy between 40% and 60% of the total volume of the soil. The pores contain air and water, and their sizes are more important than their total volume. Clay soils, for example, have more pore space than sandy soils, but the pores are so small that water and air are restricted in their movement through the soil. When the pores are full of water, air is kept out, and there is not enough oxygen for root growth. Sandy soils have large pores, which drain by gravity soon after they are filled. The water is replaced by air, but too much air speeds up nitrogen release by microorganisms. Plants can't use the nitrogen that quickly, and much of it is lost.

Too much water can be harmful. This leads to anaerobic conditions (lacking oxygen), which leach mineral nutrients and inhibit mineralization. Too much water also slows the release of nitrogen, interferes with plant growth, and accelerates the breakdown of nitrates to the extent that virtually all the nitrates may be lost in as little as half an hour.

Water in the Soil

Water in the soil occurs in three forms. **Hygroscopic water** is physically bound to the soil particles and is unavailable to plants. **Gravitational water** drains out of the pore spaces after a rain. If drainage is poor, this water interferes with normal plant growth. Plants are mainly dependent on the third type, **capillary water,** which is water held against the force of gravity in pores of the soil. The structure and organic matter of the soil—which enable the soil to hold

KEY THEME 5.2 ecology

Metal-Munching Plants

Years ago, prospectors looking for promising sites to mine silver or gold often noticed that certain plants grew on old mine tailings. They reasoned that these plants might be indicators of the precious metals entrapped in the soil. It turns out that these prospectors had it right.

Certain plants not only grow in heavy, metal-laden soils but are able to extract these metals through their root systems and accumulate them in their tissues without being damaged. Plants absorb metals because they require certain ones such as zinc and copper as components of their proteins and enzymes for normal growth and development. In certain soils, plant mineral uptake makes no distinction between heavy metals such as cadmium or selenium and these required elements. In either case, these metals are absorbed by the plant's extensive root system, which may extend a meter or more in depth. Scientists have coined a term to describe this process: **phytoremediation**—the use of plants to facilitate the removal of toxic compounds from ground water and soil. The plants that "munch metal" so well are called "hyperaccumulating" plants.

What makes a good hyperaccumulator? Researchers are attempting to answer this question, but so far, nobody knows for sure. One clue may be metal-binding polypeptides called **phytochelatins** that sequester and detoxify heavy metals in plant tissue. A survey of numerous plants has shown that phytochelatins are produced when these plants are exposed to heavy metals in soil. Interestingly, a wide range of plants—from the most advanced flowering plants (even orchids) to red, green, and brown algae—produce these detoxifying polypeptides. One possible method of accumulating heavy metals is for the plant to transport them into the cell's vacuoles, a sort of waste disposal dump.

Phytoremediation is attracting the increased attention not only of scientists but also of regulators who see it as a low-cost alternative for cleaning up contaminated sites across the country. The conventional process of soil excavation and reburial in a landfill is very expensive, typically costing about $1.5 million per acre, depending on the pollutant. The price tag opens the door for many alternative ideas. These plants can be harvested and disposed of—and in certain instances, the metals can even be recovered by sending the plants to a smelter. But can hyperaccumulators effectively get the job done?

Alpine pennycress (*Thlaspi caerulescens*) is a remarkable hyperaccumulator, able to accumulate 4% of its dry body weight in zinc. This translates into 10 metric tons of zinc per hectare. The problem, however, is that *Thlaspi* is a small and slow-growing plant. For phytoremediation to become practical, plants with metal uptake rates comparable to those of *Thlaspi*, but with faster growth rates and larger tissue mass, must be found. Screening of other plants is being done in several laboratories around the world. One plant that has been identified so far is Indian mustard, *Brassica juncea*, a relative of some highly nutritious vegetables—cabbage, cauliflower, broccoli, collard, kale, and mustard greens. Indian mustard can accumulate 3.5% of its dry body weight in lead. It also can absorb cadmium, chromium, nickel, selenium, zinc, and copper. Cattails (*Typha* spp.), which also are known to accumulate heavy metals, are already in use in some areas in the final stages of the treatment of human sewage.

Despite the advantages of phytoremediation, one drawback is that multiple crops must be planted over several years to reduce contamination to acceptable levels, while removal of the soil provides an immediate resolution. Additionally, there is concern about increasing the accumulation of these metals in the food chain as wildlife and insects eat the plants, accumulating toxicity in their bodies. In this way, toxic metals could work their way up the food chain and pose a new set of problems.

Phytoremediation was put to a big test at Chernobyl, the site of the largest environmental disaster in modern history. In 1986, the meltdown of the nuclear reactor left radioactive wastes scattered over the Ukrainian countryside. Indian mustard is being used to remove radioactive strontium from the soil. Scientists continue to test this promising natural process, and environmental cleanups of the future could be as simple as letting the flowers grow.

D. C. Scheirer

---

water against the force of gravity—the density and type of vegetational cover, and the location of underground water tables largely determine the amount of capillary water available to the plant.

The ancient Incas of Peru knew that water in the soil could rise through capillary action. Where the water table was close to the surface, they removed the upper 0.6 meter (2 feet) of soil and planted their crops down in the hollowed-out areas so that the roots would be able to reach the capillary water. They knew that in some areas having sandy soils and low annual precipitation, soils could be compacted by a heavy roller to create finer capillaries to raise water from below. This technique is effective only if the available water is within 1.5 to 3.0 meters (5 to 10 feet) of the surface and if the soils do not contain too much silt or clay.

After rain or irrigation, water in the soil drains away by gravity. The water remaining after such draining is referred to as the *field capacity* of the soil. Field capacity is mainly governed by the texture of the soil, but the structure and organic content also influence it to a certain extent. Plants readily absorb water from the soil when it is at, or near, field capacity. As the soil dries, the film of water around each soil particle becomes thinner and more tightly bound to the soil particle and less likely to enter the root. If water is not added to the soil, eventually a point is reached at which the rate of water absorption by the plant is insufficient for its needs, and the plant wilts permanently. The soil is then said to be at the *permanent wilting point*. In clay soils, the permanent wilting point is reached when the water content drops below 15%, while in sandy soils, the permanent wilting point may be as low as 4%. *Available water* is soil water between field capacity and the permanent wilting point.

Soil pH

The **pH** (acidity or alkalinity) of a soil affects both the soil and the plants growing in it in various ways. Cranberries, for example, thrive under acidic conditions, but a soil that is unusually acid or alkaline may be toxic to the roots of other plants, and mycorrhizae do not survive in soils having pH extremes. These conditions, however, do not normally affect plants directly nearly as much as they affect nutrient availability. For example, alkalinity causes minerals, such as copper, iron, and manganese, to become less available to plants, while acidity, if high enough, inhibits the growth of nitrogen-fixing bacteria. Acid soils tend to be common in areas of high precipitation where significant amounts of bases are leached from the topsoil.

It is a common agricultural practice to counteract soil acidity by adding compounds of calcium or magnesium in a process known as *liming*. Alkaline soils can be made more acidic by the addition of sulfur, which is converted by bacteria to sulfuric acid. The addition of some nitrogenous fertilizers may have the same effect.

SUMMARY

5.1 Introduction to Roots and Soils
- Roots anchor plants and absorb water and minerals in solution.

5.2 How Roots Develop
- A germinating seed radicle becomes the first root. Taproots with branch roots, or adventitious roots that become a fibrous root system, develop from the radicle or the stem just above it. Many plants have combinations of both systems.

5.3 Root Structure
- Four zones, or regions, of young roots are recognized: (1) A protective root cap that also aids in the perception of gravity. (2) A region of cell division. Its apical meristem subdivides into a protoderm, which produces the epidermis; a ground meristem, which produces the cortex; and a procambium, which produces primary xylem and primary phloem. (3) A region of elongation in which the cells produced by the apical meristem become considerably longer and slightly wider. (4) A region of maturation in which the cells mature into the distinctive cell types of primary tissues.
- Some of the epidermal cells in the region of maturation develop root hairs; the root hairs greatly increase the absorptive surface of the root. The tissues that mature in this region are similar to those of stem tips, but pith is absent in most dicot roots and originates from the procambium in monocot roots.
- At its inner boundary, the cortex has an endodermis with Casparian strips made of lignin and suberin.
- Next to the endodermis toward the center of the root are parenchyma cells constituting the pericycle. Branch roots and the vascular cambium arise in the pericycle.
- In dicot roots, the primary xylem usually first forms a solid core with two to several arms in the center of the root; a pith may be present in monocot roots.
- Primary phloem is first produced in discrete patches between the primary xylem arms, but the tissues eventually appear as concentric cylinders. In woody plants, a cork cambium usually arises in the pericycle and produces cork tissues similar to those of stems. Roots may graft together naturally. There are no leaves in roots.

5.4 Specialized Roots
- Specialized roots include those for food or water storage; pneumatophores; aerial roots (velamen roots, prop roots, photosynthetic roots, and adventitious roots); contractile roots; and buttress roots. Haustoria are peglike, parasitic roots.

5.5 Mycorrhizae
- Mycorrhizae are mutualistic associations between roots and fungi.

5.6 Root Nodules
- Some plants have nitrogen-fixing bacteria in nodules on their roots.

5.7 Human Relevance of Roots
- Root crops include sugar beets, beets, turnips, rutabagas, parsnips, carrots, sweet potatoes, yams, and cassava. Several spices are obtained from roots. Other uses of roots include the production of alcohol and the extraction of dyes, drugs, insecticides, and poisons.

5.8 Soils
- Soils contain a mixture of ingredients, including sands, rocks, silt, clay, humus, dead organic matter, plant roots, small animals, microorganisms, plus air and water, within pore spaces of various sizes.
- A vertical column of soil exhibits horizons; the topsoil is divided into an upper *A* horizon and a lower

E horizon. The *B* horizon (subsoil) usually contains more clay and is lighter in color than the topsoil. The *C* horizon (bottom portion) constitutes the weathered soil parent material.

- Living organisms in the soil decompose organic matter, the source of most important plant nutrients. Animals also cultivate the soil. Soil erosion is affected by topography.
- Soil texture refers to the relative proportions of sand, silt, and clay. More than half of the composition by weight of mineral matter is oxygen.
- Soil structure is the arrangement of the soil particles into aggregates. Good soils are highly granular and have pore spaces that constitute about half the total volume.
- Water in the soil occurs as hygroscopic water, gravitational water, and capillary water.
- The field capacity of the soil is the amount of water that remains after the rest of the water has drained away by gravity. Soil reaches the permanent wilting point when plants wilt permanently because they can no longer extract enough water from the soil for their needs. Available water is soil water between field capacity and the permanent wilting point.

REVIEW QUESTIONS

1. Distinguish between a tiny root and a root hair. What is the function of a root hair?
2. What is the difference between parasitic roots and mycorrhizae?
3. If you were shown cross sections of a young root and a young stem from the same dicot plant, how could you tell them apart?
4. What is the function of the root cap, and from which meristem does it originate?
5. How do endodermal cells differ from other types of cells?
6. Where do branch roots originate?
7. List some spices and drugs obtained from roots.
8. What is soil parent material? Where does it come from, and how does it become soil?
9. What is the difference between soil texture and soil structure?
10. What types of soil water are recognized? What is available soil water?

DISCUSSION QUESTIONS

1. Japanese gardeners regularly trim away parts of the root system to assist in dwarfing a Bonsai plant. A plant's food is obtained through photosynthesis in the leaves; can you suggest why trimming the roots can cause dwarfing?
2. It was suggested that roots perceive gravity through the root cap. Would it really matter if roots grew randomly in the soil instead of responding to gravity?
3. From the viewpoint of the plant, can you suggest a practical reason for branch roots originating internally instead of at the surface?
4. When you eat a yam or a sweet potato, what kinds of compounds and cells are you consuming?
5. Persons associated with commercial nurseries and greenhouses often sterilize their soil by heating it to get rid of pests, but then they have to wait for a short time after it has cooled to use it. Why?

CHAPTER

6

Stems

6.1 Introduction to Stems
6.2 External Form of a Woody Twig
6.3 Origin and Development of Stems
   Key Theme 6.1: Standing in Fields of Stone
6.4 Tissue Patterns in Stems
   Steles
   Herbaceous Dicotyledonous Stems
   Woody Dicotyledonous Stems

   Key Theme 6.2: Dendroclimatology
   Monocotyledonous Stems

6.5 Specialized Stems
   Rhizomes
   Runners and Stolons
   Tubers
   Bulbs
   Corms
   Cladophylls
   Other Specialized Stems

6.6 Wood and Its Uses
   Properties of Wood

OVERVIEW

After a brief introduction, this chapter discusses the origin and development of stems. Structures such as the apical meristem and the tissues derived from it, including leaf gaps, cambia, secondary tissues, and lenticels, are included. This general discussion is followed by notes on the distinctions between herbaceous and woody dicot stems and monocot stems. This section covers annual rings, rays, heartwood and sapwood, resin canals, bark, laticifers, and vascular bundles.

Next, there is a survey of specialized stems (rhizomes, stolons, tubers, bulbs, corms, cladophylls, and others). The chapter concludes with a discussion of the economic importance of wood and stems.

Tree trunks of giant sequoia (*Sequoiadendron giganteum*) found only in Sequoia National Park within the Sierra Nevada mountain range of California; sequoias are the bulkiest trees in the world. A member of a related species, the coastal redwood (*Sequoia sempervirens*), which grows only along the Pacific Coast, is the tallest tree in the world. Together, these two tree species share the title of State Trees of California. © Doug Sherman/Geofile

6.1 Introduction to Stems

Learning Outcome

a. Explain indeterminate growth and how this characteristic makes some plants different from animals.

A brief check of items around the house and garage—or even the house itself—soon reveals that much of what we use daily and take for granted, from pencils and pianos to newspapers and brooms, has some wood content. Most of that wood comes from plant stems. In fact, stems have been an integral part of human life ever since cave dwellers first used wooden clubs to kill for food.

Grafting, which usually involves artificially uniting stems or parts of stems of different but related varieties of plants, has been practiced by humans for hundreds of years. The careful matching of certain tissues is critical to its success, as is seen in the discussion of grafting in Chapter 14 and in Appendix 4. To understand how and why grafts may or may not be successful and to identify which parts of stems are useful, we first need to examine the structure of stems and learn the basic functions of the various tissues.

Unlike animals, some plants have *indeterminate growth* (i.e., they can grow indefinitely), with the meristems at their tips increasing their length and other meristems increasing their girth for hundreds or even thousands of years. In stems, the cells produced by the meristems usually become the familiar, erect, aerial *shoot system* with branches and leaves. In certain plants, such as ferns or perennial grasses, this shoot system may develop horizontally beneath or at the surface of the ground; in other plants, the stem may be so short and inconspicuous as to appear nonexistent. In a number of plants, stems are modified in ways that allow specialized functions, such as climbing or the storage of food or water.

6.2 External Form of a Woody Twig

Learning Outcomes

a. Define the following terms: *node, internode, blade, petiole,* and *stipule.*
b. Explain what is meant by deciduous growth, and explain how woody plants may be identified in the winter.

A woody twig consists of an axis with attached leaves (Fig. 6.1). If the leaves are attached to the twig alternately or in a spiral around the stem, they are said to be *alternate,*

Figure 6.1 A woody twig. (a) The twig in its winter condition. (b) The twig as it appeared the summer before.

or *alternately arranged.* If the leaves are attached in pairs, they are said to be *opposite,* or *oppositely arranged,* or if they are in whorls (groups of three or more), their arrangement is *whorled.* The area, or region (not structure), of a stem where a leaf or leaves are attached is called a **node,** and a stem region between nodes is called an **internode.** A leaf usually has a flattened **blade** and in most cases is attached to the twig by a stalk called the **petiole.**

Each angle between a petiole and the stem contains a bud. The angle is called an *axil,* and the bud located in the axil is an *axillary bud.* In flowering plants (angiosperms), axillary buds may become branches, or they may contain tissues that will develop into the next season's flowers. Most buds are protected by one to several *bud scales,* which fall off when the bud tissue starts to grow.

There is often (but not always) a *terminal bud* present at the tip of each twig. A terminal bud usually resembles an axillary bud, although it is often a little larger. Unlike axillary buds, terminal buds do not become separate branches, but, instead, the meristems within them normally produce tissues that make the twig grow longer during the growing season. The bud scales of a terminal bud leave tiny scars around the twig when they fall off in the spring. Counting the number of groups of *bud scale scars* on a twig can reveal the age of the twig.

Sometimes other scars of different origin also occur on a twig. These scars come from a leaf that has **stipules** at the base of the petiole. Stipules are paired, often somewhat leaflike, appendages that may remain throughout the life of the leaf. In some plants, they fall off as the buds expand in the spring, leaving tiny *stipule scars.* The stipule scars may resemble a fine line encircling the twig, or they may be very inconspicuous small scars on either side of the petiole base.

Deciduous trees and shrubs (those that lose all their leaves annually) characteristically have dormant axillary buds with **leaf scars** left below them after the leaves fall. Tiny **bundle scars,** which mark the location of the water-conducting and food-conducting tissues, are usually visible within the leaf scars. There may be one to many bundle scars present, but more often than not, there are three. The shape and size of the leaf scars and the arrangement and numbers of the bundle scars are characteristic for each species. One can often identify a woody plant in its winter condition by means of scars and buds.

6.3 Origin and Development of Stems

Learning Outcomes

a. Name the tissues that develop from shoot apices and the meristems from which various tissue types are derived.

b. State the differences between primary tissues and secondary tissues.

c. List and describe the function of each of the following: vascular cambium, cork cambium, stomata, and lenticels.

There is an *apical meristem* (tissue in which cells actively divide) at the tip of each stem, and it is this meristem that contributes to an increase in the length of the stem. The apical meristem is dormant before the growing season begins. It is protected by bud scales of the bud in which it is located and to a certain extent by leaf **primordia** (singular: **primordium**), the tiny embryonic leaves that will develop into mature leaves after the bud scales drop off and growth begins. The apical meristem in the embryonic stem of a seed is also dormant until the seed begins to germinate.

When a bud begins to expand or a seed germinates, the cells of the apical meristem undergo mitosis, and soon three *primary meristems* develop from it (see Figs. 4.1 and 6.2). The outermost of these primary meristems, the **protoderm,** gives rise to the *epidermis.* Although there are exceptions, the epidermis is typically one cell thick and usually becomes coated with a thin, waxy, protective layer, the *cuticle.* A cylinder of strands constituting the **procambium** appears to the interior of the protoderm. (The procambium produces water-conducting *primary xylem* cells and *primary phloem* cells that have several functions, including the conduction of food.)

The remainder of the meristematic tissue, called **ground meristem,** produces two tissues composed of parenchyma cells. The parenchyma tissue in the center of the stem is the **pith.** Pith cells tend to be very large and may break down shortly after they are formed, leaving a cylindrical, hollow area. Even if they do not break down early, they may eventually be crushed as new tissues produced by other meristems add to the girth of the stem, particularly in woody plants. The other tissue produced by the ground meristem is the **cortex.** The cortex may become more extensive than the pith, but in woody plants, it, too, eventually will be crushed and replaced by new tissues produced from within. The parenchyma of both the pith and the cortex function in storing food or sometimes, if chloroplasts are present, in manufacturing it.

All five of the tissues produced by this apical meristem complex (epidermis, primary xylem, primary phloem, pith, and cortex) arise while the stem is increasing in length and are called *primary tissues.* As these primary tissues are produced, the leaf primordia and the *bud primordia* (embryonic buds in the axils of the leaf primordia) develop into mature leaves and buds (Fig. 6.2). As each leaf and bud develop, a strand of xylem and phloem, called a *trace,* branches off from the cylinder of xylem and phloem extending up and down the stem and enters the leaf or the bud. As the traces branch from the main cylinder of xylem and phloem, each trace leaves a little, thumbnail-shaped gap in the cylinder of vascular tissue. These gaps, called **leaf gaps** and **bud gaps,** are filled with parenchyma tissue (Fig. 6.3).

A narrow band of cells between the primary xylem and the primary phloem may retain its meristematic nature and become the **vascular cambium,** one of the two lateral meristems. The vascular cambium is often referred to simply as the *cambium.* The cells of the cambium continue to divide indefinitely, with the divisions taking place mostly in a plane parallel to the surface of the plant. The secondary tissues

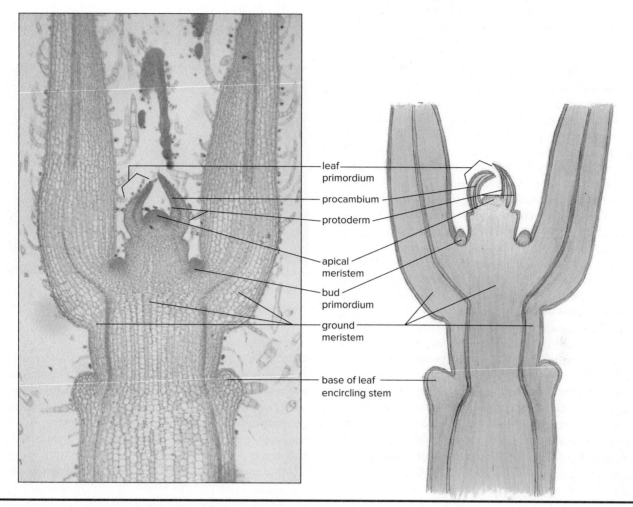

Figure 6.2 A longitudinal section through the tip of a *Coleus* stem. ×800. © G. S. Ellmore

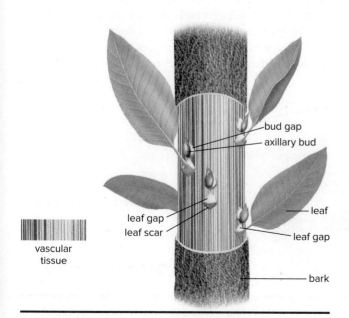

Figure 6.3 A portion of a stem showing leaf gaps and bud gaps in the cylinder of vascular tissue.

produced by the vascular cambium add to the girth of the stem instead of to its length (Fig. 6.4).

Cells produced by the vascular cambium become *tracheids, vessel elements, fibers,* or other components of *secondary xylem* (*inside* of the meristem, toward the center), or they become *sieve tube members, companion cells,* or other components of *secondary phloem* (*outside* of the meristem, toward the surface). The functions of these secondary tissues are the same as those of their primary counterparts—secondary xylem conducts water and soluble nutrients, while secondary phloem conducts, in soluble form, food manufactured by photosynthesis throughout the plant.

In many plants, especially woody species, a second cambium arises within the cortex or, in some instances, develops from the epidermis or phloem. This is called the **cork cambium,** or **phellogen.** The cork cambium produces boxlike **cork cells,** which become impregnated with **suberin,** a waxy substance that makes the cells impervious to moisture. The cork cells, which are produced annually in cylindrical layers, die shortly after they are formed. The cork cambium may also produce parenchyma-like **phelloderm** cells to the inside. Cork tissue makes up the outer bark of woody plants; it functions in reducing water loss and in

KEY THEME 6.1 ecology
Standing in Fields of Stone

The combination of high altitude; dry, cool air; low rainfall; high winds; and poor soil has provided sustenance for the oldest known living species on the earth! These ancient warriors, whose great age was unknown until 1953, are the **bristlecone pines** that flourish atop the arid mountains of the Great Basin from Colorado to California. The oldest—almost 5,000 years old—is located in the Ancient Bristlecone Pine Forest high in the White Mountains of California. Some of the trees standing today were seedlings when the great pyramids were built, were middle-aged trees during the time of Christ, and today are hoary patriarchs standing in fields of stone.

There are two species of bristlecones (Box Figure 6.1A), one living in the westernmost regions (*Pinus longaeva*) and the other inhabiting the eastern regions (*Pinus aristata*). The trees do not grow very tall, with none over 60 feet and many much shorter. Typical of their girth is the bristlecone named the "Patriarch," which is just over 36.5 feet around, but it is a relative youngster at 1,500 years. With the short summer growing season high in these mountains, bristlecones typically grow 1/100th of an inch or less in diameter in any given year. The trees stand in isolation, each appearing as a sentinel, overlooking an otherwise barren rock-strewn landscape. Their needles are remarkable. While occurring five per bundle and about 1 to 1.5 inches in length, they can live

Box Figure 6.1A An ancient bristlecone pine (*Pinus aristata*) from the White Mountains of California. © *Stephen Saks/Getty Images*

for 20 to 30 years before being cast off. This extraordinary leaf longevity gives the trees a stable photosynthetic output and sustains the tree during years of unusual stress when producing new leaf tissue would be difficult. The trees must generate new leaves and cones as well as produce enough reserves for the long winter months, all on scant annual precipitation of about 10 inches.

---

protecting the stem against mechanical injury (see the discussion of *periderm* in Chapter 4).

Cork tissue cuts off water and food supplies to the epidermis, which soon dies and is sloughed off. In fact, if the cork were to be formed as a solid cylinder covering the entire stem, vital gas exchange with the interior of the stem would not be possible. In young stems, such gas exchange takes place through the stomata, located in the epidermis (see Figs. 7.6 and 9.13). As woody stems age, **lenticels** (see Fig. 4.14) develop beneath the stomata. As cork is produced, the parenchyma cells of the lenticels remain, so that exchange of gases (e.g., oxygen, carbon dioxide) can continue through spaces between the cells. Lenticels occur in the fissures of the bark of older trees and often appear as small bumps on younger bark. In birch and cherry trees, the lenticels form conspicuous horizontal lines.

Differences between the activities of the apical meristem and those of the cambium and cork cambium become apparent

Figure 6.4 An illustration of how a cell of the vascular cambium produces new secondary phloem cells to the outside and new secondary xylem cells to the inside. Note, in cross section, that the cambium gradually becomes shifted away from the center as new cells are produced. Phloem is produced before xylem in secondary growth.

Ecologists have long noted the peculiar distribution and growth habits of certain plants. Why do plants grow where they do? What adaptations permit them to survive in life-threatening environments? The bristlecones' age and habitat offer several insights. These trees grow in places on earth where no other plant can grow. One answer appears to be the type of soil in which they are anchored. Stands of bristlecone pine grow on outcrops of dolomite, an alkaline limestone substrate low in nutrients but of higher moisture content than the surrounding sandstone. The granite and sandstone formations surrounding the dolomite outcroppings support sagebrush and Limber pine, but not bristlecones. At these altitudes, the radiant sunlight is extreme. Dolomite reflects more sunlight than other rocks and thus keeps the root zones cooler and more moisture-laden during the important growing season.

Another survival tactic is revealed by taking small core samples of the wood. These trees have large amounts of dieback (deadwood that is no longer functional), reducing the amount of tissue that the leaves need to supply with food. For example, one bristlecone over 4,000 years old is nearly 4 feet in diameter, but its functional conducting tissue is a ring only 10 inches wide.

Other characteristics provide bristlecones with survival advantages. Because their dense, resin-filled wood renders their sites inhospitable for colonization by pathogenic fungi or bacteria, they are relatively free from these attacks. These trees are often struck by lightning, but with the absence of ground cover and decaying leaf litter, fire rarely spreads from tree to tree.

As with other trees, the age of these living trees is determined by an instrument called an **increment borer,** a type of drill that is inserted into the tree trunk at its widest girth. Using a hollow drill bit, a linear core of wood is obtained that can be "read" for the number of **annual rings.** In this way, the age of a tree can be obtained without cutting the tree down to examine the annual rings revealed in the cross section of the stump. Each year, a tree will add a layer of wood to its trunk, and these layers become the annual rings that can be observed in a cross section of the trunk. During spring growth of wood, large-diameter water-conducting cells are formed; later in the summer, the water-conducting cells produced have a smaller diameter. This difference in appearance between early (spring) wood and late (summer) wood is sufficient to make each growth increment distinctive, and thus counting of the rings possible.

In 1957, the bristlecone pine "Methuselah" was discovered and determined to be 4,723 years old. Methuselah remains today as the world's oldest living organism. But what about tomorrow? After surviving nearly five millennia, Methuselah is being protected against a more insidious enemy. Standing in its field of stone without a marker because of fear of vandalism, Methuselah serves as a reminder that the human species can be just as destructive as any microbe.

D. C. Scheirer

---

if one drives a nail into the side of a tree and observes it over a period of years. The nail may eventually become embedded as the stem increases in girth, but it will always remain at the same height above the ground because the cells that increase the length of a stem are produced only at the tips.

6.4 Tissue Patterns in Stems

Learning Outcomes

a. State the meaning and origin of the terms *dicot* and *monocot.*
b. Describe how herbaceous and woody dicotyledonous stems differ.
c. Provide a sketch of a woody stem, and show the location of heartwood, sapwood, and bark.
d. Draw a generalized monocot stem cross section to show how vascular bundles are arranged.

Steles

Primary xylem, primary phloem, and the pith, if present, make up a central cylinder called the **stele** in most younger and a few older stems and roots. The simplest form of stele, called a *protostele,* consists of a solid core of conducting tissues in which the phloem usually surrounds the xylem. Protosteles were common in primitive seed plants that are now extinct, and they are also found in whisk ferns, club mosses (see Chapter 21), and other relatives of ferns. *Siphonosteles,* which are tubular with pith in the center, are common in ferns.

Most present-day flowering plants and conifers have *eusteles* in which the primary xylem and primary phloem are in discrete *vascular bundles,* as discussed in the section "Herbaceous Dicotyledonous Stems."

Flowering plants develop from seeds that have either one or two "seed leaves," called **cotyledons,** attached to their embryonic stems (see Chapters 8 and 23). The seeds of pines and other cone bearing trees have several (usually eight) cotyledons. The cotyledons usually store food needed by the young seedling until its first true leaves can produce food themselves.

Flowering plants that develop from seeds having two cotyledons are called **dicotyledons** (usually abbreviated to **dicots**), while those developing from seeds with a single cotyledon are called **monocotyledons** (abbreviated to **monocots**). Dicots and monocots differ from one another in several other respects; the differences in stem structure are noted in the following sections, and a summary of these and other differences in these two groups of flowering plants is given in Table 8.1.

Herbaceous Dicotyledonous Stems

In general, plants that die after going from seed to maturity within one growing season (**annuals**) have green,

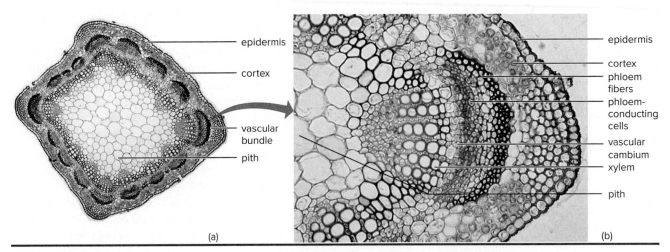

Figure 6.5 (a) A cross section of an alfalfa (*Medicago*) stem. ×40. The tissue arrangement is typical of herbaceous dicot stems. (b) An enlargement of a small portion of the outer part of the stem. ×400. *(a-b)* © *James E. Bidlack*

herbaceous (nonwoody) stems. Most monocots are annuals, but many dicots (discussed next) are also annuals.

The tissues of annual dicots are largely primary, although *cambia* (plural of *cambium*) may develop some secondary tissues. Herbaceous dicot stems (Fig. 6.5) have discrete **vascular bundles** composed of patches of xylem and phloem. The vascular bundles are arranged in a cylinder that separates the cortex from the pith, although in a few plants (e.g., foxgloves), the xylem and the phloem are produced as continuous rings (cylinders) instead of in separate bundles.

The procambium produces only primary xylem and phloem, but later, a vascular cambium arises between these two primary tissues and adds secondary xylem and phloem to the vascular bundles. In some plants, the cambium extends between the vascular bundles, appearing as a narrow ring, producing not only the conducting tissues within the bundles but also the parenchyma cells between them. In other plants, the cambium is not in an uninterrupted cylinder but is instead confined to the bundles, each of which has its own small band of cambium between the xylem and phloem.

Woody Dicotyledonous Stems

In the early stages of development, the primary tissues of stems of young herbaceous dicots, woody dicots, and cone-bearing trees are all arranged in a similar fashion. In woody plants, however, obvious differences begin to appear as soon as the vascular cambium and the cork cambium develop. The most conspicuous differences involve the secondary xylem, or wood, as it is best known (Fig. 6.6). Some tropical trees (e.g., ebony), in which both the vascular cambium and the cork cambium are active all year, produce an ungrained, uniform wood. The wood of most trees, however, is produced seasonally. In trees of temperate climates, virtually all growth takes place during the spring and summer and then ceases until the following spring.

When the vascular cambium of a typical broadleaf tree first becomes active in the spring, it usually produces

Figure 6.6 A cross section of a portion of a young linden (*Tilia*) stem. ×300. © *Kingsley Stern*

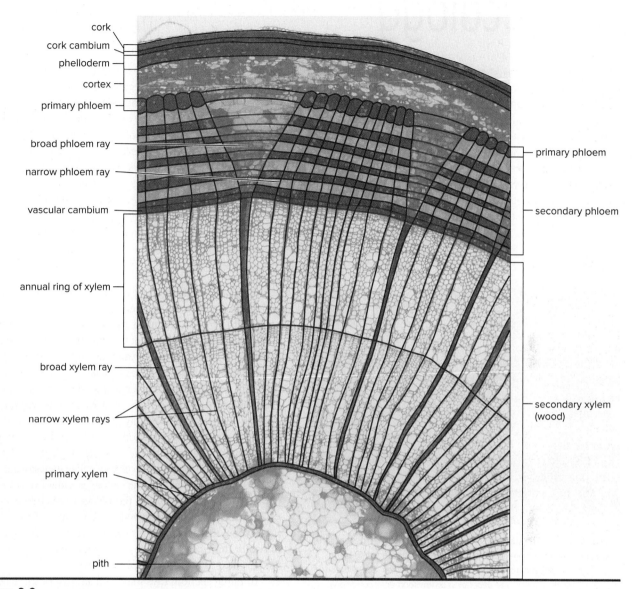

Figure 6.6 Continued.

relatively large vessel elements of secondary xylem; such xylem is referred to as *spring wood*. As the season progresses, the vascular cambium may produce vessel elements whose diameters become progressively smaller in each succeeding series of cells produced, or there may be fewer vessel elements in proportion to tracheids produced until tracheids (and sometimes fibers) predominate.

The xylem that is produced after the spring wood, and which has smaller or fewer vessel elements and larger numbers of tracheids, is referred to as *summer wood*. Over a period of years, the result of this type of switch between the early spring and the summer growth is a series of alternating concentric rings of light and dark cells. One year's growth of xylem is called an annual ring. In conifers, the wood consists mostly of tracheids, with vessels and fibers being absent. Annual rings are still visible, however, because the first tracheids produced in the spring are considerably larger and lighter in color than those produced later in the growing season. Note that an annual ring normally may contain many layers of xylem cells, and it is all the layers produced in one growing season that constitute an annual ring—not just the dark layers.

The vascular cambium produces more secondary xylem than it does phloem. Xylem cells also have stronger, more rigid walls than those of phloem cells, and they are less likely to collapse under tension. As a result, the bulk of a tree trunk consists of annual rings of wood. The annual rings not only indicate the age of the tree (because, normally, only one is produced each year), but they can also tell something of the climate and other conditions that occurred during the tree's lifetime (Fig. 6.7). For example, if the rainfall during a particular year is higher than normal, the annual ring for that year will be wider than usual because of increased growth. Sometimes, caterpillars or locusts will strip the leaves of a tree shortly after they have appeared. This usually results in a narrow annual ring, because very little growth can take place after defoliation.

If there is a fire that doesn't result in the death of the tree, it may be possible to determine when the fire occurred, because the burn scar may appear next to a given ring.

KEY THEME 6.2 ecology — Dendroclimatology

Dendroclimatology is the science of determining climate history by examining annual growth rings of trees. These annual rings provide a temporal record of environmental conditions and can be used to date important events in history. For example, Alaskan Inuit oral history tells of a disaster described as "the time that summertime did not come." Many people were said to have died as a result of this cold period and the famine that followed. Dendroclimatology data have shown that, in fact, a dramatic climate event occurred in 1783, probably as a result of a large volcanic eruption in Iceland in June of that year. During a good growing season, a tree lays down a wide ring of growth, with dense wood in late summer. During the summer of 1783, annual rings were normal at the beginning of the summer, but the density of the wood was unusually low in late summer. This indicates that summer ended quickly, before the trees could produce dense late summer wood. The Laki volcanic eruption in Iceland resulted in low temperatures throughout the Northern Hemisphere in 1783 and apparently had an especially severe effect on the people of northwest Alaska.

Figure 6.7 Colored scanning electron micrograph of a cross section through a small stem. Note the series of concentric rings radiating out from the center (top right) of the stem. The apparent "holes" are tracheary elements, and they vary in size depending on environmental conditions each year, and hence form the annual rings. © Clouds Hill Imaging Ltd./Science Source

The most recent season's growth is next to the vascular cambium, and one need only count the rings back from the cambium to determine the actual year of the fire.

It is not necessary to cut down a tree to determine its age. Instead, botanists and foresters use an *increment borer* to find out how old a woody plant is. This device, which resembles a piece of pipe with a handle on one end, removes a plug of wood from the tree perpendicular to the axis, and the annual rings in the plug can then be counted. The small hole left in the tree can be treated with a disinfectant to prevent disease and covered up without harm to the tree.

A count of annual rings has produced some red faces on at least one occasion. The Hooker Oak, which was named in honor of Sir Joseph Hooker, a famous British botanist who once examined it, was located in the community of Chico, California. Until its demise in 1977, thousands of visitors from all over the world visited the huge oak, which provided enough shade for 9,000 people on a summer day. A plaque indicating the tree to be over 1,000 years old was located beneath the tree. A count of rings after its death, however, revealed that the Hooker Oak was less than 300 years old.

When a tree trunk is examined in transverse, or cross section, fairly obvious lighter streaks or lines can be seen radiating out from the center across the annual rings (see Figs. 6.6 and 6.8). These lines, called **vascular rays,** consist of parenchyma cells that may remain alive for 10 or more years. Their primary function is the lateral conduction of nutrients and water from the stele, through the xylem and phloem, to the cortex, with some cells also storing food. Any part of a ray within the xylem is called a *xylem ray,* while its extension through the phloem is called a *phloem ray.* In trees such as basswood (*Tilia*), some of the phloem rays, when observed in cross section, flare out from a width of two or three cells near the cambium to many cells wide in the part next to the cortex (see Fig. 6.6).

In radial section, rays may be from two or three cells to 50 or more cells deep, but the majority of rays in both xylem and phloem are one or two cells wide. Ray cells can be seen in cross section if a woody stem is cut or split lengthwise along a ray (Fig. 6.8). Another view of rays (in tangential section) is obtained when the stem is cut at a tangent (i.e., cut lengthwise and off center) (see Fig. 6.17).

As a tree ages, the protoplasts of some of the parenchyma cells that surround the vessels and tracheids grow through the pits in the walls of these conducting cells and balloon out into the cavities. As the protoplasm continues to expand, much of the cavity of the vessel or tracheid becomes filled. Such protrusions, called *tyloses* (singular: *tylosis*), prevent further conduction of water and dissolved substances. When this occurs, resins, gums, and tannins begin to accumulate, along with pigments that darken the color of the wood.

This older, darker wood at the center is called **heartwood,** while the lighter, still-functioning xylem closest to the cambium is called **sapwood** (Fig. 6.9). Except for giving strength and support, the heartwood is not of much use to the tree, because it can no longer conduct materials. A tree may live and function perfectly well after the heartwood has

Figure 6.8 A three-dimensional, magnified view of a block of a woody dicot.

Figure 6.9 (a) A cross section of a 137-year-old black walnut tree (*Juglans nigra*) showing the darker heartwood to the inside and the lighter sapwood to the outside, along with the vascular cambium, phloem, and cork. (b) Longitudinal section showing heartwood and sapwood in relation to other tissues of a tree trunk. © *James E. Bidlack*

rotted away and left the interior hollow. It is even possible to remove part of the sapwood and other tissues and apparently not affect the tree very much, as has been done with giant trees, such as the coastal redwoods of California, where holes big enough to drive a car through have been cut out without killing the trees (Fig. 6.10).

Sapwood forms at roughly the same rate as heartwood develops, so there is always enough "plumbing" for the vital conducting functions. The relative widths of the two types of wood, however, vary considerably from species to species. For example, in the golden chain tree (a native of Europe and a member of the Legume Family), the sapwood is usually only one or two rings wide, while in several North American trees (e.g., maple, ash, and beech), the sapwood may be many rings wide.

Pines and other cone-bearing trees have xylem that consists primarily of tracheids; no fibers or vessel elements are produced. Because it has no fibers, the wood tends to be softer than wood with fibers and is commonly referred to as *softwood*, while the wood of woody dicot trees is called *hardwood*.

In many cone-bearing trees, resin canals are scattered not only through the xylem but throughout other tissues as well. These canals are tubelike and may or may not be branched; they are lined with specialized cells that secrete resin (discussed in Chapter 22) into their cavities (Fig. 6.11). Although resin canals are commonly associated with cone-bearing trees, they are not confined to them. Tropical flowering plants, such as olibanum and myrrh trees, have resin ducts in the bark that produce the soft resins frankincense and myrrh of biblical note.

While the vascular cambium is producing secondary xylem to the inside, it is also producing secondary phloem to the outside. The term **bark** is usually applied to all the tissues

Figure 6.10 This coastal redwood is thriving despite the removal of its lower heartwood and a little of its sapwood. Cork cells of bark are dead and their removal would not affect growth. © *Kingsley Stern*

Figure 6.11 Resin canals in a portion of a pine (*Pinus*) stem. ×100. © *James E. Bidlack*

outside the cambium, including the phloem. Some scientists distinguish between the *inner bark,* consisting of primary and secondary phloem, and the *outer bark* (*periderm*), consisting of cork tissue and cork cambium. Despite the presence of fibers, the thin-walled conducting cells of the phloem usually are not able to withstand for many seasons the pressure of thousands of new cells added to their interior, and the older layers become crushed and functionless.

The parenchyma cells of the cortex to the outside of the phloem also function only briefly because they, too, become crushed or sloughed off. Before they disappear, however, the cork cambium begins its production of cork, and because new xylem and phloem tissues produced by the vascular cambium arise to the inside of the older phloem, the mature bark may consist of alternating layers of crushed phloem and cork.

The younger layers of phloem nearest to the cambium transport, via their sieve tubes, sugars and other substances in solution from the leaves where they are made to various parts of the plant, where they are either stored or used in the process of *respiration* (discussed in Chapter 10). The presence of sugar in the phloem was recognized in the past by Native Americans. Some stripped the young phloem and cambium from Douglas fir trees and used the dried strips as food for winter and in emergencies.

Specialized cells or ducts called **laticifers** are found in about 20 families of herbaceous and woody flowering plants. These cells are most common in the phloem but occur throughout all parts of the plants. The laticifers, which resemble vessels, form extensive branched networks of latex-secreting cells originating from rows of meristematic cells. Unlike vessels, however, the cells remain living and may have many nuclei.

Latex is a thick fluid that is white, yellow, orange, or red in color and consists of gums, proteins, sugars, oils, salts, alkaloidal drugs, enzymes, and other substances. Its function in the plant is not clear, although some believe it aids in closing wounds. Some forms of latex have considerable commercial value (see the discussion under "The Spurge Family" in Chapter 24). Of these, rubber is the most important. Members of indigenous Amazon tribes utilized rubber for making balls and containers hundreds of years before Pará rubber trees were cultivated for their latex. The chicle tree produces a latex used in the making of chewing gum. Several poppies, notably the opium poppy, produce latex-containing, important drugs, such as morphine and its derivative, known as heroin. Other well-known latex producers include milkweeds, dogbanes, and dandelions.

Monocotyledonous Stems

Most monocots (e.g., grasses, lilies) are herbaceous plants that do not grow tall. The stems have neither a vascular cambium nor a cork cambium and thus produce no secondary vascular tissues or cork. As in herbaceous dicots, the surfaces of the stems are covered by an epidermis, but the xylem and phloem tissues produced by the procambium

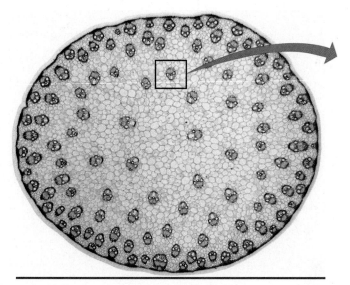

Figure 6.12 A cross section of a monocot (corn—*Zea mays*) stem. ×20. © *James E. Bidlack*

Figure 6.13 A single vascular bundle of corn (*Zea mays*) enlarged. ×400. © *Kingsley Stern*

appear in cross section as discrete vascular bundles scattered throughout the stem instead of being arranged in a ring (Fig. 6.12).

Each bundle, regardless of its specific location, is oriented so that its xylem is closer to the center of the stem and its phloem is closer to the surface. In a typical monocot, such as corn, a bundle's xylem usually contains two large vessels with several small vessels between them (Fig. 6.13). The first-formed xylem cells usually stretch and collapse under the stresses of early growth and leave an irregularly shaped air space toward the base of the bundle; the remnants of a vessel are often present in this air space. The phloem consists entirely of sieve tubes and companion cells, and the entire bundle is surrounded by a sheath of thicker-walled sclerenchyma cells. The parenchyma tissue between the vascular bundles is not separated into cortex and pith in monocots, although its function and appearance are the same as those of the parenchyma cells in cortex and pith.

In a corn stem, there are more bundles just beneath the surface than there are toward the center. Also, a band of sclerenchyma cells, usually two or three cells thick, develops immediately beneath the epidermis, and parenchyma cells in the area develop thicker walls as the stem matures. The concentration of bundles, combined with the band of sclerenchyma cells beneath the epidermis and the thicker-walled parenchyma cells, all contribute to giving the stem the capacity to withstand stresses resulting from summer storms and the weight of the leaves and the ears of corn as they mature.

In wheat, rice, barley, oats, rye, and other grasses, there is an *intercalary meristem* (discussed in Chapter 4) at the base of each internode; like the apical meristem, it contributes to increasing stem length. Although the stems of such plants elongate rapidly during the growing season, growth is columnar (i.e., there is little difference in diameter between the top and the bottom) because there is no vascular cambium producing tissues that would add to the girth of the stems.

Palm trees, which differ from most monocots in that they often grow quite large, do so primarily as their parenchyma cells continue to divide and enlarge without a true cambium developing. Several popular houseplants (e.g., ti plants, *Dracaena, Sansevieria*) are monocots in which a secondary meristem develops as a cylinder that extends throughout the stem. Unlike the vascular cambium of dicots and conifers, this secondary meristem produces only parenchyma cells to the outside and secondary vascular bundles to the inside.

Several commercially important cordage fibers (e.g., broomcorn, Mauritius and Manila hemps, sisal) come from the stems and leaves of monocots, but the individual cells are not separated from one another by *retting* (a process that utilizes the rotting power of microorganisms thriving under moist conditions to break down the thin-walled parenchyma cells) as they are when fibers from dicots are obtained. Instead, during commercial preparation, entire vascular bundles are scraped free of the surrounding parenchyma cells by hand; the individual bundles then serve as unit "fibers." If such fibers are treated with chemicals or bleached, the cementing middle lamella between the cells breaks down. Monocot fibers are not as strong or as durable as most dicot fibers.

6.5 Specialized Stems

Learning Outcome

a. Provide descriptions of rhizomes, stolons, tubers, bulbs, corms, cladophylls, and tendrils.

Although most higher plants have an erect shoot system, many species have specialized stems that are modified for various functions (Fig. 6.14). The overall appearance of specialized stems may differ markedly from that of the stems discussed

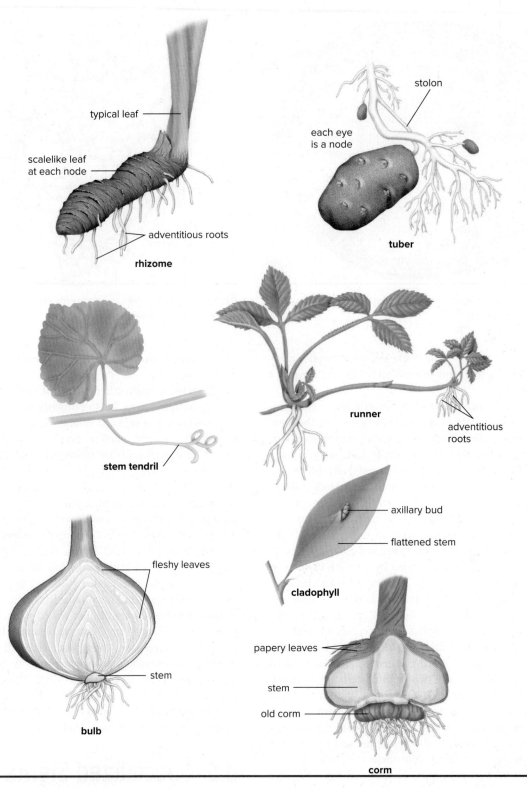

Figure 6.14 Types of specialized stems.

so far, but all stems have *nodes, internodes,* and *axillary buds;* these features distinguish them from roots and leaves, which do not have them. The leaves at the nodes of these specialized stems are often small and scalelike. They are seldom green, but full-sized, functioning leaves may also be produced. Descriptions of some of the specialized stems follow.

Rhizomes

Rhizomes (Fig. 6.14) are horizontal stems that grow below ground, often near the surface of the soil. They superficially resemble roots, but close examination will reveal scalelike leaves and axillary buds at each node, at least during some stage of development, with short to long internodes in between. Adventitious roots are produced all along the rhizome, mainly on the lower surface. As indicated in Chapter 5, the word *adventitious* refers to structures arising at unusual places, such as roots growing from stems, or leaves or buds appearing at places other than leaf axils and tips of stems. A rhizome may be a relatively thick, fleshy, food-storage organ, as in irises, or it may be quite slender, as in many perennial grasses or some ferns.

Runners and Stolons

Runners are horizontal stems that differ from rhizomes in that they grow aboveground, generally along the surface; they also have long internodes (Fig. 6.14). In strawberries, runners are usually produced after the first flowers of the season have appeared. Several runners may radiate out from the parent plant and within a few weeks may grow 1 meter (3 feet) or more long. Adventitious buds appear at alternate nodes along the runners and develop into new strawberry plants, which can be separated and grown independently. In saxifrages and some other houseplants, runners may produce new plants at intervals as they grow out and hang over the edge of the pot.

Stolons are similar to runners but are produced beneath the surface of the ground and tend to grow in different directions but usually not horizontally. In Irish potato plants, tubers are produced at the tips of stolons.

Some botanists consider stolons and runners to be variations of each other and prefer not to make a distinction between them.

Tubers

In potato plants, several internodes at the tips of stolons become **tubers** as they swell from the accumulation of food (Fig. 6.14). The mature tuber becomes isolated after the stolon to which it was attached dies. The "eyes" of the potato are actually nodes formed in a spiral around the thickened stem. Each eye consists of an axillary bud in the axil of a scalelike leaf, although this leaf is visible only in very young tubers; the small ridges seen on mature tubers are leaf scars.

Bulbs

Bulbs (Fig. 6.14) are actually large buds surrounded by numerous fleshy leaves, with a small stem at the lower end. Adventitious roots grow from the bottom of the stem, but the fleshy leaves comprise the bulk of the bulb tissue, which stores food. In onions, the fleshy leaves usually are surrounded by the scalelike leaf bases of long, green, aboveground leaves. Other plants producing bulbs include lilies, hyacinths, and tulips.

Corms

Corms resemble bulbs but differ from them in being composed almost entirely of stem tissue, except for the few papery, scalelike leaves sparsely covering the outside (Fig. 6.14). Adventitious roots are produced at the base, and corms, like bulbs, store food. The crocus and the gladiolus are examples of plants that produce corms.

Cladophylls

The stems of butcher's broom plants are flattened and appear leaflike. Such flattened stems are called **cladophylls** (or *clad-odes* or *phylloclades*) (Fig. 6.14). There is a node bearing very small, scalelike leaves with axillary buds in the center of each butcher's broom cladophyll. The feathery appearance of asparagus is due to numerous small cladophylls. Cladophylls also occur in greenbriers, certain orchids, prickly pear cacti (Fig. 6.15), and several other lesser-known plants.

Other Specialized Stems

The stems of many cacti and some spurges are stout and fleshy. Such stems are adapted for storage of water and food. Other stems may be modified in the form of *thorns,* as in the honey locust, whose branched thorns may be more than 3 decimeters (1 foot) long, but not all thornlike objects are necessarily modified stems. For example, at the base of the

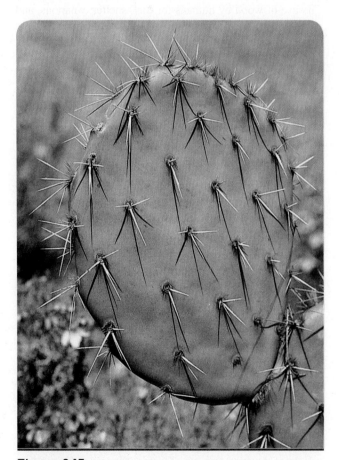

Figure 6.15 The flattened stems of prickly pear cacti (*Opuntia*) are cladophylls on which the leaves have been reduced to spines.
© Kingsley Stern

petiole of most leaves of the black locust is a pair of *spines* (modifed *stipules;* stipules were mentioned in the discussion of twigs and are discussed further in Chapter 7). The *prickles* of raspberries and roses, both of which originate from the epidermis, are neither thorns nor spines. Tiger lilies produce small, aerial *bulblets* in the axils of their leaves.

Climbing plants have stems modified in various ways that adapt them for their manner of growth. Some stems, called *ramblers,* simply rest on the tops of other plants, but many produce **tendrils** (see Fig. 6.14). These are specialized stems in the grape and Boston ivy, but are modified leaves or leaf parts in plants like peas and cucumbers. In Boston ivy, the tendrils have adhesive disks. In English ivy, the stems climb with the aid of adventitious roots that arise along the sides of the stem and become embedded in the bark or other support material over which the plant is growing.

6.6 Wood and Its Uses

Learning Outcomes

a. Describe how the density and durability of wood can affect its use in commercial products.

b. List five types of wood and what products may be derived from each.

The use of wood by humans for fuel, shelter, weapons, and other purposes dates back into antiquity, and present uses are so numerous that it would be impossible to list in a work of this type more than the most important ones. Before discussing the economic importance of wood, let's take a brief look at its properties.

In a living tree, up to 50% of the weight of the wood is water. Before the wood can be used, seasoning reduces the moisture content to 10% or less, either by air-drying it in ventilated stacks or by drying it in special ovens known as kilns. The seasoning has to be done gradually and under carefully controlled conditions, or the timber may warp and split along the rays, making it unfit for most uses. The dry part of wood is composed of 60% to 75% *cellulose* and about 15% to 25% *lignin,* an organic substance that makes the walls of xylem cells tough and hard. Other substances present in smaller amounts include resins, gums, oils, dyes, tannins, and starch. The proportions and amounts of these and other substances determine how various woods will be used (Fig. 6.16).

Properties of Wood

Density

The density of wood is among its most important physical properties. Technically, the density is the weight per unit volume. The weight is compared with that of an equal volume of water. If the weight of a volume of wood equals the weight of the same volume of water, it is said to have a specific gravity of 1.0. The specific gravity is less than 1.0 if the wood weighs less than the water and greater than 1.0 if the wood weighs more than water. Because of the considerable air space within the cells, most woods have a specific gravity of less than 1.0. The range of specific gravities of known woods varies from 0.04 to 1.40, the lightest commercially used wood being balsa with a specific gravity of about 0.12. Woods with specific gravities of less than 0.50 are considered light; those with specific gravities of above 0.70 are considered heavy. Among the heaviest woods are the South American ironwood and lignum vitae, with specific gravities of over 1.25. Lignum vitae, obtained from West Indian trees, is extremely hard wood and has been used instead of metal in the manufacture of main bearings for drive shafts of submarines because it is self-lubricating and less noisy.

Durability

A wood's ability to withstand decay caused by organisms and insects is referred to as its *durability.* Moisture is needed for the enzymatic breakdown of cellulose and other wood substances by decay organisms, but the seasoning process usually reduces the moisture to a level below that necessary for the fungi and other decay organisms to survive. Other natural constituents of wood that repel decay organisms include tannins and oils. Wood with a tannin content of 15% or more may survive on a forest floor for many years after diseases of the phloem and other causes have toppled it. Among the most durable of American woods are cedar, catalpa, black locust, red mulberry, and Osage orange. The least durable woods include cottonwood, willow, fir, and basswood.

Types of Sawing

Logs are usually cut longitudinally in one of two ways: along the radius or perpendicular to the rays (Fig. 6.17). Radially cut, or *quartersawed,* boards show the annual rings in side view; they appear as longitudinal streaks and are the most conspicuous feature of the wood. Only a few perfect quartersawed boards can be obtained from a log, making them quite expensive. Boards cut perpendicular to the rays (tangentially cut boards) are more common. In these, the annual rings appear as irregular bands of light and dark alternating streaks or patches, with the ends of the rays visible as narrower and less conspicuous vertical streaks. Lumber cut tangentially is referred to as being *plain-sawed,* or *slab cut.* Slabs are the boards with rounded sides at the outside of the log; they are usually made into chips for pulping.

Knots

Knots are the bases of lost branches that have become covered over time by new annual rings of wood produced by the cambium of the trunk. They are found in greater concentration in the older parts of the log toward the center, because in the forest, the lowermost branches of a tree (produced while the trunk was small in girth) often die from insufficient light. When a branch dies and falls off, the cambium

Hemlock. This relatively soft, light, straight-grained, resin-free wood, with its uniformly long fibers, is becoming one of the most important species for paper pulp. It is also used for structural lumber and plywood, and for boxes, barrels, and concrete forms.

White Pine. A softwood widely used in home construction and for virtually everything from masts and matches to boxes and crates. Its soft, uniform texture and straight grain cuts easily in every direction, polishes well, and warps or swells little.

Red Spruce. A favorite for violin sounding boards because of its high resonant qualities. A softwood, it is easy to work, and is light in relation to its strength and stiffness. These qualities also make it eminently suitable for ladder rails, canoe paddles, and oars.

Bald Cypress. Because it is weather-resistant without treatment, this wood was widely used for cross ties in the early days of railroading. Today it is used for water tanks and other applications requiring prolonged contact with water.

White Oak. Makes good barrels because the wood is resilient, durable, and impermeable to liquids. This hardwood, which is about twice as dense as white pine, has many other uses ranging from flooring to fine cabinet work.

White Ash. Perfect for baseball bats, tennis racquets, oars and long tool handles. This hardwood's major virtues are straight grain, stiffness, strength, moderate weight, good bending qualities, and capacity for wearing smooth.

Figure 6.16 Uses of common North American woods.

Hard Maple. In bowling pins and flooring for bowling alleys its uniform texture and hardness result in resistance to abrasion. The Romans used it for spears and lances. We turn it on lathes to make spools, bobbins, cue sticks, and croquet balls.

Hickory. A hardwood unsurpassed for the handles of impact tools like axes and hammers, and for skis, because of its hardness, strength, toughness, and resiliency. In horse-and-buggy days it was widely used for wheel spokes and rims, singletrees, and buggy shafts.

Black Walnut. A choice hardwood for fine furniture and interior paneling, because of the beauty of the heartwood grain, its ability to stay in place after seasoning, and its good machining properties. It is harder than oak and shock-resistant.

Figure 6.16 Continued.

Figure 6.17 How the surfaces of plain-sawed, quartersawed, and transversely sawed wood appear. (See also Fig. 6.8.)

at its base also dies, but the cambium of the trunk remains alive and increases the girth of the tree, slowly enveloping the dead tissue of the branch base until it may be completely buried and not visible from the surface. Knots usually weaken the boards in which they occur.

Wood Products

In the United States and Canada, about half of the wood produced is used as lumber, primarily for construction; the sawdust and other waste formed in processing the boards are converted to particle board and pulp. A considerable amount of lumber goes into the making of furniture, which may be constructed of solid wood or covered with a *veneer*. A veneer is a very thin sheet of desirable wood that is glued to cheaper lumber; it is carefully cut so as to produce the best possible view of the grain.

The next most extensive use of wood is for *pulp*, which among other things is converted by various processes to paper, synthetic fibers, plastics, and linoleum. In recent years, it has been added as a filler to commercial ice cream and bread. Some hardwoods are treated chemically or heated under controlled conditions to yield a number of chemicals, such as wood alcohol and acetic acid, but other sources of these products are now usually considered more economical. Charcoal, excelsior, cooperage (kegs, casks, and barrels), railroad ties, boxes and crates, musical instruments, bowling pins, tool handles, pilings, cellophane, and Christmas trees are but a few of the additional wood products worth billions of dollars annually on the world market (see Fig. 6.16).

In developing countries, approximately half of the timber cut is used for fuel, but in the United States and Canada, a little less than 10% is currently used for that purpose. In colonial times, wood was the almost exclusive source of heating energy. In the early 2000s, Brazil's major cities were still using scrub timber from the surrounding forests to energize their utilities, but rapidly depleting supplies and problems related to the greenhouse effect (discussed in Chapter 25) have pointed to the need for alternate sources of energy. Many types of coal are wood that has been compressed for millions of years until nearly pure carbon remains. The formation of coal and other fossils is discussed in Chapter 21. Although the world's supply of coal is still plentiful, the rate at which this fossil fuel is being consumed makes it obvious that resources will eventually be exhausted unless we find renewable or less destructive alternatives to meet our energy demands.

Some of the vast array of secondary products from stems, including dyes, medicines, spices, and foods, are discussed in later chapters and in the appendices.

SUMMARY

6.1 Introduction to Stems
- Stem structure and function need to be examined to understand practical uses of stems.
- The shoot system of plants is usually erect, but some stems may be horizontal or modified for climbing or food/water storage.

6.2 External Form of a Woody Twig
- Leaves of woody twigs may be arranged alternately, oppositely, or in a whorl. Nodes are stem regions where leaves are attached; internodes occur between nodes. Most leaves have petioles and blades. Axillary buds occur in leaf axils. Most buds are protected by bud scales. Terminal buds occur at twig tips. Terminal bud scales, when they fall, leave bud scale scars that help one determine the age of the twig.
- Stipules are paired appendages present at the bases of some leaves; when they fall off, they leave small scars on the twig. When whole leaves fall, they leave leaf scars on the twig, with tiny bundle scars within the leaf-scar surfaces.

6.3 Origin and Development of Stems
- Each stem has an apical meristem at its tip that produces tissues that increase the length of the stem. Leaf primordia develop into mature leaves when growth begins. Three primary meristems develop from an apical meristem: the protoderm gives rise to the epidermis; the procambium produces primary xylem and primary phloem; and the ground meristem produces pith and cortex.
- As leaves and buds develop from primordia, traces of xylem and phloem branch off from the main cylinder, leaving leaf gaps or bud gaps.

- A vascular cambium, producing secondary tissues, may arise between primary xylem and phloem. Secondary xylem cells include tracheids, vessel elements, and fibers. Secondary phloem cells include sieve-tube members and companion cells.
- In many plants, a cork cambium producing cork and phelloderm cells develops near the surface of the stem. Cork cells, which are part of the outer bark (periderm), have suberin in their walls. Suberin is impervious to moisture, and the outer bark, therefore, aids in protection. Lenticels in the bark permit gas exchange.

6.4 Tissue Patterns in Stems
- Primary vascular tissues and the pith, if present, constitute the stele. Protosteles have a solid core of xylem, usually surrounded by phloem; siphonosteles are tubular, with pith in the center; eusteles have the vascular tissues in discrete bundles.
- Dicotyledons (dicots) are plants whose seeds have two seed leaves (cotyledons), while monocotyledons (monocots) have seeds with one seed leaf. Herbaceous dicots have vascular bundles arranged in a ring in the stem.
- Woody dicots have most of their secondary tissues arranged in concentric layers. The most conspicuous tissue is wood (secondary xylem). In broadleaf trees, spring wood usually has relatively large vessel members, while summer wood has smaller vessels and/or a predominance of tracheids.
- An annual ring is one year's growth of xylem. A tree's age and other aspects of its history can be determined from annual rings. Rays, which function in lateral conduction, radiate out from the center of the trunk. Older wood toward the center (heartwood) ceases to function when its cells become plugged with tyloses. Younger, functioning wood (sapwood) is closer to the surface. A tree's functions are not particularly affected by the rotting of its heartwood.
- The wood of cone-bearing trees consists primarily of tracheids, and resin canals are often present. The wood of conifers has no fibers or vessels and is called softwood, while the wood of woody dicots is called hardwood. In woody plants, older tissues composed of thin-walled cells become crushed and functionless, and some are sloughed off.
- Laticifers are latex-secreting cells or ducts found in various flowering plants. The latex of some plants has considerable commercial value.
- Monocot stems have scattered vascular bundles and no cambia. The parenchyma tissue is not divided into pith and cortex. Each vascular bundle is surrounded by a sheath of sclerenchyma cells. Numerous bundles and a band of sclerenchyma cells and thicker-walled parenchyma cells just beneath the surface of monocot stems aid in withstanding stresses.

- Palm trees are monocots that become large because their parenchyma cells continue to divide. Other monocots develop a secondary meristem that produces parenchyma cells and secondary vascular bundles. Grasses have at the base of each internode intercalary meristems that contribute to rapid increases in length. Several commercially important cordage fibers are obtained from monocots.

6.5 Specialized Stems

- Specialized stems include rhizomes, stolons, tubers, bulbs, corms, cladophylls, and tendrils. Such stems may have adventitious roots.
- The dry part of wood consists primarily of cellulose and lignin. Resins, gums, oils, dyes, tannins, and starch are also present. Properties of wood that play a role in its use include density, specific gravity, and durability.
- Logs are usually cut longitudinally along the radius (quartersawed) or perpendicular to the rays (tangentially cut, plain-sawed, or slab cut). Knots are bases of lost branches that have become covered over by new wood; they usually weaken the boards in which they occur.

6.6 Wood and Its Uses

- About half the timber produced in the United States and Canada is used as lumber. Sawdust and waste are converted to particle board and pulp for paper, synthetics, and linoleum. Other timber is used for cooperage, charcoal, railroad ties, boxes, tool handles, and so forth. Developing countries use a greater proportion of their timber for fuel.

REVIEW QUESTIONS

1. What is the function of bud scales?
2. How can you tell the age of a twig?
3. Distinguish among procambium, vascular cambium, and cork cambium.
4. How can you tell, when you look at a cross section of a young stem, whether it is a dicot or a monocot?
5. What are laticifers?
6. An Irish or white potato is a stem, but a sweet potato is a root. How can you tell?
7. Distinguish among corms, bulbs, and tubers.
8. If you were looking at the top of a wooden desk, how could you tell if the wood had been radially or tangentially cut (quartersawed or plain-sawed)?
9. What differences are there between heartwood and sapwood?
10. What is meant by the specific gravity of wood?

DISCUSSION QUESTIONS

1. If the cambium of a tropical tree were active all year long, how would its wood differ from that of a typical temperate climate tree?
2. A nail driven into the side of a tree will remain at exactly the same distance from the ground for the life of the tree. Why?
3. Do climbing plants have any advantages over erect plants? Any disadvantages?
4. If two leaves are removed from a plant and one is coated with petroleum jelly, while the other is not, the uncoated leaf will shrivel considerably sooner than the coated one. Would it be helpful to coat the stems of young trees with petroleum jelly? Explain.
5. Suggest some reasons for heartwood being preferred to sapwood for making furniture.

CHAPTER 7

Leaves

7.1 Introduction to Leaves
7.2 Leaf Arrangements and Types
7.3 Internal Structure of Leaves
7.4 Stomata
7.5 Mesophyll and Veins
7.6 Specialized Leaves
   Shade Leaves
   Leaves of Arid Regions

   Key Theme 7.1: More on Leaf Structure
   Leaves of Aquatic Areas
   Tendrils
   Spines, Thorns, and Prickles
   Storage Leaves
   Flower-Pot Leaves
   Window Leaves
   Reproductive Leaves
   Floral Leaves (Bracts)
   Insect-Trapping Leaves
7.7 Autumnal Changes in Leaf Color
7.8 Abscission
7.9 Human and Ecological Relevance of Leaves
   Key Theme 7.2: Glass Cuts from Grass?

OVERVIEW

This chapter introduces leaves by comparing them with solar panels and by discussing their general functions, morphology, and dimensions. This is followed by descriptive information on basic leaf types and specific forms and arrangements. The chapter next discusses the internal structure of leaves, including epidermis and cuticle, stomata, glands, mesophyll, and veins.

Specialized leaves, including tendrils, spines, flower-pot leaves, window leaves, reproductive leaves, floral leaves, and insectivorous leaves, are then examined. The chapter concludes with an explanation of autumnal color changes and abscission and some observations on the human and ecological relevance of leaves.

Fall colors of autumn leaves on a pond in New England. © *Doug Sherman/Geofile*

7.1 Introduction to Leaves

Learning Outcome

a. Distinguish among simple, pinnately compound, and palmately compound leaves.

When trying to solve complex problems, we often turn to the natural world for inspiration. For example, engineers are working to optimize the production and storage of renewable energy. **Leaves** have much to teach us about this process. Their position on the plant maximizes light interception, they harvest a range of the solar spectrum, and they are able to store the energy they capture so the plant can use it later. Engineers use these design concepts when building solar panels to provide us with renewable energy.

Leaves originate as **primordia** in buds. In early spring, a leaf primordium may consist of fewer than 200 cells, but in response to changes in temperature, day length, and availability of water, hormones are produced that stimulate these cells to begin dividing. Within a few days or weeks, the original 200 cells have multiplied, differentiated, and expanded into a structure consisting of millions of cells. In some plants (e.g., *Eucalyptus*), the first leaves produced (*juvenile leaves*) may appear quite different in form from those produced later.

At maturity, most leaves have a stalk, called the **petiole,** and a flattened **blade,** or *lamina,* which has a network of veins (*vascular bundles*). A pair of leaflike, scalelike, or thornlike appendages, called *stipules,* are sometimes present at the base of the petiole. Occasionally, leaves may lack petioles; when they do, they are said to be **sessile.** Leaves of **deciduous** trees normally live through only one growing season, and even those of evergreen trees rarely function for more than a few years.

Leaves of flowering plants are associated with *leaf gaps* (as illustrated in Chapter 6), and all have an *axillary bud* at the base. Leaves may be *simple* or *compound.* A **simple leaf** has a single blade, while the blade of a **compound leaf** is divided into *leaflets* (see Fig. 7.4). Regardless of the number of leaflets, a compound leaf still has a single axillary bud at its base, with the leaflets having no such buds. **Pinnately compound** leaves have the leaflets in pairs along an extension of the petiole called a **rachis,** while **palmately compound** leaves have all the leaflets attached at the same point at the end of the petiole. Sometimes, the leaflets of a pinnately compound leaf may be subdivided into still smaller leaflets, forming a *bipinnately compound* leaf (Fig. 7.1).

The flattened surfaces of leaves, which are completely covered with a transparent protective layer of cells, the *epidermis,* admit light to the interior. Many leaves twist daily on their petioles, so that their upper surfaces are inclined at right angles to the sun's rays throughout daylight hours (Fig. 7.2).

Green leaves capture the light energy available to them using the most important process for life on earth. This process, called *photosynthesis* (discussed in Chapter 10), traps

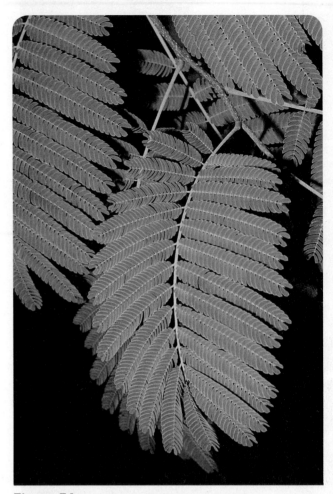

Figure 7.1 A bipinnately compound leaf of *Albizia* sp.
© *Kingsley Stern*

Figure 7.2 Hyacinth bean (*Lablab purpureus*) growing on the side of a gazebo. Note how each leaf is oriented to receive the maximum amount of light. © *James E. Bidlack*

and stores energy in sugar molecules that are constructed from ordinary water and from carbon dioxide present in the atmosphere. All the energy needs of living organisms ultimately depend on photosynthesis.

The lower surfaces of leaves (and in some plants, the upper surfaces as well) are dotted with tiny pores (**stomata**), which not only allow entry for the carbon dioxide gas needed for photosynthesis but also play a role in the diffusion out of the leaf of oxygen produced during photosynthesis. Water vapor evaporating from moist interior cell surfaces can also escape via the stomata. The evaporation of water cools the leaf, but excessive water loss can result in damage to the plant. The *stomatal apparatus*, which consists of a pore bordered by a pair of sausage-shaped *guard cells*, controls water loss when the guard cells inflate or deflate, opening or closing the pore.

Leaves also perform other functions. For example, all living cells *respire* (respiration is also discussed in Chapter 10), and in the process of this and other metabolic activities, waste products are produced. These wastes accumulate in the leaves and are disposed of when the leaves are shed, mostly in the fall. Before dropping from the plant, the leaves are sealed off at the bases of their petioles (see the discussion of leaf colors and *abscission* later in this chapter). The next season, the discarded leaves are replaced with new ones.

Leaves play a major role in the movement of water absorbed by roots and transported throughout the plant. Most of the water reaching the leaves evaporates in vapor form into the atmosphere by a process known as **transpiration** (discussed in Chapter 9). In some plants, there are special openings called *hydathodes* at the tips of leaf veins. *Root pressure* forces liquid water out of hydathodes, usually at night when transpiration is not occurring. The loss of water through hydathodes is called **guttation**. The expelled water may contain ions secreted by root cells. Other functions of leaves are discussed throughout this chapter.

Figure 7.3 Floating leaves of a giant water lily (*Victoria amazonica*), which may be up to 2 meters (6.5 feet) in diameter. The larger leaves are capable of supporting the weight of a child, without sinking.
© Kingsley Stern

7.2 Leaf Arrangements and Types

Learning Outcome

a. Describe three types of leaf arrangements.

The variety of shapes, sizes, and textures of leaves seems to be almost infinite. The leaves of some of the smaller duckweeds are less than 1 millimeter (0.04 inch) wide. The mature leaves of the Seychelles Island palm can be 6 meters (20 feet) long, and the floating leaves of a giant water lily, which reach 2 meters (6.5 feet) in diameter (Fig. 7.3), can support, without sinking, weights of more than 45 kilograms (100 pounds) distributed over their surface. In lilies, pines, ferns, and many other plants, different forms of leaves (e.g., tiny, papery scales; colored leaves called *bracts;* spines) may be produced, along with typical photosynthetic leaves, on the same plant. Genetic studies are revealing how plants produce such a broad array of leaves. However, we still have much to learn about the adaptive significance of differences in leaf shape.

Common types of leaves and leaf arrangements are illustrated in Figure 7.4. Leaves are attached to stems at regions called **nodes,** with stem regions between nodes being known as **internodes.** The arrangement of leaves on a stem (*phyllotaxy*) in a given species of plant generally occurs in one of three ways. In most species, leaves are attached alternately or in a spiral along a stem, with one leaf per node, in what is called an **alternate** arrangement. In some plants, two leaves may be attached at each node, providing an **opposite** arrangement. When three or more leaves occur at a node, they are said to be **whorled.**

The arrangement of veins in a leaf or leaflet blade (*venation*) may also be either pinnate or palmate. In **pinnately veined** leaves, there is one primary vein called the **midvein,** which is included within an enlarged **midrib;** secondary veins branch from the midvein. In **palmately veined** leaves, several primary veins fan out from the base of the blade. The primary veins are *parallel* to one another in monocots (Fig. 7.5) and diverge from one another in various ways in dicots. The branching arrangement of veins in dicots is called *netted,* or *reticulate venation* (see Fig. 7.9). In a few leaves (e.g., those of *Ginkgo*), no midvein or other large veins are present. Instead, the veins fork evenly and progressively from the base of the blade to the opposite margin. This is called *dichotomous venation* (see Fig. 7.4k).

7.3 Internal Structure of Leaves

Learning Outcome

a. Describe where the epidermis, mesophyll, and veins are found in a leaf.

Figure 7.4 Types of leaves and leaf arrangements. (a) Palmately compound leaf of a buckeye. (b) Opposite, simple leaves of a dogwood. (c) Pinnately compound leaf of a black walnut. (d) Alternate, simple but lobed leaves of a tulip tree. (e) Parallel-veined leaf of a grass. (f) Palmately veined leaf of a maple. (g) Whorled leaves of a bedstraw. (h) Pinnately veined, lobed leaf of an oak. (i) Linear leaves of a yew. (j) Globe-shaped succulent leaves of string-of-pearls. (k) Fan-shaped leaf of a *Ginkgo* tree, showing dichotomous venation.

If a typical leaf is cut transversely and examined with a microscope, three regions stand out: *epidermis, mesophyll,* and *veins* (referred to as *vascular bundles* in our discussion of roots and stems) (Fig. 7.6). The epidermis is a single layer of cells covering the entire surface of the leaf. The epidermis on the lower surface of the blade often has more *stomata,* which are discussed in the next section.

When seen from the top, the wavy, undulating walls of epidermal cells often resemble pieces of a jigsaw puzzle fitted together. Except for *guard cells,* the upper epidermal cells for the most part do not contain chloroplasts.

Their function is to protect the delicate tissues inside the leaf. A coating of waxy **cutin** (the **cuticle**—see Chapter 4) is normally present, although it may not be visible with ordinary light microscopes without being stained. In addition to the cuticle, many plants produce other waxy substances on their surfaces (Fig. 7.7). Wax production can be impacted by air pollution in as little as 24 hours after exposure. Presumably, the wax provides added protection to the leaves.

In some plants, waste materials occasionally accumulate and crystallize in epidermal cells. Different types of **glands** may also be present in the epidermis. Glands occur in the form of depressions, protuberances, or appendages either directly on the leaf surface or on the ends of hairs (see Chapter 4). Glands often secrete sticky or smelly substances, often to ward off insect pests.

7.4 Stomata

Learning Outcome

a. Explain how guard cells open and close.

The lower epidermis of most plants generally resembles the upper epidermis, but it typically has a thinner layer of cutin and is perforated by numerous tiny pores called **stomata** (singular: **stoma**) (Fig. 7.8). Some plants (e.g., alfalfa, corn) have stomata in both leaf surfaces, while others (e.g., water lilies) have them exclusively on the upper epidermis;

Figure 7.5 A portion of a monocot leaf showing parallel veins.
© Kingsley Stern

Figure 7.6 A stereoscopic view of a portion of a typical leaf.

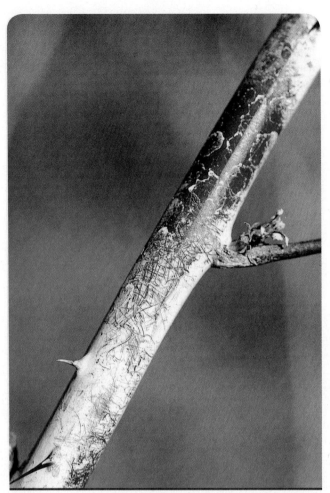

Figure 7.7 Waxes are sometimes produced on the surfaces of leaves, stems, and fruits, giving them a whitish appearance, as seen on this black raspberry cane. © *Kingsley Stern*

they are absent altogether from the submerged leaves of aquatic plants. Stomata are very numerous, ranging from about 1,000 to more than 1.2 million per square centimeter (6,300 to 8 million per square inch) of surface. An average-sized sunflower leaf has about 2 million of these pores throughout its lower epidermis. Each pore is bordered by two sausage- or dumbbell-shaped cells that are usually smaller than most of the neighboring epidermal cells. These **guard cells,** which originate from the same parental cell, are part of the epidermis, but they, unlike most of the other cells of either epidermis, contain chloroplasts.

The functioning of guard cells is aided by the photosynthesis that takes place within them. The primary functions include (1) regulating gas exchange between the interior of the leaf and the atmosphere, and (2) regulating evaporation of the water entering the plant at the roots. Guard cell walls are distinctly thickened but quite flexible on the side adjacent to the pore. As the guard cells inflate or deflate with changes in the amount of water within the cells, their unique construction causes the stomata to open or close. When the guard cells are inflated, the stomata are open; when the water content of the guard cells decreases, the cells deflate, and the stomata close. (For more detailed discussions of this stomatal mechanism, see "Regulation of Transpiration" in Chapter 9 and "Turgor Movements" in Chapter 11.)

7.5 Mesophyll and Veins

Learning Outcome

a. Distinguish between palisade and spongy mesophyll.

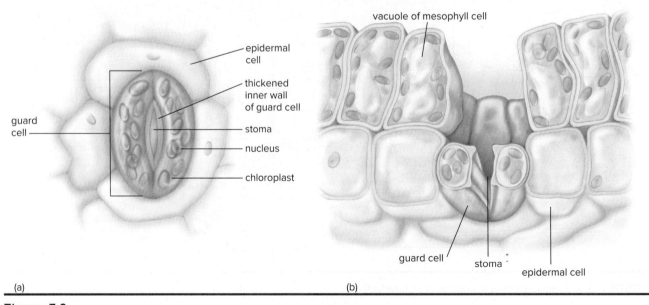

Figure 7.8 A dicot stoma. (a) Surface view. (b) View in transverse section.

Most photosynthesis takes place in the **mesophyll** between the two epidermal layers, with two regions often being distinguishable. The uppermost mesophyll consists of compactly stacked, column-shaped parenchyma cells, often in two or more layers. This region is called the **palisade mesophyll** and may contain more than 80% of the leaf's chloroplasts. This dense arrangement of chloroplast-packed cells maximizes the capture of light hitting the upper leaf surface. The lower region, consisting of loosely arranged spherical parenchyma cells with abundant air spaces separating them, is called the **spongy mesophyll.** Its cells also have numerous chloroplasts. The loose arrangement of cells in the shady side of the leaf allows for the free movement of carbon dioxide taken in through stomata.

Parenchyma tissue with chloroplasts is called *chlorenchyma*. Chlorenchyma tissue is found in the outer parts of the stems of herbaceous plants as well as in leaves. Inside the leaf, the surfaces of mesophyll cells in contact with the air are moist. If the moisture level decreases below a certain point, the stomata close, significantly reducing further drying.

Veins (*vascular bundles*) of various sizes are scattered throughout the mesophyll (Fig. 7.9). They consist of xylem and phloem tissues surrounded by a jacket of thicker-walled parenchyma cells called the **bundle sheath.** The veins give the leaf its "skeleton." The phloem transports sugars and other carbohydrates from the leaves to other parts of the plant. Water, sometimes located more than 100 meters (330 feet) above the ground, is brought into the leaf by the xylem, which, like the phloem, is part of a vast network of "plumbing" throughout the plant. Because veins run in many directions in the network, particularly in dicots, it is common, when examining a cross section of a leaf under the microscope, to see veins cut transversely, lengthwise, and at a tangent, all in the same section.

Monocot leaves, besides having *parallel veins,* usually do not have the mesophyll differentiated into palisade and spongy layers. Some monocot leaves (e.g., those of grasses) have large, thin-walled *bulliform cells* on either side of the main central vein (midrib) toward the upper surface (Fig. 7.10). Under dry conditions, the bulliform cells partly collapse, causing the leaf blade to fold or roll; the folding or rolling reduces transpiration (see also Fig. 11.21).

Forty million years were required for plants with broad leaves to evolve from the leafless vascular plants that first colonized land. No one knows why it took so long for this common feature of plants to evolve. One suggestion is that carbon dioxide levels in the early atmosphere were so high that few stomata were needed to take in an adequate supply. Because stomata are also necessary for transpirational cooling, leaves with a sparse distribution of stomata would have overheated and died. As atmospheric carbon dioxide levels dropped dramatically during the late Paleozoic era, more stomata evolved on leaves, and they were consequently able to cool themselves. This allowed for the development of larger leaves.

Figure 7.9 Net venation in an *Aristolochia* leaf. ×50. © *Kingsley Stern*

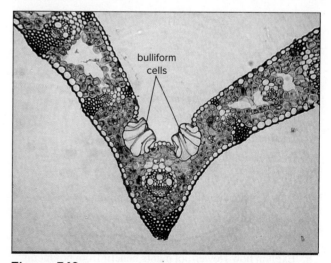

Figure 7.10 Part of a cross section of a grass leaf. When the bulliform cells are turgid, the leaf is expanded. When insufficient water is available, the bulliform cells collapse, causing the leaf blade to roll inward, reducing water loss through the stomata. ×100. © *James E. Bidlack*

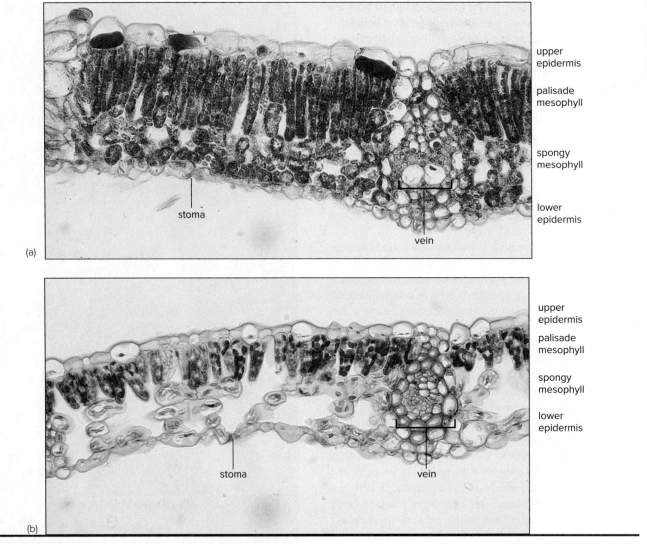

Figure 7.11 Portions of cross sections of maple (*Acer*) leaves. The normally green chloroplasts are stained red. (a) A leaf exposed to full sun. (b) A leaf exposed to shade. Note the reduction in mesophyll cells and chloroplasts in the shade leaf. ×400. *(a-b) © Kingsley Stern*

7.6 Specialized Leaves

Learning Outcomes

a. Explain why leaf structure is so variable.
b. List main features of shade leaves.
c. Describe unique features of the leaves of arid plants.
d. Describe adaptations of aquatic leaves.
e. Describe how plants benefit from producing tendrils.
f. Distinguish among spines, thorns, and prickles.
g. Describe water storage adaptations in leaves.
h. Explain how flower-pot plants benefit from producing pouches in their leaves.
i. Describe the environment in which window leaves are produced.
j. Describe the unique features of walking fern leaves.
k. Distinguish between bracts and leaves.
l. Explain how plants benefit from trapping insects.

Leaves have evolved to allow plants to thrive in a wide range of ecological niches. Consequently, they are highly variable. By looking at leaf structure, you can often see how the plant is adapted to its environment. A few of these leaf specializations will be described in this section.

Shade Leaves

A single tree may have leaves that appear similar, but they are actually quite different. For example, because leaves in the shade receive less light, they tend to be larger than their counterparts in the sun. In addition, because they receive less intense sunlight and heat, they are thinner and have fewer well-defined mesophyll layers and fewer chloroplasts (Fig. 7.11). They also do not have as many hairs.

> ### KEY THEME 7.1 ecology
>
> #### More on Leaf Structure
>
> Leaves, the main sites of photosynthesis in most plants, function in light interception and gas exchange. Because the uptake of carbon dioxide through leaf stomata must be balanced against water loss, leaf structure usually shows clear connections to prevailing environmental conditions. For instance, plants in arid environments often have thick, leathery leaves, with fewer stomata, and a dense covering of hairs that reflect light and may reduce water loss from the leaf surface. On an individual plant, leaves that develop in the shade (shade leaves) are generally thinner and have fewer leaf hairs than leaves that develop in the sun (sun leaves). Leaves are also often modified chemically or physically for defense against herbivores.

Leaves of Arid Regions

Leaf modifications are generally more pronounced in different climatic zones or habitats. Because of the limited availability of water, wide temperature ranges, and high light intensities, plants growing in arid regions have developed adaptations that allow them to thrive under such conditions. Many have thick, leathery leaves and fewer stomata, or stomata that are sunken below the surface in special depressions, all of which reduce the loss of water through transpiration. They also may have succulent, water-retaining leaves or no leaves at all (with the stems taking over the function of photosynthesis), or they may have dense, hairy coverings. Pine trees, whose water supply may be severely restricted in the winter when the soil is frozen, have some leaf modifications similar to those of desert plants. They include sunken stomata, a thick cuticle, and a layer of thick-walled cells (the **hypodermis**) beneath the epidermis (Fig. 7.12). The leaves of compass plants face east and west, with the blades perpendicular to the ground, so that when the sun is overhead, it strikes only the thin edge of the leaf, minimizing moisture loss (Fig. 7.13).

Leaves of Aquatic Areas

The submerged leaves of plants that grow in water usually have considerably less xylem than phloem because water cannot be pulled through the xylem via transpiration. The mesophyll, which is not differentiated into palisade and spongy layers, has large air spaces. Other modifications are described in the sections that follow.

Tendrils

There are many plants whose leaves are modified as **tendrils.** These modified leaves, when curled tightly around more

Figure 7.12 A pine needle in cross section. ×400. © G.S. Ellmore

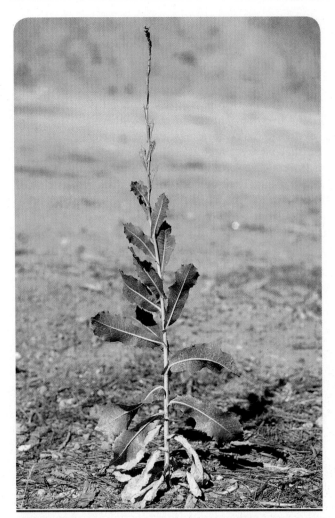

Figure 7.13 A compass plant (*Lactuca serriola*). The leaves of this and other compass plants such as *Silphium laciniatum* have their blades parallel to the sun and are oriented toward the east and west. © *Kingsley Stern*

Figure 7.14 The terminal leaflets of this garden pea plant leaf are modified as tendrils. © *Kingsley Stern*

The tendrils of many other plants (e.g., grapes) are not modified leaves but develop instead from stems.

Spines, Thorns, and Prickles

The leaves of many cacti and other desert plants are modified as **spines.** This reduction in leaf surface correspondingly reduces water loss from the plants, and the spines also tend to protect the plants from browsing animals. In such desert plants, photosynthesis, which would otherwise take place in leaves, occurs in the green stems. Most spines are modifications of the whole leaf, in which much of the normal leaf tissue is replaced with sclerenchyma, but in a number of woody plants (e.g., mesquite, black locust), it is the stipules at the bases of the leaves that are modified as short, paired spines. Like grape and other tendrils, many spinelike objects arising in the axils of leaves of woody plants are modified stems rather than modified leaves. Such modifications should be referred to as **thorns** to distinguish them from true spines. The **prickles** of roses and raspberries, however, are neither leaves nor stems, but are outgrowths from the epidermis or cortex (Fig. 7.15).

Storage Leaves

As previously mentioned, desert plants may have *succulent leaves* (i.e., leaves that are modified for water retention). The adaptations for water storage involve large, thin-walled parenchyma cells without chloroplasts just beneath the epidermis. These nonphotosynthetic cells contain large vacuoles that can store substantial amounts of water. If removed from the plant and set aside, the leaves will often retain much of the water for up to several months. Many plants with succulent leaves carry on a special form of photosynthesis called CAM photosynthesis, which is discussed in Chapter 10.

The fleshy leaves of onion, lily, and other bulbs store large amounts of carbohydrates, which are used by the plant during rapid growth early in the subsequent growing season.

rigid objects, help the plant in climbing or in supporting weak stems. The leaves of garden peas are compound (Fig. 7.14), and the terminal leaflets are reduced to whiplike strands that, like all tendrils, are very sensitive to contact. If you lightly stroke a healthy tendril, there is a sudden, rapid growth of cells on the opposite side, and it starts curling in the direction of the contact within a minute or two. If the contact is very brief, the tendril reverses movement and straightens out again. If, however, the tendril encounters a suitable solid support (e.g., a twig), the stimulation is continuous, and the tendril coils tightly around the support as it grows.

As the tendrils develop, they become coiled like a spring. When contact with a support is made, not only does the tip curl around it but the direction of the coil reverses (Chapter 11); sclerenchyma and collenchyma cells then develop in the vicinity of contact. The sclerenchyma cells provide rigid support, while the collenchyma cells impart flexibility. This makes a very strong but flexible attachment that protects the plant from damage during high winds.

Figure 7.15 (a) The *spines* of this barberry (*Berberis*) are modified leaves. (b) *Thorns* (modified stems) on a branch of the African *Acacia* tree. (c) Although referred to by some as rose "thorns," these are actually *prickles* that are lateral outgrowths from the epidermis or the cortex just beneath it. *(a) © Kingsley Stern; (b) © EggImages/Alamy; (c) © Steven P. Lynch*

Flower-Pot Leaves

Some leaves of *Dischidia* (Fig. 7.16), an *epiphyte* (a plant that grows non-parasitically on other plants) from tropical Australasia, develop into pouches that become the home of ant colonies. The ants carry in soil and add nitrogenous wastes, while moisture collects in the pouches. This creates a good growing medium for roots, which develop adventitiously from the same node as the leaf and grow down into the soil contained in the pouch. In other words, this extraordinary plant not only reproduces itself by conventional means but also, with the aid of ants, provides its own fertilized growing medium and flower pots and then produces special roots, which grow in the pots.

Window Leaves

In the Kalahari Desert of Botswana and South Africa, there are at least three plants belonging to the Carpetweed Family (Aizoaceae) that have unique adaptations to living in dry, sandy areas. Their leaves, which are shaped like ice-cream cones, are about 3.75 centimeters (1.5 inches) long (Fig. 7.17) and are buried in the sand; only the dime-sized wide end of a leaf is exposed at the surface. This exposed end is covered with a relatively transparent, thick epidermis with few stomata and a waxy cuticle. There is a mass of tightly packed, transparent water-storage cells below the exposed end; these allow light coming through the "windows" to penetrate to the chloroplasts in the mesophyll, located all around the inside of the shell of the leaf. This arrangement, which keeps most of the plant buried and away from drying winds, allows the plant to thrive under circumstances that most other plants could not tolerate. Window leaves also occur in succulent plants of a few other families.

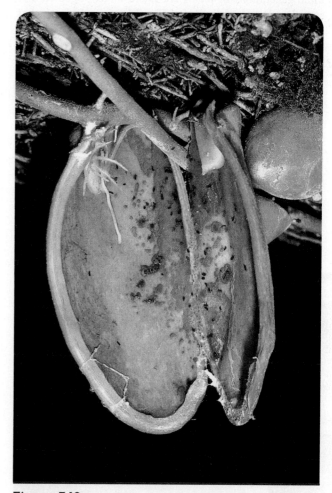

Figure 7.16 A flower-pot leaf of *Dischidia*. The hollow leaf has been sliced lengthwise to reveal the roots inside that have started to grow down from the top. *© Kingsley Stern*

Figure 7.17 A window plant (*Fenestraria*). Note the transparent tips of the window leaves. © *Kingsley Stern*

Figure 7.19 A poinsettia (*Euphorbia pulcherrima*) "flower." There are several flowers without petals in the center. The most conspicuous parts of the "flower" are the red-colored bracts (modified leaves) surrounding the true flowers. © *Kingsley Stern*

Reproductive Leaves

Some of the leaves of the walking fern are most unusual in that they produce new plants at their tips. Occasionally, three generations of plants are found linked together. The succulent leaves of air plants (Fig. 7.18) have little notches along the leaf margins in which tiny plantlets are produced, complete with roots and leaves, even after a leaf has been removed from the parent plant. Each of the plantlets can develop into a mature plant if given the opportunity to do so.

Floral Leaves (Bracts)

Specialized leaves known as **bracts** are found at the bases of flowers or flower stalks. In the poinsettia, the flowers themselves have no petals, but the brightly colored floral bracts that surround the small flowers function as petals in attracting pollinators (Fig. 7.19). In dogwoods and a few other plants, the tiny flowers in their buttonlike clusters do have inconspicuous petals. However, the large white-to-pink bracts that surround the flower clusters, which appear to the casual observer to be petals, are actually modified leaves. In Clary's annual sage (*Salvia viridis*), large, colorful bracts are produced at the top of flowering stalks, well above the flowers (Fig. 7.20).

Figure 7.18 A leaf of an air plant (*Kalanchoë*), showing plantlets being produced along the margins. © *Kingsley Stern*

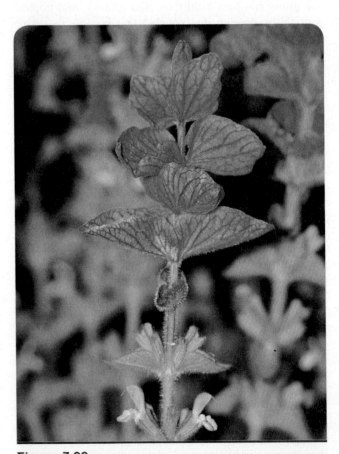

Figure 7.20 Clary's annual sage (*Salvia viridis*). Note the small flowers along the lower half of the stem and the large, colorful bracts toward the top. © *Kingsley Stern*

Insect-Trapping Leaves

Highly specialized *insect-trapping leaves* are intriguing. Almost 200 species of flowering plants are known to have these leaves. Insectivorous plants grow mostly in swampy areas and bogs of tropical and temperate regions. In such environments, certain nutrients, particularly nitrogen, may be deficient in the soil, or they may be in a form not readily available to the plants. Some of these elements are furnished when the soft parts of insects and other small organisms trapped by the specialized leaves are broken down and digested. All the plants have chlorophyll and are able to make their own food. It has been demonstrated that they can develop normally without insects if they are given the nutrients they need. The following plants represent four types of insect-trapping mechanisms.

Pitcher Plants

The blades of leaves of many *pitcher plants* are flattened and function like those of any other leaves. Some of the leaves of these curious plants, however, are larger and cone-shaped or vaselike. In some species, these larger pitcher leaves have umbrella-like flaps above the open ends (Fig. 7.21), but the flaps don't prevent some rain water from accumulating at the bottom. Some Asian pitcher plants are vines with leaves whose long petioles are twisted around branches for support. Their pitchers are formed at the tips of the leaves.

Pitcher leaves have nectar-secreting glands around the rim. The distinctive odor produced by these glands attracts insects, which, while foraging, often fall into the watery fluid at the bottom. If the insects try to climb out, they find the walls highly polished and slippery. In fact, the walls of some pitcher plant leaves are coated with wax, and as the insects struggle up the surface, their feet become coated with the wax, which builds up until the victims seem to have acquired heavy, clodlike boots. Most insects never make it up the walls, but even if they do, they still face a formidable barricade of stiff, downward-pointing hairs near the rim. Eventually, they drown, and their soft parts are digested by bacteria and by enzymes secreted by the plant's digestive glands near the bottoms of the leaves.

In North America, the pitcher plants produce their pitchers in erect clusters on the ground, but the mechanisms for trapping insects are similar to those of Asian species. Malaysian tree frogs, which have sticky pads on their feet, are undaunted by pitcher-plant leaves and lay their eggs in them. The eggs contain a chemical that neutralizes the pitcher's digestive enzymes.

Merchandisers have been shameless in their wholesale collection of these plants for sale, and pitcher plants may become extinct in the wild. Several species of pitcher plants have been placed on threatened and endangered species lists.

Sundews

The tiny plants called *sundews* (Fig. 7.22) often do not measure more than 2.5 to 5.0 centimeters (1 to 2 inches) in diameter. The roundish to oval leaves are covered with up to 200 upright, glandular hairs that look like miniature clubs. There is a clear, glistening drop of sticky fluid containing digestive enzymes at the tip of each hair. As the droplets sparkle in the sun, they may attract insects, which find themselves stuck if they alight. The hairs are exceptionally sensitive to contact, responding to weights of less than one-thousandth of a milligram, and bend inward, surrounding any trapped insect within a few minutes. The digestive enzymes break down

Figure 7.21 Insect-trapping leaves of pitcher plants (*Sarracenia*). © *Kingsley Stern*

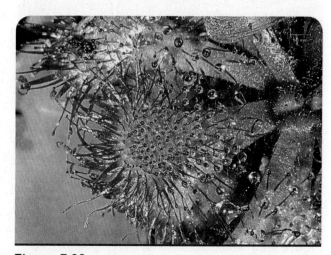

Figure 7.22 Leaves covered with glands of the sundew (*Drosera*) attract and capture insects. © *Steven P. Lynch*

the soft parts of the insects, and after digestion has been completed (within a few days), the glandular hairs return to their original positions. If bits of nonliving debris happen to catch in the sticky fluid, the hairs barely respond, showing they can distinguish between protein and something "inedible." Some sundew owners regularly feed their plants tiny bits of hamburger and boiled egg white.

In Portugal, relatives of sundews with less specialized leaves are used in houses as flypaper. In response to contact by living insects, the edges of specialized leaves of similar plants called *butterworts* rapidly curl over and trap unwary victims.

Venus's Flytraps

The leaves of the *Venus's flytrap* (Fig. 7.23) look like steel traps. This plant is found in nature only in wet areas of North Carolina and South Carolina. The two halves of the blade appear to be hinged along the midrib, with stiff, hairlike projections along their margins. There are three tiny trigger hairs on the inner surface of each half. If two trigger hairs are touched simultaneously or if any one of them is touched twice within a few seconds, the blade halves suddenly snap together, trapping the insect or other small animal. As the creature struggles, the trap closes even more tightly. Digestive enzymes secreted by the leaf break down the soft parts of the insect, which are then absorbed. After digestion has been completed, the trap reopens, ready to repeat the process. Venus's flytrap leaves, like those of sundews, do not normally close for bits of debris that might accidentally fall on the leaf, because nonliving material does not move around and stimulate the trigger hairs.

Bladderworts

Bladderworts (Fig. 7.24), which are found submerged and floating in the shallow water along the margins of lakes and streams, have finely dissected leaves with tiny bladders. The stomach-shaped bladders are between 0.3 and 0.6 centimeter (0.125 to 0.25 inch) in diameter and have a trapdoor over the opening at one end. The trapping of aquatic insects and other small animals takes place through a complex mechanism. Four curled but stiff hairs at one end of the trapdoor act as

Figure 7.23 Mouthlike leaves of the Venus's flytrap (*Dionaea muscipula*) capture insects and secrete enzymes to kill insects and obtain nutrients from them. © *Steven P. Lynch*

Figure 7.24 (a) A bladderwort (*Utricularia*). (b) A single bladder. ×40. *(a)* © *Kingsley Stern; (b)* © *James E. Bidlack*

triggers when an insect touches one of them. The trapdoor springs open, and water rushes into the bladder. The stream of water propels the victim into the trap, and the door snaps shut behind it. The action takes place in less than one-hundredth of a second and makes a distinct popping sound, which can be heard with a sensitive underwater microphone. The trapped insect eventually dies and is broken down by bacteria, and the breakdown products are absorbed by cells in the walls of the bladder.

Science-fiction writers have contributed to superstitions and beliefs that deep in the tropical jungles there are plants capable of trapping humans and other large animals. No such plants have ever been found. The largest pitcher plants known hold possibly 1 liter (roughly 1 quart) of fluid in their pitchers, and small frogs have been known to decompose in them, but the trapping of anything larger than a mouse or possibly a small rabbit seems very unlikely.

7.7 Autumnal Changes in Leaf Color

Learning Outcome

a. Explain why leaves change color in the fall.

The leaves of many oaks and several other plants generally turn some shade of brown or tan when their cells break down and die, due to a reaction between leaf proteins and tannins stored in the cell vacuoles. This is similar to the formation of leather when tannins react with animal hides. Leaves of many other deciduous plants, however, exhibit a variety of colors and drop before turning brown.

The chloroplasts of mature leaves contain several groups of pigments, such as green *chlorophylls* and *carotenoids*, which include yellow *carotenes* and pale yellow *xanthophylls*. Each of these groups plays a role in photosynthesis. Usually, considerably more chlorophyll than other pigments is present, and the intense green color of the chlorophylls masks or hides the presence of the carotenes and xanthophylls. In the fall, however, the photosynthesis ceases as chlorophylls break down. This allows the other colors to become visible. These changes occur as the day length shortens and nights become cooler.

Water-soluble *anthocyanin* and *betacyanin* pigments may also accumulate in the vacuoles of the leaf cells in the fall. Anthocyanins, the more common of the two groups, are red if the cell sap is slightly acidic, blue if it is slightly alkaline, and of intermediate shades if it is neutral. Betacyanins are usually red; they apparently are restricted to several plant families, such as the cacti (Cactaceae); the Goosefoot Family (Chenopodiaceae), to which beets belong; the Four-o'clock Family (Nyctaginaceae); and the Portulaca Family (Portulacaceae).

Some plants (e.g., birch trees) consistently exhibit a single shade of color in their fall leaves, but many (e.g., maple, ash, sumac) vary considerably from one locality to another or even from one leaf to another on the same tree, depending on the combinations of carotenes, xanthophylls, and other

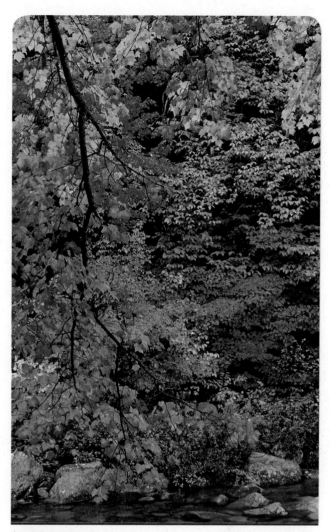

Figure 7.25 Sugar maple branches showing fall coloration above the Swift River in New Hampshire. © *Doug Sherman/Geofile*

pigments present. Some of the most spectacular fall colors in North America occur in the Eastern Deciduous Forest, particularly in New England (Fig. 7.25) and the upper reaches of the Mississippi Valley. In parts of Wisconsin and Minnesota, one can observe the brilliant reds, oranges, and golds of maples, the deep maroons (and yellows) of ashes, the bright yellows of aspen, and the seemingly glowing reds of sumacs and wahoos, all in a one place. Some fall coloration is found almost anywhere in temperate zones where deciduous trees and shrubs exist.

7.8 Abscission

Learning Outcome

a. Describe the process of abscission in leaves of temperate zone trees in the fall.

In addition to changes in leaf color, **deciduous** trees undergo other changes in the fall. In preparation for winter, they must lose their leaves through a process called **abscission.**

Figure 7.26 The abscission zone of a leaf. © *Kingsley Stern*

Abscission occurs as a result of changes that take place in an *abscission zone* near the base of the petiole of each leaf (Fig. 7.26). Sometimes the abscission zone can be seen externally as a thin band of slightly different color on the petiole. Hormones that apparently inhibit the formation of the specialized layers of cells to facilitate abscission are produced in young leaves. As the leaf ages, hormonal changes take place, and at least two layers of cells become differentiated. Closest to the stem, the cells of the *protective layer*, which may be several cells deep, become coated and impregnated with fatty *suberin*. On the leaf side, a *separation layer* develops in which the cells swell, sometimes divide, and become gelatinous. In response to any of several environmental changes (such as lowering temperatures, decreasing day lengths or light intensities, lack of adequate water, or damage to the leaf), the pectins in the middle lamella of the cells of the separation layer are broken down by enzymes. All that holds the leaf onto the stem at this point are some strands of xylem. Wind and rain then easily break the connecting strands, leaving tiny bundle scars within a leaf scar (see Fig. 6.1), and the leaf falls to the ground.

In temperate climates, new leaves are produced in the spring and are shed in the fall, but in the tropics, the cycles coincide with wet and dry seasons rather than with temperature changes. Even evergreen trees shed their leaves; they do so a few at a time, however, so that they never have the bare look of deciduous trees in winter.

7.9 Human and Ecological Relevance of Leaves

Learning Outcome

a. List 10 human uses for leaves.

Humans use shade trees in landscaping for cooling as well as for aesthetic effects. The leaves of shade plants planted next to a building can make a significant difference in energy costs to the homeowner. Humans also eat the leaves of cabbage, parsley, lettuce, spinach, kale, and chard and the petioles of celery and rhubarb, to mention a few. Many herbs and spices are derived from leaves, including thyme, marjoram, oregano, tarragon, peppermint, spearmint, wintergreen, basil, dill, sage, cilantro, and savory.

Various dyes (e.g., a yellow dye from bearberry, a reddish dye from henna, and a pale blue dye from blue ash) can be extracted from leaves (see Appendix 3), although nearly all commercial dyes are now derived from coal tar. Many cordage fibers for ropes and twines come from leaves, with various species of *Agave* (century plants) accounting for about 70% of the world's production. Bowstring fibers are obtained from a relative of the common houseplant *Sansevieria* (see Chapter 24), and Manila hemp fibers, which are used both in fine-quality cordage and in textiles, are obtained from the leaves of a close relative of the banana. Panama hats are made from the leaves of the panama hat palm, and palms and grasses are used in the tropics as thatching material for huts and other buildings.

In the high mountains of Chile and Peru, the leaves of the yareta plant are used for fuel. They produce a resin that causes the leaves to burn with an unusually hot flame. Leaves of many plants produce oils. Petitgrain oil, from a variety of orange tree leaves, and lavender, for example, are used for scenting soaps and colognes. Patchouli and lemongrass oils are used in perfumes, as is citronella oil, which is used as a mosquito repellent. Eucalyptus oil, camphor, cajeput, and pennyroyal (Fig. 7.27) are all used medicinally.

Leaves are an important source of drugs used in medicine and of narcotics and poisons. Cocaine, obtained from plants native to South America, has been used medicinally and as a local anesthetic, but its use as a narcotic has, in recent years, become a major problem in Western cultures. Andean natives chew coca leaves while working and are reported capable of performing exceptional feats of labor with little or no food while under the influence of cocaine. Apparently, the drug, which is highly addictive in forms such as "crack," anesthetizes the nerves that convey hunger pangs to the brain.

Belladonna is a drug complex obtained from leaves of the deadly nightshade, a native of Europe. The plant has been used in medicine for centuries, and several drugs are now isolated from belladonna. Included among the isolates is atropine, which is used in shock treatments, to dilate eyes, to relieve pain locally, and to slow secretions during surgery. Scopolamine, also a belladonna derivative, is used in tranquilizers and sleeping aids. Another European plant, the foxglove, is the source of digitalis. This drug has been used for centuries in regulating blood circulation and heartbeat.

Tobacco is another widely used leaf. More than 5.7 million tonnes (6.3 million tons) of tobacco for smoking, chewing, and use as snuff are produced annually around the world. At present, about 130,000 U.S. citizens die annually

KEY THEME 7.2 anatomy

Glass Cuts from Grass?

We all know that if we drop a glass bottle or dish, we need to pick up the pieces, so that someone won't get hurt on the sharp edges of the broken glass. But did you know that something as soft as a blade of grass can also give you a nasty cut if you don't handle it carefully?

Many leaves, including those of grasses, are produced with tiny teeth along the margins. In some instances when a person brushes against the edge of the leaf, the teeth alone have enough rigidity to act as a miniature saw that can produce a cut in human skin. One of the reasons grasses are generally not palatable to humans is that they accumulate silica—the primary ingredient of glass. In some grasses, including rice cutgrass (*Leersia oryzoides*—Box Figure 7.2A), the tiny teeth along the margins are covered with a film of glass that gives the downward-pointing teeth additional rigidity. This tiny saw instantly cuts if human skin is brushed against it.

If you have the opportunity to do so, examine the leaf margins of several grasses with a microscope and see if any of them have teeth encased in glass.

Box Figure 7.2A Part of the margin of a rice cutgrass (*Leersia oryzoides*) leaf, showing the "glass"-encased teeth that easily cut human skin. ×40. © Kingsley Stern

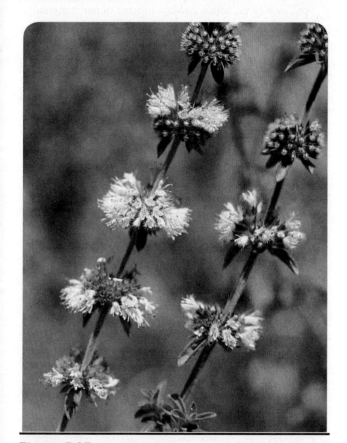

Figure 7.27 Flowers of a pennyroyal plant (*Hedeoma*). The leaves are the source of a pungent oil. © Kingsley Stern

from lung cancer attributable to smoking. Cigarette smoking is also a principal contributor to cardiovascular diseases. Heart disease causes one in five smoking-related deaths. The use of chewing tobacco increases the risk of mouth and throat cancers. Federal law forbids concentrations of more than five parts per billion of nitrosamines (cancer-causing chemicals) in cured meats, but the levels of nitrosamines in the five most popular brands of chewing tobacco range from 9,600 to 289,000 parts per billion. Though humans have long used tobacco, it has been in only the last several decades that its health threat has become appreciated.

Similarly, marijuana (*Cannabis sativa*) has been used in various ways for thousands of years. Marijuana plants produce a group of chemical compounds called cannabinoids. One of these compounds, tetrahydrocannabinol (THC) is an intoxicant. Other cannabinoids have therapeutic benefits, easing nausea caused by chemotherapy and muscle spasms in adults with multiple sclerosis. While there may be some benefits in treating certain ailments with cannabis-based products, the jury is still out on whether or not these are suitable alternatives to modern medicines. Research is underway to determine whether these compounds provide an effective pain management strategy for some people. Side effects of medical marijuana include headaches, sleepiness, and dizziness. Research is also underway to determine the effects of recreational marijuana use. Many of the cancer-causing compounds in tobacco smoke are also present in marijuana smoke. In addition, the use of marijuana products has been found to impair learning, memory and attention. Cannabidiol (CBD) is a nonintoxicating cannabinoid extracted from the flowers and buds of cannabis. Strong scientific evidence supports the value of CBD for the treatment of certain forms of childhood epilepsy. The efficacy of CBD for other uses, including treating anxiety or reducing inflammation, is not yet well characterized in humans.

The drug lobeline sulfate is obtained from the leaves of a close relative of garden lobelias. It is used in compounds taken by those who are trying to stop smoking. The leaves of several species of *Aloe*, especially *Aloe vera*, yield a juice that is used to treat various types of skin burns.

Several beverages are extracted by the brewing of leaves. Numerous teas have been obtained from a wide variety of plants, but most now in use come from a close relative of the garden camellia. Maté, the popular South American tea, is brewed from the leaves and twigs of a relative of holly. The alcoholic beverages pulque and tequila are produced from mashed leaves of *Agave* plants, and absinthe liqueur receives its unique flavor from the leaves of wormwood, a relative of western sagebrush, and from other flavorings, such as anise.

Insecticides of various types are also derived from leaves. A type of rotenone and a substance related to nicotine are obtained from tropical plants; both are effective against a variety of insects. Mexico's cockroach plant has leaves that when dried are highly effective in killing cockroaches, flies, fleas, and lice. Water extracts of the leaves (as well as other parts) of India's neem tree are reported to control more than 100 species of insects, mites, and nematodes, with little effect on useful predator insects.

Carnauba and caussu waxes are obtained from the leaves of tropical palms. Leaves are used extensively by florists in floral arrangements and bouquets, and they have many other aesthetic uses.

Leaves may find more extensive use in the future as a direct source of food. It has been shown that a curd obtained by coagulating juice squeezed from alfalfa leaves contains more than 40% protein, and juices from other leaves have yielded better than 50% protein. Experiments are underway to make the leaf curd palatable for human consumption. Additional information on past and present uses of leaves is provided in Chapter 24 and in Appendices 2, 3, and 4. The scientific names of the plants that are the sources of the various substances discussed are listed in Appendix 1.

SUMMARY

7.1 Introduction to Leaves
- Leaves are similar to solar panels in that they are covered with a transparent epidermis that admits light to the interior, and many twist on their petioles so that their flat surfaces are inclined to the sun throughout the day.
- All leaves originate as primordia. Most leaves consist of a blade and a petiole that may have paired stipules at the base. Leaves may be simple or compound, and all are associated with leaf gaps.
- The lower and often the upper surfaces of leaves have pores (stomata) that permit air circulation and facilitate photosynthesis. Leaves also respire, accumulate wastes, and eliminate excess moisture via transpiration. The shapes, sizes, and textures of leaves vary greatly.

7.2 Leaf Arrangements and Types
- The venation and the compounding of leaves may be either pinnate or palmate. The main vein is the midvein in simple leaves and the rachis in compound leaves. Monocot leaves have parallel veins. Dichotomous venation occurs in a few leaves.

7.3 Internal Structure of Leaves
- The epidermis is a surface layer of cells that is coated with a cuticle. Waxes, glands, hairs, and an occasional cellular crystal may also be present.

7.4 Stomata
- The lower epidermis usually contains numerous stomata, each formed by a pair of guard cells that regulate both evaporation of water vapor from the leaf and gas exchange between the interior and the atmosphere.

7.5 Mesophyll and Veins
- The mesophyll between the upper epidermis and the lower epidermis may be divided into an upper palisade layer, which consists of rows of parenchyma cells containing numerous chloroplasts, and a lower spongy layer in which the parenchyma cells are loosely arranged. Veins traverse the mesophyll.

7.6 Specialized Leaves
- Leaves may be specialized, usually in adaptation to specific combinations of environmental factors. Specializations include thick leaves, fewer stomata, water-storage leaves, tendrils, spines, food-storage leaves, flower-pot leaves, window leaves, reproductive leaves, bracts, and insect-trapping leaves.

7.7 Autumnal Changes in Leaf Color
- Leaves change color as green chloroplast pigments break down, revealing pigments of other colors, and different pigments accumulate in cell vacuoles.

7.8 Abscission
- Environmental factors and hormonal changes that occur in autumn in the abscission zone at the petiolar base of each deciduous leaf cause leaves to drop.

7.9 Human and Ecological Relevance of Leaves
- Leaves are a source of many useful products, including food, beverages, dyes, fuel, spices, flavorings, oils, medicinal and narcotic drugs, insecticides, and waxes.

REVIEW QUESTIONS

1. Leaves have no secondary xylem and phloem. Why not?
2. What are bracts?

3. Identify or define hydathode, transpiration, guard cells, mesophyll, venation, glands, and compound leaf.
4. How can one distinguish between the upper and lower epidermis in most leaves?
5. What is the function of bundle sheaths?
6. How do leaves in shaded areas differ from leaves in sunlit areas on the same plant?
7. What leaf modifications are associated with dry areas, wet areas (e.g., lakes), climbing, and reproduction?
8. Why do leaves turn different colors in the fall?

DISCUSSION QUESTIONS

1. Describe the advantages and disadvantages to a plant in having very tiny or very large leaves.
2. In Chapter 3, it was noted that living cells are connected to one another by plasmodesmata that extend through tiny holes in the walls. If this is true, does a leaf really need veins for the acquisition and distribution of materials?
3. Is there any advantage to a leaf in having palisade mesophyll on the upper half and spongy mesophyll on the lower half?
4. Pines and other plants that grow in very cold climates have sunken stomata just like plants of the desert, yet the annual precipitation where they grow may be very abundant. Is there any advantage to their having such a modification?
5. Because tendrils and spines can be either modified leaves or modified stems, how might you determine the origin of a given specimen?

CHAPTER

8

Flowers, Fruits, and Seeds

8.1 Introduction to Flowers, Fruits, and Seeds
8.2 Differences between Dicots and Monocots
8.3 Structure of Flowers
8.4 Fruits
  Introduction
  Fruit Regions
  Key Theme 8.1: Goober Peas
  Kinds of Fruits
8.5 Fruit and Seed Dispersal
  Dispersal by Wind
  Dispersal by Animals
  Dispersal by Water
  Other Dispersal Mechanisms and Agents
8.6 Seeds
  Structure
  Germination
  Longevity
  Key Theme 8.2: The Seed That Slept for 1,200 Years

OVERVIEW

The structure, parts, and some modifications of flowers are described in this chapter. Dicots and monocots are briefly compared. The nature and development of fruits are discussed next, and fruit structure and parthenocarpy are described. A listing and discussion of types of dry and fleshy fruits follows, and then fruit and seed dispersal are explored. An examination of seed structure and germination and some observations on dormancy and longevity of seeds, stratification, and vivipary conclude the chapter.

Sulfur butterflies (*Phoebis sennae*) visit this nectar-rich *Hibiscus* spp. flower in Cedar Key, Florida. *Courtesy of Cliff Pelchat*

8.1 Introduction to Flowers, Fruits, and Seeds

Learning Outcomes

a. Provide some examples of variations in flower structure, including size and fragrance, as well as flowering plant habitat.
b. Explain the differences among annual, biennial, and perennial plants.

> **NOTE to the Reader**
>
> This chapter emphasizes the form and structure of flowers, fruits, and seeds, with brief reference to reproduction. Chapter 23 covers more details of the life cycle and evolution of flowering plants. Even if your course does not include Chapter 23 in the curriculum, it is recommended that you read it as an adjunct to Chapter 8.

A number of years ago, an Australian farmer, while plowing a field, glanced back and was startled to see what looked like flowers being tossed to the surface of the furrows. He climbed off his tractor to take a closer look and found that what he saw were, indeed, flowers. Furthermore, the flowers were attached to pale plants that had no chlorophyll. He reported his find to a university, where botanists determined that the farmer had stumbled upon the first known underground flowering plant. The plant proved to be an orchid that lived on organic matter in the soil and was pollinated by tiny flies that found the below-ground flowers via mud cracks that developed in the dry season (Fig. 8.1).

The underground-flowering orchid is only one of nearly 300,000 known species of flowering plants. Unknown numbers of undescribed species grow in remote areas, particularly in the tropics. The flowering plants are vital to humanity, providing countless useful products, with just 11 species—10 of them members of the Grass Family (Poaceae)—furnishing 80% of the world's food. This subject is discussed in more detail in Chapter 24.

Flowers range in size from the minute flowers of the duckweed, *Wolffia columbiana,* whose entire plant body is only 0.5 millimeter (0.02 inch) wide, with flowers little more than 0.1 millimeter long, to the enormous *Rafflesia* flowers (Fig. 8.2a) of Indonesia that are up to 1 meter (3 feet, 3 inches) in diameter and may weigh 9 kilograms (20 pounds) each.

A unique species known as the corpse flower (*Amorphophallus titanum*) produces an interesting bloom (also called an *inflorescence,* discussed later in this chapter) that deserves mention for its large size and offensive odor. The bloom itself (Fig. 8.2b), which can be over 2 meters (6 feet, 6 inches) tall, protects small interior flowers and is derived from an underground corm that can remain hidden for many years. Carrion beetles are attracted to an odor of rotting flesh emitted by the bloom and act as pollinators when they travel from one flower to the next. The rarity and impressive display of corpse flower blooms (less than 300 in recorded history) inspire curious spectators to visit and enjoy botanical gardens around the world.

(a)

(b)

Figure 8.1 (a) Painting of Australian orchids (*Rhizanthella gardneri*) that complete their entire life cycle underground. (b) Photograph of the flowers after temporarily removing the soil. *(a)* © Walter Linsenmaier/National Geographic/Getty Images; *(b)* © Fred and Jean Hort

Figure 8.2 (a) A *Rafflesia* flower. These flowers may be as much as 1 meter (3 feet, 3 inches) in diameter. (b) The rare bloom of a corpse flower (*Amorpho phallus titanum*), which is known for its large size and offensive odor. *(a) Courtesy of Troy Davis; (b) Courtesy of Myriad Botanical Gardens, Oklahoma City*

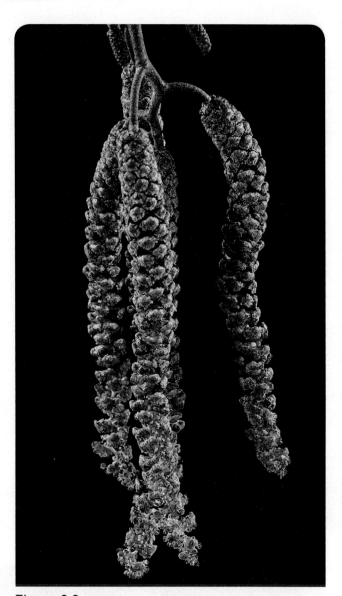

Figure 8.3 Male catkins of an alder (*Alnus*). Each catkin consists of numerous tiny, inconspicuous, wind-pollinated flowers that have no petals. © *Kingsley Stern*

and garden weeds are quite inconspicuous and lack odor (Fig. 8.3), but other flowers are strikingly beautiful, particularly when viewed with a dissecting microscope. Their fragrances, which can range from exhilarating to seductive or even putrid, are the basis of international perfume and animal-repellent industries.

Flowering plant habitats are as varied as their form. Besides the underground habitat mentioned in the chapter introduction, some flowering plants (*epiphytes*) go through their life cycles dangling from wires or other plants (Fig. 8.4). They also occur in both fresh and salt water, in the cracks and crevices of rocks, in deserts and jungles, in frigid arctic regions, and in areas where the temperatures regularly soar to 45°C (113°F) in the shade. In fact, they can be found almost anywhere they receive their basic needs of light, moisture, and a minimal supply

Flowers may be any color or combination of colors of the rainbow, as well as black or white; they may have virtually any texture, from filmy and transparent to thick and leathery, from spongy to sticky, hairy, prickly, or even dewy wet to the touch. Flowers of many trees, shrubs,

Flowers, Fruits, and Seeds   123

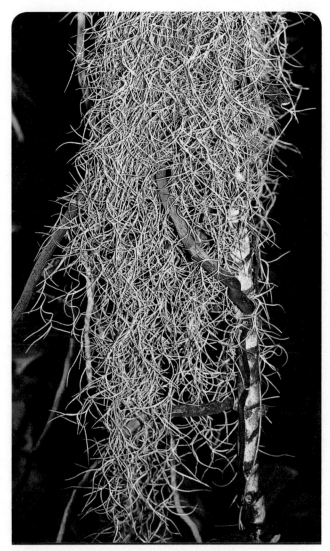

Figure 8.4 Spanish moss (*Tillandsia*). Spanish moss is a non-parasitic flowering plant that goes through its life cycle suspended from other plants or objects such as tree limbs or wires. This plant should not be confused with *lichens,* which consist of an alga and a fungus (see Chapter 19) and which may hang suspended from other objects. © *Kingsley Stern*

of minerals. One species of chickweed survives at an altitude of 6,135 meters (over 20,000 feet) in the Himalaya Mountains, and fumitory plants in the same area flower even when the night temperatures plummet to −18°C (0°F).

Flowering plants can go from the germination of a seed to a mature plant producing new seeds in less than a month, or the process may take as long as 150 years. In **annuals,** the cycle is completed in a single season and ends with the death of the parent plant. **Biennials** take two growing seasons to complete the cycle; **perennials,** however, may take several to many growing seasons to go from a germinated seed to a plant producing new seeds. However, some perennials may produce flowers in their first year of growth or on new growth that dies back each winter, while other parts of the plant may persist indefinitely.

There are two classes of flowering plants, the *Magnoliopsida* and *Liliopsida,* previously known as the *Dicotyledonae* and the *Monocotyledonae.* Despite the revised class names, the two groups are still commonly referred to as **dicots** and **monocots.** Members of the two classes are usually distinguished from one another based on the features listed in Table 8.1.

8.2 Differences between Dicots and Monocots

Learning Outcome

a. List five characteristics that can be used to distinguish dicots from monocots.

Approximately three-fourths of all flowering plant species are dicots, including many annual plants and virtually all flowering trees and shrubs. Monocots, which are primarily herbaceous, are believed to have evolved from primitive dicots. They include species that produce bulbs (e.g., lilies), grasses and related plants, orchids, irises, and palms.

Table 8.1 Some Differences between Dicots and Monocots

Dicots	Monocots
1. Seed with two cotyledons (seed leaves)	1. Seed with one cotyledon (seed leaf)
2. Flower parts mostly in fours or fives or multiples of four or five	2. Flower parts in threes or multiples of three
3. Leaf with a distinct network of primary veins	3. Leaf with more or less parallel primary veins
4. Vascular cambium, and frequently cork cambium, present	4. Vascular cambium and cork cambium absent
5. Vascular bundles of stem in a ring	5. Vascular bundles of stem scattered
6. Pollen grains mostly with three *apertures* (thin areas in the aperture wall—see Chapter 23)	6. Pollen grains mostly with one *aperture*

Palms, like other monocots, do not have a vascular cambium but become large through a *primary thickening* of cells that occurs just below the apical meristem. Further discussion on the differences between dicots and monocots, as well as some recent revisions in plant taxonomy, are provided in Chapter 16.

8.3 Structure of Flowers

Learning Outcomes

a. List the nonreproductive parts of a flower, and explain how they may contribute to protection and/or pollination.

b. Describe the male and female parts of a flower as well as some of the variations in these reproductive structures.

Regardless of form, all flowers share certain basic features. A typical flower develops several different parts, each with its own function (Fig. 8.5). Each flower, which begins as an embryonic *primordium* that develops into a *bud* (discussed in Chapter 6), occurs as a specialized branch at the tip of a stalk called a **peduncle,** which may in some instances have branchlets of smaller stalks called **pedicels.** In the case of a single flower, the stalk is typically referred to as a pedicel. A peduncle or pedicel swells at its tip into a small pad known as a **receptacle.** The other parts of the flower, some of which are in *whorls,* are attached to the receptacle (a whorl consists of three or more plant parts—such as leaves—encircling another plant part—such as a twig—at the same point on an axis).

The outermost whorl typically consists of three to five small, usually green, somewhat leaflike **sepals.** The sepals of a flower, which are collectively referred to as the **calyx,** may, in some flowers, be fused together. In many species, the calyx protects the flower while it is in the bud.

The next whorl of flower parts consists of three to many **petals;** the petals are collectively known as the **corolla.** Showy corollas attract pollinators, such as bees, moths, or birds. Bees, which can see ultraviolet light, detect in some corollas special markings invisible to humans. The petals are distinct units in peach flowers, but in some flowers (e.g., petunias), the petals are fused together into a single, flared, trumpetlike sheet of tissue (Fig. 8.6). The corolla may not be showy or conspicuous in many tree and weed species and is often missing altogether or highly modified in wind-pollinated plants such as grasses. The calyx and the corolla together are referred to as the **perianth.** *Bracts* are specialized leaves that may be as colorful as petals and can attract pollinators the way petals do; they are discussed in Chapter 7.

The male part of the flower includes the **stamens,** which produce pollen, and the female part of the plant includes the **pistil,**

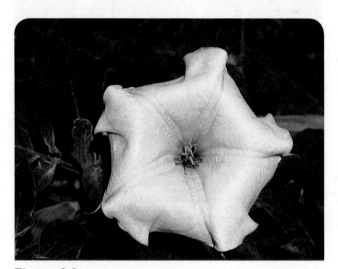

Figure 8.6 A flower of jimson weed (*Datura*). The petals are united in a single, flared sheet of tissue. © *Kingsley Stern*

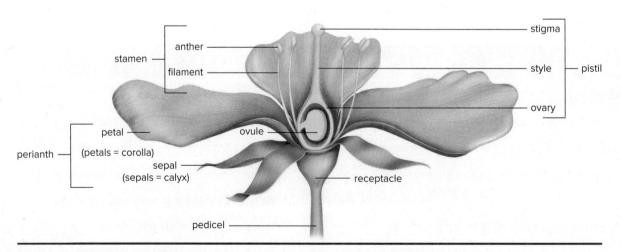

Figure 8.5 Parts of a generalized flower. The interior structure of the ovule and the sexual processes involved are discussed in Chapter 23.

which produces the egg. Several to many stamens are attached to the receptacle around the base of the often greenish pistil in the center of the flower. Each stamen consists of a semi-rigid but otherwise usually slender **filament** with a sac called an **anther** at the top. The development and dissemination of **pollen grains** in anthers are described in Chapter 23. In most flowers, the pollen is released through lengthwise slits that develop on the anthers, but in members of the Heath Family (Ericaceae) and those of a few other groups, the pollen is released through anther pores.

The *pistil*, which often is shaped like a tiny vase that is closed at the top, consists of three regions. At the top is the **stigma,** which is usually connected by a slender, stalklike **style** to the swollen base called the **ovary**. The stigma varies in form among plant species, from a narrow point to a bulb to a set of broad arms or branches. The ovary later develops into a *fruit*.

There is evidence that ovaries first evolved when the margins of leaves bearing ovules rolled inward. Such ovule-bearing leaves were called **carpels.** In some instances, two or more carpels eventually fused together, producing *compound* ovaries, consisting of two to several carpels. The number of carpels is often reflected by the number of lobes or divisions of the stigma. Each segment of a mature ovary, such as a tomato or an orange, for example, represents a carpel. Carpels are discussed further in Chapter 23. Some texts refer to pistils as carpels, but because a pistil can consist of one to several carpels, such references are incorrect.

The ovary is said to be **superior** if the calyx and corolla are attached to the receptacle at the base of the ovary, as in pea and primrose flowers. In other instances, the ovary is **inferior** when the receptacle grows up around (and in some cases fused to) the ovary, so that the calyx and corolla appear to be attached at the top, as in apple, cactus, and carrot flowers. A cavity containing one or more egg-shaped **ovules** lies within the ovary; short stalks attach ovules to the wall of the cavity. After *fertilization* each ovule becomes a **seed.** Details of fertilization and ovule development are given in Chapter 23.

Peach flowers are produced singly, each on its own stalk, referred to as a pedicel, but many other flowers, such as lilac, grape, and bridal wreath, are produced in **inflorescences.** Inflorescences are groups of several to hundreds of flowers that may all open at the same time, or they may follow an orderly progression to maturation (Fig. 8.7). The main branch of these inflorescences is referred to as the peduncle, and the little stalks that arise

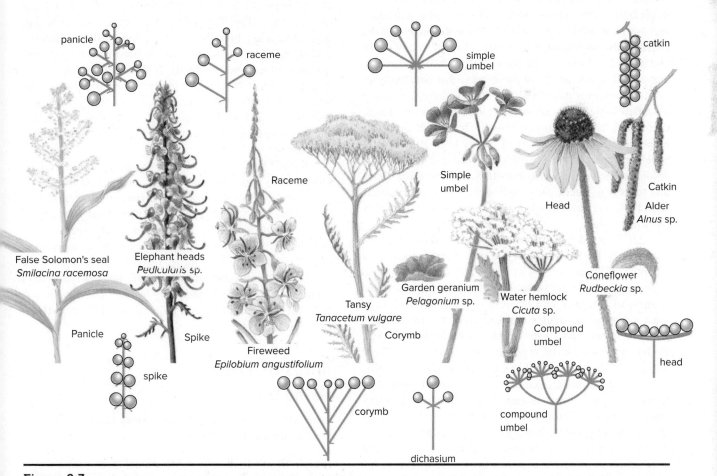

Figure 8.7 Inflorescence types. Each ball represents a flower. In all inflorescences shown, except for the dichasium and catkin, the lowermost or outermost flowers open first. The flowers then open in succession upward or inward. In a dichasium, the central flower opens first, and the side flowers open simultaneously. In a catkin, all the flowers open simultaneously.

from the peduncle are called pedicels, so there is one pedicel for each flower.

A discussion of primitive and specialized flowers, along with their evolutionary development and ecological adaptations, is given in Chapter 23.

8.4 Fruits

Learning Outcomes

a. Provide the botanical definition of a fruit, and explain why some people may use a different definition in everyday life.
b. Identify the regions of mature fruits.
c. State five types of fleshy and dry fruits, and explain how simple, aggregate, and multiple fruits are derived from the flowers.

Introduction

In 1893, the U.S. Supreme Court, in the case of *Nix* v. *Hedden*, ruled that a tomato was legally a vegetable rather than a fruit. This was in keeping with the public's general conception of fruits as being relatively sweet and dessert-like. Vegetables, on the other hand, were considered more savory and useful as salad or main-course foods. Regardless of the court's decision, a **fruit**, botanically speaking, is any ovary and its accessory parts that has developed and matured. It also usually contains one or more seeds that develop from fertilized ovules within the ovary. By this definition, many so-called vegetables, including tomatoes, string beans, cucumbers, and squashes, are really fruits. On the other hand, vegetables can consist of leaves (e.g., lettuce, cabbage), leaf petioles (e.g., celery), specialized leaves (e.g., onion), stems (e.g., potato), roots (e.g., sweet potato), stems and roots (e.g., beets, carrots), flowers and their peduncles (e.g., broccoli), flower buds (e.g., globe artichoke), or other parts of the plant.

All fruits develop from ovaries, so they are produced only by the flowering plants. *Fertilization* (see Chapter 23) usually determines whether or not the ovary or ovaries (and sometimes the receptacle or other tissues) of a flower will develop into a fruit. Unless at least a few of the ovules are fertilized, the flower normally withers and drops without developing further. Pollen grains contain specific stimulants called *hormones* (discussed in detail in Chapter 11) that may initiate fruit development, and sometimes a small amount of dead pollen is all that is needed to stimulate an ovary into becoming a fruit. It is the hormones produced by the developing seeds, however, that promote the greatest fruit growth. These hormones, in turn, stimulate the production of more fruit growth hormones by the ovary wall. In a few instances (e.g., the cultivated banana), fruits develop without fertilization. These fruits are said

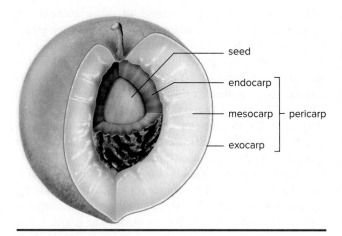

Figure 8.8 Regions of a mature peach fruit.

to be *parthenocarpic*. Parthenocarpy is discussed in Chapter 23.

Fruit Regions

Most of a mature fruit has three regions, which sometimes merge and can be difficult to distinguish from one another (Fig. 8.8). The skin forms the **exocarp,** while the inner boundary around the seed(s) forms the **endocarp.** The endocarp may be hard and stony (as in a peach pit around the seed). It also may be papery (as in apples), or it may not be distinct from the **mesocarp,** which is the often fleshy tissue between the exocarp and the endocarp. The three regions collectively are called the **pericarp.** In dry fruits, the pericarp is usually quite thin.

Some fruits consist of only the ovary and its seeds. Others have adjacent flower parts, such as the receptacle or calyx fused to the ovary or different parts modified in various ways. Fruits may be either fleshy or dry at maturity, and they may split, exposing the seeds. They may be derived from a single ovary or from more than one. Traditionally, all these features have been used to classify fruits, but unfortunately, not all fruits lend themselves to neat pigeonholing by such characteristics. Some of these problems are pointed out in the classification that follows.

Kinds of Fruits

Fleshy Fruits

Fruits whose mesocarp is at least partly fleshy at maturity are classified as fleshy fruits.

Simple fleshy fruits develop from a flower with a single pistil. The ovary may be superior or inferior, and it may be *simple* (derived from a single **carpel**), or it may consist of

KEY THEME 8.1 ecology

Goober Peas

The peanut is often ranked by the food industry as the most popular "nut" in the United States, occupying two-thirds of the nut market. Other nuts, such as walnuts, almonds, and pecans, grow on trees. Peanut plants, however, are legumes, so they are more closely related to peas than to nut trees. In the southern United States peanuts are sometimes called "goober peas." This term was derived from the native language of African slaves, who referred to peanuts as "nguba" and relied on them as an important source of nutrition.

Peanut plants are unusual because they produce their fruits underground. Small, yellow flowers (Box Figure 8.1A) are formed aboveground, where they are self-pollinated. This causes the ovary, or "peg," to grow. It turns downward and penetrates the soil, where it matures into what we recognize as a peanut (Box Figure 8.1B). This reproductive strategy may have evolved to protect the fruit from herbivores and prepare for the next generation, because the seed is buried in the soil, ready to germinate when the environment is acceptable.

(A) (B)

Box Figures 8.1A and 8.1B (A) Small, yellow flower of a peanut plant. (B) Peanuts exposed after excavation from the soil. (A-B) © *Shelley H. Jansky*

---

two or more carpels and be *compound* (for a discussion of the derivation of carpels and compound ovaries, see Chapter 23). The ovary alone may develop into the fruit, or other parts of the flower may develop with it.

A **drupe** is a simple fleshy fruit with a single seed enclosed by a hard, stony endocarp, or pit (Fig. 8.9). It usually develops from flowers with a superior ovary containing a single ovule. The mesocarp is not always obviously fleshy, however. In coconuts, for example, the *husk* (consisting of the mesocarp and the exocarp), which is usually removed before the rest of the fruit is sold in markets, is very fibrous (the fibers, incidentally, are used in making mats and brushes). The seed ("meat") of the coconut is hollow and contains a watery *endosperm* (see Chapter 23) commonly referred to as "milk." It is surrounded by the thick, hard endocarp typical of drupes. Other examples of drupes include the stone fruits (e.g., apricots, cherries, peaches, plums, olives, and almonds). In almonds, the husk, which dries somewhat and splits at maturity, is removed before marketing, and it is the endocarp that we crack to obtain the seed.

Berries usually develop from a compound ovary and commonly contain more than one seed. The entire pericarp is fleshy, and it is difficult to distinguish between the mesocarp and the endocarp (Fig. 8.10). Three types of berries may be recognized.

A *true berry* is a fruit with a thin skin and a pericarp that is relatively soft at maturity. Although most contain more than one seed, notable exceptions are dates and avocados, which have only one seed. Typical examples of true berries include tomatoes, grapes, persimmons, peppers, and eggplants. Some fruits that include the word *berry* in their common name (e.g., strawberry, raspberry, blackberry) botanically are not berries at all.

Some berries are derived from flowers with inferior ovaries, so that other parts of the flower also contribute to

Figure 8.9 Representative drupes. (a) Peaches. (b) Almonds. (c) Olives. *(a) © Ingram Publishing/SuperStock; (b-c) © Kingsley Stern*

the flesh. They can usually be distinguished by the remnants of flower parts or their scars that persist at the tip. Examples of such berries include gooseberries, blueberries, cranberries, pomegranates, and bananas. Because fruit development in the cultivated banana is parthenocarpic, there are no seeds. Several other species of banana produce an abundance of seeds.

Pepos are berries with relatively thick rinds. Fruits of members of the Pumpkin Family (Cucurbitaceae), including pumpkins, cucumbers, watermelons, squashes, and cantaloupes, are pepos.

The *hesperidium* is a berry with a leathery skin containing oils. Numerous outgrowths from the inner lining of the ovary wall become saclike and swollen with juice as the fruit develops. All members of the Citrus Family (Rutaceae) produce this type of fruit. Examples include oranges, lemons, limes, grapefruit, tangerines, and kumquats.

Pomes are simple fleshy fruits for which most of the flesh comes from the enlarged floral tube or receptacle that grows up around the ovary. The endocarp around the seeds is papery or leathery. Examples include apples, pears, and quinces. In an apple, the ovary consists of the core and a little adjacent tissue. The remainder of the fruit has developed primarily from the floral tube, with a little tissue contributed by the receptacle (Fig. 8.11). Botany texts often refer to pomes, pepos, some berries, and other fruits derived from more than an ovary alone as *accessory fruits* or as fruits having *accessory tissue*.

Figure 8.10 Representative berries. (a) Grapes. (b) Tomatoes. *(a-b) © Kingsley Stern*

Dry Fruits

Fruits whose mesocarp is definitely dry at maturity are classified as *dry fruits*.

Dry Fruits That Split at Maturity (Dehiscent Fruits) The fruits in this group are distinguished from one another by the way they split.

The **follicle** splits along one side or seam (*suture*) only, exposing the seeds within (Fig. 8.12). Examples include larkspur, columbine, milkweed, and peony.

The **legume** splits along two sides or seams (Fig. 8.13). Thousands of members of the Legume Family (Fabaceae) produce this type of fruit. Examples include peas, beans, garbanzo beans, lentils, carob, kudzu, and mesquite.

Figure 8.11 Apples (representative of pomes). The bulk of the flesh is derived from the floral tube that grows up around the ovary. *© Kingsley Stern*

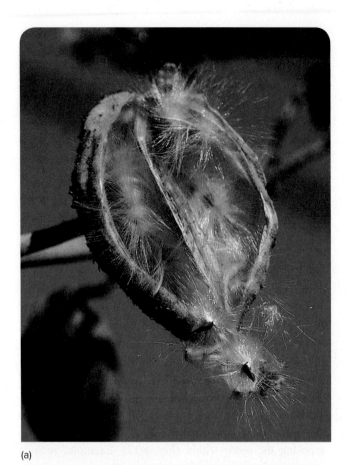

(a)

Peanuts are also legumes, but they are atypical in that the fruits develop and mature underground (discussed in Box 8.1). The seeds are usually released in nature by bacterial breakdown of the pericarp instead of through an active splitting action.

Siliques also split along two sides or seams, but the seeds are borne on a central partition, which is exposed when the two halves of the fruit separate (Fig. 8.14a). Such fruits, when they are less than three times as long as they are wide, are called *silicles* (Fig. 8.14b). Siliques and silicles are produced by members of the Mustard Family (Brassicaceae), which includes broccoli, cabbage, radish, shepherd's purse, and watercress.

Capsules are the most common of the dry fruits that split (Fig. 8.15). They consist of at least two carpels and split in a variety of ways. Some split along the partitions between the carpels, while others split through the cavities (*locules*) in the carpels. Still others form a cap toward one end that pops off and releases the seeds, or they form a row of pores through which the seeds are shaken out as the capsule rattles

(b)

Figure 8.12 Follicles. (a) Milkweed. (b) *Magnolia*. The fruit of the magnolia is actually an aggregate fruit consisting of approximately 40 to 80 individual, one-seeded follicles on a common axis. The follicles and axis fall from the magnolia tree as a unit. *(a-b)* © *Kingsley Stern*

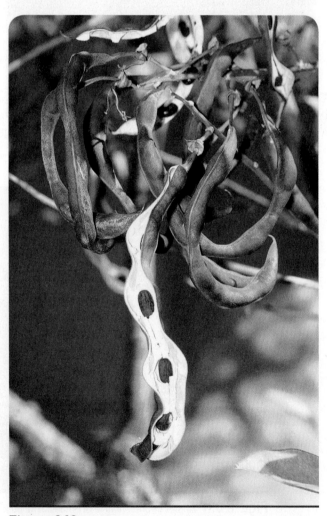

Figure 8.13 Legumes of a coral tree (*Erythrina*). © *Kingsley Stern*

Flowers, Fruits, and Seeds 131

in the wind. Examples include irises, orchids, lilies, poppies, violets, and snapdragons.

Dry Fruits That Do Not Split at Maturity (Indehiscent Fruits) In this type of dry fruit, the single seed is united with the pericarp.

Only the base of the single seed of the **achene** is attached to its surrounding pericarp. Accordingly, the husk (pericarp) is relatively easily separated from the seed. Examples include sunflower "seeds" (the edible kernel plus the husk constitute the achene) (Fig. 8.16), buttercup, and buckwheat.

Figure 8.14 (a) A silique after it has split open. The seeds are borne on a central, membranous partition. (b) Silicles of dollar plant (*Lunaria*). *(b) © Kingsley Stern*

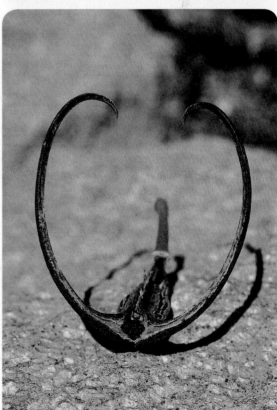

Figure 8.15 Capsules. (a) Butterfly iris (*Moraea*). (b) *Bletilla* orchid. (c) Autograph tree (*Clusia rosea*). (d) Unicorn plant (*Proboscidia*). *(a–d) © Kingsley Stern*

Figure 8.16 Dry fruits that do not split at maturity. (a) Achene of a sunflower sliced open. (b) Grain (caryopsis) of corn, cut lengthwise. (c) Nuts (acorns) of an oak. The acorn was cut lengthwise above the cup. *(a-c)* © Kingsley Stern

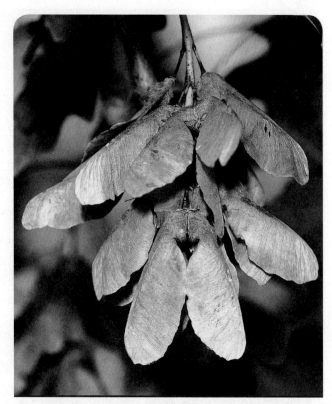

Figure 8.17 Samaras of a big-leaf maple (*Acer macrophyllum*). Maple samaras are produced in pairs that separate at maturity.
© Kingsley Stern

Nuts are one-seeded fruits similar to achenes, but they are generally larger, and the pericarp is much harder and thicker. They develop with a cup, or cluster, of bracts at their base. Examples include acorns (Fig. 8.16), hazelnuts (filberts), and hickory nuts. Botanically speaking, many nuts in the popular sense are not nuts. We have already seen that peanuts are atypical legumes and that coconuts and almonds are drupes. Walnuts and pecans are also drupes, whose "flesh" withers and dries after the seed matures. Brazil nuts are the seeds of a large capsule, and a cashew nut is the single seed of a unique drupe. It appears as a curved appendage at the end of a swollen pedicel, which is eaten raw in the tropics or made into preserves or wine. Pistachio nuts are also the seeds of drupes.

The pericarp of the **grain** (**caryopsis;** plural: **caryopses**) is tightly united with the seed and cannot be separated from it (Fig. 8.16). All members of the Grass Family (Poaceae), including corn, wheat, rice, oats, and barley, produce grains.

In **samaras,** the pericarp surrounding the seed extends out in the form of a wing or membrane, which aids in dispersal (Fig. 8.17). In maples, samaras are produced in pairs, but in ashes, elms, and the tree of heaven, they are produced singly.

The twin fruit called a **schizocarp** (Fig. 8.18) is well represented by the Parsley Family (Apiaceae). Members of this family include parsley, carrots, anise, caraway, and dill. Upon drying, the twin fruits break into two one-seeded segments called *mericarps*.

Aggregate Fruits

An **aggregate fruit** is derived from a single flower with several to many pistils. The individual pistils develop into tiny drupes or other fruitlets, but they mature as a clustered unit on a single receptacle (Fig. 8.19). Examples include raspberries, blackberries, and strawberries. In a strawberry, the cone-shaped receptacle becomes fleshy and red, while each pistil becomes a little *achene* on its surface. In other words, the strawberry, while being an aggregate fruit, is also partly composed of accessory tissue.

Multiple Fruits

Multiple fruits are derived from several to many individual flowers in a single inflorescence. Each flower has its own receptacle, but as the flowers mature separately into fruitlets, they develop together into a single larger fruit, as in aggregate fruits. Examples of multiple fruits include mulberries,

Figure 8.18 Schizocarps of carrots. The schizocarp is comprised of a pair of seeds which can separate into two one-seeded mericarps upon drying. © *Shelley H. Jansky*

(a)

(b)

Figure 8.19 (a) A blackberry flower. Note the numerous green pistils. (b) Blackberries, representative of aggregate fruits. *(a) © Kingsley Stern; (b) © Pixtal/age fotostock*

Osage oranges (Fig. 8.20), pineapples, and figs. Pineapples, like bananas, usually develop parthenocarpically (see Chapter 23), and there are no seeds. The individual flowers are fused together on a fleshy axis, and the fruitlets coalesce into a single fruit.

Figs mature from a unique "outside in" inflorescence. The individual flowers of the inflorescence are enclosed by the common receptacle, which has an opening to the outside at the tip (Fig. 8.21). Such a multiple fruit arrangement is referred to as a *syconium*. Some fig varieties develop parthenocarpically, but others are pollinated by tiny wasps that crawl in and out through the opening. Some multiple fruits, such as those of the sweet gum, are dry at maturity.

8.5 Fruit and Seed Dispersal

Learning Outcome

a. Describe fruit and seed adaptations that are used to promote their dispersal.

Why are so many species of orchids rare, while dandelions, shepherd's purse, and other weeds occur all over the world? Why do some plants occupy entire mountain ranges, while others are confined to small niches? The answers to these questions involve many different factors, including climate, soil, the adaptability of the plant, and its means of seed dispersal. It is important for seeds to be carried away from the

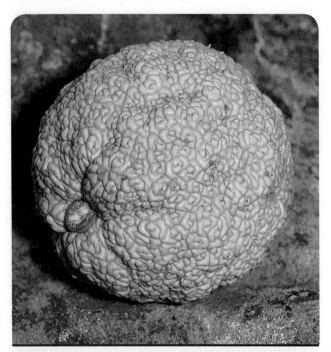

Figure 8.20 An Osage orange (*Maclura*), representative of multiple fruits. The mature ovaries of many flowers become united in a single unit. © *Kingsley Stern*

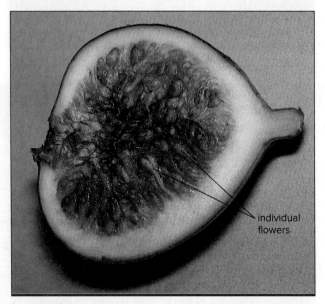

Figure 8.21 Section through a developing fig. © *Kingsley Stern*

released from the tree. In a brisk wind, samaras may be carried up to 10 kilometers (6 miles) away from their source, although most are relatively evenly distributed within a few meters of the tree. In hop hornbeams, the seed is enclosed in an inflated sac that gives it some buoyancy in the wind. In some members of the Buttercup and Sunflower Families (Ranunculaceae and Asteraceae), the fruits have plumes, and in the Willow Family (Salicaceae), the comose (hairy) seeds are well-designed for wind dispersal. In button snakeroots and Jerusalem sage, the fruits are too large to be airborne, but they are spherical enough to be rolled along the ground by the wind.

Seeds themselves may be so tiny and light that they can be blown great distances by the wind. Orchids and heaths, for example, produce seeds with no endosperm that are as fine as dust and equally light in weight. In catalpa and jacaranda

Figure 8.22 Types of seeds and fruits dispersed by wind.

mother plant before they germinate. This prevents competition with the mother plant and avoids inbreeding. The following sections describe several effective seed dispersal mechanisms.

Dispersal by Wind

Evolution has created fruits and seeds with a variety of adaptations for wind dispersal (Fig. 8.22). The samara of a maple has a curved wing that causes the fruit to spin as it is trees, the seeds rather than the fruits are winged. Dandelion fruitlets have plumes that radiate out at the ends like tiny parachutes; these catch even a slight breeze. In tumble mustard and other tumbleweeds, the whole above-ground portion of the plant may *abscise* (separate from the root) and be blown away by the wind, releasing seeds as it bumps along.

Dispersal by Animals

Birds, mammals, and ants all act as seed dispersal agents (Fig. 8.23). Shore birds may carry seeds great distances in mud that adheres to their feet. Other birds and mammals eat fruits whose seeds pass unharmed through their digestive tracts. Some bird-disseminated fruits contain laxatives that speed their passage through the birds' digestive tracts. In blackbirds, the seeds may remain in the tract as little as 15 minutes, but in mammals, the period is more commonly about 24 hours. In the giant tortoises of the Galápagos Islands, seeds do not pass through the tract for 2 weeks or more, and the seeds usually will not germinate unless they have been subjected to such treatment (see Chapter 11). Some seeds and fruits are gathered and stored by rodents, such as squirrels and mice, which do not recover all the seeds they bury. Blue jays, woodpeckers, and other birds carry away nuts and other fruits, which they may drop in flight.

Many fruits and seeds adhere to the fur or feathers of animals and birds. Bedstraw and bur clover fruits are covered with small hooks that catch in fur (or a hiker's socks). The large capsules of unicorn plants have two giant, curved extensions about 15 centimeters (6 inches) long. These can catch on the fetlock of a deer or other animal that happens to step on the fruit, and the seeds are scattered as the animal moves along. Twinflowers and flax have fruits with sticky appendages that adhere to fur on contact, and those of the puncture vine penetrate the skin and stick by means of hard, little prickles (Fig. 8.24).

Bleeding hearts, trilliums, and several dozen other plants have appendages on their seeds that contain oils attractive to ants (Fig. 8.25). The Swedish botanist Johan Rutger Sernander once estimated that more than 36,000 such seeds were carried by members of a single ant

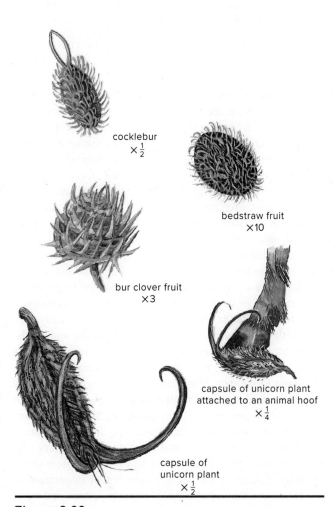

Figure 8.23 Types of seeds and fruits dispersed by animals and birds.

Figure 8.24 Fruits of a puncture vine (*Tribulus terrestris*) clinging to a bicycle tire. © *Kingsley Stern*

Figure 8.25 Seeds of the Pacific bleeding heart (*Dicentra formosa*). The white appendages are *elaiosomes*, which are removed from the seeds by ants and used for food. © Kingsley Stern

colony to their nest, where the ants stripped off the appendages for food but did not harm the seeds themselves.

Dispersal by Water

Some fruits contain trapped air, adapting them to water dispersal. Many sedges, for example, have seeds surrounded by inflated sacs that enable the seeds to float (Fig. 8.26). Others have waxy material on the surface of the seeds, which temporarily prevents them from absorbing water while they are floating. Sometimes, a heavy downpour will create a torrent of water that dislodges masses of vegetation along a stream bank, carrying whole plants and their fruits to new locations. Large raindrops themselves may splash seeds out of their opened capsules.

Seeds and fruits of a few plants have thick, spongy pericarps that absorb water very slowly. Such fruits are adapted to dispersal by ocean currents, even though salt water eventually may penetrate enough to kill the delicate embryos. Enough fruits are beached before this occurs to ensure the survival of the species. Contrary to popular belief, coconuts that fall into water usually become waterlogged and sink in a few days. Rarely, if ever, are they carried out to sea.

Other Dispersal Mechanisms and Agents

Fruits of some legumes, touch-me-nots, and members of other families mechanically eject seeds, sometimes with considerable force. For example, the splitting action of drying witch hazel capsules may fling the seeds over 12 meters (40 feet) away. Fruits of dwarf mistletoes may be violently released in response to the heat of a warm-blooded animal coming close

(a)   (b)

Figure 8.26 Sedge adaptation to water dispersal. (a) A sedge plant. (b) A sedge fruit. The seed is enclosed within an inflated covering that enables it to float on water.

to the plants. In fact, small welts have been raised by dwarf mistletoe fruits on the skin of humans who ventured close to the plants. In manroots and a few other members of the Pumpkin Family (Cucurbitaceae), the seed release resembles a geyser eruption as a frothy substance containing the seeds squirts out of one end of the melonlike fruits.

In filarees and other members of the Geranium Family (Geraniaceae), each carpel of the fruit splits away and curls back from a central axis. Each fruitlet consists of a single, pointed seed with a long, slender tail that is sensitive to changes in humidity (Fig. 8.27). At night when the humidity increases, the tail is relatively straight, but in the sun, it coils up like a corkscrew, literally drilling the pointed seed into the ground as it does so and effectively planting it in the process.

Humans, both intentionally and unintentionally, are by far the most efficient transporters of fruits and seeds. Travelers and explorers have carried many noxious weeds along with valuable food and medicinal plants from one continent to another. Most countries now have strict regulations barring the importation of plant materials, except by special permit, and carrying some plants across borders is not legal under any circumstances. In the United States, for example, certain fruits and seeds that might carry diseases harmful to local agriculture may be barred from entry. Arizona, California, and Hawaii have border inspections to try to prevent the importation of popcorn, citrus, and other fruits; such plants may be carriers of diseases or pests that are presently under control.

Figure 8.27 Filaree (*Erodium*) fruitlet. (a) Under humid conditions. (b) Under dry conditions. Alternate coiling and uncoiling causes the fruitlet to be "screwed" into the ground.

8.6 Seeds

Learning Outcomes

a. Diagram and label a mature dicot seed (e.g., bean) and a monocot seed (e.g., corn) in section to show the parts and regions.

b. Explain the changes that occur when a seed germinates, and state the environmental conditions essential to germination.

c. List the types of factors that control dormancy. Explain how dormancy may be broken both naturally and artificially.

Structure

Remember that ovules develop into seeds. The concave side of a kidney bean (the seed of a dicot plant) has a small, white scar called the *hilum*. The hilum marks the point at which the ovule was attached to the ovary wall. A tiny pore called the *micropyle* is located next to the hilum. If this bean is placed in water for an hour or two, it may swell enough to split the seed coat. Once the seed coat is removed, the two halves, called **cotyledons,** can be distinguished (Fig. 8.28). The cotyledons, which have a tiny, immature plantlet along one edge between them, are food-storage organs that also function as the "seed leaves" of the seedling plant. The cotyledons, and the tiny, rudimentary bean plant to which they are attached, constitute the *embryo*. Some seeds (e.g., those of grasses and all other monocots) have only one cotyledon.

The tiny embryo plantlet has undeveloped leaves and a meristem at the upper end of the embryo axis. This embryo shoot is called a **plumule.** The cotyledons are attached just below the plumule. The very short part of the stem above the cotyledons is called the **epicotyl,** while the stem below the attachment point is the **hypocotyl.** In an embryo, it is often difficult to tell where the stem ends and the root begins, but the tip that will develop into a root is called a **radicle.** When a kidney bean germinates, the hypocotyl lengthens and bends, becoming hook-shaped. The top of the hook emerges from the ground, pulling the cotyledons above the ground. Once the cotyledons have emerged, the hook straightens out. In lima beans and peas, however, the hypocotyl remains short, so that the cotyledons do not emerge above the surface (see Chapter 11). Kidney beans exhibit epigeous germination, while lima beans undergo hypogeous germination.

In other seeds, the cotyledon(s) may not play a significant role in food storage. In corn, for example, the bulk of the food-storage tissue is *endosperm* (see Chapter 23). Corn "seeds" (Fig. 8.29) also display other features not seen in beans. The plumule and the radicle are enclosed in tubular, sheathing structures called the **coleoptile** and the **coleorhiza,** respectively. These protect the delicate tissues within the seeds during germination. After the coleoptile and coleorhiza have grown to several millimeters in length, their development ceases, and the plumule and radicle burst through the tips.

Germination

Germination is the beginning or resumption of growth of a seed, and it depends on the interplay of a number of internal and external factors. In order to germinate, a seed must first be *viable* (alive). Many seeds also require a period of **dormancy** before they germinate. Dormancy is brought about by mechanical and/or physiological circumstances. In the Legume Family (Fabaceae) and others, seeds may have seed coats so thick or tough that they prevent the absorption of water or oxygen. Some seeds even have a one-way valve that lets moisture out but prevents its uptake. Dormancy in such seeds may sometimes be broken artificially by *scarification,* in which the seed coat is broken by nicking, cracking, or dipping in a concentrated acid for a few seconds to a few minutes. In nature, such seeds may remain dormant until the seed coat is damaged by the mechanical abrasion of rock particles in the soil, alternate thawing and freezing, or in some cases, bacterial action.

Dormancy may also be brought about by growth-inhibiting substances in the seed coat, the interior of the seed, or tissues of the fruit surrounding it. Many desert plants have inhibitors in the seed coat. These have to be washed away by soaking rains before germination will occur. The inhibitors have evolved for survival of the species by preventing germination unless there has been sufficient rainfall for a seedling to become established.

Apples, pears, citrus fruits, tomatoes, and other fleshy fruits contain inhibitors that prevent germination of the seeds within the fruits. Once the seeds are removed and washed,

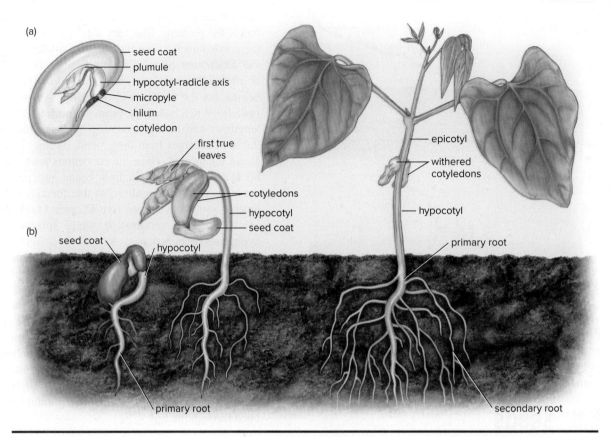

Figure 8.28 A common garden bean. (a) Seed structure. (b) Germination and development of the seedling.

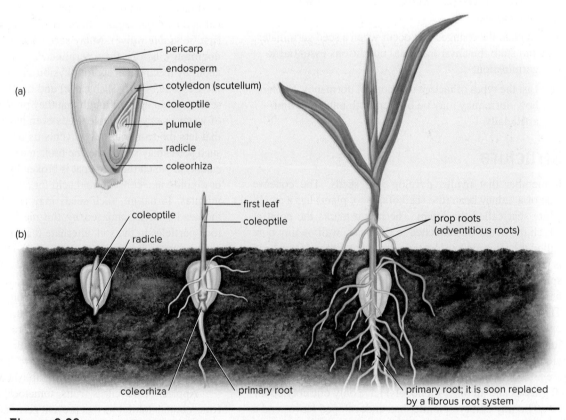

Figure 8.29 Corn. (a) Grain structure. (b) Germination and development of the seedling.

they germinate readily. The embryos of some seeds, such as those of the American holly, consist of only a few unspecialized cells when the fruit ripens. After the fruit has dropped, the seeds will not germinate until the embryo has developed fully with the aid of food materials stored in its endosperm. Such a process of development is called *after-ripening*.

In many woody plants of temperate areas, germination stimulators need to be present to initiate growth. These normally do not develop unless the seeds encounter a wet period accompanied by cold temperatures. Usually, this period needs to be a minimum of 4 to 6 weeks. The dormancy of such seeds can be broken artificially by placing them in a refrigerator, preferably in damp sand, for a few weeks.

Even when mechanical and physiological barriers to germination are not present, a seed will not normally germinate unless environmental factors are favorable. Water and oxygen are essential. Many seeds imbibe 10 times or more their total weight in water before the radicle emerges. Some seeds, such as those of castor beans (Fig. 8.30) and certain spurges, have appendages that absorb water to speed up germination.

After water has been imbibed, enzymes begin to function in the cytoplasm, which has now been rehydrated. Some enzymes convert stored proteins to amino acids, others convert fats and oils to soluble compounds, and still other enzymes aid in the conversion of starch to sugar. The soluble substances can then be transported to the embryo, and respiration, which in a dormant seed is almost imperceptible, can be greatly accelerated. Aerobic respiration provides the energy for germination in virtually all species, although in some plants, such as rice and barnyardgrass, the seeds have developed the ability to survive long periods of alcoholic fermentation when oxygen levels are low. When oxygen levels are sufficient, the cascade of events for germination are initiated, and a new plant begins to develop as mitosis and cell elongation take place. The process of cellular respiration and alternative pathways are discussed in Chapter 10.

If seeds are kept waterlogged after planting, available oxygen is greatly reduced, and the seeds or seedlings may die. Most seeds require temperatures within certain ranges to germinate. These usually need to be above freezing but below 45°C (113°F). Germination percentages tend to be low near either extreme, however. Most crop plants have an optimum (ideal) germination temperature of between 20°C and 30°C (68°F and 86°F).

The role of light in germination varies among plant species. Seeds of some varieties of lettuce will not germinate in the dark (see the discussion of *phytochrome* in Chapter 11), while other seeds, such as the California poppy, germinate only in the dark. In lettuce seeds, the light apparently inactivates germination inhibitors, while in the California poppy, it stimulates inhibitor formation. (See the additional discussion of dormancy in Chapter 11.)

Longevity

From time to time, one reads or hears of seeds of wheat or other edible plants germinating after lying dormant in Egyptian pyramids or Native American tombs for thousands of years, but none of these reports has been confirmed. In fact, there is evidence in a few instances that rats or rodents in recent times carried the seeds concerned to their nests. However, a report of seeds of the aquatic lotus plant germinating after about 1,200 years and another documenting the germination of Arctic tundra lupine seeds that were frozen for an estimated 10,000 years have been confirmed.

Seeds remain viable for periods that vary greatly, depending on the species and the storage conditions. Some seeds, such as those of certain willows, cottonwoods, orchids, and tea, remain viable for only a few days or weeks, regardless of how they are stored, but the period of viability of most seeds is extended by months or even years when they are stored at low temperatures and kept dry.

By law, packets of vegetable and flower seeds sold in stores are dated, giving the buyer a rough idea of how long a significant number of the seeds might be expected to remain viable. Generally, seeds of Pumpkin Family (Cucurbitaceae) members (e.g., squash, cantaloupe, cucumber) retain a relatively high percentage of viability for several years, while those of members of the Lily Family (Liliaceae) (e.g., onion, leek, chives) retain a good percentage of viability for only 2 or 3 years. Properly stored wheat seeds have been reported to retain more than 30% viability for more than 30 years, and some weed seeds stored under conditions of low oxygen, high humidity, and cool temperatures have remained viable for even longer periods.

In 1879, William J. Beal buried 20 pint-sized bottles of weed seeds on the campus of what is now Michigan State University in East Lansing, Michigan. Each bottle contained 1,000 seeds of 20 different species of weeds. Every 5 years, a bottle of seeds was dug up and the seeds were planted, until the schedule was changed in 1920 to every 10 years.

Figure 8.30 Castor bean (*Ricinus*) seeds. Note the small, water-absorbing appendage (caruncle) at the end of each seed.
© Kingsley Stern

KEY THEME 8.2 physiology — The Seed That Slept for 1,200 Years

Rip Van Winkle would appreciate this. An Oriental Sacred Lotus (*Nelumbo nucifera*) seed collected from the sediment of a dry lake bottom near a small village in northeastern China has germinated after being dormant for over 1,200 years. It is one of the oldest living seeds ever found. The Beijing Institute of Botany donated Sacred Lotus seeds to a team of UCLA scientists, who dated the seeds with a nondestructive method called accelerator mass spectroscopy. Small amounts of tissue (less than 10 milligrams) were sampled for radiocarbon dating prior to germination studies. Before the use of this newer dating method, whole seeds had to be destroyed in the process of dating, thereby eliminating the possibility of testing the seeds for viability. After lying dormant in a bed of black clay at depths of 0.5 to 2.8 meters (1.5 to 9 feet), the germinated 1,200-year-old seed (1,288 ± 271 years) was the oldest, but not the only, survivor from the subterranean tomb. Three other ancient lotus seeds found at various depths germinated, and the UCLA team determined that one was more than 600 years old and another was more than 300.

While reports have claimed seed germination of more ancient seeds recovered from dry archaeological sites in Egypt (such as King Tut's tomb in the pyramids of Giza and from the tombs of other pharaohs), experts now agree that these reports are unreliable. Apart from these lotus seeds, the oldest documented viable seeds are lupine seeds that were frozen in Arctic tundra and seeds from Professor Beal's germination study (see text section "Longevity").

These Sacred Lotus seeds have managed to ward off the ravages of time. Existing in an impenetrable seed coat and mired in an oxygen-deficient mud, the seeds have intact genetic and enzymatic systems that reactivated when the seeds were split open and soaked in water. (Enzymes are proteins that speed up chemical reactions in the cell.) A key enzyme that repairs proteins was present during germination; this enzyme has been found in similar quantities in modern-day Sacred Lotus seeds. The repair enzyme functions in converting damaged amino acids back to their naturally occurring functional form. This is especially important in "repairing" the proteins of the cell membrane. Without intact membranes, cells are not able to function, and such damage will lead to the death of the cell and eventually the organism.

The architecture of the fruit no doubt plays a key role in the longevity of these seeds. The fruits (also called pods) of the Sacred Lotus (Box Figures 8.2A and 8.2B) are round to oblong, 10 to 13 centimeters (4 to 5 inches) long, and 8 to 10 centimeters (3 to 4 inches) in diameter, and each contains a single seed. The fruit wall, or pericarp, which is impervious to water and airtight, is initially green and turns purplish brown and becomes dry and notably hard. Chinese botanists who first investigated similar ancient lotus seeds collected from the same deposits were unable to get them to germinate, even after 20 months of soaking in water. It wasn't until they scarified (filed open to permit water absorption) the seeds, as prescribed in a 1,400-year-old Chinese manuscript Ch'i Min Yao Su (*Important Technology for People in Harmony*), that they were ultimately successful in germinating the seeds. The hard, airtight fruit walls are the most significant of the structural features that contribute to the exceptional longevity of the seeds.

The Sacred Lotus was introduced to China following the introduction of Buddhism from India in the first century B.C. It is regarded as a symbol of purity and strength, emerging from the mire of lake waters and opening its crimson flowers to the heavens. The earliest dated depiction of Buddha (in A.D. 240) shows him surrounded with a halo and seated cross-legged on a lotus throne. From old Chinese manuscripts, we learn that lotus has been cultivated as a crop plant (most parts are edible) for the past 4,000 years, and traditional Chinese herbal medicine considered the Sacred Lotus a mainstay of its pharmaceutical collections. The large number of lotus seeds collected from the site suggests that the plant was under cultivation in this now dried-up lotus lake. It is likely that the 1,200-year-old seed was derived from a plant cultivated by Buddhists at this site in northeastern China.

Unlike Rip Van Winkle, who, following a drink of liquor, slept for only 20 years, these Sacred Lotus seeds are remarkable in their capacity to revive after more than 1,000 years. Lessons learned from these plants can shape our thinking about our own aging and the possibilities that may exist in the future for extending life at the margins.

D. C. Scheirer

(A)

(B)

Box Figures 8.2A and 8.2B (A) Green lotus fruit (also referred to as a "pod") before drying. (B) Dried lotus pod showing seeds. *(A) Kingsley Stern; (B) © James E. Bidlack*

A few species of both dicots and monocots produce seeds that have no period of dormancy at all. In some instances, the embryo, which develops from the zygote, continues to grow in a phenomenon known as *vivipary*. In the red mangrove, a tropical tree associated with coastal waters and estuaries, each fruit contains a single seed in which the embryo continues to grow while the fruit is still on the tree, reaching a length of 25 centimeters (10 inches) or more before the seedling becomes detached and essentially plants itself in the mud below (Fig. 8.31).

SUMMARY

8.1 Introduction to Flowers, Fruits, and Seeds

- Flowers occur in a wide variety of sizes, colors, textures, and habitats. Annuals complete their life cycles in one growing season; biennials take two seasons; and perennials may take several to many years to complete their cycles.

8.2 Differences between Dicots and Monocots

- Dicots and monocots are distinguished from one another by the differences in numbers of flower parts and cotyledons, venation, presence or absence of cambia, vascular bundle arrangement, and pollen grain apertures.

8.3 Structure of Flowers

- A typical flower consists of a peduncle and a receptacle to which are attached sepals (calyx), petals (corolla), stamens, and pistil. Flowers may have no petals, or the petals may be fused together. Many flowers may be in an inflorescence.
- Stamens consist of a pollen-bearing anther and a stalk or filament.
- A pistil consists of a stigma, a style, and an ovary. The ovary may be superior or inferior and contains one or more ovules.

8.4 Fruits

- A fruit, which is unique to flowering plants, is a mature ovary. Hormones promote the greatest fruit growth. Parthenocarpic fruits are seedless and develop without fertilization occurring, but not all seedless fruits are parthenocarpic.
- A mature fruit has an outer exocarp, an inner endocarp around the seed(s), and a mesocarp between the exocarp and endocarp; the three regions may be fused together as a pericarp. At maturity, fruits may be fleshy or dry. A fruit may be derived from the ovary alone or from adjacent flower parts as well.
- Fleshy fruits may develop from a flower with a single pistil (simple fruit); aggregate fruits develop from a single flower with more than one pistil; multiple fruits are derived from flowers in an inflorescence.
- Simple fleshy fruits include drupes and berries; berries may be true berries, pepos, or hesperidiums. Pomes have flesh derived from both the receptacle and the ovary.

Figure 8.31 Young seedlings of red mangrove (*Rhizophora mangle*), whose seeds have no dormant period and germinate while the fruit is still on the tree. The seedlings grow to lengths of up to 25 centimeters (10 inches) before falling and becoming planted in the mud below. Ocean currents and tides also distribute mangrove seeds throughout tropical tidal zones. The dispersal has been so effective that there are 60,000 square kilometers (23,000 square miles) of mangroves in Southeast Asia alone. Mangrove wood is harvested for fuel, and in the past 40 years, the groves in some localities have been reduced in area by more than 50%. The fruit is sweet and edible. © *Kingsley Stern*

When the first bottle was dug up in 1884, seeds of most of the weeds germinated; in 1960, seeds of evening primrose, curly dock, and moth mullein still germinated; and in 1980, 29 moth mullein seeds, 1 mullein seed, and 1 mallow seed germinated—101 years after they were placed in the bottles. It is of interest to note that a mallow seed previously had not germinated since 1899. Only six of the original bottles now remain; they are not scheduled to be unearthed until the year 2040. Recent evidence indicates that the timing of the digging up of the seed bottles did not take into account the fact that certain temperature patterns are critical to the germination of many weed seeds and that if Beal's experiment had been conducted in a different way, the germination results would have been quite different.

- Accessory fruits consist of more than the ovary alone; some aggregate fruits (e.g., strawberries) are largely composed of accessory tissue. The individual fruitlets of a multiple fruit develop together into a single larger fruit.
- Some dry fruits split as they mature; such fruits include follicles, legumes, siliques or silicles, and capsules.
- Non-splitting dry fruits include achenes, nuts, grains (caryopses), samaras, and schizocarps.

8.5 Fruit and Seed Dispersal

- Fruits and seeds may have wings, plumes, and other adaptations for wind dispersal. Some fruits and seeds have adaptations for animal, bird, or water dispersal. Some fruits eject seeds with force, and some have modifications that drill seeds into the ground.
- Humans disperse many seeds, and most countries and a few states have strict regulations governing the importation of plant materials, primarily to control the spread of pests and diseases.

8.6 Seeds

- A bean seed has a hilum, a micropyle, a seed coat, two cotyledons, and an embryonic bean plant consisting of a plumule and a radicle.
- In grains, the plumule and the radicle are protected by a coleoptile and a coleorhiza, respectively.
- Germination of a seed depends on the cessation of dormancy. Dormancy may be sustained by mechanical circumstances; scarification may break dormancy in such seeds.
- Dormancy may also be induced by growth or germination inhibitors, or after-ripening may need to occur. Cold temperatures may be necessary for the germination of some seeds; stratification may break the dormancy of such seeds.
- A seed will not germinate unless environmental factors, including water, oxygen, light, and certain temperature ranges, are favorable.
- Seeds remain viable for a few days to more than 100 years. The viability of most seeds is extended by storage at low temperatures under dry conditions, but some weed seeds have their viability extended by storage under humid, cool conditions that include little oxygen.
- Some plants produce seeds that undergo no dormancy at all. The growth of the embryo while the seed and fruit are still on the plant is termed vivipary.

REVIEW QUESTIONS

1. Define calyx, corolla, receptacle, peduncle, pedicel, pistil, filament, ovary, and carpel.
2. Indicate the features by which dicots are distinguished from monocots.
3. What is the difference between a fruit and a vegetable?
4. What causes an ovary to develop into a fruit?
5. What are the various parts of a fruit?
6. How do fleshy fruits differ from dry fruits?
7. Distinguish among simple, aggregate, and multiple fruits.
8. Distinguish among achenes, grains, samaras, and nuts.
9. What adaptations do seeds and fruits have for dispersal by water and animals?
10. Define plumule, radicle, coleoptile, coleorhiza, hypocotyl, after-ripening, stratification, and vivipary.

DISCUSSION QUESTIONS

1. Most wind-pollinated flowering plants have inconspicuous, nonfragrant flowers. How might nature be affected if all flowers were that way?
2. Do you believe the botanical distinction between fruits and vegetables is a good one? If not, how would you change it?
3. In discussing pomes, it was observed that the bulk of the flesh in an apple comes from the floral tube. What could you do to prove or disprove this?
4. Seed and fruit dispersal is achieved with the aid of wind, water, animals, mechanical means, and humans. If you were "designing" a new plant, can you think of any new way in which it might be dispersed?
5. When volcanic activity or coral polyps cause new islands to appear in the oceans, they eventually acquire some vegetation. Would you expect the types of dispersal mechanisms for the flowering plants on these islands to be the same as they were for ancient continents?

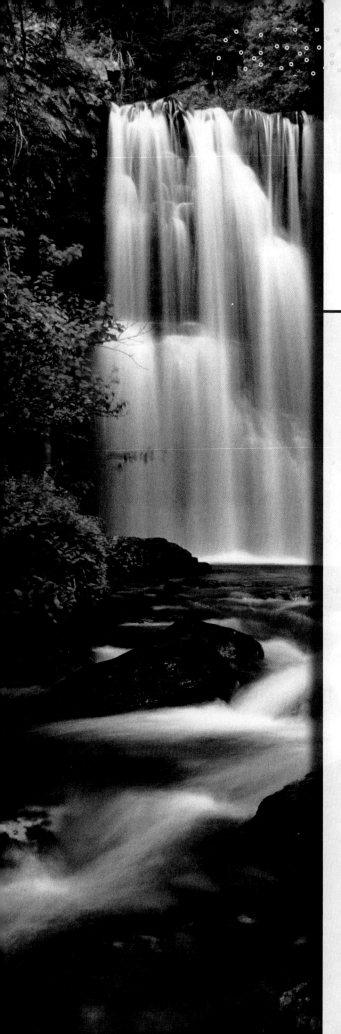

CHAPTER 9

Water in Plants

9.1 Introduction to Water in Plants
9.2 Molecular Movement
   Diffusion
   Osmosis
   Plasmolysis
   Imbibition
   Key Theme 9.1: Measuring Water Potential and Psychrometry
   Active Transport
9.3 Water and Its Movement through the Plant
   The Cohesion-Tension Theory
9.4 Regulation of Transpiration
9.5 Transport of Food Substances (Organic Solutes) in Solution
   The Pressure-Flow Hypothesis
   Key Theme 9.2: Xylem and Phloem Synopsis
9.6 Mineral Requirements for Growth
   Macronutrients and Micronutrients

OVERVIEW

This chapter begins by introducing molecular movement through a comparison between balls in motion in a room and molecular activity. This is followed by a discussion of diffusion, osmosis, turgor, plasmolysis, imbibition, and active transport.

The entry of water into the plant is then explored; this is followed by a discussion of the movement of water through the plant, the evaporation of water into leaf air spaces, and transpiration. A discussion of mineral requirements for growth concludes the chapter.

Rockwell Falls and foliage at Glacier National Park in Montana.
© Design Pics/Natural Selection John Reddy

9.1 Introduction to Water in Plants

Learning Outcome

a. Explain what happens to the distribution of constantly moving tennis balls in two rooms, connected by a hole, to allow passage of the balls.

Nearly everyone has had the experience of driving along a highway or city street when someone in the car says, "What's that smell?" Soon a bakery, or a paper mill, or perhaps a dead skunk comes into view, and the smell gets stronger. Then it fades away as the source is left behind. We take for granted that there is a correlation between our proximity to an odor source and the intensity of the odor—but how does the odor reach us (Fig. 9.1)?

Imagine two adjacent rooms identical in height, width, and length, with no windows, doors, fixtures, or furniture. Now suppose we lift a small flap in the ceiling of one of them and drop in 100 tennis balls. These are ordinary tennis balls except they have perpetual-motion motors in them that cause them to travel in any direction at 30 mph. The tennis balls bounce off the walls, floor, and ceiling, and collide with one another, each time being deflected at a different angle. Shortly after they are introduced, the tennis balls become randomly distributed throughout the room.

Now what would you expect to happen if we opened a small hole in the wall between the two rooms? Eventually, a tennis ball will travel at just the right angle to go through the hole into the other room. Theoretically, it could then bounce straight back into the first room, but it seems unlikely that it would. Reason tells us that long before it might happen to strike the exact angle it needed to return, several more balls would come in from the first room. Given enough time, some balls might bounce back into the first room, but in the long run, each of the two rooms would end up with roughly 50 tennis balls, with an occasional ball going between the rooms in either direction.

The situation just described is analogous to what takes place in nature on a molecular level.

9.2 Molecular Movement

Learning Outcome

a. Define diffusion, osmosis, plasmolysis, imbibition, and active transport.

Molecules and ions (discussed in Chapter 2) are constantly in random motion. Visual evidence of this can be seen with a light microscope. If a drop of ink is diluted with water and observed through the microscope under high power, the tiny carbon particles of the ink appear to be in constant motion. This motion, known as *Brownian movement,* is the result of the bombardment of the visible particles by invisible water molecules, which are in constant motion.

Diffusion

The differing intensity of odors discussed earlier involves molecules behaving somewhat like the tennis balls. Through their random motion, molecules tend to become distributed throughout the space available to them. For example, if perfume molecules are kept in a bottle, they will become distributed throughout the bottle, but if the stopper is removed, they will eventually become dispersed throughout the room, even if there is no air movement.

This movement of molecules or ions from a region of higher concentration to a region of lower concentration is called **diffusion** (Fig. 9.2). Molecules that are moving from a region of higher concentration to a region of lower concentration are said to be moving *along a diffusion gradient,* while molecules going in the opposite direction are said to be moving *against a diffusion gradient.* When the molecules, through their random movement, have become distributed throughout the space available, they are considered to be in a state of *equilibrium.* The rate

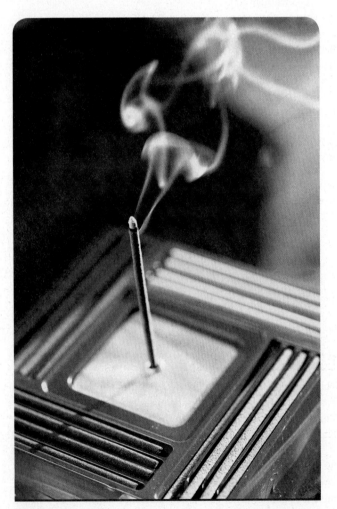

Figure 9.1 Even with no air current, the odor from this incense will eventually fill the room through the process of diffusion. © *Image Source/Getty Images*

Water in Plants

Osmosis

Solvents are liquids in which substances dissolve. Although the cytoplasm of living cells is bounded by membranes, it is now well known that water (a solvent) moves freely from cell to cell. This has led scientists to believe that plasma, vacuolar, and other membranes have tiny holes or spaces in them, even though such holes or spaces are invisible to the instruments presently available. It also has led to the construction of models of such membranes (see Fig. 3.7). Membranes through which different substances diffuse at different rates are described as **semipermeable.** All plant cell membranes appear to be semipermeable.

In plant cells, **osmosis** is essentially the *diffusion of water through a semipermeable membrane from a region where the water is more concentrated to a region where it is less concentrated.* Osmosis ceases if the concentration of water on both sides of the membrane becomes equal.

A demonstration of osmosis can be made by tying a membrane over the mouth of a thistle tube that has been filled with a solution of 10% sugar in water (i.e., the solution consists of 10% sugar and 90% water). Fluid rises in the narrow part of the tube as osmosis occurs when the thistle tube is immersed in water (Fig. 9.3).

Although the previous simple definition of osmosis serves our purposes, plant physiologists prefer to define and discuss osmosis more precisely in terms of *potentials*. It is possible to prevent osmosis by applying pressure. Just enough pressure to prevent fluid from moving as a result of osmosis is referred to as the osmotic pressure of the solution. In other words, osmotic pressure is the pressure required to prevent osmosis. The **osmotic potential** (represented by ψ_s) of a solution is a measure of the potential of water to move from one cell to another as influenced by solute concentration.

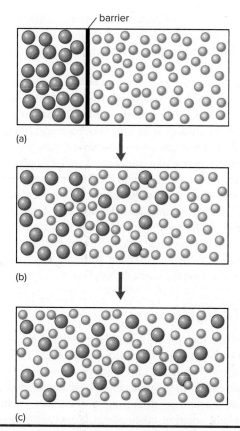

Figure 9.2 Simple diffusion. (a) A barrier separates two kinds of molecules. (b) When the barrier is removed, random movement of individual molecules results in both kinds moving from a region of higher concentration to a region of lower concentration. (c) Eventually, equilibrium (even distribution) is reached. The rate of diffusion gradually slows down as equilibrium is approached.

of diffusion depends on several factors, including pressure, temperature, and the density of the medium through which it is taking place.

Except within the area immediately surrounding the source, unaided diffusion requires a great deal of time because molecules and ions are infinitesimally small. Something that is less than a millionth of a millimeter in diameter is going to take a long time to move just 1 millimeter, even though the amount of movement may be great in proportion to the size of the particle concerned. In gases, there is a great deal of space between the molecules and correspondingly less chance of the molecules bumping into each other and thus being slowed down. Accordingly, gas molecules occupy a space that becomes available to them relatively rapidly, while liquids do so more slowly, and solids are slower yet.

Large molecules move much more slowly than small molecules. If you added a tiny drop of a dye (which has relatively large molecules) to one end of a bathtub of water without disturbing the water in any way, it would take years for the dye molecules to diffuse throughout the tub and reach a state of equilibrium. In nature, however, wind and water currents distribute molecules much faster than they ever could be distributed by diffusion alone.

Figure 9.3 A simple osmometer, made by tying a semipermeable membrane over the mouth of a thistle tube.

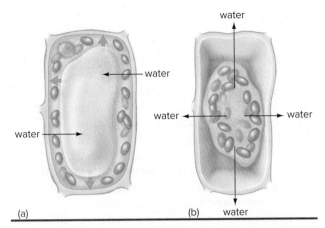

Figure 9.4 (a) A turgid cell. Water has entered the cell by osmosis, and turgor pressure is pushing the cell contents against the cell walls. (b) Water has left the cell, and turgor pressure has dropped, leaving the cell flaccid. The vacuole has disappeared. ×200.

Water enters a cell by osmosis until the osmotic potential is balanced by the resistance to the expansion of the cell wall. Water gained by osmosis may keep a cell firm, or **turgid,** and the **turgor pressure** that develops against the walls as a result of water entering the vacuole of the cell is called **pressure potential** (represented by ψ_p).

The release of turgor pressure can be heard each time you bite into a crisp celery stick or the leaf of a young head of lettuce. When we soak carrot sticks, celery, or lettuce in water to make them crisp, we are merely increasing the turgor of the cells (Fig. 9.4).

The **water potential** (represented by ψ_w) of a plant cell is essentially its osmotic potential and pressure potential combined ($\psi_w = \psi_s + \psi_p$). If we have two adjacent cells of different water potentials, water will move from the cell having the higher water potential to the cell having the lower water potential.

Osmosis is the primary means by which water enters plants from their surrounding environment. In land plants, water from the soil enters the cell walls and intercellular spaces of the epidermis and the root hairs and travels along the walls until it reaches the endodermis. Here it crosses the differentially permeable membranes and cytoplasm of the endodermal cells on its way to the xylem. Water flows from the xylem to the leaves, evaporates within the leaf air spaces, and diffuses out (*transpires*) through the stomata into the atmosphere. The movement of water takes place because there is a water potential gradient from relatively high soil water potential to successively lower water potentials in roots, stems, leaves, and the atmosphere.

Plasmolysis

If you place turgid carrot and celery sticks in a 10% solution of salt in water, they soon lose their rigidity and become limp enough to curl around your finger. The water potential inside the carrot cells is greater than the water potential outside, so diffusion of water out of the cells into the salt solution takes place. If you were to observe such cells with a microscope, you would see that the vacuoles, which are largely water, had disappeared and that the cytoplasm had pulled away from the walls and had concentrated in the middle of the cell. Such cells are said to be *plasmolyzed*. This loss of water through osmosis, which is accompanied by the shrinkage of protoplasm away from the cell wall, is called **plasmolysis** (Fig. 9.5).

Figure 9.5 A portion of a leaf of the water weed *Elodea*. (a) Normal cells. (b) Plasmolyzed cells. ×100. *(a-b)* © *Kingsley Stern*

KEY THEME 9.1 physiology — Measuring Water Potential and Psychrometry

Water is essential in plants to maintain cell turgor, for solute transport, as a medium for chemical reactions, as a raw material for photosynthesis and many other reactions, and for cooling plant surfaces. It is for these reasons, and more, that water potential is studied, and scientists have developed instruments to determine the status of water in plants. One popular instrument for measuring water potential is a thermocouple leaf psychrometer (Box Figure 9.1A), which can be used to determine total water potential of intact plant tissue. Other sensors can be used to measure the osmotic potential of sap within plant tissue after the sample has been treated. The psychrometer measures water potential based upon the relative humidity of the air that has been allowed to equilibrate with a small area of the leaf surface. A small thermocouple within this air space is cooled for a short period of time, which causes a small droplet of water to form. Once cooling is stopped, the droplet evaporates from the thermocouple and produces a voltage that is measured and used to calculate water potential. Total water potential is usually measured first and then the sample is frozen in liquid nitrogen to destroy cells so a second measurement can be made to determine osmotic potential. The difference in measurements can then be used to determine pressure potential because water potential is the sum of osmotic and pressure potentials ($\psi_w = \psi_s + \psi_p$).

Water potential measurements, along with the components of water potential, are very important in determining how water will move in plants and hence the physiological processes essential to plant growth and development.

Box Figure 9.1A Water potential of a leaf being measured with a thermocouple psychrometer. © ELITechGroup Biomedical Systems-formerly Wescor

---

If plasmolyzed cells are placed in fresh water before permanent damage is done, water reenters the cell by osmosis, and the cells become turgid once more.

Imbibition

Osmosis is not the only force involved in the absorption of water by plants. *Colloidal* materials (i.e., materials that contain a permanent suspension of fine particles) and large molecules, such as cellulose and starch, usually develop electrical charges when they are wet. The charged colloids and molecules attract water molecules, which adhere to the internal surfaces of the materials. Because water molecules are polar, they can become both highly adhesive to large organic molecules such as cellulose and cohesive with one another. As discussed in Chapter 2, polar molecules have slightly different electrical charges at each end due to their asymmetry. This process, known as **imbibition,** results in the swelling of tissues, whether they are alive or dead, often to several times their original volume. Imbibition is the initial step in the germination of seeds (Fig. 9.6).

The physical forces that develop during germination can be tremendous, even to the point of causing a seed to split a rock weighing several tons (Fig. 9.7). It has been found, for example, that a pressure of 42.2 kilograms per square centimeter (600 pounds per square inch) is needed to break the seed coat of a fresh walnut and that water being imbibed by a cocklebur seed develops a force of up to 1,000 times that of normal atmospheric pressure. Yet, when water and oxygen reach walnut and cocklebur seeds, they germinate readily, as do seeds that fall into the crevices of rocks or have boulders roll over on them.

The huge stone blocks used in the construction of the pyramids of Egypt are believed to have been quarried by hammering rounded wooden stakes into holes made in the face of the stone and then soaking the stakes with water. As the stakes swelled, the force created by imbibition split the rocks.

Active Transport

Return for a moment to our two rooms with the tennis balls. Suppose that, besides the 100 tennis balls, we drop in 50 slightly underinflated basketballs; these basketballs are

Figure 9.6 Black-eyed pea seeds before (top) and after (bottom) imbibition of water. © *Kingsley Stern*

Figure 9.7 A live oak that grew from an acorn lodged in a small crack in the rock. When it rained, the acorn imbibed water, and the force of the swelling split the rock. A root is now slowly widening the split. © *Kingsley Stern*

also extraordinary in having perpetual-motion motors that propel them in any direction at 12 mph. They should also become randomly distributed throughout the room shortly after they are introduced. Assume, however, that the hole in the wall (which is large enough for the passage of a tennis ball) is not quite large enough to allow a basketball to pass through freely. The basketballs will then remain in the first room. However, if we were to install a mechanical arm next to the hole in the second room, and if this arm could grab basketballs that come near the hole and squeeze them through in one direction, basketballs would be transported into the second room *through the expenditure of energy*. The basketballs obviously would gradually accumulate in the second room in greater numbers.

Plants expend energy to move substances, too. Plant cells generally have a larger number of mineral molecules and ions than exist in the soil immediately next to the root hairs. If it were not for the barriers imposed by the semipermeable membranes, these molecules and ions would move from a region of higher concentration in the cells to a region of lower concentration in the soil.

Most molecules needed by cells are polar, and those of solutes may set up an electrical gradient across a semi-permeable membrane of a living cell. To pass through the membrane, molecules require special embedded transport proteins (see Chapter 3). The transport proteins are believed to occur in two forms: one facilitating the transport of specific ions to the outside of the cell and the other facilitating the transport of specific ions into the cell.

The plants absorb and retain these solutes against a diffusion (or electrical) gradient through the expenditure of energy. This process is called **active transport.** It requires an enzyme complex that controls a *proton pump*. The pump helps regulate which ions pass through the cell membrane and this is often accompanied by the expenditure of energy-rich ATP molecules (discussed in Chapter 10).

Figure 9.8 A mangrove tree. Mangroves flourish in tropical tidal zones where the salt content of the water is high enough to plasmolyze the cells of most other plants. The mangroves still obtain water via osmosis, which takes place because the mangrove cells accumulate an unusually high concentration of organic solutes; some are also able to excrete excess salt. © *Kingsley Stern*

Mangroves, saltbush, and certain algae thrive in areas where the water or soil contains enough salt to kill most vegetation. Such plants accumulate large amounts of organic solutes, including the carbohydrate *mannitol* and the amino acid *proline*. The organic solutes facilitate osmosis, despite the otherwise adverse environment (Fig. 9.8). The leaves of some mangroves also have salt glands through which they excrete excess salt.

9.3 Water and Its Movement through the Plant

Learning Outcomes

a. Provide at lease two reasons why plants need water.
b. Explain the cohesion-tension theory and how it contributes to water movement in plants.

If you were to cover the soil at the base of a plant with foil, place the pot where it receives light, and then put the potted plant under a glass bell jar, you would notice moisture accumulating on the inside of the jar within an hour or two. Because of the foil barrier, the water could not have come directly from the soil; it had to have come through the plant. More than 90% of the water entering a plant passes through and evaporates—primarily into leaf air spaces and then through the stomata into the atmosphere (see Fig. 9.10)—with usually less than 5% of the water escaping through the cuticle. This process of water vapor loss from the internal leaf atmosphere is called *transpiration* (Fig. 9.9).

The amount of water transpired by plants is greater than you might suspect. For example, mature corn plants each transpire about 15 liters (4 gallons) of water per week,

Figure 9.9 A potted plant sealed under a bell jar. The surface of the soil has been covered with foil so that no water could evaporate from it. Note the accumulation of moisture on the inside of the glass. The moisture came through the plant by transpiration. © *James E. Bidlack*

while four-tenths of a hectare (1 acre) of corn may transpire more than 1,325,000 liters (350,000 gallons) in a 100-day growing season. A hardwood tree uses about 450 liters (120 gallons) of water while producing 0.45 kilogram (1 pound) of wood, and the 200,000 leaves of an average-sized birch tree can transpire more than 3,785 liters (1,000 gallons) per day during the growing season. Humans recycle much of their water via the circulatory system, but if they were to have requirements similar to those of plants, each adult would have to drink well over 38 liters (10 gallons) of water per day.

Why do living plants require so much water? Water constitutes about 90% of the weight of young cells. The thousands of enzyme actions and other chemical activities of cells take place in water and, although relatively negligible, some water is used for photosynthesis. The exposed surfaces of the mesophyll cells within the leaf have to be moist at all times, for it is through this film of water that the carbon dioxide molecules needed for the process of photosynthesis enter the cell from the air. Water is also needed for cell turgor, which gives rigidity to herbaceous plants.

Consider also what it must be like in the mesophyll of a leaf that is exposed to the midsummer sun in areas where the air temperature soars to well over 38°C (100°F). If it were not for the evaporation of water molecules from the moist surfaces, which provides cooling, and reradiation of energy by the leaf, the intense heat would damage the plant. Sometimes, the transpiration is so fast that more water is lost than is taken in. The stomata may then close, preventing wilting. The role of abscisic acid in reducing water loss is discussed in Chapter 11.

How does water travel through the roots in the soil and then up the trunk to the topmost leaves of a tree? We know that interconnected tubes of xylem extend throughout the plant, from the young roots up through the stem and branches to the tiny veinlets of the leaves. We also know that the water, following a water potential gradient, travels to the start of this "plumbing system" by osmosis. Water then rises through the columns by a combination of factors in a process that has been the subject of much debate for the past 200 years (Fig. 9.10).

One of the earliest explanations for the rise of water in a living plant was given in 1682 by the English scientist Nehemiah Grew. He suggested that cells surrounding the xylem vessels and tracheids performed a pumping action that propelled the water along. This was questioned,

Figure 9.11 Capillarity in narrow tubes. The smaller the diameter of the tube, the greater the rise of the fluid.

however, when it was found that water will also rise in dead stems. Then, after Marcello Malpighi suggested it, the belief that capillary action moved the water became popular.

It is well known that the height water will rise in a narrow tube is inversely proportional to the diameter of the tube. It is also known that this rise occurs through the forces involved in the forming of a concave *meniscus* (curved surface) at the top of the water column (Fig. 9.11). Even though water can rise 1 meter (3 feet) or more in a very narrow tube, air must be present above the column for the forces to work, which is not the case in a plant. In fact, any air introduced into a water column in xylem interferes with the rise of water. Also, while capillarity might produce enough force to raise water a meter or two, the diameter of the tubes is not small enough to raise it more than that.

The pioneer plant physiologist Stephen Hales discovered and measured *root pressure* as one means by which water moves through plants. When some plants are pruned after growth has begun in the spring, water will exude from the cut ends. This is the result of root pressure. Some plants do not "bleed" when they are pruned, and the force exerted by root pressure in these plants has been shown to be less than 30 grams per square centimeter (a few pounds per square inch). This is considerably less than what is needed to raise water to the tops of tall trees. Furthermore, root pressure seems to drop to negligible amounts in the summer, when the greatest amounts of water are moving through the plant.

The Cohesion-Tension Theory

Stephen Hales also identified a pulling force due to evaporation of water from leaves and stems. This has led to the cohesion-tension theory, the most satisfactory explanation for the rise of water in plants thus far suggested. Water molecules are electrically neutral, but they are asymmetrical in shape (see Chapter 2). This results in the molecules having very slight positive charges at one end and very slight negative charges at the other end. Such molecules are said to be

Figure 9.10 Pathway of water through a plant.

polar. When the negatively charged end of one water molecule comes close to the positively charged end of another water molecule, weak hydrogen bonds hold the molecules together.

We know that water molecules adhere to capillary walls (e.g., those of xylem tracheids and vessels) and cohere to each other, creating a certain amount of tension. It is possible, for example, to fill a small glass with water, place a thin, smooth sheet of cardboard over the mouth, and invert the glass without the water spilling. This is because the adhesion of the water molecules to the cardboard and the cohesiveness of the water molecules to one another hold the cardboard against the rim of the glass.

When water evaporates from the mesophyll cells in a leaf and diffuses out of the stomata (*transpires*), these cells develop a lower water potential than the adjacent cells. Because the adjacent cells then have a correspondingly higher water potential, replacement water moves into the first cells by osmosis. This continues across rows of mesophyll cells until a small vein is reached. Each small vein is connected to a larger vein, and the larger veins are connected to the main xylem in the stem, and that, in turn, is connected to the xylem in the roots that receive water, via osmosis, from the soil. As transpiration takes place, it creates a "pull," or tension, on water columns, drawing water from one molecule to another all the way through an entire span of xylem cells. The cohesion required to move water to the top of a tall tree is considerable, but the cohesive strength of the water columns is usually more than adequate. Any breaking of the tension through the introduction of a gas bubble results in a temporary or permanent blocking of water transport. This is seldom a problem, however, because small bubbles may be redissolved and larger gas bubbles rarely block more than a few of the numerous capillary tubes of xylem at any time the tissue is functioning.

Water molecules first enter the plant through root hairs and cortical cells via plasmodesmata. The flow of water continues as it moves partly through cell cytoplasm and partly through spaces between cells. Water can also move between cellulose fibers in the walls and through spaces in the centers of dead cells. Most water and solutes can travel across the epidermis and cortex via the cell walls until they reach the endodermis. There, the water and solutes are forced by *Casparian strips* to cross the plasma membrane and enter the cytoplasm of the endodermal cells on their way to the vessels or tracheids of the xylem (Fig. 9.12; see also Chapter 5).

If significant transpiration is occurring, the roots are likely to grow rapidly toward available water. In corn plants, for example, the main roots may grow at a rate of more than 6 centimeters (2.3 inches) a day. Solutes, as well as water, may move so rapidly during periods of rapid transpiration that there is little osmosis taking place across the endodermis. Scientists believe that at such times water

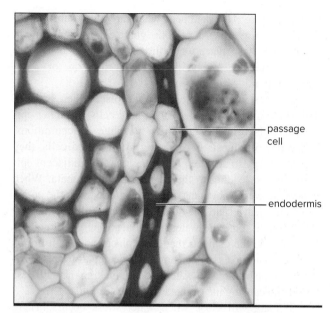

Figure 9.12 Part of the center of a buttercup root, showing endodermal cells with Casparian strips. ×600. © *Kingsley Stern*

may be pulled through the roots by bulk flow, which is the passive movement of a liquid from higher to lower water potential.

In summary, "columns" of water molecules are pulled through the plant from roots to leaves, and the abundant water of a normally moist soil supplies these "columns" as the water continues to enter the root by osmosis (see Fig. 9.10); simply put, the difference between the water potentials (water "concentrations") of two areas (e.g., soil and the air around stomata) generates the force to transport water in a plant.

9.4 Regulation of Transpiration

Learning Outcome

a. Draw a representative stomatal apparatus, and explain how this helps to regulate transpiration.

Two guard cells and an opening called the **stoma** (plural: **stomata**) constitute the stomatal apparatus. These stomatal apparatuses, which often occupy 1% or more of the surface area of a leaf, *regulate transpiration and gas exchange.* Control of transpiration is, however, strongly influenced by the water-vapor concentration of the atmosphere. The guard cells bordering each stoma have relatively elastic walls with radially oriented microfibrils, making them analogous to pairs of sausage-shaped balloons joined at each end, each with a row of rubber bands around it. The part of the wall adjacent to the hole itself is considerably thicker than the remainder of the wall (Fig. 9.13). This thickness allows each stoma to be opened and closed by means of changes in the turgor of the guard cells. The stoma is closed when turgor pressure is low and open when turgor pressure is high. Changes in turgor pressures in the guard cells, which contain chloroplasts, take place when they are exposed to changes in light intensity, carbon dioxide concentration, or water concentration.

Changes in turgor pressure take place when osmosis and active transport between the guard cells and other epidermal cells bring about shifts in solute concentrations. While photosynthesis is occurring in the guard cells, they expend energy to acquire potassium ions from adjacent epidermal cells, leading to the opening of the stomata. When photosynthesis is not occurring in the guard cells, the potassium ions leave, and the stomata close. With an increase in potassium ions, the water potential in the guard cells is lowered, and the osmosis that takes place as a result brings in water that makes the cells turgid. The departure of potassium ions also results in water leaving, making the cells less turgid and causing the stomata to close (see Fig. 9.13).

Stomata will close when water stress occurs. There is evidence that the hormone abscisic acid is produced in leaves subject to water stress and that this hormone causes membrane leakages, which induce a loss of potassium ions from the guard cells and cause them to deflate.

The stomata of most plants are open during the day and closed at night. However, the stomata of a number of desert plants are open only at night when there is less water stress on the plants. This conserves water but makes carbon dioxide needed for photosynthesis inaccessible during the day.

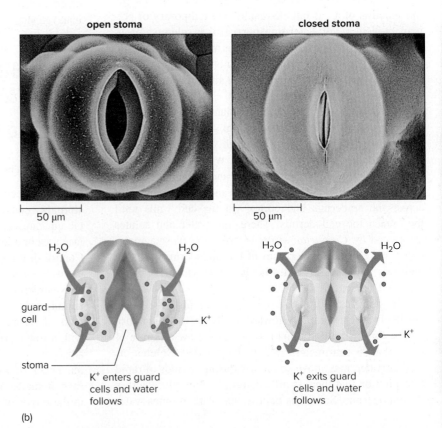

Figure 9.13 (a) A small portion of the leaf epidermis of spiderwort (*Tradescantia* sp.) with several stomata interspersed among ordinary epidermal cells. Each stoma is bordered by a pair of guard cells, and each guard cell is flanked to the outside by an epidermal cell called a *subsidiary cell*. ×100. (b) *Left*. An open stoma. The guard cells swell when turgor pressure in them increases and the stoma opens as the thinner outer walls stretch more than the thicker inner walls. *Right*. The stoma closes when the turgor pressure in the guard cells decreases. The reasons for the changes in turgor pressure are discussed in the text. ×400. (a) © Science Source; (b) © Jeremy Burgess/SPL/Photo Researchers, Inc.

Figure 9.14 A barrel cactus (*Ferrocactus*). The stems store, in the form of organic acids, carbon dioxide taken in at night; the carbon dioxide is then released inside the plant during the day for use in photosynthesis. © *Kingsley Stern*

Such plants convert the carbon dioxide available at night to organic acids, which are stored in cell vacuoles. The organic acids are then converted back to carbon dioxide during the day when photosynthesis occurs (Fig. 9.14). A specialized form of photosynthesis called *CAM photosynthesis* uses the carbon dioxide released from the organic acids. CAM photosynthesis is discussed in Chapter 10.

Other desert plants have their stomata recessed below the surface of the leaf or stem in small chambers. These chambers, called stomatal crypts, often are partially filled with epidermal hairs, which further reduce water loss by slowing down air movement. Similar recessed stomata are found in the leaves of pine trees, which have little water available to them in winter when the soil is frozen (see Fig. 7.12). A few tropical plants that occur in damp, humid areas (e.g., ruellias; see Chapter 4) have stomata that are raised above the surface of the leaf, while plants of wet habitats generally lack stomata on submerged surfaces.

Although light and carbon dioxide concentration affect transpiration rates, several other factors play at least an indirect role. For example, air currents speed up transpiration as they sweep away water molecules emerging from stomata.

Humidity plays an inverse but direct role in transpiration rates: high humidity reduces transpiration, and low humidity accelerates it. Temperature also plays a role in the movement of water molecules out of a leaf. The transpiration rate of a leaf at 30°C (86°F), for example, is about twice as great as it is for the same leaf at 20°C (68°F). The adaptive modifications of leaves and their surfaces and the availability of water to the roots also play important roles in influencing the amount of water transpired. Leaf modifications are discussed in Chapter 7.

If a cool night follows a warm, humid day, water droplets may be produced through structures called **hydathodes** at the tips of veins of the leaves of some herbaceous plants. This loss of water in liquid form is called **guttation** (Fig. 9.15). Minerals absorbed at night are pumped into the intercellular

Figure 9.15 Droplets of guttation water at the tips of leaves of young barley plants. © *Kingsley Stern*

spaces surrounding the vessels and tracheids of the xylem. As a result, the water potential of the xylem elements is lowered, and water moves into them from the surrounding cells. In the absence of transpiration at night, the pressure in the xylem elements builds to the point of forcing liquid water out of the hydathodes in the leaves. Although the droplets resemble dew, the two should not be confused. Dew is water that is condensed from the air, while guttation water is literally forced out of the plant by root pressure. As the sun strikes the droplets in the morning, they dry up, leaving a residue of salts and organic substances, one of which is used to manufacture commercial flavor enhancers such as monosodium glutamate. In the tropics, the amount of water produced by guttation can be considerable. In taro plants, used by the Polynesians to make poi, a single leaf may produce as much as a cupful (about 240 milliliters) of water overnight through guttation.

9.5 Transport of Food Substances (Organic Solutes) in Solution

Learning Outcome

a. Explain the pressure-flow hypothesis and the steps involved in moving sugars from the source to the sink.

One of the most important functions of water in the plant is the *translocation* (transportation) of dissolved sugars by the phloem. Many of the studies that led to our present knowledge of the subject used aphids (small, sucking insects) and organic compounds designed as radioactive tracers.

Most aphids feed on phloem by inserting their tiny, tubelike mouthparts (*stylets*) through the leaf or stem tissues until a sieve tube is reached and punctured. The turgor pressure of the sieve tube then forces the fluid in the tube through the aphid's digestive tract, and it emerges at the rear as a droplet of "honeydew." In some studies, research workers anesthetized feeding aphids and cut their stylets, so that much of the tiny tube remained where it had been inserted. Fluid exuded (sometimes for many hours) from the cut stylets and was then collected and analyzed (Fig. 9.16).

Carbon dioxide, a basic raw material for photosynthesis, can be synthesized with radioactive carbon. By exposing a photosynthesizing leaf to radioactive carbon dioxide, the pathway of manufactured food substances can be traced. The radioactive substances produce an image on film corresponding to the food pathway. Data obtained from such studies reveal that dissolved organic molecules, such as sugars, are confined entirely to the sieve tubes while they are being transported. At one time, it was believed that ordinary diffusion and cyclosis (discussed in Chapter 3) were responsible for the movement of the substances from one sieve-tube member to the next, but it is now known that the substances move through the phloem at approximately 100 centimeters (almost 40 inches) per hour—far too rapid a movement to be accounted for by diffusion and cyclosis alone.

The Pressure-Flow Hypothesis

The most widely accepted theory for movement of substances in the phloem is called the **pressure-flow** (or **mass-flow**) **hypothesis**. According to this theory, food

(a)

(b)

Figure 9.16 (a) Color enhanced scanning electron micrograph of an aphid sucking sap from a young cabbage plant. X50. (b) Photomicrograph showing a cross section of an aphid probing a leaf vein for phloem sap. X10. *(a) © P. Dayanandan/Science Source; (b) © Steven P. Lynch*

Figure 9.17 The pressure-flow hypothesis.

substances in solution (organic solutes) flow from a *source,* where water enters by osmosis (e.g., food-producing tissue, such as the mesophyll tissue of a leaf, or food-storage tissue, such as the cortex of a root or rhizome). The water exits at a *sink,* which is a place where food is utilized, such as the growing tip of a stem or root. The organic solutes are then moved along concentration gradients from sources to sinks (Fig. 9.17).

First, in a process called *phloem-loading,* active transport carries sugars into the sieve tubes of the smallest veinlets. This decreases the water potential in the sieve tubes, and water then enters these phloem cells by osmosis. Turgor pressure, which develops as a result of osmosis, forces the fluid through the sieve-tube network toward the sinks.

As the food substances (mainly sucrose) in solution are actively removed at the sink, water also exits the sieve tubes, and the pressure in these sieve tubes is lowered, causing a mass flow from the higher pressure at the source to the lower pressure at the sink. Most of the water diffuses back to the xylem, where it then returns to the source and is transpired or recirculated. The pressure-flow hypothesis explains how viruses introduced by aphids are carried through the phloem.

KEY THEME 9.2 physiology
Xylem and Phloem Synopsis

In general, xylem transports the majority of water (Box Figure 9.2A) passively up the plant, from the roots to the shoots. This water movement is facilitated by a relatively higher water potential in the roots and a gradient of lower water potentials, with some fluctuations, from the stems to the leaves and ultimately the atmosphere. However, there is also water that dissolves sugars in the phloem, and this can move some water in a downward direction. Water movement in plants is generally in an upward direction, but it can also move around the plant, depending on pressure potential and the osmotic drive of water from a higher concentration to a lower concentration.

A similar generalization can be made in that phloem transports the majority of sugars (Box Figure 9.2B), actively, from the leaves to other plant parts that are growing, such as fruits and roots. Because leaves are the main photosynthetic organ of most plants, they are the origin of sugar synthesis, and logically, the first place from which sugar is transported from the source to a sink where sugar is needed. That is not to say that sugar always moves from the leaves to other plant parts, because sugar can be transported and stored for later use, or it can even end up in fruits to promote seed disbursement. For instance, sugars can be stored in roots during winter months and then move up to the shoots or other plant parts as needed. Hence, sugar movement is inevitably from where it is synthesized or stored to a place where there is a physiological or other need for energy.

Box Figure 9.2A Depiction of water droplets. Water is generally transported passively through the xylem of plants. © fotointeractiva/123RF

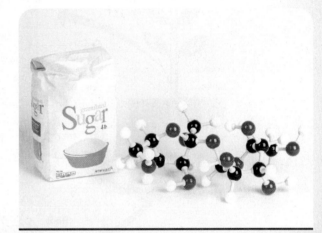

Box Figure 9.2B Sugar and a representation of the sucrose molecule. Sugar is generally transported actively through the phloem of plants. © McGraw-Hill Education

9.6 Mineral Requirements for Growth

Learning Outcome
a. List the macronutrients and micronutrients required by plants, and describe a function of each.

Growth phenomena are not entirely controlled by internal means. Light, temperature, soil structure, minerals, and other external factors all play a role. In fact, growth depends on a complex, interrelated combination of chemical and physical forces, both internal and external, which are in delicate balance with one another.

Plants may take up many elements from the soil, but besides the carbon and oxygen obtained from carbon dioxide and the hydrogen obtained from water, only 15 elements are essential to most plants as building blocks for the many compounds they synthesize. Sodium, a comparatively abundant element, is apparently required by few plants. The **essential elements** can be remembered by a sentence (Fig. 9.18) that includes the symbols of the elements involved.

Macronutrients and Micronutrients

The mineral elements (also called inorganic nutrients) are usually put into two categories: (1) *macronutrients,* which are used by plants in greater amounts and constitute from 0.5% to 3.0% of the dry weight of the plant, and (2) *micronutrients,* which are needed by the plant in very small amounts, often constituting only a few parts per million of the dry weight.

The macronutrients are nitrogen, potassium, calcium, phosphorus, magnesium, and sulfur, with the first four usually making up about 99% of the nutrient total. The remaining elements, the micronutrients, are present in amounts ranging from bare traces—as in the case of sodium and cobalt, neither of which may actually be essential for some

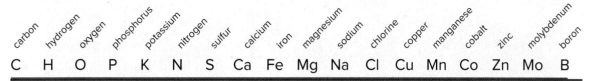

Figure 9.18 Elements essential as building blocks for compounds synthesized by plants.

Table 9.1	Uses of Essential Elements in Plants	
Element	**Some Functions**	**Deficiency Symptoms**
Nitrogen	Part of proteins, nucleic acids, chlorophyll	Relatively uniform loss of color in leaves, occurring first on the oldest ones
Potassium	Activates enzymes; concentrates in meristems	Yellowing of leaves, beginning at the margins and continuing toward center; lower leaves mottled and often brown at the tip
Calcium	Essential part of middle lamella; involved in movement of substances through cell membranes	Terminal bud often dead; young leaves often appearing hooked at tip; tips and margins of leaves withered; roots dead or dying
Phosphorus	Necessary for respiration and cell division; high-energy cell compounds such as ATP	Plants stunted; leaves darker green than normal; lower leaves often purplish between veins
Magnesium	Part of the chlorophyll molecule; activates enzymes	Veins of leaves green but yellow between them, with dead spots appearing suddenly; leaf margins curling
Sulfur	Part of some amino acids	Leaves pale green with dead spots; veins lighter in color than the rest of the leaf area
Iron	Required for chlorophyll-protein complexes and some steps of electron transport in cellular respiration	Larger veins remaining green while rest of leaf yellows; mainly in young leaves*
Manganese	Activates some enzymes	Dead spots scattered over leaf surface; all veins and veinlets remain green; effects confined to youngest leaves
Boron	May function in cell cycle regulation, hormone responses, membrane function, nucleic acid synthesis	Petioles and stems brittle; bases of young leaves break down

*NOTE: The symptoms of iron deficiency may be caused by several factors, such as overwatering, cold temperatures, and nematodes (small roundworms) in the roots. The iron may be relatively abundant in the soil, but its uptake may be prevented or sharply reduced by these environmental conditions. Iron becomes more soluble under acid conditions—so much so that it can produce toxic conditions for most plants. Acid soil plants, on the other hand, have a much higher iron requirement than plants that require more non-acid conditions; accordingly, azaleas and other plants having high iron requirements can achieve normal growth in a non-acid soil.

All micronutrients are harmful to plants when present in excessive quantities. Copper will kill algae in concentrations of one part per million, and boron has been used in weed killers. Even macronutrients are harmful if present in heavy amounts, although non-essential elements sometimes can counteract their toxicity.

plants—to 1,500 parts per million of iron and manganese and up to 10,000 parts per million of chlorine.

In addition to these elements, specific organisms may require others. For example, certain algae apparently require the elements vanadium, silicon, or iodine, while some ferns utilize aluminum. Several loco weeds absorb and accumulate selenium in amounts constituting up to 5 micrograms per gram of dry weight. Selenium, which is often fatally poisonous to livestock, appears to enhance the growth of these plants by reducing toxic effects of phosphates, but there is no direct evidence it is essential to them.

When any of these elements are deficient in the soil, the plants exhibit characteristic symptoms of the deficiency, which disappear after the problem has been corrected (Fig. 9.19). **Hydroponics** is a technique that can be used to grow plants using mineral nutrient solutions without soil. Through the use of specially formulated nutrient solutions that lack certain elements, deficiency symptoms can be detected. Table 9.1 shows some of the uses of essential elements in plants and describes the symptoms of deficiency for each element. Commercial hydroponics has become popular because it offers some important advantages over traditional farming. Hydroponics uses less space, enables better control of nutrients, and uses fewer pesticides. There is much potential for hydroponics in space travel or even life on other planets.

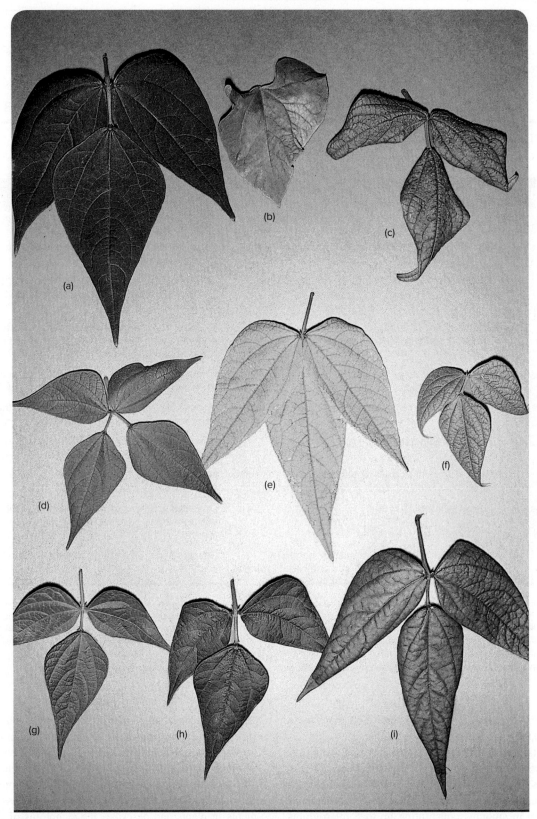

Figure 9.19 Leaves of bean plants grown in media deficient in various elements to show deficiency symptoms. (a) A normal plant that has been furnished with all the essential elements. The other plants were grown in media deficient in specific elements, as follows: (b) Potassium. (c) Phosphorus. (d) Calcium. (e) Nitrogen. (f) Sulfur. (g) Micronutrients. (h) Magnesium. (i) Iron. © *Doug Sherman/McGraw-Hill Companies, Inc*

SUMMARY

9.1 Introduction to Water in Plants

- Molecules and ions are in constant random motion and tend to distribute themselves evenly in the space available to them. They move from a region of higher concentration to a region of lower concentration by simple diffusion along a diffusion gradient; they may also move against a diffusion gradient. Evenly distributed molecules are in a state of equilibrium. Diffusion rates are affected by pressure, temperature, molecule size and density, and other factors.

9.2 Molecular Movement

- Osmosis is the diffusion of water through a semipermeable membrane. It takes place in response to concentration differences of dissolved substances.
- Osmotic potential is a measure of the potential of water to move from one cell to another as influenced by solute concentration. The pressure that develops in a cell as a result of water entering it is called turgor. Water moves from a region of higher water potential (osmotic potential and pressure potential combined) to a region of lower water potential when osmosis is occurring. Osmosis is the primary means by which plants obtain water from their environment.
- Plasmolysis is the shrinkage of the cytoplasm away from the cell wall as a result of osmosis taking place when the water potential inside the cell is greater than the potential outside.
- Imbibition is the attraction and adhesion of water molecules to the internal surfaces of materials; it results in swelling and is the initial step in the germination of seeds.
- Active transport is the expenditure of energy by a cell that results in molecules or ions entering or leaving the cell against a diffusion gradient.

9.3 Water and Its Movement through the Plant

- Water that enters a plant passes through it and mostly transpires into the atmosphere via stomata. Water retained by the plant is used in photosynthesis and other metabolic activities.
- Water moves from high potential to low potential. The cohesion-tension theory also postulates that water rises through plants because of the adhesion of water molecules to the walls of the capillary-conducting elements of the xylem, cohesion of the water molecules, and tension on the water columns created by the pull developed by transpiration.

9.4 Regulation of Transpiration

- Transpiration is regulated by humidity and the stomata, which open and close through changes in turgor pressure of the guard cells. These changes, which involve potassium ions, result from osmosis and active transport between the guard cells and the adjacent epidermal cells.
- Aquatic, desert, tropical, and some cold-zone plants have modifications of stomatal apparatuses or specialized forms of photosynthesis that adapt them to their particular environments.
- Guttation is the loss of water at night in liquid form through hydathodes at the tips of leaf veins.

9.5 Transport of Food Substances (Organic Solutes) in Solution

- The translocation of food substances takes place in a water solution, and according to the pressure-flow hypothesis, such substances flow along concentration gradients between their sources and sinks.

9.6 Mineral Requirements for Growth

- Growth phenomena are controlled by both internal and external means, and by chemical and physical forces in balance with one another. Besides carbon, hydrogen, and oxygen, 15 other elements are essential to most plants. When any of the essential elements are deficient in the plant, characteristic deficiency symptoms appear.

REVIEW QUESTIONS

1. Distinguish among diffusion, osmosis, active transport, plasmolysis, and imbibition.
2. Why do living plants need a great deal of water for their activities?
3. Explain how a tall tree gets water to its tips without the aid of mechanical pumps.
4. What is the difference between transpiration and guttation?
5. Explain the pressure-flow hypothesis.
6. What are macronutrients? List them.
7. When nutrients are deficient in the soil, how are the deficiencies manifested in plants?

DISCUSSION QUESTIONS

1. Why is salted meat less likely to spoil than unsalted meat?
2. Why would dye molecules in a bathtub of water take a long time to diffuse completely throughout the tub but perfume molecules released in a closed room take considerably less time to do the same thing?
3. Why does osmosis not cause submerged water plants to swell up and burst?
4. Some bodies of water, such as the Dead Sea, have considerably higher salt concentrations than those of the human body. If you were swimming in such water, would you expect your cells to become plasmolyzed? Why?

CHAPTER 10

Plant Metabolism

10.1 Introduction to Plant Metabolism

10.2 Enzymes and Energy Transfer
Oxidation–Reduction Reactions

10.3 Photosynthesis
The Essence of Photosynthesis
Introduction to the Major Steps of Photosynthesis
A Closer Look at Photosynthesis

Key Theme 10.1: Photosynthesis and Pizza
Other Significant Processes That Occur in Chloroplasts

10.4 Respiration
The Essence of Respiration
Introduction to the Major Steps of Respiration
Factors Affecting the Rate of Respiration
A Closer Look at Respiration

10.5 Additional Metabolic Pathways

10.6 Assimilation and Digestion
Key Theme 10.2: Photosynthesis, Global Warming, and Tropical Rain Forests

OVERVIEW

Enzymes and energy transfer are explained first to introduce general metabolic concepts. Photosynthesis and respiration are then presented at three different levels: the essence of the process is examined, the major steps are briefly introduced, and the processes are explored in greater detail. One or two levels may be sufficient for some readers; others will want to explore all three. The chapter discusses the importance of the main features of each process and summarizes the light-dependent reactions, light-independent reactions, glycolysis, the citric acid cycle, and the electron transport system. It concludes with a tabular comparison between photosynthesis and respiration and makes a few brief observations on additional metabolic pathways.

Light shining through a canopy of the Snoqualmie National Forest near Greenwater, Washington. © *Rob A. Johnston/Walkabout Wolf Photography /Getty Images*

10.1 Introduction to Plant Metabolism

Learning Outcome

a. Define *metabolism, photosynthesis,* and *respiration.*

One of the most beautiful and breathtaking scenes in nature is the appearance of sunlight rays beaming through a canopy of trees. It can remind us of the essential role of the sun in creating the building blocks of life and generating oxygen for organisms to breathe through **photosynthesis.** Through photosynthesis, which requires little more than air, water, green pigment, and light—parts of the water and air are combined in cells and stored as sugar. Also, as long as any cell remains alive, stored energy is released by **respiration** to fuel the cell's activities. Without the proper balance between photosynthesis and respiration found in nature, life on earth as we know it would not exist (Fig. 10.1). Photosynthesis and respiration form the basis for most of our discussion in this chapter.

Each living cell of a plant contains genetic information that programs all of its metabolic activities. Energy is required for all forms of **metabolism,** which may be defined as all the interrelated biochemical processes that take place in a living organism. When rosebushes flower, or when leaves appear on a maple tree in the spring, or any other form of life activity occurs, the dynamic process of acquiring, releasing, and transferring energy from one form to another takes place. Plants and animals release energy during their life cycles; the energy is then recycled or used by other living organisms.

Photosynthetic cells can convert light energy to a usable form, and that energy may then be released during respiration, facilitating growth, development, and reproduction.

Although photosynthetic organisms carry on both *photosynthesis* and *respiration,* most animals, including humans, carry on only respiration, and they rely upon green plants for oxygen, food, shelter, and many other items.

10.2 Enzymes and Energy Transfer

Learning Outcomes

a. Explain how enzymes function in anabolism and catabolism.

b. Describe what happens to electrons during oxidation and reduction reactions.

Enzymes (proteins that speed up chemical reactions in cells without being used up in the reactions) regulate just about every metabolic activity. In biochemical reactions, one or more enzymes catalyze the many forms of energy conversion that take place within cells. In some cases, these enzymes help form the chemical bonds needed to build molecules through a process called **anabolism.** Enzymes also break chemical bonds through the process of **catabolism.** Most reactions of photosynthesis are anabolic because they result in the construction of molecules that are stored for energy, whereas reactions in **cellular respiration** (referred to as "respiration" in the text) are generally catabolic, because they release energy held in chemical bonds. Photosynthesis builds organic compounds by combining carbon dioxide and water, forming carbohydrates. Respiration, on the other hand, breaks down those carbohydrates, producing carbon

Figure 10.1 Summary of photosynthesis and aerobic respiration. Photosynthesis uses the energy from the sun to form carbohydrates from carbon dioxide and water. Aerobic respiration uses oxygen to break down carbohydrates to release energy, carbon dioxide, and water. © Ingram Publishing

dioxide and water, which may be used once again in photosynthesis. As we will see in the following sections this photosynthesis–respiration cycle relies on an enzyme complex that splits water molecules and releases electrons that function in temporarily storing biochemical energy. These electrons transfer energy from one form to another through *oxidation–reduction reactions*.

Oxidation-Reduction Reactions

Both photosynthesis and respiration include many **oxidation–reduction reactions**. **Oxidation** is the *loss of one or more electrons;* it removes electrons from a compound. **Reduction** is the *gain of one or more electrons;* it adds electrons to a compound. In most oxidation–reduction reactions, oxidation of one compound is coupled with reduction of another compound catalyzed by the same enzyme or enzyme complex. When an electron is removed, a proton may follow, with the result that a hydrogen atom is often removed during oxidation and added during reduction. Oxygen is usually the oxidizing agent (i.e., the final acceptor of the electron), but oxidations can occur without oxygen being involved.

10.3 Photosynthesis

Learning Outcomes

a. Write the generalized reaction for photosynthesis.
b. Explain how carbon dioxide, water, light, and chlorophyll contribute to photosynthesis.
c. Distinguish between light-dependent and light-independent reactions.
d. Provide a generalized diagram for the light-dependent reactions of photosynthesis, and label the photosystems as well as the steps that produce oxygen, ATP, and NADPH.
e. Provide a generalized diagram of the Calvin cycle, and explain how rubisco functions to "fix" carbon dioxide and why ATP and NADPH are needed in the cycle.
f. Distinguish between C_3 and C_4 plants.

Oil and coal today provide about 90% of the energy needed to power trains, trucks, ships, airplanes, factories, computers, communication systems, and a multitude of electrically energized appliances. The energy within that oil and coal was originally captured from the sun by plants and algae growing millions of years ago and then transformed into fossil fuels by geological forces.

The energy needs of transportation, industry, and homes seem insignificant, however, when compared with the combined energy requirements of all living organisms. Every living cell requires energy just to remain alive, and more energy is needed for the cell to reproduce, grow, or do physical work as part of an organism. In addition, oxygen is vital to nearly all life in processes that release stored energy.

Photosynthesis, at least indirectly, is not only the principal means of keeping all forms of humanity functioning but also the sole means of sustaining life at any level—except for a few bacteria that derive their energy from sulfur salts and other inorganic compounds. This unique manufacturing process of green plants furnishes raw material, energy, and oxygen. In photosynthesis, energy from the sun is harnessed and, with the aid of chlorophyll, is transformed from light energy to biochemical energy in the bonds between the atoms in a sugar molecule. Oxygen is given off as a by-product of the process.

It has been estimated that all of the world's green organisms (including those in the oceans) together produce between 100 billion and 200 billion metric tons (between 110 billion and 220 billion tons) of sugar each year. To visualize that much sugar, consider that it is enough to make about 300 quadrillion sugar cubes, with a total volume exceeding that of 2 million Empire State Buildings.[1]

Much of the sugar produced by plants is converted to wood, fibers (such as cotton and linen), and other structural materials. The first products of photosynthesis may also be converted to disaccharides, such as sucrose; polysaccharides, such as starch; and other storage forms of carbohydrates. The digestive activities of living organisms break down the carbohydrates to smaller molecules.

Sugars produced by photosynthesis are also needed for the synthesis of amino acids for proteins and other cell constituents. In fact, photosynthesis produces more than 94% of the dry weight of green organisms, with the remainder coming from the soil or dissolved matter.

The capacity of plants to meet our energy needs may well determine the ultimate size of the global human population. In some heavily populated parts of the world, the food supply already is falling short of providing enough energy to sustain life, and starvation is widespread. Meanwhile, in the Western world, many people consume too much food and are spending large sums of money on weight reduction. We will, however, eventually approach a point at which the human population will need to stabilize, or even those in the most affluent areas could exceed the capacity of the plants to sustain them.

Much of photosynthesis worldwide occurs in organisms living in the oceans. It is estimated that between 40% and 50% of the oxygen in the atmosphere originates in oceans and lakes. Laboratory tests have shown, however, that pollutants—such as the PCBs (polychlorinated biphenyls) used in electrical insulators—are capable of stopping many delicate algae from carrying on photosynthesis, even at concentrations as low as 20 parts per billion. The concentration of such substances in ocean waters at present is considerably less than one part per billion, but PCB concentrations of up to five or more parts per billion have been reported in some estuaries. The use of PCBs and related

---

1. One quadrillion is 1,000,000,000,000,000. New York City's Empire State Building has 102 stories and is 381 meters (1,250 feet) tall.

> **NOTE to the Reader**
>
> Photosynthesis is undoubtedly the most important process on earth to life as we know it. It is also a complex process that can be summarized briefly or examined in detail. What follows is a discussion of the subject at three different levels: (1) the essence of photosynthesis; (2) a brief introduction to the major steps of photosynthesis; and (3) a closer look at photosynthesis. The process of respiration, which is discussed after photosynthesis, is treated in similar fashion. The third level, in particular, contains detail that may or may not be discussed in your course.

chemicals has been curtailed in the United States, but other countries are still using them, and residues are still washing off into rivers and on into the oceans. It is important to make the world community fully aware of the dangers of allowing pollutant concentrations to build up to the point of creating significant problems that could ultimately adversely affect all life.

1. The Essence of Photosynthesis

Cells need energy to do their work and to reproduce. The energy for most of this cellular activity comes from energy-storing molecules commonly known as **ATP** (*adenosine triphosphate*).[2] ATP doesn't last more than a few seconds and has to be produced constantly. Instead of ingesting food to obtain energy as animals do, plants can make ATP, using light as the source of energy, just like a solar-powered engine. Without light, however, ATP production stops, and the cells could quickly die. In a solar-powered engine, this problem is avoided by using some of the generated electricity to charge batteries before the sun goes down. Plants also accumulate energy for later use by building sugar molecules for short-term energy storage or starch for longer-term energy storage.

The energy captured by light during photosynthesis is stored in a simple sugar molecule produced from carbon dioxide (CO_2) present in the air and water (H_2O) absorbed by the plant. When the carbon dioxide and water are combined and ultimately a sugar molecule ($C_6H_{12}O_6$, glucose) is produced in a chloroplast, oxygen gas (O_2) is released as a by-product. The oxygen diffuses out into the atmosphere. The overall process can be depicted by the equation that follows. This is a simplified equation; there are many intermediate steps to the process, and glucose is not the first product of photosynthesis.

---

[2]. There are millions of ATP molecules in living cells. When an ATP molecule releases its energy, it gives up the terminal phosphate group in the row of three attached to its base and becomes ADP (adenosine diphosphate). An ADP molecule becomes an ATP molecule again when it regains its third phosphate group and stores energy. ATP is an important participant in many reactions that require the transfer of energy.

$$6CO_2 + 12H_2O + \text{light energy} \xrightarrow[\text{enzymes}]{\text{chlorophyll}}$$

$$C_6H_{12}O_6 + 6O_2 + 6H_2O$$

Photosynthesis takes place in chloroplasts (see Chapter 3) or in cells with membranes in which chlorophyll is embedded. The principals in the process are *carbon dioxide, water, light,* and *chlorophyll;* a brief examination of each follows.

Carbon Dioxide

Our atmosphere consists of approximately 78% nitrogen and 21% oxygen. The remaining 1% is made up of a mixture of less common gases, including 0.040% (400 parts per million) carbon dioxide and hydrogen, helium, argon, and neon. The carbon dioxide in the air surrounding the leaves of plants reaches the chloroplasts in the mesophyll cells by diffusing through the stomata into the leaf interior. The carbon dioxide goes into solution in a thin film of water on the outside walls of cells. It then diffuses through the cell walls, across cell membranes, and into the cytoplasm, where it finally reaches the chloroplasts.

The amount of carbon dioxide constantly being taken from the atmosphere during daylight hours by all green plants is enormous. Just four-tenths of a hectare (1 acre) of corn (10,000 plants) accumulates more than 2,500 kilograms (5,512 pounds) of carbon from the atmosphere during a growing season. Over 10 metric tons (11 tons) of carbon dioxide are needed to furnish this much carbon. It also has been calculated that the total present atmospheric supply of carbon dioxide (more than 2.2 billion metric tons, or about 50 metric tons over each hectare of the earth's surface) would be completely used up in about 22 years if it were not constantly being replenished.

A large reservoir of carbon dioxide in the oceans has helped maintain atmospheric carbon dioxide levels throughout most of recorded history, but in the last 50 years, those levels have begun to climb. Plant and animal respiration, decomposition, natural fires, volcanoes, and similar sources replace carbon dioxide at roughly the same rate at which it is removed during photosynthesis. The use of fossil fuels, pollution, deforestation, and other human activities, however, have disrupted this cycle by adding more carbon dioxide to the atmosphere than is removed.

Measurements of atmospheric carbon dioxide at the Mauna Loa research station on the big island of Hawaii have demonstrated this steady increase. Increased carbon dioxide levels can cause global increases in temperatures because, in the atmosphere, this gas acts much like a glass panel in a greenhouse by trapping solar radiation from the sun. Climatic models project that the earth's average atmospheric temperature will rise 1.0°C to 3.5°C (1.8°F to 6.3°F) by the year 2100. Since the 1980s, global warming, presumably as a result of this increase, has become a major political and scientific issue. See the section "Global Warming" in Chapter 25 for a discussion of this problem.

Although increases in carbon dioxide levels may enhance photosynthesis and increase food production, insects, bacteria, and viruses that proliferate with warmer temperatures can offset these potential gains. Under carefully monitored conditions, commercial greenhouses pump carbon dioxide through pipes placed over plant beds to supplement their natural supply and have found that fertilizing plants with this gas has increased yields by more than 20%. Indeed, this indicates that plants have the potential to limit elevated carbon dioxide levels in the atmosphere. However, recent evidence suggests that many plants develop fewer stomata when carbon dioxide levels increase, thereby adapting them to such changes and reducing photosynthetic efficiency. The abundance of photosynthetic organisms that may or may not proliferate under these conditions is a key factor affecting global warming.

Increased carbon dioxide levels are causing temperatures to rise globally, which, in turn, will result in longer growing seasons in middle and high latitudes and, hence, will increase global photosynthesis. Nonetheless, the same temperature increases may accelerate plant and animal respiration and decomposition, which would add carbon dioxide to the environment. Respiration at higher temperatures may also diminish benefits to be anticipated from an increase in numbers of photosynthetic organisms during global warming. Through careful evaluation of global trends and physiological processes, scientists should be better able to predict what is in store for our planet's future and how to improve management of the environment.

Water

Less than 1% of all the water absorbed by plants is used in photosynthesis; most of the remainder is transpired or incorporated into cytoplasm, vacuoles, and other materials. Water supplies electrons in photosynthesis. Oxygen is released as a by-product. This has been demonstrated by conducting photosynthetic experiments using either carbon dioxide or water containing isotopes of oxygen. When the isotope is present only in the water, it appears in the oxygen gas released. If, however, it is present only in the carbon dioxide, it is confined to the sugar and water produced and never appears in the oxygen gas, demonstrating clearly that the water is the sole source of the oxygen released.

If water is in short supply, it may indirectly become a limiting factor in photosynthesis; under such circumstances, the stomata usually close and sharply reduce the carbon dioxide supply.

Light

Light exhibits properties of both waves and particles. Energy reaches the earth from the sun in waves of different lengths, the longest waves being radio waves and the shortest being gamma rays. About 40% of the radiant energy we receive

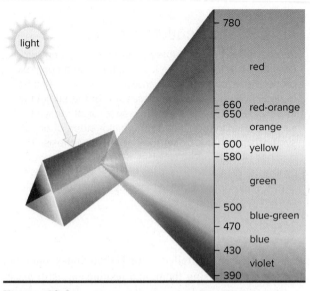

Figure 10.2 Visible light that is passed through a prism is broken up into individual colors with wavelengths ranging from 390 nanometers (violet) to 780 nanometers (red).

is in the form of visible light. If this visible light is passed through a glass prism, it splits into its component colors. Reds are on the longer wavelength end and violets on the shorter wave end, with yellows, greens, and blues between (Fig. 10.2). Although nearly all of the visible light colors can be used in photosynthesis, those in the violet to blue and red-orange to red wavelengths are used most extensively. Light in the green range is reflected; this is why leaves are green. Leaves commonly absorb about 80% of the visible light that reaches them.

The intensity of light varies with the time of day, season of the year, altitude, latitude, and atmospheric conditions. On a clear summer day at noon in a temperate zone, sunlight reaches an intensity of about 2,000 μmols[3] per square meter of surface per second. In contrast, consider that a room with fluorescent lighting produces only about 40 μmols per square meter of surface per second.

Plants vary considerably in the light intensities they need for photosynthesis to occur at optimal rates. Factors such as temperature and the amount of carbon dioxide available can also be limiting (Fig. 10.3). For example, an increase in photosynthesis will not occur in some plants receiving more than 670 μmols of light per square meter unless supplemental carbon dioxide is provided. With supplements, however, rates of photosynthesis will continue to increase up to about 1,000 μmols. Herbaceous plants on a forest floor can survive with less than 2% of full daylight, and some mosses are reported to thrive on intensities as low as 0.05% to 0.01%. Most land plants that grow naturally in the open need at least 30% of full daylight to thrive. The optimal amount for some species of trees approaches full daylight, while shade plants often do well in 10% of full daylight.

---

3. A μmol is a measurement of the number of photons striking a specified area such as a leaf surface.

Figure 10.3 Effects of light and temperature on two forms of photosynthesis. Both forms of photosynthesis, known as C_3 and C_4, respectively, are discussed in this chapter. (a) In C_3, the rate of photosynthesis will not increase beyond a certain intensity of light. In C_4 plants, when additional carbon dioxide is available, photosynthetic rates undergo up to a 30% increase in light intensity. (b) In C_3 plants, quantum yield of photosynthesis decreases as temperatures increase, whereas in C_4 plants, the quantum yield of photosynthesis is not significantly affected by temperature fluctuations between 10°C and 40°C.

Light that is too intense may change the way in which some of a cell's metabolism takes place. For example, higher light intensities and temperatures may change the ratio of carbon dioxide to oxygen in the interiors of leaves, which, in turn, may accelerate *photorespiration*. Photorespiration is typically considered to be a wasteful process that uses oxygen and releases carbon dioxide, although it may help some plants survive under adverse conditions. It differs from common aerobic respiration in its chemical pathways.

Photooxidation, which results in the destruction ("bleaching") of chlorophyll by light, may also occur. High light intensities may cause chlorophyll molecules to go to a different, excited state. The energy released from the excited chlorophyll is passed to oxygen molecules, which become highly reactive and bleach the chlorophyll. During the fall season in temperate regions, photooxidation plays a significant role in the breakdown of chlorophyll in leaves, resulting in the autumn colors discussed in Chapter 7. High light intensities may also cause an increase in transpiration, resulting in the closing of stomata. A sharp reduction in the available carbon dioxide supply inevitably follows.

Chlorophyll

There are several different types of **chlorophyll** molecules, all of which contain one atom of magnesium. They are very similar in structure to the heme of hemoglobin, the iron-containing red pigment that transports oxygen in blood. Each molecule has a long lipid tail, which anchors the chlorophyll molecule in the lipid layers of the thylakoid membranes of chloroplasts (Fig. 10.4).

Chloroplasts of most plants contain two major types of chlorophyll. Chlorophyll *a* is blue-green in color and has the formula $C_{55}H_{72}MgN_4O_5$. Chlorophyll *b* is yellow-green in

Figure 10.4 The structure of a molecule of chlorophyll *a*, the most important of the pigments involved in photosynthesis. The boxlike ring structure on the left, with magnesium and nitrogen inside, functions in capturing light energy. The tail, which extends into the interior of a thylakoid membrane, is insoluble in water; all chlorophyll molecules are, however, fat soluble.

color and has the formula $C_{55}H_{70}MgN_4O_6$. Usually, a chloroplast has about three times more chlorophyll *a* than *b*. The more chlorophyll *a* there is in a cell, the brighter green the cell and the tissue of which it is a part appear to be. When a molecule of chlorophyll *b* absorbs light, it transfers the energy to a molecule of chlorophyll *a*. Chlorophyll *b*, then, makes it possible for photosynthesis to take place over a broader spectrum of light than would be possible with chlorophyll *a* alone (see Fig. 10.7).

Other photosynthetic pigments include carotenoids (yellowish to orange pigments), phycobilins (blue or red pigments found in cyanobacteria and red algae), and several other types of chlorophyll. Chlorophylls *c, d,* and possibly *e* take the place of chlorophyll *b* in certain algae, and several other photosynthetic pigments are found in bacteria. The various chlorophylls are all closely related and differ from one another only slightly in their molecular structure.

In chloroplasts, about 250 to 400 pigment molecules are grouped as a light-harvesting complex called a **photosynthetic unit,** with large numbers of these units in each granum. Two types of these photosynthetic units function together in the chloroplasts of green plants to carry out the first phase of photosynthesis, the *light-dependent* reactions, which are discussed in the next section.

2. Introduction to the Major Steps of Photosynthesis

Photosynthesis takes place in two series of steps called the **light-dependent reactions** and the **light-independent reactions.** Although the light-independent reactions use products of the light-dependent reactions, both processes occur simultaneously.

The Light-Dependent Reactions

The *light-dependent reactions* are the first major steps in the conversion of light energy to biochemical energy. The reactions are initiated when units of light energy (*photons*) strike chlorophyll molecules embedded in the thylakoid membranes of chloroplasts.

Our knowledge of the light-dependent reactions essentially began in the 1930s in England through a discovery by Robin Hill, a biochemist. He found that a solution of fragmented and whole chloroplasts, isolated from leaves that had been ground up and centrifuged, could briefly produce oxygen if an electron acceptor were present to receive electrons from water. In 1951, it was shown that **NADP** (*nicotinamide adenine dinucleotide phosphate*), which is derived from the B vitamin *niacin,* was a natural electron acceptor in this reaction. In honor of its discoverer, the process became known as the *Hill reaction.*

During the light-dependent reactions,

1. water molecules are split apart, releasing electrons and hydrogen ions, and oxygen gas is released;
2. the electrons from the split water molecules are passed along an electron transport system;
3. energy-storing ATP molecules are produced;
4. some hydrogen from the split water molecules reduces NADP to form **NADPH** (*reduced nicotinamide adenine dinucleotide phosphate*), which carries hydrogen and is used in the second phase of photosynthesis, the *light-independent reactions.*

The Light-Independent Reactions

The *light-independent reactions* (or *carbon-fixing and reducing reactions*) complete the conversion of light energy to chemical energy by utilizing ATP and NADPH to form sugars. Some scientists refer to the light-independent reactions as the *dark reactions* because they don't directly require light, but darkness has nothing to do with their functioning. In fact, even though light is not directly required in the same sense as it is for the light-dependent reactions, light is nevertheless required for the activation of the enzymes involved, and the processes normally occurs only in daylight.

The light-independent reactions are a series of reactions that take place outside of the grana in the stroma of the chloroplast, if the products of the light-dependent reactions are available. They may initially proceed in different ways, depending on the particular kind of plant involved, but they all go through the **Calvin cycle,** discovered and elucidated by Dr. Melvin Calvin of the University of California. In 1961, Calvin received a Nobel Prize for unraveling how this most common type of light-independent reactions takes place.

In this cycle, carbon dioxide (CO_2) from the air is combined with a 5-carbon sugar (*RuBP, ribulose bisphosphate*), and then the combined molecules are converted, through several steps, to sugars, such as glucose ($C_6H_{12}O_6$). Energy and electrons involved in these steps are furnished by the ATP molecules and NADPH produced during the light-dependent reactions. Some of the sugars that are produced during the light-independent reactions are recycled, while others are stored as starch or other polysaccharides (simple sugars strung together in chains). A summary of simplified photosynthetic reactions is shown in Figure 10.5. More detailed diagrams of the light-dependent and light-independent reactions are shown in Figures 10.8, 10.10, and 10.13.

Two molecules of a 3-carbon sugar compound (*3PGA—3-phosphoglyceric acid*) are shown as the first stable substance produced when carbon dioxide from the air and RuBP are combined and then converted during the light-independent reactions (Fig. 10.5). Some grasses and many plants of arid regions fix carbon differently. They produce a 4-carbon acid as the first product, followed by the Calvin cycle. This 4-carbon pathway is discussed, along with another variation found mostly in desert plants, in the next section.

Figure 10.5 A simplified summary of photosynthetic reactions.

3. A Closer Look at Photosynthesis

A great deal has been learned about photosynthesis since 1772, when Joseph Priestley (1733–1804), a naturalist in England, received a medal for demonstrating that a sprig of mint "restored" oxygen, so that a mouse could live in air that had been used up by a burning candle. Seven years later, Jan Ingen-Housz (1730–1799) of Holland, who visited England to treat the royal family for smallpox, carefully repeated Priestley's demonstrations. He showed that the air was restored only when green parts of plants were receiving sunlight.

In 1782, Jean Senebier (1742–1809), a Swiss pastor, discovered that photosynthesis requires carbon dioxide, and in 1796, Ingen-Housz showed that carbon is a plant nutrient. The final component of the overall photosynthetic reaction was explained in 1804 by another Swiss, Nicholas Theodore de Saussure (1767–1845), who showed that water is required.

The Light-Dependent Reactions Reexamined

As noted earlier, light has characteristics of both waves and particles. Sir Isaac Newton, while experimenting with a prism over 300 years ago, produced a spectrum of colors from visible white light and postulated that light consisted of a series of discrete particles he called "corpuscles."

Newton's theory only partially explained light phenomena, however, and by the middle of the 19th century, James Maxwell and others showed that light and all other parts of the extensive electromagnetic spectrum travel in waves.

By the late 1800s, with the discovery that a *photoelectric effect* can be produced in all metals, the wave theory also became inadequate to explain certain attributes of light.

When a metal is exposed to radiation of a critical wavelength, it becomes positively charged, because the radiation forces electrons out of the metal atoms. The ability of light to force electrons from metal atoms depends on its particular wavelength—its energy content—and not on its intensity or brightness. The shorter the wavelength, the greater the energy, and vice versa.

In 1905, Albert Einstein proposed that the photoelectric effect results from discrete particles of light energy he called **photons.** In 1921, he received the Nobel Prize in physics for this work. Both waves and particles (photons) are now recognized as aspects of light. The energy (quantum) of a photon is not the same for all kinds of light; those of longer wavelengths have lower energy, and those of shorter wavelengths have proportionately higher energy.

Chlorophylls, the principal pigments of photosynthetic systems, absorb light primarily in the violet to blue and in the red wavelengths. This was first ingeniously demonstrated in 1882 by T. W. Engelmann. He focused a tiny spectrum of light on a filament (single row of cells) of *Spirogyra*, a freshwater alga. The alga had been mounted in a drop of water on a microscope slide containing bacteria that move toward an oxygen source. As shown in Figure 10.6, the bacteria assembled in greatest numbers along the algal filament in the blue and red portions of the spectrum, demonstrating that oxygen production is directly related to the light the chlorophyll absorbs. An analysis using living material is called a *bioassay*.

Each pigment has its own distinctive pattern of light absorption, which is referred to as the pigment's *absorption spectrum* (Fig. 10.7). When a pigment absorbs light, the energy levels of some of the pigment's electrons are raised. The energy from an excited electron is released when it drops back to its ground state. When this occurs, the energy

Figure 10.6 Engelmann's experiment. A tiny spectrum of light was focused on a microscope slide with a filament (row) of algal cells suspended in water containing bacteria that move toward an oxygen source. The bacteria assembled mostly in the areas exposed to the violet to blue as well as the red portions of the spectrum.

Figure 10.7 The absorption spectra of chlorophyll *a*, chlorophyll *b*, and a carotenoid. The maximum absorption of the chlorophylls is in the blue and red wavelengths. The maximum absorption of the carotenoids is in the blue-green to green parts of the visible spectrum.

may be emitted immediately as light (a phenomenon called *fluorescence*). In chlorophyll, the emitted light is characteristically in the red part of the visible light spectrum, so an extract of chlorophyll placed in light (especially ultraviolet or blue light) will appear red. The absorbed energy may also be emitted as light after a delay (a phenomenon called *phosphorescence*); it may otherwise be converted to heat. The most important use of the absorbed energy, however, is its storage in chemical bonds for photosynthesis.

Photosystems The two types of photosynthetic units present in most chloroplasts make up **photosystems** known as *photosystem I* and *photosystem II* (Fig. 10.8). Photosystem II received its "II" designation because it was discovered after photosystem I, but we know now that the events of photosynthesis that take place in photosystem II come *before* those of photosystem I.

While both photosystems produce ATP, only organisms that have photosystems I *and* II can produce NADPH and oxygen as a consequence of electron flow. At least 2.8 billion years ago, photosynthetic forms of bacteria (cyanobacteria) are believed to have evolved from primitive bacteria with the development of chlorophyll *a* and photosystem II. As a consequence of the oxygen-generating steps in photosystem II that have evolved, humans and other organisms are dependent upon photosynthetic organisms that generate oxygen in the air we breathe today. Because photosynthetic organisms can generate oxygen from water, this very process in the future could be exploited to sustain human life during space travel and for the colonization of other planets.

Each photosynthetic unit of photosystem I consists of 200 or more molecules of chlorophyll *a*, small amounts of chlorophyll *b*, carotenoid pigment with protein attached,

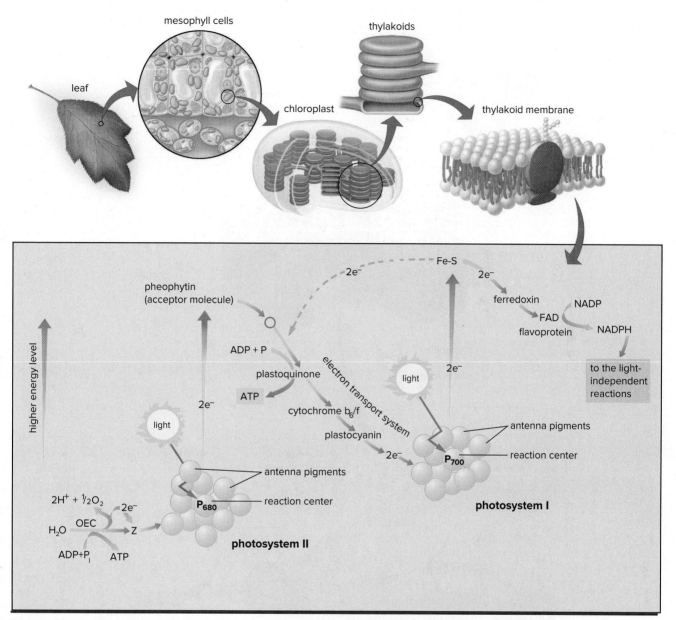

Figure 10.8 The light-dependent reactions of photosynthesis, which occur in more than one way. In noncyclic photophosphorylation, involving photosystems I and II, which convert light energy to biochemical energy in the form of ATP and NADPH, water molecules are split, releasing electrons, protons, and oxygen gas. The electrons are subsequently used to produce NADPH, whereas the protons are used, in part, to enable production of ATP. Oxygen gas is a by-product of this noncyclic photophosphorylation, although aerobic organisms rely upon this gas for respiration. The ATP and NADPH are used in the carbon-fixing reactions that convert CO_2 to sugars (see Fig. 10.10). Only photosystem I is involved in cyclic photophosphorylation. In this relatively simple system, electrons boosted from a photosystem I reaction-center molecule are shunted back into the reaction center via the electron transport system. ATP is produced from ADP, but no NADPH or oxygen is produced.

and a special reaction-center molecule of *chlorophyll* called P_{700}. Although all pigments in a photosystem can absorb photons, the *reaction-center* molecule is the only one that can actually use the light energy. The remaining photosystem pigments are called *antenna pigments* because together they function like an antenna in gathering and passing light energy to the reaction-center molecule (Fig. 10.9). Iron-sulfur proteins are the primary electron acceptors for photosystem I (i.e., iron-sulfur proteins first receive electrons from P_{700}).

A photosynthetic unit of photosystem II consists of chlorophyll *a*, ß-carotene (the precursor of vitamin A) attached to protein, chlorophyll *b*, and a reaction-center molecule of chlorophyll *a*; these special molecules are called P_{680}.

The letter *P* stands for pigment and the numbers *700* and *680* of the reaction-center molecules of chlorophyll *a* refer to peaks in the absorption spectra of light with wavelengths of 700 and 680 nanometers, respectively. These peaks differ slightly from those of the otherwise identical

Figure 10.9 Organization of the thylakoid membrane showing the relative positions of photosystems and protein complexes. Some hydrogen ions (protons) from the stroma are pumped into the thylakoid space (lumen), producing a hydrogen gradient. ATP is produced when these hydrogen ions and those from water flow from the thylakoid space into the stroma through the ATP synthase complex.

chlorophyll *a* molecules of the photosynthetic units. A primary electron acceptor called *pheophytin* (or *Pheo*) is also present in photosystem II. One reaction-center molecule was found by Johann Deisenhofer, Robert Huber, and Hartmut Michel of Germany to have over 10,000 atoms. They received a Nobel Prize in 1988 for their work in determining the atomic structure.

Water-Splitting (Photolysis) When photons of light are absorbed by P_{680} molecules of a photosystem II reaction center (located near the inner surface of a thylakoid membrane), the light energy boosts electrons to a higher energy level. This is referred to as *exciting* electrons. Excited electrons are unstable and can rapidly revert to their lower energy level, releasing absorbed energy, perhaps as heat. However, during photosynthesis, the excited electrons are passed to an acceptor molecule, called *pheophytin*, at the beginning of an electron transport system.

From pheophytin, electrons are shuttled to another acceptor, **PQ** (*plastoquinone*), within the thylakoid membrane.[4] PQ is mobile and moves through the lipid bilayer toward the inner side of the thylakoid membrane, unloading electrons to the *cytochromes* that are next in line. Electrons extracted from water by a manganese-containing *oxygen-evolving complex* (OEC) replace the electrons lost by the P_{680} molecule. It has been suggested that there is an oxidation–reduction system, usually designated Z, operating between water and P_{680}. The transfer of electrons from Z to P_{680} reoxidizes Z and prepares it for accepting additional electrons from the OEC. Recent investigations indicate that for each two water molecules that are split, one molecule of oxygen is produced, along with four protons and four electrons.

This metabolic pathway eventually evolved in photosynthetic bacteria (cyanobacteria). The abundance of water, as an electron source, probably facilitated the mechanism that generated oxygen as a by-product of photosynthesis. This process increased the supply of oxygen in earth's atmosphere and made it possible for energy-efficient aerobic respiration to evolve.

Electron Flow and Photophosphorylation The high-energy acceptor molecule PQ releases electrons that originated from photosystem II to an electron transport system, which functions something like a downhill bucket brigade. This electron transport system consists of iron-containing pigments called **cytochromes** and other electron transfer molecules, plus *plastocyanin*—a protein that contains copper. When electrons pass along the electron transport system and protons subsequently move across the thylakoid membrane by *chemiosmosis* (see next section, "Chemiosmosis"), ATP molecules are formed from ADP in the process of *photophosphorylation*.

A somewhat similar series of events occurs in photosystem I. When photons of light are absorbed by P_{700} molecules in a photosynthetic unit, the energy excites electrons, which are transferred to an iron-sulfur acceptor molecule designated Fe-S. This then passes electrons to another iron-sulfur acceptor molecule, **Fd** (*ferredoxin*), which, in turn, releases them to a carrier molecule designated as **FAD** (*flavin adenine dinucleotide*). FAD contains flavoprotein, which assists in the reduction of NADP to NADPH. The electrons removed from the P_{700} molecule are replaced by electrons from photosystem II via the electron transport system outlined previously. This overall movement of electrons from water to photosystem II to photosystem I to NADP is called *noncyclic* electron flow because it goes in one direction only. The production of ATP that correspondingly occurs is designated as *noncyclic photophosphorylation*.

Photosystem I can also work independently of photosystem II. When it does, the electrons boosted from P_{700} reaction-center molecules (of photosystem I) are passed from ferredoxin to plastoquinone (instead of NADPH) and back into the reaction center of photosystem I. This process is *cyclic* electron flow. ATP generated by cyclic electron flow is called *cyclic photophosphorylation*, but water molecules are not split and neither NADPH nor oxygen is produced in this process.

Chemiosmosis The oxygen-evolving complex on the inside of a thylakoid membrane catalyzes the splitting of water molecules, producing protons and electrons, as well as oxygen gas. These electrons, used to replenish those excited in chlorophyll, are then transferred in bucket-brigade fashion through an electron transport system, ultimately reducing NADP to NADPH. As electrons travel through this transport system, additional protons move from the stroma to the inside of the thylakoid membrane in oxidation–reduction reactions when electrons pass from photosystem II to PQ. These protons join with the protons from the split water molecules and thereby contribute to an accumulation of four protons toward the inside of the thylakoid membrane (an area also known as the thylakoid *lumen*).

Although some protons are used in the production of NADPH on the stroma side of the thylakoid membrane, there is still a net accumulation of protons in the thylakoid lumen from the splitting of water molecules and electron transport. This establishes a proton gradient, giving special proteins in the thylakoid membrane the potential for moving protons from the thylakoid lumen back to the stroma. Movement of protons across the membrane is thought to be a source of energy for the synthesis of ATP. The action has been described as similar to the movement of molecules during osmosis and has been called **chemiosmosis,** or the *Mitchell theory*, after its author, Peter Mitchell. In this model, protons move across a thylakoid membrane through protein channels called ATPase. With the proton movement, ADP and phosphate (P) combine, requiring an input of energy and producing ATP (Fig. 10.9). This chemiosmotic mechanism for connecting electron flow with the conversion of ADP to ATP

---

4. Except for the high speed (trillionths of a second) at which it usually carries out its function, an acceptor molecule could be compared to a pickup order sent to a clerk in a store, in the sense that an item is picked up from a source, held temporarily, and then transferred elsewhere.

KEY THEME 10.1 physiology — Photosynthesis and Pizza

The next time you order pizza, think of all the things that went into making that pizza (Box Figure 10.1A). Of course, the very bread used to make the crust came from wheat seeds, produced as a result of photosynthesis. You can probably ascertain that any fruits or vegetables, including tomatoes, green peppers, onions, and spices, also were produced because photosynthesis took place. The cheese, pepperoni, sausage, bacon, and, perhaps, even anchovies on the pizza came to exist as a result of animals that ate plants produced through photosynthesis. Even the pizza box itself was made as a direct result of photosynthesis, which produces the cellulose we need for paper products, clothing, and wood. As you turn the pages (also made from plants) of this chapter, try to envision the remarkable processes that enabled the conversion of light energy into biological energy. It is this energy, derived from the photosynthetic light reactions, that fuels the light-independent reactions of photosynthesis and, hence, the synthesis of organic molecules needed for life on earth. If this remarkable process of photosynthesis could be replicated on another habitable planet, it is possible that life could be sustained extraterrestrially. Because photosynthesis produces oxygen, as well as food and resources for clothing and shelter, it is likely that plants will be used in the future for space exploration. Imagine how much fun astronauts will have, eating pizza produced from plants grown on another planet!

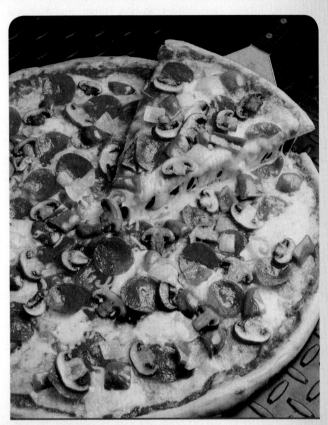

Box Figure 10.1A Next time you eat pizza, think of the photosynthetic processes that helped produce your delicious meal!
© Brand Z Food/Alamy

is essentially the same as that of oxidative phosphorylation in mitochondria (see Fig. 10.14), except that in mitochondria, oxygen (instead of NADP) is the terminal electron acceptor.

The Light-Independent Reactions Reexamined

We have seen how the synthesis of ATP and NADPH is set in motion during the *light-dependent reactions*. Both of these substances play key roles in the synthesis of carbohydrate using carbon dioxide from the atmosphere, which during the *light-independent reactions* reaches the interior of chlorenchyma tissues via stomata. As indicated earlier, the light-independent reactions are really a series of reactions, each mediated by an enzyme in this major phase of photosynthesis.

The light-independent reactions take place in the stroma of the chloroplasts and, as long as the products of the light-dependent reactions are present, they do not directly need light in the same sense that the light-dependent reactions do. However, they normally take place only during daylight hours because some of the enzymes involved in the light-independent reactions require light for their activation or conversion to a form in which they can actively catalyze steps of the light-independent reactions.

The Calvin Cycle Melvin Calvin studied the light-independent reactions using a round, flat container to grow green algae, to which he introduced radioactive carbon. By tracing the radioactive carbon in metabolites as the algae carried out certain reactions, he was able to show the steps through which carbon dioxide is fixed and converted to carbohydrates. The carbohydrate produced during these reactions facilitates growth, including the development of leaves, stems, roots, flowers, and other plant structures. From an ecological standpoint, this process is essential because it is the main biosynthetic pathway through which carbon enters the web of life. As discussed next, the **Calvin cycle,** or *photosynthetic carbon reduction (PCR)* pathway (Fig. 10.10), runs in five main steps:

1. Six molecules of carbon dioxide (CO_2) from the air combine with six molecules of ribulose 1,5-bisphosphate (RuBP, the 5-carbon sugar continually

Figure 10.10 The Calvin cycle (light-independent reactions) of photosynthesis. The photograph on top of the cycle shows the apparatus that Melvin Calvin used to study photosynthesis. The cycle takes place in the stroma of chloroplasts, where each step is controlled by a different kind of enzyme. Carbon dioxide molecules from the air enter the cycle one at a time, making six turns of the cycle necessary to produce one molecule of a 6-carbon sugar such as glucose ($C_6H_{12}O_6$). *Photo courtesy of Lawrence Berkeley National Laboratory Copyright Notice: © 2010 The Regents of the University of California Lawrence Berkeley National Laboratory Person/Principle Investigator*

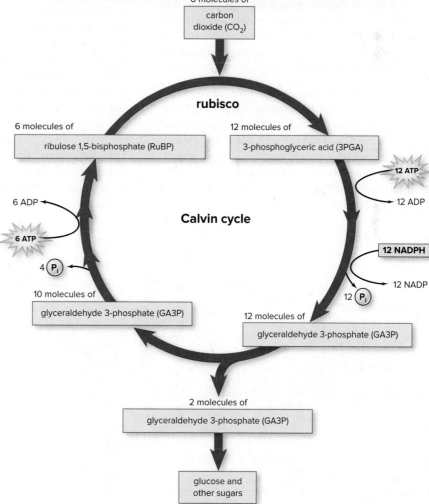

being formed while photosynthesis is occurring), with the aid of the enzyme rubisco (RuBP carboxylase/oxygenase).[5]

2. The resulting six 6-carbon unstable complexes are immediately split into twelve *3-carbon* molecules known as 3-phosphoglyceric acid (3PGA), the first stable compound formed in photosynthesis.

3. The NADPH (which has been temporarily holding the hydrogen and electrons released during the light-dependent

---

5. Rubisco is a huge, complex enzyme that may constitute up to 30% of the protein present in a living leaf. It is the most abundant protein known.

reactions) and ATP (also from the light-dependent reactions) supply energy and electrons that chemically reduce the 3PGA to twelve molecules of glyceraldehyde 3-phosphate (GA3P, 3-carbon sugar phosphate).

4. Ten of the twelve glyceraldehyde 3-phosphate molecules are restructured, using another six ATPs, and become six 5-carbon molecules of RuBP, the sugar with which the Calvin cycle was initiated.
5. This leaves a net gain of two GA3P molecules, which either can contribute to an increase in the carbohydrate content of the plant (glucose, starch, cellulose, or related substances) or can be used in pathways that lead to the net gain of lipids and amino acids.

Because rubisco catalyzes the formation of the 3-carbon compound 3PGA as the first isolated product in these light-independent reactions, plants that carry out this process are called **C_3 plants.** However, as indicated by its name, the enzyme RuBP carboxylase/oxygenase has the potential to fix both CO_2, through its carboxylase activity described by the Calvin cycle, and O_2, through its oxygenase activity. The oxygenase activity of rubisco enables C_3 plants to undergo a process called **photorespiration.**

Photorespiration While the carboxylase activity of rubisco ultimately results in the accumulation of carbohydrates, the same enzyme can exhibit an oxygenase activity that catalyzes the combination of oxygen with RuBP and a subsequent pathway that releases carbon dioxide. This *photorespiration* may be interpreted as an alternate process that competes with the carbon-fixing role of photosynthesis. However, the oxygenase activity of rubisco responsible for photorespiration is not necessarily an ominous alternative, because it provides a salvage pathway to allow C_3 plants to survive under hot, dry conditions. Evidence suggests that photorespiration helps dissipate ATP and accumulated electrons from the light reactions, thereby preventing photooxidative damage.

Factors that determine the extent of photorespiration in C_3 plants include temperature, $CO_2:O_2$ ratio, and kinetic properties of rubisco. Hot and dry climates generally promote photorespiration because stomata are closed under these conditions, leading to a decreased $CO_2:O_2$ ratio within the leaf and near rubisco complexes. Oxygen gas can accumulate from photolysis in a leaf with closed stomata, and photorespiration is more likely to occur, particularly when the carbon dioxide concentration drops roughly below 50 parts per million. When temperatures are cooler and moisture is present, the stomata are more likely to open, which leads to an increased $CO_2:O_2$ ratio. Under milder climatic conditions, C_3 plants are more efficient at fixing carbon dioxide.

Photorespiration requires cooperation among chloroplasts, peroxisomes, and mitochondria to facilitate shuttling of intermediates along the photorespiratory pathway. The products of photorespiration are the 2-carbon phosphoglycolic acid, which is processed to some extent in peroxisomes and eventually released as carbon dioxide in mitochondria, and the 3-carbon phosphoglyceric acid that can reenter the Calvin cycle. No ATP is produced by photorespiration.

The 4-Carbon Pathway Sugarcane, corn, sorghum, and at least 1,000 other species of tropical grasses or arid region plants subject to high temperature stresses have a distinctive leaf anatomy called *Kranz anatomy* (Fig. 10.11).

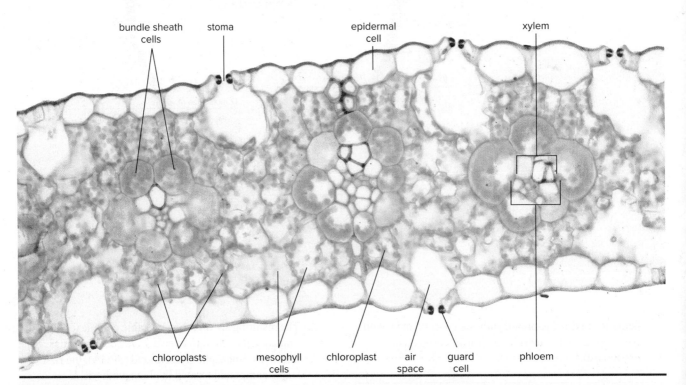

Figure 10.11 A portion of a cross section of a leaf of corn (*Zea mays*), a C_4 plant with Kranz anatomy leaves. Compare this with leaves of typical C_3 plants, as illustrated in Chapter 7. © *Kingsley Stern*

Kranz anatomy leaves have two forms of chloroplasts. In the bundle sheath cells surrounding the veins, there are large chloroplasts, often with few to no grana. The large chloroplasts have numerous starch grains, and the grana, when present, are poorly developed. The chloroplasts of the mesophyll cells are much smaller, usually lack starch grains, and have well-developed grana.

Plants with Kranz anatomy produce a *4-carbon* compound, *oxaloacetic acid,* instead of the 3-carbon 3-phosphoglyceric acid during the initial steps of the light-independent reactions. Oxaloacetic acid is produced when a 3-carbon compound, *phosphoenolpyruvate (PEP)*, and carbon dioxide are combined in mesophyll cells with the aid of a different carbon-fixing enzyme, **PEP carboxylase.** Depending on the species, the oxaloacetic acid may then be converted to aspartic, malic, or other acids. PEP carboxylase is not sensitive to oxygen and, hence, has a greater enzyme affinity for carbon dioxide, so there is no photorespiratory loss of carbon dioxide captured in the organic acids. The carbon dioxide is transported to the bundle sheath cells, where it is released and enters the Calvin cycle, just as in C_3 plants. The carbon dioxide concentration can be kept high in relation to the oxygen concentration in the bundle sheath cells, thereby keeping the reaction of rubisco with oxygen very low (Fig. 10.12).

Because the PEP system produces 4-carbon compounds, plants having this system are called 4-carbon, or **C_4 plants,** to distinguish them from plants that have only the 3-carbon, or C_3, system. Besides Kranz anatomy, C_4 plants have other characteristic features:

1. High concentrations of PEP carboxylase are found in the mesophyll cells. This is significant because PEP carboxylase allows the conversion of carbon dioxide to carbohydrate at much lower concentrations than does rubisco (found only in the bundle sheath cells), the corresponding enzyme in the Calvin cycle.
2. The optimum temperatures for C_4 photosynthesis are much higher than those for C_3 photosynthesis, allowing C_4 plants to thrive under conditions that would adversely affect C_3 plants.

Obviously, the C_4 pathway furnishes carbon dioxide to the Calvin cycle in a more roundabout way than does the C_3 pathway, but the advantage of this extra pathway is a major reduction of photorespiration in C_4 plants. Also, in C_4 plants, the C_4 pathway in the mesophyll cells results in carbon dioxide being picked up even at low concentrations (via the enzyme PEP carboxylase) and in carbon dioxide being concentrated in the bundle sheath, where the Calvin cycle takes place almost exclusively. The Calvin cycle enzyme, rubisco, catalyzes the reaction in which RuBP reacts with carbon dioxide rather than oxygen. Consequently, C_4 plants, which typically photosynthesize at higher temperatures, have photosynthetic rates that are two to three times higher than those of C_3 plants. However, at lower temperatures, C_3 photosynthesis is more efficient than

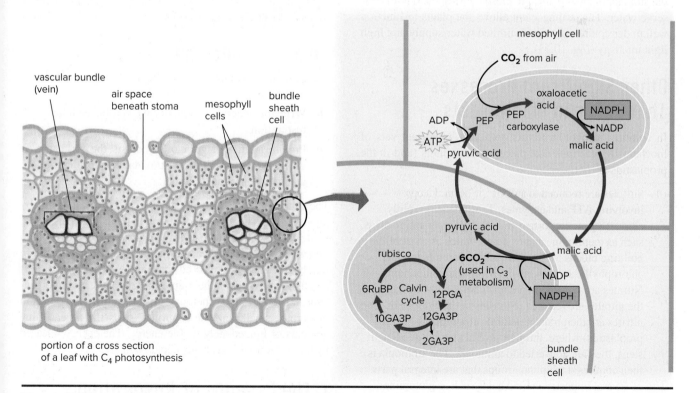

Figure 10.12 An illustration of the C_4 photosynthesis pathway. Carbon dioxide is converted to organic acids in mesophyll cells. After the acids move into bundle sheath cells, some carbon dioxide is released and enters the Calvin cycle, where it becomes a 3-carbon compound that moves back to a mesophyll cell; there it is converted to PEP, which accepts carbon dioxide from the air.

that of C_4 plants because the cost of photorespiration at those temperatures is usually less than the two extra ATPs required for the C_4 pathway.

CAM Photosynthesis *Crassulacean acid metabolism (CAM)* photosynthesis is found in plants of about 30 families, including cacti, stonecrops, orchids, bromeliads, and other succulents that are often stressed by limited availability of water. A few succulents do not have CAM photosynthesis, however, and several nonsucculent plants do. Many CAM plants are facultative C_3 plants that can switch to C_3 photosynthesis during the day after a good rain or when night temperatures are high. Plants with CAM photosynthesis typically do not have a well-defined palisade mesophyll in the leaves, and, in contrast to the chloroplasts of the bundle sheath cells of C_4 plants, those of CAM photosynthesis plants resemble the mesophyll cell chloroplasts of C_3 plants.

CAM photosynthesis is similar to C_4 photosynthesis in that 4-carbon compounds are produced during the light-independent reactions. In these plants, however, the organic acids (mainly malic acid) accumulate at night and break down during the day, releasing carbon dioxide. The enzyme PEP carboxylase is responsible for converting the carbon dioxide plus PEP to organic acids at night when the stomata are open. During daylight, the organic acids diffuse out of the cell vacuoles in which they are stored and are converted back to carbon dioxide for use in the Calvin cycle. A much larger amount of carbon dioxide can be converted to carbohydrate each day than would otherwise be possible, because the stomata of such plants are closed during the day to conserve water. This arrangement allows the plants to function well under conditions of both limited water supply and high light intensity (Fig. 10.13).

Other Significant Processes That Occur in Chloroplasts

In addition to photosynthesis, two very important sets of biochemical reactions take place in chloroplasts (and in the proplastids of roots).

1. Sulfates are reduced to sulfide via several steps involving ATP and enzymes. The sulfide is rapidly converted into important sulfur-containing amino acids, such as methionine and cysteine, which are part of the building blocks for proteins, anthocyanin pigments, chlorophylls, and several other cellular components.
2. Nitrates are reduced to organic compounds. Initially, the nitrates are reduced to nitrites in the cytoplasm. The nitrites are then transported into chloroplasts (or root proplastids) where, through several enzyme-mediated steps, they are converted to ammonia. The ammonia is then converted to amino groups that are integral parts of several important amino acids such as glutamine and aspartic acid. Glutamine is an important form of nitrogen storage in roots or specialized stems such as those of carrots, beets, and potato tubers.

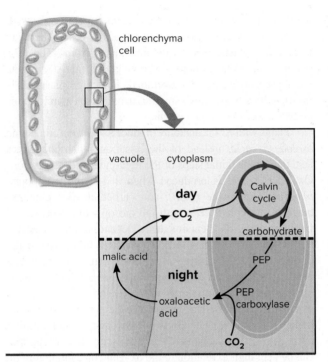

Figure 10.13 Crassulacean acid metabolism (CAM) photosynthesis found in orchids, pineapples, and many desert plants such as cacti. CAM is similar to C_4 photosynthesis, but the plants have their stomata closed during daylight heat, thus conserving water. The organic acids accumulate at night and break down during the day, releasing carbon dioxide, which then enters the Calvin cycle and C_3 metabolism while the stomata are closed.

10.4 Respiration

Learning Outcomes

a. Explain the difference between aerobic and anaerobic respiration.

b. Write the generalized reaction for aerobic cellular respiration.

c. Provide a crude diagram of glycolysis, the citric acid cycle, and electron transport/oxidative phosphorylation, and list the products of each.

The solar energy that is converted into biochemical energy by photosynthesis is stored in organic compounds such as wood, while coal and oil contain energy originally captured by green organisms in the geological past. If the organic compounds are burned, the energy is released very rapidly in the form of heat and light, and much of the usable energy is lost. Living organisms, however, "burn" their energy-containing compounds in numerous, small, enzyme-controlled steps that release tiny amounts of immediately usable energy. The energy released by these reactions is typically captured in ATP.

1. The Essence of Respiration

Respiration is essentially the release of energy from glucose molecules that are broken down to individual carbon dioxide molecules. The process takes place in all active cells 24 hours a day, regardless of whether or not photosynthesis is occurring simultaneously in the same cells. It is initiated in the cytoplasm and completed in the mitochondria. The energy, stored in chemical bonds containing high-energy electrons, is released from simple sugar molecules that are broken down during a series of steps controlled by enzymes. No oxygen is needed to initiate the process, but in **aerobic respiration** (the most widespread form of respiration), the process cannot be completed without oxygen gas (O_2). The controlled release of energy is the significant event; carbon dioxide (CO_2) and water (H_2O) are the by-products. Aerobic respiration can be summed up in the following equation, but bear in mind that respiration, like photosynthesis, is a complex process that involves many steps not reflected in a simplified equation.

$$C_6H_{12}O_6 + 6O_2 \xrightarrow{\text{enzymes}} 6CO_2 + 6H_2O + \text{energy}$$
glucose, oxygen, carbon dioxide, water

Anaerobic respiration and **fermentation** are two forms of respiration that were probably carried on in the geological past when there was no oxygen in the atmosphere. These forms of respiration are still carried on today by certain bacteria and other organisms in the absence of oxygen gas. Anaerobic respiration and fermentation release less than 6% of the energy released from glucose by aerobic respiration. The two forms differ from one another in the manner in which hydrogen released from the glucose is combined with other substances. Fermentation is very important industrially, particularly in the brewing and baking industries. Two well-known forms of fermentation are illustrated by the following equations:

$$C_6H_{12}O_6 \xrightarrow{\text{enzymes}} 2C_2H_5OH + 2CO_2 + \text{energy (ATP)}$$
glucose, ethyl alcohol, carbon dioxide

$$C_6H_{12}O_6 \xrightarrow{\text{enzymes}} 2C_3H_6O_3 + \text{energy (ATP)}$$
glucose, lactic acid

The relatively small amount of energy released during these inefficient forms of respiration is partly stored in two ATP molecules. The actual amount of energy stored is roughly only 29% of the approximately 48 kcals[6] of energy released in anaerobic respiration.

2. Introduction to the Major Steps of Respiration

Glycolysis

In most forms of carbohydrate respiration, the first major phase takes place in the cytoplasm and requires no oxygen gas (O_2). This phase, called **glycolysis,** involves three

---

6. A kcal is the energy required to raise the temperature of 1 kilogram of water from 14.5°C to 15.5°C.

main steps and several smaller ones, each controlled by an enzyme. During the process, a small amount of energy is released, and some hydrogen atoms are removed from compounds derived from a glucose molecule. The essence of this complex series of steps is as follows:

1. In a series of reactions, the glucose molecule becomes a fructose molecule carrying two phosphates (P).
2. This sugar (fructose) molecule is split into two 3-carbon fragments called glyceraldehyde 3-phosphate (GA3P).
3. Some hydrogen, energy, and water are removed from these 3-carbon fragments, leaving pyruvic acid (Fig. 10.14).

Two ATP molecules supply the energy needed to start glycolysis. By the time pyruvic acid has been formed, however, four ATP molecules have been produced from the energy released along the way, for a net gain of two ATP molecules. A great deal of the energy originally in the glucose molecule remains in the pyruvic acid. The hydrogen ions and high-energy electrons released during glycolysis are picked up and temporarily held by an acceptor molecule, **NAD** (*nicotinamide adenine dinucleotide*). What happens to them next depends on whether they are used for *aerobic respiration, true anaerobic respiration,* or *fermentation*.

Aerobic Respiration

In aerobic respiration (the most common type of respiration), glycolysis is followed by two major stages: the *citric acid cycle* and *electron transport*. Both stages occur in the mitochondria and involve many smaller steps, each of which is controlled by enzymes (see Fig. 10.14).

The Citric Acid (Krebs) Cycle The **citric acid cycle** was originally named the *Krebs cycle* after Hans Krebs, a British biochemist who received a Nobel Prize in 1953 for unraveling many of the complex reactions that take place in respiration. The name *citric acid cycle,* or *tricarboxylic acid* (*TCA*) *cycle,* reflects the important role played by organic acids during the process.

Before entering the citric acid cycle, which takes place in the fluid matrix in the compartments formed by the cristae of mitochondria (see Chapter 3), carbon dioxide is released from pyruvic acid that was produced by glycolysis. What remains is restructured to a 2-carbon acetyl group. This acetyl group combines with an acceptor molecule called *coenzyme A* (*CoA*). This combination (called *acetyl CoA*) then enters the citric acid cycle, which is a series of biochemical reactions that are catalyzed by enzymes. Little of the energy originally trapped in the glucose molecule is released during glycolysis. As the citric acid cycle proceeds, however, high-energy electrons and hydrogen are successively removed. This removal takes place from a series of organic acids and, after transfer, ultimately produces compounds such as **NADH** (*reduced nicotinamide adenine dinucleotide*) and **FADH₂** (*reduced flavin adenine dinucleotide*), as well as a small amount of ATP. Carbon dioxide is produced as a by-product during the reaction.

Figure 10.14 A summary of cellular respiration. In the first phase, glycolysis, which takes place in the cytoplasm, a sugar molecule is converted to two pyruvic acid molecules. The subsequent phases take place within the matrix of a mitochondrion. In aerobic respiration, the pyruvic acid is broken down in the citric acid cycle, and energy is transferred to compounds such as NADH, ATP, and $FADH_2$. Carbon dioxide and hydrogen are also released. The released hydrogen is carried by an electron transport system and combined with oxygen, forming water. In anaerobic respiration and fermentation, pyruvic acid is converted in the absence of oxygen gas to ethyl alcohol or lactic acid with little release of energy.

Electron Transport Much of the energy originally in the glucose molecule has now been transferred to the acceptors NAD and FAD, which became NADH and FADH$_2$, respectively. NADH and FADH$_2$ are electron donors to an *electron transport system* consisting of special acceptor molecules arranged in a precise sequence on the inner membranes of mitochondria. The electrons flow through a series of carrier molecules, many of which are part of protein complexes, down an energy gradient. Some of these electron carriers also accept protons and release them to the intermembrane space of the mitochondrion. Shuttling of protons in this way causes a buildup of protons outside the mitochondrial matrix, thereby establishing an electrochemical gradient. Through chemiosmosis, additional protein complexes couple the transport of protons back into the matrix with phosphorylation of ADP to form ATP. The production of ATP stops if there are no electron donors or electron acceptor oxygen molecules.

The acceptor molecules include iron-containing proteins called *cytochromes*. Energy is released in small increments at each step along the system, and ATP is produced from ADP and P. As the final step in aerobic respiration (see Fig. 10.14), oxygen acts as the ultimate electron acceptor, producing water as it combines with hydrogen.

By the time electron transport is complete, the recoverable energy locked in a molecule of glucose has been used to create high-energy ATP molecules. The energy in ATP is then available for use in the synthesis of other molecules and for growth, active transport, and a host of other metabolic processes. Aerobic respiration produces a net gain of 36 ATP molecules from one glucose molecule, using up six molecules of oxygen and producing six molecules of carbon dioxide and a net total of six molecules of water. For each mole (180 grams) of glucose aerobically respired, 686 kcal of energy is released, with about 39% of it being stored in ATP molecules and the remainder being released as heat.

Anaerobic Respiration and Fermentation

In living organisms, glucose molecules may often undergo glycolysis without enough oxygen to complete aerobic respiration. In such cases, the hydrogen released during glycolysis is simply transferred from the hydrogen acceptor molecules back to the pyruvic acid after it has been formed, creating ethyl alcohol in some organisms, and lactic acid or similar substances in others. Some energy is released during either fermentation or true anaerobic respiration, but most of it remains locked up in the alcohol, lactic acid, or other compounds produced.

In *true anaerobic respiration*, the hydrogen removed from the glucose molecule during glycolysis is typically combined with an inorganic ion, as, for example, when sulfur bacteria (discussed in Chapter 17) convert sulfate (SO_4) to sulfur (S) or another sulfur compound, or when certain cellulose bacteria produce methane gas (CH_4) by combining the hydrogen with carbon dioxide.

Oxygen gas is not required to make these compounds, but few organisms can live long without oxygen, and many that carry on fermentation can also respire aerobically.

If oxygen becomes available, the remaining energy can be released by further breakdown of these compounds. About 7% of the total energy in a glucose molecule is removed during anaerobic respiration or fermentation. So much of that energy goes into the making of the alcohol or the lactic acid or is dissipated as heat that there is a net gain of only two ATP molecules (compared with 36 ATP molecules produced in aerobic respiration). The forms of anaerobic respiration are adaptive to the organisms that have them in that they recycle NAD and allow glycolysis to continue.

Living cells can tolerate only certain concentrations of alcohol. In media in which yeasts are fermenting sugars, for example, once the alcohol concentration builds up beyond 12%, the cells die and fermentation ceases. This is why most wines have an alcohol concentration of about 12% (24 proof).

Many bacteria carry on both fermentation and true anaerobic respiration simultaneously, making it difficult to distinguish between the two processes. Some texts use the terms *anaerobic respiration* and *fermentation* interchangeably to designate respiration occurring in the presence of little or no oxygen gas.

3. Factors Affecting the Rate of Respiration

Temperature

Temperature plays a major role in the rate at which the respiratory reactions occur. For example, when air temperatures rise from 20°C (68°F) to 30°C (86°F), the respiration rates of plants double and sometimes even triple. The faster respiration occurs, the faster the energy is released from sugar molecules, with an accompanying decrease in weight. In growing plants, this weight loss is more than offset by the production of new sugar by photosynthesis. In harvested fruits, seeds, and vegetables, however, respiration continues without sugar replacement, and some water loss occurs. Respiring cells convert energy stored as starch or sugar primarily to ATP, but much of the energy is lost in the form of heat, with only 39% being stored as ATP. Most fresh foods are kept under refrigeration, not only to lower the respiration rate and retard water loss but also to dissipate the heat. Keeping the temperatures down is also important to prevent the growth and reproduction of food-spoiling molds and bacteria, which may thrive at warmer temperatures.

Heat inactivates most enzymes at temperatures above 40°C (104°F), but a few organisms, such as cyanobacteria and algae in the hot springs of Yellowstone National Park and similar places, have adapted in such a way that they are able to thrive at temperatures exceeding 60°C (140°F)—heat that would kill other organisms of comparable size almost instantly.

Water

Water inside cells and their organelles acts as a medium in which the enzymatic reactions can take place. Living cells often have a water content of more than 90%, but

Table 10.1 Summary of ATP Synthesized from One Glucose Molecule during Respiration

METABOLIC PATHWAY (Aerobic Respiration)	CELLULAR LOCATION			ATP Produced (Substrate-Level and Oxidative Phosphorylation)
	Cytoplasm	Mitochondrial Matrix	Inner Mitochondrial Membrane	
Glycolysis	2 ATP			2 ATP
	2 NADH		2 (2 ATP)	4 ATP
Conversion of pyruvate to acetyl CoA / electron transport and oxidative phosphorylation		2 (1 NADH)	2 (3 ATP)	6 ATP
Citric acid (Krebs) cycle / electron transport and oxidative phosphorylation		2 (1 ATP)		2 ATP
		2 (3 NADH)	2 (9 ATP)	18 ATP
		2 (1 FADH$_2$)	2 (2 ATP)	4 ATP
Total				36 ATP

Table 10.2 Summary Comparison of Photosynthesis and Respiration

Photosynthesis	Respiration
1. Stores energy in sugar molecules	1. Releases energy from sugar molecules
2. Uses carbon dioxide and water	2. Releases carbon dioxide and water
3. Increases weight	3. Decreases weight
4. Occurs only in light	4. Occurs in either light or darkness
5. Occurs only in cells containing chlorophyll	5. Occurs in all living cells
6. Produces oxygen in green organisms	6. Utilizes oxygen (aerobic respiration)
7. Produces ATP with light energy	7. Produces ATP with energy released from sugar

the plant with unique colors, aromas, poisons, and other compounds that may attract or deter other organisms or give them a competitive edge in nature. Humans have exploited many secondary compounds from plants for medicinal, culinary, or other purposes. It has been estimated that 50,000 to 100,000 such compounds exist in plants, with only a few thousand of these thus far having been identified. Secondary metabolic products may be derived from modification of amino acids and related compounds to produce alkaloids, or through specialized conversions such as the *shikimic acid pathway* (phenolics) and *mevalonic acid pathway* (terpenoids). Examples of these compounds are shown in Table 10.3. Lignin, which is a component of secondary cell walls, is, for example, synthesized through the shikimic acid pathway. Because it is hard to digest and is toxic to some predators, it protects plants from herbivorous animals.

10.6 Assimilation and Digestion

Learning Outcome

a. Provide definitions to distinguish between assimilation and digestion.

Sugars produced through photosynthesis may undergo many transformations. Some sugars are used directly in respiration, but others not needed for that purpose may be transformed into lipids, proteins, or other carbohydrates. Among the most important carbohydrates produced from simple sugars are sucrose, starch, and cellulose. Much of the organic matter produced through photosynthesis is eventually used to build protoplasm and cell walls. This conversion process is called **assimilation.**

Table 10.3 Examples of Plant Secondary Compounds

Compound	Example Source	Human Use
Alkaloids		
Codeine	Opium poppy	Narcotic pain reliever; cough suppressant
Nicotine	Tobacco	Narcotic; stimulant
Quinine	Quinine tree	Used to treat malaria
Phenolics		
Lignin	Woody plants	Hardwood furniture and baseball bats
Salicin	Willow tree	Aspirin precursor
Tetrahydrocannabinol	Marijuana	Glaucoma treatment, nausea suppressant, and relaxant
Terpenoids		
Camphor	Camphor tree	Component of medicinal oils and disinfectants
Menthol	Mints and eucalyptus tree	Strong aroma; used in cough medicines
Rubber	Rubber tree	Rubber tires, rubber bands, and other commercial products

When photosynthesis is taking place, sugar may be produced faster than it can be used or transported away to other parts of the plant. When this happens, the excess sugar may be converted to large, insoluble molecules, such as starch or oils, temporarily stored in the chloroplasts, and then later changed back to a soluble form that is transported to other cells. The conversion of starch and other insoluble carbohydrates to soluble forms is called **digestion** (Fig. 10.15). The process is nearly always one of *hydrolysis,* in which water is taken up and, with the aid of enzymes, the links of the chains of simple sugars that make up the molecules of starch and similar carbohydrates are broken by the addition of water. The disaccharide malt sugar (maltose), for example, is transformed to two molecules of glucose, with the aid of an enzyme (maltase), by the addition of one molecule of water, as follows:

$$\underset{\text{maltose}}{C_{12}H_{22}O_{11}} + \underset{\text{water}}{H_2O} \xrightarrow{\underset{\text{(enzyme)}}{\text{maltase}}} \underset{\text{glucose}}{2C_6H_{12}O_6}$$

Fats are broken down to their component fatty acids and glycerol, and proteins are digested to their amino acid building blocks in similar fashion. Digestion is carried on in any cell where there may be stored food, with very little energy being released in the process. In animals, special digestive organs also play a role in digestion, but plants have no such additional "help" in the process. In both plants and animals, digestion within cells is similar and is a normal part of metabolism.

Except in insect-trapping plants, digestion takes place within plant cells where the carbohydrates, fats, or proteins are stored, while in animals, digestion usually occurs outside of the cells in the digestive tract. Apart from the location, the process is essentially similar in plants and animals.

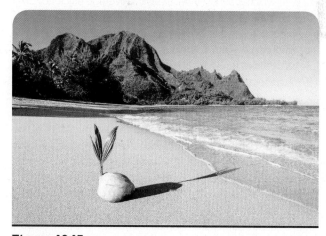

Figure 10.15 Germination requires an intricate series of events, including digestion of stored starches and oils. Depicted here is a germinating coconut (*Cocos nucifera*) at the water's edge on Tunnels Beach (Makua Beach) on the north shore of Kauai, Hawaii. © *M Swiet Productions/Getty Images*

KEY THEME 10.2 ecology

Photosynthesis, Global Warming, and Tropical Rain Forests

In much the same way that the windows of your car trap heat during the summer, carbon dioxide in our atmosphere traps radiation from the sun, which helps keep our planet warm. Without CO_2 and other greenhouse gases, the temperature on earth would be about 33°C (60°F) cooler than it is now. Conversely, increasing the concentration of these gases in our atmosphere can cause escalating global temperatures. For the past 1,000 years, a careful balance maintained by the oceans and photosynthetic organisms has kept atmospheric CO_2 levels fairly constant at 0.028%. Since the 1850s, fossil fuel combustion and deforestation have contributed to increasing atmospheric CO_2 to a current level of 0.040%. Increases in greenhouse gases such as CO_2 have caused a global warming of 0.6°C (1.1°F) since 1900, and predictions suggest an overall global warming of 1.0 to 3.5°C (1.8 to 6.3°F) (or even more, according to recent estimates) by the end of the 21st century.

Billions of metric tons of carbon are converted to organic matter through carbon fixation by continental and marine ecosystems each year. Some of this carbon balance is regulated by the ability of oceans and seas to equilibrate CO_2, and most scientists agree that a large amount of the CO_2 is absorbed by terrestrial ecosystems. Among these ecosystems, tropical rain forests (Box Figure 10.2A) are the most productive, with photosynthetic efficiencies that average 2.2 kilograms (4.8 pounds) of dry organic matter produced per square meter per year, which is almost three times the average efficiency of all other continental ecosystems. Although this superior efficiency may provide a substantial contribution to global carbon fixation, the amount of earth's landmass covered with tropical rain forests declined from 30% in the 1950s to about 5% in 2010.

In general, as you move toward the tropics, there is an increase in the diversity of all organisms. This includes plants. Even though tropical rain forests currently occupy only 5% of the world's land surface, they contain over half of the world's plant and animal species. For example, Costa Rica is about one-fourth the size of Great Britain, but it has almost six times the number of plant species. Experts estimate that deforestation contributes to the extinction of over 100 plant, animal, and insect species each day. For many reasons it is unfortunate to suffer such a rapid loss of life. For example, 25% of Western drugs are derived from rain-forest ingredients, and some may contain possible cures for life-threatening diseases.

While the photosynthetic capacity of tropical rain forests may increase as a consequence of high CO_2 levels in the atmosphere, this does not necessarily mean that these ecosystems are a panacea to the looming threat of global warming. Between 10 and 30 million hectares (25 and 74 million acres) of rain forest are lost every year to ranching, logging, mining, and otherwise developing areas for human needs. For each hectare (2.5 acres) of forest that is cleared and burned, about 220 metric tons (242 tons) of carbon are released, which directly increases the amount of CO_2 in the atmosphere. Each year, deforestation in tropical rain forests accounts for 20% to 30% of all carbon dioxide introduced into the atmosphere. The consequences are magnified because the hectares of tropical rain forest that are lost to deforestation no longer remain to absorb excess CO_2.

Some evidence suggests that, as the CO_2 content in the air rises, tropical rain forests will likely increase photosynthetic capacity, thereby removing excess carbon from the atmosphere. However, increasing temperatures that accompany the accumulation of greenhouses gases may, at some point, hinder the ability of plants to increase photosynthetic efficiency. This is because plant respiration, which releases CO_2, increases exponentially with increasing temperature, whereas photosynthetic rates generally increase to a temperature optimum and then decline.

The results of a study conducted from 1984 through 2000 in La Selva, Costa Rica, have shed new light on possible scenarios for changes in atmospheric CO_2 concentrations during global warming. In this 16-year study, the investigators measured annual diameter increments of tree girth along with daily temperatures and atmospheric CO_2 levels. Even though the tree species studied were adapted to tropical rain-forest ecosystems, the data collected demonstrated relatively lower forest productivity at higher temperatures. These findings suggest that, as temperatures rise, increased plant respiration in tropical rain forests may add to ongoing atmospheric CO_2 accumulation and accelerate global warming. These studies provide a compelling argument to do what we can now to slow the process of global warming before it gets out of hand.

Some countries are doing what they can to combat the problem of deforestation. In the mid-1970s, Costa Rica established a system of national parks and reserves to protect 12% of the country's land area from degradation. The current Costa Rican government continues to promote conservation programs through the preservation of rain forests and has a goal of expanding protected areas to 25% in the near future. Similar efforts may, indeed, slow the ever-increasing threat of global warming.

Box Figure 10.2A Lush vegetation in a Costa Rican rain forest. *Courtesy of Joe Vogt*

SUMMARY

10.1 Introduction to Plant Metabolism
- Metabolism is defined as the sum of all interrelated biochemical processes that take place in a living organism.

10.2 Enzymes and Energy Transfer
- Enzymes catalyze reactions of metabolism. Many of these include oxidation–reduction reactions. Oxidation is a loss of electrons; reduction is a gain of electrons.

10.3 Photosynthesis
- Photosynthesis is an anabolic process that combines carbon dioxide and water in the presence of light with the aid of chlorophyll; oxygen is a by-product. All life depends on photosynthesis, which takes place in chloroplasts.
- Carbon dioxide constitutes 0.040% of the atmosphere, but the percentage has been rising in recent years. Increased carbon dioxide levels have the potential to elevate global temperatures through the "greenhouse effect."
- Chlorophyll b and carotenoids are antenna pigments that direct light energy to chlorophyll a. Photosynthetic units containing chlorophylls and accessory pigments absorb units of light energy, become excited, and pass this energy to acceptors during the light-dependent reactions of photosynthesis.
- During the light-dependent reactions of photosynthesis, which occur in thylakoid membranes of chloroplasts, water molecules are split, and oxygen gas is released. Hydrogen ions and electrons are released from water and transferred to produce NADPH and ATP.
- The two types of photosynthetic units present in most chloroplasts are photosystems I and II. The events that take place in photosystem II come before those of photosystem I. Each photosystem has a reaction-center molecule of chlorophyll a that boosts electrons to a higher energy level when the molecule is excited by light energy.
- Photosystem II boosts electron excitation to a level that, when it encounters photosystem I, has the potential to reduce NADP to NADPH through noncyclic electron flow. Photosystem I, by itself, can cycle electrons for the generation of ATP. ATP production involves both electron transport while the photosystems are operating and proton movement across thylakoid membranes.
- The light-independent reactions occur through a series of reactions known as the Calvin cycle, which takes place in the stroma of chloroplasts. In the first step, carbon dioxide is combined with RuBP through catalytic action of the enzyme rubisco to form two molecules of the 3-carbon compound GA3P. The ATP and NADPH from the light-dependent reactions furnish energy to eventually convert GA3P to 6-carbon carbohydrates. This cycle also regenerates RuBP to enable continued carbon fixation.
- In the light-independent reactions of C4 plants, 4-carbon oxaloacetic acid is initially produced instead of 3-carbon PGA. In the leaf mesophyll of C4 plants, there are large chloroplasts, which contain rubisco in the bundle sheaths, and small chloroplasts, which contain higher concentrations of PEP carboxylase that facilitate the conversion of carbon dioxide to carbohydrate at much lower concentrations than is possible in C3 plants.
- CAM photosynthesis occurs in succulent plants whose stomata are closed and admit little CO_2 during the day. Regular photosynthesis occurs as the 4-carbon compounds that accumulate at night are converted back to carbon dioxide during the day.

10.4 Respiration
- Respiration is a catabolic process that takes place in the cytoplasm and mitochondria of cells. The energy is released, with the aid of enzymes, from simple sugar and organic acid molecules.
- In aerobic respiration, stored energy release requires oxygen; CO_2 and water are by-products of the process.
- Anaerobic respiration and fermentation do not require oxygen gas, and much less energy is released. The remaining energy is in the ethyl alcohol, lactic acid, or other such substances produced. Some released energy is stored in ATP molecules. Temperature, available water, and environmental oxygen affect respiration rates.
- Glycolysis, which occurs in the cytoplasm, requires no molecular oxygen; two phosphates are added to a 6-carbon sugar molecule, and the prepared molecule is split into two 3-carbon sugars (GA3P). Some hydrogen, energy, and water are removed from the GA3P, producing pyruvic acid. There is a net gain of two ATP molecules. Hydrogen ions and electrons released during glycolysis are picked up by NAD, which becomes NADH.
- In aerobic respiration, which occurs in the mitochondria, pyruvic acid loses some CO_2, is restructured, and becomes acetyl CoA. Energy, CO_2, and hydrogen are removed from the acetyl CoA in the citric acid cycle, which involves enzyme-catalyzed reactions of a series of organic acids.
- NADH passes the hydrogen gained during glycolysis and the citric acid cycle along an electron transport system; small increments of energy are released and partially stored in ATP molecules, and the hydrogen is combined with oxygen gas, forming water in the final step of aerobic respiration.
- Hydrogen removed from glucose during glycolysis is typically combined with an inorganic ion in anaerobic respiration. The hydrogen is combined with the pyruvic acid or one of its derivatives in fermentation. Both processes occur in the absence of oxygen gas, with only about 7% of the total energy in the glucose molecule being released, for a net gain of two ATP molecules.

- Two molecules of NADH and two ATP molecules are gained during glycolysis when two 3-carbon pyruvic acid molecules are produced from a single glucose molecule. Another molecule of NADH is produced when the pyruvic acid molecule is restructured and becomes acetyl CoA prior to entry into the citric acid cycle.
- In the citric acid cycle, acetyl CoA combines with 4-carbon oxaloacetic acid, producing first a 6-carbon compound, next a 5-carbon compound, and then several 4-carbon compounds. The last 4-carbon compound is oxaloacetic acid. Two CO_2 molecules are also released during this process.
- Some hydrogen removed during the citric acid cycle is picked up by FAD and NAD; one molecule of ATP, three molecules of NADH, and one molecule of $FADH_2$ are produced during one complete cycle. Energy associated with electrons and/or with hydrogen picked up by NAD and FAD is gradually released as the electrons are passed along the electron transport system; some of this energy is transferred to ATP molecules during oxidative phosphorylation.
- Energy used in ATP synthesis during oxidative phosphorylation is believed to be derived from a gradient of protons formed across the inner membrane of a mitochondrion, while electrons are moving in the electron transport system by chemiosmosis.
- Altogether, 38 ATP molecules are produced during the complete aerobic respiration of one glucose molecule; two are used to prime the process, so there is a net gain of 36 ATP molecules.

10.5 Additional Metabolic Pathways

- In addition to photosynthesis and respiration, other metabolic pathways are required for growth, development, reproduction, and survival. Essential products of additional pathways include nucleotides, proteins, chlorophylls, and fatty acids. Secondary metabolites include alkaloids, phenolics, and terpenoids.

10.6 Assimilation and Digestion

- Conversion of sugar produced by photosynthesis to fats, proteins, complex carbohydrates, and other substances is termed assimilation. Digestion takes place within plant cells with the aid of enzymes. During digestion, large, insoluble molecules are broken down by hydrolysis to smaller, soluble forms that can be transported to other parts of the plant.

REVIEW QUESTIONS

1. Why are enzymes important, and how do they contribute toward metabolism?
2. What happens during the light-dependent and light-independent reactions of photosynthesis?
3. What roles do water, light, carbon dioxide, and chlorophyll play in photosynthesis?
4. Which gases influence rubisco activity, and how does this activity affect photosynthesis?
5. Explain how products of photosynthesis relate to those used by respiration.
6. How does glycolysis contribute toward aerobic respiration, anaerobic respiration, and fermentation?
7. What do cells get out of aerobic respiration?
8. How do temperature, water, and oxygen affect respiration?
9. How is chemiosmosis in photosynthesis similar to oxidative phosphorylation in respiration?
10. What are cytochromes, and how do they function in photosynthesis and respiration?
11. What additional metabolic pathways are demonstrated by plants, and why are they necessary?

DISCUSSION QUESTIONS

1. If space exploration revealed a planet that had the water, sunlight, and climate necessary to maintain plant life, would it be possible to produce oxygen and food necessary for human life?
2. How would you genetically engineer a C_3 plant to achieve the high efficiency of C_4 photosynthesis?
3. Can growing more plants and increasing the abundance of other photosynthetic autotrophs regulate the "greenhouse effect"?
4. How does cellular respiration benefit life on earth if it is a catabolic process?
5. Explain why secondary metabolites, originally designed by plants to deter pests or attract pollinators, have commercial value to humans.

CHAPTER

11

Growth and Development

11.1 Introduction to Growth and Development
11.2 Nutrients, Vitamins, and Hormones
    Plant Hormones
    Key Theme 11.1: Plant Hormones beyond "The Classic Five"
    Other Hormones or Related Compounds
11.3 Hormonal Interactions
    Apical Dominance
    Senescence
    Other Hormonal Interactions
11.4 Plant Movements
    Growth Movements
    Turgor Movements
    Taxic Movements
    Miscellaneous Movements
11.5 Photoperiodism
    Phytochromes and Cryptochromes
11.6 Temperature and Growth
11.7 Dormancy and Quiescence

OVERVIEW

This chapter introduces growth phenomena by providing distinctions among growth, differentiation, and development. This is followed by a discussion of plant hormones (auxins, gibberellins, cytokinins, abscisic acid, ethylene) and their roles in plant growth and development. The chapter explores plant movements, including nutations and twining, contraction, nastic, tropic, turgor, and solar tracking movements. The discovery and functions of photoperiodism and phytochrome are briefly surveyed. The chapter concludes with a discussion of the relationship of temperature to growth and dormancy.

Plants respond to many stimuli, including gravity, which allows roots to grow downward and shoots to grow upward. © *Design Pics*

11.1 Introduction to Growth and Development

Learning Outcomes

a. Distinguish between determinate and indeterminate growth.
b. Describe the components of development.

If you have ever grown squash in a garden, you may have noticed that one day the fruits are small and then only a week or so later they are huge. That may have left you wondering how they grew so big so fast. The words *grow, growing,* and *growth* are used in several ways. If, for example, you see a rubber balloon being inflated, you may say it is growing in size. Or if there is a leak in the roof, you might say that the puddle it forms is growing bigger. In the biological realm, however, growth is always associated with cells. It may be defined simply as an *irreversible increase in mass due to the division and enlargement of cells,* and may be applied to an organism as a whole or to any of its parts.

Plants go through a predictable sequence of growth stages. Plants or plant parts may exhibit determinate growth, in which the meristem stops producing new cells at a certain developmental stage, producing a structure of predictable size. In contrast, meristems responsible for indeterminate growth produce parts that continue to grow until the plant is killed by an environmental factor, such as frost. Products of indeterminate meristems, then, are variable in size.

All living organisms begin as a single cell and increase in mass through cell division and enlargement until a body consisting of up to billions of cells is formed. After cells have enlarged, **differentiation** occurs; that is, the cells develop different forms adapted to specific functions, such as conduction, support, or secretion. If differentiation did not occur, the result would be a shapeless blob with no distinct tissues and little, if any, coordination. The coordination of growth and differentiation of a single cell into multicellular tissues and organs is called **development.**

What regulates development? What determines where and when growth and differentiation occur? The answers to these questions lie in the organism's genes and its environment. In the 1950s, F. C. Steward and his colleagues at Cornell University illustrated this concept with a tissue culture experiment. They isolated cells from a carrot root and placed them in jars containing nutrients and growth hormones. The cells divided, grew, and developed into embryo-like structures. The embryos eventually grew into new carrot plants, demonstrating that every cell has all the information necessary to create an entire plant. This exciting discovery laid the foundation for modern cell biology. Genes, the basic units of heredity, control the synthesis and development of *enzymes,* which are proteins that catalyze every metabolic step within cells. Certain genes, called housekeeping genes, are programmed to synthesize specific enzymes continually in every living cell, while other genes produce enzymes only in specific tissues or at specific times during a plant's life cycle. The environment determines which genes will be expressed. In an organ or an organism, the environment includes both external and internal factors. The internal environment includes nutrients, vitamins, and hormones. The external environment includes water, minerals, gases, light, and temperature, among many factors.

11.2 Nutrients, Vitamins, and Hormones

Learning Outcomes

a. Distinguish between vitamins and hormones.
b. Describe the effects of the five major plant hormones on growth.
c. Explain why oligosaccharins and brassinosteroids are considered to be plant hormones.

Nutrients furnish the elements and energy for plant growth and maintenance. A lack of nutrients restricts the normal development of plants. Plants typically obtain all of their nutrients from the air and soil.

Vitamins play an important role in reactions catalyzed by enzymes. Most vitamins are *coenzymes* or parts of coenzymes, which are organic molecules that participate in catalyzed reactions, mostly by functioning as electron acceptors or donors. Vitamins are required in relatively small amounts for the normal growth and development of all organisms. Most vitamins essential to all living organisms are produced by plants, although a few are also synthesized by animals. Vitamin A, for example, does not occur in plants. Carotene pigments in chloroplasts act as *precursors* for vitamin A in animals.

Hormones are produced mostly in actively growing regions of plants and are active in far smaller amounts than vitamins and enzymes. Hormones are transported from their point of origin to another part of the plant where they have specific effects, such as causing stems to bend, initiating flowering, or even inhibiting growth.

Every plant growth response results from a series of biochemical steps, called a **signal transduction pathway** (Fig. 11.1). Commonly, a signaling molecule induced by an environmental cue binds to a receptor in the plasma membrane. The interaction of the signaling molecule with the receptor causes the release of a messenger molecule in the cytoplasm. This results in a cascade of events that transfers the information from one molecule to another. Ultimately, an enzyme is produced or activated, catalyzing a reaction in the cell. This results in a detectable response to the environmental cue.

Plant Hormones

Only five classes of plant hormones regulate most plant development. Each type of hormone can have multiple effects. Hormones act by binding to receptors, which

Figure 11.1 The three main stages of cell signaling are illustrated here. On the left, the receptor (tan) embedded in the plasma membrane is activated when it receives a signal (green). Then, the information is carried through a series of molecules (red, purple, and yellow) resulting in a cellular response. *Source: © The McGraw-Hill Companies, Inc.*

triggers a signal transduction pathway. The outcome may include changes in the types of enzymes produced in a tissue or changes in transport across membranes. Hormones may promote or inhibit growth. The concentration of a hormone and its interaction with other hormones may influence the effect it has on plant growth.

The major known types of plant hormones are auxins, gibberellins, cytokinins, abscisic acid, and ethylene.

Ethylene is a simple two-carbon molecule, while the other major plant hormones are ring structures, as shown in Figure 11.2.

Auxins

In 1881, Charles Darwin and his son Francis were fascinated by the fact that *coleoptiles* (tubular sheaths protecting the shoots of grass seedlings) bend toward a light source.

Figure 11.2 Chemical structures of the five major classes of plant hormones.

They noticed that the bending did not occur if they placed little metal foil caps over the coleoptile tips, but the bending resumed when the caps were removed. They concluded that the coleoptile tips must be sensitive to light. Charles Darwin died in 1882, but others following up on his work demonstrated that a water-soluble "influence" was produced in the coleoptile tips.

In 1926, Frits Went, a young plant physiologist in Holland, cut off the tips of oat coleoptiles and placed the cut ends on a gelatin-like support medium called **agar** (a substance obtained from marine algae). After a few hours, he removed the tips and cut the agar into little square blocks, which he then placed on decapitated coleoptiles. He found that if he placed the block squarely over the cut top of a coleoptile, it grew straight up, but if he placed the block off center, the tip bent away from the side on which the block had been placed (Fig. 11.3). He also found that the more coleoptile tips he placed on the agar to begin with, the more pronounced was the response. His experiments demonstrated that something produced in the coleoptile tips moved out into the agar and that this substance caused the coleoptiles to bend. Went named the substance **auxin** (from the Greek word *auxein*, meaning "to increase"), a term now in widespread use.

Auxin molecules are similar in structure to the amino acid tryptophan, which is found in both plant and animal cells. Auxin is typically synthesized from tryptophan in plants.

Auxin production occurs mainly in apical meristems, buds, young leaves, and other active young parts of plants. It is sometimes difficult to predict how cells will respond to auxins because responses vary according to concentration, location, and other factors. For example, if an auxin of a specific concentration promotes shoot growth in a certain plant, the same auxin of identical concentration will usually inhibit its root growth. At appropriate concentrations, auxins normally stimulate the enlargement of cells by increasing the plasticity (irreversible stretching) of cell walls.

Hormone concentrations were first measured using *bioassays*, which evaluate the effect of a substance on living tissues. For example, a fixed number of shoot tips can be placed on agar for a certain time period, and a block of the agar can then be placed off center on a decapitated coleoptile. The concentration of agar in that block can be determined by the angle of bending of the coleoptile that occurs when compared with the bending brought about by a similar block containing a known amount of auxin (Fig. 11.4).

Hormone concentrations today are more frequently determined by vaporizing a sample and moving it through a tube of liquid or powdered material by a technique known as *gas chromatography*. As the sample moves through the column, the hormone and other components of the sample separate, and the amount of hormone present then can be determined relatively precisely.

Auxins also may have many other effects, including triggering the production of other hormones or growth

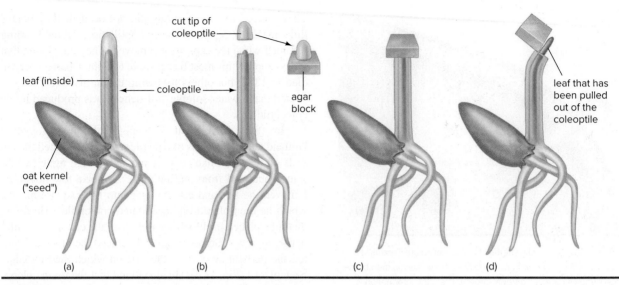

Figure 11.3 Went's experiment with oat coleoptiles focused on germinated oat "seeds" with intact coleoptiles, as shown in (a). In (b), the tip of a coleoptile was cut off, placed on a small agar block, and left for an hour. When this agar block was placed on a decapitated coleoptile, growth was vertical, as shown in (c). When the agar block was placed off center so that only half of the decapitated coleoptile was in contact with it, the tip bent away from it (d). This experiment demonstrated that something affecting growth diffused from the coleoptile tip into the agar and from the agar to the part of the decapitated coleoptile touching the block.

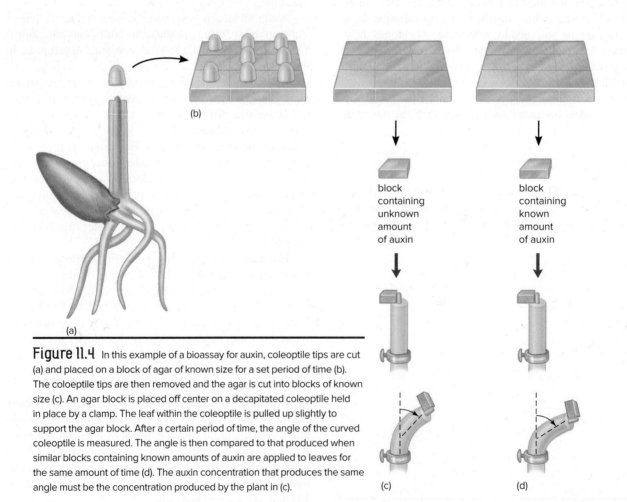

Figure 11.4 In this example of a bioassay for auxin, coleoptile tips are cut (a) and placed on a block of agar of known size for a set period of time (b). The coleoptile tips are then removed and the agar is cut into blocks of known size (c). An agar block is placed off center on a decapitated coleoptile held in place by a clamp. The leaf within the coleoptile is pulled up slightly to support the agar block. After a certain period of time, the angle of the curved coleoptile is measured. The angle is then compared to that produced when similar blocks containing known amounts of auxin are applied to leaves for the same amount of time (d). The auxin concentration that produces the same angle must be the concentration produced by the plant in (c).

regulators (especially ethylene), causing dictyosomes to increase rates of secretion, controlling some phases of respiration, and influencing many developmental aspects of growth. Auxins promote cell enlargement and stem growth, cell division in the cambium, initiation of roots, and differentiation of cell types. Auxins delay developmental processes such as fruit and leaf abscission, as well as fruit ripening. They also inhibit lateral branching. Compared to dicots, monocots are less sensitive to auxin. Higher concentrations, however, will kill almost any living plant tissues.

Movement of auxins from the cells where they are produced requires the expenditure of energy from ATP molecules. Auxin movement is relatively slow (about 1 centimeter per hour), although an active transport "pumping" mechanism carries the hormone up to 10 times faster than it would travel by simple diffusion alone. The movement is *polar,* which means the auxins move away from their source, usually from a stem tip toward the roots. This polar movement occurs even when a stem is inverted. Auxin can move through a plant in the phloem or through diffusion from cell to cell.

In the past, it was believed that several auxins occurred in nature, but until relatively recently, most scientists thought that *indoleacetic acid* (*IAA*) was the only active auxin. Current research, however, indicates that plants do actually produce three other growth regulators that are not converted to IAA but bring about many of the same responses as IAA. These compounds, called endogenous auxins, are *phenylacetic acid* (*PAA*), *4-chloroindoleacetic acid* (*4-chloroIAA*), and *indolebutyric acid* (*IBA*). The roles of these additional auxins are still being determined. One possibility is that they contribute to the pool of auxins that a plant must maintain to be able to respond quickly to environmental cues.

Auxins are used widely in agriculture and horticulture. At proper concentrations, auxin can stimulate the formation of roots on almost any plant organ. In plant nurseries, these substances are applied in powdered, paste, or liquid form to the bases of cut segments of stems to stimulate root growth (Fig. 11.5). Synthetic auxins presently in use include NAA (naphthalene acetic acid), 2,4-D (2,4-dichlorophenoxyacetic acid), and MCPA (2-methyl-4-chlorophenoxyacetic acid).

Orchardists spray fruit trees with auxins to promote uniform flowering and fruit set, and they later spray the fruit to prevent the formation of abscission layers and subsequent premature fruit drop. A substantial saving in labor expenses results from being able to go through an orchard or a pineapple plantation and pick all of the fruit at one time. If auxins are applied to flowers before pollination occurs, seedless fruit can be formed and developed. Some orchardists use auxins for controlling the number of fruits that will mature in order not to have too many small ones, and some even control the shapes of plants for sales appeal.

Weeds were controlled by hand labor and by the use of caustic or otherwise poisonous substances until it was discovered that synthetic auxins, including ichlorophenoxyacetic acid (2,4-D) and, richolorphenoxyacetic acid (2,4,5-T), when sprayed at low concentrations, kill some weeds. Broad-leaved

Figure 11.5 Four weeks after treating the *Gardenia* cutting on the left with auxin, more roots and healthier foliage were observed, compared with the untreated cutting on the right. © *Kingsley Stern*

plants such as dandelions, plantains, and dicots are more susceptible to low concentrations of auxins than narrow-leaved plants such as grasses. They have been used on lawns to kill weeds without noticeably affecting the grass.

2,4,5-T was banned in 1979 for most uses in the United States by the Environmental Protection Agency. The controversy over the use of 2,4,5-T began in the 1960s during the Vietnam War, when Agent Orange, a 1-to-1 mixture of 2,4-D and 2,4,5-T, was used to defoliate jungles. Subsequent tests and experiments with 2,4,5-T have produced leukemia, miscarriages, birth defects, and liver and lung diseases in laboratory animals. The diseases and defects are apparently caused by TCDD (2,3,7,8-tetrachlorodibenzoparadioxin), a dioxin contaminant that evidently is unavoidably produced in minute amounts during the manufacture of 2,4,5-T. TCDD harms or even kills laboratory animals in doses as low as a few parts per billion.

Gibberellins

In 1926, Eiichi Kurosawa of Japan reported the discovery of a new substance that caused "foolish seedling" disease in rice. The stems of rice seedlings infected with a certain fungus grew twice as long as those of uninfected plants, but the stems were weakened so that they eventually collapsed and died. Kurosawa found that extracts of the fungus applied to uninfected plants brought about the same growth stimulations as the fungal disease. Nine years later, other Japanese scientists were able to crystallize the substance, which was named **gibberellin,** after the scientific name of the fungus that produced it (*Gibberella fujikuroi*). It was not until the 1950s, however, that the true chemical structure of gibberellin was elucidated by a British group headed by Brian Cross and an American group led by Frank Stodola.

There are well over 100 gibberellins, customarily abbreviated as **GA**. Each form of GA is identified by a subscript (e.g., GA_6). They have been isolated from immature seeds (especially those of dicots), root and shoot tips, young leaves, and fungi. Most forms of GA produced by plants are inactive, apparently functioning as precursors to active forms. GA moves through xylem and phloem and, unlike that of auxin, the movement is not polar.

Acetyl coenzyme A, which is vital to respiration, functions as a precursor to GA. Interest in the hormonal properties of GA is high, for their ability to increase growth in plants is far more impressive than that of auxin alone, although traces of auxin apparently need to be present for GA to produce its maximum effects.

Most dicots and a few monocots grow faster with an application of GA, but coniferous trees such as pines and firs generally show little, if any, response. If GA, at the appropriate concentration, is applied to a cabbage plant, it may become 2 meters (6 feet) tall, and bush bean plants can become pole beans with a single application (Fig. 11.6). Many genetically dwarf plants grow as tall as their normal counterparts after an application of the appropriate GA. Interestingly, GA stimulates foliage height without significant changes in root growth.

Gibberellins not only dramatically increase stem growth, but they are also involved in nearly all of the same regulatory processes in plant development as auxins. In some kinds of plants, flowering can be brought about by applications of GA, and the dormancy of buds and seeds can be broken. Some GA appears to lower the threshold of growth; that is, plants may start growing at lower temperatures than usual after an application of GA. For example, an application to a lawn could cause it to turn green 2 or 3 weeks earlier in the spring.

In some varieties of lettuce and cereals, the seeds normally germinate only after being subjected to a period of cold temperatures. An application of GA can eliminate the cold requirement for germination. Application of GA often hastens the conversion of juvenile foliage to adult foliage. Juvenile plants are incapable of producing flowers, while adult plants are able to produce flowers if provided with appropriate conditions.

Figure 11.6 Effect of gibberellins on flowering and plant height. Cabbage plants on the right were grown in a warm greenhouse but did not grow tall and flower until treated with gibberellins. The short cabbage plants on the left were not treated with gibberellins.
© Omikron/Science Source

Several *growth retardants* available commercially inhibit or block the synthesis of GA. Applications of growth retardants result in stunted plants because stem elongation is inhibited. When applied to certain commercially grown crops such as chrysanthemums, flowers with thicker, stronger stalks are produced.

The relatively high cost of GA has limited the extent of its use in horticulture and agriculture. However, GA has been used experimentally to increase yields of sugarcane and hops, and it has revolutionized the production of seedless grapes by increasing the size of the fruit and lengthening fruit internodes, which results in slightly wider spaces between grapes in the bunches. Better air circulation between grapes reduces their susceptibility to fungal diseases, and the need for hand thinning is eliminated. GA has been used in navel orange orchards to delay the aging of the fruit's skin and in celery to increase the length and crispness of petioles (the edible part). GA is now used to increase seed production in conifers and to increase the rate of malting in breweries by enhancing starch digestion.

Cytokinins

By the close of the 19th century, botanists suspected that there must be something that regulates cell division in plants. In 1913, Gottlieb Haberlandt of Germany discovered an unidentified chemical in the phloem of various plants. This chemical stimulated cell division and initiated the production of cork cambium. In 1941, Johannes van Overbeek discovered that coconut "milk" (a liquid *endosperm;* endosperm is discussed in Chapter 23) contained something that increased the growth rate of tissues and embryos, and within a few years, several substances that accelerated cell division were isolated from coconut "milk." In 1955, a substance named *kinetin* was found to stimulate the proliferation of parenchyma cells in tobacco pith when auxin was also present. Kinetin was isolated from heat-treated herring sperm and is not a naturally occurring substance. Its discovery was important, however, because it demonstrated that a simple chemical could promote cell division. In 1964, the identity of a substance very similar to kinetin was determined in kernels of corn. These stimulants to cell division came to be known as **cytokinins.**

The several types of cytokinins differ somewhat in their molecular structure and possibly in origin, but they are similar in composition to adenine. You'll recall that adenine is a building block of one of the four nucleotides found in DNA. Cytokinins are synthesized in root tips and germinating seeds.

If auxin is present during the cell cycle, cytokinins promote cell division by speeding up the progression from the G_2 phase to the mitosis phase, but no such effect takes place in the absence of auxin. Cytokinins also play a role in the enlarging of cells, the differentiation of tissues, the development of chloroplasts, the stimulation of cotyledon growth, and the delay of aging in leaves, as well as in many of the growth phenomena also brought about by auxins and gibberellins. Cytokinins move throughout plants via the xylem, phloem, and parenchyma cells.

As discussed earlier in this chapter, it is possible to induce a few cells to develop into complete new plants. By controlling cytokinin and auxin concentrations in the growth medium, scientists can induce tissue and organ development. Similarly, certain bacteria stimulate the growth of galls (tumors) on plants, either by producing cytokinins and auxins that promote the unorganized growth of tumor cells or by transferring bacterial hormone synthesis genes to the DNA of the host plant. The incorporation of the transferred genes results in host tissue that produces more auxin and cytokinin than normal, which, in turn, results in tumor growth.

Despite their role in cell division and enlargement, however, there is no evidence that cytokinins initiate or promote animal cancers or have any other effect on animal cells. In fact, some natural cytokinins act as anticancer agents by killing human cancer cell lines.

Experiments have shown that cytokinins prolong the life of vegetables in storage. Related synthetic compounds have been used extensively to regulate the height of ornamental shrubs and to keep harvested lettuce and mushrooms fresh. They also have been used to shorten the straw length in wheat, so as to minimize the chances of the plants blowing over in the wind, and to lengthen the life of cut flowers. Many have not yet been approved for general agricultural use.

Abscisic Acid

Although the promotion of growth by auxins and gibberellins had been amply demonstrated by the 1940s, plant physiologists began to suspect that something else produced by plants could have opposite (inhibitory) effects. In 1949, Torsten Hemberg, a Swedish botanist, showed that substances produced in dormant buds (i.e., buds whose development is temporarily arrested) blocked the effects of auxins. He called these growth inhibitors *dormins.*

In 1963, three groups of investigators working independently in the United States, Great Britain, and New Zealand discovered a growth-inhibiting hormone, which was in 1967 officially named **abscisic acid (ABA).** Later, it was shown that ABA and dormins are the same compound.

ABA is synthesized in plastids, apparently from carotenoid pigments. It is found in many plant materials but is particularly common in fleshy fruits, where it evidently prevents seeds from germinating while they are still on the plant. When ABA is applied to seeds outside of the fruit, germination usually is delayed. Because the stimulatory effects of other hormones are inhibited by ABA, cell growth is usually also inhibited. Like the movement of gibberellins and cytokinins, that of ABA throughout plants is nonpolar (nondirectional).

ABA was originally believed to promote the formation of abscission layers in leaves and fruits. However, the evidence suggesting that ethylene (discussed in the following section, "Ethylene") is far more important than ABA in abscission is overwhelming, and, despite its name, ABA has little, if any, influence on the process.

ABA apparently helps leaves respond to excessive water loss. When the leaves wilt, ABA is produced in amounts several times greater than usual. This interferes with the transport or retention of potassium ions in the guard cells, causing the stomata to close. When the uptake of water again becomes sufficient for the leaf's needs, the ABA breaks down, and the stomata reopen. ABA produced in times of drought is transported from the shoot to the root and causes an increase in both root growth and water uptake. Despite the growth-inhibiting effects of ABA, there is no evidence that it is toxic to plants.

Ethylene

In 1901, Dimitry Neljubow, a Russian student at the St. Petersburg Botanical Institute, discovered that pea seedlings grown in the dark in a laboratory showed reduced stem elongation, increased swelling of stems, and abnormal horizontal growth. When the seedlings were placed outside in fresh air, however, they resumed normal growth. It was determined that the simple gas **ethylene,** produced by gas lamps in the laboratory, had produced the abnormal growth.

In 1934, R. Gane discovered that ethylene was produced naturally by fruits, although it had been known for some time that the ripening of green fruits could be accelerated artificially by placing them in ethylene. It is now known that ethylene is produced not only by fruits but also by flowers, seeds, leaves, and even roots. Several fungi and a few bacteria are also known to produce it, and its regulatory effects make it a hormone in the broad sense of the word. Ethylene is produced from the amino acid methionine; oxygen is required for its formation.

The production of ethylene by plant tissues varies considerably under different conditions. A surge of ethylene lasting for several hours becomes evident after various tissues, including those of fruits, are bruised or cut, and applications of auxin can cause an increase in ethylene production of 2 to 10 times. As pea seeds germinate, the seedlings produce a surge of ethylene when they meet interference with their growth through the soil. This apparently causes the stem tip to form a tighter crook, which may aid the seedling in pushing to the surface.

When wind blows across a green field or animals walk across it, plants in the field respond to these and other mechanical stresses with an increase in the production of fibers and collenchyma tissue. Cell elongation is also inhibited, resulting in shorter, sturdier plants. The responses, called *thigmomorphogenesis,* are under the control of genes that are activated by touch. Several substances, including enzymes, ethylene, and a protein called *calmodulin,* are involved. Calmodulin, which may play a role in several plant growth responses, constitutes up to 2% of a plasma membrane and is activated when it binds to calcium. The calmodulin-calcium complex activates enzymes in membranes, which then activate or deactivate other enzymes, and plays a role in several kinds of plant reactions to stimuli, including thigmomorphogenesis.

Ethylene apparently can trigger its own production. If minute amounts are introduced to the tissues that produce the gas, a tremendous response by the tissues often results. This can lead to harmful effects. Flowers, for example, may fade in the presence of excessive amounts, and leaves may abscise (Fig. 11.7).

Figure 11.7 In this illustration of the effect of ethylene on holly twigs, two similar twigs were placed under glass jars for a week; at the same time, a ripe apple was placed under the jar on the right. Ethylene produced by the apple caused abscission of the holly leaves. © *Kingsley Stern*

KEY THEME 11.1 physiology
Plant Hormones Beyond "The Classic Five"

While plant hormones are typically defined as natural compounds secreted by plants to regulate growth, the broader term "plant growth regulator (PGR)" may be applied to many organic substances that, in minute amounts, affect plant processes. The "classic five" plant hormones include auxins, gibberellins, cytokinins, abscisic acid, and ethylene. However, there are other compounds, whether they be naturally secreted or exogenously applied, that can promote, inhibit, or otherwise modify plant processes. A few of the recently discovered PGRs are involved with pest resistance and include jasmonates, nitric oxide, salicylic acid, and systemin, but there are others, such as brassinosteroids that influence cell development and expansion, as well as polyamines that help regulate mitosis and meiosis (Box Table 11.1A). As new methods for isolating, identifying, and characterizing plant compounds are refined, there will surely be more PGRs revealed, demonstrating the remarkable ways in which plants respond to the environment. Visit the website for the Plant Growth Regulation Society of America at http://www.pgrsa.org/ to learn more about scientists and their research in this fascinating field of plant biology.

Box Table 11.1A Plant Growth Regulators and Some of Their Known Actions

Plant Growth Regulator	Plant Processes Affected
The "Classic Five"	
• Auxins	Cell enlargement
• Cytokinins	Cell division
• Gibberellins	Internode elongation
• Abscisic acid	Abscission
• Ethylene	Senescence
Additional Plant Growth Regulators	
• Brassinosteroids	Cell development and expansion
• Jasmonates	Insect and disease resistance
• Nitric oxide	Signaling in defense
• Polyamines	Regulation of mitosis and meiosis
• Salicyclic acid	Systemic acquired resistance
• Systemin	Defense against herbivores

In ancient China, growers ripened fruits in rooms where incense was being burned, and citrus growers ripened their fruits in rooms equipped with a kerosene stove. In houses where gas heating is used, occupants often experience great difficulty in growing houseplants, and greenhouse heaters that use gas create a dilemma for owners attempting to promote plant growth.

In the days before electric street lights became common, gas lights were used, and in some cities, leaves fell from the shade trees if they were located near a gas line that leaked. In these instances, minute amounts of ethylene gas resulting from the fuel combustion or leakage influenced nearby plants. In fact, as little as 1 part of ethylene per 10 million parts of air may be sufficient to trigger responses.

Today, the commercial use of ethylene is extensive. It has been used for many years to ripen harvested green fruits, such as bananas, mangoes, and honeydew melons, and to cause citrus fruits to color up before marketing. Since ethylene production almost ceases in the absence of oxygen, apple and other fruit growers have found that if they place unripe fruit in sealed warehouses after harvest, reduce temperatures to just above freezing, and replace the air with inert nitrogen gas or carbon dioxide, the fruit will remain metabolically inactive for long periods. The growers can then remove the fruit in batches throughout the year, add as little as one part per million of ethylene, and have ripe fruit at any time. This is why apples are always available in grocery stores, even though harvesting is usually confined to a few weeks in the fall.

Fruits that respond to ethylene usually have a major increase in respiration just before ripening occurs. The increase in ethylene production at that time is often up to 100 times greater than it was a day or two earlier. The accompanying major increase in respiration is called a *climacteric,* and fruits that exhibit such phenomena are called *climacteric fruits.* Some fruits, such as grapes, are *nonclimacteric* and do not respond to ethylene.

A few growers still use natural ethylene to ripen pears and peaches by wrapping each fruit individually in tissue paper. The paper retards the escape of ethylene from the fruit and hastens ripening. "Resting" potato tubers may be induced to sprout following brief applications of ethylene, and seeds may be stimulated to germinate if given a short exposure to the gas just before sprouting, although treatment after sprouting inhibits growth. While ethylene production is normal and necessary for fruit ripening, it has been a problem for the production of food in space. On earth, ethylene is dissipated by air movement. In spaceships, though, it accumulates and causes food to ripen too quickly. NASA developed a filtering system to remove excess ethylene from the atmosphere in space shuttles. This technology is being used on earth to extend the shelf life of perishable foods in storerooms and display units.

Ethylene is used in Hawaii to promote flowering in pineapples, and it causes members of the Pumpkin Family (Cucurbitaceae) to produce more female flowers and, thus, more fruit. Many trees in nurseries are grown in containers and are usually crowded. Consequently, they are often tall and spindly, but applications of small amounts of ethylene to the trunks while they are enclosed in plastic tubes causes a marked thickening, making the trees sturdier and less likely to break.

Other Hormones or Related Compounds

A number of compounds called *oligosaccharins*, which are released from cell walls by enzymes, influence cell differentiation, reproduction, and growth in plants and therefore must be considered hormones (Box Table 11.1A). However, oligosaccharins produce their effects at concentrations of up to 1,000 times less than those of auxins, and not only are their effects (e.g., growth promotion, inhibition of flowering) highly specific, but the responses to them are essentially identical in all species. *Brassinosteroids* bind to receptors on cell surfaces, affecting apical dominance, gravitropism, seed germination, root growth, xylem formation, and fruit abscission.

11.3 Hormonal Interactions

Learning Outcomes

a. Define *apical dominance*.
b. Describe factors that lead to plant senescence.
c. Explain how gibberellins and auxins work together to regulate seed germination.

Apical Dominance

For centuries, gardeners and nursery workers have deliberately removed terminal buds of plants to promote bushier growth, knowing that the buds are involved in *apical dominance*. Apical dominance is the suppression of the growth of the lateral buds (also called axillary buds) by the tips of branches. Apical dominance is believed to be brought about by an auxin-like inhibitor in a terminal bud. It is strong in trees with conical shapes and little branching toward the top (e.g., many pines, spruces, firs) (Fig. 11.8) and weak in trees that branch more often (e.g., elms, ashes, willows).

As the tissues produced by the apical meristem of the terminal bud increase in length, the source of the inhibitor moves farther away and the concentration of the inhibitor gradually drops in the vicinity of the lateral buds. We presume that the inhibitory effect eventually falls below a threshold, allowing the lateral buds to grow. Each lateral bud then itself becomes a branch with a terminal bud that

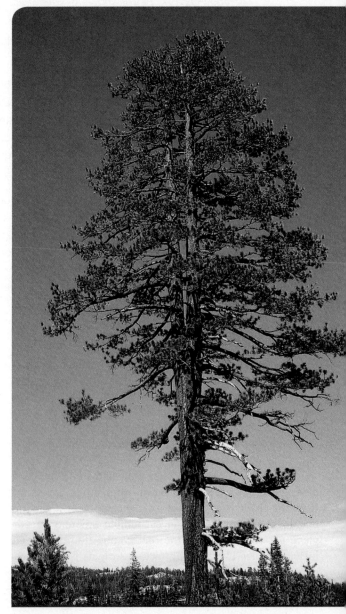

Figure 11.8 A Jeffrey pine tree. The trunk of this tree, which would normally be single, is forked because of the earlier removal of the terminal bud, allowing a lateral bud to grow. © *Kingsley Stern*

produces its own inhibitor, which, in turn, suppresses the new lateral buds that develop beneath it. Experiments with pea plants have shown that the axillary buds will begin to grow as little as 4 hours after a terminal bud has been removed. If cytokinins are applied in appropriate concentration to axillary buds, however, they will begin to grow, even in the presence of a terminal bud, thereby offsetting apical dominance. The extent of the suppression depends on how close the axillary buds are to the terminal bud and on the amount of inhibitor involved. Cytokinins are produced in the roots and then transported to axillary buds, where they break dormancy. Auxin is also involved in controlling cytokinin concentration.

Senescence

The breakdown of cell components and membranes that eventually leads to the death of the cell is called **senescence.** As mentioned in Chapter 7, every year, the leaves of deciduous trees and shrubs senesce and drop through the process of abscission. Even evergreen species typically retain their leaves for only 2 or 3 years (with the notable exception of bristlecone pines, which hold their leaves for up to 30 years), and the aboveground parts of many herbaceous perennials senesce and die at the close of each growing season.

Why do plant parts senesce? As plants grow, new organs develop in part from nutrients that are recycled from senescing organs. Senescence is an orderly process, regulated by the actions of many molecular players, including cytokinins, ethylene, auxins, salicylic acid, jasmonic acid, and abscisic acid. There is much yet to learn about the ways in which these compounds interact to systematically move plant tissues to their death.

Other Hormonal Interactions

Many plants can be grown in test tubes in which scientists can manipulate the components of the substrate and evaluate their effects on growth. The development of roots and shoots in tissue culture has been found to be mainly regulated by a combination of auxins and cytokinins. Experiments with living pith cells of tobacco plants have shown that such cells will enlarge when supplied with auxin and nutrients, but they will not divide unless small amounts of cytokinin are added. By varying the amounts of cytokinin, it is possible to stimulate the pith cells to differentiate into roots or into buds from which stems will develop.

Gibberellins and ABA are important in the regulation of seed germination. Gibberellins promote cell elongation, thereby enabling the developing root to break through the seed coat. After water has been imbibed by cereal seeds, gibberellins released by the embryo stimulate the secretion of enzymes that digest endosperm. ABA has been shown to inhibit the synthesis and release of these enzymes. The relative amounts of gibberellins and ABA in seeds other than cereal grains also influence whether the seeds germinate or remain dormant. In some plants, soaking the seeds in gibberellin breaks dormancy and allows them to germinate.

11.4 Plant Movements

Learning Outcomes

a. Distinguish among the major types of plant growth movements.

b. Describe situations in which plants rely on phototropism and gravitropism.

c. List examples of turgor movements in plants.

d. Define taxis.

e. Describe how plants benefit from dehydration movements.

Growth Movements

Growth movements result from different growth rates across an organ. They occur mainly in young parts of a plant and, as a rule, are quite slow, usually taking at least 2 hours to become apparent, although the plant may have begun microscopic changes within minutes of receiving the stimulus.

Movements Resulting Primarily from Internal Stimuli

Nutations Charles Darwin once attached a tiny sliver of glass to the tip of a plant growing in a pot. Then he suspended a piece of paper blackened with carbon over the tip, and as the plant grew, he raised the paper just enough to allow the tip to touch the paper without hurting the plant. He found that the growing point traced a spiral pattern in the blackened paper (Fig. 11.9). We know now that such nutations (also referred to as *spiraling* movements) are common to many plants and are displayed by most plant organs, and they result from unequal growth rates on the inner and outer surfaces of organs.

Figure 11.9 Charles Darwin demonstrated nutation (spiraling growth) by attaching a sliver of glass to the tip of a plant and suspending a piece of paper covered with carbon over the plant so it touched the glass sliver. As the plant grew, it etched a spiral pattern on the paper.

Nodding Movements Germinating seeds of legumes, such as garden beans, develop a thickened crook in the hypocotyl and exhibit a slow, oscillating movement (i.e., the bent hypocotyl "nods" from side to side like an upside-down pendulum) as the seedling pushes up through the soil. This nodding movement apparently helps the seedling to push through the soil.

Twining Movements Although twining movements are mostly stimulated internally, external forces, such as gravity and contact, may also play a role. These movements occur when cells in the stems of climbing plants, such as morning glory, differentially elongate, causing visible spiraling in growth (in contrast with the spiraling movements previously mentioned, which are not visible to the eye). Tendril twining, which is initiated by contact, results from an elongation of cells on one side of the stem and a shrinkage of cells on the opposite side, followed by differences in growth rates (Fig. 11.10). Gelatinous fibers in tendrils play a role in coiling. Plants with a cylinder of these fibers are able to coil in different directions. In contrast, plants that produce fibers on the inner surface of the tendril can coil in only one direction.

Contraction Movements In Chapter 5, we noted that the bulbs of a number of dicots and monocots have contractile roots, which pull them deeper into the ground where the environment is more constant. For example, Easter lily bulblets are pulled into the ground by root contraction. The detection of light by the bulblets stimulates this activity. The underground portion of the plant above the bulb may be responsible for depth perception. In other plants, roots contract when day/night temperature differences are detected.

Nastic Movements When flattened plant organs, such as leaves or flower petals, first expand from buds, they characteristically bend up and down as the cells in the upper and lower parts of the leaf alternate in enlarging faster than those in the opposite parts. Such nondirectional movements (i.e., movements that do not result in an organ being oriented toward or away from the direction of a stimulus) are called **nastic**. Nastic movements may involve differential growth or turgor changes in special cells. *Epinasty* is the permanent downward bending of an organ, often the petiole of a leaf, in response either to an unequal flow of auxin through the petiole or to ethylene. Nastic movements that result from changes in turgor pressure include sleep movements and contact movements, which are discussed later.

Movements Resulting from External Stimuli

Growth toward or away from a stimulus is called a **tropism**. Tropic movements can be divided into three phases:

1. In the initial *perception* phase, the organ receives a greater stimulus on one side.
2. Then, *transduction* occurs, during which one or more hormones become unevenly distributed across the organ.
3. Finally, *asymmetric growth* occurs as a result of greater cell elongation on one side.

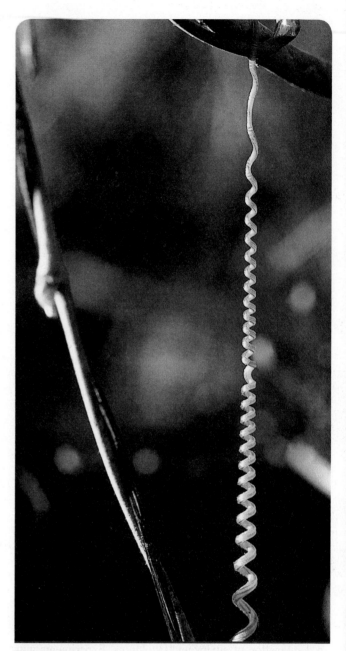

Figure 11.10 Typical twining of a tendril produced by a manroot plant. Note that the direction of coiling reverses near the midpoint.
© *Kingsley Stern*

Phototropism The main shoots of most plants growing in the open tend to develop vertically, although the branches often grow horizontally. If a box is placed over a plant growing vertically and a hole is cut to admit light from one side, the tip of the plant will begin to bend toward the light within a few hours. If the box is later removed, a compensating bend develops, causing the tip to grow vertically again. Such a growth movement toward light is called a *positive phototropism* (Fig. 11.11). A similar bending away from light is called a *negative phototropism*. The shoot tips of most plants are positively phototropic, while roots are either insensitive to light or negatively phototropic.

Growth and Development 199

Figure 11.11 This tulip plant received light from the left for several weeks. Note how all visible plant parts are oriented on the side that received light. This is an example of positive phototropism. © Brand X Pictures/Getty Images

the light, causing the cells there to elongate more and produce a bend. Careful experiments have shown, however, that stem tips growing in the open have the same total amount of auxin present as stem tips from the same species receiving light from one side. Other experiments indicate that the auxin migrates away from the light, accumulating in greater amounts on the opposite side, promoting greater elongation of cells on the "dark" side. Apparently, an active transport system enables the auxin to migrate against a diffusion gradient. Photoreceptor proteins called phototropins initiate the phototropism response. They are embedded in the plasma membrane and unfold when exposed to blue light. This initiates a cascade of events inside the cell, leading to cell elongation.

Different intensities of light may bring about different phototropic responses. In Bermuda grass, for example, stems tend to grow upright in the shade and parallel with the ground in the sun. In other plants, such as the European rock rose, which grows among rocks or on walls, the flowers are positively phototropic, but once egg cells in the flowers are fertilized, they become negatively phototropic. As the pedicels (stalks) elongate, the developing fruits are buried in cracks and crevices, where the seeds then may germinate. Phototropic responses are not confined to flowering plants. A number of mushrooms, for example, show marked positive responses to light, and certain fungi that grow on horse dung have striking phototropic movements, which are discussed in Chapter 19.

We have already noted that if the tip of a coleoptile is covered or removed, the structure will not bend toward light (see Fig. 11.3). We have also noted that auxin at certain concentrations promotes the elongation of cells. For some time, it was believed that stem tips bent toward light because auxin was destroyed or inactivated on the exposed side, leaving more growth-promoting hormone on the side away from

Gravitropism Growth responses to the stimulus of gravity are called **gravitropisms.** The primary roots of plants tend to be positively gravitropic, while shoots forming the main axis of plants are typically negatively gravitropic (Fig. 11.12). Gravity is sensed by root cap cells, which contain amyloplasts (starch-containing plastids).

Figure 11.12 This *Coleus* plant was placed on its side the day before the photograph was taken. The stems bent upward within 24 hours of the pot being tipped over. This is an example of negative gravitropism. © Kingsley Stern

Figure 11.13 In this root cap of a tobacco plant, amyloplasts (the spherical dark objects) have settled to the bottom of each cell. This triggers the gravitropic response. ×2,000. *Courtesy of John Z. Kiss*

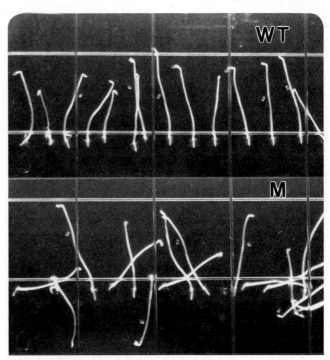

Figure 11.14 These tobacco seedlings have been grown in the dark. Normal "wild type" seedlings in the top row are growing as expected. The seedlings in the bottom row are mutants with much less starch than normal plants. The mutant seedlings are disoriented, suggesting that normal amyloplasts are important to perceive gravity. ×0.5. *Courtesy of John Z. Kiss*

These gravity-sensing cells are large, with organelles and cytoskeleton elements mostly confined to the periphery. This allows amyloplasts, which are more dense than the cytoplasm, to settle to the bottom of each root cap cell (Figs. 11.13 and 11.14). This results in the accumulation of auxin on the lower side of the root cap, ultimately leading to an auxin gradient in the elongation zone of the root (above the root cap) and a downward curvature. There is likely a secondary mechanism that also contributes to gravitropism, but it is not yet understood.

The stimulus of gravity can be negated by rotating a plant placed in a horizontal position. A simple device called a *clinostat* uses a motor and a wheel to rotate a potted plant slowly about a horizontal axis (Fig. 11.15). As the plant rotates, both the stem and the roots continue to grow horizontally instead of exhibiting characteristic gravitropic responses. Starch grains cannot settle while the plant is moving, and this apparently prevents transport of auxin, which would bring about cell elongations that produce curvatures of the root and stem.

Other Tropisms A response to contact with a solid object is called a *thigmotropism*. One of the most common thigmotropic responses is seen in the coiling of tendrils and in the twining of climbing plant stems (see Figs. 7.14 and 11.10). Such responses can be relatively rapid, with some tendrils wrapping around a support two or more times within an hour. The coiling results from cells in contact becoming slightly shorter while those on the opposite side elongate.

Roots often enter cracked water pipes and sewers. In fact, roots have been known to grow upward for considerable distances in response to water leaks. Some scientists have called such growth movements *hydrotropisms*, but most plant physiologists today doubt that responses to water and several other "stimuli" are true tropisms. Other external stimuli that produce tropic responses designated as tropic by some scientists include chemicals (*chemotropism*),

Figure 11.15 This clinostat is a tool used by plant biologists to negate the effects of gravity. Growing plants or seedlings are slowly rotated so that amyloplasts do not settle to the bottom and typical growth or bending of stems away from or roots toward gravity does not occur.

temperature (*thermotropism*), wounding (*traumotropism*), electricity (*electrotropism*), darkness (*skototropism*), and oxygen (*aerotropism*).

Greater concentrations of roots tend to occur on the north and south sides of wheat seedlings, and it has been suggested that magnetic forces may be involved; the term *geomagnetotropism* has been proposed for this phenomenon. Some of these tropic or tropic-like responses have been artificially induced, but others take place regularly in nature. For example, germinating pollen grains produce a long tube that follows a diffusion gradient of a chemical released within a flower; this is considered a chemotropism. Thermotropic responses to cold temperatures may be seen in the shoots of many common weeds, which grow horizontally when cold temperatures prevail and return to erect growth when temperatures become warmer.

Turgor Movements

Turgor movements result from changes in internal water pressures and are often, but not always, initiated by contact with objects outside of the plant. The cells involved may be in normal parenchyma tissue of the cortex, or they may be in special swellings called *pulvini* located at the bases of leaves or leaflets. Some turgor movements may be quite dramatic, taking place in a fraction of a second. Others may require up to 45 minutes to become visible.

The sudden movements of bladderworts (discussed in Chapter 7) result from turgor changes apparently triggered by electric charges released upon contact or as a result of variations in light or temperature. The springing of the trap of the Venus's flytrap (also discussed in Chapter 7) was thought to be brought about in similar fashion, but recent research has shown that the trap closes when its outer epidermal cells expand rapidly, and it reopens when the inner epidermal cells expand in the same way. About one-third of the ATP available in the cells is used in each movement, so repeated stimulation of the trap by touching the trigger hairs readily fatigues the trap if sufficient time for ATP replenishment is not given between stimulations.

The sensitive plant (*Mimosa pudica*) has well-developed pulvini at the bases of its many leaflets and a large pulvinus at the base of each leaf petiole. When the leaf is stimulated by touch, heat, or wind, there is a type of chain reaction in which potassium ions migrate from one half of each pulvinus to the other half. This is followed by a rapid shuttling of water from half of the pulvinar parenchyma cells to those of the other half. The loss of turgor results in the folding of both the leaflets and the leaf as a whole (Fig. 11.16).

If a part of the stem of a sensitive plant is cut off and then immediately reattached with a water-filled piece of rubber tubing, it can be shown that something is transmitted through the water from above the cut. Within a few minutes after a leaf above the cut is stimulated to fold, a leaf or two below the cut will also fold. Although potassium ions have been shown to leave cells in pulvini that are losing turgor, it is not known if the reaction is transmitted across water in the rubber tubing by these

(a)

(b)

Figure 11.16 In part *a*, the upper drawing shows the turgid pulvinus of a sensitive plant (*Mimosa pudica*) before being touched. The lower drawing shows how the leaves fold after being touched as cells in the pulvinus loses turgor. Part *b* shows a sensitive plant in which the leaflets of the leaf toward the bottom have folded upward in response to being bumped. The other leaves have remained fully expanded. *(b) © Kingsley Stern* ions, by electrical charges, or by something else. A normal response to a stimulus by a sensitive plant takes from a few seconds to less than 2 minutes.

The redwood sorrel (*Oxalis oregana*), whose leaflets also have pulvini, flourishes on the floor of redwood forests, where the light may be only 1/200th that of full sunlight. Occasionally, light of sufficient intensity to damage delicate leaves may temporarily penetrate the overhead canopy. When this occurs, the leaflets begin to fold downward within 10 seconds and are completely folded in about 6 minutes, unfolding once again when the brighter light is gone.

Turgor contact movements are not confined to leaves. Many flowers exhibit movements of stamens and other parts, with most such movements facilitating pollination. The pollen-receptive stigmas found in flowers of the African sausage tree have two lobes. These lobes fold inward, enclosing pollen grains that have landed on them. The stigmas of monkey flowers (Fig. 11.17) and Asian cone flowers exhibit similar turgor movements upon contact, while the pollen-bearing stamens of barberry and moss rose flowers snap inward suddenly upon contact, dusting visiting insects with pollen (Fig. 11.18).

Figure 11.18 In this barberry flower, the stamens are spread away from the pistil before contact (a), and they pull inward in response to contact (b).

Orchid flowers exhibit some of the most spectacular of all turgor contact movements, including those by which little sacs of pollen are forcibly attached to the bodies of visiting insects. In a few instances, the benefactors even receive dunkings in small "pans" of fluid from which they escape via a trap door. Brief discussions and illustrations of a few examples are given in Figure 23.18.

"Sleep" Movements (Circadian Rhythms)

Members of several flowering plant families, particularly the Legume Family (Fabaceae), which includes peas and beans, exhibit movements in which leaves or petals fold as though "going to sleep." The folding and unfolding usually takes place in regular daily cycles, with folding most frequently taking place at dusk and unfolding occurring in the morning. Such cycles, which have been more extensively documented in the Animal Kingdom, have come to be known as **circadian rhythms.**

In prayer plants (*Maranta*—Fig. 11.19) and others where the leaves or flowers fold up and reopen daily, turgor movements are involved, as are the external stimuli of light and temperature. These circadian rhythms are controlled by a biological "clock" on cycles of approximately 24 hours. In a few instances, nastic movements apparently are initiated by external stimuli alone.

Circadian rhythms appear to be controlled internally by the plants, although they are also geared to changing day lengths and seasons. They do not generally accelerate when temperatures increase. The actual timing of circadian rhythms varies with the species, although most plants exhibiting sleep movements do so at dusk and at dawn. The flowers of several species pollinated by night-flying moths, however, open at night.

About 200 years ago, the famous Swedish botanist Linnaeus planted wedge-shaped segments of a circular garden with plants that exhibited sleep movements. The plants were arranged in successive order of their sleep

Figure 11.17 The white, two-lobed structure in the center of this monkey flower is the stigma. Before pollination, the stigma is open (a). After the flower has been touched by a pollinator, the lobes quickly fold together (b). © *Kingsley Stern*

Species of certain warmer marine-water algae called *dinoflagellates* (discussed in Chapter 18) glow in the dark through *bioluminescence*, a process by which chemical energy is converted to light energy. One species always glows brightly within 2 or 3 minutes of midnight, even if it is maintained in culture in continuous dim light. This dinoflagellate also glows when culture containers in which it is suspended are jarred, and it displays two other types of circadian rhythms. One rhythm involves cell division, with peak mitotic activity occurring just before dawn, and the other rhythm pertains to photosynthesis, which reaches a maximum around noon.

Another type of rhythm is seen in the giant bamboos of Asia, which send up huge, flowering stalks every 33 or 66 years, even if the plants have been transplanted to other continents and are growing under different conditions. These flowering stalks use up all the energy reserves of the bamboo, and they die shortly after appearing. This has especially been a problem in cities where nearly all of the bamboo plantings were propagated from a single source and then all died simultaneously after flowering.

Solar Tracking

Leaves often twist on their petioles and, in response to illumination, become perpendicularly oriented to a light source. In fact, many plants have solar-tracking leaves with the blades oriented at right angles to the sun throughout the day. Some scientists have referred to solar-tracking movements as *heliotropisms*, but, unlike phototropic responses of stems and roots, growth is not involved.

It has been widely reported that sunflowers exhibit heliotropic movements and face the sun throughout the day. This may be true when the plants are very young. However, when sunflowers are grown commercially for their seeds, the flowers always face east as they develop and remain facing in that direction until the seeds mature and the plant dies.

Strictly speaking, the twisting of petioles that facilitates heliotropic movements should be called *phototorsion*, because motor cells (in pulvini) at the junction of the blade and the petiole control the movement (see turgor movements in this chapter). If one views a tree from directly overhead or observes a vine growing on a fence, it is perhaps surprising to note how little overlap of the leaves occurs. Each leaf is oriented so that it receives the maximum amount of light available (Fig. 11.20).

Water Conservation Movements

Leaves of many grasses have special, thin-walled cells (*bulliform cells*) below parallel, lengthwise grooves in their surfaces. During periods when sufficient water is not available, these cells lose their turgor, and the leaf rolls up or folds (Fig. 11.21). Experiments with certain prairie grasses have shown that the rolling effectively reduces transpiration to as low as between 5% and 10% of normal.

(a)

(b)

Figure 11.19 Sleep movements are illustrated by this prayer plant (*Maranta*). At noon, the leaves are open (a), but at 10 P.M., they have folded inward (b). © *Kingsley Stern*

movements throughout a 24-hour day. One could tell the approximate time simply by noting which part of the garden was "asleep" and which was "awake." A few others copied the garden clock idea, but the expense and difficulty of obtaining all the plants from different parts of the world and replanting them each year proved to be too great for the practice to be continued.

The movements are produced by turgor changes caused by the passage of water into and out of cells at the bases of the leaves. The function of these movements is not clear, and the rhythms are also not confined to sleep movements.

Figure 11.20 Leaves of this wild grape (*Vitis* sp.) are oriented so there is little overlap, and consequently, leaf area for photosynthesis is maximized. © *Herbert Kehrer/Getty Images*

(a) (b)

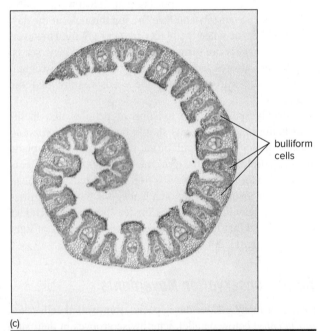

(c)

Figure 11.21 Grass leaves can roll inward under drought stress in order to conserve water. Under normal water regimes, the leaf is flat and open (a), but the leaf rolls when water is limited (b). When viewed under a microscope, the collapsed thin-walled bulliform cells can be seen. *(c) © Kingsley Stern*

Taxic Movements

Taxis is a type of movement that occurs in several groups of plants and fungi but not flowering plants. In response to a stimulus, the cell or organism, either propelled or pushed by **flagella** (whiplike appendages) or **cilia** (short, whiplike appendages), moves toward or away from the source of the stimulus. In contrast to tropism, in which a plant grows toward or away from a stimulus, taxis requires motility.

Stimuli for taxic movements include chemicals, light, oxygen, and gravitational fields. In ferns, for example, the female reproductive structures produce a chemical that prompts a *chemotaxic* response in the male reproductive cells (sperms)—that is, the sperms swim toward the source of the chemical. Certain one-celled algae exhibit *phototaxic* responses, swimming either toward or away from a light source. Other similar organisms exhibit *aerotaxic* movements in response to changes in oxygen concentrations.

Miscellaneous Movements

Slime molds, discussed in Chapter 18, inhabit damp logs and debris. During their active stages, their protoplasm, which has no rigid cell walls, migrates over dead leaves and other substrates in a crawling-flowing motion, somewhat like thin, slowly moving gelatin. Certain cyanobacteria (e.g., *Oscillatoria*) wave slowly back and forth or slide up and down against each other in *gliding movements,* and diatoms (one-celled algae with thin glass shells) give the appearance of being jet-propelled. It is not certain just how these movements occur, but there is evidence that submicroscopic fibrils produce rhythmic waves that bring about the motion. In the case of diatoms, some materials are apparently forced out of the cells, and the friction set up by the process may propel the organisms through the water in which they are found.

Dehydration movements do not require living cells or hormones, because the forces are purely physical. They are caused by imbibition or by the drying out of tissues or membranes. The individual fruitlets of filarees and other members of the Geranium Family (Geraniaceae) have long, stiff, pointed extensions, which are sensitive to changes in humidity. As humidity decreases during the day, they coil up, and then at night, when humidity increases, the coils relax. This alternating coiling and uncoiling results in the pointed fruitlets planting themselves into the ground in a corkscrew-like fashion (see Fig. 8.27).

A number of fruits that are podlike and dry at maturity split in various ways with explosive force, flinging seeds as far as 12 meters (39 feet) from the plant. Examples of such fruits include those of the garbanzo bean, witch hazel, vetches, and Mexican poppies. In ice plants and stonecrops, rain or dew causes the mature fruits to open due to the swelling of membranes. Pressures inside the squirting cucumber build up to the point where, upon abscission of the fruit, the seeds are expelled from the stalk in distances of up to 10 or more meters (33 feet). Dwarf mistletoes, which are parasitic on coniferous trees, produce tiny, sticky fruits that are explosively released and adhere to the trunks and branches of trees in the vicinity.

11.5 Photoperiodism

Learning Outcomes

a. Describe the adaptive advantage of the photoperiodic control of flowering.
b. Summarize the role of phytochrome in plant responses to the environment.

In the early 1900s, two American plant physiologists, Wightman W. Garner and Henry A. Allard, became curious about a tobacco plant growing at a research center in Beltsville, Maryland. The plant grew 3 to 4 meters (10 to 13 feet) tall during the summer but failed to flower in August when the normal-sized tobacco plants flowered. They took the giant plant into a greenhouse for protection during the winter and were surprised when it flowered in December.

Garner and Allard decided to conduct some experiments with tobacco plants. If the plants were started in pots in the fall in a greenhouse, they grew only about a meter (3 feet) tall before flowering. They wondered what would happen if they kept tobacco plants, as well as some soybean seedlings that would not bloom in Beltsville before September, in complete darkness from 4 P.M. until 9 A.M., thereby allowing them only 7 hours of daylight. They found that all the plants flowered, and further investigations showed that the length of day (actually, the length of the night) was directly related to the onset of flowering in many plants. They published the results of their investigations in 1920 and later called the phenomenon they had discovered **photoperiodism** (Fig. 11.22). Photoperiod is the number of hours of light in a 24-hour period. Near the equator, the photoperiod is 12 hours throughout the year. In temperate regions, the photoperiod is longer in the summer than in the winter. Consequently, photoperiodism has tremendous survival value, because it allows plants to sense changes in the seasons and produce flowers when reproduction is most likely to be successful.

The critical length of day (i.e., the maximum or minimum length of day) for the initiation of flowering is often about 12 to 14 hours, although it can vary considerably. Plants that will not flower unless the day length is shorter than the critical length (e.g., in the fall or spring) are called **short-day plants.** These plants actually measure the length of uninterrupted darkness in a 24-hour cycle. So short-day plants actually need long nights. They include asters, chrysanthemums, dahlias, goldenrods, poinsettias, ragweeds, sorghums, salvias, strawberries, and violets. Plants that will not flower unless periods of light are longer than the critical length are called **long-day plants.** These include garden beets, larkspur, lettuce, potatoes, spinach, and wheat. Such plants usually flower in the summer but will also flower when left under continuous artificial illumination. Accordingly, leafy vegetables, such as lettuce and spinach, need to be harvested in the spring and grown again in the fall in temperate latitudes if *bolting* (producing flowering stalks) is to be avoided.

The photoperiodic control of flowering has been intensively studied in *Arabidopsis*, a model plant with a facultative long-day photoperiod response (it flowers earlier in long days than in short days). This control mechanism requires a complex interaction among photoreceptors (such as phytochromes), the circadian clock, and genes that regulate flowering in response to photoperiod. Photoperiod and the circadian clock regulate the expression of the *CONSTANS* (*CO*) gene so that levels of the CO protein increase across daylight hours. (Note that gene names are in italics but proteins are not.) CO then induces the expression of the *FT* (*FLOWERING LOCUS T*) gene, which, in combination with additional gene products, leads to the production of flower buds.

Indian grass and several other grasses have two critical photoperiods; they will not flower if the days are too short, and they will not flower if the days are too long. Such species are referred to as **intermediate-day plants.** Other plants, particularly those of tropical origin, will flower under any length of day, providing, of course, they have received the minimum amount of light necessary for normal growth. Such plants are called **day-neutral plants** and include garden beans, calendulas, carnations, cyclamens, cotton, nasturtiums, roses, snapdragons, sunflowers, and tomatoes, as well as many common weeds, such as dandelions.

With some plants, small differences in day length may be critically important. Some varieties of soybeans, for example, will not flower when days are 14 hours long but will flower if the day length is increased to 14½ hours. This difference could amount to less than 320 kilometers (200 miles) of latitude, with certain varieties grown in the southern states not producing fruit in the northern states, and vice versa.

Commercial florists and nursery owners have made extensive use of photoperiods, manipulating with artificial

Figure 11.22 Photoperiodism is critical in many plants. In poinsettia, plants must receive less than 8.5 hours of light per day in order to produce red bracts and flowers. This is particuarly important to prepare pointsettias for the holidays. Not only is the poinsettia a popular Christmas plant, it is the number-one flowering potted plant in the United States, even though its traditional sales period is just 6 weeks. *Provided by Scott Bauer/USDA*

light the flowering times of poinsettias, some lilies, and other plants so they will flower at times of the biggest demand, such as Christmas or Mother's Day. The light intensities used to lengthen the days artificially can be very low—often less than 2 μmols per square meter per second (i.e., one-thousandth the intensity of full sunlight)—and can come from incandescent bulbs, which have more red wavelengths than fluorescent lamps and therefore have more effect on phytochromes, which are involved in photoperiodism and are discussed in the next section, "Phytochromes and Cryptochromes."

A number of vegetative activities of plants are also affected by photoperiods. In the shortening days of the fall, for example, many woody plants will begin undergoing the changes that lead to the dormancy of buds, regardless of temperature. In the spring, certain seeds respond to photoperiods in their germination, both long-day and short-day species having been discovered. Plants that produce tubers (e.g., Jerusalem artichoke) may develop them only under short-day conditions, even though they are long-day plants with respect to flowering. Usually, these photoperiodically controlled responses are valuable to the plants in that they prepare them for changes in the seasons and thus ensure their survival and perpetuation.

Phytochromes and Cryptochromes

Experiments that were conducted after Garner and Allard's discoveries prompted researchers to look for a pigment that controlled photoperiodism. They had already found that the initiation of flowering can be inhibited if plants are exposed to even a brief period of red light during the night, which suggested the existence of a light-sensitive pigment. Within a few years, such a pigment was discovered. It was not visible, however, and a special pigment-analysis instrument had to be constructed to detect it. In 1959, after the pigment had been isolated, it was named **phytochrome**. Since 1959, several different phytochromes have been identified.

Phytochromes are pale blue, proteinaceous pigments that apparently occur in all higher plants and, like all pigments, absorb light. Only minute amounts are produced, mostly in meristematic tissues. Phytochromes occur in two stable forms, either of which can be converted to the other: P_{red}, or P_r, is a form that absorbs red light; $P_{far-red}$, or P_{fr}, is an active, related form that absorbs the far-red light found at the edge of the visible light spectrum. When either form absorbs light, it is converted to the other form, so that P_r becomes P_{fr} when it absorbs red light, and P_{fr} becomes P_r when it absorbs far-red light. P_r is stable indefinitely in the dark. The normal effect of light in nature is to cause more P_r to become P_{fr} than vice versa. P_{fr} converts back to P_r in the dark over a period of several hours or else becomes inactivated, but its conversion in the presence of appropriate light is instantaneous (Fig. 11.23). The CO protein mentioned in the previous section is stabilized

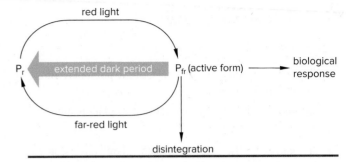

Figure 11.23 Phytochrome interconversions.

by far-red light and destabilized by red light through a phytochrome B-dependent mechanism.

Phytochrome pigments play a role in a great variety of plant responses: in addition to photoperiodism, they are involved with several aspects of plant development, changes in plastids, the production of anthocyanin pigments, and the detection of shading by other plants. To account for the vast diversity of morphological and physiological responses to light, studies investigating light-signal transduction pathways are currently being conducted (Fig. 11.24).

One of the most studied phenomena associated with phytochrome pigments involves the germination of seeds. Some seeds, for example, do not germinate in darkness, but the red part of the spectrum in sunlight converts P_r to P_{fr}, which, in turn, unblocks the germinating mechanism. If such seeds (e.g., those of "Grand Rapids" lettuce) are given suitable moisture and oxygen conditions and then exposed only to red light, they will germinate readily, but if they are given only far-red light, they will not. Furthermore, if they are given alternating flashes of red and far-red light, the type of light in the final flash determines whether or not they will germinate. If the last flash, which may be of only a few seconds' duration, is red light, the seeds will germinate, but if it is far-red light, they will not.

When a seedling first emerges from the soil, light changes the P_r to P_{fr}, triggering a reduction in the production of ethylene by the cells. This, in turn, allows the crooks in the hypocotyls to relax, and the plant straightens up. Elongation of stems appears to be inhibited by P_{fr}. The mechanism is so sensitive that as little as 3 seconds of clear moonlight may produce a shortening of internodes of **etiolated** oat seedlings (those that are spindly and pale from having been grown in the dark). More far-red light reaches young, growing stems of a tree that is shaded, however, causing the P_{fr} to be converted to P_r, thus lowering the inhibition so that the young stems grow longer and out from under the shade.

A second group of blue, light-sensitive pigments known as **cryptochromes** also play a role in circadian rhythms and interact with phytochromes in plants in controlling reactions to light. Cryptochromes, like phytochromes, are important for maintaining the stability of CO proteins involved in flower induction.

Figure 11.24 This experiment illustrates the effect of subjecting one leaf of a short-day plant (cocklebur) to short days while the rest of the plant is exposed to long days. (a) The short-day plant exposed to long days. No flowers were produced. (b) The same plant exposed to the same long days while one leaf was covered with black paper 16 hours a day for a few weeks (mimicking short days). The plant flowered, presumably because some substance that initiates flowering was produced in the shaded leaf and then diffused or was carried to the stem tip where flower buds are produced.

11.6 Temperature and Growth

Learning Outcome

a. Explain the concept of growing degree accumulation and be able to calculate growing degree days for prediction of development and harvest dates.

Each species of plant has an optimum temperature for growth—although the optimum may vary throughout the life of the plant—and a minimum temperature below which growth will not occur. Each species also has a maximum temperature above which injury may result (or, at least, growth will cease).

Dr. Frits Went, the discoverer of auxin, experimented with the growth of tomato plants under conditions in which night temperatures were lowered to 17°C (63°F) while day temperatures were maintained at 25°C (77°F). These plants were found to grow better than plants kept at a steady temperature around the clock. Went applied the term **thermoperiodism** to this phenomenon. It has since been shown that the optimal thermoperiod may change with the growth stage of the plant. Young pepper plants have been shown to develop best when night temperatures are approximately 25°C (77°F), but as they grow older, the optimum night temperature for their development drops more than 11°C (20°F). Lower night temperatures often result in a higher sugar content in plants and may produce greater root growth, although some plants, such as peas, seem to be unaffected in this regard.

Growers of crop plants often calculate **growing degree days** (GDD) to predict development and harvest dates. Each crop species has a minimum, or base, temperature for growth. The simplest way to calculate degree day accumulation across the season is to subtract the base temperature from the average temperature for each day and then sum those values across the season. For example, if a crop has a base temperature of 12°C and the high/low temperatures across a three-day period are 18/12, 15/13, and 16/14, then the degree day accumulation will be 15 (18 + 12/2) − 12 = 3, plus 14 − 12 = 2 plus 15 − 12 = 3, for a total accumulation of 8. When a certain GDD value is reached, then the crop will be ready for harvest. This concept is also used to predict when disease or insect pressure will be severe and control measures need to be implemented.

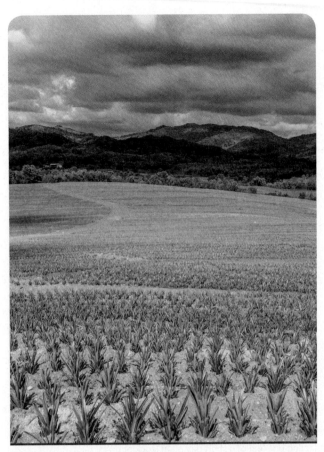

Figure 11.25 The harvest date for this pineapple field can be predicted based on daily temperatures during the growing season.
© Mario De Moya F/Shutterstock.com

In Hawaii, Growing Degree Days (GDD) are used to predict fruit harvest dates. This is important when scheduling labor for harvest and in planning marketing efforts. Rather than using air temperature throughout the growing season, though, two growth periods are analyzed in different ways. Fruit induction is stimulated with a growth regulator so that all plants in a field initiate fruits at the same time. From the time of fruit induction until the first flower opens, air temperature is used to calculate GDD. Then, from the first flower to harvest, GDD is based on fruit temperature. The GDD model also takes into account reductions in growth when temperatures are above optimum (Fig. 11.25).

11.7 Dormancy and Quiescence

Learning Outcome

a. Describe the adaptive significance of seed dormancy.

As the days grow shorter in the fall, cells of perennial plants in regions with cold winters produce sugars and amino acids that lower the point at which cold damage will occur. Protective bud scales develop on buds, and leaves become senescent. As cell metabolism slows down, many plants that are now prepared for winter become *dormant*. **Dormancy** is defined as a period of growth inactivity in seeds, buds, bulbs, and other plant organs even when temperature, water, and day length would typically cause plant growth.

Quiescence is a state in which a seed cannot germinate unless environmental conditions normally required for growth are present. Abscission layers develop in leaf petioles in the fall as plants undergo a change from an active condition to either a dormant or a quiescent one. In temperate and cold climates, most plants and their seeds go through a dormant period lasting from a few days to several months, and in each case, the dormancy nearly always has some survival value for the species. Many of the stone fruits (e.g., cherries, peaches, and plums) need at least several weeks of rest at temperatures below 7°C (45°F) before the buds will develop into flowers or new growth, and in some instances, the buds will not develop before the tree has encountered a period of freezing temperatures. This dormancy allows the plants to reproduce under the conditions prevailing in the cooler temperate zones of the world. Seeds of desert plants germinate only after appreciable rain has fallen, and seeds of plants native to vernal pools and other wet areas may germinate only if they are immersed in water.

Depending on the species, the breaking of seed dormancy may be controlled by a variety of factors referred to as *after-ripening*. Some plants become sensitive to photoperiods only after they have been subjected to a period of chilling, either while they are still seeds or shortly after germination. Growers often place moistened seeds of such plants in a refrigerator for several weeks. This process is known as *stratification* and is one of several practices employed to break dormancy. In some biennial plants, the cold treatment can be replaced by an application of gibberellins. Natural stratification occurs when winter grains are sown in the fall, and it has been shown that the extent of flowering and yield in such grains is profoundly reduced in the absence of a period of natural chilling.

Some plants have their dormancies broken only after they have received a series of similar stimulations, so a single stimulation, such as an unseasonal rainstorm, will not trigger growth. Other plants, such as poor man's pepper (*Lepidium virginicum*), require both certain temperatures and certain photoperiods for germination, and some weeds require specific thermoperiods plus some mechanical abrasion of the seed coat to break dormancy. Still other seeds may require specific enzyme action or a combination of conditions acquired only by passing through the digestive tracts of birds, bats, or other animals before they will germinate.

When a mechanical tomato picker was being developed in California during the early 1960s, tomato breeders wanted to have fruits that would stand up to the bumping they would receive in the machinery. They tried, therefore, to produce suitable varieties. Their search for new genes to use in their breeding programs led them to the wild tomatoes of the Galápagos Islands. When they took the seeds back to California, they experienced difficulty in getting them to germinate, even after subjecting them to a wide variety of stratification treatments. Then they remembered that tomatoes are eaten by the giant tortoises of the Galápagos Islands. They tried feeding the tomatoes to giant tortoises in a zoo. After passing through the animals, the seeds germinated normally. Apparently, an enzyme in the tortoises' digestive tracts broke the dormancy of the seeds.

Germination is prevented by a thick or restrictive seed coat in the seeds of plants such as cocklebur, and the seeds of some legumes are almost stonelike because of the exceptional hardness of the outer covering or seed coat. Growers often scarify (nick or file) seed coats to allow water and gases to enter. In nature, various decay organisms corrode the seed coats, or the seeds themselves may secrete enzymes that aid in this process. The seeds of certain lupines have remarkable valves that restrict the uptake of water, preventing germination before the seed coat is eroded by other means. Many plants have seeds that will not germinate until they have been exposed to fire, and one such Australian legume with a thick seed coat will not readily germinate until heat from a fire causes a plug in the seed to pop out.

Dormancy may be restricted to certain parts of the plant rather than occurring in the plant as a whole. In coffee, for example, only the flower buds may show dormancy. In other woody plants, the onset of dormancy progresses from axillary or lateral buds toward the base of a branch on to the terminal bud at the tip. Some seeds go through two cycles of dormancy before full growth occurs. In wake-robins (*Trillium* spp.), the radicle (embryo root) of the seed emerges and becomes established after being subjected to cold, but the epicotyl does not emerge until a second period of cold has been encountered.

In many seeds, growth inhibitors, such as ABA, are present in the seed coat, often in abundance. Before germination can occur, these inhibitors may need to be leached out by rain, broken down by enzymes, or have their effects overcome by other hormones. Studies have shown that gibberellin, which has the capacity to break dormancy when applied artificially, is produced by the germinating seeds of both cone-bearing and flowering plants. Ethylene, which is also effective in breaking dormancy, is produced by various germinating seeds, including those of oats, peanuts, and clover. Cytokinins do not generally appear to be involved in breaking dormancy, but a few instances are known of their doing so artificially. Although auxins are formed during the early stages of germination, they probably are not involved in the control of dormancy.

Many aspects of growth discussed in this chapter are applied to, or are involved in, vegetative propagation, which is discussed in Chapter 14 and Appendix 4.

SUMMARY

11.1 Introduction to Growth and Development

- Growth is defined as an irreversible increase in volume due to the division and enlargement of cells. As cells mature, they differentiate into forms adapted to specific functions.
- Development is a change in form as a result of both growth and differentiation. Cells themselves assume different shapes and forms that allow volume to increase at a greater rate than surface area.

11.2 Nutrients, Vitamins, and Hormones

- Many growth phenomena are influenced by hormones, which are produced in minute amounts in one part of an organism and transported to another part, where they have specific effects on growth, flowering, and other plant activities.
- Vitamins are organic molecules that function in activating enzymes; they are sometimes difficult to distinguish from hormones.
- Darwin and his son noted in 1881 that coleoptiles bend toward a light source. Frits Went, in 1926, followed up on the Darwinian observations and demonstrated that something in coleoptile tips moved out into agar when decapitated tips were placed on it. He called the substance auxin and showed that it could cause coleoptiles to bend.
- Auxins stimulate the enlargement of cells and are involved in many other growth phenomena. Synthetic auxin-like compounds have been used as weed killers and defoliants.
- Gibberellins promote stem growth but not root growth
- Cytokinins promote cell division and can be used to stimulate the growth of axillary buds.
- Abscisic acid causes buds to become dormant and apparently helps leaves respond to excessive loss of water.
- Ethylene gas hastens the ripening of fruits and is used commercially to ripen green fruits.

11.3 Hormonal Interactions

- Apical dominance is the suppression of the growth of lateral buds by the tips of branches.
- Senescence is the breakdown of cell parts that leads to the death of the cell.

11.4 Plant Movements

- Plant movements, such as nutations and nodding, twining, contractile movements, and nastic moments, are growth movements that result primarily from internal stimuli.
- Tropisms are permanent, directed growth movements that result from external stimuli, such as light (phototropism), gravity (gravitropism), contact (thigmotropism), and chemicals (chemotropism).
- Turgor movements result from changes in internal water pressures; they may be very rapid or may take up to 45 minutes to become visible. Turgor movements include contact movements and sleep movements in which leaves or flowers fold daily in what is known as a circadian rhythm. Other types of plant movements include water conservation movements, taxes, and dehydration movements.

11.5 Photoperiodism

- Photoperiodism is a response of plants to the number of hours of light in a 24-hour period. Day length terminology refers to the length of the light period, even though the plant measures the length of the dark period. Short-day plants will not flower unless the photoperiod (hours of light) is shorter than a critical day length, and long-day plants will not flower unless the photoperiod is longer. Intermediate-day plants have two critical photoperiods; the flowering of day-neutral plants is independent of photoperiod.

- The photoperiodic control of flowering requires a complex interaction among photoreceptors, the circadian clock, and genes that regulate flowering in response to photoperiod.
- Phytochromes are light-sensitive pigments in all higher plants that play a role in many different plant responses. Phytochromes occur in two forms, each of which can be converted to the other by the absorption of light. Daylight generally results in more P_r being converted to P_{fr} (the active form) than vice versa. P_r absorbs red light and P_{fr} absorbs far-red light. P_{fr} will convert back to P_r over a period of several hours in the dark; P_r is stable indefinitely in the dark.
- Cryptochromes have functions similar to those of phytochromes.

11.6 Temperature and Growth

- Each species of plant has an optimum temperature for growth.
- Estimates of accumulation of growing degree days can be used to predict crop plant growth and maturity.

11.7 Dormancy and Quiescence

- Dormancy is a period of growth inactivity in seeds, buds, bulbs, and other plant organs even when appropriate environmental conditions are met.
- Quiescence is a state in which a seed cannot germinate unless appropriate environmental conditions exist.
- The change from dormancy to a state in which germination will occur in seeds is controlled by a variety of factors, including temperature, moisture, photoperiod, thickness of seed coat, enzymes, and growth inhibitors.
- Stratification is the breaking down of dormancy in seeds by keeping them refrigerated and moist for several weeks.

REVIEW QUESTIONS

1. What are the differences among hormones, enzymes, and vitamins?
2. Auxins, gibberellins, and cytokinins all promote growth. How does one distinguish among them?
3. How do hormonal herbicides function?
4. List several commercial applications of ethylene gas.
5. What role do amyloplasts play in gravitropism?
6. Distinguish among internally and externally induced growth movements, turgor movements, and dehydration movements.
7. How do day-neutral and intermediate-day photoperiod plants differ in their requirements?
8. What is the difference between dormancy and quiescence? How may dormancy be artificially broken?
9. What are the requirements for a seed to germinate?
10. How does phytochrome differ from other plant pigments?

DISCUSSION QUESTIONS

1. If green plants require light in order to produce food needed for growth, why are seedlings that are germinated in the dark taller than those of the same age grown in the light?
2. If you remove a terminal bud from a stem, will growth in length stop altogether? Explain.
3. Would it be technically correct to say that some plants go to sleep at night? Explain.
4. Many plants produce seeds that require a period of dormancy before they will germinate. Of what value is the dormancy to the plant? Where might you expect to find plants with seeds that do not undergo dormancy?
5. The rapid turgor movements of the sensitive plant (*Mimosa pudica*) are rare in the Plant Kingdom. If all plants evolved such movements, would the phenomenon have possible survival value or other value to the plants? Explain.

CHAPTER

12

Meiosis and Alternation of Generations

12.1 Introduction to Meiosis and Alternation of Generations
12.2 The Phases of Meiosis
    Key Theme 12.1: Why Plants Have Sex Lives
        Division I (Meiosis I, or Reduction Division)
        Division II (Meiosis II, or Equational Division)
12.3 Alternation of Generations
    Key Theme 12.2: Ordering Complexity

OVERVIEW

After a comparison of asexual and sexual reproduction, this chapter explores the necessity for reducing the chromosome number at some stage in the life cycle of sexually reproducing organisms. The phases of meiosis are systematically covered. The chapter also presents a discussion of alternation of generations and general rules for understanding the process.

Microscopic view of *Fuchsia* pollen, a product of meiosis and mitosis.
© *Image Source/Getty Images*

12.1 Introduction to Meiosis and Alternation of Generations

Learning Outcomes

a. Compare and contrast sexual and asexual reproduction.
b. Explain why meiosis is necessary for sexual reproduction.

For centuries, farmers have grown crops for food and saved some seeds for the next year's crop. Sometimes these seeds develop into plants similar to those from the previous year, but sometimes the seeds develop into plants that are very different, even when growing conditions stay the same. Differences in color, size, taste, and texture of fruits and vegetables can be attributable to the parents that produced the seed. The processes in which each parent produces egg and/or sperm are elegantly designed to result in different progeny. The cell divisions that make egg and sperm, known as meiosis, are the basis of variation in seed-propagated plants and a major contributor to evolutionary change.

In Chapter 3, we noted that plants grow through an increase in the number of cells brought about by mitosis and cell division. Mitosis ensures that the number of chromosomes and the DNA content in the daughter cells will be identical with those of the parent cell. Plants, however, do not live forever. In order to perpetuate the species they must reproduce, or they will become extinct.

Reproduction may take place through natural vegetative propagation, which is discussed in Chapter 14. **Asexual reproduction** (the prefix *a* of *asexual* means "without"; asexual reproduction, therefore, simply means

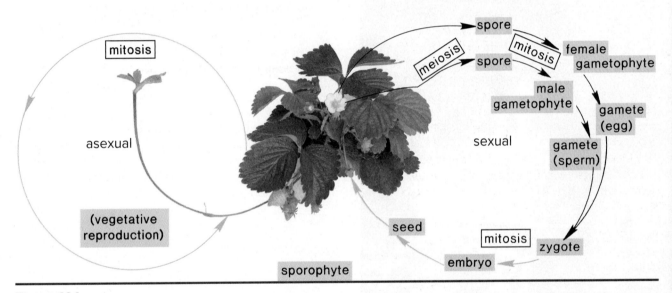

Figure 12.1 This image shows an overview of asexual reproduction (on the left) and sexual reproduction (on the right) using strawberry as an example. © Shelley H. Jansky

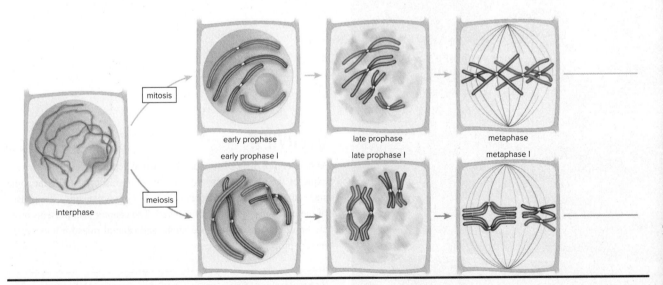

Figure 12.2 A comparison of mitosis and meiosis in diploid cells.

reproduction without sex) produces offspring that are genetically identical to their parents. Although products of asexual reproduction are typically genetically uniform, mutations occasionally provide genetic variability, which is necessary for evolution.

Nearly all plants, however, also undergo sexual reproduction, which in flowering and cone-bearing plants ultimately results in the formation of **seeds.** In sexual reproduction, sex cells called **gametes** are produced. Two gametes, called **egg** and **sperm** in higher plants and animals, unite to form a single cell called a **zygote.** The zygote is the first cell of a new individual (Fig. 12.1).

If gametes were not reduced in chromosome number, the zygote, which results from the union of two gametes, would have twice as many chromosomes as each of the gametes. If such zygotes were then to develop into mature organisms, each of the new organism's cells would have double the original number of chromosomes, and when they produced gametes, the zygotes resulting from the union of this next generation of gametes would have four times the number of chromosomes of the grandparents. If that were to continue for 20 generations, a species having 16 chromosomes in each cell to begin with would end up with 8,388,608 chromosomes per nucleus! The problem does not actually arise, however, because at a certain point in the life cycle of all sexually reproducing organisms, the unique process of **meiosis** occurs.

Meiosis results in gametes that have only half the number of chromosomes of the parent plant. When the gametes form zygotes as they unite in pairs, the original chromosome number is restored.

Since some aspects of meiosis are similar to those of mitosis (the two processes are compared in Fig. 12.2), it is strongly recommended that you review mitosis in Chapter 3 before going through the phases of meiosis outlined in the following section.

Keep in mind that all living cells, before undergoing meiosis, have two sets of chromosomes—one from the male parent, the other from the female parent. Generally, the members of each pair of chromosomes are identical with each other in length, in the amount of DNA present, in the genes they carry, and in the possession of a centromere at precisely the same location. However, as discussed in Chapter 13, their genes may produce contrasting characteristics. Such chromosome pairs are called *homologues,* or **homologous chromosomes.**

When a cell undergoes meiosis, four cells are produced from two successive divisions, which generally take place without pausing. Because of the remarkable events that occur during meiosis, the four cells are never identical with the original cell or with one another.

12.2 The Phases of Meiosis

Learning Outcomes

a. Contrast metaphase 1 with anaphase 1 of meiosis.

b. Differentiate between metaphase I and metaphase II.

As in mitosis, DNA must make a copy of itself before meiosis actually begins, so that each chromosome initially has two identical DNA molecules held together by a centromere. Like mitosis, meiosis is a continuous process that has been divided into arbitrary phases for convenience. In reality, the lines between one phase and another are indistinct.

Four main phases are recognized in each of the two divisions, and the first phase is further subdivided. Since the main phases of the first division bear the same names as those of the second division, it has become customary to designate first division phases with a Roman numeral I and those of the second division with a Roman numeral II.

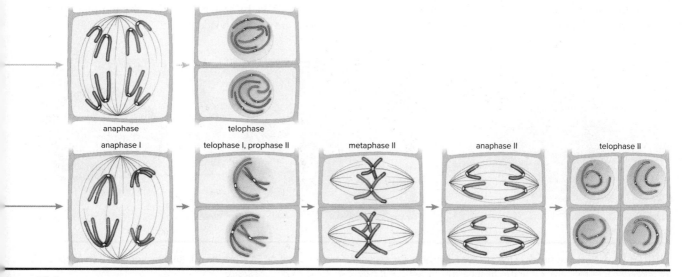

gure 12.2 Continued.

KEY THEME 12.1 evolution

Why Plants Have Sex Lives

Imagine a population of plants on a hillside in southern California. Over the centuries, the environment on that hillside will vary due to climate change and local disturbances. Perhaps the plants are exposed to warmer temperatures and higher ozone levels than they experienced 200 years ago. If the plants have no mechanism to adapt to change, then the population will become extinct (Box Figure 12.1A). How do the plants keep up with a changing environment? Sexual reproduction is the key. First, it brings together new combinations of genes from different parents. Second, the production of gametes by meiosis reshuffles existing gene combinations in the parents via recombination and segregation. Homologous chromosomes pair up during meiosis, and at least one crossover is needed to maintain each pair. The recombination of alleles as a result of crossing-over adds to the shuffling of genetic information. Genomics studies are revealing new insights into recombination events. Sites of crossing-over are not random; so-called recombination hotspots have up to 70 times more recombination than average. In addition, recombination is generally reduced near centromeres. So, depending on chromosomal location, some genes are shuffled much more frequently at meiosis than others. This observation has interesting implications for our opening scenario. It seems, then, that genes are probably not randomly arranged on

Box Figure 12.1A This population of wildflowers carries genetic variation that allows it to adapt to changes in its environment.
© Shutterstock/Leena Robinson

chromosomes. Are "housekeeping" genes (those required for day-to-day maintenance of basic life functions) located near centromeres, where they are not likely to be reshuffled from generation to generation? Conversely, are genes that allow adaptation to changing environments located in recombination hotspots? Genomics studies will soon provide answers to those questions. Stay tuned.

The number of the chromosomes is reduced to half during the first division, and no further reduction in number takes place as a result of the second division. Accordingly, some have referred to division I as a *reductional division* and to division II as an *equational division.*

Meiosis generally takes much longer to complete than mitosis. In lilies, for example, mitosis runs its full course in about 24 hours, while meiosis takes about 2 weeks. In other organisms, meiosis may take months or even years. After the process has begun in human females, for example, it takes as long as 50 years to reach its conclusion, since it remains in a state of arrest for most of that time.

Division I (Meiosis I, or Reduction Division)

Prophase I

At prophase I, the chromosomes coil, becoming shorter and thicker; the chromosomes also become aligned in homologous pairs; the nuclear envelope and the nucleolus disassociate; and parts of each closely associated pair of chromosomes are exchanged with each other (Fig. 12.2; see also Fig. 12.3a, b). Two homologous chromosomes paired at meiosis are called a bivalent.

(a)

(b)

(c)

Figure 12.3 Featured stages of meiosis in a *Lilium* anther, in sequential order, include (a) Early prophase I, (b) Late prophase I, (c) Anaphase I, (d) Telophase I, (e) Metaphase II, and (f) Telophase II. ca. ×400. *(a–f)* © Biophoto Associates/Science Source

As with prophase of mitosis, the beginning of this phase is marked by the appearance of chromosomes as faint threads in the nucleus when viewed under a microscope. Each chromosome gradually coils like a spring, so it becomes as much as 100 times shorter in length, and correspondingly, it becomes thicker. As the chromosomes become shorter and thicker, homologues become aligned in pairs, and eventually, two **chromatids** can be distinguished for each chromosome. Each pair of chromosomes, therefore, has four chromatids; each chromatid is a double-stranded DNA molecule with its associated proteins. The four chromatids of a homologous pair each have their own **centromere** (a dense, constricted area of a chromosome to which *spindle fibers,* consisting of microtubules that look like thin threads, become attached; see Figs. 12.2 and 12.3b).

As prophase I progresses, parts of the chromatids of the homologous chromosomes break and are exchanged with each other. Depending on the length of the chromosomes, this exchange may occur at one to several points along the length of the paired homologous chromosomes. The evidence for this exchange of parts appears a little later in prophase I, when the chromosomes of each homologous pair appear to push each other apart. It can then be seen that adjacent chromatids have *crossed over* at one to several points.

Crossing-over results in an exchange of some of the DNA contributed by the two parents, which is the basis for some of the variability seen in the offspring. An X-shaped figure called a **chiasma** (plural: **chiasmata**) results from each crossover (Fig. 12.4). Chiasmata are necessary for the attachment of homologous chromosomes to opposite spindle poles at meiosis and their subsequent separation at meiosis I. They result in the recombination of alleles on chromosomes.

After the chiasmata have appeared, the chromatids separate, each remaining the same length as it was originally and with the same amount of DNA, but now possessing "traded" material. The process is something like swapping cell phone cases with a friend. When you are through with the exchange, you still have structurally identical phones, but each would have a different combination of parts.

Figure 12.4 Chiasmata are visible points of crossing-over between homologous chromosomes; they are shown here where the dark green chromosomal segments meet the light green regions.

By the end of prophase I, the nuclear envelope and the nucleolus have disassociated and disappeared, and *spindle fibers* (microtubules having the appearance of fine threads) are beginning to form. As in mitosis, some spindle fibers are attached to the centromeres of the chromosomes, while others extend in arcs from pole to pole.

Metaphase I

At metaphase I, the pairs of chromosomes become aligned at the equator of the cell. The now complete spindle becomes more apparent (Fig. 12.2).

Metaphase I resembles metaphase of mitosis, except that when the chromosomes move to the imaginary, circular, platelike *equator,* homologous chromosomes are lined up directly opposite one another on each side of the equator (Fig. 12.5). As in mitosis, at this stage, two chromatids of a chromosome are held together at the centromere, and each homologous chromosome functions as a single unit.

Anaphase I

At anaphase I, each chromosome, consisting of two chromatids, migrates to a **pole** (the region at each end of the cell, analogous to the poles of the earth). That is, the two chromosomes migrate to opposite poles; see Figs. 12.2 and 12.3c.

In anaphase of mitosis, the two chromatids of each chromosome, which are being held together at their centromere, separate as the sister centromeres move apart and a chromatid of each pair migrates to an opposite pole.

(d)

(e)

(f)

Figure 12.3 Continued.

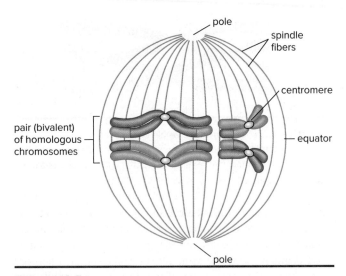

Figure 12.5 A diagram of paired homologous chromosomes at metaphase I of meiosis.

Anaphase I of meiosis is fundamentally different from anaphase of mitosis in that the chromatids of each chromosome remain together at their centromeres and do not separate from one another. Instead, a whole chromosome from each pair migrates to an opposite pole. Consequently, when the chromosomes arrive at the poles, each still consists of two chromatids, but only half the total number of chromosomes is at each pole. Since each chromosome is involved in at least one crossover in prophase I, one of the chromatids will consist of a mixture of original DNA and DNA from a homologous chromosome.

Telophase I

Depending upon the species, chromosomes in telophase I now either partially revert back to interphase, becoming longer and thinner as they do so, or proceed directly to division II; two new cells will eventually form, each with half the chromosome number of the original cell (see Figs. 12.2 and 12.3d). Normally, no new nuclear envelopes are organized around the chromosomes, but the nucleoli usually do reappear. This phase has been completed when the original cell has become two cells or two nuclei.

Division II (Meiosis II, or Equational Division)

The events of the second division of meiosis are similar to those of mitosis, except that there is only one set of chromosomes. In some organisms where there is no interphase, the process proceeds directly from telophase I to prophase II or metaphase II.

Prophase II

At prophase II, chromosomes become shorter and thicker, and their two-stranded nature once more becomes apparent (see Fig. 12.2).

Metaphase II

At metaphase II, the centromeres of the chromosomes become aligned along the equator and new spindles become conspicuous and complete.

The two spindles may be formed either perpendicular to the ones that had formed in metaphase I or along the same plane. A spindle fiber is attached to each centromere, and other spindle fibers extend in arcs from pole to pole (see Figs. 12.2 and 12.3e).

Anaphase II

At anaphase II, the centromeres and chromatids of each chromosome separate and migrate to opposite poles (see Fig. 12.2).

Telophase II

At telophase II, the coils of the chromatids (now called chromosomes again) relax, so that the chromosomes become longer and thinner. Note that, at this point, each chromosome consists of one DNA molecule, in contrast to two at the beginning of meiosis II. New nuclear envelopes and nucleoli appear for each group of chromosomes (see Fig. 12.2).

Telophase II is accompanied by the formation of new cell walls between each of the four groups of chromosomes (see Figs. 12.2 and 12.3f). Each of the four products of telophase II carries half of the chromosomes that were present at the beginning of meiosis, and none of the four cells will have the same combination of DNA. In plants, these four cells develop into haploid structures from which gametes are produced by mitosis.

12.3 Alternation of Generations

Learning Outcomes

a. Define haploid and diploid.

b. Explain the meaning of alternation of generations using the terms sporophyte and gametophyte.

As we have noted, meiosis occurs at some point in the life cycle of all organisms that reproduce sexually. In diploid organisms (see below) every product of meiosis contains one copy of each chromosome. The chromosomal complement consisting of two complete sets of chromosomes is restored when gametes unite, forming a zygote.

Any cell having one set of chromosomes is said to be **haploid,** and any cell with two sets of chromosomes is said to be **diploid.** By the time meiosis is complete, four *haploid* cells have been produced from one *diploid* cell. The gametes are haploid, while a zygote is diploid.

Many plant species contain more than two sets of chromosomes and are said to be polyploid. Cells having three sets of chromosomes are said to be *triploid* ($3x$). Examples of triploid plants include navel oranges and seedless watermelons. Since the homologous chromosomes of a triploid cell undergoing meiosis cannot pair properly, gametes are typically inviable and seeds are not produced. Bread wheat is hexaploid, with six sets of chromosomes, while strawberries are octaploid, with eight sets. The products of meiosis in bread wheat would have three copies of chromosomes (half of the original complement of six) and those of strawberry would contain four copies of each chromosome.

(a)

(b)

Figure 12.6 When going for a walk, you may be treading on gametophyte tissue if you are stepping on simple plants such as mosses (a). If more advanced plants, such as grasses, are underfoot, then you are walking on sporophytes (b). *(a) © McGraw-Hill Education/ Richard Gross, photographer; (b) © Shutterstock/jaroslava V*

In most animals, the only haploid cells (i.e., cells with a single set of chromosomes) are the gametes (egg and sperm). In plants and other organisms, however, this is generally not so. In the life cycle of a diploid plant species that undergoes sexual reproduction, there is an alternation between a multicellular diploid ($2x$) **sporophyte** ($2n$) phase and a multicellular haploid (x) **gametophyte** (n) phase. This is commonly referred to as **alternation of generations.**

It is important to understand the terminology described here: "$2n$" refers to the sporophyte generation, or the stages in the life cycle that result from fertilization; "n" refers to the gametophyte generation, or the stages in the life cycle that result from meiosis; "x" refers to the number of sets of chromosomes at any stage. In diploid plants, n corresponds to x. That is, the sporophyte generation ($2n$) is $2x$, while the gametophyte generation (n) is x. However, as mentioned previously, many plant species are polyploid, with more than two sets of chromosomes. In polyploid species, n does not equal x. For example, we mentioned that strawberries contain eight sets of chromosomes. Consequently, strawberry plants, which constitute the sporophyte generation ($2n$), are $8x$. Pollen and eggs in strawberry flowers constitute the gametophyte generation (n) and are $4x$.

The diploid plant body is called a sporophyte. It develops from a zygote and eventually produces **sporocytes,** each of which undergoes meiosis, producing four **spores.** The spores undergo mitosis to produce multicellular haploid bodies called gametophytes. Specialized cells within these gametophytes function as gametes.

In chapters to follow, we will see that the gametophytes of many primitive plant species are more conspicuous than the sporophytes. However, as members of the Plant Kingdom become more complex, the gametophytes become smaller (Fig. 12.6). In the most advanced plants, the gametophyte generation is microscopic, while the sporophyte is the dominant generation. The switch from one generation to the other takes place as spores are produced by meiosis (sporophyte to gametophyte) and again when a zygote is produced through fusion of gametes, or **fertilization** (gametophyte to sporophyte).

Alternation of generations can be seen in the Protistan, Fungal, and Plant Kingdoms. It becomes most conspicuous, however, in the Plant Kingdom.

If you will note, along with the accompanying diagram (Fig. 12.7), the following six rules for the alternation

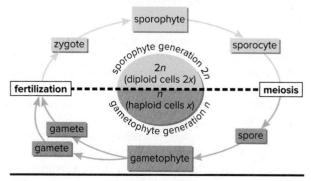

Figure 12.7 A typical life cycle of plants or other organisms that undergo sexual reproduction with an alternation of generations.

KEY THEME 12.2 evolution — Ordering Complexity

Biologists continually struggle with the task of classifying organisms. It would be ideal to find a system in which all organisms fit into well-defined categories. However, the array of living organisms on the planet is very complex, with an incredibly broad range of lifestyles. Life cycles provide one classification option. But even with this one method, several systems have been defined. They are outlined in Box Table 12.2A. Organisms with a haplobiontic life cycle have a multicellular haploid or diploid phase, but not both. Therefore, they do not exhibit an alternation of generations, and it is not appropriate to say that they have gametophyte and sporophyte generations. In contrast, an alternation of generation is observed in diplobiontic organisms because both the haploid (gametophyte) and diploid (sporophyte) generations are multicellular. Another classification system separates the haplobiontic category into two types, those in which the haploid generation is multicellular (haplontic) and those in which the diploid generation is multicellular (diplontic). An alternation of generations (multicellular haploid and diploid) is seen only in organisms with a haplodiplontic life cycle. Finally, with organisms that exhibit an alternation of generations, the sporophyte and gametophyte can look similar (isomorphic) or different (heteromorphic). As you examine life cycles throughout the remainder of this book, think about how they fit into these classification schemes.

Box Table 12.2A Classification Schemes Based on Life Cycles

Life Cycle	Features	Example
Haplobiontic	Multicellular haploid or diploid phase	Some algae (*Chlamydomonas*); most fungi
Diplobiontic, dominant haploid	Multicellular gametophyte (haploid) and sporophyte (diploid) phases; gametophyte is larger than sporophyte	Bryophytes
Diplobiontic, dominant diploid	Multicellular gametophyte (haploid) and sporophyte (diploid) phases; gametophyte is larger than sporophyte	Some algae (*Laminaria*), vascular plants
Haplontic	Multicellular haploid, but not diploid phase	Some algae (*Chlamydomonas*); most fungi
Diplontic	Multicellular diploid, but not haploid phase	Some algae (*Vaucheria, Acetabularia*)
Haplodiplontic	Multicellular gametophyte (haploid) and sporophyte (diploid) phases	Some algae (*Ulva*), bryophytes, vascular plants
Isomorphic	Sporophyte and gametophyte generations look identical	Some algae (*Ulva*)
Heteromorphic	Sporophyte and gametophyte generations look different from each other	Some algae (*Laminaria*); bryophytes, vascular plants

of generations, you should have little trouble following the life cycle of any sexually reproducing organism discussed in this book.

1. The first cell of the *gametophyte generation* is a *spore*, and the last cell is a *gamete*.
2. Any cell of the gametophyte generation (*n*) contains half as many chromosomes as the sporophyte generation (*2n*).
3. The first cell of the *sporophyte generation* is a *zygote*, and the last cell is a *sporocyte*.
4. Any cell of the sporophyte generation (*2n*) contains twice as many chromosomes as the gametophyte generation.
5. The change from the sporophyte to the gametophyte generation occurs as a result of *meiosis*.
6. The change from the gametophyte to the sporophyte generation occurs as a result of the fusion of gametes at *fertilization*.

The word *generation* as used in *alternation of generations* simply means phase of a life cycle and should not be confused with the more widespread use of the word pertaining to time or offspring. Sexual reproduction produces new gene combinations in three ways: (1) Recombination as a result of crossing over in meiosis I combines pieces of maternal chromosomes with pieces of paternal ones. (2) Segregation at meiosis I randomly assorts maternal and paternal chromosomes into daughter cells. (3) The union of egg and sperm cells at fertilization brings together unique combinations of genes.

Consequently, meiosis and fertilization ensure that every individual in a sexually reproducing population is unique. In fact, while genes are always passed from one generation to another, every individual contains a combination of genes that will last only one generation and will never occur again.

SUMMARY

12.1 Introduction to Meiosis and Alternation of Generations

- Asexual reproduction may take place through vegetative propagation, and it produces genetically identical offspring. Sexual reproduction results from the union of two haploid gametes to produce a zygote. The products of sexual reproduction are genetically different from their parents and each other.
- Meiosis ensures that gametes will have half the chromosome number of the parent and that offspring will not be identical to parents.

12.2 The Phases of Meiosis

- Meiosis requires two successive divisions, each of which, like mitosis, is divided into arbitrary phases even though the process is continuous. In prophase I, the chromosomes become paired and exchange parts. The similar chromosomes of each pair are referred to as homologous. Exchange of parts of chromatids occurs by crossing over and formation of chiasmata as they break and reattach to chromatids of homologous chromosomes.
- In metaphase I, the chromosomes become aligned at the equator in pairs, and in anaphase I, whole chromosomes from each pair migrate to opposite poles. In telophase I, the chromosomes either partially revert to their interphase state or initiate the second division, which is similar to mitosis.
- In prophase II, the chromosomes of each of the two groups become shorter and thicker again, with both groups becoming aligned at their respective equators in metaphase II. In anaphase II, the chromatids of each chromosome separate and migrate to opposite poles, and in telophase II, the chromatids (now called chromosomes again) lengthen, and new nuclear envelopes and nucleoli appear for all four cells. New cell walls are produced between each of the four cells.
- Since the chromatids exchanged parts earlier, none of the four cells have identical combinations of DNA. In addition, each cell has half the original number of chromosomes. Each of the four cells constitutes a spore, which may develop into a structure within which gametes may be produced by mitosis.

12.3 Alternation of Generations

- Any cell having one set of chromosomes is said to be haploid, or to have x chromosomes; any cell having two sets of chromosomes is said to be diploid, or to have $2x$ chromosomes.
- In the life cycle of an organism that undergoes sexual reproduction, there is an alternation between a gametophyte (n) phase and a sporophyte ($2n$) phase.
- The change from the gametophyte phase to the sporophyte phase occurs when two gametes (each n) unite, forming a zygote ($2n$) in the process of fertilization. The change from the sporophyte to the gametophyte phase occurs as a result of a sporocyte undergoing meiosis. This switching of phases is referred to as alternation of generations.

REVIEW QUESTIONS

1. During which phase of meiosis does each of the following events occur? (a) crossing-over (b) chromatids separate at their centromeres and migrate to opposite poles (c) chromosomes become aligned in pairs at the equator
2. Is there any difference between mitosis and the second division phases of meiosis?
3. What is the significance of meiosis with respect to sexual reproduction?
4. What is the difference between $2x$ and $2n$?
5. At what stage of a life cycle does the chromosome number of cells switch from x to $2x$?

DISCUSSION QUESTIONS

1. If an organism reproduces very freely by asexual means, is there any advantage to its also reproducing sexually?
2. Would it make any difference if the events of the second division of meiosis took place before the events of the first division?
3. Would it be correct to say that a bivalent has four times the amount of DNA present in the chromosomes of a cell in anaphase II?
4. Would you assume that the length of a chromosome might have anything to do with the number of crossovers it might form with its homologue? Explain.
5. Two mitotic divisions may take place in as little as a few hours, whereas meiosis may take much longer. Can you suggest any reasons for this?

CHAPTER

13

Genetics and Molecular Biology

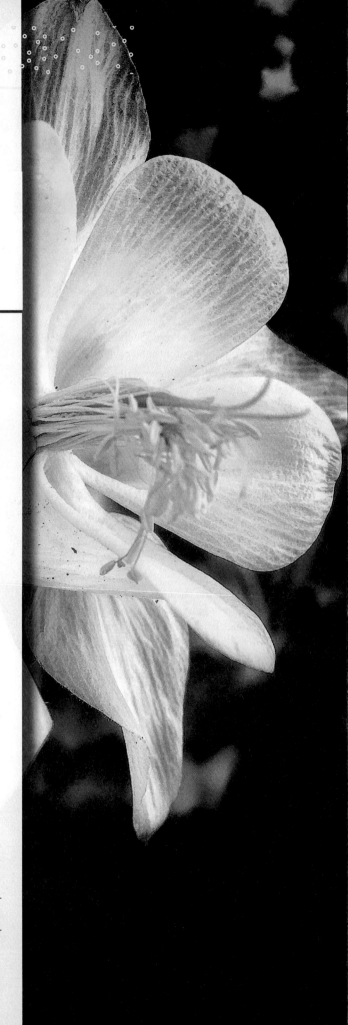

13.1 Introduction to Genetics and Molecular Biology

13.2 Molecular Genetics
   Structure of DNA
   DNA Functions
   Key Theme 13.1: Massive DNA Sequencing
   Key Theme 13.2: The Polymerase Chain Reaction (PCR)

13.3 Cytogenetics
   Changes in Chromosome Structure
   Changes in Chromosome Number

13.4 Mendelian Genetics
   Mendel's Studies
   The Monohybrid Cross
   The Dihybrid Cross
   The Backcross
   The Testcross
   Incomplete Dominance
   Interactions among Genes
   How Genotype Controls Phenotype

13.5 Quantitative Traits

13.6 Extranuclear DNA

13.7 Linkage and Mapping

13.8 The Hardy-Weinberg Law

OVERVIEW

In this chapter, the topic of genetics is introduced with the story of Barbara McClintock's discovery of transposable elements ("jumping genes") in corn. The chapter progresses from genetics at the molecular level to the cellular level (cytogenetics) to the organismal level (Mendelian genetics and related topics) and then finishes with population genetics (the Hardy-Weinberg law).

A single flower of Columbine (*Aquilegia* spp.), a wild flower in temperate regions of the world. © Crystal Neitz/Stockimo/Alamy

13.1 Introduction to Genetics and Molecular Biology

Learning Outcome

a. Explain why Barbara McClintock's work was groundbreaking.

Every so often, someone comes along who is able to push beyond our limits of understanding in science and change the way we view the natural world. Barbara McClintock was one of the most significant scientists in 20th-century biology because she caused a major shift in the way we view gene organization. She was a geneticist who carried out most of her work at the Cold Spring Harbor laboratory in New York in the 1940s and 1950s. While looking at corn seedlings for one of her genetic studies, she noticed that patches of color often appeared where they did not seem to belong. There were patches of yellow in green leaves or patches of green in white leaves. She also noticed spots of dark color in corn kernels that should have been light-colored, based on her genetic studies (Fig. 13.1). Instead of ignoring these plants as genetic oddities, McClintock undertook years of research to identify the cause of these patterns.

In the early 1950s, McClintock published work introducing the concept of **transposition,** or the movement of a piece of a chromosome to another chromosomal location. A **transposable genetic element** (or *jumping gene* in popular literature) is a DNA fragment that can move to a new location on the same chromosome or even another chromosome. If it moves into an existing gene, then the function of that gene is disrupted. A transposable element can also move out of a gene, restoring its original function. So, the patches of yellow in green leaves were regions where transposable elements were inserted into a gene involved in chlorophyll synthesis. Green areas on yellow leaves resulted when a transposable element moved out of a chlorophyll synthesis gene and allowed it to function again.

McClintock was a highly respected figure in the field of genetics and was able to describe in detail the actions of several transposable elements in corn. Accordingly, even though the concept of movable genes was difficult to fit into the framework of conventional genetics, McClintock's ideas were generally accepted by her peers. Before long, transposable elements were found in other organisms as well.

The discovery of the structure of DNA by James Watson and Francis Crick in 1953 ushered in the new field of **molecular genetics.** At that time, geneticists thought of McClintock's work as a remnant of a previous era and not relevant to modern genetics. However, studies of DNA and gene structure transformed the concept of a gene. Genes were no longer abstract entities but were actual sequences of nucleic acids that could be isolated and characterized. The concept of mutation within a gene became understandable.

McClintock's work was formally recognized in 1983 when she was awarded a Nobel Prize. We now know that transposable elements are widespread and have probably been in organisms for a long time. Amazingly, 85% of the DNA in corn is composed of transposable elements. We do not know, however, if they represent genetic parasites causing mutations as they move about an organism's genetic material or if they perform valuable functions. One theory is that they allow nature to tinker with chromosomes much as human genetic engineers do. It may be evolutionarily beneficial to copy, move, and rearrange pieces of chromosomes, creating new and occasionally better combinations of genes within an organism.

Transposable elements (transposons) have become powerful tools for molecular geneticists. Researchers can use transposons to create mutations. When a transposable element inserts into a gene, it causes a mutation and often alters the gene product. Conversely, function is restored when the transposon excises itself from the gene. Transposable elements, then, are capable of causing reversible, often directed, mutations. By comparing the phenotype of a normal plant with one that contains a transposon inserted into a specific gene, scientists can determine the function of that gene.

It is interesting that transposons are often activated under genomic stress. Examples of genomic stress include chromosomal damage by radiation, creation of hybrids between divergent species, and attack by pathogens.

Figure 13.1 This region of an ear of corn shows the effects of transposable elements. The speckles in the center kernel are clusters of cells in which a transposable element has removed itself from a purple pigment gene, so purple pigment is produced. Yellow areas in that kernel do not express purple pigment because the transposable element has interrupted that pigment gene. © *Kingsley Stern*

The activation of transposons in these situations results in changes in gene expression that might allow the plant to tolerate the stress. This is not a targeted response to the stressful event, but is instead a random attempt to deal with a challenge that the plant may not have encountered before.

13.2 Molecular Genetics

Learning Outcomes

a. Describe the arrangement of the DNA components in a double helix.
b. List the four major functions of DNA.

In Chapter 3, we mentioned that chromosomes in the nucleus of each cell carry genetic information that is passed on from one generation to the next. Chromosomes are composed of two types of large molecules: DNA and protein. Until the early 1950s, most scientists believed that proteins carried the genetic information because they thought that the 20 amino acids in proteins could provide more diversity than the four bases in DNA. We now know, of course, that DNA is responsible for carrying genetic information.

Structure of DNA

Chromosomes are composed of chromatin, which is DNA and associated proteins. A DNA molecule is a simple, elegant chain of building blocks called **nucleotides.** Each nucleotide consists of three parts: (1) a nitrogen-containing compound, called a **base;** (2) a 5-carbon sugar, named **deoxyribose;** and (3) a phosphate group. Both the nitrogenous base and the phosphate group are bonded to the sugar (Fig. 13.2). Four types of nucleotides occur in DNA. Each has a unique nitrogenous base, but all have the same phosphate group and deoxyribose sugar. Two of the bases—**adenine** (A) and **guanine** (G)—are called **purines.** Each has a molecular structure that resembles two linked rings. The other two bases—**cytosine** (C) and **thymine** (T)—are called **pyrimidines.** They each have a molecular structure consisting of a single ring.

Biochemists assign numbers to carbon atoms in order to keep track of them. The five carbons in the sugar molecule in a nucleotide are numbered 1' through 5', starting with the carbon after the oxygen atom and proceeding clockwise around the molecule. The carbon that is not in the ring, therefore, is the 5' carbon. The numbers contain the prime symbol (') to distinguish them from carbons in the bases, which are also components of nucleotides. When nucleotides join together in long chains, the phosphate component links the 5' carbon in one nucleotide to the 3' carbon in the next nucleotide via a phosphodiester bond. When a DNA double helix forms, one strand is oriented so the 5' carbon is at the top of the nucleotide; the parallel strand is in the other orientation, with the 5' carbon at the bottom of the nucleotide. Strand orientation is important in nucleic acids because a new chain can only grow in the 5' to 3' direction. The nucleotides in a DNA molecule are bonded to each other

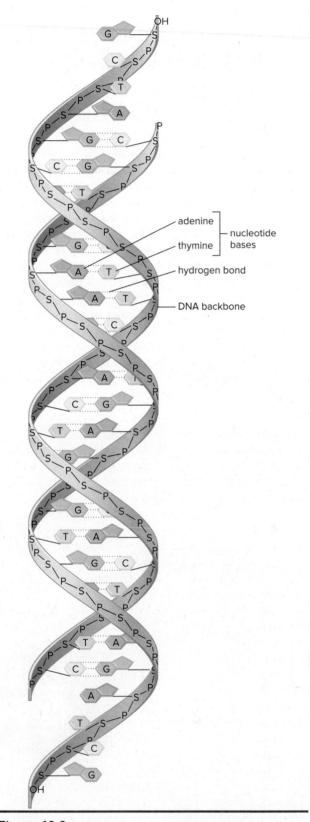

Figure 13.2 In this enlargement of a small portion of a DNA molecule, the rungs of the twisted ladder formed by the two entwined spiraling strands consist of nitrogen-containing bases supported by alternating units of sugar (S) and phosphate (P) molecules. The purine adenine or guanine (A or G) occurs opposite the pyrimidine thymine or cytosine (T or C). The purines and pyrimidines opposite each other are held together by hydrogen bonds linking the nitrogenous bases of the paired molecules.

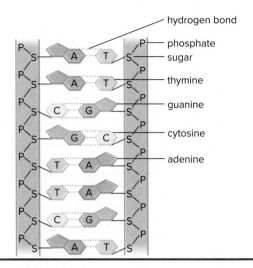

Figure 13.3 This tiny portion of a DNA molecule shows eight base pairs. Since a chromosome is millions of base pairs long, the number of variations in sequences of pairs is virtually unlimited.

end-on-end through phosphodiester bonds, but also side-by-side through hydrogen bonds. This creates a structure that looks like a spiral staircase. The handrails for the stairs are composed of alternating sugar and phosphate groups held together by phosphodiester bonds (Fig. 13.3). Each sugar is also bound to a base. Hydrogen bonds hold each base on one side of the helix to another base on the other side, making the steps of the staircase. Although the structure of a DNA molecule is very simple—alternating sugar and phosphate groups on the outside and pairs of bases in the middle—it provides all the genetic diversity on the planet. The variation comes in the form of base pairs. A single chromosome contains a DNA molecule that is often hundreds of millions of base pairs in length. The sequence of base pairs on a DNA molecule determines whether, for example, that DNA will direct the synthesis of a chlorophyll molecule or a hemoglobin molecule.

As mentioned earlier, in the early 1950s, James Watson and Francis Crick began to unravel the mystery of DNA structure. Much of their work involved piecing together existing information to develop a model. They knew Linus Pauling had postulated that the structure of DNA might be similar to the structure of protein. Pauling had shown part of the structure of some proteins to be helical and maintained by hydrogen bonds between the amino acids. Watson and Crick also knew from X-ray work by Rosalind Franklin and Maurice Wilkins that DNA molecules are composed of regularly repeating units in a helical arrangement. Finally, studies by Erwin Chargaff indicated that the number of pyrimidine nucleotides (cytosine + thymine) in DNA equals the number of purine nucleotides (adenine + guanine). This suggested to Watson and Crick that purines pair with pyrimidines.

When they had put together all the facts, Watson and Crick concluded that a DNA molecule consists of a double helix whose two strands appear to be wrapped around an invisible pole—one strand spiraling in one direction and the other strand spiraling at the same angle in the opposite direction. They also concluded that nucleotides are linked in ladderlike fashion between the two strands. Furthermore, in order for the nucleotides to fit precisely in the DNA molecule, two linked purines or two linked pyrimidines would be either too wide or too narrow. However, a purine and a pyrmidine linked together would fit perfectly. Accordingly, they concluded that the steps of the staircase had to consist exclusively of purine-pyrimidine pairs. In order for hydrogen bonds to form correctly between base pairs, guanine must pair with cytosine, and thymine must pair with adenine. Watson and Crick's model is now universally accepted as an authentic representation of a DNA molecule.

DNA Functions

The DNA in a cell must perform four major tasks. It must (1) store genetic information; (2) copy that information for future generations of cells; (3) express that information; and (4) occasionally change its message, or **mutate**.

Storage of Genetic Information

Most DNA in plants resides in the nucleus. The nucleus is a compartment in which proteins that interact with DNA are concentrated. This compartmentalization is important because it creates an environment in which biochemical reactions are optimized by high concentrations of substrates and the enzymes that act on them.

The genetic information in a DNA molecule resides in the sequence of nucleotides. The sequence GAATCC, for example, codes for a different set of amino acids than CTTAGA. Early scientists were fooled by the simplicity of a DNA molecule and did not believe it could provide information for the survival of all organisms on earth. However, the DNA in a typical plant cell contains millions or billions of base pairs. A barley plant with 5 billion base pairs of DNA in its nucleus could produce $4^{5,000,000,000}$ different DNA sequence combinations! This provides more than enough genetic variability for the plant.

At this point, we need to introduce some important concepts, beginning with that of the gene. We are all familiar with the idea that genes control physical traits, such as eye color in people and height in plants. How do they do that? From a molecular perspective, a **gene** is a segment of DNA (several thousand base pairs long) that directs the synthesis of a protein. That protein is then used by the cell as structural or storage material, or it influences the activities of the cell by acting as an **enzyme**. As you recall from Chapter 2, enzymes are organic catalysts. Any activity in the cell, from photosynthesis to cell-wall construction, absolutely depends on enzymes to facilitate chemical reactions. Therefore, all activities of the cell are controlled by enzymes, which require genes for their synthesis. Each kind of organism has many different genes in its **genome**, which is a term for *the sum total of DNA in an organism's chromosomes.*

KEY THEME 13.1 molecular — Massive DNA Sequencing

Technological breakthroughs in DNA sequencing in the early 21st century rival those in computer electronics. The Human Genome Project of the 1990s stimulated an interest in large-scale DNA sequencing for the diagnosis of human diseases. The investment in research and development of new methods for high throughput sequencing has benefited not only the medical community but also many other fields of science, including forensics, livestock breeding, and pathogen epidemiology. Botanists are now also reaping the benefits of these new technologies. So-called "next generation sequencing" allows DNA samples to be sequenced without the complex, expensive, and time-consuming cloning and mapping steps used in the Human Genome Project and other early sequencing efforts. The cost of DNA sequencing is 10,000 times less now than it was 10 years ago, and it continues to drop. Box Figures 13.1A and 13.1B show a flow cell and a DNA sequencer, respectively. DNA samples are added to the flow cell, which is then placed in the loading dock (the black tray) on the left side of the machine. Optical scanners (upper part of the left side) scan the samples and feed sequence information into the computer. Reagents for the sequencing reactions are found in the right side of the machine. To put the power of "next gen" sequencing into perspective, in 2008, the total amount of DNA sequence stored in GenBank (the US repository for DNA sequence data) totaled about 100 billion base pairs. That represented all the DNA data that had ever been generated and published from all US research labs. Just 4 years later, using next gen sequencing technology, a single lab could generate that much information in just 2 days! The ability to sequence massive amounts of DNA is especially important for plant scientists, since the DNA sequences of many plant species are very large. For example, the entire DNA complement of a pine tree is 40 billion base pairs. That is 13 times the size of the human genome. It took 10 years to sequence the human genome, but now an entire genome can be sequenced in months. Once the sequence of a species is determined, it is relatively easy to sequence other members of that species for comparison. Next generation sequencing technologies are changing the questions that scientists can answer. The challenge to scientists now is not the ability to generate DNA sequence data but the ability to analyze it. As you can imagine, computers with enormous storage and processing capabilities are needed to sort through billions of bits of data. More importantly, bioinformaticists, scientists with backgrounds in computer science, statistics, and biology, are needed to make sense out of these huge data sets.

(A)

(B)

Box Figures 13.1A and 13.1B (A) A flow cell, which holds the sample for DNA sequencing. (B) A high throughput DNA sequencer.
(A-B) © Shelley H. Jansky

A relatively simple organism such as a bacterium typically has several thousand genes in its genome. On the other hand, more complex eukaryotic organisms usually have between 10,000 and 50,000 genes per genome. The differences between organisms lie in the differences in the composition of their genomes. Organisms that are very dissimilar—for example, bacteria and higher plants—have many differences in their DNA sequences. The genomes of similar organisms, such as domestic plants and their wild relatives, are very similar, yet different enough to account for dramatic differences in appearance.

Replication (Duplication) of Information

DNA must duplicate itself precisely in order to pass along its information from generation to generation. DNA replication occurs during the S phase of the cell cycle (see Chapter 3). Differences between the DNA of one organism and that of another lie in nucleotide sequence and number. If the four nucleotides are synthesized in living cells, then what tells the cell exactly how to put them together to form DNA molecules with proper nucleotide sequences?

Watson and Crick predicted that if the two strands of DNA are "unzipped" by an enzyme breaking the hydrogen bonds between the nucleotides down the middle of the ladder, then each separated strand provides all the information needed to put together a new double-stranded helix (Fig. 13.4). In other words, since guanine can pair only with cytosine, and vice versa, a guanine nucleotide in a DNA strand can only bond to a cytosine nucleotide. If the two strands of a DNA molecule "unzip," then individual nucleotide "building blocks" can line up next to each of the single chains in precise sequence and be bonded together to form a new chain; each single chain acts as a template for the creation of a new double-stranded DNA molecule. When replication is complete, two double helices have been created from a single one. Each new DNA molecule consists of one strand from the original molecule and another built using that parental strand as a template. This is called **semiconservative replication.**

Replication is a remarkably rapid and precise process. During DNA replication, nucleotides are added by the enzyme **DNA polymerase** at an amazing rate of approximately 833 per second in prokaryotes and 33 per second in eukaryotes. However, even at this rate, a plant chromosome that is 200 million base pairs long would take 70 *days* to replicate. Because a cell must replicate its DNA before it undergoes mitosis, this replication rate is far too slow for normal plant growth. How is this problem resolved? In two ways: (1) Replication proceeds simultaneously in both directions on the chromosome, cutting the replication time in half, and (2) Replication begins at many points on a chromosome simultaneously.

As you might imagine, mistakes during DNA replication can have serious consequences. Fortunately, the nucleus contains enzymes that are capable of "proofreading" during replication. They detect and correct mismatched base pairs, such as guanine paired with adenine or thymine paired with cytosine. Consequently, about only one in every million nucleotides added during replication is incorrectly paired. However, each nucleus of a corn plant contains 4 billion base pairs of DNA. Therefore, about 4,000 mutations (changes in DNA sequence) occur with each cell division.

Expression of Information

Since the nature and structure of DNA are of fundamental importance to its ability to control the form and function of organisms, there appears to be a paradox. If every cell in an organism contains the same genetic information, then why do cells within a plant differ from each other? That is, why is a bark cell so different from a leaf cell?

While each cell in a plant contains the same DNA information, different subsets of this information are read in each cell type. For example, a plant has epidermal cells that secrete a protective waxy cuticle, and it has mesophyll cells that are highly specialized for photosynthesis. The epidermal cells express genes (read DNA sequences) that encode enzymes necessary to build a waxy cuticle. Mesophyll cells, on the other hand, will not express cuticle genes but do express a set of genes that encode enzymes for chlorophyll synthesis. One of the most active areas of biology research today is directed toward understanding how organisms control gene expression.

A cell's environment can influence the set of genes expressed by that cell. Many cellular changes occur as a result of the expression of new sets of genes induced by hormones (see Chapter 11). We know, for example, that an epidermal cell expresses genes for cell elongation if the hormone auxin is present; when mesophyll cells are moved from dark into light, many dramatic changes in gene expression occur as genes for the photosynthetic machinery are turned on.

Figure 13.4 This DNA molecule is replicating from left to right. The DNA molecule replicates by "unzipping" along the hydrogen bonds, which link the pairs of nucleotides in the middle, and then each half serves as a template for new nucleotides to be brought in. As the new nucleotides fit into the appropriate places and are linked together, two double helices form.

KEY THEME 13.2 molecular — The Polymerase Chain Reaction (PCR)

How many times have you said to yourself, "Why didn't I think of that?" Many scientists must have had that thought when Kary Mullis described the polymerase chain reaction (PCR) in 1984. While Mullis was driving to his cabin in northern California one evening, he was thinking about a method for determining the arrangement of nucleotide sequences in DNA molecules. The technique used DNA replication enzymes to synthesize short strands of DNA in a test tube. After running through several scenarios in his head, he realized that he could create a chain reaction in a test tube. Every reaction would double the amount of DNA synthesized, causing an exponential growth in the number of DNA molecules created. Twenty cycles of DNA replication would create from a single molecule a million identical DNA molecules. For years scientists had worked with replication enzymes in DNA synthesis experiments, but none of them had conceived the power of a chain reaction in a test tube.

The polymerase chain reaction allows scientists to replicate (duplicate) specific DNA sequences in small test tubes. The reaction requires DNA polymerase, nucleotides (actually, nucleoside triphosphates, the high-energy precursors to nucleotides), short (approximately 20 nucleotide) single-stranded DNA molecules called *primers,* and DNA from an organism to serve as a template. The reaction follows these steps:

1. The primer is made with the aid of a DNA synthesizer. The primer DNA sequence is critical because it determines what piece of an organism's DNA will be replicated during the PCR reaction.
2. DNA polymerase, nucleotide building blocks, primers, and DNA template are mixed together in a small test tube.
3. The test tube is placed in a *thermal cycler.* This machine is capable of rapidly heating and cooling samples.
4. The tube is heated to approximately 94°C to separate the template DNA into single-stranded molecules. In a living cell, enzymes carry out this process. In an experimental setting, however, heat provides a simple way to make the DNA single-stranded.
5. The tube is then cooled down to between 35°C and 60°C. This allows the primers to locate and bind to complementary base sequences on the template DNA. Remember that DNA molecules are typically double helices with adenine-thymine pairs and guanine-cytosine pairs.
6. The tube is heated to 72°C, which is the optimal temperature for the DNA polymerase used in PCR reactions. Because the enzyme for PCR is derived from thermophilic (heat-loving) bacteria, it is not destroyed by heating (see Chapter 17). DNA polymerase locates the short, double-stranded regions where primers have bound to template DNA; it then uses nucleotide building blocks to make double-stranded DNA near the primer.
7. Steps 4 through 6 are repeated 20 to 30 times.

After completion of the PCR reaction, millions of copies of a double-stranded DNA molecule are present in the test tube. The real power of the reaction, though, comes from its ability to amplify only the DNA sequence adjacent to the primer binding site. This allows a researcher to make an endless supply of a specific segment of DNA. Box Figure 13.2A shows how a target DNA sequence is amplified through cycles of PCR.

The polymerase chain reaction has become a standard tool in many research laboratories. Countless applications for this technique have been developed and more are on the way. PCR is an important tool because (1) large quantities of DNA can be synthesized from very small samples and (2) only the target DNA sequence is amplified.

How is the PCR reaction used today? Here is a short list of examples:

1. **Forensics.** Samples of hair, blood, and skin contain enough DNA to be amplified by PCR. A comparison of amplified DNA samples from a crime scene with DNA from a suspect can provide compelling evidence.
2. **Disease detection.** Primers have been developed specifically to amplify DNA from pathogens. For example, a person's blood could provide template DNA for the PCR reaction with a primer that binds only to a sequence from the human immunodeficiency virus (HIV). Then, even low levels of DNA produced by the virus would be amplified by PCR. This makes early disease detection possible. PCR is also being used to identify pathogens in plants.
3. **Food safety.** Pathogens in food can be detected quickly and accurately using PCR. Primers specific for pathogenic strains of *Escherichia coli* can detect very low levels of the pathogen in hamburger. Similar tests are available for detection of the organisms that cause *Salmonella* poisoning and botulism, among others.

The expression of genetic information requires two processes. First, the information coded by DNA is *transcribed.* During **transcription,** a copy of the gene message is made using RNA (ribonucleic acid) building blocks. RNA is similar to DNA with a few important differences. First, RNA nucleotides contain **ribose** sugars instead of deoxyribose sugars. Ribose contains one more oxygen atom than deoxyribose. Second, thymine is replaced by **uracil,** another nitrogenous base. Finally, RNA molecules are typically single-stranded. After transcription has been completed in the nucleus, the RNA

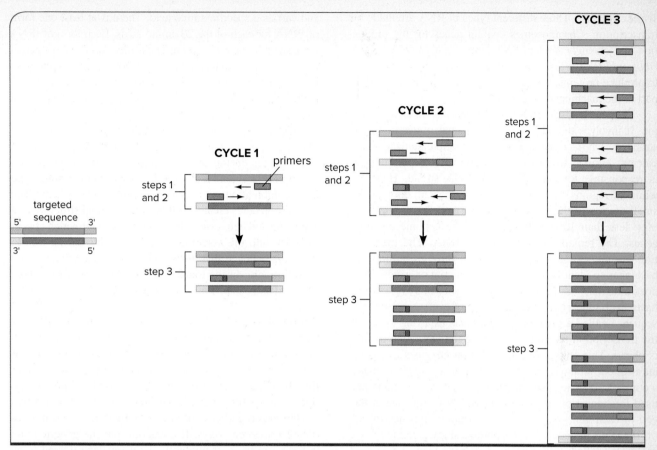

Box Figure 13.2A Three cycles of the polymerase chain reaction.

4. **Fetal testing.** A small sample of fetal tissue can be amplified using primers specific to DNA regions that, in mutant form, cause human diseases. Primers are available to detect sickle-cell anemia, phenylketonuria, cystic fibrosis, hemophilia, and Huntington's disease, to name a few. Of course, a moral dilemma is created if the fetus tests positive.
5. **Evolutionary relationships.** DNA from extinct animals can be amplified and compared with the DNA of modern relatives. With the use of PCR, DNA from a 120-million-year-old weevil and from a 250-million-year-old bacterium have already been extracted and amplified.
6. **Conservation efforts.** DNA from tissue samples collected at poaching sites can be amplified and compared to the DNA of pelts from suspected poachers. PCR also allows scientists to determine whether commercial products containing ivory, meat, or feathers were taken from endangered species.
7. **Genetics of dung.** Even DNA from dung can be amplified with PCR! Scientists trying to understand why the giant ground sloth became extinct have located 20,000-year-old dung in caves. PCR amplification of the dung is being used for comparison with related present-day species. In addition, careful choice of primers allows researchers to amplify DNA from the animal's food and its parasites.

Only a few examples have been presented here. Application of PCR seems to be limited only by our imagination. We are capable of performing this amazingly simple yet powerful reaction because we understand the elements responsible for DNA replication.

is modified and then travels to the cytoplasm, where *translation* occurs. During **translation,** the RNA message provides the information necessary to construct proteins. It might help to remember the following when you are trying to distinguish between transcription and translation. During transcription, a nucleic acid, RNA, is made using another nucleic acid, DNA, as a template. This is analogous to a stenographer transcribing shorthand to longhand. During translation, a protein is made using a nucleic acid (RNA) as a template. This is analogous to a translator converting an English text to Spanish.

Transcription Three different types of RNA are made by transcription. Chromosomes contain genes for the production of *messenger RNA (mRNA), transfer RNA (tRNA),* and *ribosomal RNA (rRNA).* Messenger RNA will be translated to produce proteins, while tRNA and rRNA will be used as part of the machinery for translation.

The mechanism of RNA synthesis is relatively simple. It involves the addition of RNA nucleotides to single-stranded DNA molecules, using complementary base pairing, similar to DNA replication. **RNA polymerase** is responsible for assembling the RNA strand. However, while all regions of the genome are replicated, only a very small portion of the genome is transcribed. In most eukaryotes, less than 10% of the DNA in the genome contains genes. The remainder is **noncoding DNA.** The transcription machinery must therefore search through millions of base pairs to identify small regions that contain genes. How does it do that? We have unraveled some of this mystery but have much to learn. At the beginning of every gene, a DNA sequence called a **promoter region** acts as a flag to signal the transcription machinery scanning the DNA that a gene is ahead. This causes the transcription enzymes to associate with one strand of the DNA molecule and begin to make a copy of the gene. The complementary DNA strand is not used for transcription. A *terminator* DNA sequence at the end of each gene signals the transcription enzymes to fall off the DNA molecule. At this point, a single-stranded RNA molecule, called a **transcript,** has been made using one strand of a gene as a template. The transcript will undergo several modifications and move to the cytoplasm before translation begins.

As we mentioned before, not every gene is expressed in every cell. In addition to identifying genes, the transcription machinery must determine which genes to transcribe, or express. This is a very complex process and involves interactions among genes and signals from the environment. In addition, cells must control the amount of each type of protein they produce. Until recently, scientists thought that the large amount of nonprotein-coding DNA in the genome represented evolutionary "junk." That is, this DNA was what was left over from failed experimentation by organisms during evolution. On the contrary, it now appears that this nonprotein-coding DNA is fundamental to the control of gene expression. Because nonprotein-coding DNA is not translated, scientists never see protein products. However, RNA transcription products from these sequences appear to carry out an amazing array of activities that control the amount of protein produced from conventional "coding" genes.

Nonprotein-coding genes are also responsible for making some of the machinery needed for translation. Chromosomes contain genes for building tRNA. When transcribed, these genes produce molecules of tRNA approximately 80 nucleotides long. The single-stranded RNA produced by transcription spontaneously forms base pairs in some regions, creating a three-dimensional molecule of a characteristic shape. Transfer RNA will act as the "translator" during translation. One end of the tRNA molecule binds to the mRNA, while the other end carries a specific amino acid. There is at least one form of tRNA for each of the 20 amino acids. Each form of tRNA has a specific *anticodon* loop. The **anticodon** is a sequence of three nucleotides that recognizes a codon on mRNA and base-pairs with it. An amino acid binds to the other end of the tRNA with the aid of an enzyme that specifically identifies both a single tRNA strand and its proper amino acid.

Genes for rRNA are also transcribed in the nucleus. This RNA is used to construct **ribosomes,** which act as workbenches during protein synthesis. A ribosome is composed of a large and a small subunit, each containing a specific combination of rRNA molecules and proteins. The rRNA molecules create the three-dimensional structure of a ribosome and help orient the proteins, most of which appear to assist with the assembly of new proteins during translation. It is believed that mRNA slides between the large and small subunits during translation. All ribosomes in a cell are identical and capable of reading any mRNA.

Translation RNA transcripts that code for proteins are called *mRNAs.* It seems puzzling that a nucleic acid with only four possible nucleotides (adenine, uracil, cytosine, and guanine) can be used to make a protein polymer that may have 20 different amino acids. One nucleotide is incapable of encoding one amino acid. Even two nucleotides together could be combined in a maximum of only 16 different ways. It turns out that the genetic code is based on trios of nucleotides we earlier referred to as *codons.* Like words and sentences, codons and genes can be read from one direction only, thereby preventing the possibility of expressing the wrong amino acid. There are 64 possible ways to combine nucleotides, taking three at a time. This is more than enough to encode all 20 amino acids. The order of nucleotides on an mRNA molecule (determined by the sequence of nucleotides in a gene during transcription) determines the sequence of amino acids added during translation. For example, the codon AUG (adenine-uracil-guanine) on an mRNA encodes the amino acid methionine, whereas the codon AGU on an mRNA encodes the amino acid serine. The genetic code (Fig. 13.5) shows the matching of each three-nucleotide codon with a specific amino acid. With few exceptions, the genetic code is universal. In other words, the same code is used in bacteria, protists, fungi, plants, and animals.

As mentioned before, tRNAs act as translators, or decoders, during translation. The anticodon part of the tRNA binds to the mRNA codon by base pairing with it, while the other end of the tRNA molecule has, according to the genetic code, the correct amino acid bound to it. In a sense, then, we can think of tRNAs as the dictionary used to decode an mRNA during translation.

The start of translation is usually signaled by a ribosome in the cytoplasm binding to the mRNA and scanning it for the first AUG codon. This sets the reading frame because each set of three nucleotides is translated in succession after the first AUG. A start tRNA with the amino acid methionine and an anticodon

Codon	Amino Acid	Codon	Amino Acid	Codon	Amino Acid	Codon	Amino Acid
UUU	phe (phenylalanine)	UCU	ser (serine)	UAU	tyr (tyrosine)	UGU	cys (cysteine)
UUC	phe	UCC	ser	UAC	tyr	UGC	cys
UUA	leu (leucine)	UCA	ser	UAA	STOP	UGA	STOP
UUG	leu	UCG	ser	UAG	STOP	UGG	trp (tryptophan)
CUU	leu	CCU	pro (proline)	CAU	his (histidine)	CGU	arg (arginine)
CUC	leu	CCC	pro	CAC	his	CGC	arg
CUA	leu	CCA	pro	CAA	gln (glutamine)	CGA	arg
CUG	leu	CCG	pro	CAG	gln	CGG	arg
AUU	ile (isoleucine)	ACU	thr (threonine)	AAU	asn (asparagine)	AGU	ser
AUC	ile	ACC	thr	AAC	asn	AGC	ser
AUA	ile	ACA	thr	AAA	lys (lysine)	AGA	arg
AUG	met (methionine) (start)	ACG	thr	AAG	lys	AGG	arg
GUU	val (valine)	GCU	ala (alanine)	GAU	asp (aspartic acid)	GGU	gly (glycine)
GUC	val	GCC	ala	GAC	asp	GGC	gly
GUA	val	GCA	ala	GAA	glu (glutamic acid)	GGA	gly
GUG	val	GCG	ala	GAG	glu	GGG	gly

Figure 13.5 The genetic code. There are 64 different codons, each of which consists of three bases and nucleotides and specifies one of the 20 amino acids plus *stop* and *start* information.

that base-pairs with the AUG on the mRNA joins the mRNA and the ribosome, forming a three-part complex (Fig. 13.6). Another tRNA that base-pairs to the next codon and has its correct amino acid attached inserts in the complex next to the start tRNA. An enzyme in the ribosome then links the two adjacent amino acids together to begin to form a protein. As translation continues, new tRNAs base-pair to each mRNA codon in order, and new amino acid linkages are made. For example, an mRNA with the sequence AUG-AAG-UGU will be translated into a polypeptide with the sequence methionine-lysine-cysteine until a stop codon (UGA, UAA, or UAG) is reached. At that point, a newly made polypeptide is released from the complex.

To summarize, the flow of information in a cell is from DNA in the nucleus, where it is transcribed into mRNA, and then transported into the cytoplasm, where mRNA is translated into polypeptides. Proteins are composed of polypeptides. Because these processes occur in all living organisms, it has come to be known as the *Central Dogma of Molecular Genetics* (Fig. 13.7). Truly, a cell's identity is determined by the genes it expresses, which, in turn, produce the myriad of proteins required for its survival and function.

Mutation

When Professor Jansky's son told a friend that his mom was a scientist, the boy asked, "Can she make mutants?" While the term *mutation* may conjure up images from science fiction movies, most mutations are not so dramatic. In fact, as mentioned earlier, **mutations**, or changes in DNA sequence, happen every time DNA replicates. In addition to DNA replication errors and other spontaneous types of changes, agents called **mutagens** can alter DNA sequences. Ultraviolet light, ionizing radiation, and certain chemicals act as mutagens. DNA repair enzymes can often find and correct damaged DNA. If left uncorrected, though, a permanent change in DNA sequence, or mutation, occurs. Most mutations are *silent*, with no visible consequence of the DNA alteration. Remember that most DNA in a cell is noncoding. Even altered coding sequences may produce functional proteins.

Mutations are either *somatic* or *germ-line*. A **somatic mutation** occurs in a body cell and will exist in all cells produced by mitosis of the mutant cell. A somatic mutation may show up as a *sport*, or a branch that looks different from the others on a plant. These types of mutations are sometimes sources of new types of horticultural plants. Navel oranges and Red Delicious apples originated from sports and have been clonally propagated by grafting (see Chapter 14).

A **germ-line mutation** occurs in tissue that will produce gametes, or sex cells. Unlike somatic mutations, germ-line mutations will be passed on to future generations through seeds. A germ-line mutation is generally not apparent until it is passed on to offspring, which will carry the altered DNA in all cells. This mutation has now become a permanent feature of that plant's lineage. Germ-line mutations are important

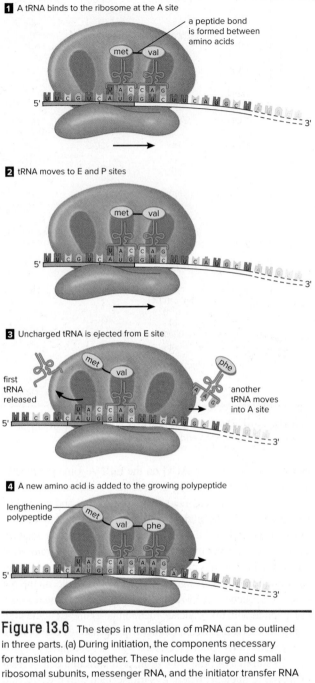

Figure 13.6 The steps in translation of mRNA can be outlined in three parts. (a) During initiation, the components necessary for translation bind together. These include the large and small ribosomal subunits, messenger RNA, and the initiator transfer RNA carrying the amino acid methionine. (b) During elongation, the ribosome moves along the messenger RNA. Transfer RNA molecules bring amino acids to the ribosome, where a polypeptide chain is constructed. (c) When the ribosome reaches a stop codon, translation terminates. The large and small ribosomal subunits, messenger RNA, transfer RNA, and polypeptide then separate from each other.

for the genetic improvement of horticultural and agronomic plants. They are responsible for variability in all traits, including flower color and fragrance, fruit taste and texture, grain yield and quality, and disease and stress tolerance.

While we tend to think of mutations as "bad," they are an essential feature of DNA. All the genetic variability in nature has arisen as a result of mutation. The accumulation of mutations is the mechanism that drives evolution of species (see Chapters 15 and 16). That variability was essential for the first plants, which lived in the ocean, to become established on land; it continues to be necessary for plants to survive and compete in an ever-changing environment.

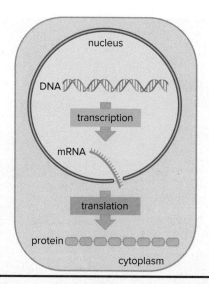

Figure 13.7 The Central Dogma of Molecular Genetics states that the flow of information in a cell proceeds from the DNA in the nucleus via mRNA to the cytoplasm, where information from the mRNA is translated into proteins.

13.3 Cytogenetics

Learning Outcomes

a. Define *cytogenetics*.
b. Distinguish between an inversion and a translocation.
c. Contrast aneuploid with polyploid.

During the middle of the 20th century, corn cytogenetics dominated the biological sciences, due in large part to the efforts of Barbara McClintock and other corn geneticists, including Roland Emerson, George Beadle, Charles Burnam, and Marcus Rhoades. **Cytogenetics** is the study of chromosome behavior and structure from a genetic point of view. This includes chromosome movement during meiosis and mitosis, chromosome pairing during meiosis, and chromosomal variation due to changes in chromosome structure and number. Chapters 3 and 12 discuss chromosome behavior during mitosis and meiosis, respectively. Therefore, this section will emphasize changes in chromosome structure and number. These changes occur at low frequency, but since their effects are often significant, they can have a dramatic influence on the evolutionary success of populations or species.

Changes in Chromosome Structure

Occasionally, a chromosome will break into two or more pieces. Enzymes may repair the damage, leaving the chromosome unaltered. However, sometimes a broken chromosome becomes reorganized before the pieces are attached together again. The most significant events from an evolutionary perspective include *inversions* and *translocations*.

An **inversion** results when a piece of a chromosome is broken and reinserted in the opposite orientation. For example, if the chromosome contains genes ABCDE in that order, then an inversion might create a chromosome with the genes in this order: **ABDCE**. The inverted chromosome can pair up with the normal homologous one at meiosis, but crossing-over within the inverted region will create gametes that do not survive because they contain some duplicated chromosomal regions and are missing other segments. Plants carrying an inverted chromosome therefore are less fertile than normal ones. The only gametes that are passed on from one generation to another are those in which crossovers occurred only outside the inverted region. Therefore, allele combinations within an inverted region are not rearranged by meiosis and are inherited together as blocks.

A **translocation** results when a piece of a chromosome breaks off and becomes attached to another one. A plant carrying a translocation, like one carrying an inversion, will not be as fertile as a normal plant. This is because some gametes from translocation plants carry unusual combinations of chromosomes.

Translocations and inversions can be important factors in **speciation** (the formation of new species) during evolution. Plants carrying chromosomes from these events may become reproductively isolated from normal plants and will eventually develop into new species. Crosses between the new species and the old one will produce offspring with low fertility and, therefore, a poor reproductive capacity. These types of events have been important in the evolution of many plants. For example, cultivated rye (*Secale cereale*) and four wild species (*S. vavilovii, S. africanum, S. montanum, S. silvestre*) each contain 14 chromosomes. However, translocations have resulted in the wild species being reproductively isolated from the cultivated species. Genetic exchange among these species is limited by translocations. See Chapter 15 for a more detailed discussion of mutations caused by chromosomal rearrangements.

Changes in Chromosome Number

Compared to animals, plants are remarkably tolerant of changes in chromosome number. Mistakes made during the pairing and separation of chromosomes during meiosis can result in gametes carrying extra or missing chromosomes. If one of these gametes is involved in fertilization, then the resulting zygote will be *aneuploid*. An **aneuploid** plant carries an extra one or more chromosomes or, less commonly, is missing one or more chromosomes. For example, the cells in a carrot plant typically contain nine pairs of homologous chromosomes, for a total of 18 chromosomes. An aneuploid carrot plant would have 19 chromosomes if it had an extra copy of one chromosome or 17 chromosomes if it were missing one of a pair. As you would expect, aneuploid plants generally differ in appearance from their normal counterparts due to the effects of extra or missing doses of many genes. Geneticists sometimes use aneuploids to determine the chromosomal locations of genes.

Occasionally, meiosis may completely fail to halve the chromosome number, producing gametes with the somatic (body cell) chromosome number. Because they carry the

Figure 13.8 Leaves of potato plants. *Left:* A tetraploid. *Right:* A diploid. © *Shelley H. Jansky*

sporophytic (2n) chromosome number, they are called $2n$ gametes. If these gametes are involved in fertilization, then the resulting offspring are *polyploid*. A **polyploid** plant has at least one complete extra set of chromosomes. Using the carrot example, a polyploid carrot could have 27 chromosomes (three sets of nine chromosomes) or 36 chromosomes (four sets) or some other multiple of nine. Polyploid plants are often larger and higher-yielding than their diploid counterparts. For this reason, many of our cultivated plants are polyploid. These include potato, cotton, peanut, wheat, oats, strawberry, and sugarcane (Fig. 13.8). The larger and longer-lasting flowers of polyploids make them attractive as ornamental plants; marigold, snapdragon, lily, and hyacinth are among the numerous examples of ornamental polyploids. In addition, triploid plants (polyploids containing three copies of each chromosome) are sterile and do not produce seeds. Seedless fruits, such as bananas and watermelons, are triploid. In Chapter 15, polyploidy is discussed as an important mechanism for the production of interspecific hybrids in nature.

13.4 Mendelian Genetics

Learning Outcomes

a. Describe Gregor Mendel's best-known research studies.
b. Distinguish among P, F_1, and F_2 generations in genetic studies.
c. Outline the steps in a monohybrid cross.
d. Explain the law of independent assortment.
e. Show the elements of a backcross.
f. Compare and contrast a testcross with a backcross.
g. Define *incomplete dominance*.
h. Explain why a dihybrid cross, such as that encountered with flower color in the blue-eyed Mary, may not follow the expected 9:3:3:1 phenotypic ratio.

Genetics as a science originated about a generation before its significance became appreciated in the scientific community as a whole. An Austrian monk, Gregor Mendel (Fig. 13.9) taught between 1853 and 1868 in what became the Czechoslovakian city of Brno. In the monastery there, he carried out a wide range of studies in physics, mathematics, and natural history. He also became an authority on bees and meteorology and kept notes on experiments involving two dozen different kinds of plants. Today, he is best known for the studies he conducted with peas.

Figure 13.9 Oil painting of Austrian botanist Gregor Johann Mendel, who founded the science of genetics. © *James Kong-Holmes/Science Source*

Mendel published the results of his studies on pea plants in a biological journal in 1866 and sent copies of his paper to leading European and American libraries. He also sent copies to at least two eminent botanists of the time. His work, nevertheless, was completely ignored or overlooked until 1900, when three botanists (Eric von Tschermak of Austria, Carl Correns of Germany, and Hugo de Vries of Holland), working independently in their own countries, reached the same conclusions as Mendel. Each, as a result of library research, came across Mendel's original paper.

Mendel's Studies

Peas, unlike many flowering plants, are *self-fertile*. In other words, a pollen grain of a pea flower can germinate on its own stigma, allowing a sperm cell to fertilize an egg and the ovule to develop into a viable pea seed. Although Mendel did not know about mitosis and meiosis, he had noticed that if a pea plant grew to a certain height, its offspring, if grown under the same conditions, would always reach approximately the same height, generation after generation. He wondered what would happen if he crossed a tall plant with a short one. Would the offspring be intermediate in height?

To make such a cross, Mendel needed to prevent self-pollination. He did so by reaching into a flower of one plant and removing the stamens before the pollen had matured. Then, he took pollen from another plant and applied it by hand to the stigma of the first flower. He also covered the experimental flowers with small bags to prevent insects from bringing pollen from other flowers after the cross had been made.

When Mendel made his crosses, the results were astonishing. All of the offspring were tall. There were no short or intermediate plants. He found, however, that if he allowed these tall plants to pollinate themselves, they produced offspring in a ratio of approximately three tall plants to one short plant (Fig. 13.10).

Mendel also tried crosses between peas with contrasting traits, such as smooth versus wrinkled seeds, and yellow versus green seeds.

Mendel referred to the original plants involved in making the crosses as the **parental generation (P)**; their offspring were the **first filial (F_1)** generation. Filial means "of or relating to a son or daughter." The offspring of F_1 plants were called the **second filial (F_2) generation**.

There are several reasons Mendel was successful in developing a model for inheritance when countless others before him had failed. He performed carefully planned and executed crosses between pure-breeding parents. Each parent was genetically uniform for the trait under study—a green-seeded parent produced only green-seeded offspring. Mendel also counted the number of offspring in each cross, something none of his predecessors had done. The use of his math background to apply statistical analyses to his data was unique, allowing him to develop and test inheritance models. Finally, he chose traits under simple genetic control. His predecessors usually had studied traits such as weight, which is controlled by many genes and strongly influenced by environment. A summary of some of Mendel's data is shown in Table 13.1.

As Mendel began to collect data in his experiments, he realized that there must be elements in the pea plant responsible for controlling traits such as seed color and stem height. He referred to this unknown agent as a *factor*. He also deduced that each plant must have two such factors for each characteristic, since even though all the offspring of an F_1 generation appeared identical, the F_2 generation revealed

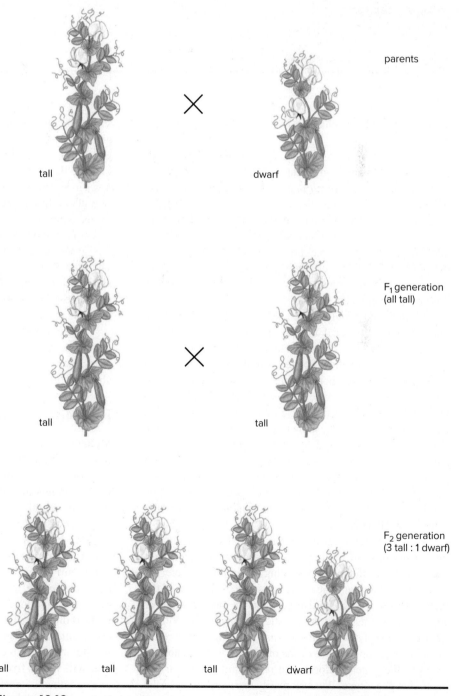

Figure 13.10 In this illustration of three generations of pea plants, one parent is tall and the other is short. Their offspring comprise the F_1 generation. All F_1 plants are tall. Self-pollination of the F_1 generation creates the F_2 generation, which segregates for height.

Table 13.1 — Summary of Mendel's Data

Parents	F₁	F₂
Yellow × green seeds	All yellow	6,022 yellow: 2,001 green
Smooth × wrinkled seeds	All smooth	5,474 smooth: 1,850 wrinkled
Green pod × yellow pod	All green	428 green: 152 yellow
Long stem × short stem	All long	787 long: 277 short
Axial flowers × terminal flowers	All axial	651 axial: 207 terminal
Inflated pods × constricted pods	All inflated	882 inflated: 299 constricted
Purple flowers × white flowers	All purple	705 purple: 224 white

some plants with one characteristic and others with the contrasting form.

These discoveries and deductions came to be known as the **law of unit characters.** Stated simply, this law says that *factors, which always occur in pairs, control the inheritance of traits.* Today, we know that the paired factors are **alleles** of genes. Alleles are alternative forms of a gene. For example, the seed color gene has an allele for yellow color and an allele for green color. Genes are always at the same position (**locus**) on homologous chromosomes (see Chapter 12). We also now know that there are typically thousands of genes on every chromosome.

From his data and analyses, Mendel also deduced that one allele may conceal the expression of the other. For example, in the F_1 generation following a cross between green-podded and yellow-podded parents, all the plants had green pods. However, both parents obviously had to contribute something to the cross, since some of the F_2 plants had green pods, while others had yellow pods. This deduction led to another principle, known as the **law of dominance.** This principle says that *for any given pair of alleles, one may mask the expression of the other.* The expressed allele is referred to as **dominant,** and the one that is not expressed (masked) is **recessive.** In the green-podded by yellow-podded cross, the green-pod allele is dominant, and the yellow-pod allele is recessive.

Mendel's crosses also made it clear that a plant's having green pods did not indicate whether or not both members of the pair of alleles controlling pod color were present, because the dominant one could mask the expression of the recessive one. To distinguish between the physical appearance of an organism and the genetic information it contains, we use the terms *phenotype* and *genotype*. **Phenotype** refers to the physical appearance of the organism, while **genotype** refers to the genetic information responsible for contributing to that phenotype. We customarily use words to describe phenotypes and letters to designate genotypes.

In Mendel's crosses, for example, seven pairs of phenotypes are shown as parents; these phenotypes include yellow seeds and green seeds, smooth seeds and wrinkled seeds, green pods and yellow pods, and so on. In designating the genotypes of each of these plants, a capital letter is used to indicate the dominant member of a pair of genes; *the same letter* in lowercase is used to indicate the corresponding recessive allele. In the green-podded × yellow-podded pea cross, for example, green is dominant over yellow. If a plant has a green phenotype, then its genotype could be either *GG* or *Gg*. The genotype of yellow-podded plants would always be *gg*. The plant is said to be **homozygous** if both alleles of a pair are identical (e.g., *GG* or *gg*) and **heterozygous** if the pair is composed of contrasting alleles (e.g., *Gg*). Mendel's true-breeding parents were homozygous for the traits he studied.

It is amazing that, nearly 150 years ago, Mendel was able to develop hereditary laws that are widely recognized today. Remember that he did his work before anyone knew about chromosomes, meiosis, or genes.

The Monohybrid Cross

Mendel was successful largely because he carried out well-planned crosses with true-breeding parents. Geneticists today perform similar experiments when trying to understand the genetic basis of uncharacterized traits. A common strategy is to generate a **monohybrid cross.** In this scenario, a single trait is studied. A cross is made between two true-breeding parents differing for that trait, producing an F_1 generation. Then, these F_1 plants are intercrossed to produce an F_2 generation. As an example, we will examine a cross between a homozygous green-podded pea plant and a homozygous yellow-podded one (Fig. 13.11). The genotype of the homozygous dominant parent (green-podded) is *GG*; the genotype of the homozygous recessive parent (yellow-podded) is *gg*. You will recall from Chapter 12 that gametes have only one member of each pair of homologous chromosomes. The gametes of the green-podded parent (*GG*), therefore, will be *G*; the gametes of the yellow-podded parent (*gg*) will be *g*. No matter which egg of one parent unites with which sperm of the other parent, all the zygotes of this cross will be heterozygous, with the genotype *Gg*. This means that all individuals in the F_1 generation will have the green-podded phenotype, and all will be heterozygous.

The next step is the actual monohybrid cross. Members of the F_1 generation are crossed with each other (or self-pollinated) to produce an F_2 generation. All of the F_1 plants have the same phenotype (green-podded) and genotype (*Gg*). Each F_1 plant produces two kinds of gametes (*G* and *g*) in roughly equal numbers. When the gametes are produced, half will be *G*, and half will be *g*. A *G* egg may unite with a *G* sperm, producing a *GG* zygote. Alternatively, purely at random, the *G* egg may unite with a *g* sperm, producing a *Gg* zygote. The same type of random combinations occur with *g* eggs, so that either *Gg* or *gg* zygotes are produced. When all the offspring of a large number of such crosses are counted, a genotypic ratio of 1 *GG* : 2 *Gg* : 1 *gg* is produced. The phenotypic ratio will be approximately three green-podded plants (*GG* and *Gg*) to one yellow-podded plant (*gg*). These four equally possible F_2 offspring are shown at the bottom of Figure 13.11. The monohybrid cross illustrates the law of segregation, which states that during meiosis, the two copies of a gene separate so that an offspring receives one copy from each parent.

The Dihybrid Cross

So far, these examples have involved single genes. However, it is often desirable to study a pair of genes using a **dihybrid cross**. This is especially valuable in gene mapping studies. That topic is discussed later in this chapter. The **law of independent assortment** states that *the factors (genes) controlling two or more traits segregate independently of each other*. This is one law that, while true for many traits, often breaks down. That is because the two genes under study may be physically close to each other on a chromosome. If this is the case, then the genes are linked together and will follow each other through meiosis. They do not segregate independently. In this situation, the genes are said to be **linked**. In the dihybrid example that follows, however, we will assume that the pair of genes is **unlinked**. That is, they are on different chromosomes or are far apart on a single chromosome.

Recall that tallness in peas is dominant and dwarfness is recessive, while green pod is dominant over yellow pod. The homozygous dominant (*GGTT*) parent will have a green-podded, tall phenotype, while the recessive parent (*ggtt*) will be yellow-podded and dwarf. All the gametes from the dominant parent will be *GT*, and those of the recessive parent will be *gt*. The F_1 generation will be composed of dihybrid (*GgTt*) plants that are green-podded and tall (Fig. 13.12).

When intercrossed to produce the F_2 generation, these dihybrid members of the F_1 generation will produce four types of gametes in equal proportions. The gamete

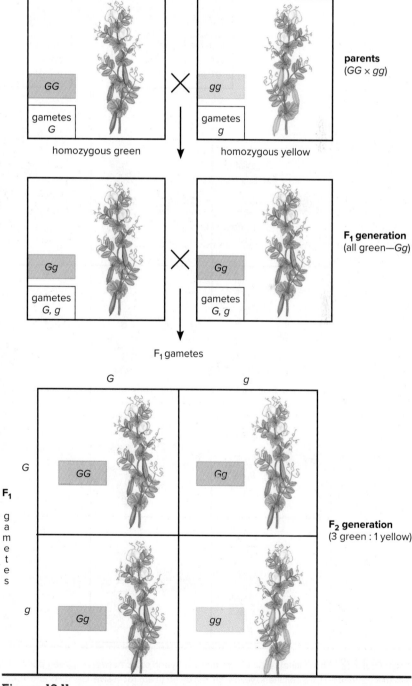

Figure 13.11 A Punnett square illustrating a monohybrid cross between a green-podded pea plant and a yellow-podded pea plant. Green (*G*) is dominant; yellow (*g*) is recessive.

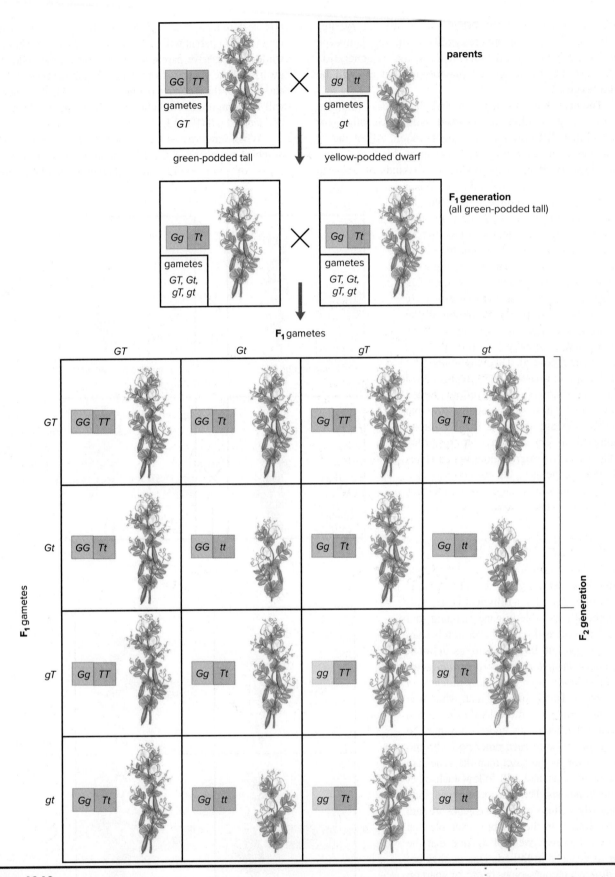

Figure 13.12 This Punnett square illustrates a dihybrid cross between a green-podded, tall pea plant and a yellow-podded, dwarf pea plant. Green (*G*) and tall (*T*) are dominant; yellow (*g*) and dwarf (*t*) are recessive. Genes *G* and *T* are not linked.

genotypes are *GT, Gt, gT,* and *gt.* Remember that each gamete will carry only one allele of each gene. Since any one of the four kinds of gametes can unite randomly with any of the four kinds of gametes of the other parent, 16 combinations are possible.

In order to systematically outline all possible combinations of gametes, a diagram called a **Punnett square** is used to determine the genotypes of the zygotes. Figures 13.11 and 13.12 illustrate Punnett squares. The Punnett square diagram looks somewhat like a checkerboard, with the gametes of one parent across the top and the gametes of the other parent on the left side. To fill in each square on the checkerboard, look at the gamete designation above the square and to the side of it. Multiply the gamete proportions (e.g., 1/4 × 1/4) and add together the gamete letters (e.g., *GT* + *GT*). In Figure 13.12, the first square would therefore be 1/16 *GGTT.*

Nine different zygote genotypes are possible in the 16-box Punnett square. They are produced in a ratio of 1 *GGTT*: 2 *GGTt*: 1 *GGtt*: 2 *GgTT*: 4 *GgTt*: 2 *Ggtt*: 1 *ggTT*: 2 *ggTt*:1 *ggtt.* Four phenotypes are possible in a ratio of 9 green-podded, tall: 3 green-podded, dwarf: 3 yellow-podded, tall: 1 yellow-podded, dwarf. A geneticist observing a 9:3:3:1 ratio from a dihybrid cross can deduce that the genes controlling the two traits are unlinked and are exhibiting dominance. Genetic ratios such as the ones previously outlined are expected when large numbers of individuals are studied. Note in Mendel's data (Table 13.1) that he looked at over 8,000 F_2 offspring following a monohybrid cross involving seed color. With small numbers, chance alone may not produce the expected ratios. Human populations, for example, are divided relatively equally into males and females (a genetically controlled trait), but in smaller groups, such as families, the general population ratio of one male to one female may not be evident.

The Backcross

If a scientific theory is valid, then one should be able to test it experimentally. Mendel tested his predictions using backcrosses. A **backcross** is a cross between a hybrid and one of its parents. Mendel had found in pea flowers that red color is dominant and that white is recessive. Therefore, all F_1 offspring of a cross between a homozygous red-flowered parent (*RR*) and a homozygous white-flowered parent (*rr*) would be red (*Rr*). The gametes of these F_1 plants would be either *R* or *r*, with each type being produced in equal numbers. If the F_1 hybrid (*Rr*) is crossed to a homozygous recessive plant (*rr*), then half of the offspring will be *Rr*. These plants received the dominant allele from the heterozygous parent and a recessive allele from the homozygous parent. The remaining offspring from the backcross received a recessive allele from the heterozygous parent, so they would be *rr*. Following backcrosses of F_1 hybrids to recessive parents, Mendel observed the expected ratio of 1 red-flowered plant (*Rr*) to 1 white-flowered plant (*rr*). This provided further confirmation for his inheritance theory.

The Testcross

In the F_2 generation of a monohybrid cross, there are two genetically different types of plants that look identical. Homozygous dominant plants look just like heterozygotes. Suppose you are interested in determining which plants are homozygous dominant for the tall allele. How would you identify them? One of the easiest ways is to cross the tall plants with short plants. A cross between a plant with the dominant phenotype and a homozygous recessive plant is called a **testcross.** This cross will determine whether the plant with the dominant phenotype is homozygous or heterozygous. The short plants are homozygous recessive, so you know their genotype. If all the offspring from a cross between a tall plant and a short plant are tall, then the tall plant must be homozygous dominant. If some (approximately 50%) of the offspring are short, then the tall plant must be heterozygous.

Incomplete Dominance

Although dominance was demonstrated by Mendel's crosses, in many other instances, one allele does not completely mask the expression of another allele. In other words, some genes exhibit **incomplete dominance**, in which a heterozygote is intermediate in phenotype to the two homozygotes. In snapdragons, for example, the F_1 plants of a cross between a red-flowered parent and a white-flowered parent are all pink-flowered. When an F_2 generation is produced from such a cross, the flowers of the offspring appear in a ratio of 1 red: 2 pink: 1 white (Fig. 13.13).

Interactions among Genes

So far, our discussion of inheritance has focused on single genes or pairs of genes acting independently of each other. This type of autonomous behavior, however, is the exception rather than the rule. Genes are generally responsible for the production of proteins that are components of complex biochemical pathways. It is often necessary to consider the genotype of an individual at more than one locus before a prediction of phenotype can be made. For example, in a plant called blue-eyed Mary (*Collinsia parviflora*), two genes control flower color in a biochemical pathway as follows:

$$\text{colorless} \xrightarrow{\text{gene } W} \text{magenta} \xrightarrow{\text{gene } M} \text{blue}$$

If a plant contains at least one dominant *W* allele, then it can proceed through the first step of the pathway and produce either magenta or blue flowers. The flowers will be magenta if the plant is homozygous recessive for the *M* gene (*mm*) or blue if it has at least one copy of the dominant *M* allele (*MM* or *Mm*), allowing it to complete the second step in the pathway. The enzyme produced by the dominant *W* allele catalyzes a pathway that synthesizes a compound used by

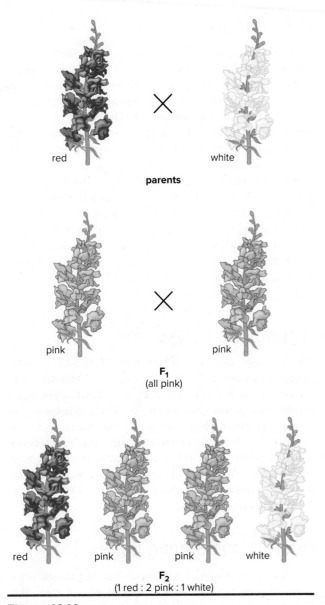

Figure 13.13 Incomplete dominance. A cross between a red-flowered and a white-flowered variety of snapdragons produces F₁ offspring with an intermediate phenotype.

the gene M enzyme. Therefore, a homozygous recessive (ww) plant cannot produce colored flowers even if it contains a dominant M allele. Of course, a plant that is homozygous recessive for both genes ($wwmm$) will also produce colorless (white) flowers.

How Genotype Controls Phenotype

Mendel had no way of knowing how his factors (genes) could produce plants that varied in phenotype. However, as our understanding of the molecular basis of genetics increases, so does our ability to explain Mendelian inheritance patterns.

Mendel discovered that the allele for smooth seeds is dominant over the wrinkled seed allele. How does one allele mask the expression of another? Often, the dominant allele codes for a protein that can effectively catalyze a reaction in a cell and produce a phenotype we recognize. The recessive allele, on the other hand, represents a mutant form that has an altered DNA sequence. The protein product that results from transcription and translation of that recessive (mutant) allele is defective. It cannot catalyze the reaction and, therefore, does not produce a functional product. One copy of the normal allele is usually enough to allow the cell to produce ample quantities of normal protein. Therefore, a heterozygous plant looks normal. A homozygous recessive plant, on the other hand, cannot make any normal protein from its two mutant alleles.

What is the difference between the smooth seed allele and the wrinkled seed one? The smooth pea phenotype (RR or Rr) produces an enzyme allowing the seeds to accumulate high levels of starch. Seeds with the rr genotype do not produce a functional form of that enzyme and so have high levels of sucrose rather than starch. The sucrose in the rr seeds causes them to absorb water during development. When they dry, they lose that water and shrivel. The RR and Rr seeds contain lower levels of sucrose and do not absorb as much water during development; consequently, they do not shrivel when they dry. Recent molecular genetic studies have revealed that the r allele produces a defective enzyme because it has had an 800-nucleotide piece of DNA inserted into the gene. As you might expect, compared with the normal allele, transcription and translation of the allele carrying this insert produce a much different (and, in this case, nonfunctional) protein. How did the insert get there? The best guess is that a transposable element inserted into the pea chromosome at this location sometime during the pea plant's evolutionary history.

Molecular genetics can also explain how other Mendelian traits are expressed. For example, as you recall, flower color in snapdragons is inherited as an incompletely dominant trait. In this case, one copy of the functional allele produces some red pigment, but two copies produce noticeably more pigment. Red-flowered plants therefore have two copies of an allele that is critical in the red pigment synthesis pathway and are RR. Pink-flowered plants have one copy of that allele, producing light red (pink) pigmentation, and are RR'. (We use superscripts instead of upper- and lowercase letters here because we are not looking at complete dominance.) Finally, white plants have two copies of the defective allele ($R'R'$) and cannot make the enzyme necessary for red pigment production.

13.5 Quantitative Traits

Learning Outcome

a. Explain how quantitative traits differ from simply inherited traits.

While simply inherited traits such as snapdragon flower color effectively demonstrate Mendel's principles, many traits are under much more complex genetic control. **Quantitative traits,** such as yield and days to flowering, exhibit a range of phenotypes, rather than the discrete phenotypes studied by Mendel. Typically, they are dramatically influenced by environment. Consider tomato fruit yield as an example of a quantitative trait. In identical environments, some tomato plants produce more fruit than others due to genetic differences. However, genetically identical plants will also produce different amounts of fruit when grown in different environments. They will produce large yields under optimal growing conditions and lower yields under stressed conditions. Therefore, phenotype (yield in this example) is determined by both genotype and environment. This is the "nature versus nurture" concept. Because quantitative traits are strongly influenced by environment and are generally controlled by many genes, they do not produce typical Mendelian ratios. Instead, statistical tools such as distributions, means, and variances are used to study quantitative traits. Molecular geneticists are able to identify chromosomal fragments, called **quantitative trait loci,** or QTLs, associated with quantitative traits. Presumably, these fragments contain genes that influence the trait and behave like Mendelian genes. Using this approach, quantitative geneticists are beginning to understand the nature of complex traits such as growth rate, fruit quality, and flowering time.

13.6 Extranuclear DNA

Learning Outcome

a. Explain how the endosymbiont theory predicts the presence of extranuclear DNA in eukaryotic organisms.

A statement earlier in this chapter suggested that *most* DNA in plants resides in the nucleus. Where is the remaining DNA found? In plants, **extranuclear DNA** is found in both mitochondria and chloroplasts. According to the **endosymbiotic theory,** mitochondria and chloroplasts were free-living bacteria at some time in their evolutionary history. They became incorporated into cells of organisms that evolved into plants and established a symbiotic association. As you might expect, then, the DNA in mitochondria and chloroplasts is similar to bacterial DNA.

Some genes involved in photosynthesis and respiration are located in the chloroplast and mitochondrial genomes, respectively. Because many copies of each organelle exist in a plant cell and multiple copies of each gene may be found in each organelle, extranuclear genes do not exhibit Mendelian segregation ratios. In addition, in most plant species, sperm cells rarely carry mitochondria and chloroplasts, so extranuclear genes are typically passed on to the next generation by only the female parent. This is called **maternal inheritance.**

Leaf variegation exhibits maternal inheritance because it results from mutations in chloroplast synthesis genes. Chloroplast genes are targeted by some herbicides. If these genes can be manipulated by genetic engineers, then crop plants can be made to survive treatment with those herbicides. Consequently, a farmer could spray the herbicide on a field and kill all plants except the crop. See Chapter 14 for a more detailed discussion of genetically engineered plants.

13.7 Linkage and Mapping

Learning Outcome

a. In the cross *AB/ab* x *ab/ab*, predict offspring ratios if genes *A* and *B* are 20 map units apart.

There are typically thousands of genes on each chromosome, and the closer the genes are to one another on a chromosome, the more likely they are to be inherited together. Genes that are together on a chromosome are said to be linked.

In 1906, just a few years after the basic details of meiosis became known, W. Bateson, R. C. Punnett (after whom the Punnett square is named), and E. B. Saunders became the first to report on linkage in sweet peas. They knew from earlier work with sweet peas that purple flower color was dominant and red was recessive. They also knew that pollen grains with an oblong shape ("long pollen") were dominant and that spherical pollen grains ("round pollen") were recessive.

When Bateson and Punnett crossed a homozygous purple long plant (*PPLL*) with a homozygous red round plant (*ppll*), all the F_1 offspring were purple and long, as expected. When they crossed the F_1 plants with one another to obtain an F_2 generation, however, the F_2 offspring phenotypes were produced in numbers that differed markedly from the expected 9:3:3:1 ratio. Puzzled by these results, they tried a backcross, crossing F_1 plants (*PpLl*) with the recessive parent (*ppll*). Again, the expected 1:1:1:1 ratio was not produced. Instead, they obtained a ratio of 7 purple long: 1 purple round: 1 red long: 7 red round. They could not adequately explain what they had observed. In 1910, however, T. H. Morgan, who had observed similarly puzzling ratios in his experiments with fruit flies, correctly theorized that linkage and crossing-over were responsible.

If the genes for purple and red and those for long and round were on separate pairs of chromosomes, then the genotypes of the F_1 would be all *PpLl*, and the homozygous recessive would be *ppll*. A backcross to the F_1 (*PpLl* × *ppll*) should have produced 1 *PpLl*: 1 *Ppll*: 1 *ppLl*: 1 *ppll* instead of the 7:1:1:7 ratio actually obtained. Apparently, the F_1 parent produced many more *PL* and *pl* gametes than *Pl* and *pL* gametes. If we assume the genes are linked, then we can use a slash (/) to separate the alleles on one chromosome from those on the homologous chromosome. The parents were *PL/PL* and *pl/pl*. The first parent would produce gametes containing the *PL* chromosome and the other would produce *pl* gametes. The F_1 plant, then, would be *PL/pl* because it received a *PL* chromosome from one parent and a homologous *pl* chromosome from the other. The F_1 plant would produce just two kinds of gametes, namely *PL* and *pl*, and the backcross (*PL/pl* × *pl/pl*)

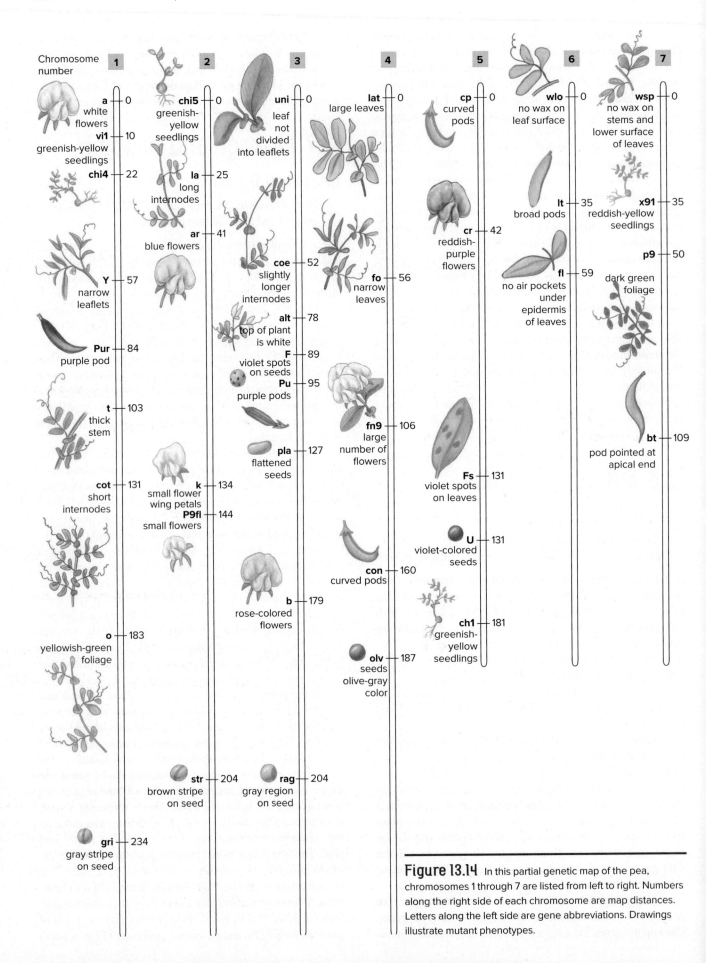

Figure 13.14 In this partial genetic map of the pea, chromosomes 1 through 7 are listed from left to right. Numbers along the right side of each chromosome are map distances. Letters along the left side are gene abbreviations. Drawings illustrate mutant phenotypes.

would have produced only two kinds of phenotypes, namely purple long (*PL/pl*) and red round (*pl/pl*) in equal numbers. But some purple round and red long phenotypes were also produced. The only plausible explanation for this seems to be that some crossing-over (see prophase I of meiosis in Chapter 12) occurs between genes *P* and *L*.

Crossing-over is a regular, seemingly random event during meiosis. Therefore, it may occur by chance between any pair of linked genes. Each gene has a specific location (locus) on a chromosome and, presumably, crossing-over is more likely to happen between two genes that are far apart on a chromosome than between two genes that are close together. We can use the frequency of crossing-over between genes to create a genetic map of each chromosome. A genetic map is similar to a map of a straight road, with genes corresponding to cities and map distances corresponding to the distances between the cities. A **map unit** equals 1% crossing-over between a pair of genes. Suppose we make a backcross (*PL/pl* × *pl/pl*), and 90% of the offspring are either *PL/pl* or *pl/pl*. These are **parental type** offspring because the gametes from the heterozygous parent (*PL* and *pl*) do not carry chromosomes for which a crossover between *P* and *L* occurred. The remaining 10% of the offspring are **recombinant types** (*Pl/pl* and *pL/pl*) because a crossover occurred between genes *P* and *L* to produce *Pl* and *pL* gametes. Remember that the heterozygous parent contains *P* and *L* on one chromosome and *p* and *l* on the homologous chromosome. The only way, then, it can produce *Pl* and *pL* gametes is through crossing-over. Because 10% of the offspring (and, therefore, gametes) were recombinant, the map distance between genes *P* and *L* is 10 map units.

In Bateson and Punnett's backcross, two of every 16 (12.5%) of the offspring were produced through recombination due to crossing-over. We can deduce from this that the genes controlling flower color and pollen shape in sweet peas are 12.5 map units apart on a chromosome. It is interesting that Mendel did not observe linkage in his studies. The traits he analyzed happened to be on different chromosomes in peas or far apart when they were on the same chromosome. Imagine the difficulty he would have encountered if he had chosen to work on genes that were linked.

Linkage maps are essential tools for geneticists because they give us a physical framework in which to organize genes. Massive mapping efforts are currently underway in all major crop plants, with new genes and DNA sequences being added to maps every day. A simplified example of a genetic map of the pea is shown in Figure 13.14.

13.8 The Hardy-Weinberg Law

Learning Outcome

a. Summarize the Hardy-Weinberg law.

Before Mendel's work became known, most biologists studied quantitative traits and believed that inherited characteristics resulted from a blending of those furnished by the parents. It was difficult to understand why unusual characteristics did not eventually become so diluted that they essentially disappeared. After Mendel, R. C. Punnett and other biologists asked why dominant genes did not eventually completely eliminate recessive ones in breeding populations. After all, a cross between two heterozygotes produces only 25% homozygous recessive individuals. G. H. Hardy, a mathematician, and W. Weinberg, a physician, pointed out the reason in 1908, and their observation became known as the Hardy-Weinberg law. The **Hardy-Weinberg law,** which essentially specifies the criteria for genetic equilibrium in large populations, states that the proportions of dominant alleles to recessive ones in a large, random mating population will remain the same from generation to generation unless there are forces that change those proportions. In small populations, for example, random losses of alleles can occur if, by chance, the individuals carrying those alleles do not mate as often as other individuals. In populations of any size, selection is the most significant cause of exceptions to the Hardy-Weinberg law and can cause dramatic changes in the proportions of dominant to recessive alleles. Discussions of artificial selection and natural selection are found in Chapters 14 and 15, respectively.

SUMMARY

13.1 Introduction to Genetics and Molecular Biology
- McClintock's work with corn changed beliefs about genes and chromosomes.

13.2 Molecular Genetics
- A DNA molecule consists of nucleotides. Each nucleotide contains a base, a sugar, and a phosphate group.
- An RNA molecule is similar to DNA but is typically single-stranded, has a different sugar, and has uracil instead of thymine.
- Watson and Crick developed a model of a DNA molecule that is a double-stranded helix resembling a spirally twisted ladder.
- DNA stores genetic information, replicates itself, expresses its information, and can mutate.
- A genome is the sum total of the information in an organism's DNA.
- A DNA molecule replicates by "unzipping," and the resulting single strands serve as templates for two new double-stranded molecules.
- With DNA as a template, mRNA, tRNA, and rRNA are made during transcription.
- During translation, mRNA from transcription is used to make polypeptides; tRNA decodes mRNA information. Ribosomes are made of rRNA and proteins.
- Mutations are changes in DNA sequence.

13.3 Cytogenetics
- Inversions and translocations may be important speciation mechanisms.
- An aneuploid has a normal chromosome number plus or minus one to several chromosomes. A polyploid has extra complete sets of chromosomes.

13.4 Mendelian Genetics
- Gregor Mendel founded the science of genetics while experimenting with peas. The pea varieties he crossed had pairs of contrasting characteristics on homologous chromosomes. Offspring of the parental (P) generation were called the F_1 generation; the F_2 generation resulted from self-pollination of F_1 plants.
- The agents controlling a plants' characteristics were called "factors." Mendel deduced that each plant had two factors (later known as alleles) for each characteristic. His law of unit characters states that "factors, which always occur in pairs, control the inheritance of various characteristics." The suppressing factor was called dominant and its counterpart recessive.
- Phenotypes (described with words) denote appearance; genotypes (shown with letters) designate genetic makeup. Capital letters designate dominant alleles; lowercase letters designate recessive alleles.
- A homozygous plant has identical alleles; a heterozygous plant has contrasting alleles.
- Monohybrid cross offspring are produced in a ratio of 3 dominant to 1 recessive phenotype. Dihybrid crosses produce a 9:3:3:1 phenotypic ratio. A backcross between an F_1 hybrid and its recessive parent produces a phenotypic ratio of 1:1 (monohybrid) or 1:1:1:1 (dihybrid).
- A testcross is used to determine if a plant with the dominant phenotype is homozygous or heterozygous.
- If a gene exhibits incomplete dominance, the heterozygote phenotype is intermediate to that of the two homozygotes.
- Genes produce enzymes, which control biochemical reactions, ultimately determining phenotypes.

13.5 Quantitative Traits
- Quantitative trait genotypes are strongly influenced by environment.

13.6 Extranuclear DNA
- Extranuclear DNA is found in the mitochondria and chloroplasts of plants.

13.7 Linkage and Mapping
- Linked genes are inherited together. Bateson, Punnett, and Saunders observed a 7:1:1:7 ratio in a dihybrid cross, and they postulated that it resulted from linkage and crossing-over.
- Chromosomal mapping requires calculating crossover percentages to determine the positions of genes on chromosomes.

13.8 The Hardy-Weinberg Law
- The Hardy-Weinberg law explains why recessive alleles in a population do not disappear.
- Selection is the most significant of the seven exceptions to the Hardy-Weinberg law.

REVIEW QUESTIONS

1. Using S, P, and B as abbreviations for sugar, phosphate, and base, respectively, draw a DNA double helix in the arrangement you would see if you untwisted the helix so you were looking at a ladder structure.
2. If a DNA molecule had a string of cytosine-thymine base pairs, how would it differ in width from a normal molecule?
3. Explain how DNA replicates.
4. Using the genetic code, determine the amino acid sequence produced by the following gene (DNA) sequence. Remember that during transcription, RNA forms complementary base pairs with DNA. For example, if the DNA sequence were TAG, then transcription would produce an RNA fragment with the sequence AUC. TACACAGCAACT

DISCUSSION QUESTIONS

1. Why do you suppose gene expression requires an intermediate RNA molecule? That is, why does the translation machinery read mRNA instead of the gene itself?
2. If proofreading enzymes became more accurate and efficient, then what do you suppose would happen to the speed at which organisms evolve?
3. Mendel's peas were self-pollinating. Describe advantages and disadvantages of self-pollination.

CHAPTER

14

Plant Breeding, Propagation, and Biotechnology

14.1 Crop Plant Evolution
   Origins of Agriculture
   Evolution under Domestication

14.2 Plant Breeding
   Breeding Methods Using Sexually Compatible Germplasm
   Breeding Methods Using Sexually Incompatible Germplasm

   Key Theme 14.1: Genome Editing

14.3 Plant Propagation
   Seed Propagation
   Asexual Plant Propagation

OVERVIEW

In this chapter, the processes that have produced and propagated domesticated plants are discussed. Proposed origins of our agricultural societies are presented along with the phenotypic changes that occurred during crop domestication. Plant breeding methods using conventional and nonconventional strategies are outlined. Then, propagation methods using seeds and asexual propagules are discussed.

Plant breeders develop new crop varieties that meet the needs of farmers and consumers. This is a field of mixed greens growing in the Salinas Valley, California. © *David Litman/Shutterstock*

14.1 Crop Plant Evolution

Learning Outcomes

a. Explain how plants change as a result of domestication.
b. Discuss hypotheses for the shift from a hunter-gatherer lifestyle to an agricultural one.
c. Describe the traits typically selected during domestication.

It is clear that the human species is currently an eater of grass seeds. We have become "canaries." It is also clear that the world's food supply depends on 12 or 15 plant species. It probably was not always so, although wheat, barley, rice, and maize have been the foundations of our high civilizations. The current trend is for the major crops to become even more major and for the lesser ones to dwindle.[1]

In fact, although there are approximately 250,000 species of flowering plants, only six species—wheat, rice, corn, potato, sweet potato, and cassava—provide 80% of the calories consumed by humans worldwide (Fig. 14.1). An additional eight plants—sugarcane, sugar beet, bean, soybean, barley, sorghum, coconut, and banana—complete the list of major crops grown for human consumption.

In the 200,000 years since modern humans have inhabited this earth, plants have been cultivated for only the most recent 10,000 years, or about 5% of human history. However, in that time, humans have dramatically changed the plant landscape. Hunter-gatherer societies have given way to agricultural societies and the development of cities and civilizations. Such concentrations of people are vulnerable to catastrophes such as drought and famine. Domesticated plants depend on humans for their survival, but humans have also become dependent on plants for our survival.

Figure 14.1 These six foods—potatoes, rice, corn, cassava, wheat, and sweet potatoes—meet most of the caloric needs of people worldwide. © *Shelley H. Jansky*

---

1. J. P. Harlan. 1992. *Crops and Man*, 2d ed. Madison, WI: Amer. Soc. Agron.

Plants have been **domesticated** by genetically altering them to meet the needs of human societies. Strictly speaking, a domesticated plant is one whose reproductive success depends on human intervention. This is an ongoing evolutionary process, and plants are found in a continuum from purely wild to fully domesticated. Crop plants continue to evolve as a result of breeding efforts. It is amazing that, although humans have eaten thousands of types of plants in the past, the human race currently relies on only a handful to supply almost all nutritional needs.

Origins of Agriculture

Domestication of crop plants, and the development of industrial societies that followed, is a recent event in our history. Of the 80 billion humans that have lived on this earth, 90% were hunter-gatherers, 6% were farmers, and only 4% were industrialized.

It is not clear why most humans shifted from a hunter-gatherer lifestyle to an agricultural one. Several hypotheses exist, but they are purely speculative. Hunter-gatherers may have had better lives than cultivators did. Their diet and health were better, their starvation rate was lower, and they worked far fewer hours. Their work week was only about 20 hours long! Because they had abundant free time, hunter-gatherers may have simply begun to grow plants in gardens as a hobby. They may have begun to cultivate a few plants to supplement lean times in the gathering schedule. Or they may have had spiritual reasons for growing plants in a protected environment. Because agricultural practices arose independently in many parts of the world over thousands of years, each of these scenarios probably explains why humans began to domesticate plants.

People began to domesticate plants in the Near East around 10,000 years ago. Many agricultural sites from this time period were located on the western edge of what is now Iran. Domesticated plants were developed in Asia and the New World 1,000 to 3,000 years later. Farming probably began in Africa and the New World about 4,000 years ago.

The first crops domesticated were cereal grains. Root crops and legumes, such as peas and beans, were domesticated 1,000 to 2,000 years later. These were followed by vegetables, and then oil, fiber, and fruit crops. Finally, plants used for forage, decoration, and drugs were first domesticated only about 2,000 years ago. Few crops have been brought into domestication in recent centuries. Plant breeders currently devote much more energy to improving existing crops than to developing new ones. Until recently, it was widely believed that agriculture originated in one region in a short period of time near the Upper Euphrates River in present-day Turkey (Fig 14.2). However, new analyses are leading scientists to conclude that the domestication of the first crops (wheat and barley) occurred in several locations over a prolonged period of time (millennia instead of centuries). Similarly, wild rice was likely brought into agriculture and domesticated multiple times in Asia. So, the idea to domesticate plants was not the result of one group's clever inspiration. Instead, domestication may have been a natural progression in the activities of many hunter-gatherer societies as they learned to manage

Figure 14.2 The Euphrates river and farmland seen from Dura Europos. © *Joel Carillet/Getty Images*

their surroundings and manipulate plants to suit their needs. For further discussion on the origin of cultivated plants and regions of crop domestication, see Chapter 24.

Evolution under Domestication

It is as if man had been appointed managing director of the biggest business of all, the business of evolution . . . Whether he is conscious of what he is doing or not, he is in point of fact determining the future direction of evolution on this earth. That is his inescapable destiny, and the sooner he realizes it and starts believing in it, the better for all concerned.[2]

As soon as people began sowing seeds of selected plants, instead of simply harvesting wild plants, dramatic genetic changes occurred in those populations in response to selection pressure. The seed dispersal mechanism was probably the first trait altered during domestication. The seeds of wild plants are typically carried away by wind, water, or animals. This minimizes the competition for light and nutrients that results from overcrowding. People most likely gathered seeds from plants that retained them instead of dispersing them. Early domesticators would have focused also on plants with high seed numbers and large seed size (Fig. 14.3). Competition in crowded cultivated plots would favor high seedling vigor and fast germination rate. In addition, while seed dormancy mechanisms are important in wild plants, plants that survived in cultivated gardens were those that germinated immediately upon planting. These changes accumulated over many generations and created the cultivated plants we know today. Many of these plants have been so dramatically altered from their wild forms that they cannot survive without the aid of humans. For example, if a cornfield is abandoned, it will cease to produce any corn plants within just a year or two.

The first serious attempts to understand the geographic origins of domesticated plants were made by the Russian geneticist Nikolai Ivanovich Vavilov in the 1930s. Based on extensive collecting expeditions carried out worldwide, Vavilov proposed that each crop plant was first domesticated in the region of the world where its genetic diversity is greatest. Tragically, Vavilov's career and life were cut short when he was imprisoned by the Stalin government in 1940 on charges of espionage and, ironically, efforts to harm agriculture. He died in prison 3 years later. Although more recent studies indicate that patterns of crop domestication cannot be adequately explained by Vavilov's *Centers of Origin*, he can be credited with initiating studies that have led to our understanding of the relationships between cultivated plants and their wild relatives. As we will see later in this chapter, wild relatives of crop plants are valuable sources of genes for modern crop improvement.

a.

b.

Figure 14.3 Modern corn (a) was probably domesticated from teosinte (b), which provided high seed numbers and large seed size. *(a) © Kingsley Stern; (b) Courtesy of John Doebley*

---

2. J. S. Huxley. 1957. *New Bottles for New Wine*. New York: Harper and Sons.

14.2 Plant Breeding

Learning Outcomes

a. Define *plant breeding*.
b. Explain why breeding strategies for self-pollinated crops differ from those for cross-pollinated ones.
c. Describe ways in which genes from sexually incompatible germplasm may be introduced into crop plants.

The systematic breeding of crop plants has been carried out only in the past 200 years. Until then, crop improvement was based on simply selecting and planting seeds from desirable plants. This method was moderately effective because plant populations were genetically diverse and could be improved gradually across generations (Fig. 14.4). However, it took an understanding of genetics and plant reproduction, and the training of scientists in these fields, to develop the sophisticated crop-breeding methods we know today.

Plant breeding is accelerated evolution guided by humans rather than nature. Breeders replace natural selection with human selection in order to modify plants genetically to meet our needs. Remember from Chapter 13 that, according to the Hardy-Weinberg law, a population of plants will remain in genetic equilibrium unless an outside force changes its genetic makeup. Plant breeders act as that outside force, molding plant populations to contain higher proportions of alleles that are desirable to humans. Typically, the primary goal of plant-breeding programs is improved yield, with disease resistance, pest resistance, and stress tolerance contributing to yield. With new sources of agricultural land dwindling, a rapidly changing climate, and the human population continuing to grow, efforts must concentrate on improving the amount of food harvested per acre and developing crops that can grow on lower-quality land.

Breeding Methods Using Sexually Compatible Germplasm

The currency of plant breeders is crop germplasm, the sum of all genes in the crop and its relatives. Breeders must distinguish between sexually compatible and incompatible germplasm. Sexually compatible germplasm can contribute new genes through crosses to the crop. Sexually incompatible germplasm (usually distantly related wild germplasm, but sometimes other organisms as well) cannot be crossed to crop cultivars, so more technically difficult strategies must be employed to access their genes.

Strategies

A plant breeder tends to divide crop plant species into two groups: **self-pollinating** and **cross-pollinating.** A self-pollinating plant is capable of fertilizing itself, while a cross-pollinating one cannot. Breeding methods for self-pollinating plants are radically different from those for cross-pollinating ones. A self-pollinating plant tends to be highly **homozygous** because all of its genes came from the same parent. The plant's mother was also its father! As you might expect, these plants have undergone a significant amount of **inbreeding.** As the species has evolved, it has adapted to and thrives on inbreeding. On the other hand, a cross-pollinating plant is likely to be highly **heterozygous,** because its mother and father were different plants. Cross-pollinating species tend to require a high level of heterozygosity in order to be productive.

Self-pollinated crop plants include wheat, rice, oats, barley, peas, beans, tomatoes, and peppers. Some fruit trees, such as apricots, nectarines, peaches, and citrus fruits are also primarily self-pollinated. The most primitive form of breeding in this group of plants is called **pure-line selection.** This method simply involves collecting seed from each of several plants, growing all the seed from each plant in a row, and selecting the most desirable row. All the plants in a row will be related to each other, since they came from the same plant. The seeds from the best row can then be propagated as a new "pure-line" variety.

A breeder with an understanding of reproductive biology can make crosses between self-pollinating plants. In fact, today, most plant breeders create hybrid populations by crossing desirable inbred parents. Then, as the hybrid plants self-pollinate, highly diverse groups of offspring will be created. These offspring of the self-pollinated hybrids provide the basis for selecting pure-line varieties. Crosses between normal wheat varieties and dwarf plants have been created to develop dwarf varieties. More of the energy in these plants is directed toward producing seeds rather than stalks, resulting in higher yields.

Norman Borlaug (1914–2009), who is known as the "Father of the Green Revolution," was awarded the Nobel Peace Prize in 1970 for increasing the world food supply,

Figure 14.4 Potato tuber diversity. Genetic variation provides the foundation for efforts to improve plants through breeding.
© Shelley H. Jansky

particularly through developing new strains of wheat in Mexico and other countries. The new high-yielding strains, which were produced from crosses with a dwarf variety from Japan, were widely planted. Before the project began in 1944, Mexico had imported wheat for many years, but the new varieties were so successful that, within 20 years, Mexico had quadrupled its wheat production and had enough to export to other countries. India and Pakistan experienced similar gains. It has been argued that, by increasing grain yields in developing countries, Borlaug has saved more lives than anyone in history (Fig. 14.5).

Although the Green Revolution has been a dramatic success, gains have not been unqualified. Irrigation and inorganic fertilizer are needed to support the rapid growth of these plants. Because many inorganic fertilizers are produced from fossil fuels, the cost of farming increases with fuel prices. On the other hand, the high-input system of agriculture has allowed food production to increase without a corresponding increase in the amount of land used for farming. This means that in areas where population growth is strong, high-yield farming prevents deforestation of wild areas.

Cross-pollinated species include corn, rye, alfalfa, and clover, as well as most fruits, nuts, and vegetables. The simplest form of selection in cross-pollinated crops is **mass selection.** With this method, many plants from a population are selected, and seeds from these plants are then used to create the next generation. Again, the seeds from the best plants are chosen and propagated, and so on, for many generations. This slowly molds the genetic makeup of the population to fit the breeder's preferences. For example, suppose variation in seed size is due to genetic differences and the breeder always collects from plants with the largest seeds. After several generations of selection, the breeder has genetically altered the population so that a larger proportion of plants carries the genes for large seed size.

Outcrossing (crossing between genetically different plants) in cross-pollinated crops often results in hybrid vigor, or **heterosis** (Fig. 14.6). These plants are large, vigorous, fertile, and high yielding. Conversely, self-pollination of cross-pollinated plants will result in **inbreeding depression** in the form of small size, poor vigor, low reproductive capacity, and a high proportion of abnormal plants, due to the expression of deleterious recessive alleles. Breeders, however, do not avoid inbreeding cross-pollinated species. In fact, a common breeding method involves forced self-pollination for several generations to create inbred lines in which deleterious alleles have been eliminated. Then, selected inbred lines are crossed to produce **hybrid seed.** Using this method, in which crosses between unrelated inbred lines often produce hybrids that dramatically outyield their parents, the most dramatic success story to date involves corn (maize). In 1908, in one of the earliest hybrid corn studies, plant breeder G. H. Shull crossed two inbred lines of corn, each of which produced 20 bushels per acre. The hybrid offspring yielded 80 bushels per acre, a quadruple increase in yield. Most corn varieties in the United States are hybrids (Fig. 14.7).

Some early American and European varieties, called **heirloom varieties,** are grown as open-pollinated populations of plants. Each variety is a mixture of genotypes, and all plants are allowed to pollinate each other, or open-pollinate, during seed production. These varieties are not as uniform as modern varieties, but their genetic variability allows them to produce a crop under many different environments. For example, a dry year and an insect pest might severely compromise all the plants in a hybrid variety if the variety is not genetically capable of surviving these conditions. An open-pollinated

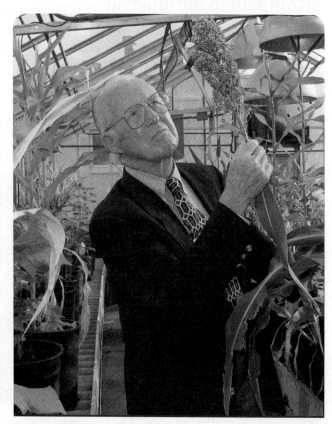

Figure 14.5 Norman Borlaug, the father of the Green Revolution. © Bill Meeks/AP Images

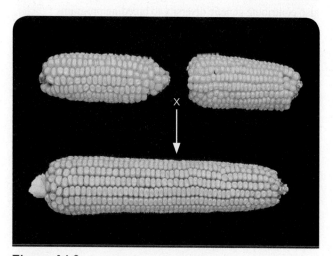

Figure 14.6 Crosses between inbred lines (top) often produce high-yielding hybrids (bottom). Such vigor is called heterosis. © Shelley H. Jansky

Figure 14.7 Hybrid corn plants. © *Kingsley Stern*

variety is likely to have some plants that are adapted to drought and others that can resist insect damage. The Seed Savers Exchange in Decorah, Iowa, has been created to preserve and distribute heirloom varieties. Heirloom varieties are important for farm market producers and growers of organic crops. They also contribute genetic diversity to breeding programs.

Germplasm Collection and Gene Banks

Progress in plant breeding is absolutely dependent on genetic variability. It is impossible to improve a population if there is no genetic variability for the trait of interest. For example, a breeder who needs to develop rust fungus-resistant wheat must begin with a population containing at least one plant with some degree of resistance. Plant breeders are, therefore, concerned with the *germplasm* resources of crop plants. As mentioned earlier, a crop plant's **germplasm** is the sum total of its genes. A breeder must have access to crop plant germplasm-containing genes for traits that are important for current breeding efforts. In the future, new traits may become important. For example, in a decade or two, we may need to develop plants with resistance to a new pathogen.

The plant varieties used in agriculture today have resulted from centuries of selection for specific traits. They are often genetically uniform and, therefore, may not be good sources of new genetic variability for further advances in breeding. In addition, their homogeneity makes them vulnerable to outbreaks of pests. The Irish potato famine in the mid-1800s occurred because people in that region relied on a few varieties of potatoes for most of their food supply. When a pathogen spread through the fields, all the plants succumbed, and the people lost their entire potato crop. As recently as 1970, 15% of the United States corn crop was wiped out by a fungal pathogen. Genetic uniformity in the corn varieties grown at that time resulted in this significant loss.

In order to meet current and future needs for plant genetic diversity, **gene banks** have been established worldwide. The International Plant Genetic Resources Institute is responsible for coordinating international efforts to collect and conserve crop plant germplasm. The institute also assists regional and national gene banks. The collaborative efforts of state and federal government agencies, along with private industries, constitute the National Plant Germplasm System in the United States. Scientists regularly conduct collecting expeditions in regions where crop wild relatives (CWR) are found (Fig. 14.8). They then bring plant samples back to gene banks, where they are catalogued, propagated, and screened for desirable traits. Some seeds or other **propagules** are put into long-term storage for future needs, while others are given to plant breeders for immediate use.

Breeding Methods Using Sexually Incompatible Germplasm

In recent years, powerful new tools have been added to the toolboxes of plant breeders. Most breeding progress today is made using crosses among crop plants and their relatives, as described in the previous section, "Breeding Methods Using Sexually Compatible Germplasm." However, in many breeding programs, species boundaries no longer exist. Through techniques such as *protoplast fusion* and *gene splicing*

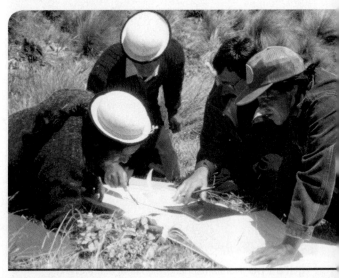

Figure 14.8 Botanists on a plant-collecting expedition in Peru. *Courtesy David Spooner*

(discussed in sections that follow), genes from virtually any living organism can be incorporated into crop plants. Everything, therefore, from an antifreeze gene in North Atlantic fish to a pesticide gene in bacteria, is now accessible to breeding programs.

Protoplast Fusion

There are times when a breeder would like to tap into the genetic diversity of a plant species that is distantly related to a crop plant. However, it is impossible to make a cross between the two species and obtain viable seeds. An alternative strategy involves **protoplast fusion.** With this method, cells of each species are grown in a liquid nutrient solution. Then, their cell walls are chemically stripped off to produce protoplasts. The protoplasts of the two species are mixed together and stimulated, with the aid of an electric current or chemical solution, to fuse with each other. When screening the fusion products, a scientist must distinguish between *autofusions,* where two protoplasts of the same species have fused together, and *hybrid fusions,* in which the protoplast of one species has fused with that of the other species. One way of doing this involves dyeing the protoplasts of each species with a different color before the fusion event. Then, the scientist can examine the fusions with a microscope and identify protoplasts that show a mixture of the colors. The selected protoplasts are then carried through a series of *tissue culture* events (see "Micropropagation" in this chapter) that cause them to grow into whole plants. These new plants carry genes from two distantly related species. They are called **somatic hybrids** because they resulted from combining (hybridizing) *somatic,* or body, cells from two different plants. In contrast, plants usually reproduce when gamete cells (eggs and sperm) are combined, producing sexual hybrids.

Although protoplast fusion appears to have value for the incorporation of new genes into crop plants, few success stories are currently available. Somatic hybrids have been created between potato cultivars and nontuber-bearing wild relatives. Some are highly resistant to a number of bacterial and fungal diseases, but varieties have not yet been created from the hybrids or their offspring. Somatic hybrids between haploid ($1x$) Clementine and diploid ($2x$) mandarin oranges have produced the seedless, easy-to-peel oranges that are readily available every winter.

Gene Splicing and Transgenic Plants

Genes from virtually any organism, from viruses to humans, can now be inserted into plants, creating **transgenic plants.** The technologies that have been developed to support this work arose from basic research to understand DNA structure and function. **Recombinant DNA** technology is based on the knowledge that DNA behaves the same no matter what organism carries it. Because the DNA code is nearly universal, a plant cell can read a DNA sequence from almost any other organism and produce a foreign gene product.

The serendipitous discovery of **restriction enzymes** paved the way for molecular genetic engineering. These enzymes act as molecular scissors, chopping up DNA into pieces small enough for genetic analysis. In nature, bacteria produce restriction enzymes, protecting themselves from viral pathogens by destroying pieces of foreign DNA. A key feature of these enzymes is their ability to cut DNA at specific sites and create DNA fragments with "sticky" ends. When DNA is cut by certain restriction enzymes, one strand is slightly longer than the other. Each fragment then is a double-stranded DNA molecule with a single-stranded tail at each end. This tail is called a sticky end because it will base-pair with the tail of any other fragment cut with the same enzyme (Fig. 14.9).

How are transgenic plants made? For example, how could a bacterial insect resistance gene be inserted into a corn plant? First, the insect resistance gene in the bacterium must be discovered. This is actually a very difficult task, and the details are beyond the scope of this book (as expected, finding a specific gene in the enormous genome of eukaryotes is even more difficult and is like looking for a needle in a haystack).

Once the gene is found in the bacterium's genome, a restriction enzyme is used to cut the bacterial chromosome on both sides of the gene, but not within the DNA sequence of the gene. After that, the bacterial DNA and a *cloning vector* are cut with the same restriction enzyme. A *cloning vector* is a piece of DNA that can copy itself in a living cell. A common example of a cloning vector is a **plasmid,** a small, circular piece of DNA naturally found in bacteria and capable of replicating independently of the bacterial chromosome. Dozens of custom-made plasmids containing restriction enzyme cutting sites have been created by genetic engineers. If the cut bacterial DNA is mixed with cut plasmids, then their sticky ends will join together and create recombinant DNA. In this case, the recombinant

Figure 14.9 Some restriction enzymes can cut DNA so that one strand is longer than the other. The DNA ends are said to be sticky because they will pair with any DNA fragment that has been cut with the same restriction enzyme.

DNA is a plasmid with an insect resistance gene attached to it. *Escherichia coli*, the bacterial workhorse of molecular genetics, can be stimulated to take up the plasmid via a process called **transformation**. When the *E. coli* cells multiply in a flask of a nutrient medium, they will replicate their recombinant plasmids. At this point, *E. coli* is cloning our insect resistance gene for us. A culture of the bacterium will contain millions of copies of the gene (Fig. 14.10).

After the insect resistance gene has been cloned by *E. coli*, there are enough copies to manipulate it and try to insert it into plant cells, again through transformation. The plasmids can be removed from a culture of *E. coli*, and the same restriction enzyme can be used to cut out the resistance gene and purify it. Plant cells will not take up foreign DNA as readily as bacterial cells. Nevertheless, a number of techniques to transform plants with foreign DNA have been developed. This section will focus on the two major ones.

The most common way to transform a plant relies on a natural genetic engineer, *Agrobacterium tumefaciens*. This bacterium infects plant cells and then inserts DNA (called **T-DNA**, or transfer DNA) from its plasmids into the plant's chromosomes. This causes the plant to produce cancer-like tumors that harbor and feed the bacterium (Fig. 14.11). Human genetic engineers, however, have removed the virulence (disease-causing) genes from the T-DNA region and replaced them with genes of human interest. For example, an insect resistance gene could be inserted into the T-DNA region of the *Agrobacterium* plasmid (Fig. 14.12). Then, the corn cells can be infected with the bacterium, which will insert its T-DNA into the chromosomes of those cells. Transformed cells can be cultured to produce whole plants carrying and expressing the insect resistance gene. The gene has now been stably incorporated into a plant chromosome and will be passed on to subsequent generations (Fig. 14.13).

When the *Agrobacterium* transformation technique was first developed, it was not effective in monocots. As a result, an alternative strategy using a **particle gun** was developed especially for use with monocots. This technique actually first used a modified shotgun, with DNA attached

Figure 14.10 In a common gene cloning technique, a plasmid from a bacterium and a gene of interest from foreign DNA are cut with the same restriction enzyme. The gene can then be stably inserted into the plasmid, which is then taken up by the bacteria. When the bacteria multiply, they replicate the inserted gene.

Figure 14.11 A crown gall, caused by the bacterium *Agrobacterium tumefaciens*, on a tomato plant. © *Central Science Laboratory/Science Source*

Figure 14.12 After a gene is cloned, it can be inserted into a T_i plasmid in *Agrobacterium tumefaciens*. When the bacterium infects plant tissue, the gene is inserted into plant DNA.

to the shot. Today's particle guns are much more refined (and expensive!) than that first prototype but still rely on shooting DNA into plant tissue. With this biolistics technique, very small tungsten or gold pellets would be coated with the cloned insect resistance gene. Then the DNA can be shot into corn cells and, if enough plant cells are bombarded, some of them will have the gene incorporated into their chromosomes. Precisely how this process works is still a mystery, but through the use of this technique, foreign genes have been permanently inserted into crop plants. In fact, transgenic corn containing a bacterial insect resistance gene is now widely grown in the United States.

It is difficult to control the number of gene copies inserted during transformation. In addition, genes are inserted at random locations in plant chromosomes during this step. Sometimes, the foreign genes are inserted in areas of the genome that are not expressed by the plant, or they may insert into a critical portion of existing plant genes. In addition, considerable effort is often needed to get the foreign gene expressed at suitable levels in appropriate tissues. Plant transformation and gene expression, therefore, are the most challenging aspects of transgenic plant production.

Transgenic corn, soybean, cotton, and canola varieties containing herbicide and insect resistance are grown extensively in North America. Traits such as disease, insect, and

Figure 14.13 Steps in making a transgenic plant.

Figure 14.14 Compared to conventional rice (left), golden rice (right) is colored due to the production of beta-carotene, which the human body converts to vitamin A. *(left)* © Sergey Molchenko/Shutterstock *(right)* © Olga Popova/Shutterstock

herbicide resistance are important for the producers of those crops, but they are not noticed by consumers. Benefits from transgenics that directly help consumers have been more elusive. The most famous example is golden rice, which was developed around the turn of the 21st century (Fig. 14.14). Conventional rice is a staple food in parts of the world where vitamin A deficiency is common, leading to blindness and other serious health problems. To address this problem, rice was transformed with genes for beta-carotene synthesis in the rice endosperm (the part of the rice plant that is consumed). Beta-carotene is converted to vitamin A in the human body. It has been technically challenging to engineer the rice plants to produce nutritionally beneficial levels of beta-carotene in high-yielding varieties. It has also been politically challenging to promote the adoption of genetically engineered rice. However, there is optimism that golden rice will become widely available in the near future.

Nonedible transgenic plants are also showing potential in some novel ways. Plants are being created to produce biodegradable polymers in order to reduce our dependence on plastics made from nonrenewable resources. Plants are also being engineered to produce industrial products, such as modified starch for packaging.

Finally, transgenic technology is providing us with ornamental plants never before seen in nature. Most popular cut flowers do not produce blue hues, and efforts are presently being directed toward the development of blue carnations, chrysanthemums, and roses. For example, a violet carnation named Moonshadow contains a petunia gene for blue color (Fig. 14.15). Genetic engineers are also trying to extend the shelf life of cut flowers, mainly by blocking genes for ethylene synthesis. In addition, "flowering-time" genes are being used to develop plants that flower under day-length conditions that would normally prevent flowering; plant architecture genes are being used to produce either compact or vine-type plants; and fragrance genes have been identified in an effort to restore fragrance quality to roses and carnations.

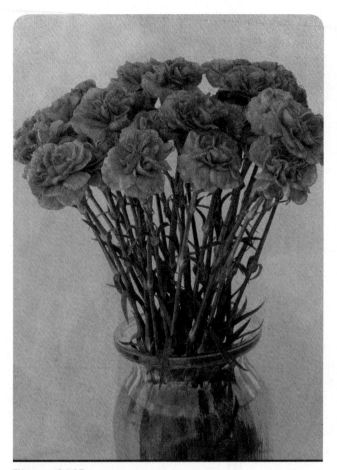

Figure 14.15 The Moonshadow variety of carnation contains a petunia gene for blue color. © Shelley H. Jansky

Pros and Cons of Transgenic Plants

There are approximately 190 million hectares (470 million acres) acres of transgenic crops grown worldwide, accounting for 10% of crop production. The most common transgenic crops are soybean, corn, cotton, and canola. Most often, these plants contain a gene making them resistant to the herbicide glyphosate, commercially sold as Roundup, or an insect resistance gene that produces a protein called Bt toxin (because it is derived from the bacterium *Bacillus thuringiensis*). In the United States, 94% of cotton, 93% of corn, 94% of soybean, and 95% of sugar beet acreage are planted as transgenic crops. It is estimated that 75% of all processed food products in the United States contain a genetically modified product.

On the positive side, proponents of transgenic crops argue that transgenic crops are environmentally friendly because they allow farmers to use fewer and less noxious chemicals for crop production. For example, a 21% reduction in the use of insecticide has been reported on Bt cotton. In addition, when glyphosate is used to control weeds, then other, more persistent herbicides do not need to be applied. In addition, farmers can reduce tilling practices, which protects the soil from erosion. However, recent studies indicate that there may be negative human health effects from exposure to glyphosate.

KEY THEME 14.1 molecular Genome Editing

The Hollywood film *Gattaca* presents a science fiction story about the use of genetic engineering to "perfect" the human species. In this movie, genes responsible for genetic disease and other human imperfections are corrected to create a human super race. How much of this film is reality and how much is fantasy? Is it really possible to genetically engineer animals and plants to perfection? Genetic engineering is a powerful technology that has been used for decades to improve plants, animals, and microbes for human use. However, the application of genetic engineering technology, such as creating transgenic plants and animals, has been limited by the requirement for extensive evaluation of the technology and strict guidelines for its use. In recent years, a rapidly developing area of genetic engineering, called genome editing, has become poised to revolutionize genetic engineering by being more effective and potentially safer. Genome editing uses special enzymes called sequence-specific nucleases (SSNs) to target regions of a genome for modification. SSNs can be engineered to target virtually any DNA sequence and are used to make breaks in double-stranded DNA. Once a break has been made, the living cell is forced to repair the break, which can also allow new DNA to be introduced into the genome. As such, genes can be knocked-out or modified, giving new functions. This provides infinite possibilities for genetic engineering. In 2013, a groundbreaking SSN called CRISPR/Cas was first used for genome editing in eukaryotic cells (Box Figure 14.1A). In just three years, CRISPR/Cas has been used to develop micro pigs with targeted mutations in growth genes, mosquitoes unable to spread malaria, and human embryos in which genetic diseases have been corrected. To date, no reports of germ-line modifications have been made in humans using CRISPR/Cas or any other SSN. However, this possibility has already alerted the scientific field, and worldwide restrictions on SSN technology have been put in place. What are the implications for CRISPR/Cas and other SSNs in agriculture? Genome editing is rapidly being used to develop new plants for research and genetically

Figure 14.1A The CRISPR-CAS9 editing complex (gray) has targeted a DNA region (blue) for editing. © *Science Photo Library/Alamy Stock Photo*

engineered crops for agriculture. The ability to target any genomic region for mutagenesis allows genes to be studied in plants in which natural mutants are not available. Gene editing has been used to mutate a sugar metabolism gene in potato, resulting in improved processing quality. In wheat, a disease susceptibility gene has been inactivated, resulting in resistance to powdery mildew. Both of these varieties carry no transgenes and are not regulated as transgenic. This is an important distinction from traditional genetic engineering, in which genes are introduced into random places in the DNA. The precision of genome editing in combination with the ability to target any genomic region creates endless possibilities for genetic engineering. Nevertheless, further research is needed to understand the effects of genome editing technology and its safety before we leap into the futuristic-world attempt to engineer the perfect human.

Nathan Butler

There is concern about whether herbicide-resistance genes will move into populations of weeds. Crop plants are sometimes grown in areas where weedy relatives also live. This is especially true outside of the United States. (Remember that most crop plants did not originate in the United States, and they do not have wild relatives here.) If the crop plants hybridize with weedy relatives, then this herbicide-resistance gene will be perpetuated in the offspring. In this way, the resistance gene can make its way into the weed population. If this happens, a farmer can no longer use glyphosate, for example, to kill those weeds. This scenario is not likely to occur in many instances because there are no weedy relatives growing near the crop plant. However,

in some cases, it may become a serious problem. For example, canola readily hybridizes with mustard weed species and could transfer its herbicide-resistance genes to those weeds. In addition, Canadian farmers have reported that herbicide-resistant canola has invaded wheat fields like a weed and, of course, cannot be killed with a major herbicide.

In addition to questions regarding environmental safety, human health issues resulting from the addition of transgenes must be addressed. The major concern over the consumption of foods derived from transgenic crops is the potential for the transgene protein product to cause an allergic reaction. For example, a Brazil nut gene was added to soybean to increase its methionine content. Methionine is an amino acid commonly added to animal feed. However, this gene produced a protein that caused an allergic reaction in some people. Although the transgenic soybean was being developed as an animal feed, there was concern that it might find its way into the human food chain, and consequently, it was dropped from production.

14.3 Plant Propagation

Learning Outcomes

a. Describe the relationship between plant breeding and plant propagation.
b. Explain how commercial seeds are produced.
c. List five methods of asexual propagation.

After creating genetically superior plants through breeding, it is necessary to have methods to perpetuate, or propagate, the improved plants. Plant propagation techniques enable us to grow thousands of acres of a superior corn variety or to have the same daylily variety growing in thousands of backyards. Some plants are most easily propagated through sexual reproduction via seed propagation. With other plants, asexual reproduction methods are used more extensively.

Seed Propagation

Hybrid varieties are often grown from the seed produced by crosses between two inbred parents. A common summer job for teenagers in corn-belt regions is detasseling corn (removing male flowers). Rows of one inbred parent are planted next to rows of a second inbred parent (Fig. 14.16). Then, the tassels of the first parent are removed to prevent self-pollination. Pollen from the second parent is carried by the wind to the female flowers (the ears) of the first parent and carries out fertilization. The mature ears of the detasseled parent are then collected, and the kernels are harvested as hybrid seed. That seed will be sold to farmers for planting their hybrid cornfields during the next growing season.

Inbred line varieties are typically grown from seed. It is especially easy to generate seed from these varieties. They are simply grown in a field and allowed to self-pollinate. Seeds collected from these plants will grow into a uniform population similar to the plants from which the seeds were collected. For example, if you collect seeds from an inbred green bean variety, you can grow plants that are nearly identical to that variety in appearance and fruit quality. In contrast, seeds collected from self-pollinating a hybrid plant will produce a highly variable population of plants.

Commercial seeds usually come from fields planted solely for seed production. The fields are isolated from other fields of the same crop in order to prevent contamination by foreign pollen and mixing of seeds by harvest equipment. Seed quality depends on the health of the plants that produce it. Growing conditions, consequently, are meticulously monitored. Often, the plants are grown in arid regions where irrigation can be controlled and seeds can dry out before harvest. Most grass and forage seed production is concentrated in the Pacific Northwest, while vegetable and flower seeds are generated in the coastal valleys of California. There is also a growing international seed production industry.

When seeds are mature, they are harvested and stored in a controlled environment. The **viability** of most seeds is best maintained in cool, dry storage conditions. Samples are periodically removed from a seed lot and tested for vigor and viability. You may sometimes see the results of these tests on packages of seeds you purchase for your garden. You will also see a date stamped on the seed package. Seeds of some species of plants retain their viability in storage much longer than others. Seeds of green beans, lettuce, onions, and peppers lose their viability after a year or two in storage, while those of beets and tomatoes may be stored for 5 years or more.

Before planting seeds, steps may be taken to ensure the growth of a vigorous stand of seedlings. In preparation for planting, seeds may be dusted with a *protectant,* such as a fungicide. The red coating on seed corn is such a fungicide. Sometimes, seeds are dusted with beneficial bacteria before planting. Legumes such as peas and beans establish mutualistic associations with bacteria that *fix* nitrogen (convert nitrogen from the air into forms the plants can use). Dusting seeds with these bacteria will encourage that relationship to develop.

It is important to plant seeds in a suitable bed. The soil should be moist enough to allow the seeds to imbibe water, which is essential to begin germination. However, if the soil is soggy, the new roots will not have enough oxygen to keep up with the respiration needed for active growth. Seeds of different species are adapted for growth at different soil temperatures. Celery, lettuce, and onion seeds will not germinate in hot summer soil, while tomato, bean, cucumber, and sweet corn seeds germinate poorly in cool spring soils. Initially, seeds rely on stored food for growth. After a seedling establishes a root system, it can be fertilized to support rapid growth.

Figure 14.16 In this field planted for hybrid corn seed production, rows that have been detasseled are adjacent to rows with tassels. Pollen from the tassels will land on ears of detasseled plants. Hybrid corn will be harvested from detasseled plants. *Courtesy of Randy Manning*

In their early stages of growth, seedlings are easily shaded and stunted by weed growth, so weed control is important until the plants can shade the spaces between rows. Seedlings may also compete with each other for light, water, and nutrients, making thinning necessary for production of a good crop.

Asexual Plant Propagation

While the cloning of animals is a relatively recent event, botanists have been cloning plants for centuries! Many plants are easy to **propagate** asexually, using vegetative parts rather than seeds. For example, **crown division** is a simple technique in which a plant is separated into several pieces, each of which contains a portion of the crown and root system (Fig. 14.17). This is commonly used for many ornamental perennial plants. A breeder who has identified a superior plant of an asexually propagated species can perpetuate that genotype indefinitely. This avoids the genetic variability and consequent unpredictability that results from seed propagation.

Cuttings

Of course, nothing lives forever. Actually, that's not exactly true. Through asexual propagation, a plant can live indefinitely. A clonal colony of a Tasmanian bush (*Lomati tasmanica*) is estimated to be 43,600 years old. Closer to home, a cutting from a favorite plant could be propagated indefinitely and would essentially live forever.

Propagation of plants from parts such as leaves is called propagation by **cuttings.** If a stem is used as a cutting, it needs to be coaxed into producing **adventitious** roots (roots produced on internodes or other parts of plant organs). Buds on the stem will grow into the shoot system. Leaf and root cuttings, though, must develop both adventitious roots and shoots (Fig. 14.18). The formation of adventitious structures requires plant cells near the wound to **dedifferentiate** (become less specialized) and create a new **meristematic** region.

After a cutting is made, it must be kept in an environment that will allow wound healing and the development

Figure 14.17 A sexual reproduction of a daylily by crown division. © *Shelley H. Jansky*

Figure 14.18 Asexual reproduction by stem cuttings of an ornamental fig (*Ficus*) (left) and leaf cuttings of *Sansevieria*. Note the adventitious roots developing from both types of cuttings.
© *Shelley H. Jansky*

of new organs. The most critical step is prevention of water loss. Because a stem or leaf cutting has lost its root system, its tissues must be kept moist until roots can once again supply water. This is easily done by keeping the cuttings in an enclosed container to maintain high humidity and to reduce **transpiration.** In commercial production systems, frequent misting or constant fog is used to keep cuttings from drying out. Sometimes, rooting is stimulated by application of a rooting powder containing a growth regulator such as **auxin.** Rooting powders also often contain fungicides to prevent pathogens from entering wounds. The cuttings must be kept in a potting mix such as perlite or vermiculite, which holds water but also drains well. This is preferable to rooting cuttings in water, because oxygen levels in water are low and are quickly depleted as young cells in developing roots undergo respiration. Temperature is also a factor in successfully generating cuttings. Ideally, young roots should be warm enough to enhance growth, while the aboveground parts should be cool enough to minimize growth and transpiration. For this reason, cuttings are often grown on heating pads or in heated beds. Once roots are established, fertilizer helps to stimulate growth. Commercial production of ornamental plants relies on large-scale propagation by both seeds and cuttings (Fig. 14.19).

Many houseplants are easily propagated by cuttings. These include African violet (*Saintpaulia*), snake plant (*Sansevieria*), *Begonia, Coleus,* and *Kalanchoë*. Outdoor plants propagated by cuttings include *Forsythia, Geranium,* rose (*Rosa*), *Spiraea,* raspberry (*Rubus*), and juniper (*Juniperus*). The main advantage of propagation by cuttings is that identical copies of a valuable plant can be made. A major disadvantage is that diseases carried by the mother plant, including those caused by viruses, fungi, and bacteria, are also propagated.

Figure 14.19 Large-scale propagation of ornamental plants in a commercial operation. © Dr. Ellen Leue, Ball Horticulture

Figure 14.21 Steps in air layering. (a) Cuts are made at an angle to the axis of a stem of a rooted plant. (b) Damp sphagnum moss is wrapped around the cut area. (c) Polyethylene film is wrapped around the moss, and the ends of the film are taped shut. (d) Aluminum foil is wrapped around the film-covered moss. Adventitious roots develop and protrude through the moss. Then the rooted portion can be cut off and planted. © Shelley H. Jansky

Layering

Layering is a modified form of cutting propagation and works well for some plants that are not easy to propagate by cuttings. This procedure allows the adventitious root system to develop before the new plant part is severed from the parent plant.

Tip layering is used with blackberries, boysenberries, and other plants with flexible stems. The canes are bent over until the tips touch the ground; the tips are then covered with a small mound of soil. Roots form on the portion of buried stem, and eventually, shoots also appear (Fig. 14.20). These new plants can then be cut from the parent stems. Variations of tip layering include forcing a stem to lie horizontally and covering it with small mounds of soil at intervals, or heaping soil around the base of a plant so that the individual stems produce roots there. Once roots have been established, the individual plantlets or pieces of stem can be cut off from the original parent and grown independently.

Air layering is sometimes used to propagate tropical trees and shrubs. It is useful for producing a few large plants from a single plant with a rigid stem. A branch or main stem is wounded with a sterile knife and then may be dusted with a rooting powder. The wound is produced by gashing the stem or, alternatively, by *girdling*. **Girdling** is the removal of a ring of bark around the stem. Damp sphagnum moss is wrapped around the wound, and the area is covered with clear plastic film to retain moisture. Then, aluminum foil is placed over the plastic film to reflect sunlight and prevent overheating (Fig. 14.21). When roots are observed in the moss through the plastic film, the layer is removed from the parent plant and transplanted.

Grafting

Grafting is a process by which segments of different plants are connected and induced to grow together as one plant. It has been performed as a horticultural art for thousands of years, dating back to around 1560 B.C. Historically, grafting has been performed mainly to clone plants that are difficult to propagate as cuttings. Today, many fruit and nut trees are grafted. Trees are bred for high-quality fruits or nuts. Then, they are grafted onto root systems selected for traits such as winter hardiness, dwarfing, and disease resistance. The top part of the graft is called the **scion.** The bottom portion, which forms the root system, is the **rootstock.** To take grafting a step further, though, it is possible to graft several different but related plants onto the same rootstock. You could, for example, have grapefruit, orange, and lemon grafted onto the same *Citrus* rootstock. Almond, plum, and apricot can be grafted onto a peach rootstock. Many nurseries

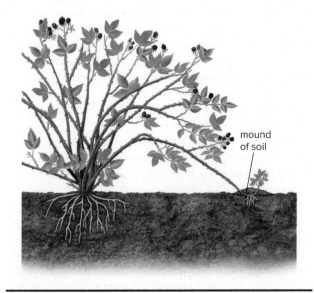

Figure 14.20 The tip layering technique involves bending the tips of canes to the ground and then covering each tip with a mound of soil. When a new plant has developed at the tip, it can be cut from the parent plant and grown independently.

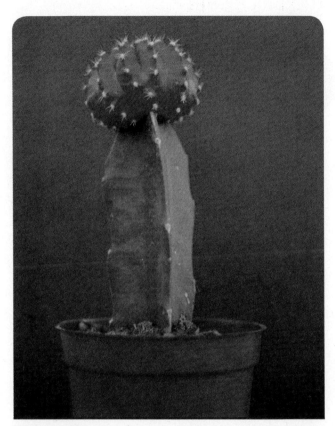

Figure 14.22 A brightly colored cactus stem has been grafted onto a green cactus plant. © *Shelley H. Jansky*

Figure 14.23 A simple graft. The root portion (*stock*) and the portion to be grafted onto the stock (*scion*) are cut so that the two parts will fit together with the cambium of both portions in close contact.

sell grafted apple trees containing two or more varieties. Brightly colored novelty cactus plants, which are not capable of photosynthesis, survive because they are grafted onto normal plants (Fig. 14.22).

To describe how grafting is performed, assume that a breeder has developed a new extra juicy apple variety, but it cannot survive winters in northern states. This new apple variety could be grafted onto a young native tree with a hardy root system. In late spring, young branches from the tree with the juicy apples (the scions) could be collected and kept in a cooler to inhibit the buds from growing. Then, later in the spring when the rootstock began to grow, the scion branches would be brought out of the cooler. Using sterilized instruments, the top of the rootstock would be cut diagonally several inches above the soil line. At the same time a diagonal cut would be made in a scion branch that is similar in diameter to that of the rootstock. Successful grafting depends on good contact between the **vascular cambium** of the scion and that of the rootstock. Therefore, the scion and rootstock must be aligned so their cambia are in contact. They are then held together with tape or string (Fig. 14.23). The graft union is often also sealed with wax to prevent the wounded tissues from drying out. If the graft is successful, then several weeks later, the vascular cambia of the scion and rootstock will have grown together. The tree now consists of a single plant with one variety of rootstock below and a different scion variety above. An expanded and illustrated discussion of grafting is given in Appendix 4.

Propagation of Specialized Stems and Roots

Herbaceous perennials produce shoots that die during the winter, but in the spring, they regrow from underground storage structures. The underground structures that survive over winter may be **bulbs**, *corms, tuberous roots,* or *rhizomes* (discussed in Chapter 6) (Fig. 14.24). These structures are typically fleshy, with abundant food-storage tissue, and contain buds that will sprout in the spring.

Micropropagation

One of the major disadvantages of most asexual propagation techniques is that they also propagate pathogens. Once a plant is diseased, there is no immune system to activate or antibiotic that can be administered as a cure. The best way to eliminate the effects of plant disease is to prevent exposure to pathogens. This is nearly impossible in greenhouse and field environments. However, it is possible to maintain plants in a disease-free state if they are grown in sterile test tubes. This is called **micropropagation.** Other advantages of micropropagation include the capacity to multiply and grow large numbers of plants in a small area. In addition, since the plants do not need to be watered or fertilized, little maintenance is required.

Propagating plants through micropropagation is similar to growing them as cuttings. The major difference is that the plants are grown *in vitro* in a sterile medium and maintained in controlled environment rooms. The medium includes a support matrix composed of agar, a gelatinous material extracted from red algae. Inorganic salts are added to the medium to provide macro- and micronutrients, such as nitrogen, phosphorus, calcium, and iron. Sucrose is added to supplement the sugars produced by the plant. In addition, vitamins such as thiamine, nicotinic acid, and inositol

Figure 14.24 (a) Propagation of specialized stems. From right to left, they include tubers, runners, bulbs, and rhizomes. (b) Examples of specialized stems include tulip bulbs (left), *Crocus* corms (middle), and ginger rhizome (right). *(b) © Shelley H. Jansky*

are generally included in the growth medium. Commonly, growth regulators are also added. After the ingredients are combined and pH is adjusted, the mixture is poured into *test tubes*. The tubes are then capped and put in an **autoclave** (a large form of pressure cooker) to sterilize them (Fig. 14.25). When the medium cools, it solidifies like Jell-O™. This acts as the "soil" in the micropropagation system, providing plants with support, nutrients, and water.

Micropropagation, like other forms of asexual propagation, relies on the property of **totipotency** (the capacity of a cell to give rise to any structure of a mature organism) of plant cells. Each living cell has the genetic information and, therefore, the capacity to develop into any cell type. Micropropagation usually begins with an excised piece of leaf or stem tissue, or **explant,** and carries it through three steps.

The first step is establishment of explants in **tissue culture.** Micropropagation requires sterile plant material as well as growth media. Plant parts must be **disinfested** to remove surface contaminants without killing the plant tissue. Common disinfestants include bleach and ethanol. This procedure does not remove internal contaminants, including pathogens, so it is important to begin with disease-free plants. After plant parts are disinfested, they are inserted into the growth medium in test tubes under sterile conditions. Often, a reach-in chamber, called a **laminar flow hood,** is used for this step (Fig. 14.26). Filtered air is blown across the work surface in the chamber to prevent the introduction of contaminants. Test tubes containing sterile plants are then placed in a clean room with artificial lighting and temperature control. The goal of this first step is to obtain sterile, viable plant tissue cultures (Fig. 14.27).

After cultures are established, they typically will be induced to develop multiple shoots in a multiplication medium. These **microshoots** can be separated and placed in a new medium by a process called **subculturing** (Fig. 14.28). This step is similar to propagation by cuttings, except it is carried out under sterile conditions. It is not unusual to subculture plants every 4 weeks, making approximately four new plants from every one in a test tube. At this rate, it is theoretically possible to produce a million plants from one plant in just 10 months. Although these multiplication rates are not

Figure 14.25 This autoclave uses pressurized steam to sterilize media, glassware, and instruments. © *Kingsley Stern*

Figure 14.27 Tissue culture plantlets. © *Shelley H. Jansky*

In other cases, plants are induced to form roots *in vitro* by transferring them to a rooting medium. Compared to the multiplication medium, the rooting medium usually contains reduced levels of cytokinins and increased auxin levels. When possible, the most economical approach is an *ex vitro* one, in which microshoots are rooted in potting mix and treated as cuttings.

The last step in micropropagation is the transfer of plants back to an outdoor environment. This is often the most difficult step. Because the humidity is high in test tubes, plants grown *in vitro* do not produce as much wax on their cuticles as do those grown outdoors. In addition, their stomata do not close as readily in response to water stress. Therefore, tissue culture plantlets must be acclimatized to an outdoor environment. First, the humidity in the growth chamber is reduced for a few weeks before the plants are

realized in commercial systems, many tissue culture laboratories have the capacity to produce millions of plants per year.

The third step in micropropagation is root formation. Some explants, such as those from African violets, will spontaneously produce roots in multiplication medium.

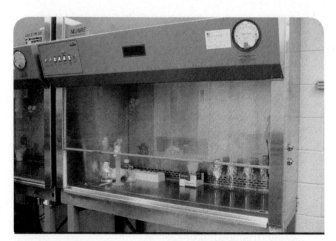

Figure 14.26 This laminar flow hood provides a sterile work area by blowing sterile air from the back of the hood to the opening in front. © *Shelley H. Jansky*

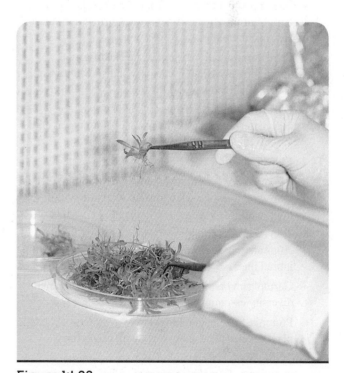

Figure 14.28 Plantlets that have developed from cultured meristematic tissue are separated and further cultured to maturity. © *Kingsley Stern*

brought out of their test tubes. Then, when they are transferred to soil in pots, they are maintained in a high-humidity environment for several weeks.

Commercial micropropagation has become a successful venture with a number of plants. Some plants that propagate slowly by other asexual means may be rapidly increased with micropropagation. These include orchids, Boston fern, African violet, and *Hosta*. In some cases, with the use of tissue culture techniques, as many plants can be produced in a month as would be produced in a year with other techniques. New cultivars can also be rapidly multiplied to meet high market demand. For example, new apple rootstocks are often propagated *in vitro* because conventional asexual reproduction methods cannot adequately supply market needs. Tissue culture protocols, using various combinations of nutrients and growth regulators, have been developed for a number of woody plants that are otherwise difficult to root.

Tissue culture is now being used to propagate endangered plant species. Small pieces of just a few plants can be used to establish tissue cultures. Then, nearly unlimited numbers of plantlets can be produced and returned to their natural habitat.

SUMMARY

14.1 Crop Plant Evolution
- Most of our food is derived from just a handful of plant species.
- We domesticate plants by genetically altering them to meet our needs.
- The impetus for the shift from hunter-gatherer to agricultural societies is not clear. Current hypotheses are speculative.
- Plant domestication began in the Near East approximately 10,000 years ago and spread to Asia, Africa, and the New World.
- During domestication, plants were selected for nonshattering seeds, high yield, seedling vigor, and absence of seed dormancy.

14.2 Plant Breeding
- Two methods for improvement of self-pollinated crops are pure-line selection and selection within self-pollinated offspring of hybrid plants.
- Two methods for improvement of cross-pollinated crops are mass selection and creation of hybrids from inbred lines.
- Germplasm banks are critical repositories of genetic diversity essential for plant-breeding progress.
- Protoplast fusion combines entire genomes of related plant species that cannot mate with each other.
- Transgenic plants contain DNA from foreign organisms. To make a transgenic plant, a gene is spliced out of the donor and inserted into a vector. The vector is inserted into a host, such as a bacterium, which clones the foreign donor DNA. The next steps, transformation of the plant with the cloned foreign gene and expression of that gene, are the most difficult ones to perform.

14.3 Plant Propagation
- Seed propagation requires production of high-quality seed, an adequate seed storage environment, and appropriate conditions for seedling growth.
- Cuttings provide a method of asexual propagation for many plant species.
- Grafting unites pieces of two different, but related, plants and can be used for cloning.
- Another form of asexual reproduction utilizes natural propagules, such as bulbs, corms, and tubers.
- Micropropagation is an asexual propagation technique

REVIEW QUESTIONS

1. Describe the traits that early humans selected during domestication of plants.
2. Explain why knowledge of a plant's reproductive system is important when choosing a breeding strategy.
3. Outline the steps you would use to create a petunia plant carrying a corn gene for red flower color.
4. Differentiate between a rootstock and a scion.
5. Explain how you would create an apple tree that produces both Granny Smith and Red Delicious apples.
6. Describe the four steps that are usually carried out in micropropagation.

DISCUSSION QUESTIONS

1. Suppose you grew a hybrid tomato plant and saved some of its seeds to plant in your garden next year. Would those seeds grow into plants that looked like the hybrid plant from which you collected them? What if you collected seeds from an inbred line of tomato and planted them?
2. Describe advantages and disadvantages of micropropagation compared with making cuttings in a greenhouse.
3. What might be some advantages of having several varieties of apple or plum on the same tree?
4. Describe advantages and disadvantages of producing a crop of potato plants using asexual (tuber) rather than sexual (true seed) propagation.
5. Many people are opposed to the production of transgenic crop plants. Do you feel that position is justified?

CHAPTER 15

Evolution

15.1 Introduction to Evolution
   Evolution in Agriculture
   Evolutionary Medicine
   Species Extinctions and Invasions
   Biotechnology and Evolution

15.2 A Brief Overview of the Early Development of Evolutionary Concepts
   The First Revolution
   The Second Revolution
   The Third Revolution

15.3 Charles Darwin

15.4 Evidence for Evolution

15.5 Microevolution—Evolution within Species
   Natural Selection
   Mutations
   Migration
   Genetic Drift
   Rates of Evolution

15.6 Macroevolution—How Species Evolve
   Geographic Isolation
   Ecological Isolation
   Mechanical Isolation

15.7 The Role of Polyploidy in Evolution
   Apomixis

15.8 Discussion

   Key Theme 15.1: Our Daily Bread

OVERVIEW

An introduction to the early development of evolutionary concepts is followed by a discussion of Charles Darwin and his contributions to the theory of evolution through natural selection. Evidence for evolution is given, and mechanisms of organic evolution, including mutations, migration, and genetic drift, are discussed. The roles of isolation and polyploidy in the evolution of species are explored. The chapter concludes with a discussion of some aspects of past and present controversy surrounding evolution.

The South African orchid (*Disa atricapilla*, also referred to as *Orthopenthea atricapilla*) has evolved to mimic the appearance of a female wasp. It is pollinated when male wasps attempt to copulate with it. *Courtesy of Marland Holderness*

15.1 Introduction to Evolution

Learning Outcomes

a. Describe the role of evolutionary principles in the study of botany.
b. Explain how Darwin's theory of evolution was influenced by plant and animal breeding.
c. Discuss the connection between insecticide resistance and antibiotic resistance.
d. Explain why evolutionary biologists are interested in species invasions.
e. Explain why genetic engineers are interested in evolution.

Few historical events have had a greater impact on society in general and the biological sciences in particular than the publication of Charles Darwin's book *On the Origin of Species by Means of Natural Selection, or the Preservation of Favoured Races in the Struggle for Life*. Darwin's theory of evolution through natural selection (the tendency of organisms with favorable adaptations to their environment to survive and produce new generations) has stimulated an enormous amount of thinking and research. It has also provided an explanation, based on natural laws, for the diversity of life around us.

There have been disagreements over the nature, the mechanisms, and even the existence of evolution. These have stemmed, in part, from a failure of people in diverse fields to define terms and to distinguish between fact and theory in matters of evolution. Evolution itself, for example, has been broadly defined by some as simply being synonymous with change. We are told that everything from cars to computers to cultures is evolving. We need, therefore, to distinguish between change in inanimate objects and *organic evolution*, which is the accumulation of genetic changes in populations of living organisms over many generations.

Although we are only two decades into the 21st century, it is already being called the "Century of Biology." The field of biology is responding to the need to address major issues in society, such as the ability to produce enough food for a large and growing population, the effects of humans on the global environment, alternatives to fossil fuels, the development of improved tools to fight disease, and the role of biotechnology in everything from food production to medicine. Evolutionary biology is in a particularly advantageous position to address these issues. Evolution provides the framework for understanding biological phenomena from DNA molecules to entire ecosystems. It explains both the unity of life, by studying common ancestors, and the diversity of life, by considering adaptive change. As such, evolution is arguably the single most important concept in all of biology. Evolutionary biology is currently one of the most exciting and rapidly developing fields of science. Our knowledge of evolutionary biology affects us in a diverse set of arenas, including medicine, agriculture, forensics, biotechnology, economics, and law.

The term *evolution* often conjures up a very specific image, such as the development of human beings from primates. However, applications of evolutionary studies permeate biology in ways that may surprise you. The following paragraphs illustrate how evolution is at the core of many advances in biology.

Evolution in Agriculture

As you will see later in this chapter, Charles Darwin became the founder of modern evolutionary thought after the publication of his book. His use of the term *natural selection* was derived from his knowledge that for millennia, humans have used **artificial selection,** or breeding, to change agricultural and domestic plants and animals. In fact, the first chapter was devoted to domestic animals. Plant and animal breeding can be thought of as evolution directed by humans. We rely on domestic plants and animals almost exclusively for our food and clothing. They are the products of the artificial selection of wild ancestors to suit our needs. Without the successful use of evolutionary principles throughout our recent history, we would still be hunter-gatherers (Fig. 15.1).

In addition to being the products of change, agricultural systems are agents of change, by influencing the evolution

Figure 15.1 (a) Wild tomatoes have small fruits. During domestication, humans altered their genetic composition so that (b) cultivated tomatoes produce large fruits. *(a)* © *Dr. Roger Chetelat/ C.M. Rick Tomato Genetics Resource Center (b)* © *Adrian Burke/Getty Images*

of native species. For example, insect populations evolve resistance to insecticides when they are exposed to high concentrations in agricultural fields. Selection pressure kills insects that do not have the genetic ability to detoxify or avoid the toxins. Most of the insects that survive do so because they have genes that allow them to live in the presence of the pesticide. These are the insects that mate and pass their genes on to the next generation. It takes only a few generations for a population of susceptible insects to evolve into a population of resistant ones. Agricultural entomologists trained in evolutionary genetics are studying ways to prevent or delay the evolution of insecticide resistance. For example, alternative pesticides from different chemical families can be used. Or, using another example from evolutionary biology, farmers can develop ways to encourage the establishment of populations of natural enemies of the pests in agricultural fields.

Evolutionary Medicine

An exciting new application of evolution is called evolutionary medicine. It applies evolutionary principles to the ways in which illnesses are treated. For example, a bacterial infection involves an interaction between a population of bacteria and the human body. If antibiotics are not used appropriately to treat the infection, then the bacteria have an opportunity to evolve, resulting in the genetic ability to overcome that biochemical weapon. This is similar to the insecticide example previously mentioned. Consequently, those antibiotics will be ineffective in all people infected by the altered bacterium. There are currently some strains of disease-causing bacteria that are resistant to all known antibiotics. The evolution of drug resistance has increased the cost of medical treatment and reduced its effectiveness. Some people fear that many infectious diseases may become untreatable, but evolutionary biologists believe that through an understanding of the evolution of drug resistance, the medical community can develop strategies to prevent this from happening.

A very different biomedical application of evolutionary research involves the discovery of natural compounds that are produced by plants and animals for survival but that may also be useful in medicine. The World Health Organization has identified over 20,000 plants that humans have used as medicines. For example, potent anticancer compounds have been identified in the Pacific yew and the rosy periwinkle of Madagascar. Other natural plant products are used as scents, emulsifiers, and food additives. Evolutionary biologists provide critical expertise in the search for new useful natural products. Based on their knowledge of family trees, they can predict where to look for new products in relatives of useful plants. In addition, if they understand the adaptive reasons for producing the valuable compounds, they can look for plants that might have undergone selection pressure for those adaptations.

Species Extinctions and Invasions

Because of human activity, species in many parts of the world are disappearing at alarming rates. Of course, the loss of these species results in the loss of potentially valuable natural products. From a broader perspective, though, the loss of species can undermine the health of entire ecosystems that depend on complex interactions among their members. Conservation biology uses evolutionary principles to understand the mechanisms by which species expand or contract in response to changing environments. Some species expand rapidly, while others move toward extinction. Studies of natural selection, migration patterns, and the effects of mutation help scientists explain and predict changes in populations over time.

In contrast to studying species extinctions, evolutionary biologists also try to understand species invasions. Newly introduced species often establish themselves at the expense of native species. For example, natural selection in plants introduced into areas where there are no natural pests or herbivores will favor the expenditure of energy for reproduction rather than defense. This allows them to establish themselves quickly and outcompete native vegetation. These non-native invasive species pose a serious threat to the survival of native species in many ecosystems (Fig. 15.2). The effect of invasive species on natural populations is especially apparent in the Hawaiian Islands, one of the most isolated groups of islands in the world. Ships and planes that visit the islands carry foreign plants, animals, fungi, and microbes. As a result of these foreign introductions, Hawaii is now home to 47 of the 100 most invasive species in the world. These invaders threaten native populations by competing for space. This is an especially alarming scenario, since, as isolated island dwellers, the native species are not found anywhere else. Therefore, they are particularly vulnerable to extinction. In order to protect native species, evolutionary biologists must understand the consequences of the interactions between established species and invaders. Then, they can intervene to try to tip the balance in favor of the native species.

Figure 15.2 Kudzu is an invasive plant introduced into the United States from Japan. It can grow as much as 20 meters (65.6 feet) per year, choking out native vegetation as shown here growing over a forest. © *Tim Mainiero/Getty Images*

Biotechnology and Evolution

The plant biotechnology industry relies heavily on evolutionary biologists to determine the fate of *transgenes* (foreign genes introduced into plants through genetic engineering) after genetically engineered crop plants are planted in agricultural fields. Some transgenic crops, for example, carry a gene that allows them to survive herbicide applications. This is useful in agricultural production systems because it allows the farmer to apply a specific herbicide, killing weeds but not the crop. What happens, though, if weedy relatives of the crop live in the area and the crop plant mates with them? Evolutionary biologists are studying the fitness of transgenes in natural systems, along with the rates of gene exchange between crop plants and their wild relatives. This will help them determine the potential for undesirable consequences, such as herbicide-resistant weeds.

15.2 A Brief Overview of the Early Development of Evolutionary Concepts

Learning Outcomes

a. Explain how Aristotle's Scale of Nature differs from our current understanding of natural systems.

b. Describe how the voyage of the *Beagle* led to the first evolution revolution.

c. List the three biological theories that intersected to create the second evolution in biology.

d. Describe how studies of the molecular biology of development have contributed to the third revolution in our understanding of evolution.

More than 2,300 years ago, Aristotle (384–322 B.C.) observed grand design in nature and arranged all organisms known to him from the simplest to the most complex in what he called the Scale of Nature. Aristotle's arrangement implied that all organisms were static and didn't evolve. These beliefs were widespread until at least the 17th century and were not extensively challenged before Darwin. *Fossils* had been found many centuries before Darwin but were not properly identified until the 15th century, when Leonardo da Vinci (1452–1519) correctly observed that they were parts of previously existing organisms that had become extinct.

Georges Cuvier (1769–1832), a French zoologist and *paleontologist* (one who studies fossils; fossils and fossilization are discussed in Chapter 21), used comparative anatomy toward the close of the 18th century to classify animals. Cuvier, however, firmly believed that organisms did not change over time. When geological finds showed apparent evolution of organisms in rock formations of certain regions, Cuvier tried to explain them away with what was called *catastrophism*. He theorized that mass extinctions or catastrophes had occurred wherever a new geological find revealed a different group of fossils, and the presence of the new fossils was due to repopulation of the region by species migrating in from surrounding areas.

By the end of the 18th century (before the principles of genetics were known), many prominent biologists had come to believe that hereditary changes in populations over long periods of time (evolution) occurred as a result of the inheritance of acquired characteristics. One of the more prominent supporters of this widespread idea was Jean Baptiste Lamarck (1744–1829). He believed, for example, that giraffes acquired their long necks over many generations as a result of the gradual increase in neck length as shorter-necked animals stretched to reach leaves on the branches of trees. Slight stretching of the neck was believed to have been passed on to the offspring as it occurred, and eventually, numerous tiny increases due to individual stretching added up to the present great neck length of giraffes (Fig. 15.3).

Molecular studies of epigenesis are revealing new insights into the concept of inheritance of acquired traits, but in a different form than proposed by Lamarck. Epigenetics is the study of heritable changes in gene expression not due to mutation. This can be accomplished by blocking gene

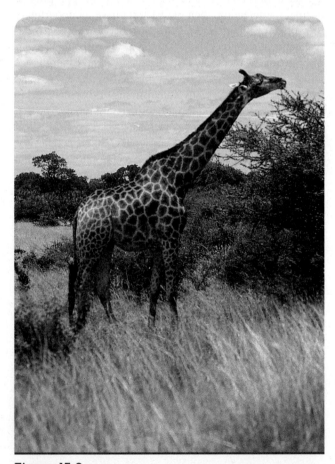

Figure 15.3 Lamarck and his contemporaries believed that the long neck of a giraffe developed over time as the animals stretched to reach higher leaves and that the little length increases were inherited and became cumulative. This theory of inheritance of acquired characteristics was experimentally disproved. © *Kingsley Stern*

expression through the temporary addition of methyl groups to the DNA molecule. When the methyl groups are removed, expression is restored. Recent studies have demonstrated that mRNA expression can also be controlled by methylation. Epigenetic changes occur in certain tissues during development and are typically reset during sexual reproduction. However, epigenetic alterations can also be carried across generations. It has been suggested that epigenetic control of gene expression allows for short-term adaptation to changing environments, with the possibility of reversing the phenotype. That is, a plant can use epigenetic alterations to switch between phenotypes, depending on environment, as an alternative to the permanent changes that result from mutation (changes in the nucleotide sequence of DNA).

The First Revolution

As discussed later in this chapter, at the age of 22, Charles Darwin set out on an adventure that would fundamentally alter the course of his life and the field of biology. The journey established the foundation for an evolutionary theory that would turn scientific thought upside-down. During the 5-year voyage around the world, the Englishman served as the ship's naturalist and had the opportunity to see the world as few others could. He hiked in tropical rain forests, discovered fossils of extinct animals, and observed the unique communities of the Galápagos Islands. When he returned to England, he spent most of his time thinking and writing about his amazing discoveries on his voyage on the *Beagle*. After more than 20 years of contemplation and writing, Darwin published *On the Origin of Species* in 1859.

There are few scientific discoveries that have shaken the foundation of human understanding as firmly as Darwin's theory of evolution of species by natural selection. It affected not only biology, but also psychology and anthropology, and even arts and literature. It gave us a new perspective of what it means to be human. Darwin's studies were significant in two ways. First, his travels allowed him to document incredible amounts of evidence to support the fact that evolution actually does occur. Second, Darwin described natural selection as a plausible mechanism for evolution.

The Second Revolution

The second revolution occurred 70 years later, in the 1930s. It is important to realize that when Darwin proposed his theory of natural selection, the scientific community did not understand the basis for the inheritance of traits. That did not come until 1900, when scientists independently discovered Gregor Mendel's inheritance theories (discussed in Chapter 13). In the early 1900s, mechanisms of inheritance were uncovered and scientists demonstrated that genes are on chromosomes. By the 1930s the new field of population genetics was developing an understanding of how genes flow through populations and generations. The theories of Darwinian natural selection, Mendelian genetics, and population genetics intersected at this time to provide a more comprehensive understanding of mechanisms of evolution. After this period, fundamental advances in evolutionary thought did not occur for another half century.

The Third Revolution

We are now in our third "evolution revolution." This is the evolution of development, known among biologists as "Evo-Devo." DNA sequencing efforts in the latter part of the 20th century revealed that vastly different organisms share many of the same genes. This was not expected. It turns out that crucial genes called homeobox genes have been passed down nearly unchanged for hundreds of millions of years. The mouse and human genomes are surprisingly similar—so why do we look so different? The differences are due to regulatory genes that act as developmental switches. These master control genes turn other target genes on and off, so their protein products are produced only at certain times and in specific places. Homeobox genes are critical in determining the body plan because proper development depends on the right genes being active at the right times.

Until recently, we believed that the diversity of forms in nature resulted from innumerable novel creations, starting from scratch with each new species. We now understand that evolution uses highly developed genetic systems in new ways when mutations alter master switches. Consequently, there is no need to reinvent the wheel when a new species is evolving. Organisms with nearly identical genomes can look very different because different developmental programs created them. Over evolutionary time, the same genes can be used for different purposes in different species. For example, plants have master switches to control flower development. Amazingly, plant master switches are genetically similar to those in animals. Even during the domestication of plants and animals, humans have inadvertently taken advantage of developmental control genes. Teosinte, the ancestor of corn, for example, produces tiny, hard kernels on a small ear (see Chapter 14). However, selection of a major control gene altered ear development dramatically, producing the ears that are familiar to us (see Fig. 14.3). So, even though teosinte and modern corn look very different, they are genetically similar.

15.3 Charles Darwin

Learning Outcome

a. Explain how Darwin was influenced by the writings of Thomas Malthus.

As a boy, Charles Darwin (1809–1882) (Fig. 15.4) showed an interest in natural history, but in deference to his physician father's wishes, he enrolled in medical school; he later studied theology at Cambridge University, without, however, excelling in either subject. In 1831, when he was 22 years old, he graduated with a degree in theology but was not really sure what he wanted to do with his life.

Figure 15.4 Charles Darwin at different ages throughout his life in the 1800s. © *Shutterstock/Everett Historica*

Shortly after Darwin's graduation, King William IV of England commissioned a sailing vessel, the HMS *Beagle,* to undertake a voyage around the world to chart coastlines, particularly those of South America. Young Darwin accepted an unpaid position as assistant naturalist and captain's companion on the voyage, which began December 27, 1831.

During the voyage, Darwin read a geology book that made a profound impression on him. The book, authored by Charles Lyell, was based on a theory of John Hutton, who believed the earth was much older than the few thousand years it previously had been thought to be. Hutton suggested that the earth had been subjected to cycles of upheaval and erosion over great periods of time. The erosion resulted in rock fragments and weathered debris being washed down rivers into the oceans, where the deposits gradually accumulated in thick layers. The thick layers, which slowly were converted into sedimentary rocks (many of which contained fossils), became land as they were elevated above sea level. Hutton believed that, given enough time, all geological changes could be attributed to slow, natural processes, and Lyell suggested that the slow changes occurred at a uniform rate. Although some modern geologists incorporate many of Hutton's ideas about gradual change over long periods of time, they point to evidence that geological changes have not always proceeded at a uniform rate. Darwin himself was not convinced that geological changes occurred at a uniform rate, but he did agree with Lyell that the earth must be very old indeed and that large changes were due to the gradual accumulation of small changes.

Today, we can estimate the age of the earth by determining in rocks the ratio of a radioactive element to the nonradioactive element it becomes. We know that a radioactive element gradually loses its radioactivity, although sometimes at a variable rate. The amount of time it takes for a radioactive element to lose half of its radioactivity is referred to as its **half-life.** The half-life for each radioactive element is unique. For example, uranium, which becomes lead as it loses its radioactivity, has a half-life of 4.5 billion years, while radioactive carbon has a half-life of 5,730 years, and radioactive nitrogen has a half-life of only 10 minutes. By checking ratios of uranium to the lead it becomes when it loses its radioactivity, we now know that the earth is probably at least 5 billion years old.

Darwin had many opportunities to collect plants and animals on both sides of South America, as well as in the Galápagos Islands and along the coasts of Australia and New Zealand. He kept a daily journal and spent countless hours alone on horseback, collecting and observing the living world around him. This gave him ample opportunity to think about the forms and distribution of the myriad new organisms he encountered. His thoughts slowly led to the development of ideas that later blossomed into his theory of evolution through natural selection.

Upon his return to England in 1836, Darwin retired to the country and began working on his collections and journal. He also carried on extensive correspondence with other biologists and made many detailed investigations into pollinating mechanisms, earthworm ecology, geographical distributions of plants and animals, and several other areas of natural history.

Throughout his activities, Darwin was guided by a concept he had adapted from an essay on human populations and food supplies written by Thomas Malthus in 1798. Malthus observed that populations grow geometrically until food supplies, disease, wars, and other factors limit their growth. Darwin realized that although humans might artificially increase their food supply through selective breeding and cultivation, wild plants and animals could not do so and were therefore vulnerable to a process of selection in nature, which would explain changes in natural populations. He was reluctant to publish his ideas, however, and did not begin putting together his book on the origin of species[1] until 1856.

Meanwhile, an English naturalist by the name of Alfred R. Wallace (1823–1913), who made major contributions to our knowledge of animal geography, independently concluded that natural selection contributed to the origin of new species and sent Darwin a brief essay on the topic in 1858. At the urging of Charles Lyell, Darwin and Wallace jointly presented a paper of their views to the Linnaean Society of London in 1858, and in 1859, Darwin's classic book *On the Origin of Species by Means of Natural Selection* was published.

---

1. Our understanding of the term species has been modified and refined since Darwin's time, but, as discussed in Chapter 16, most biologists today think of a species in general terms as a population of individuals of similar form and structure with a common ancestry, capable of interbreeding freely with one another in nature but not generally interbreeding with individuals of other dissimilar populations. A shorter definition is simply to refer to a species as a group of organisms with a common gene pool. Note that "species" is like the word "sheep," in that it is spelled and pronounced the same in either singular or plural usage. There is no such thing as a plant or animal "specie."

15.4 Evidence for Evolution

Learning Outcome

a. Explain how two plant species may be similar in form even if they are not closely related to each other.

Similarities in the forms of living organisms often provide evidence for evolution. Homologies (characteristics shared by different organisms) point to common ancestry. For example, all members of the Mustard Family (Brassicaceae), which includes broccoli, cabbage, radish, and stocks, produce a pungent, watery juice that has the smell of sulfur. However, similarities are not always due to common ancestry. For example, a number of African spurges (species of *Euphorbia*) and American cacti have similar, succulent stems even though they are not closely related (Fig. 15.5). The plants do share similar arid habitats to which they have independently become adapted (an *adaptation* is a characteristic that makes an organism better suited to its environment). In other words, plants of very different ancestry have adapted in similar ways to common environmental conditions in different parts of the world. This type of evolution is called *convergent evolution*.

Other evidence comes from the structure and relationships of proteins, DNA, and other molecules, as well as the common use of ATP by living organisms. Some fundamental metabolism genes, for example, are found universally in all living organisms, which suggests they appeared very early, probably in a single organism, and that they have been successively inherited by the myriad of organisms in existence today.

Fossils (Fig. 15.6) provide compelling evidence for descent with modification. The simplest fossils are generally found in the oldest geological strata, while more complex forms tend to be found in younger strata.

Still further evidence is drawn from the geographical distribution of organisms. Many groups are confined to a single continent or island. In some instances where similar organisms occur on more than one landmass, there is evidence that the landmasses were once linked, which would have permitted terrestrial migration.

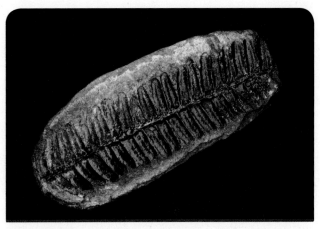

Figure 15.6 A fossil fern. © ElementalImaging/Getty Images

Figure 15.5 The barrel cactus on the left (*Jasminocereus howellii*) and the barrel spurge on the right (*Euphorbia obesa*) are unrelated but have evolved similar forms in adaptation to arid habitats. This is an example of convergent evolution. © Kingsley Stern

15.5 Microevolution–Evolution within Species

Learning Outcomes

a. List the four principles on which the theory of natural selection is based.
b. Define *mutation*.
c. Explain how migration affects gene flow.
d. Describe a situation in which genetic drift may occur.
e. Define *punctuated equilibrium*.

Natural Selection

Darwin observed that animal breeders could change populations of domestic animals by retaining and mating those with desirable traits. Darwin also noted that, in nature, a limited supply of food and other resources prevents some individuals from reproducing. In addition, the competitive abilities of individuals vary within populations. He reasoned that those best adapted to use available resources would increase in number in succeeding generations, and those less well suited would decrease. Darwin also used comparative anatomy and embryology in developing his theory of descent with modification, which he called **natural selection,** and based it on four principles:

1. *Overproduction.* Many organisms produce enormous numbers of offspring. For example, a single tree produces thousands of viable seeds each year. A single mushroom can produce millions of spores, each with the potential to become a new mushroom-bearing fungus.
2. *Struggle for existence.* All the germinating seeds, spores, and other reproductive structures of organisms compete for moisture, light, nutrients, and space. In nature, the amounts of these elements are insufficient to support all of these organisms, and many die.
3. *Inheritance and accumulation of favorable variations.* All organisms are genetically variable. Hereditary variations that have survival value are inherited from generation to generation and accumulate with time, while other variations that do not promote survival and reproduction are gradually eliminated.
4. *Survival and reproduction of the fittest (differential survival and reproduction).* Individuals best adapted to the environment (the "fittest") have the best chance to survive and reproduce. A plant that produces insect-repellent compounds, for example, may have a higher probability of surviving to reproductive age and producing more offspring. The offspring of that insect-resistant plant may then bear its inherited features.

One of the criticisms of Darwin's theory when it was published was that it did not explain how hereditary variations originated. It should be remembered, however, that Mendel's findings were not published until later, and the details of mitosis and meiosis did not become known until the 20th century. Today, with our knowledge of genetics, we have come to understand the mechanisms of evolution in populations much better than was possible in Darwin's time.

Mutations

All organisms, from aardvarks to zinnias, occur in populations composed of a few to billions of genetically variable individuals. The genetic composition of a population will change over generations if the environment favors certain hereditary variations and disfavors others.

In some instances, the environment may alter the phenotype without affecting the genetic constitution of an organism. Plants that grow relatively tall at sea level may be dwarfed when they are transplanted to cooler areas in the mountains or drier areas near deserts. However, their genetic composition has not changed. They are still capable of growing tall in an appropriate environment. Phenotypic changes in populations, however, often result due to genetic changes.

A **mutation** is a DNA sequence change. It is the basis for genetic variability among and within species. The smallest mutation is a change in a single nucleotide pair (see Chapter 13) and is called a point mutation. If it occurs in a structural gene, it could code for a different protein. Mutations also occur in the regulatory genes that control the expression of structural genes. They may cause the plant to generate a larger or smaller amount of the protein product. On a much larger scale, mutations can occur at the whole chromosome level. Changes within chromosomes may occur in several ways. A part of a chromosome may break off and be lost (*deletion*), or a piece of a chromosome may become attached to another (*translocation*). In some instances, a part of a chromosome may break off and then become reattached in an inverted position (*inversion*) (Fig. 15.7). Like point mutations, chromosomal changes may affect gene expression levels. Most mutations are harmful. However, some either are silent (have no effect on the survival of the phenotype) or produce a characteristic that may help the organism survive changes in its environment.

Migration

Gene flow between populations occurs when individuals or gametes migrate from one population to another. The amount of gene flow depends on population size and the

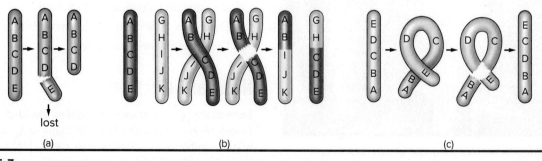

Figure 15.7 Examples of mutations caused by chromosomal changes include: (a) *Deletion*. Part of a chromosome breaks off and is lost. (b) *Translocation*. Part of one chromosome becomes attached to another. (c) *Inversion*. Part of a chromosome breaks off, becomes inverted from its original position, and then is reattached.

extent to which populations are isolated from one another. If interbreeding occurs over wide distances, as might occur in wind-pollinated plants, a single individual's genes can quickly be spread from one population to another. On the other hand, gene flow occurs at a much slower rate when interbreeding is restricted to small, isolated populations. This is often the case with plants that occur only in specialized habitats, such as near springs or on certain soil types. A small amount of gene flow over longer distances can be expected to occur, even though most of the interbreeding takes place between closely associated individuals in a population.

Genetic Drift

As observed in Chapter 13, the Hardy-Weinberg law states that alleles of genes tend to remain at a constant frequency from generation to generation. However, by chance alone, the genetic makeup of a population may change due to random events. This is called *genetic drift*. In a large population, genetic drift is unlikely to make any significant difference. In a small population, however, the success or failure of a single individual to multiply may cause a marked change in its frequency and, in some cases, cause it to disappear altogether or, on the other hand, increase to a frequency of 100%. In other words, in a small population, the random shift in gene frequencies by chance may bring about evolutionary changes.

Rates of Evolution

Darwin believed that evolution by natural selection is a slow and gradual process. Contemporary biologists, however, favor the *punctuated equilibrium* model, which holds that major changes have taken place in spurts of maybe 100,000 years or less, followed by periods of millions of years during which changes have been minor. They base their hypotheses on fossils, which, when arrayed according to their ages, reveal large gaps in the record. Although missing-link fossils are occasionally discovered, the record does little to support Darwin's concept of gradual, long-term change, even though it is believed that possibly as few as one organism in a million may have become a fossil. Others opposed to hypotheses of evolution through sudden change argue that because probably only a tiny percentage of organisms became fossilized and usually only the harder parts of organisms (e.g., bones, teeth, wood) are preserved, drawing definite conclusions from fossil evidence about evolution through either gradual or sudden change may be speculative.

The conditions necessary for an organism to become a fossil are very specialized and limited in occurrence (see Chapter 21). This makes it quite improbable that large numbers of missing-link fossils will ever be found. Proponents of evolution through periods of rapid change argue that under conditions of changing climates or other situations exerting strong selection pressures on forms with adaptive mutations, new species of organisms could arise in

less than 100 generations, making 100,000 years ample time for considerable evolution to occur.

15.6 Macroevolution—How Species Evolve

Learning Outcomes

a. Distinguish between prezygotic and postzygotic isolation barriers.
b. Describe how geographic isolation can lead to speciation.
c. Describe an ecological isolation mechanism.
d. Describe a mechanical isolation mechanism.

If new forms of genes are introduced into an interbreeding population, they may gradually spread throughout the population, and the genetic composition of the whole population will change in time. If some barrier divides the population, however, the two new populations eventually may become distinct from each other, sometimes in a relatively short period of time. Isolation barriers can be classified as prezygotic (occurring before fertilization) or postzygotic (occurring after fertilization). Examples of prezygotic barriers include geographic isolation, ecological isolation, and mechanical isolation, while postzygotic barriers include zygote death, sterility in hybrid plants, and low fitness in hybrids. Experimental data indicate that prezygotic barriers are more significant isolating mechanisms than postzygotic barriers. The point of fertilization is considered to be important because it is wasteful for the mother plant to devote energy to the development of hybrid seeds if they are not going to grow into successful plants. Therefore, it is better to avoid hybridization altogether than to produce unsuccessful hybrid plants.

Geographic Isolation

How can one population evolve into two populations that are so different they cannot interbreed? Imagine, for example, that a population of white-flowered daisies occurs throughout a wide valley that is bounded on both sides by high mountains. If a mutation occurs that results in red flowers, the new color may, in time, spread throughout the entire population. If, however, before the mutation occurred, there was a volcanic eruption that blocked off part of the valley, we would then have two populations of white-flowered daisies isolated from each other. If the same mutation for red color should then occur, it would spread through only one population. Meanwhile, if a mutation for hairy leaves occurred in the other population, that characteristic could spread throughout the second population but the geographic barrier prevents it from spreading to the population of red-flowered daisies. In time, the genetic changes could become so great that even if the barrier were removed, plants from the two populations could not interbreed, and two distinct species of daisies would exist.

Figure 15.8 Two species of redbud, both native to North America are (a) Eastern redbud (*Cercis canadensis*) and (b) Western redbud (*Cercis occidentalis*). They look similar but are geographically isolated. *(a) © Holly Hildreth/McGraw-Hill Education (b) © Kingsley Stern*

In the United States, there are two closely related species of small trees called redbuds (*Cercis*) that look very much alike. The eastern redbud (*Cercis canadensis*) occurs on the borders of streams, mostly east of the Mississippi River between the Canadian border and Florida, where precipitation occurs throughout the year. The western redbud (*Cercis occidentalis*) is native to stream areas in California, Utah, Nevada, and Arizona, where most of the precipitation occurs in the winter and spring. The two species can be artificially hybridized, but each is so adapted to its own wild habitat and climate that specimens of either species die when transplanted to the other's wild habitat (Fig. 15.8). Presumably, a single species of redbud once grew along river margins throughout much of temperate North America. When the eastern populations became geographically isolated from those of the west, however, random mutations arose independently in the east and the west; with gene flow between the two populations no longer possible, two species now exist where formerly there was only one.

Several other factors contribute to the development of new species from geographically isolated populations. When separation first occurs, it is unlikely that both populations will have genes that are identical in all respects, and each of the two new populations may have only a subset of the genetic variation that was present in the original population. In addition, geographically isolated populations will be subjected to selection pressures from the environment.

Ecological Isolation

Isolation leading to the development of new species is not limited to physical barriers, such as mountains or oceans. Ecological factors such as climate or soils may play a role, as do time and mechanical isolating factors. As a result, related species can be *sympatric* (occupy overlapping ranges of territory) without exchanging genes.

A mutant form within a population may flower at a different time, preventing an exchange of genes between it and nonmutant forms. In the temperate deciduous woods of eastern North America and in the Columbia River basin in the Pacific Northwest, there are many populations of early spring–flowering herbs called *Dutchman's breeches* (*Dicentra cucullaria*), which are discussed further in Chapter 16. Dutchman's breeches and *squirrel corn* (*Dicentra canadensis*), a close relative, often grow together in the eastern deciduous forest. Both species have highly dissected leaves that are so similar in appearance that many people assumed Dutchman's breeches and squirrel corn were the same species (Fig. 15.9).

It is believed, however, that at some point in geological history, mutations occurred that caused some plants to begin flowering after other plants had already set seed. As a result, one group became reproductively isolated from the other group. In due course, other mutations affecting the form and fragrance of the flowers and the shape and pigmentation of food-storage bulblets beneath the surface also occurred, but the plants continued to occupy the same habitats. In short, we now have two closely related but distinct species growing together without interbreeding, simply because ecological isolation has led to hybridization failure.

Mechanical Isolation

Other isolating mechanisms may be mechanical. In orchids, for example, the pollen is usually produced in little sacs called *pollinia* (Fig. 15.10), which stick to the heads or bodies of visiting insects. If pollination is to occur, the pollinia must be inserted within a concave stigma. Each species of orchid is constructed so that it is highly unlikely that a pollinium of one species will be inserted within the stigma of another species. As a result, many species of related orchids can be sympatric without genes being exchanged.

Four closely related sympatric species of Peruvian *Catasetum* orchids, which can be artificially hybridized very easily, have no known natural hybrids, despite their being pollinated by a single species of bee. Microscopic examination of the pollinators has shown that the pollinium of one

Evolution 271

Figure 15.9 Two woodland wildflowers, (a) Dutchman's breeches plant (*Dicentra cucullaria*), and (b) squirrel corn (*Dicentra canadensis*), are sympatric but ecologically isolated because they flower at different times. *(a-b) © Carol Wolfe*

Figure 15.10 A pair of *pollinia* (sacs of pollen) produced by members of the Orchid Family (*Orchidaceae*). The pollinia become attached to the bodies of visiting insects, which transport them to other orchid flowers. This is an example of mechanical isolation. ×10. *© Kingsley Stern*

species is attached to the insect's head, that of another is attached to the insect's back, that of a third to the abdomen, and that of the fourth only to the left front leg. Even after visits to hundreds of flowers, none of the pollinia are misplaced.

15.7 The Role of Polyploidy in Evolution

Learning Outcomes

a. Explain how polyploidy can restore fertility to interspecific hybrids.
b. Define *apomixis*.

Recent genomics studies have revealed that nearly all flowering plant species have polypoidy in their history even if they are not currently polyploid. Species that we consider to be diploid are actually products of an ancient polyploidization event, followed by evolution of a diploid genetic system. Polyploidy is much more common in plants than in animals for a few reasons. First of all, animals require highly tuned regulation of gene expression in tissues and cannot tolerate the imbalances that result from doubling the chromosome complement. Second, reproductive systems in animals are typically severely disrupted following polyploidization. Finally, polyploidy often results from interspecific hybridization, which is common in plants but not in animals.

When a diploid hybrid is formed between two related species, the plant carries one copy of each chromosome from its maternal parent and one copy of each chromosome from its paternal parent. When this hybrid plant undergoes meiosis to form gametes, the chromosomes may be too dissimilar to pair with each other, resulting in a sterile plant. This is an evolutionary dead-end. However, if the chromosome complement of each parent is doubled in the hybrid, then the chromosomes from the maternal parent can pair with each other; chromosomes from the paternal parent will also pair with each other. The hybrid is fertile and can produce offspring. A polyploid generated from an interspecific cross is called an *allopolyploid*. Alternatively, a polyploid formed from a single species is called an *autopolyploid*. An autotetraploid has four copies of every chromosome, which can associate in groups of two, three, or four at meiosis. Fireweed is an example of an autopolyploid (Fig. 15.11). Autopolyploidy can lead to unbalanced gametes and a consequent reduction in fertility. Evolution in autopolyploids, therefore, often leads to mechanisms that promote the pairing of chromosomes at meiosis. For example, corn is diploid with a sporophytic chromosome number of 20 ($2n = 2x = 20$). A large number of gene duplications have been found in corn, suggesting that sometime in its distant past, it was tetraploid ($2n = 4x = 20$). Subsequent divergent evolution of the duplicated genes occurred over millions of years, so corn now behaves as a diploid.

Evolution can act in novel ways on plants containing extra sets of chromosomes. Sometimes, genes on one pair of chromosomes become expressed only in certain tissues or at certain times in the life cycle, while the same genes on the other pair of chromosomes have a different expression pattern. In other cases, mutations in the extra genes can lead to new functions. Newly formed polyploids are often able to expand their range considerably because of the flexibility imparted by extra genomes. Many domesticated crops are polyploid, and new evidence indicates that gene duplication was important because it provided extra copies of genes to act as raw material for the changes required to convert wild species to cultivated forms.

Apomixis

Sterile hybrids also may reproduce asexually. One form of asexual propagation, called **apomixis,** includes the production of seeds without fertilization. For example, a cell in an ovule may divide by mitosis to form an embryo. Asexual propagation also includes other forms of vegetative reproduction, which are discussed in Chapters 4 through 7, Chapter 14, and Appendix 4.

Species that have the option to reproduce by both apomixis and sexual reproduction can be highly successful in nature. Dandelions, for example, are among the most successful plants known. They reproduce apomictically, as well as sexually through fertilization (Fig. 15.12).

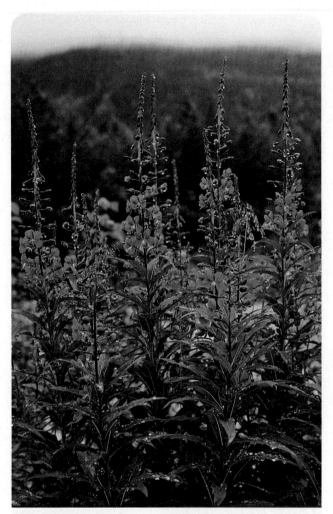

Figure 15.11 Fireweed (*Epilobium angustifolium*) is an autopolyploid found primarily in North American mountainous areas below 3,050 meters (10,000 feet). One subspecies has four sets of chromosomes, with a chromosome number of $2n = 4x = 36$. A second subspecies has eight sets of chromosomes, with a chromosome number of $2n = 8x = 72$. © *Kingsley Stern*

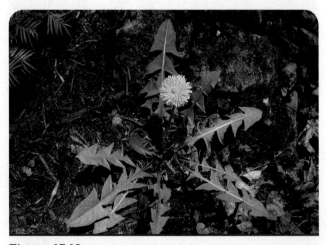

Figure 15.12 In addition to reproducing by ordinary vegetative and sexual means, dandelions such as this one reproduce apomictically. Apomixis is a form of sexual reproduction through which seeds are produced without fertilization. © *Shelley H. Jansky*

KEY THEME 15.1 evolution — Our Daily Bread

Cereal crops, such as wheat and rye, have provided the foundation for agriculture since they were first domesticated 10,000 years ago. Wheat was domesticated from a wild allotetraploid that was more vigorous, higher-yielding, and more broadly adapted than its wild diploid parents. Through breeding, this species evolved into emmer wheat, which is used today for pasta production. When tetraploid wheat hybridized with a third diploid wild relative, an allohexaploid was created. This new species became the basis for the development of modern bread wheat. The extra genome in hexaploid bread wheat allowed it to occupy a broader range than tetraploid emmer wheat. The third genome also carried genes for proteins that contribute to bread quality. Today, hexaploid cultivars account for 90% of world wheat production (Box Figure 15.1A). Extensive breeding efforts in the past century, however, have led to a dramatic narrowing of the genetic base of cereals. Efforts are now underway to broaden the gene pool with wild relatives. One strategy is to recreate "synthetic" wheat by crossing the wild relatives and then doubling their chromosomes to create new polyploid wheats. Another option is to cross wheat cultivars with wild relatives. Both strategies have

Box Figure 15.1A This wheat field is being harvested with a combine that separates the grain from the chaff. © *Aleksandar Dickov/Shutterstock*

been successful. With continued human population growth and increasing impacts of climate change, it will be critical to access the genetic diversity needed to breed highly productive cereal crops.

15.8 Discussion

Learning Outcome

a. Describe evidence to support the theory of evolution.

Darwin's theory of evolution through natural selection initially caused a great deal of controversy because, even though Darwin believed to his death in a divine Creator, he also believed that the Creator had used natural laws to bring all living things into being gradually over long periods of time. Most of his contemporaries, however, were guided by a literal interpretation of the biblical account of creation and were convinced that all living things had been created in six days and had existed unchanged since the beginning.

If we now know and understand the mechanisms of organic evolution, then why are there still disagreements about the concept of evolution? Obviously, a lack of objective analysis of the evidence is a factor, but it is not the sole reason. Science deals with tangible facts and evidence that can be measured or experimentally tested; beliefs stemming from metaphysics or religion are outside the realm of science to prove or disprove.

Virtually no objective thinkers will deny the extraordinary impact that the theory of evolution has had on our concept of the living world. Nearly all who have studied the evidence are led to the conclusion that evolutionary processes are the sole plausible explanation for the unity of life, at the molecular and cellular levels, as well as for the extraordinary diversity of the organisms on our planet. There is little unanimity of thought as to the precise pathways of evolution. One authority may be convinced that a certain group evolved from another, while another equally eminent authority may maintain that the reverse occurred. Part of the reason for such paradoxes is that the historical record is quite incomplete. Fossils represent a very small fraction of organisms that once existed. With incomplete evidence, scientists can deal only in probabilities, and it is inevitable that different interpretations result.

SUMMARY

15.1 Introduction to Evolution
- Evolutionary concepts are important in all disciplines of biology.
- Humans have used artificial selection to breed plants and animals.
- Evolutionary medicine applies principles of evolution to the treatment of illness.
- Evolutionary principles are used to predict species extinctions and invasions.
- Biologists are studying the fate of transgenes in the environment.

15.2 A Brief Overview of the Early Development of Evolutionary Concepts

- The theory of organic evolution received its greatest impetus from Charles Darwin's *On the Origin of Species*.
- Before 1900, many biologists, notably Lamarck, believed that organic evolution occurred as a result of the inheritance of acquired characteristics. This theory was discredited experimentally, but new studies of epigenesis are revealing that environment can affect gene expression.
- Evolutionary thought has undergone three major revolutions. The first was created by the publication of Darwin's book. The second resulted from combining concepts from theories of Darwinian natural selection, Mendelian genetics, and population genetics.

15.3 Charles Darwin

- Charles Darwin spent five years as a naturalist on the HMS Beagle, where he observed a wide range of plant and animal life worldwide.
- When Darwin returned to England, using thoughts formulated during his journey, he developed his theory of evolution through natural selection.

15.4 Evidence for Evolution

- Evidence in support of organic evolution is drawn from fossils and from the form, ecology, geographical distributions, and relationships of living organisms, homologies, molecular structures, and analogies.

15.5 Microevolution—Evolution within Species

- Darwin's natural selection theory is based on four principles: (1) overproduction; (2) struggle for existence; (3) variation and inheritance; and (4) survival and reproduction of the fittest.
- The primary mechanisms of organic evolution include natural selection, mutation, migration, and genetic drift.
- Darwin believed that evolution through natural selection was a gradual process over great periods of time. Some contemporary biologists believe that organic evolution has taken place in spurts between long periods of little change, based on evidence from the fossil record.

15.6 Macroevolution—How Species Evolve

- If new forms of genes are introduced into a freely interbreeding population, they will gradually spread throughout the population, and the genetic composition of the population will change over time. If a population is divided by a barrier, then genetic variants occurring in one of the divided populations will not spread to the other. In time, because of the isolation, each new population may develop into separate species incapable of breeding with one another.

15.7 The Role of Polyploidy in Evolution

- Isolation and polyploidy play major roles in the evolution of species, especially in plants.
- Allopolyploids arise from interspecific hybridization, while autopolyploids form from intraspecific hybrids. The extra genetic material in polyploids provides opportunities for range expansion, gene evolution, and domestication.

15.8 Discussion

- Opinions and convictions on origins vary, but few can deny the major impact that theories of evolution have had on modern peoples and on their concepts of life.
- Biologists generally feel evolution is the only plausible explanation for the unity of life at the molecular and cellular levels and for the great diversity of life, but there is little agreement among them as to the precise pathways of evolution in the past.

REVIEW QUESTIONS

1. How does organic evolution differ from other forms of evolution?
2. How did Darwin's theory of evolution differ from Lamarck's?
3. What basic modifications have been made in evolutionary theory since Darwin's time?
4. What is meant by a mechanism of evolution?
5. What are some advantages and disadvantages of allopolyploidy compared to autopolyploidy?

DISCUSSION QUESTIONS

1. One of Darwin's principles was that there is a struggle for existence among living organisms. Do plants struggle with one another to survive? If so, how do they do it?
2. Some populations change noticeably in form within a hundred years. If only one gene in every 200,000 mutates and if most mutations are harmful, how is such change possible?
3. Do you think there might be scientifically supportable alternatives to organic evolutionary theory to account for the diversity of life around us? Explain.

CHAPTER

16

Plant Names and Classification

16.1 Introduction to Plant Classification
16.2 Development of the Binomial System of Nomenclature
    Linnaeus
    International Codes of Botanical Nomenclature
16.3 Development of the Kingdom Concept
16.4 Classification of Major Groups
16.5 Species Concepts
    Morphological Species Concepts
    Interbreeding Species Concepts
    Ecological Species Concepts
    Phylogenetic Species Concepts
    Eclectic Species Concepts
    Nominalistic Species Concepts
16.6 The Future of Plant Classification

OVERVIEW

This chapter begins with a discussion of the problems involved with the use of common names for plants, as illustrated by a survey of such names for two related species of American spring-flowering perennials. It continues with a brief historical account of the events that led to the development and acceptance of Linnaeus's Binomial System of Nomenclature. The history of the development of a six-kingdom classification is presented, along with a list of the phyla and classes included in the kingdoms covered in this text. A dichotomous key to the kingdoms and phyla of organisms is provided. New information about the use of DNA sequence data for plant classification is also presented. The chapter concludes with a brief discussion of species concepts.

Like many wild plants, *Castilleja indivisa* is known by more than one common name. It is called Texas paintbrush or entireleaf Indian paintbrush, and it is an annual flower native to Texas. One goal of plant taxonomists is to assign only one Latin name to each species. © *Steven P. Lynch*

16.1 Introduction to Plant Classification

Learning Outcome

a. Discuss problems associated with the use of common names in science.

Biologists are sometimes thought by the general public to be pompous when they refer to peanuts as *Arachis hypogaea* or an African hoopoe bird as *Upupa epops*. They aren't, however, merely showing off or being difficult. Rather, they're identifying organisms by their scientific names through a system that has evolved over centuries. It has become vital to us to be able to consistently distinguish among existing organisms (with current global estimates of 8.7 million species), as well as those that have become extinct.

All living organisms are given a single, two-word Latin scientific name, and many also have common names. Only one correct scientific name applies to all individuals of a species, no matter where they're found, but many common names may be given to the same species, and the same common name may refer to a number of different species.

The scientific name *Dicentra cucullaria,* for example, was given to a spring-flowering plant native to eastern North America and the Columbia River basin in the West. Its unique flower shape, which looks like the baggy pants of a traditional Dutch costume, has resulted in its being given the common name *Dutchman's breeches*. It also, however, has the common names of *little-boys' breeches, kitten's breeches, breeches-flower, boys-and-girls, Indian boys-and-girls, monkshood, white eardrops, soldier's cap, colicweed, little blue stagger, white hearts, butterfly banners, rice roots,* and *meadow bells*. In addition to these English common names, the plant has Native American names, and in the Canadian province of Quebec, it has French names.

Often growing with *Dicentra cucullaria* is a related plant, with similar leaves but with slightly different flowers, having the scientific name *Dicentra canadensis*. Because of the similarities between the two plants and their close association in the woods where they occur, some people in the past assumed that they were merely two different forms of the same plant. This is reflected in the fact that some of the common names of *Dicentra cucullaria* (e.g., colicweed, Indian boys-and-girls, little blue stagger) have also been applied to *Dicentra canadensis*, in addition to the names *squirrel corn, turkey corn, turkey pea, wild hyacinth, fumitory, staggerweed,* and *trembling stagger*. However, the problem of the diversity and the overlapping of common names for these two plants doesn't stop there. The name *monkshood*, for example, is widely used for *Aconitum* species, which are in the Buttercup Family (Ranunculaceae)—a family only distantly related to the Fumitory Family (Fumariaceae) of *Dicentra*. The names *soldier's cap, rice roots, meadow bells,* and *turkey corn* have also been applied to completely unrelated plants. Both species of *Dicentra* are shown in Chapter 15.

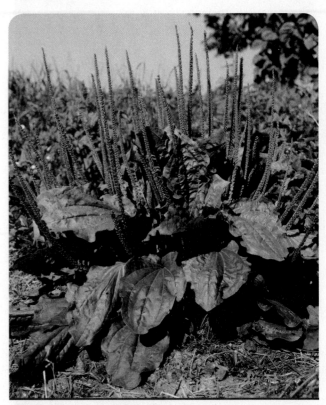

Figure 16.1 Plantain (*Plantago major*) has at least 300 common names. © Arco Images GmbH/Alamy

In Europe, with its many languages, common names can become especially numerous. The widespread weed with the scientific name *Plantago major,* for example, is often called *broad-leaved plantain* in English, but it also has no fewer than 45 other English names, 11 French names, 75 Dutch names, 106 German names, and possibly as many as several hundred more names in other languages, with literally dozens of these common names also applying to quite different plants (Fig. 16.1). If it were not for the early recognition by biologists and others of the urgent need for worldwide uniformity in naming and classifying all organisms, utter chaos eventually might have prevailed.

16.2 Development of the Binomial System of Nomenclature

Learning Outcomes

a. Explain why scientific names are in Latin.
b. Describe the Binomial System of Nomenclature.
c. Discuss the steps taken to name a newly discovered plant.

The first person known to attempt to organize and classify plants was Theophrastus, the brilliant student of Plato and Aristotle. His 4th-century B.C. classification of nearly 500 plants into trees, shrubs, and herbs, along with his distinctions among plants on the basis of leaf characteristics, was used for hundreds of years. It was not until the 13th century A.D. that a distinction was made between *monocots* and *dicots* on the basis of stem structure.

Herbalists added many plants to Theophrastus's list, and by the beginning of the 18th century, details of fruit and flower structure, in addition to form and habit, were used in classification schemes. European scholarly institutions were beginning to bulge with thousands of plants that explorers had brought back from around the world, and confusion over scientific and common names was multiplying.

The use of Latin in schools and universities had become widespread, and it was then customary to use descriptive Latin phrase names for both plants and animals. All organisms were grouped into **genera** (singular: **genus**), with the first word of the Latin phrase indicating the particular genus to which the organisms belonged. For example, all known mints were given phrase names (*polynomials*) beginning with the word *Mentha*, the name of the genus. Likewise, the phrases for lupines began with *Lupinus,* and those for poplars began with *Populus.* A complete phrase name for spearmint read *Mentha floribus spicatis, foliis oblongis serratis.* Roughly translated, it means "Mentha with flowers in a spike (an elongated but compact flower cluster); leaves oblong, saw-toothed."

Linnaeus

At this point, the Swedish naturalist Carolus Linnaeus (1707–1778) (Fig. 16.2) began improving the way organisms were named and classified. The system he established worked so well that it has persisted to the present. In fact, Linnaeus's system is now used throughout the world.

Linnaeus inherited his passion for plants from his father, who was a minister and an amateur gardener. He is said to have been impressed at the age of 4 by his father's remarks about the uses of neighborhood plants in his home community of Råshult, located in southern Sweden. After a brief tenure as a student at the University of Lund, he spent most of his time making excursions to Lapland, Holland, France, and Germany. Eventually, he became the professor of botany and medicine at Uppsala, where he inspired large numbers of students, 23 of whom became professors themselves. He frequently led large field trips into the countryside, accompanied by a musical band.

When Linnaeus began his work, he set out to classify all known plants and animals according to their genera. In 1753, he published a two-volume work entitled *Species Plantarum,* which is important as the first book used by plant taxonomists to assign names. In this book, Linnaeus not only included a referenced list of all the Latin phrase names previously given to the plants but, when necessary, also changed some of the phrases to reflect relationships, placing one to many similar organisms called **species**[1] in

---
1. As noted in Chapter 15, *species* is like the word *sheep* in that it is spelled and pronounced the same in both singular and plural usage. There is no such thing as a plant or animal "specie." Currently there are many different concepts for species, as discussed later in this chapter.

Figure 16.2 Carolus Linnaeus. © *Photo Researchers, Inc.*

each genus. He limited each Latin phrase to a maximum of 12 words, and in the margin next to the phrase, he listed a single word, which, when combined with the generic name, formed a convenient abbreviated designation for the species. The word in the margin for spearmint was *spicata,* and the word for peppermint was *piperita.* Accordingly, the abbreviated name for spearmint was *Mentha spicata* and for peppermint, *Mentha piperita.*

Although Linnaeus originally considered the phrase names the official names of the plants, he and those who followed him eventually replaced all the phrase names with abbreviated ones. Because of their two parts, these abbreviated names became known as *binomials,* and the method of naming became known as the *Binomial System of Nomenclature.* Today, all species of organisms are named according to this system, which lists the genus name first, followed by the species name. The genus name is capitalized, while the species name is in lowercase. Both genus and species names are italicized. The author of the name, either in full or abbreviated, follows the genus and species names. For example, the full scientific name for spearmint is written *Mentha spicata* L., the L. standing for Linnaeus; the full scientific name for the wild dandelion native to Scandinavia is written *Taraxacum officinale* Wiggers, after Fredericus Henricus Wiggers, who was the first (in his *Flora of Holstein,* published in 1780) to describe the species.

Besides establishing the Binomial System of Nomenclature, Linnaeus tried to do more than publish just a long

Figure 16.3 Cover page of *Species Plantarum* by Linnaeus.
© Science Source/Science Source

list of plant names. After all, such lists aren't very useful if they can't be used to identify plants. Linnaeus organized all known plants into 24 classes,[2] which were based mainly on the number of *stamens* (pollen-bearing structures) in flowers. All plants with five stamens per flower, for example, were placed in one class, while those with six stamens were placed in another class. Plants and other organisms that don't produce flowers (e.g., mosses, fungi) were put in a class of their own. His arrangement was for convenience and was artificial because it did not necessarily reflect natural relationships. For the first time, however, it became possible for workers to identify plants previously unknown to them. This classification was used in *Species Plantarum* and other works by Linnaeus (Fig. 16.3) and for a short period of time was adopted by some, but not all, botanists.

---

2. Note that Linnaeus used the word *class* for a strictly artificial category of classification, resulting in unrelated organisms being grouped together simply for convenience. Modern use of the word differs from that of Linnaeus in that organisms are now assigned to a class on the basis of natural relationships.

International Codes of Botanical Nomenclature

International Code of Botanical Nomenclature (ICBN)

In 1867, more than 100 years after *Species Plantarum* was published, about 150 European and American botanists met in Paris to try to standardize rules governing the naming and the classifying of plants. Currently, botanists meet every 6 years for this purpose at the International Botanical Congress. At the first meeting, the botanists agreed to use the works of Linnaeus as the starting point for all scientific names of plants and decided that his binomials, or the earliest ones published after him, would have priority over all others.

Today, the modified rules constitute what is known as the *International Code of Botanical Nomenclature,* which is a single book with a common index to its English, French, and German translations of the various rules and recommendations. It now specifically recognizes Linnaeus's *Species Plantarum* as the starting point for scientific names of plants (except for mosses, fossil plants, and some groups of algae, as previously mentioned) and spells out details of naming and classifying, which are followed by botanists of all nationalities. A similar code for animals has been developed and established by zoologists. It, too, uses a work of Linnaeus as the starting point for scientific names.

If a botanist finds a plant that is new to science, the *International Code of Botanical Nomenclature* (ICBN) requires that he or she take at least two steps to have the plant officially recognized. (1) DNA sequence data must be deposited in a public database such as GenBank. (2) An annotated herbarium specimen of the plant, designated by the author as a *type specimen,* must be deposited in an herbarium to which the public may have access. A discussion of herbaria and herbarium specimens is given in Chapter 23.

International Code of Nomenclature for Cultivated Plants (ICNCP)

A cultivated plant (cultigen) is one whose origin or selection is primarily due to intentional human activity. Such a plant may arise by deliberate or accidental hybridization or by selection from existing cultivated stock; alternatively, it may be a selection from a wild population. Often, cultivated plants are maintained clonally (for example, apple cultivars) or as inbred lines (for example, corn inbreds). All of these factors lead to patterns of variation often not suitable to the use of the ICBN. In 1952, a proposal was published by Joseph Lanjouw and colleagues for an independent set of nomenclature rules for cultivated plants, and in 1953 the first edition of the *International Code of Nomenclature for Cultivated Plants* (ICNCP) was published. The latest (eighth) edition was published in 2009.

A fundamental difference between the ICNCP and the ICBN is their approach to classification. Groups of plants named and classified in the ICBN are collectively designated as **taxa** (singular: **taxon**) and are grouped in nested ranks (with species contained within genera, and genera within families, etc.). The ICNCP uses two classification terms for cultigens, the terms *cultivar* and *cultivar-group*. Most cultigens have names consisting of a Latin name (either a genus or a genus and species name) that is governed by the *International Code of Botanical Nomenclature*—for example, *Primula,* to which is added a cultigen epithet, enclosed in single quotes (e.g., *Primula* 'Old Irish Scented'). Cultigens can then be grouped by categories of use to humans, and these groups imply no statement of relationships, unlike ranks in the ICBN. For example, the *Primula* cultivars 'Old Irish Scented,' 'MacEatt's Blue,' and 'Osborne Green' are best cultivated under outdoor conditions and have been assembled into the *Primula* Border Auriculata Group. Group names can be formed by any trait, including flower color, flowering time, growth habit, or even common chromosome numbers.

PhyloCode

Knowledge of phylogenetic (evolutionary) relationships of all life is changing rapidly due to the explosion of new data arising from molecular systematics. Many long-accepted taxa are being shown to represent groups that are not "natural" in the sense that they are non-monophyletic (not descended from a common ancestor); see the section "Phylogenetic Species Concepts," later in this chapter). Traditional nomenclature as governed by the ICBN encounters problems when naming newly discovered clades, or branches of a phylogenetic tree (see Fig. 16.6). The problem is caused by the dependence of names on their taxonomic rank in this system. Under the ICBN, every species is considered to be part of a genus, all genera are part of families, all the way up the taxonomic hierarchy, and taxa at every rank need to be named. Rules in the ICBN dictate that the correct name for a taxon is generally the earliest name published for that taxon at the rank at which it is classified. As a consequence, if a clade is reclassified at a different rank, its name is usually changed. Furthermore, it is difficult to name clades one at a time as they are discovered without having to change the names of other clades. However, under a different system called phylogenetic nomenclature, these problems don't exist because priority operates independent of rank. The earliest phylogenetically defined name for a clade, regardless of its rank, has priority over later-defined names for the same clade. Phylogenetic definitions (e.g., the smallest clade containing species X and Y) link names directly to clades. In the early 1990s, Kevin de Queiroz and Jacques Gauthier proposed this new nomenclatural system to name clades, in which the use of ranked categories such as the genus and family is not mandatory, and these categories (if used) have no bearing on names. A set of rules called the PhyloCode has been drafted to govern phylogenetic nomenclature.

At present, the draft PhyloCode is being used by some taxonomists to name clades, and publications are currently in preparation by an international group to formalize this system.

16.3 Development of the Kingdom Concept

Learning Outcome

a. Explain why it is important to recognize more than two kingdoms.

If you were to ask the average person to describe the differences between plants and animals, you would probably be told that plants are green, don't eat each other, and don't move. A distinction between plants and animals in general has probably always existed in the minds of intelligent beings, and it was natural when classification schemes were first developed that all living organisms would be placed, according to the highest category of "kingdom," in either the *Plant Kingdom* or the *Animal Kingdom*.

While this distinction still works well for complex plants and animals, it breaks down for the so-called simpler organisms. For example, there are more than 300 species of single-celled organisms called *euglenoids* (see Fig. 18.23) inhabiting a variety of freshwater habitats. These microscopic creatures have small, whiplike structures, called *flagella,* that propel them through the water, and they can ingest food particles through a groove called a *gullet*. Both of these features would be considered animal-like, and so euglenoids in the past have been treated as animals in some textbooks. Many of these organisms also, however, have chloroplasts, and if light is present, they can carry on photosynthesis efficiently enough to eliminate the need for ingesting food. Accordingly, they have often been treated as plants in botany books.

A similar problem has existed with slime molds, which resemble masses of protoplasm slowly creeping or flowing across dead leaves and debris, engulfing and feeding on bacteria and other substances as they go, and looking somewhat like amoebae, which botanists and zoologists alike have traditionally called animals. When slime molds reproduce, however, dramatic changes take place. They become stationary and develop reproductive bodies that are distinctly funguslike, causing some to consider them fungi (Fig. 16.4).

In an attempt to overcome this problem, biologists John Hogg and Ernst Haeckel proposed a third kingdom in the 1860s. All organisms that did not develop complex tissues (e.g., algae, fungi, and sponges) were placed in a kingdom called *Protoctista*. This third kingdom included such a heterogeneous variety of organisms, however, that in 1938, another biologist by the name of Herbert F. Copeland proposed that it again be divided. He assigned the name *Monera* to all single-celled organisms with prokaryotic cells (the differences between prokaryotic and eukaryotic cells

Figure 16.4 Slime molds are similar to animals in (a) their vegetative state, which is motile and feeds as it moves, but slime molds also become (b) fungus-like when they reproduce. Consequently, they are difficult to classify. *(a) © McGraw-Hill Education/ Richard Gross; (b) © Alexander62/Getty Images*

are discussed in Chapters 3 and 17), leaving the algae, fungi, and single-celled organisms with eukaryotic cells in the Kingdom Protoctista.

Although many biologists thought Copeland's four-kingdom system of classification was a definite improvement, it, too, had problems—particularly the fact that basic differences in the mode of nutrition existed among organisms within the Kingdom Protoctista. As a result, during the 1970s, many biologists adopted variations of a five-kingdom system proposed by R. H. Whittaker in 1969 (Table 16.1).

In Whittaker's system, three kingdoms of more complex organisms based on three basic forms of nutrition (photosynthesis, ingestion of food, and absorption of food in solution) were recognized, along with two kingdoms of protists, which were distinguished on the basis of differences in cellular structure.

The five-kingdom arrangement also has had critics and problems, particularly since the early 1980s, when Carl Woese, a microbiologist and leading authority on bacteria, argued that Kingdom Monera should be divided into two kingdoms. Woese based his contention on previously unrecognized fundamental differences between two major groups of bacteria (discussed in Chapter 17). His proposal for two kingdoms of bacteria gained increasing acceptance in the 1990s and is now generally recognized.

After acceptance of Woese's kingdom classification, James Lake and his colleagues proposed creating yet another kingdom by dividing the bacterial Kingdom Archaea into two kingdoms, making three kingdoms of prokaryotes in all. The slime molds, which have no cell walls during their active state but do develop walls when they reproduce, still do not fit well into the six-kingdom system, and other organisms traditionally regarded as fungi may be more closely related to protists. Much information still needs to be accumulated to understand natural relationships, and until this occurs, taxonomic categories will remain subject to different interpretations.

16.4 Classification of Major Groups

Learning Outcome

a. List the following in order, from the largest to the smallest group: order, family, kingdom, phylum, class.

Since Linnaeus's time, a number of classification categories have been added between the levels of kingdom and genus. Genera are now grouped into *families*, families into *orders*, orders into *classes*, classes into *phyla (divisions)*,[3] and phyla (divisions) into *kingdoms*. Fundamental differences in RNA, cell wall content, and other features observed within two distinct groups of bacteria in Kingdom Archaea are considered by many systematists to be great enough to warrant recognizing three superkingdoms called **domains.** Most prokaryotes are placed in Domain Bacteria, while other prokaryotes are placed in Domain Archaea. All organisms with eukaryotic cells are placed in Domain Eukarya.

Because the recognition of domains does not otherwise affect classification at and beneath the kingdom level, the classification in this text emphasizes a hierarchy with kingdoms at the top. A classification of major groups of living organisms into six kingdoms, based primarily on a modification of Whittaker's five-kingdom system, is found in Table 16.2. It is the classification followed throughout this book. Figure 16.5a outlines the relationships among organisms, as understood by taxonomists until recently. Figure 16.5b focuses on the evolutionary history and current relationships among plants.

Viruses, which don't have cellular structure or most of the other attributes of living organisms, are not included in

---

3. *Division* is equivalent to the term *phylum*, which has been used universally in animal classification. Until 1993, the *International Code of Botanical Nomenclature,* which governs the naming and classification of plants, required the use of *division* for plants. In 1993, an International Botanical Congress revised the rules and permitted the use of *phylum* interchangeably with the term *division* for plants and fungi. The term *division* has been dropped by most botanists, who have adopted *phylum* in its place.

this classification; they are discussed in Chapter 17 after the bacteria. Lichens, each of which is a combination of an alga and a fungus, are discussed with the fungi in Chapter 19. If we were to give a complete classification of the common onion, it would look like this:

Kingdom: Plantae
   Phylum: Magnoliophyta
      Class: Liliopsida
         Order: Liliales
            Family: Liliaceae
               Genus: *Allium*
                  Species: *Allium cepa* L.

The second part of the scientific name of a species, called the *specific epithet,* is followed by the name of the author, usually in abbreviated form. The author is the person (or persons) who originally named the plant or placed the species in a particular genus. Many plants have one author's name in parentheses, followed by the name of another author outside the parentheses. For example, the scientific name of mountain hemlock is *Tsuga mertensiana* (Bong.) Carr. In this instance, August Bongard (abbreviated to Bong.), a professor of botany at St. Petersburg in the early 19th century, first described the species under the genus *Pinus.* Half a century later, Élie Carrière (abbreviated to Carr.), an authority on conifers at the Paris Natural History Museum, transferred the species to the genus *Tsuga.* Because Bongard was the first to describe the species and Carrière was the first to place the species in a different genus, both names are included with the scientific name. Binomials are printed in italics. The first letter of a generic (genus) name is capitalized, whereas all of the specific epithet that follows is lowercase. In the case of *Allium cepa,* the L. following the name stands for Linnaeus, because this plant was one of the 7,300 species he included in his *Species Plantarum.*

In addition to these categories of classification, various "in-between" categories, such as *subphylum, subclass,* and *suborder,* have been used, and species themselves are sometimes further divided into *subspecies, varieties,* and *forms.* A few authors also recognize "super" categories, such as *superkingdom* and *superorder.* One classification widely used in the past recognized only two divisions (phyla) in Kingdom Plantae (Division [Phylum] Bryophyta for nonvascular plants and Division [Phylum] Tracheophyta, with a number of subdivisions [subphyla], for all vascular plants). These subphyla are treated as full phyla in this and a number of other texts.

Because human judgment has to enter into the compilation of classifications, there undoubtedly will never be complete agreement on them, but they are nevertheless useful, indeed necessary, for making some order of the diversity of life around us.

Those who specialize in identifying, naming, and classifying organisms are called *taxonomists.* Today, taxonomists who use DNA sequences and evolutionary processes in trying to sort out natural relationships refer to themselves as *systematists.* Taxonomists and systematists construct keys to help others identify organisms with which they may not be familiar. Most such keys are *dichotomous;* that is, they give the reader pairs of statements based on features of the organisms. By carefully examining an organism and choosing from each pair the statements that most closely apply to the organism, we can arrive at an identification. If, for example, we agree with the first of the two statements given for each number in the key, then we either use the identification at the end of the line or proceed to the next indented statement beneath. If we disagree with the first of the paired statements, then we look for the matching alternate statement and proceed from there.

The key that follows this discussion deals with major groups of organisms and illustrates hypothetical natural relationships. It covers only general features and does not indicate various details or occasional exceptions that would appear in keys for lesser groups. All organisms in any given group are presumed to be more closely related to each other than to organisms in another group.

Once again, it must be emphasized that a key of this type, while calling attention to basic differences between major groups of organisms, may also oversimplify the differences, partly because it does not list all the characteristics of each member of a given group. Moreover, it does not necessarily identify exceptions to the rule or some of the intermediate forms that frequently are transferred back and forth as research uncovers new evidence or as new interpretations develop.

Keys at this level of classification also are not completely practical because to be able to arrive at an identification, one sometimes has to have available specific stages in the life cycles of the plants. Furthermore, chemical tests that are not easily performed or details that may be difficult to see with a light microscope may be required. Nevertheless, such a key is generally preferable to a simple listing of groups and their characteristics for purposes of distinguishing among them.

16.5 Species Concepts

Learning Outcomes

a. Explain why it is difficult to place plants into well-defined categories.

b. Describe the morphological species concept.

c. Describe the interbreeding species concept.

d. Describe the ecological species concept.

e. Describe the phylogenetic species concept.

f. Describe the eclectic species concept.

g. Describe the nominalistic species concept.

Because species are the basic units of classification, they receive much attention from taxonomists. It seems as if the identification of fundamental characters unique to each species should be a relatively straightforward process. However, nature is not so easily partitioned into neat compartments, especially with plants that frequently hybridize, with species

Table 16.1 Five Classifications of Organisms into Kingdoms

Two Kingdoms (Traditional)	Three Kingdoms (Hogg and Haeckel)	Four Kingdoms (Copeland)	Five Kingdoms (Whittaker)	Six Kingdoms (Woese et al.)	Features
		Monera Bacteria	**Monera** Bacteria	**Archaea** Archaebacteria	Cells prokaryotic; lack muramic acid
	Protoctista Bacteria	**Protoctista** Algae	**Protista** Algae	**Bacteria** True bacteria	Cells prokaryotic; have muramic acid
	Algae	Slime molds	Slime molds	**Protista**	Cells eukaryotic
	Slime molds	Flagellate fungi	Flagellate fungi	Algae	
	Flagellate fungi	True fungi	Protozoa	Slime molds	
	True fungi	Protozoa	Sponges	Water molds	
	Protozoa	Sponges		Protozoa	
	Sponges			Sponges	
			Fungi True fungi	**Fungi** True fungi	Absorb food in solution
Plantae Bacteria	**Plantae** Bryophytes	**Plantae** Bryophytes	**Plantae** Bryophytes	**Plantae** Bryophytes	Produce food via photosynthesis
Algae	Vascular plants	Vascular plants	Vascular plants	Vascular plants	
Slime molds					
Flagellate fungi					
True fungi					
Bryophytes					
Vascular plants					
Animalia Protozoa	**Animalia** Multicellular animals	**Animalia** Multicellular animals	**Animalia** Multicellular animals	**Animalia** Multicellular animals	Ingest food
Sponges					
Multicellular animals					

Table 16.2 Classification of Organisms in Six Kingdoms

Domain Archaea

  Kingdom Archaea
    Phylum Archaebacteria (methane, salt, and sulfolobus bacteria)

Domain Bacteria

  Kingdom Bacteria
    Phylum Eubacteria
        Class Eubacteriae (unpigmented, purple, and green sulfur bacteria)
        Class Cyanobacteriae (cyanobacteria)
        Class Prochlorobacteriae (prochlorobacteria)

Domain Eukarya

  Kingdom Protista
    Phylum Chlorophyta (green algae)
    Phylum Chromophyta (yellow-green, golden-brown, and brown algae)
    Phylum Rhodophyta (red algae)
    Phylum Euglenophyta (euglenoids)
    Phylum Dinophyta (dinoflagellates)
    Phylum Cryptophyta (cryptomonads)
    Phylum Prymnesiophyta (haptophytes)
    Phylum Charophyta (stoneworts)
    Phylum Myxomycota (plasmodial slime molds)
    Phylum Dictyosteliomycota (cellular slime molds)
    Phylum Oomycota (water molds)
    [Phylum Protozoa—protozoans]*
    [Phylum Porifera—sponges]

  Kingdom Fungi
    Phylum Chytridiomycota (chytrids)
    Phylum Zygomycota (coenocytic fungi)
    Phylum Ascomycota (sac fungi)
    Phylum Basidiomycota (club fungi)
    Phylum Deuteromycota (imperfect fungi)

  Kingdom Plantae
    Phylum Hepaticophyta (liverworts)
    Phylum Anthocerotophyta (hornworts)
    Phylum Bryophyta (mosses)
    Phylum Psilotophyta (whisk ferns)
    Phylum Lycophyta (club mosses)
    Phylum Equisetophyta (horsetails)
    Phylum Polypodiophyta (ferns)
    Phylum Ginkgophyta (*Ginkgo*)
    Phylum Cycadophyta (cycads)
    Phylum Pinophyta (conifers)
    Phylum Gnetophyta (*Gnetum, Ephedra, Welwitschia*)
    Phylum Magnoliophyta (flowering plants)

  [(Kingdom Animalia (multicellular animals))]

*Groups in brackets are not discussed in this book.

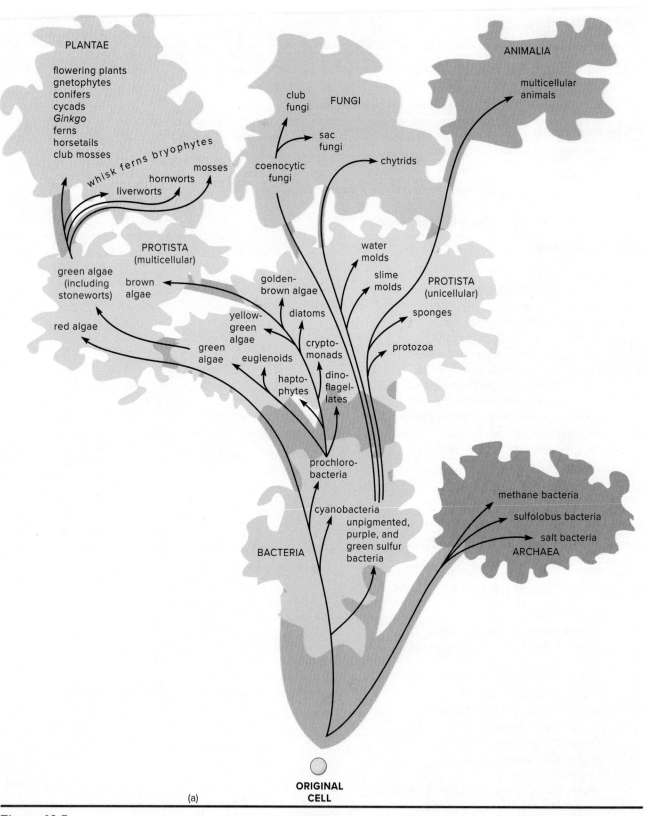

Figure 16.5a Putative relationships among kingdoms and the major groups of organisms. *Source:* Cantino, P. D., J. A. Doyle, S. W. Graham, W. S. Judd, R. G. Olmstead, D. E. Soltis, P. S. Soltis, and M. J. Donoghue. 2007. Towards a phylogenetic nomenclature of Tracheophyta. Taxon 56: 822–846.

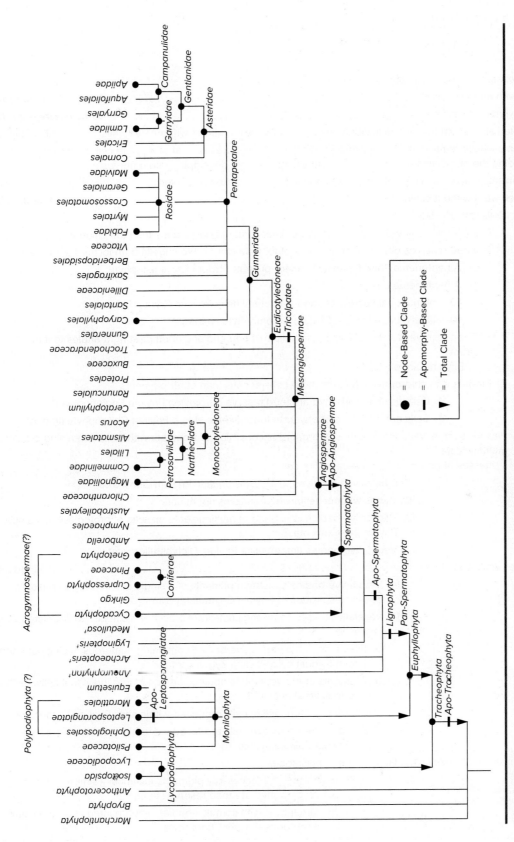

Figure 16.5b Hypothesis of phylogenetic relationships of 53 clades of land plants, based on diverse phylogenetic hypotheses, summarized by Cantino and coworkers, illustrating the nature of new classifications arising out of recent phylogenetic projects, most of them based on DNA sequence data.

A Key to Major Groups of Organisms (Exclusive of Kingdom Animalia)

1. Unicellular, prokaryotic organisms with cell walls
   2. Cell walls with muramic acid . . . . **Domain (Kingdom) Bacteria**
   2. Cell walls without muramic acid. . . . . **Domain (Kingdom) Archaea**
1. Unicellular, colonial, filamentous, or multicellular eukaryotic organisms, with or without cell walls . . . . **Domain Eukarya**
   3. Organisms whose female (and usually male) reproductive structures consist of a single cell or with sterile cells surrounding the one-celled reproductive structures; zygotes not developing into embryos
      4. Organisms unicellular, filamentous, or plasmodial (i.e., with naked protoplasm)
         5. Cell walls without chitin . . . . **Kingdom Protista**
            6. Cells with plastids
               7. Plastids with yellow, brown, or orange pigments more conspicuous than the chlorophyll pigments
                  8. Food reserves oils or carbohydrates other than starch; two flagella both located at one end of the cell
                     9. A haptonema (third flagellum) usually present Phylum Prymnesiophyta
                     9. A haptonema absent
                        10. The unequal flagella bearing stiff lateral hairs Phylum Cryptophyta
                        10. Flagella not as above Phylum Chromophyta (in part)
                  8. Food reserve starch; cells with a flagellum at one end and another at right angles to it in a central groove
                     Phylum Dinophyta
               7. Plastids with chlorophyll pigments more conspicuous than other pigments
                  11. Cells flexible; carbohydrate food reserve *paramylon* Phylum Euglenophyta[4]
                  11. Cells not flexible; carbohydrate food reserve starch Phylum Chlorophyta (in part)
            6. Plastids absent; vegetative bodies of organisms consisting of filaments with walls containing cellulose, multinucleate
               masses of protoplasm, or amoeba-like cells
               12. Cell walls with cellulose; organisms aquatic Phylum Oomycota
               12. Cell walls without cellulose; organisms not aquatic
                  13. Vegetative bodies consisting of multinucleate masses of protoplasm (plasmodia) Phylum Myxomycota
                  13. Vegetative bodies amoeba-like Phylum Dictyosteliomycota
         5. Cell walls containing chitin . . . . **Kingdom Fungi**
            14. Vegetative bodies consisting primarily of a single cell, often with rhizoids; zoospores produced Phylum Chytridiomycota
            14. Vegetative bodies consisting primarily of branched filaments; no flagellated cells produced
               15. Filaments of the vegetative bodies containing numerous nuclei; not partitioned into individual cells Phylum Zygomycota
               15. Filaments of the vegetative bodies partitioned into individual cells, each with one to several nuclei
                  16. Sexual reproductive cells produced within sacs Phylum Ascomycota
                  16. Sexual reproductive cells produced externally on club-shaped structures Phylum Basidiomycota
      4. Organisms multicellular, not filamentous or plasmodial
         17. Organisms with accessory pigments essentially similar to those of higher plants; carbohydrate food reserve starch
            18. Organisms with complex bodies differentiated into nodes and internodes Phylum Charophyta

---

4. Only about a third of euglenoid species develop chloroplasts.

18. Organisms with bodies not differentiated into nodes and internodes Phylum Chlorophyta (in part)
17. Organisms with some accessory pigments differing from those of higher plants; food reserves carbohydrates other than ordinary starch
19. Organisms brownish in color due to presence of brown pigments; carbohydrate food reserve *laminarin* Phylum Chromophyta (in part)
19. Organisms reddish in color due to presence of red pigments; carbohydrate food reserve *floridean starch* Phylum Rhodophyta
3. Organisms with multicellular reproductive structures.................................... **Kingdom Plantae**
20. Plants without true xylem or phloem Phylum Bryophyta
20. Plants with true xylem and phloem
21. Plants with true leaves absent; *enations*[5] present; stems branching dichotomously Phylum Psilotophyta
21. Plants with true leaves present; enations absent; stems branching in various ways or unbranched
22. Plants with leaves having a single vein (*microphylls*)
23. Stems not ribbed and not containing silica; leaves photosynthetic Phylum Lycophyta
23. Stems ribbed and containing silica; leaves reduced to scales and non-photosynthetic Phylum Equisetophyta
22. Plants with leaves usually having more than one vein (*megaphylls*)
24. Plants reproducing by means of spores produced on the leaves Phylum Polypodiophyta
24. Plants reproducing by means of seeds developed from ovules
25. Plants without flowers; seeds not produced in enclosed ovaries
26. Leaves pinnate and large, resembling those of palms Phylum Cycadophyta
26. Leaves not pinnate or palmlike
27. Leaves fan-shaped, with numerous dichotomously forking veins Phylum Ginkgophyta
27. Leaves not fan-shaped
28. Wood containing no vessels Phylum Pinophyta
28. Wood containing vessels Phylum Gnetophyta
25. Plants with flowers; seeds produced in enclosed ovaries Phylum Magnoliophyta
29. Flowers with parts mostly in fours and fives; cotyledons two[6] Class Magnoliopsida
29. Flowers with parts mostly in threes; cotyledon one[6] Class Liliopsida

---

5. See Chapter 21 for a discussion of *enations*.
6. Note that two groups of dicots are now recognized.

often closely related, and only partially diverged from one another through speciation. An additional complicating factor is that many plants reproduce asexually or with combined sexual and asexual methods. Taxonomists often disagree about which characters to use when identifying species. There are at least 22 different ways of classifying species, and they can be grouped into the following six groups of concepts.

Morphological Species Concepts

Morphological species concepts define species based on their morphology (growth form). This has been an important system historically and relies heavily on the examination of herbarium specimens (discussed in Chapter 23). Typically, a taxonomist will sort a set of herbarium sheets based on an overall impression of morphological similarities. Then, microscopic examination of, for example, trichomes, and of a wide range of other morphological features, may allow for additional separation into species. This is a very subjective treatment, and the traits that lead to distinctions among groups may not be clear or consistently used among taxonomists. In recent years, computers have allowed taxonomists to analyze large numbers of morphological traits and use computer programs to form clusters, which are then used to help define species. However, it is still unclear how to interpret these computer-generated clusters in terms of species, and subjective decisions often still must be made.

Interbreeding Species Concepts

Many biology textbooks focus on this concept to describe species. A species is a population capable of interbreeding and is reproductively isolated from other groups. This sounds like a logical grouping strategy, but it has serious problems, especially in the plant world. Plants exhibit different degrees of crossability, and it is difficult to predict which groups are crossing with each other in natural environments. Greenhouse studies using artificial pollinations may determine which groups are capable of intercrossing, but this may not reflect the natural world, where these groups may rely on different pollinators or flower at different times. For example, it may be possible to intercross *Iris* species, but in the wild they do not hybridize because they do not flower at the same time. Also, it would take many years to intercross and assess crossing success for the many pairwise crosses of a large group of species.

Ecological Species Concepts

An ecological species is a group of related individuals that occupy a unique ecological niche. Taxonomists using this method assume that the evolution of species is governed mainly by ecological differences. For example, oak species capable of interbreeding may be found in geographic proximity. However, one species may occupy swampy, wet areas, while another is found in rich upland wooded areas. Ecological niches, therefore, maintain species identity.

Phylogenetic Species Concepts

With this grouping method, phylogenetic history determines species boundaries. Phylogenetics is the study of evolutionary similarities among groups of organisms. Therefore, individuals with a common evolutionary background are considered to be a species. Cladistic methods are used to determine the evolutionary history of a group of individuals. Cladistics is a method of examining natural relationships among organisms, based on features shared by the organisms. The relationships are portrayed in branching diagrams (evolutionary trees) called **cladograms** (Fig. 16.6). In cladistics, the value or form of a feature is referred to as a **character state**. In a feature such as flower color, for example, the character could be flower color, and the state could be purple or red. After the specific chemical or other nature of a character is determined, hypotheses (assumptions) are made about which states are primitive (ancestral) and which are derived (have evolved from something else). Obviously, the hypotheses need to be tested, and to test them, we often try to find additional characters that may support one hypothesis more than they support others. Those hypotheses that are not supported can be eliminated, and those that are supported can be subjected to further testing.

In trying to choose the best of several to many cladograms, taxonomists often use the principle of *parsimony*. Parsimony, as it applies to cladistics, is based on a principle of logic called *Occam's razor*, which states that "one should not make more assumptions than the minimum needed to explain anything." The best cladogram is usually interpreted as that which requires the fewest evolutionary changes in the taxa involved to arrive at the present situation.

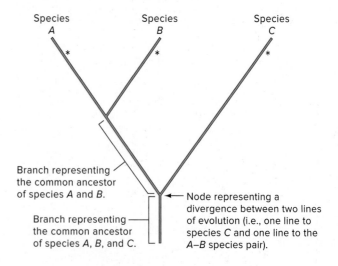

Figure 16.6 A simple cladogram showing relationships of three species derived from a common ancestor. The cladogram indicates that Species A and Species B are more closely related to each other than either is to Species C.

Cladistics has been used to infer evolutionary relationships among many different groups of organisms in all the kingdom survey chapters that follow in this book.

Eclectic Species Concepts

Proponents of this grouping strategy argue that a single criterion is not sufficient to identify species. They suggest that morphological, geographical, biological, and ecological criteria must be considered when defining species. By using so many criteria, species designations are by nature more subjective, but taxonomists using this method consider it critical to use these many criteria to have accurate assessments of what a species is.

Nominalistic Species Concepts

Some biologists believe that species do not exist. They suggest that only individuals are real and any attempt to determine higher-level groupings, such as species, genera, or families, only results in artificial classifications. The evolutionary unit of importance is actually the local interbreeding population. A larger grouping, such as a species, is subjected to a diversity of selection pressures, so it is not considered a cohesive unit.

16.6 The Future of Plant Classification

Learning Outcome

a. Explain how new DNA sequencing technologies are influencing knowledge about interrelationships among organisms.

In 1998, a group of 28 plant taxonomists (the Angiosperm Phylogeny Group) published a paper entitled "An Ordinal Classification of Flowering Plants." The classification was based on a considerable accumulation of DNA sequence data and its cladistic analysis, and it involved an extensive reorganization of flowering plant families. This reorganization was adopted by Walter Judd and his colleagues in their taxonomy text, first published in 1999 and now in its third (2008) edition. The rapid development of new DNA sequence generation techniques ("next-generation sequencing") and DNA sequence analysis is increasing the pace of the generation and refinement of cladograms and will continue to provide new insights into the interrelationships of the world's organisms.

Because this text covers an introduction to the subject of botany as a whole and does not get into the details of plant taxonomy, the widely used traditional approach has been followed throughout this edition. In this approach, all flowering plants (Phylum Magnoliophyta) have been separated into two large classes (monocots and dicots), primarily on the basis of the number of cotyledons ("seed leaves"). In addition to the number of seed leaves, monocots and dicots have a number of distinguishing features, such as the arrangement of leaf veins and numbers of flower parts. However, not all flowering plants fit neatly into these two groups. The relationships among these other groups are still being determined. In the meantime, the term *eudicot* is often used to refer to the group of dicots that share all features typical of dicots. Most dicots are eudicots.

SUMMARY

16.1 Introduction to Plant Classification
- Scientific names are used by biologists to provide unified points of reference.

16.2 Development of the Binomial System of Nomenclature
- All organisms are identified by a two-word Latin name (binomial); many also have one to many common names. Biologists recognized the need for worldwide uniformity in naming and classifying organisms.
- Theophrastus first classified plants in the 4th century B.C. Monocots and dicots were distinguished in the 13th century; herbalists later added many plants to Theophrastus's original list. Descriptive Latin phrase names were used for all organisms by the 18th century.
- Linnaeus compiled a comprehensive list of all known Latin phrase names for plants according to their genera in *Species Plantarum*, published in 1753. Linnaeus placed specific kinds of related organisms (species) in each genus; he also listed a single word next to the Latin phrase which, when combined with the name of the genus, formed an abbreviation for each species.
- The abbreviated binomial, plus the authority for it, eventually replaced the phrase name; this method of naming plants became known as the Binomial System of Nomenclature. Linnaeus organized all known plants into 24 classes, based primarily on the number of stamens in flowers.
- At a meeting in Paris in 1867, European and American botanists agreed to use Linnaeus's 1753 publication and binomials as the starting point for all scientific names of plants. The rules *(International Code of Botanical Nomenclature, ICBN)* drawn up at that meeting and periodically revised are now followed for naming and classifying plants by botanists of all nationalities.
- Cultivated plants are frequently treated by another classification system, the *International Code of Nomenclature of Cultivated Plants (ICNCP)*.

- In 1998, a group of plant taxonomists revised the traditional system of plant classification, particularly at the family level, based on DNA sequence and cladistic information. Some forms of the new classification are becoming widely adopted and frequently use yet another classification system, the *PhyloCode*.

16.3 Development of the Kingdom Concept

- At first, two kingdoms (Plant and Animal) were recognized. Hogg and Haeckel proposed a third kingdom (Protoctista) in the 1860s. In 1938, H. F. Copeland, on the basis of cellular differences, divided the Protoctista into Kingdoms Monera and Protoctista.
- In 1969, Whittaker proposed five kingdoms based on both forms of nutrition and cellular structure. In the 1980s, Carl Woese divided Whittaker's Kingdom Monera into two kingdoms, and six kingdoms are now generally recognized.

16.4 Classification of Major Groups

- Since Linnaeus's time, several classifications have been added between the levels of kingdom and genus.
- The second part of a binomial, the specific epithet, is followed by the name of the author, usually in abbreviated form.
- Subcategories (e.g., subspecies, suborder) are sometimes used in classification.
- Taxonomists construct keys to aid in the identification of organisms. Most keys are dichotomous. Keys to major groups of organisms deal only in generalities and do not always indicate occasional exceptions or allow for intermediate forms that may be transferred back and forth between groups as new knowledge about them is gained.

16.5 Species Concepts

- Cladistics is a system of classifying and inferring evolutionary relationships based on an examination of shared features and differences.
- There are many ways of classifying species, and these can be grouped into morphological, interbreeding, ecological, phylogenetic, eclectic, and nominalistic species concepts.

16.6 The Future of Plant Classification

- Taxonomists continue to organize groups of organisms based on relationships. DNA sequence data are critical in making those decisions.
- Recent data suggest that not all dicots belong in the same group. Eudicots exhibit a unified set of features. These represent most of the dicots.

REVIEW QUESTIONS

1. What are the advantages and drawbacks of using scientific names as compared with common names?
2. What is the *International Code of Botanical Nomenclature*?
3. What is meant by the term *binomial*?
4. Which phyla include simple aquatic organisms?
5. List some characteristics by which the following may be distinguished from one another and then construct a simple dichotomous key that could aid someone from another planet in identifying them: cups, oranges, hammers, bicycles, rosebushes, automobiles, plates, cats, rats, raspberry bushes, dogs, wrenches, peaches.

DISCUSSION QUESTIONS

1. Because phrase names are generally more descriptive of organisms than binomials, do you think they should be revived?
2. After Linnaeus organized all the genera known to him into classes, a number of other categories of classification between the level of kingdom and genus were added. Are these other categories really useful? Explain.
3. Do you think biologists were justified in dividing the traditional Plant and Animal Kingdoms into six kingdoms? Would it make any significant difference to the world at large if biologists decided to simply classify organisms as either plants or animals? Explain.

CHAPTER 17

Domain (Kingdom) Bacteria, Domain (Kingdom) Archaea, and Viruses

17.1 Introduction to Domain (Kingdom) Bacteria, Domain (Kingdom) Archaea, and Viruses

17.2 Features of Domains (Kingdoms) Bacteria and Archaea
    Cellular Detail and Reproduction
    Size, Form, and Classification

17.3 Domain (Kingdom) Bacteria—the True Bacteria
    Phylum Bacteriophyta

17.4 Human Relevance of the Unpigmented, Purple, and Green Sulfur Bacteria
    Composting and Compost
    Key Theme 17.1: The Social Life of Prokaryotes
    True Bacteria and Disease
    Koch's Postulates
    True Bacteria Useful to Humans

17.5 Class Cyanobacteriae—the Cyanobacteria (Blue-Green Bacteria)
    Introduction
    Distribution
    Form, Metabolism, and Reproduction
    Cyanobacteria, Chloroplasts, and Oxygen
    Human Relevance of the Cyanobacteria
    Key Theme 17.2: Cyanobacteria and Algae Blooms

17.6 Class Prochlorobacteriae—the Prochlorobacteria

17.7 Domain (Kingdom) Archaea—the Archaebacteria
    The Methane Bacteria
    The Salt Bacteria
    The Sulfolobus Bacteria
    Human Relevance of the Archaebacteria

17.8 Viruses
    Introduction
    Size and Structure
    Key Theme 17.3: Plant Viruses
    Bacteriophages
    Viral Reproduction
    Human Relevance of Viruses

17.9 Viroids and Prions

A single cell of *Escherichia coli*, a common gram-negative bacterium. Note the filaments extending from the cell surface. The more abundant and shorter filaments are common pili, used by the bacterium for colonization of eukaryotic cells, such as those in the human gastrointestinal or urinary tract. The two longer filaments are conjugative pili, required for the direct transfer of DNA from one bacterial cell to another (conjugation). The small, green spheres are RNA bacteriophage particles that adhere to the conjugative pili as the first step in infection.
© Dr. Linda M. Stannard, University of Cape Town/Science Source

OVERVIEW

Following an introduction describing luminescent bacteria associated with flashlight fish, the chapter gives an overview of the features of Domains (Kingdoms) Bacteria and Archaea. A brief investigation of bacteria and their reproduction is followed by a discussion of true bacteria, surveying representatives and the human relevance of heterotrophic and autotrophic true bacteria. Sections on composting, disease transmission, bacterial diseases, and the Koch's Postulates are included. Cyanobacteria and prochlorobacteria are explored next, and the human relevance of cyanobacteria is examined. Archaebacteria are then discussed. The chapter concludes with an overview of the nature, reproduction, and human relevance of viruses, viroids, and prions.

17.1 Introduction to Domain (Kingdom) Bacteria, Domain (Kingdom) Archaea, and Viruses

Learning Outcome

a. Explain how tiny microscopic bacteria can produce more biomass than all other organisms.

The astonishing variety and beauty of undersea life, with representatives from all kingdoms, has attracted the attention of snorkelers and scuba divers in ever-increasing numbers. The shallow tropical waters of the Great Barrier Reef of Australia, the Virgin Islands National Park, Hanauma Bay of Hawaii, and countless other reefs teem with brilliantly colored marine life. Scuba divers who explore the oceans at night in areas such as Indonesia's Banda Sea; the region just north of Darwin, Australia; or Israel's Gulf of Elat are frequently rewarded with a spectacular sight not seen by those who explore by day. They find themselves in the midst of a shimmering, constantly changing light display.

Countless numbers of lights, bright enough to read by, flicker, dart, glide slowly, or even congregate in glimmering groups. Some of the lights come from large pouches beneath the eyes of small fish, which can turn them off at will by covering them with special folds of tissue. Unlike their deep-sea relatives, these *flashlight fish,* as they are called, do not produce the light within their own bodies. Instead, the light comes from thousands of microscopic, luminescent bacteria that enjoy a *symbiotic,* or specifically, a mutualistic, relationship with the fish.

The fish furnish some food and oxygen to the bacteria from blood vessels in the pouches, while the bacteria produce the light used by the fish not only to search for food at night but also to attract live food and to confuse predators by flashing the lights on and off while swimming in erratic patterns. Both organisms derive benefit from the relationship.

Luminescent bacteria are minor representatives of the most numerous and geologically ancient of the earth's living organisms. Fossils of bacteria have been found in geological strata that are estimated to be 3.5 billion years old. Since fossils of the first organisms with eukaryotic cells date back "only" 2.7 billion years, bacteria have existed on this planet nearly a billion years longer than any other organism.

Bacteria occur in astronomical numbers in almost every conceivable natural habitat. Each gram of garden soil, for example, is estimated to contain 2 billion bacteria representing 1 million species. Bacteria are found on and in all plants and animals, in all types of soils, throughout both fresh and salt waters, in polar ice caps and bubbling hot springs, in coal and petroleum, throughout the atmosphere, in radioactive waste, and almost anywhere else one may care to look for them. In your body, bacterial cells are 10 times more abundant than human cells. The biomass of bacteria on earth exceeds that of all plants and animals combined.

Just how many species of bacteria there are is uncertain, because most have yet to be discovered. A current estimate suggests that there are likely 2 billion species on Earth.

Although bacteria that cause disease and spoilage receive most of the publicity, more than 90% of bacterial species are either harmless or useful to humans. They are used in industry for a wide variety of purposes and are so important ecologically that life as we know it today would be vastly different without the activities of these tiny organisms. Both industrial uses and the role of bacteria in ecology are discussed in later sections.

17.2 Features of Domains (Kingdoms) Bacteria and Archaea

Learning Outcomes

a. Describe major differences between prokaryotes and eukaryotes.
b. Summarize the main steps in binary fission.
c. Draw and label the three shapes exhibited by bacteria.

All members of Domains (Kingdoms) Bacteria and Archaea have prokaryotic cells. Such cells have no nuclear envelopes, but each has a long strand of DNA, ribosomes, and membranes. In addition, there usually are small DNA molecules called *plasmids* present in bacterial cells. Membrane-bound organelles, such as plastids, mitochondria, dictyosomes, and endoplasmic reticulum are lacking (Table 17.1; see Fig. 17.13).

In a number of members of these two kingdoms, cells may occur in a common matrix of gelatinous material as colonies or **filaments** (threads), but each cell is completely independent, with no protoplasmic connections between them. Some species are **motile** (can propel themselves), usually using simple **flagella** (whiplike tails, which often occur in pairs) that push or pull the cell through the water. Several filamentous species exhibit a gliding motion in which the filaments glide back and forth, but most species are nonmotile.

Bacteria typically absorb food through their walls. A few true bacteria, including *cyanobacteria* (blue-green bacteria)

Table 17.1	Some Differences between Eukaryotic and Prokaryotic Cells
EUKARYOTIC CELLS	**PROKARYOTIC CELLS**
1. Have a nuclear envelope	1. Lack a nuclear envelope
2. Have two to hundreds of chromosomes per cell; DNA is double stranded	2. Have a single, closed loop of double-stranded DNA plus (usually) several to 40 plasmids
3. Have membrane-bound organelles (e.g., plastids, mitochondria)	3. Lack membrane-bound organelles
4. Have 80S ribosomes[1]	4. Have 70S ribosomes[1]
5. Have asexual reproduction by mitosis	5. Have asexual reproduction by fission
6. Have sexual reproduction by fusion	6. Sexual reproduction unknown

[1]S is a Svedberg unit, which is used to measure the rate at which a particle suspended in a fluid settles to the bottom when centrifuged. An 80S ribosome is larger than a 70S ribosome.

and *chloroxybacteria,* can make their own food through photosynthesis.

Bacteria typically reproduce asexually through fission, a form of cell division in which one cell essentially pinches in two. Fission does not involve mitosis, because there are no nuclei or other organelles. The strand of DNA duplicates itself, however, and the DNA is distributed to the two new cells from the parent cell. Sexual reproduction is unknown, but genetic recombination occurs in several groups. It is often facilitated by *pili* (minute tubes) that form connections between cells, but even without pili, genetic recombination can occur between cells in close contact.

Cellular Detail and Reproduction

Prokaryotic cells have no membrane-bound nuclear envelopes or organelles; folds of the plasma and other membranes perform some of the functions of the organelles of eukaryotic cells. In addition, microcompartments surrounded by protein shells provide some separation of metabolic activities in bacterial cells. Ribosomes that are about half the size of those in the cytoplasm of eukaryotic cells are also present. Bacterial cells have a *nucleoid,* which is a single circular chromosome. The nucleoid is usually attached at one point to the plasma membrane. In addition, up to 30 or 40 small, circular DNA molecules called *plasmids* may be present. The plasmids *replicate* (produce duplicate copies of themselves) independently of the chromosome, and the entire complement of plasmids in a bacterial cell often consists of multiple copies of one or a few different DNA molecules. The chromosome and sometimes all or some of the plasmids replicate before the cell divides (Fig. 17.1).

Unlike eukaryotic cells, bacteria do not undergo mitosis. Instead, there is an internal reorganization of material during which the two copies of the duplicated chromosome migrate to opposite ends of the cell. Then, at approximately the middle of the cell, transverse (perpendicular) walls and cell membranes are formed, dividing the cell in two. Finally, the two new cells separate and enlarge to their original size, or in some bacteria, the cells remain attached to each other in chains. This simple form of asexual reproduction is called *fission* (Fig. 17.2).

Under ideal conditions of moisture, food supply, and temperature, a bacterium may undergo fission every 10 to 20 minutes. If a single bacterium of average size were to replicate itself every 20 minutes for 36 hours, a mass of bacteria numbering 322,981,536,679,200,000,000,000,000,000,000 and weighing 126,464,618,590 metric tons (137,438,953,472 tons) would be produced. This body of cells would be larger than the earth. Of course, bacteria do not continue to reproduce at their maximum rate for very long due to the exhaustion of food supplies and the accumulation of toxic wastes.

Bacteria do not produce gametes or zygotes; nor do they undergo meiosis. However, at least three forms of genetic recombination occur.

1. In *conjugation,* a plasmid and/or part of the DNA strand is transferred from a donor cell to a recipient cell, usually through a tiny, hollow, tubelike **pilus,** while the two cells are in contact (Fig. 17.3). Once in the recipient cell, the DNA becomes part of the new cell. Bacterial cells are genetically different from each other, and usually only cells of different mating types undergo conjugation.
2. In *transformation,* a living cell acquires fragments of DNA released by dead cells in its surroundings and incorporates them into its own cell.
3. In *transduction,* fragments of DNA are carried from one cell to another by viruses (discussed toward the end of this chapter in the section "Viruses").

Size, Form, and Classification

Except for the cyanobacteria, which are discussed separately, and a few giant bacteria, which attain lengths of up to 60 micrometers, most bacteria are less than 2 or 3 micrometers in diameter, and a few of the smaller species approach 0.15 micrometer. The latter are so small that 6,500 of them

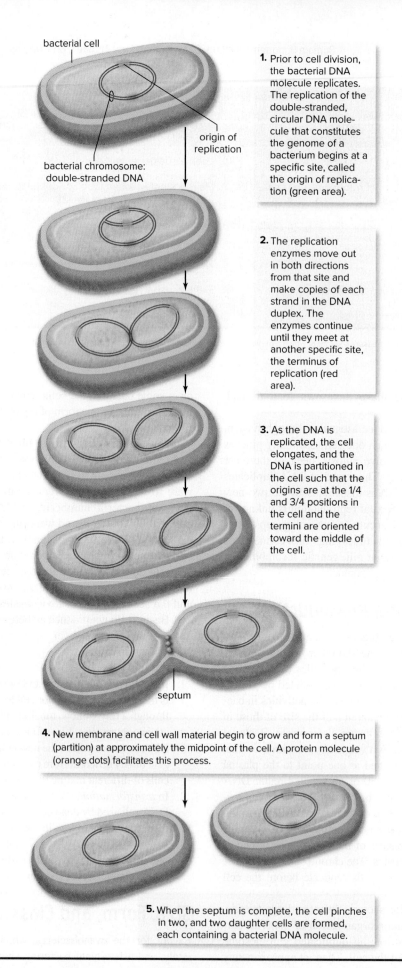

Figure 17.1 In this illustration of binary fission in bacteria, replication of DNA begins at the origin of replication. Once DNA is replicated, the cell elongates and two daughter cells are formed, each with one copy of the bacterial DNA molecule.

Figure 17.2 Scanning electron microscope image of *Escherichia coli*, showing near completion of binary fission and the point at which two new cells have been produced from a single cell. × 21,674. © CDC/Janice Haney Carr

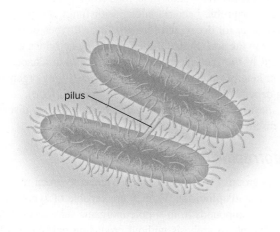

Figure 17.3 When bacteria undergo conjugation, part of the DNA strand of the donor cell migrates through a hollow, tubelike *pilus* to the recipient cell, where it is incorporated into the DNA of its new cell. Shorter pili, which cover the surfaces of both cells, are not involved in conjugation.

arranged in a row would not quite extend across the head of a pin. Because of their small size, bacteria are not visible individually to the unaided eye, and they are studied in the laboratory with electron microscopes or with the highest magnifications of light microscopes.

Millions of bacteria have yet to be described and given a name. The different kinds of bacteria occur primarily in three forms: *cocci* (singular: *coccus*), which are spherical or sometimes elliptical; *bacilli* (singular: *bacillus*), which are rod shaped or cylindrical; and *spirilla* (singular: *spirillum*), which are in the form of a helix, or spiral (Fig. 17.4). Further classification is based on other visible features, including, for example, development of slimy or gummy, capsulelike sheaths around cells; color; presence of hairlike or budlike appendages; development of internal, thicker-walled *endospores* and other cell inclusions; mechanisms of movement and features of colonies. Biochemical characteristics, including energy sources, cell-wall components, fermentation products, luminescence, optimum ranges of pH and temperature, oxygen relationships, amino acid sequences of proteins, and salt tolerance are also used in classification. Only about half of the phyla of bacteria can be cultured in the lab and, consequently, characterized by these methods. DNA sequence information is becoming increasingly important for the classification of bacteria.

Some bacteria have slender flagella, usually about 5 to 10 micrometers in length, which propel them through liquid. Others have somewhat shorter, tubelike pili, which resemble flagella but do not provide locomotion. The pili apparently enable bacteria to attach themselves to surfaces or to each other, as in conjugation, which was discussed previously in the section "Cellular Detail and Reproduction."

True bacteria are also grouped into two large categories based on the reaction of their cell walls to a dye. After a heat-fixed smear of cells has been stained blue-black with a violet dye, a dilute iodine solution, alcohol, or acetone is added. When so treated, some species rapidly lose their color (but absorb pink safranin dye) and are called *gram-negative*. Others, called *gram-positive*, retain most of the blue-black color. Gram's stain, named after Christian Gram, who discovered it in 1884, has many variations, but all the variations produce similar results. Other features used in classifying bacteria are indicated in the discussions that follow.

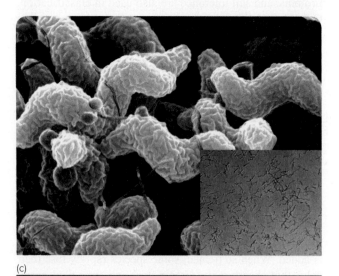

Figure 17.4 The three basic shapes of bacteria are: (a) Cocci. (b) Bacilli. (c) Spirilla. Images are computer renditions based on scanning electron micrographs. *(a) © CDC/Janice Haney Carr; (b) © Shutterstock/Tatiana Shepeleva; (c) Provided by De Wood, digital colorization by Chris Pooley, USDA-ARS*

17.3 Domain (Kingdom) Bacteria—the True Bacteria

Learning Outcomes

a. Compare and contrast saprobes and parasites.
b. Describe ways in which autotrophic bacteria synthesize food.

Phylum Bacteriophyta
Class Bacteriae—the Unpigmented, Purple, and Green Sulfur Bacteria

The true bacteria differ from archaebacteria in several ways. True bacteria have peptidoglycan in their cell walls, their genes lack introns, and fatty acids are used to build their membranes. Despite these unifying features, the true bacteria constitute a heterogeneous assemblage, with most being **heterotrophic** (organisms that cannot synthesize their own food).

The majority of heterotrophic bacteria are *saprobes* (living organisms that obtain their food from nonliving organic matter). Saprobic bacteria, along with fungi, are primarily responsible for decay and for recycling all types of organic matter in the soil. Some of their recycling activities are discussed in the section on the *nitrogen cycle* in Chapter 25. Other heterotrophic bacteria are *parasites* (living organisms that depend on other living organisms for their food).

Several parasites that cause important human diseases are discussed later in the chapter.

Autotrophic Bacteria

A few groups of true bacteria are similar to green plants in being **autotrophic;** that is, they are capable of synthesizing organic compounds from simple inorganic substances. Some carry on photosynthesis without producing oxygen as a by-product. Included in the photosynthetic bacteria are the *purple sulfur bacteria,* the *purple nonsulfur bacteria,* and the *green sulfur bacteria.* The *cyanobacteria* (blue-green bacteria) and the *chloroxybacteria,* which do produce oxygen as a product of their photosynthesis, are discussed separately later in this chapter. Cells of the first two groups appear purplish, or occasionally red to brown, because of the presence of a mixture of greenish, yellow, and red pigments. Their greenish *bacteriochlorophyll* pigments (there are several closely related ones) are very similar to the chlorophyll *a* of higher plants. No plastids are present in bacteria, and their pigments are, instead, located in folds in the plasma membrane or in small, spherical bodies. During photosynthesis, the purple bacteria substitute hydrogen sulfide, the bad-smelling gas given off by rotten eggs, for the water used by higher plants. The generalized equation for the process is as follows:

$$CO_2 + 2H_2S \xrightarrow[\text{light}]{\text{bacteriochlorophyll}} (CH_2O)n + H_2O + 2S$$

carbon dioxide + hydrogen sulfide → carbohydrate + water + sulfur

The equation for the purple nonsulfur bacteria is similar, but hydrogen from organic molecules is used instead of hydrogen sulfide. Green sulfur bacteria do use hydrogen sulfide, but their chlorophyll, called *chlorobium chlorophyll,* differs significantly in its chemistry from the chlorophylls of higher plants.

Other groups of bacteria are *chemoautotrophic;* that is, they are capable of obtaining the energy they require from various compounds or elements through chemical reactions involving the oxidation of reduced inorganic groups, such as NH_3, H_2S, and Fe^{++}, or the oxidation of hydrogen gas (as discussed in the section on oxidation-reduction reactions in Chapter 10).

Chemoautotrophic bacteria include *iron bacteria,* which transform soluble compounds of iron into insoluble substances that accumulate as deposits (e.g., in water pipes); *sulfur bacteria,* which can convert hydrogen sulfide gas to elemental sulfur and sulfur to sulfate; and *hydrogen bacteria,* which flourish in soils where they use molecular hydrogen derived from the activities of anaerobic and nitrogen-fixing bacteria. Nitrifying and nitrogen-fixing bacteria (some of which are heterotrophic) are a group of ecologically important bacteria that are discussed, along with their roles in the nitrogen cycle, in Chapter 25.

17.4 Human Relevance of the Unpigmented, Purple, and Green Sulfur Bacteria

Learning Outcomes

a. Describe optimum conditions for composting.
b. Discuss ways in which bacteria can enter our bodies and cause disease.
c. List the steps in Koch's postulates.
d. Describe ways in which bacteria are useful to humans.

Composting and Compost

Long before the existence of bacteria was known or even suspected, primitive agriculturists made rough piles of weeds, garbage, manure, and other wastes. They watched as the piles became reduced to a fraction of their original volume and changed into **compost,** a dark, fluffy material that conditions and enriches soil when mixed with it. Today, the compost pile is at the heart of the activities of many gardeners and small farmers. With many communities instituting ordinances forbidding the burning of leaves and refuse, and with space for waste disposal becoming scarce, a significant number of cities and towns have turned to the composting of street leaves and other materials. They have found that they not only save space but also are producing a useful, ecologically compatible product (Fig. 17.5).

Because bacteria are ubiquitous, plant litter in a pile of any size will eventually be decomposed. Ideally, however, a

Figure 17.5 Kitchen waste before and after composting.
© Shutterstock/Marina Lohrbach

compost pile is 2 meters wide, 1.5 meters deep, and at a minimum, 2 meters long (6 feet wide, 4.5 feet deep, and at least 6 feet long). If the pile is of lesser length or breadth, or greater depth, the conditions that favor the breakdown of materials (created by the bacteria themselves) develop at a slower pace.

Any accumulation of household garbage, leaves, weeds, grass clippings, and/or manure may be heaped together (Fig. 17.5), and it has been demonstrated that no so-called starter culture of microorganisms is needed to initiate decomposition. Once the ever-present decay organisms have material to decompose, their numbers increase rapidly, and much heat is generated. In fact, the temperature in the center of the pile often rises to 70°C (158°F). As it rises, populations of organisms adapted to higher temperatures replace those not so well adapted. The remains of the latter are then added to the compost. The high temperatures also kill many weed seeds and most disease organisms.

The proportion of carbon to nitrogen present in a compost pile largely determines the pace at which the microbial activities proceed, with a ratio of 30 carbon to one nitrogen being optimal. Microbial growth stops if the proportion of carbon gets much greater, and nitrogen is lost in the form of ammonia if the ratio drops lower.

If the materials are kept moist and turned occasionally to aerate the pile, composting may be completed in as little as 2 weeks. Shredding the materials exposes much more surface area for the microorganisms to work on and speeds up the process. Composters generally avoid or keep to a minimum a few materials, such as *Eucalyptus* and walnut leaves, which contain substances that inhibit the growth of other plants, when the compost is intended for use as a soil conditioner or crop fertilizer. Bamboo and some types of fern fronds are particularly resistant to decay bacteria and will decompose much more slowly than other materials. Domestic cat manure may contain stages of a parasite that can infect humans, although the organism should be killed if proper composting techniques are employed. Generally, however, almost any organic materials are suitable. Although compost adds humus to the soil and improves soil structure and moisture-holding capacity, its value as

KEY THEME 17.1 ecology — The Social Life of Prokaryotes

We tend to think of bacteria as simple organisms and don't expect them to have a complex social structure. However, we are learning that some bacteria are capable of communicating with each other, forming a mass of cells that act as a coordinated unit. One example is the production of biofilms. Bacteria can form a complex arrangement of cells embedded in extracellular secretions, forming a film in soils or on plant surfaces. The matrix in which the bacteria are embedded provides protection from environmental hazards. In human medicine, biofilms may be found in patients with chronic bacterial infections or on the surfaces of implants. These biofilms are a problem because it is more difficult to kill bacteria embedded in biofilms. Dental plaque is another example of a biofilm (Box Figure 17.1A). A second social activity carried out by some bacteria is quorum sensing. Bacteria secrete signal molecules into their environment and can determine, by monitoring the external concentration of these molecules, whether others are present in high density. The presence of a sufficient concentration of cells induces the bacteria to function as a colony and perform a task that contributes to the survival of the species. Myxobacteria, for example, can aggregate to form fruiting bodies when nutrients become scarce. Cells within the fruiting body will differentiate into myxospores, which may survive until conditions become suitable for growth again. Other examples of quorum sensing are found in bacteria that infect plants as pathogens and those that establish symbiotic relationships with roots.

Box Figure 17.1A Dental plaque containing the bacterium *Streptococcus mutans* (green) and an extracellular matrix (blue).
© STEVE GSCHMEISSNER/SPL/Getty Images

a fertilizer has certain limitations. The nitrogen content of good compost is about 2% to 3%, phosphorus 0.5% to 1%, and potash 1% to 2%, as compared with five to 10 times those amounts in chemical fertilizers, which also require less labor to produce. Nevertheless, with the spiraling costs of chemical fertilizers, an increasing awareness of the problems accompanying their use, and greater public enlightenment concerning the accumulation and disposal of solid wastes, compost is undoubtedly destined to play a major role in pertinent agricultural practices of the future.

True Bacteria and Disease

It has been calculated that plant diseases alone cause American farmers losses of more than $4 billion per year. Although many plant diseases are caused by fungi, a number involve bacteria. Bacteria are involved in diseases of pears, potatoes, tomatoes, squash, melons, carrots, citrus, cabbage, and cotton. Bacteria also cause huge food losses after they have been harvested or processed. They are even better known, however, as the culprits in many serious diseases of animals and humans (Table 17.2).

Because they are so tiny and often incapable of independent movement, how can bacteria infect and kill animals billions of times their size? We have learned much about their activities in the past century and know now that they can gain access to human tissues in a variety of ways.

Modes of Access of Disease Bacteria

Access from the Air Every time a person coughs, sneezes, or just speaks, an invisible cloud of tiny saliva droplets is produced (Fig. 17.6). Each droplet contains bacteria along with other microorganisms and a tiny amount of protein. The moisture usually evaporates almost immediately, leaving minute protein flakes to which live bacteria adhere. As normal breathing takes place, these may soon find their way into the respiratory tracts of other humans or animals, particularly if they have been confined within the air of a room. Fortunately, the natural resistance of most of those who acquire the bacteria in this way prevents the bacteria from multiplying to the point of causing a disease. When resistance is not there, however, a number of diseases, including diphtheria, whooping cough, some forms of meningitis and pneumonia, and strep throat, can develop. Some bacteria, such as anthrax, require inhalation of a minimum of about 8,000 spores for the disease to develop internally in humans.

Another serious disease spread through the air is tuberculosis (TB), caused by *Mycobacterium tuberculosis*. It typically affects the lungs and can lead to chest pain and coughing up blood. One in four people in the world is infected with TB bacteria. A person carrying the bacterium does not become ill, though, unless that person's immune system is compromised by, for example, HIV,

Table 17.2 Some Diseases of Humans Caused by Bacteria or Viruses

BACTERIAL DISEASES		VIRAL DISEASES	
Anthrax	Mastitis	AIDS	Meningitis (some)
Botulism	Meningitis (some)	Bird flu	Mononucleosis
Brucellosis	Pneumonia (some)	Chicken pox/shingles	Mumps
Bubonic plague	Strep throat	Common colds	Pneumonia (some)
Cholera	Syphilis	Diabetes (some)	Poliomyelitis
Diphtheria	Tetanus	Encephalitis	Rabies
Dysentery (some)	Toxic shock syndrome	Genital herpes	Rubella
Gonorrhea	Tuberculosis	Infectious hepatitis	SARS
Leprosy	Typhoid fever	Influenza	Smallpox
Lyme disease	Whooping cough	Measles	Yellow fever

Figure 17.6 Stop-action photograph of a sneeze. © *Custom Medical Stock Photo/Alamy Stock Photo*

other illnesses, or advancing age. Worldwide, TB kills nearly 2 million people each year. The TB infection rate in the United States has been falling for decades. In 2016, it hit a low of 9,272 cases, or nearly 3 cases per 100,000 Americans. However, the misuse of antibiotics has produced some bacterial strains with resistance to these drugs. Multi-drug-resistant TB strains can withstand the antibiotics isoniazid and rifampin. Extensively drug-resistant TB strains (XDR-TB) resist most anti-TB drugs. Health officials are concerned that effective treatment options may not be available for future epidemics. They are researching ways to properly manage TB and develop new tools to prevent, treat, and diagnose the disease.

In 2007, a tuberculosis scare occurred when a man from Atlanta, Georgia, flew to Paris, France, and then returned on a flight from Prague, Czech Republic, to Montreal, Canada, while infected with multi-drug-resistant tuberculosis. The incident sparked a debate in Congress on the failure of federal customs agents to stop him. Upon his return to the United States, the Centers for Disease Control and Prevention (CDC) placed him under involuntary isolation. He became the first individual subjected to CDC isolation since 1963. This incident illustrates both the seriousness with which health officials regard infectious diseases such as tuberculosis and the potential ease with which these diseases can be spread globally.

Access through Contamination of Food and Drink
In the past, open sewers and unsanitary conditions for food preparation caused a number of bacterial diseases, including cholera, dysentery, *Staphylococcus* and *Salmonella* food poisoning, and typhoid and paratyphoid fevers, to reach epidemic proportions all over the world. Although *Staphylococcus* food poisoning, which is seldom fatal, is still fairly common today, developed countries now rarely see epidemics caused by other bacteria unless a natural disaster, such as a flood or a typhoon, disrupts normal sewage disposal. For example, after Hurricane Maria hit Puerto Rico in 2017, dozens of people died of leptospirosis. This bacterial disease can be spread to humans from animals through contaminated water. The diseases are more often spread by carriers who handle food, or by houseflies.

The United States and other countries with ocean shorelines ban the harvesting of shellfish at certain times because *Salmonella* bacteria may multiply enough in clams and mussels to cause illness in humans who eat them—particularly if they are not well cooked. The *Salmonella* bacteria multiply in the intestinal tract or spread from there to other parts of the body. In 2017, an outbreak of salmonellosis caused over 170 cases of food poisoning in Virginia. The source of contamination was one chowder in a large chili and chowder cook-off. Large, temporary events such as this may not provide sufficient safety oversight during food preparation.

An illness often referred to as C. diff., which is caused by *Clostridium difficile*, is becoming increasingly common, especially in older patients in hospitals and long-term care facilities. The bacterium causes diarrhea and, in severe cases, life-threatening colon inflammation. It occurs most often after an antibiotic treatment, which kills the beneficial bacteria normally found in the gut. This allows a population explosion of *C. difficile* to occur.

Some bacteria are very common on algae in freshwater streams, lakes, and reservoirs. One such bacterium, *Legionella pneumophila,* causes *Legionnaire disease,* which killed 34 members of the American Legion attending a convention in a Philadelphia hotel in 1976. An estimated 50,000 Americans are infected annually by Legionnaire disease bacteria, which nearly always pass through the human digestive tract harmlessly without multiplying. On rare occasions, however, something unknown triggers their reproduction, mostly in older males who are heavy smokers or alcoholics. The results are often fatal.

The most deadly of all known biological toxins is produced by a bacterium with the scientific name of *Clostridium botulinum.* The name comes from *botulus,* the Latin word for "sausage," because the bacterium was first discovered after people had died from eating some contaminated sausage at a picnic.

Unlike *Salmonella* food poisoning, *botulism* is not an infection but is poisoning from a substance produced by bacteria that can grow and multiply anaerobically (in the absence of oxygen) in improperly processed or stored foods. Home-canned beans, beets, corn, and asparagus, in particular, have been known to permit the development of botulism bacteria. Just 1 gram (0.035 ounce) of the toxin is enough to kill 14 million adults, and a little more than half a kilogram (1.1 pound) could eliminate the entire human race.

The bacteria, which are present in most soils, produce unusually heat-resistant spores and are likely to be present on any soil-contaminated foods. They may not be destroyed during canning unless the food is heated for 30 minutes at 80°C (176°F) or boiled for 10 to 15 minutes. They ordinarily will not grow in foods that have been preserved in brines containing at least 10% salt (sodium chloride) or in fairly acid media, such as the juices produced by most stone fruits. The toxin is absorbed directly from the stomach and the intestines, affecting nerves and muscles and causing paralysis. While some antidotes are available, they are ineffective after symptoms have become advanced.

There is evidence that some deaths of infants due to unknown causes are actually due to botulism. For reasons that are not clear, botulism bacteria, which pass harmlessly through the digestive tracts of humans over the age of 1 year, may germinate in the intestines of infants less than a year old. About 100 cases of infant botulism occur each year in the United States. Spores from the bacterium are found in honey, so honey should not be given to infants. Botulism spores are also found in water, soil, and household dust.

Escherichia coli is a bacterium that lives harmlessly in the digestive tracts of mammals. However, some strains,

Figure 17.7 Scanning electron micrograph of *Escherichia coli*, strain O157:H7. ca. ×10,000. © Science Photo Library/Alamy Stock Photo

such as *E. coli* O157:H7 (Fig. 17.7) cause foodborne illness. Infection often leads to bloody diarrhea, and occasionally to kidney failure, especially in young children and elderly people. Most reported illnesses have resulted from eating undercooked ground beef, drinking unpasteurized milk, drinking contaminated water, and eating contaminated vegetables. In 2018, 17 people in four states were sickened and one died after eating *E. coli* O26–tainted beef. The meat packing company recalled potentially contaminated meat, and the outbreak was over two months after the illnesses were reported. Also in 2018, contaminated romaine lettuce caused five deaths and illness in 210 people in the United States. The outbreak was caused by *E. coli* O157:H7 with resistance to several antibiotics. The increasing frequency of *E. coli*–tainted fruits and vegetables may be due in part to the use of irrigation water contaminated by the bacterium.

Access through Direct Contact Bacteria responsible for diseases such as syphilis, gonorrhea, chlamydia, anthrax, and brucellosis enter the body through the skin or mucous membranes (i.e., those membranes lining tracts with openings to the exterior of the body). *Syphilis, gonorrhea,* and *chlamydia* are transmitted through sexual intercourse or other forms of direct contact, rarely through the use of public washroom towels or toilets. The symptoms of sexually transmitted diseases, which include persistent sores or discharges from the genitalia, sometimes disappear after a few weeks, leading victims to believe the body has healed itself. It is estimated that about 50% of females infected with gonorrhea at first exhibit no symptoms at all. More often than not, however, symptoms appear or reappear in different parts of the body at a later date. Because gonorrheal eye infections often occur in newborns as they pass through an infected birth tract, the United States, by law, requires that a drop of antibiotic or silver nitrate solution be placed in the eyes of infants at birth. Both syphilis and gonorrhea are curable when treated promptly, but it is very important that such treatment be sought, because failure to do so can lead to sterility, blindness, and even death.

Anthrax, which is primarily a disease of cattle and other farm animals in addition to wild animals, is sometimes transmitted to humans, particularly to workers in the wool and hide industries. Since September 2001, anthrax has received much publicity as an agent of bioterrorism. Like syphilis and gonorrhea, it can be effectively eliminated if treated early enough, but may be fatal if allowed to progress. Brucellosis, another disease of farm animals, is occasionally transmitted to humans through direct contact or through the consumption of contaminated milk. It is sometimes called *undulant fever* because of a daily rise and fall of temperature apparently associated with the release of toxins by the bacteria.

Access through Wounds When one steps on a dirty nail or is wounded in such a way that dirt is forced into body tissues, *tetanus* ("lockjaw") bacteria, which are common soil organisms, may enter dead or damaged cells. There they can multiply and produce a deadly toxin so powerful that 0.00025 gram (0.00000088175 ounce) is enough to kill an adult. For comparison, about 150 times that amount of strychnine is needed to achieve the same result. Control of tetanus through immunization is very effective and has become widespread. About 30 cases of tetanus are reported in the United States each year. Most victims are people who have not received a tetanus vaccine or who have not stayed up to date on their 10-year booster shots. Several related bacteria that gain access to the body in the same way are responsible for potentially fatal *gas gangrene,* which was feared on battlefields in the past but is now controlled through the use of antibiotics and aseptic techniques.

Access through Bites of Insects and Other Organisms
Bubonic plague (the "Black Death") and tularemia are two bacterial diseases transmitted by fleas, deerflies, ticks, or lice that have parasitized infected animals, particularly rodents.

Rat fleas, found on infected rats that inhabit dumps, sewers, barnyards, and ships acquire the plague bacteria and then pass them on to humans through their bites. The disease has been found in ground squirrels and other rodents in the United States, particularly in the West, since 1900.

In the past, bubonic plague spread with great speed and reached devastating epidemic proportions. In 1665, in London, hundreds of thousands perished from the disease, and between 1347 and 1349, it is believed to have killed one-fourth of the entire population of Europe (about 25 million persons). Although the plague is rare in North America, it can be deadly. In 2015, an adult and a teenager in Colorado died after contracting the disease. Control depends on control of rodents and fleas, which are virtually impossible to

eradicate entirely, and the use of vaccines, which produce immunity for about 6 to 12 months.

Tularemia is primarily a disease of animals, but infected ticks or deerflies may transmit the disease to humans through bites. It is an occupational disease of meat handlers and is fatal in 5% to 8% of the cases. Ticks, lice, and fleas may also transmit *rickettsias,* which cause typhus and spotted fevers. Rickettsias are extremely tiny bacteria that live within eukaryotic cells.

Mycoplasmas, referred to in the past as pleuropneumonialike organisms (PPLOs), are also minute bacteria that may be transmitted in various ways. These have no cell walls and therefore are quite variable in shape. They are found in many plants, in hot springs, and in the moist surfaces of the respiratory and intestinal tracts of animals and humans. They are responsible for a form of human pneumonia and for numerous plant diseases. Like an increasing number of bacteria, they are resistant to penicillin.

Lyme disease, which was known in Europe before 1900, has spread rapidly throughout the United States since 1975, when an outbreak occurred at Lyme, Connecticut. Its arthritis-like effects are caused by a bacterium (*Borrelia burgdorferi*) hosted by deer and field mice. It is injected into the human bloodstream by deer ticks.

Koch's Postulates

Because there are so many bacteria present everywhere, how can we be certain that a given bacterium obtained from an infected person is actually the organism responsible for the observed disease? Robert Koch, who was a German physician, became known during the latter half of the 19th century for his investigations of anthrax and tuberculosis. As a result of his work, Koch formulated rules for proving that a particular microorganism is the cause of a particular disease. His rules, with minor modifications, are still followed today. They have come to be known as *Koch's Postulates,* and their essence is as follows:

1. The microorganism must be present in all cases of the disease.
2. The microorganism must be isolated from the victim in pure culture (i.e., in a culture containing only that organism).
3. When the microorganism from the pure culture is injected into a susceptible host organism, it must produce the disease in the host.
4. The microorganism must be isolated from the experimentally infected host and grown in pure culture for comparison with that of the original culture.

True Bacteria Useful to Humans

For many years, we controlled insect pests of food plants mostly through the use of toxic sprays. Residues of the sprays remaining on the fruits and vegetables have accumulated in human tissues, often with adverse effects, while at the same time, many organisms have become immune or resistant to the toxins. In addition, the sprays kill useful organisms, and precipitation runoff washes the toxins into streams, lakes, and oceans, harming or killing aquatic organisms.

As we have become aware of the undesirable effects of the use of toxic sprays, we have looked for alternative means of controlling crop pests. Today, many harmful pests and even weeds can be significantly limited through the use of biological controls, which are discussed in Appendix 2, and with transgenic plants, discussed in Chapter 14.

Bacillus thuringiensis and Bacillus popilliae

The bacterium *Bacillus thuringiensis* (often referred to as Bt), has been remarkably effective as a biological control agent against a wide range of caterpillars and worms, including peach tree borers, European corn borers, bollworms, cabbage worms and loopers, tomato and fruit hornworms (Fig. 17.8), tent caterpillars, fall webworms, leaf miners, alfalfa caterpillars, leaf rollers, gypsy moth larvae, and cankerworms.

The bacteria, which are easily mass-produced by commercial companies, are sold in the form of a stable, wettable dust containing millions of spores. When the spores are sprayed on food plants, they are harmless to humans, birds, animals, earthworms, or any living creatures other than moth or butterfly larvae. When a caterpillar ingests any tissue with Bt spores on it, the bacteria quickly become active and multiply within the digestive tract, soon paralyzing the gut. The caterpillar stops feeding within 2 or 3 hours and slowly turns black, dropping off the plant in 2 to 4 days.

The toxin-producing gene from *Bacillus thuringiensis* was introduced into another bacterium, *Pseudomonas fluorescens,* which is used on corn to control black cutworms. The gene has also been introduced into crop plants through genetic engineering (see Chapter 14). These transgenic plants produce their own Bt toxin to protect against insect damage. This strategy is especially useful for controlling insects that live inside plants, such as cutworms, because they are not affected by insecticide sprays.

In 1983, a variety of *Bacillus thuringiensis* (var. *israelensis*) was introduced into the commercial market for the control of mosquitoes. Called BtI, the bacterium attacks only mosquito larvae ("wigglers") and one or two other pests. Six years of experiments performed on more than 70 species of fish, snails, shrimp, and insects demonstrated no adverse effects on either plants or animals other than mosquitoes (and a couple of lesser pests), even at dosages 100 times more powerful than those needed to kill mosquito larvae. BtI can be purchased as "dunks," which look like compressed donuts. The dunks are placed in water sources, such as bird baths and animal watering troughs, where the bacterium is released and kills mosquito larvae developing in the water.

Another bacterium, *Bacillus popilliae,* is also marketed in a powder form. When it is applied to soil where grubs of the highly destructive Japanese beetle are present, it causes what is known as "milky spore disease" in the grubs, which die in a few days. The spores are carried throughout the topsoil by rain water, by foraging grubs, and by organisms that feed on the grubs.

Bioremediation

Naturally occurring bacteria have shown much potential in the developing science of *bioremediation,* which involves the study of the use of living organisms in the cleanup of toxic wastes and pollution. One bacterium produces enzymes that break down nitroglycerin and trinitrotoluene, which are contaminants in the soil around some munitions factories and explosive sites.

Preliminary tests indicate that the bacteria can decompose such waste and contaminant materials into harmless residues within 6 months. Other bacteria, such as *Pseudomonas cepacia,* can perform similar feats in oil spills and chemical dumps containing degreasers, such as trichloroethylene (TCE), creosote, and even 2,4,5-T, the Agent Orange defoliant of Vietnam fame. The pollution-fighting bacteria may, however, need a few nutrients added to the contaminants to stimulate their activities. Some scientists believe bacteria may also eventually be used to break down nuclear wastes. With the probability of the development of bacteria specifically engineered (see "Gene Splicing and Transgenic Plants" in Chapter 14) to deal with a host of pollution problems, their use for this purpose may become widespread in the future.

(a)

(b)

Figure 17.8 A tomato hornworm (a) before and (b) 3 days after spraying with *Bacillus thuringiensis.* (a-b) © Kingsley Stern

Other Useful Bacteria

Human eyes contain rhodopsin, a protein that is so sensitive it reacts to light in less than one millionth of a second. Certain bacteria that contain a form of rhodopsin have proved to be invaluable in research on the chemistry of vision and have led to our understanding of how eyes convert light energy into vision.

Bacteria play a major role in the dairy industry. Milk, which is composed of proteins, carbohydrates, fats, minerals, vitamins, and about 87% water, has exceptional nutritive value for animals and is an excellent medium for the growth of many kinds of bacteria. Milk is sterile when secreted within the udder of the cow, but it picks up bacteria as it leaves the cow's body. It spoils rapidly if it is not obtained and stored under strictly sanitary conditions. Even after it has been pasteurized and refrigerated, the number of bacteria in it will increase the longer it stands. If milk is left in open containers in household refrigerators, for example, bacterial growth will not be kept in check for much longer than 24 hours.

Except for milk itself, either alone or in mixtures (e.g., ice cream), all dairy products are manufactured from raw or pasteurized milk by the controlled introduction of various bacteria. Such products include buttermilk, acidophilus milk, yogurt, sour cream, kefir, and cheese. Whey, the watery part of the milk separated from curd during cheese-making, is used, along with starches and molasses, for the commercial production of lactic acid. Lactic acid is used extensively in the manufacture of textile and laundry products, in the leather tanning industry, as a solvent in lacquers, and in the treatment of calcium and iron deficiencies in humans.

Beneficial bacteria, such as *Lactobacillus acidophilus* (a common organism in healthy digestive tracts), aid in digestion, reduce the risk of cancer, and may even reduce cholesterol levels (Fig. 17.9). In recent years, probiotics have received attention for their health promoting activities. These bacteria promote gut and immune system health. Fermented foods such a yogurt, kefir, and kimchee are good sources of probiotics. In another example of beneficial bacteria, acidophilus bacteria have been widely used, along with antibiotics, to control or eliminate human female yeast infections.

In their metabolism of various sugars, proteins, and other organic substances, bacteria also produce waste products that have important industrial uses. Such products are often produced in large quantities by culturing bacteria in huge vats. These products include acetone; explosives; solvents (e.g., nail polish remover); butyl alcohol, used in the manufacture of synthetic lacquers; dextran, used as a food stabilizer and as a blood plasma substitute; sorbose, used in the manufacture of ascorbic acid (vitamin C); and citric acid, which is the principal citruslike, tart flavoring of soft drinks, candies, and other foods. Some vitamin and medicinal preparations also involve bacterial synthesis.

Bacteria are used in the curing of vanilla pods, cocoa beans, coffee, and black tea and in the production of vinegar, sauerkraut, and dill pickles. Fibers for linen cloth are separated from flax stems by bacteria, and green plant materials are fermented in silos to produce silage for cattle feed. In recent years, the production of several important amino acids by bacteria has been exploited commercially.

Brevibacterium, the genus of bacteria responsible for the odor of Limburger and Brie cheeses, also is the source of foot odor in certain people. Bactericides are being tested to improve foot pads that merely absorb odor.

17.5 Class Cyanobacteriae– the Cyanobacteria (Blue-Green Bacteria)

Learning Outcomes

a. Outline the main differences between bacteria and cyanobacteria.

b. Describe where cyanobacteria are found.

c. Explain the significance of color variation among the cyanobacteria.

d. Discuss the proposed connection between cyanobacteria and chloroplasts.

e. Describe ways in which cyanobacteria impact humans.

Introduction

In the past, algae have been distinguished from other organisms in being photosynthetic, in having relatively simple structures, and, for those reproducing sexually, in having sex structures consisting of a single cell. As more became known about cellular details, however, it became apparent that the differences between the *cyanobacteria* (*blue-green bacteria*) (formerly known as the blue-green algae) and true algae are quite basic. Like all other members of Kingdom Bacteria, cyanobacteria have prokaryotic cells, while all

Figure 17.9 *Lactobacillus,* or lactic acid bacteria, are part of the normal flora of the human intestine and are used as probiotics. ×2,000. © Kateryna Kon/Shutterstock

algae that are assigned to Kingdom Protista have eukaryotic cells. Cyanobacteria, in fact, are so much like other true bacteria that biologists have abandoned the reference to them as algae and consider them true bacteria.

The main distinctions between organisms traditionally regarded as bacteria and cyanobacteria are (1) cyanobacteria have chlorophyll *a*, which is found in higher plants, and oxygen is produced when they undergo photosynthesis; (2) cyanobacteria also have blue *phycocyanin* and red *phycoerythrin* pigments known as *phycobilins;* (3) cyanobacteria are the only organisms that can both fix nitrogen and produce oxygen—a paradox, because nitrogen fixation is essentially an anaerobic process. Except for the *prochlorobacteria,* other bacteria capable of carrying out photosynthesis do not produce oxygen, and they do not have chlorophyll *a.*

Distribution

Cyanobacteria are found in almost as diverse a variety of habitats as other true bacteria. They are common in temporary pools or ditches, particularly if the water is polluted. They are not found in acidic waters, but they are abundant in other fresh and marine waters around the globe, from the frozen lakes of Antarctica to warm tropical seas. In the open oceans, cyanobacteria are the principal photosynthetic organisms in plankton, the tiny cells of the cyanobacterium *Synechococcus. Trichodesmium* is a marine, filamentous, nitrogen-fixing cyanobacterium that forms extensive mucilage-producing colonies in some tropical waters. Other bacteria multiply in the mucilage and become food for protozoa.

A different species of cyanobacterium is found in each temperature range of the hot springs of Yellowstone National Park, where water temperatures approach 85°C (185°F) (Fig. 17.10). There the bacteria precipitate chalky, insoluble carbonate deposits, which become a rocklike substance called *travertine.* The deposits accumulate at the rate of up to 2 to 4 millimeters per week, with other cyanobacteria often forming brilliantly colored streaks in the travertine.

Cyanobacteria are often the first photosynthetic organisms to appear on bare lava after a volcanic eruption, and they thrive in the tiny fissures of desert rocks. Some are found in jungle soils or on the shells of turtles and snails, while others live symbiotically in various types of organisms, including amoebae and other protozoans, diatoms, certain sea anemones and their relatives, some fungi, and the roots of tropical palmlike plants called *cycads.* They also flourish in tiny pools of water formed at the bases of the leaves of tropical grasses and other plants, and they are well-known components of "compound" organisms called *lichens,* which consist of a fungus and a photosynthetic partner.

Form, Metabolism, and Reproduction

The cells of cyanobacteria often occur in chains or in hairlike filaments, which are sometimes branched (Fig. 17.11). Several species occur in irregular, spherical, or platelike colonies, the individual cells being held together by the gelatinous sheaths they secrete. The sheaths may be colorless or pigmented with various shades of yellow, red, brown, green, blue, violet, or blue-black, which makes some colonies quite striking in appearance.

The cells themselves appear blue-green in color in about half of the approximately 2,700 known species. This color, which is due to the presence of green chlorophyll *a* and blue phycocyanin, is often masked by the presence of other pigments. Several yellow or orange carotenoid pigments similar to those of higher plants are usually present, and varying amounts of phycoerythrin (a red phycobilin pigment in a form unique to the cyanobacteria) may give the cells a distinct red color. The periodic appearances of large numbers of cyanobacteria with considerable amounts of phycoerythrin are believed to have given the Red Sea its name.

Cyanobacteria produce a nitrogenous food reserve called *cyanophycin.* The production of such food reserves is atypical for prokaryotic organisms. Cyanobacteria also produce and store carbohydrates and lipids. Flagella are unknown in the cyanobacteria, but some of these organisms are nevertheless capable of movement. *Oscillatoria* filaments, for example, seem to rotate on axes and move in a gliding fashion, apparently by the twisting of minute fibrils inside the cell walls while secreting mucilage that reduces friction. New cells are formed through fission, while new colonies or filaments may arise through fragmentation (breaking up) of older ones.

In the common genera *Nostoc* and *Anabaena,* which form chains of cells, fragmentation often occurs at special, larger, colorless, nitrogen-fixing cells called **heterocysts,** which are produced at intervals in the chains. Members of these two genera also may produce thick-walled cells called *akinetes,* which can resist freezing and other adverse conditions. When favorable conditions return, this survival feature enables the cells to germinate and become new

Figure 17.10 Bacteria growing in hot springs at Yellowstone National Park. Species differ in color and are found at different optimal temperatures. © Robert Glusic/Getty Images

Cyanobacteria, Chloroplasts, and Oxygen

Cyanobacteria that occur symbiotically in other organisms commonly lack a cell wall and appear to function essentially as chloroplasts. When a eukaryotic cell containing chloroplasts divides, all the chloroplasts divide at the same time. The cells of cyanobacteria occurring within the cells of other organisms divide in similar fashion, leading to speculation that chloroplasts originated as cyanobacteria or *prochlorobacteria* (discussed after the cyanobacteria), living within other cells.

Fossils of photosynthetic organisms believed to be 3.5 billion years old and closely resembling present-day cyanobacteria have been found in Australia. It was not until half a billion years later, about 3 billion years ago, that these organisms began producing oxygen as a by-product of photosynthesis. The oxygen slowly began to accumulate, becoming substantial about 1 billion years ago. At the same time that the oxygen was accumulating, other photosynthetic organisms appeared, and forms of aerobic respiration developed.

Within the last half billion years, enough *ozone,* which is a by-product of oxygen, accumulated to become an effective shield against most of the harmful ultraviolet radiation coming from the sun. Photosynthetic organisms, which had been protected from the radiation by their watery environments, were then able to survive on land.[1] Accordingly, cyanobacteria appear to have played a fundamental role in almost the entire history of living organisms.

Human Relevance of the Cyanobacteria

Cyanobacteria are included among the many aquatic and photosynthetic organisms at the bottom of many food chains. They store energy in the sugar they produce through photosynthesis. Small fish and crustaceans feed on them, only to be eaten by larger fish, which then become food for other aquatic organisms or humans.

During warmer months, algae and cyanobacteria may become abundant in bodies of fresh water, especially if the water is nutrient-rich. A floating scum or mat called a *bloom* is most likely to develop in late summer (Fig. 17.12). Cyanobacteria, in particular, tend to become abundant in blooms when agricultural runoffs are high in nitrogen and phosphorus. The water in the vicinity of a bloom often acquires a "fishy" or otherwise objectionable odor or taste.

Anabaena, Microcystis, and other genera of cyanobacteria produce toxic substances that can kill both domestic and wild animals. Cattle, hogs, horses, sheep, rabbits, dogs, and even poultry have been poisoned through drinking

Figure 17.11 Examples of cyanobacteria include (a) *Spirulina,* (b) *Gloeocapsa* (×500), and (c) *Anabaena* (×400). *(a)* © MELBA PHOTO AGENCY/Alamy Stock Photo; *(b)* © McGraw-Hill Education/Al Telser; *(c)* © Stephen Durr

chains or filaments. Some akinetes have been known to germinate after lying dormant for more than 80 years.

Cyanobacteria do not produce gametes or zygotes and do not undergo meiosis. Extensive genetic recombination has, however, been reported in a marine cyanobacterium. The method of recombination has yet to be elucidated.

---

1. See also the discussion in Chapter 18 of the role of certain green algae in the transition of aquatic organisms to land, and the discussion in Chapter 25 of the effects of ozone depletion on our environment.

KEY THEME 17.2 ecology — Cyanobacteria and Algae Blooms

Many Chammoro villagers living in Guam suffer from a disease with symptoms similar to dementia and Parkinson's. Upon closer examination, they have been found to have brain tangles and accumulations of amyloid deposits. A recent study found that these symptoms were caused by the consumption of flour from the roots of cycads that grow on the island. It turns out that cyanobacteria live in these roots and are consumed by the islanders who eat the flour. The cyanobacteria produce a toxin that causes brain damage. You may think that you will never be at risk of poisoning by cyanobacteria because you do not consume cycad flour from Guam. However, if you ever go for a swim in a lake or pond with a green paint-like scum growing in it (Box Figure 17.2A), you may be exposed. These blue-green algae blooms are common during the warm days of summer and can grow rapidly. Swimming in or around these blooms can cause skin irritation, and swallowing the water can lead to vomiting, diarrhea, and muscle pain. Dogs are especially vulnerable as they wade along the shoreline and lap up the lake water. They can die within an hour of drinking contaminated water. The short-term solution to poisoning by cyanobacteria is to monitor lakes in the summer and post warnings. In the long term, efforts to reduce the introduction of excess nutrients such as phosphorus into lakes can be effective in preventing blooms.

Box Figure 17.2A An algae bloom on a pond. © *Aaron Roeth Photography*

Figure 17.12 Aerial view of an algae bloom in Lake Erie.
Source: NASA Earth Observatory images by Joshua Stevens, using Landsat data from the U.S. Geological Survey. Caption by Kathryn Hansen.

water that had an algal-bacterial bloom. Many poisonous fish that are immune to the toxins of certain cyanobacteria become poisonous to their predators only after feeding on the bacteria.

While the cyanobacteria are alive, they produce oxygen, and the oxygen content of the water is temporarily increased. Later, however, decay bacteria decomposing the bodies of organisms that have died may deplete the available oxygen so much that fish and other organisms are killed.

Cyanobacteria can be attacked by viruses called *cyanophages*, and there has been considerable interest in using the viruses to control blooms in lakes and ponds. Many of the cyanobacteria responsible for the blooms, however, have thick, mucilaginous sheaths that may prevent viruses (and, incidentally, fungi) from penetrating the cells. Blooms that develop in calm ocean waters in the tropics can be even more extensive than those of fresh waters. A marine bloom once reported between the shore and the Great Barrier Reef of Australia was 1,600 kilometers (almost 1,000 miles) long and covered 52,000 square kilometers (20,000 square miles) of ocean.

Most cyanobacteria are not palatable to humans, but there are exceptions. Species of *Spirulina* have been used for food by the natives of the Lake Chad region of central Africa and areas around Mexico City for centuries. *Spirulina*, which has a significant vitamin content, is now cultured commercially and sold in health-food stores. The Japanese use several cyanobacteria, including two species of *Nostoc*, as side dishes, and the Chinese treat *Nostoc commune* as a delicacy. A few colorless forms are mild parasites in humans and animals.

Some strains of certain *Nostoc* species produce antibiotics that kill related strains of the same species. *Scytonema hofmannii* and other cyanobacteria produce antibiotics called *cyanobacterins* that kill many different forms of both cyanobacteria and eukaryotic algae. These cyanobacteria undoubtedly play a role in their own survival by inhibiting the growth of competing organisms.

Swimmers in Hawaii occasionally suffer from "swimmers' itch," a severe skin inflammation that is caused by a species of *Lyngbya* that sometimes becomes abundant. Ironically, the toxin produced by these organisms has been demonstrated to suppress leukemia and several other types of cancer.

In human water supplies, cyanobacteria frequently clog filters, corrode steel and concrete, cause natural softening of water, and produce undesirable odors or coloration in the water. Many communities control cyanobacteria in reservoirs through the addition of very dilute amounts of copper sulfate.

More than 40 species of cyanobacteria are known to fix nitrogen from the air at roughly the same rates as the nitrogen-fixing bacteria of the leguminous plants discussed in Chapter 25. They may be more important than originally thought in this regard. In Southeast Asia, so much usable nitrogen is produced in the rice fields by naturally occurring cyanobacteria that rice is often grown for many years on the same land without the addition of fertilizer.

17.6 Class Prochlorobacteriae– the Prochlorobacteria

Learning Outcome

a. Describe features of prochlorobacteria.

In 1976, Ralph A. Lewin of the Scripps Institute of Oceanography announced the discovery of unicellular, prokaryotic organisms with bright green cells that were living on marine animals called *sea squirts* found in shallow marine waters of Baja California. These organisms, which were given the name *Prochloron,* have the chlorophylls *a* and *b* of higher plants but no trace of the phycobilin accessory pigments associated with cyanobacteria. Instead, their accessory pigments were confined to the carotenoid pigments found in higher plants. Also, unlike the single membranes of cyanobacteria, those of the new organisms are double.

Lewin considered the pigment differences between cyanobacteria and the bright green cells of *Prochloron* to be basic enough to warrant recognition at the division level, and he proposed a new division to be known as the *Prochlorophyta.* Many microbiologists are reluctant to recognize these organisms as belonging to a separate bacterial division because the prokaryotic cell structure and chemistry are similar to those of cyanobacteria and other true bacteria (Fig. 17.13), but others agree with Lewin's assessment of the significance of the pigment system. While the pigment system is indeed significant, they are treated here as a class of true bacteria because their remaining structure and features are essentially indistinguishable from those of other true bacteria.

In 1984, Dutch biologists discovered a similar organism, which they named *Prochlorothrix,* in shallow lakes in the Netherlands. It differs from *Prochloron* in being free-living and filamentous. In the late 1980s, Sally Chisholm of the Massachusetts Institute of Technology identified yet another marine prochlorobacterium that flourishes in dim light at a depth of about 100 meters (328 feet). This organism now appears to be one of the two most numerous bacteria living in ocean waters.

The discovery of prochlorobacteria adds weight to the theory that chloroplasts may have originated from such cells living

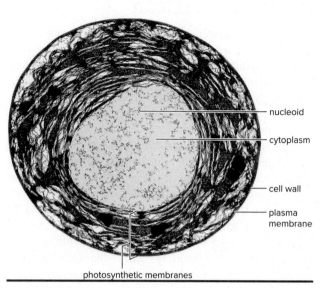

Figure 17.13 Colorized transmission electron micrograph of *Prochloron*. Note the absence of a nucleus and other organelles, and the concentric layers of photosynthetic membranes shown in green. ca. ×12,500. © *T.D. Pugh & E.H. Newcomb, University of Wisconsin-Madison*

within the cells of other organisms, especially because the pigments involved are identical with those of higher plants.

17.7 Domain (Kingdom) Archaea–the Archaebacteria

Learning Outcomes

a. List the features that differ between true bacteria and archaebacteria.
b. Explain why methane bacteria are named as such.
c. Explain how salt bacteria can live in a salty environment.
d. Describe where sulfolobus bacteria are found.

The archaebacteria represent one of two quite distinct lines of the most primitive known living organisms. They are fundamentally different in their metabolism from the other line of bacteria, the true bacteria, and they differ in their cell wall composition, gene organization, and membrane structure.

In the early 1980s, these basic differences led University of Illinois microbiologist Carl Woese and his colleagues, who have conducted considerable research on the archaebacteria, to suggest that the organisms should be separated from the true bacteria in a kingdom of their own, a suggestion that was widely adopted. Subsequently, phylogenetic studies have revealed that the differences between true bacteria and archaebacteria are so fundamental that they should be placed in two superkingdoms called *domains.* Three distinct groups of bacteria are included in the archaebacteria.

The Methane Bacteria

The *methane bacteria,* which are the largest group of the archaebacteria, are killed by oxygen and are active only under anaerobic conditions found in swamps, ocean and lake sediments, hot springs, animal intestinal tracts, sewage treatment plants, and other areas not exposed to air. Their energy is derived from the generation of methane gas from carbon dioxide and hydrogen.

Methane, or "marsh gas," is a principal component of natural gas and may have been a major part of the earth's atmosphere in early geologic times. It still is present in the atmospheres of the planets Jupiter, Saturn, Uranus, and Neptune and is the main ingredient of firedamp, which causes serious explosions in mines. Methane will burn when it constitutes only 5% to 6% of the air, and a flitting, dancing light called *ignis fatuus,* or "will-o'-the-wisp," which is occasionally seen at night over swamps and marshy places, is said to be due to the spontaneous combustion of the gas.

The Salt Bacteria

Commercial salt evaporation ponds and other shallow areas in bodies of water with high salt content often have a unique appearance from above. They can be strikingly red due to the presence of a distinctive group of archaebacteria. These are the *salt bacteria,* whose metabolism enables them to thrive under conditions of extreme salinity that instantly kills other living cells. The bacteria carry out a simple form of photosynthesis with the aid of a membrane-bound red pigment called *bacterial rhodopsin.* The concentration of salt inside the cells is much lower than in their surroundings, but their metabolism is so closely tied to their environment that the bacteria die if placed in water with lower salt concentrations (Fig. 17.14).

The Sulfolobus Bacteria

The *sulfolobus bacteria* constitute a third group of archaebacteria whose members occur in sulfur hot springs. The extraordinary metabolism of these bacteria allows them to thrive at very high temperatures—mostly in the vicinity of 80°C (176°F), with some doing very well at only 10°C below the boiling point of water. One genus (*Pyrodictium*), discovered in superheated ocean-floor areas, has a minimum temperature requirement of 82°C (179°F) and an optimum growth temperature of 105°C (220°F), and it can tolerate 110°C (230°F). The environment of species in one order of these thermophilic bacteria is also exceptionally acidic, their hot-springs habitats often having a pH of less than 2 (the neutral point on the 14-point pH scale is 7). One genus (*Thermoplasma*) in this group is bounded by only a plasma membrane and has no cell wall. It is found only in the embers of coal tailings. Another genus (*Thermoproteus*) appears to be confined to the geothermal areas of Iceland.

Figure 17.14 The north end of Utah's Lake Bonneville, as seen from the air. The water has a very high salt content. The pinkish areas are due to a red pigment produced by salt bacteria. The pigment, called *bacterial rhodopsin,* is involved in a form of photosynthesis.
© Kingsley Stern

James Lake and his colleagues discovered that the shape of the ribosomes of sulfolobus bacteria is significantly different from those of other archaebacteria, true bacteria, and eukaryotes and that the chemistry of sulfur-dependent bacteria also distinguishes them from other archaebacteria. They have proposed a third kingdom of prokaryotes named *Eocytes* (dawn cells). Other microbiologists are hesitant to base kingdom status on ribosome shape, but the controversy over the significance of these discoveries will undoubtedly continue as further research into the matter is pursued.

Human Relevance of the Archaebacteria

Archaebacteria are significant to humans in several ways. In the future, methane bacteria may be used on a large scale to furnish energy for engine fuels and for heating, cooking, and light, because the methane gas they produce can be used to replace the methane in natural gas. Methane has an octane number of 130 and has been used as a fuel for cars in Italy for over 40 years. It is clean, nonpolluting, safe, and nontoxic, and it prolongs the life of automobile engines, also making them easier to start. The methane is given off by the bacteria as they "digest" organic wastes in the absence of oxygen. Nine kilograms (20 pounds) of horse manure or 4.5 kilograms (10 pounds) of pig manure fed daily into a methane digester will produce all the gas needed for the average American adult's cooking needs. Considerably less green plant material is required to produce the same amount of methane.

The digester basically consists of an airtight drum connected by a pipe to a storage tank with a means of drawing off the sludge left after the gas has been produced. The sludge itself makes an excellent fertilizer, although many sludges that originate from municipal and large-scale agricultural plants carry with them toxic levels of metals.

In the United States, anaerobic digesters are used on farms to generate methane from animal manure. Wastewater treatment plants also produce methane as a byproduct. Although some plants use the biogas, most simply burn it. Methane is the primary source of hydrogen in the commercial production of ammonia.

17.8 Viruses

Learning Outcomes

a. Describe how the smallpox vaccine was developed.
b. List the components of a virus.
c. Define *bacteriophage*.
d. Explain how viruses reproduce.
e. List human diseases caused by viruses.

Introduction

Smallpox is a communicable disease that apparently was widespread for thousands of years, periodically killing large numbers of people. As it developed in its victims, it appeared as blisterlike lesions on the skin, which later often became permanent pits or depressions. In 1901, an outbreak of smallpox in New York caused 720 deaths. Since that time, however, it has been eliminated in the United States through vaccination. Vaccination involves the introduction of a weakened form of the disease agent into the body. The body's natural defenses, in fighting the agent, build up an immunity to the disease. The first known vaccinations were performed in England by Benjamin Jesty, a farmer, and Edward Jenner, a country physician. They both had noticed that farmhands working with cows having cowpox, a comparatively mild disease related to smallpox, did not contract smallpox during the devastating epidemics of the disease that occurred in Europe from time to time.

In 1796, Jenner scratched the skin of a boy with fluid he had obtained from a cowpox blister on the hand of a milkmaid. Six weeks later, he deliberately inoculated the boy with fluid from a blister of a smallpox victim, but the boy developed no symptoms of the disease. Jenner thus had performed a successful vaccination against a dread disease more than 50 years before Louis Pasteur developed the germ theory of disease.

The World Health Organization believes that as a result of vaccinations and vigilance, smallpox has been eradicated throughout the world. This is the only infectious disease that has ever been eradicated. However, there has been speculation that it may have been cultured by rogue nations and could reappear as an agent of bioterrorism.

Size and Structure

We know now that smallpox was caused by something considerably smaller than bacteria. During Pasteur's time, virtually all infectious agents, including bacteria, protozoans, and yeast, were called **viruses.** One of Pasteur's associates, Charles Chamberland, discovered that porcelain filters would block out bacteria but would not keep an unseen agent from passing through. The agent caused rabies, another serious disease of both animals and humans. Agents of disease that could pass through filters became known as *filterable viruses,* although the word *filterable* is no longer used. Today, we know that not only smallpox and rabies are caused by these viruses but also measles, mumps, chicken pox, polio, yellow fever, influenza, fever blisters, warts, and the common cold.

Only organisms with certain unique features are now called viruses. These features, which include a complete lack of cytoplasm or cellular structure, make viruses quite different from anything else in the six kingdoms of living organisms we now recognize. In fact, viruses seem to represent the interface between biochemistry and life. In 1946, Wendell Stanley, an American chemist, received a Nobel Prize for demonstrating that a virus causing tobacco mosaic, a common plant disease, could be isolated, purified, and crystallized and that the crystals could be stored indefinitely but would always produce the disease in healthy plants at any time they were placed in contact with them. We also know that viruses do not grow by increasing in size or dividing, nor do they respond to external stimuli. They cannot move on their own, and they cannot carry on independent metabolism. However, when inside living cells, they express their genes and use cellular machinery to produce more virus particles.

Viruses are incredibly numerous. In 1989, for example, marine biologists at the University of Bergen in Norway discovered that a teaspoon of sea water typically contains more than 1 billion viruses. They are about the size of large molecules, varying in diameter from about 15 to 300 nanometers

KEY THEME 17.3 molecular — Plant Viruses

The book *Hot Zone* and the movie *Contagion* created an awareness of emerging viruses and their dangers to the human population. The ebola, hanta, and HIV viruses are now everyday words that have become synonymous with death. There is another group of viruses that also has a significant impact: plant viruses, which cause an estimated $60 billion worth of crop loss per year worldwide. They infect plants and cause hundreds of diseases, such as tomato spotted wilt disease, tobacco mosaic disease, maize stripe disease, and apple chlorotic leaf spot disease.

Surprisingly, the first viruses ever identified were in plants. In 1898, a Dutch professor of microbiology, Dr. Martinus Beijerinck, was working to identify the disease that caused tobacco leaves to become mottled with light green and yellow spots. He demonstrated that the condition was caused not by a bacterium, as was commonly thought at the time, but rather by some other unknown pathogen in the sap of the tobacco plant. He proved this by collecting sap from a diseased plant, which was then passed through a filter capable of straining out any bacteria. When the filtered solution was reinjected into the leaf veins of healthy plants and the disease was transmitted, he had made his point. He called this filtered sap a *contagium vivum fluidium* (a contagious living fluid) and introduced the term *virus* to describe its property of being able to reproduce itself within living plants. Dr. Beijerinck's virus was later named tobacco mosaic virus (TMV), consistent with the now-established practice of naming plant viruses both by the plants they infect and by the major disease symptom (e.g., mosaic, wilting, spotted, etc.).

Not only were the first viruses discovered in plants, but the understanding of their biochemical nature was first recognized through research on tobacco mosaic virus. Today, we know that viruses are submicroscopic, infectious particles that are composed of a protein coat and a nucleic acid center. They can be seen only with an electron microscope. As obligate parasites, they can reproduce themselves only with a living cell. The biochemical nature of viruses remained unknown until 1935, when Dr. Wendell Stanley, an organic chemist in the United States, succeeded in crystallizing the protein coat of TMV. Stanley, however, did not recognize the nucleic acid content of the virus, which was later shown to be RNA. The fact that RNA could exist separately from DNA was a discovery that has had great influence on the development of molecular biology thought.

Today, we know that tobacco mosaic virus is a rigid rod, 300 nanometers by 15 nanometers, composed of a protein coat of approximately 2,100 helically arranged protein subunits surrounding an axial canal that contains a single-stranded RNA molecule consisting of 6,400 nucleotides (Box Figure 17.3A). Like all plant viruses, it is classified according to the type of nucleic acid that it contains, either DNA or RNA but never both; whether the nucleic acid is single- or double-stranded; and the shape of the virus particle (spheres, stiff rods, flexible rods).

TMV is highly contagious, so much so that it can be transmitted to healthy plants merely from the fingers of smokers of

---

(Fig. 17.15). Thousands of the smallest ones could fit inside a single bacterium of average size.

A virus consists of a nucleic acid core surrounded by a protein coat. The architecture of the protein coats varies considerably, but many have 20 sides and resemble tiny geodesic domes, while others have distinguishable head and tail regions. The nucleic acid core consists of either DNA or RNA—never both. Viruses have been classified in several ways. Originally, they were grouped according to their hosts and the types of tissues or organs they affected. Now they are separated first according to the DNA or RNA in their cores. Then they are grouped according to size and shape, the nature of their protein coats, and the number of identical structural units in their cores.

Bacteriophages

Viruses that attack bacteria have been studied extensively. These are called **bacteriophages,** or simply **phages** (Fig. 17.16). Some resemble the space exploration vehicles portrayed in science fiction movies. Each phage consists of a head on top of a thin, cylindrical core, which is surrounded by a sheathing coat. At the base of the core are six spider-like, fibril "legs," which anchor the virus in place.

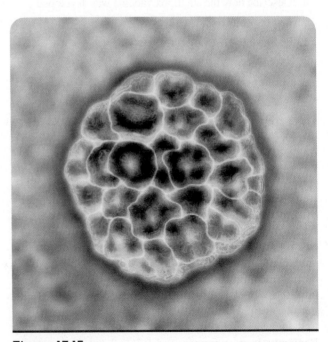

Figure 17.15 Human *papillomavirus* causes warts that can lead to cervical cancer. This image was produced using high-dynamic-range imaging with a transmission electron microscope. The viral diameter is about 55 nanometers. © Universal Images Group/BSIP/Getty Images

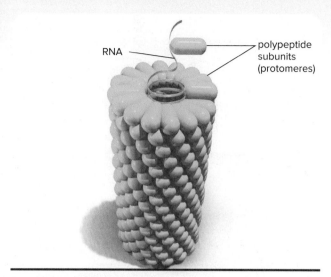

Box Figure 17.3A Structure of tobacco mosaic virus.

cigarettes that were made from infected tobacco. This is unlike most other plant viruses that can survive no more than a few hours outside their living host. Plant viruses can gain entry into a plant only through an open wound or a puncture and are typically transmitted by insect vectors such as aphids, leafhoppers, white flies, and mites. Aphids are the most important vectors, infecting healthy plants when they insert their mouth parts, called stylets, into phloem tubes for feeding. During feeding, they inject salivary secretions containing the virus particles into the plant's sieve tubes.

Once injected, viruses are transmitted within the phloem and move throughout the plant. However, viruses cannot move directly through cell walls. Rather, cell-to-cell movement of viral particles occurs via the *plasmodesmata* (singular: *plasmodesma*), which are membrane-lined, cylindrical pores through cell walls. Plasmodesmata create cytoplasmic bridges that cross cell walls to connect adjacent cells, and this transport route explains why many viral infections are systemic, affecting the entire organism.

Viruses seldom kill the plant outright but rather weaken it by causing abnormalities in leaves (such as mottling or changes in leaf color, shape, or vein patterns); changes in flower color; or irregularities in fruit size, shape, or color. Viruses can also cause fruits to ripen prematurely and to have an unpleasant taste or reduced sugar content. Crop yields of fruits and vegetables, as well as quality, can be reduced.

Few options exist for controlling plant viral diseases. The most effective control is achieved by sanitation—removing and burning diseased plants and thus killing the virus-carrying insects. Additionally, naturally resistant varieties of some plants have been developed. Chemicals remain an ineffective treatment for plant viruses because of the cost and environmental concerns.

Viral diseases affect many important agricultural crops in addition to tobacco. Crop losses worldwide are enormous each year. With the world's human population increasing at about 1.1% yearly, any disease that threatens agricultural productivity and the ability of the human population to feed itself must be taken seriously. Although not as spectacular or newsworthy as ebola or HIV, plant viruses are silent killers because they rob humanity by directly affecting the food supply.

D. C. Scheirer

Viral Reproduction

Viruses can replicate (reproduce themselves) only at the expense of their host cells. In doing so, they first become attached to a susceptible cell. Then they penetrate to the interior, some types leaving their coats on the outside. Inside the cell, their DNA or RNA directs the synthesis of new virus molecules, which are then assembled into complete viruses. These are released from the host cell, usually as it dies (Fig. 17.17).

Some viruses (e.g., those causing influenza) can mutate (see Chapter 15) very rapidly, allowing them to attack organisms that previously had been immune to them. As a result, new vaccines constantly have to be developed to combat new strains of viruses. Some viruses greatly affect the metabolism of their host cells. For example, botulism bacteria produce toxins only if specific phages are present and active. Evidence is mounting that many forms of cancer, which usually involves abnormal cell growth, are caused by viruses. Scientists also suspect that all living organisms carry viruses in an inactive form in their cells, and they are trying to discover what causes the inactive viruses to become active.

Cells of higher animals that are invaded by viruses produce a protein called *interferon*, which is released into the fluid around the cells or into the bloodstream. Minute amounts of interferon in contact with cells cause the cells to produce a protective protein that prevents or inhibits the propagation of many types of viruses within the protected cells and inhibits viruses from causing tumors that transform normal cells into tumor cells.

Because of these properties of interferon, it is now being produced in large quantities for use in controlling certain cancers and many other viral infections. The use of bacteria as hosts for donor DNA is especially promising for the future. This process, in effect, turns the bacteria into interferon synthesis centers. In 1981, scientists at the University of Washington and the Genentech Corporation of San Francisco announced that they had succeeded in producing a form of interferon by splicing interferon genes into yeast cells. Because yeast cells are larger than bacteria, the process can potentially produce much larger quantities of interferon at considerably lower cost than is possible using bacteria.

Human Relevance of Viruses

The economic impact of viruses in both developing and industrialized countries is enormous. The annual loss in work time due to common cold and influenza viruses alone

Figure 17.16 A phage virus consists of a head carrying nucleic acid, a tail, and tail fibers (a). A digitally generated stylized scanning electron microscopic view shows Enterobacteria phage T4, which infects *Escherichia coli* (b). ×25,000. *(b) © MedicalRF.com*

Figure 17.17 Stages in the development of a phage virus within a bacillus bacterium. (a) The virus becomes attached to the bacterium. (b) The DNA of the virus enters the cell. (c) Various components of the virus are synthesized by the DNA of the bacterium. (d) The viral components are assembled into units. (e) The assembled viruses are released as the bacterial wall breaks down.

amounts to millions of hours. While discomfort, adverse effects on employment, and even deaths due to viral diseases, such as chicken pox, measles, German measles, mumps, and yellow fever, have declined dramatically because immunizations against the diseases became widespread, they still take their toll. Another viral disease, infectious hepatitis, periodically still claims victims. Guillain-Barré syndrome and Epstein-Barr are debilitating diseases caused by viruses that are apparently carried by nearly everybody, but what triggers them into action is as yet still unknown. Avian flu is caused by the bird flu virus. This virus is typically confined to birds, but transmission to humans has been documented. Symptoms can range from fever and cough to life-threatening respiratory distress. Currently, human-to-human spread of the disease is rare. However, health officials are concerned that a mutation in the virus could allow it to spread rapidly, causing a pandemic (worldwide disease outbreak).

The virus associated with AIDS (acquired immune deficiency syndrome) is caused by a *retrovirus* (an RNA virus that inserts a copy of its genome into the DNA of the host cell) that is related to those that cause cancer. It was detected in a blood sample collected in 1959 from Zaire (now Democratic Republic of the Congo). AIDS was not adequately described, however, until 1981, and the virus itself was not identified until 1983. Research has shown

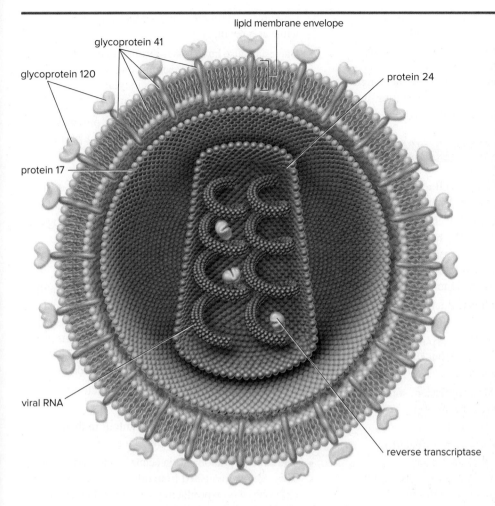

Figure 17.18 A virus is a nucleic acid coated with protein. The human immunodeficiency virus (HIV), which causes AIDS, consists of RNA surrounded by several layers of proteins. Once inside a human cell (usually a T cell, part of the immune system), the virus uses an enzyme to convert its RNA to DNA, which then inserts into the host DNA. HIV damages the human body's protection against disease by killing T cells and by using these cells to make more of itself.

that two forms of the AIDS virus, HIV-1 and HIV-2, diverged from a common ancestor in the early 1950s (Fig. 17.18). Thirty years later, the AIDS virus had spread all over the world. Retroviruses mutate so rapidly that they are capable of evolving about a million times faster than cellular organisms. Since the initial appearance of AIDS, increasing research is being applied to the development of a vaccine with the use of a genetically engineered virus. If successful, the vaccine will cause the human body to develop a defense against AIDS viruses without the disease itself developing.

The production of vaccines for a number of other diseases is undertaken by a flourishing worldwide multimillion-dollar industry. Other mass-produced viruses are used to infect ticks, insects, and other disease organisms of both animals and plants. Some viruses cause great losses when they infest creamery vats during the manufacture of dairy products or culture vats during the production of antibiotics. One phage attacks nitrogen-fixing bacteria in the roots of leguminous plants, while other phages attack diphtheria and tuberculosis bacteria. One natural virus called *Abby* (an abbreviation of Abington) is found only in the caterpillars of gypsy moths. Gypsy moth caterpillars have been particularly destructive in forests of both North America and Europe, and it is hoped that the destruction can be greatly reduced by the dissemination of this virus to infect caterpillars in the forests under attack.

17.9 Viroids and Prions

Learning Outcome

a. Explain how viroids and prions differ from viruses.

More than a dozen plant diseases (e.g., chrysanthemum stunt disease, potato spindle-tuber disease) are caused by infectious agents called *viroids*. Viroids, which are even smaller than viruses, consist of circular strands of RNA that occur in the nuclei of infected plant cells. They may remain latent in the host nuclei without any outward manifestation of the disease in some plants but cause severe disease symptoms in others. Viroids are transmitted from plant to plant via pollen or ovules or by machinery, and it is not yet known how they produce symptoms of a disease. They do appear, however, to be synthesized from RNA templates by an enzyme. In addition, unlike viruses, viroids may be seed-transmitted.

Prions, which are also smaller than viruses, appear to be particles of protein that cause diseases of animals and humans. Cruetzfeldt-Jakob disease and chronic wasting disease, its counterpart in deer, elk, and moose, are caused by prions. Prions are believed to cause disease by inducing abnormal folding of the proteins in the brain, resulting in brain damage. Prion diseases are always fatal.

SUMMARY

17.1 Introduction to Domain (Kingdom) Bacteria, Domain (Kingdom) Archaea, and Viruses

- Domains (Kingdoms) Bacteria and Archaea consist of prokaryotic organisms (bacteria). The bacteria occur as single cells, in colonies, or in the form of chains or filaments. Some cells may be motile, or they may exhibit a gliding motion; most are nonmotile.

17.2 Features of Domains (Kingdoms) Bacteria and Archaea

- Bacterial nutrition is primarily by absorption of food in solution through the cell wall, but some are photosynthetic or chemosynthetic.
- Reproduction is asexual by means of fission; some genetic recombination occurs by means of pili between cells.
- Bacteria occur as spheres (cocci), as rods (bacilli), and in spiral forms (spirilla); they are further classified on the basis of several visible features, such as sheaths, appendages, and motion. They are also classified by their chemistry and as gram-positive or gram-negative.
- Prokaryotic cells have no nuclear envelopes or organelles. Each cell has a single, closed loop of double-stranded DNA and sometimes small, circular DNA molecules called plasmids, which replicate independently of the large DNA molecule or chromosome.
- Neither meiosis nor mitosis occurs, but fission takes place with the development of a transverse wall that forms near the middle of the cell; gametes and zygotes are not produced. Conjugation facilitates genetic recombination. Transformation involves the incorporation of fragments of DNA released by dead cells; transduction involves the viral transfer of fragments of DNA from one cell to another.

17.3 Domain (Kingdom) Bacteria—the True Bacteria

- Heterotrophic bacteria are saprobes or parasites. Autotrophic bacteria are photosynthetic but do not produce oxygen; chemoautotrophic bacteria obtain their energy through oxidation of reduced inorganic groups.

17.4 Human Relevance of the Unpigmented, Purple, and Green Sulfur Bacteria

- Any nonliving organic material will eventually be decomposed to compost by bacteria and fungi. Compost is definitely good for the soil but has limited value as a fertilizer.
- Bacteria cause huge losses through plant diseases and food spoilage and many serious diseases in animals and humans. They gain access to their hosts by various means.
- Koch formulated postulates (rules) for proving that a particular microorganism is the cause of a particular disease.
- Bacteria useful to humans include *Bacillus thuringiensis, Bacillus thuringiensis* var. *israelensis,* and *Bacillus popilliae.* Bacteria also play a major role in the manufacture of dairy products, such as yogurt, sour cream, kefir, and cheese.
- Bacteria are used in the manufacture of industrial chemicals, vitamins, flavorings, food stabilizers, and a blood plasma substitute; they play a role in the curing of vanilla, cocoa beans, coffee, and tea and in the production of vinegar and sauerkraut; they aid in the extraction of linen fibers from flax stems, in the production of ensilage for cattle feed, and in the production of several important amino acids.

17.5 Class Cyanobacteriae—the Cyanobacteria (Blue-Green Bacteria)

- Cyanobacteria are virtually ubiquitous in their occurrence.
- The cells of cyanobacteria occur in a variety of forms. They are distinguished from other bacteria in having chlorophyll *a,* in producing oxygen, and in having blue and red phycobilin pigments. They produce cyanophycin, a nitrogenous food reserve.
- Cyanobacteria have no flagella, but some species have gliding movements. Fragmentation may occur at heterocysts. Akinetes may also be produced.
- Cyanobacterial cells may have been the origin of chloroplasts, because they divide as chloroplasts do when their host cells divide.
- Cyanobacteria may become very abundant in bodies of polluted fresh water. Toxic substances are produced when the bacteria die and are decomposed. At least 40 species of cyanobacteria are known to fix nitrogen.

17.6 Class Prochlorobacteriae—the Prochlorobacteria

- Prochlorobacteria are similar in form to cyanobacteria but have pigmentation similar to that of higher plants, and they lack phycobilins.

17.7 Domain (Kingdom) Archaea—the Archaebacteria

- The three groups of archaebacteria are the methane bacteria, the salt bacteria, and the sulfolobus bacteria.

17.8 Viruses

- Vaccination was first performed by Jesty and Jenner in connection with smallpox. Smallpox is believed to have been eradicated but may reappear as an agent of bioterrorism.

- Viruses, which have no cellular structure, are about the size of large molecules. Some can be isolated, purified, and crystallized yet remain virulent. They cannot grow or increase in size, and they cannot be replicated outside of a living cell. They depend upon DNA for their raw materials.

- Viruses consist of a core of nucleic acid surrounded by a protein coat. They are classified on the basis of the DNA or RNA in their cores, their size and shape, the number of identical structural units in their cores, and the nature of their protein coats.

- Bacteriophages are viruses that attack bacteria. In replicating, they become attached to a susceptible cell, which they penetrate, with their DNA or RNA directing the synthesis of new virus molecules from host material; the assembled new viruses are released when the host cell dies.

- Cells of higher animals being invaded by viruses produce interferon, a protein that causes cells to produce a protective substance that inhibits replication of viruses and inhibits the capacity of viruses to transform normal cells into tumor cells.

17.9 Viroids and Prions

- Viral diseases, such as chicken pox, measles, mumps, and yellow fever, have declined because immunizations against the diseases have become widespread. Mass-produced viruses are used to infect insects and other pests of both plants and animals. Some viruses cause losses of dairy products in creamery vats.

- Viroids and prions are disease-causing particles that are smaller than viruses.

REVIEW QUESTIONS

1. What is symbiosis? Give examples other than those mentioned in the text.
2. Why are bacteria not classified on the basis of visible features alone?
3. How do bacteria exchange DNA?
4. How does fission differ from mitosis?
5. Is photosynthesis in bacteria the same as it is in higher plants? Explain.
6. How do chemoautotrophic bacteria differ from photosynthetic bacteria?
7. What is the difference between nitrification and nitrogen fixation?
8. If decay bacteria use nitrogen, how does composting accumulate any nitrogen?
9. How are disease bacteria transmitted?
10. What are Koch's Postulates?
11. Why are many bacteria considered useful?
12. What do cyanobacteria and other bacteria have in common? How do they differ?
13. How do cyanobacteria survive freezing and desiccation?
14. What is an algal-bacterial bloom?
15. How do viruses differ from bacteria?
16. What is a vaccination?
17. What is a phage?
18. How do viruses multiply?

DISCUSSION QUESTIONS

1. If a virulent phage were to eliminate all the bacteria in North America for 1 year, how would our lives be affected?
2. What would be the feasibility and the advantages or disadvantages of using only cyanobacteria and other nitrogen-fixing bacteria for our agricultural nitrogen needs?
3. Methane gas produced by bacteria is proving to be sufficient to meet all the fuel needs of villages in India. Do you think we could produce and use methane in a similar fashion in the United States?
4. If cyanobacteria are capable only of asexual reproduction, does this mean that they can never change in form?
5. As long as viruses can multiply, why should there be any question as to whether or not they are living?

CHAPTER 18

Kingdom Protista

18.1 Introduction to Kingdom Protista
18.2 Features of Kingdom Protista
　　Key Theme 18.1: Green Plant Phylogeny
18.3 Algae
18.4 Phylum Chlorophyta—the Green Algae
　　Chlamydomonas
　　Ulothrix
　　Spirogyra
　　Oedogonium
　　Other Green Algae
18.5 Phylum Chromophyta—the Yellow-Green Algae, Golden-Brown Algae, Diatoms, and Brown Algae
　　Yellow-Green Algae (Xanthophyceae)
　　Golden-Brown Algae (Chrysophyceae)
　　Diatoms (Bacillariophyceae)
　　Brown Algae (Phaeophyceae)
18.6 Phylum Rhodophyta—the Red Algae
18.7 Phylum Euglenophyta—the Euglenoids
18.8 Phylum Dinophyta—the Dinoflagellates
18.9 Phylum Cryptophyta—the Cryptomonads
18.10 Phylum Prymnesiophyta (Haptophyta)—the Haptophytes
18.11 Phylum Charophyta—the Stoneworts
18.12 Human and Ecological Relevance of the Algae
　　Diatoms
　　Key Theme 18.2: Biofuels from Algae
　　Other Algae
　　Algin
　　Minerals and Food
　　Agar
18.13 Other Members of Kingdom Protista
18.14 Phylum Myxomycota—the Plasmodial Slime Molds
18.15 Phylum Dictyosteliomycota—the Cellular Slime Molds
　　Human and Ecological Relevance of the Slime Molds
18.16 Phylum Oomycota—the Water Molds
　　Human and Ecological Relevance of the Water Molds

Underwater kelp beds provide habitat for fish and other animals. © *Corbis*

OVERVIEW

After an introduction and a summary of the features of members of Kingdom Protista, this chapter discusses the phyla of algae, slime molds, and water molds. Brief discussions of the life cycles of representatives of the principal phyla are included.

First, the Chlorophyta (green algae) are discussed, with *Chlamydomonas, Ulothrix, Spirogyra,* and *Oedogonium* being explored in some detail; mention is also made of *Chlorella,* desmids, *Acetabularia, Volvox,* and *Ulva.* The Chromophyta (yellow-green algae, golden-brown algae, diatoms, and brown algae) are taken up next. Then the Euglenophyta (euglenoids) are briefly examined, and Phylum Dinophyta (dinoflagellates) is explored. The role of members of Phylum Dinophyta in red tides and bioluminescence is noted. After a brief overview of Phyla Cryptophyta, Prymnesiophyta, and Charophyta, the chapter summarizes differences between the phyla of algae in Table 18.1, which lists food reserves, chlorophyll pigments, and flagella. This is followed with a digest of the human and ecological relevance of the algae. The chapter concludes with a brief examination of the slime molds and water molds.

18.1 Introduction to Kingdom Protista

Learning Outcome

a. Describe features of *Coleochaete* that are similar to those of higher plants.

If you were to travel on the open ocean at night, you may be fortunate enough to see millions of tiny twinkling lights on the surface of the water. The lights are the product of *bioluminescence* carried out by algae called dinoflagellates. When the dinoflagellates are disturbed, as when a boat passes through water, they produce a flash of blue light that lasts about one-tenth of a second. The production of light startles predators, so bioluminescence is a defense response. High concentrations of bioluminescent dinoflagellates are found in some quiet areas, such as Mosquito Bay, Puerto Rico; Montego Bay, Jamaica; Castille, Maine; and Indian River Lagoon, Florida. Dinoflagellates are only one of several phyla of algae and other simple eukaryotes discussed in this chapter.

The fossil record indicates that less than 1 billion years ago, all living organisms were confined to the oceans, where they were protected from dehydration, ultraviolet radiation, and large fluctuations in temperatures. They also absorbed the nutrients they needed directly from the water. The fossil record also suggests that many times, beginning about 400 million years ago, green algae, which are important members of Kingdom Protista, began making the transition from water to land, eventually giving rise to green land plants.

Coleochaete (Fig. 18.1), a tiny, freshwater green alga that grows as an **epiphyte** (it grows on plants without harming them) on the stems and leaves of submerged plants, was probably an indirect ancestor of today's land plants. It shares several features with higher plants, including cells that resemble parenchyma, the development of a cell plate and a phragmoplast during mitosis, a protective covering for the

Figure 18.1 *Coleochaete,* a green alga that has several features of higher plants and is believed to be an indirect ancestor of land plants. ×500. © *Kingsley Stern*

zygote, and the production of a ligninlike compound. Lignin adds mechanical strength to cell walls and provides protection against microbes. The ligninlike compound was present 100 million years earlier than the presumed evolution of plant life from water to land, and its presence in algal cells could explain how such organisms adapted to land habitats.

18.2 Features of Kingdom Protista

Learning Outcome

a. Describe general features of protists.

As indicated in Chapter 16, all organisms with eukaryotic cells are considered by many to constitute a superkingdom known as *Domain Eukarya,* which consists of Kingdoms Protista, Plantae, Fungi, and Animalia. We begin our study of this domain with Kingdom Protista. This kingdom consists of eukaryotes that do not fit into any of the other kingdoms. Therefore, they lack a set of unifying features. Many, including the euglenoids, protozoans, and some algae, consist of a single cell, whereas other algae are multicellular or occur as colonies or filaments. Nutrition is equally varied, with the algae being photosynthetic, the slime molds and protozoans ingesting their food, the euglenoids either carrying on photosynthesis or ingesting their food, and the oomycetes and chytrids absorbing their food in solution.

Individual life cycles vary considerably, but reproduction is generally by cell division and sexual processes. Many single-celled protists are *motile* (capable of spontaneous movement), usually by means of flagella, but other members of Kingdom Protista, especially those that are multicellular,

Hemisphere. By 1990, however, the blooms were also occurring with increasing frequency throughout both temperate and tropical waters of the Southern Hemisphere. As mentioned in the introduction to this chapter and discussed in Chapter 11, dinoflagellates also play a major role in oceanic bioluminescence.

Reproduction is by cell division. Sexual reproduction appears to be rare.

18.9 Phylum Cryptophyta— the Cryptomonads

Learning Outcome

a. Describe characteristic features of cryptomonads.

The **cryptomonads** constitute a small group of asymmetrical, somewhat flattened, unicellular, marine and freshwater algae with two flagella, both flagella bearing short hairs throughout their length. Each cryptomonad cell has one or more plates on the inside of the plasma membrane, with a few also having plates outside of the plasma membrane. The plates may be green, blue, yellowish green, red, or brown. There is a single, two-lobed chloroplast with starch granules surrounding a central pyrenoid. In addition to the nucleus, there is a distinctive *nucleomorph* that possibly represents a vestigial nucleus of a primitive organism that may have, in ancestral forms, lived symbiotically within the larger cell. In most species, the *gullet,* which is analagous to that of a euglenoid, is lined with two or more rows of *ejectosomes.* Ejectosomes are structurally different from the stinging, harpoonlike projectiles (trichocysts) of protozoans and some dinoflagellates but unreel as long threads similar to trichocysts. Asexual reproduction is by mitosis; sexual reproduction is unknown (Fig. 18.26).

18.10 Phylum Prymnesiophyta (Haptophyta)—the Haptophytes

Learning Outcome

a. Describe characteristic features of haptophytes.

Most *haptophytes* are unicellular, with two smooth flagella of similar length inserted at the apex. Their pigments and carbohydrate reserves are similar to those of members of the Chromophyta, from which the majority differ in having a third flagellum—the *haptonema.* Unlike true flagella, however, the haptonema, which is located between the true flagella, does not function in propulsion; instead, it has a sticky tip that aids in capturing food. Haptophyte cells may be naked, but more often they are covered with scales that are sloughed off into the water as they are replaced from

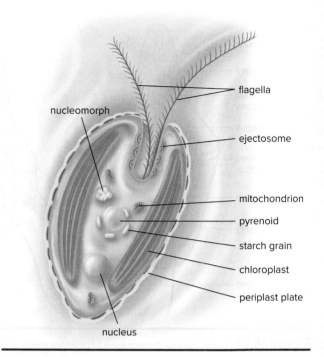

Figure 18.26 Cellular features of *Cryptomonas* sp., a cryptomonad. ×2,000.

beneath, where they are produced by dictyosomes. One group of haptophytes, the *coccolithophores,* has scales formed of calcite crystals. Two disc-shaped chloroplasts are present in each cell. Haptophytes, which occur in both salt and fresh water, are smaller than most diatoms and dinoflagellates but occur in huge numbers, especially in the tropics. Haptophytes are a major component of marine plankton.

18.11 Phylum Charophyta— the Stoneworts

Learning Outcome

a. Describe characteristic features of stoneworts.

Stoneworts (Fig. 18.27) are primarily aquatic organisms of shallow, freshwater lakes and ponds; they often carry calcium precipitate on their surfaces. A stonewort body consists of an axis with short, lateral branches in whorls, so it looks somewhat like a small horsetail plant (discussed in Chapter 21). Many scientists place stoneworts in their own phylum (Charophyta), but others consider them a class of green algae because of their pigmentation and a few other green algal features. Their sexual reproduction is oogamous (nonmotile female gamete and motile male gamete), and, unlike other algae discussed to this point, male gametes of stoneworts are produced by multicellular structures. Other features of both their vegetative growth and reproduction are more complex than those of any of the green algae. Some botanists consider them more closely related to mosses than to algae.

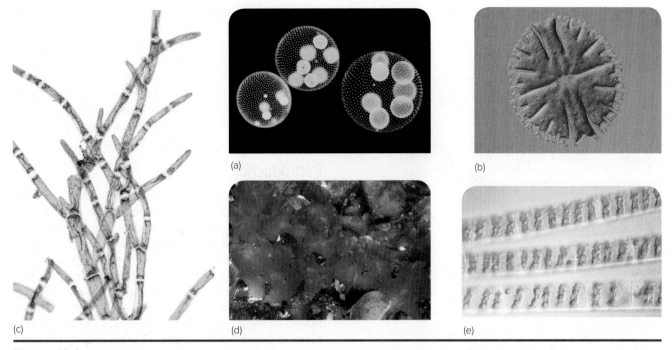

Figure 18.2 Representative green algae. (a) *Volvox*. The cells form hollow, spherical colonies that spin as the flagella of each cell beat in a coordinated motion. New colonies are produced within the older ones. (b) *Micrasterias,* a desmid. These algae consist of single cells that often have a constriction in the center. (c) *Cladophora,* a branched filamentous alga. (d) *Ulva* (sea lettuce), a multicellular alga. (e) *Spirogyra,* an unbranched filamentous alga. *(a-b)* © *Lebendkulturen.de/Shutterstock; (c)* © *Buccaneer/123RF; (d)* © *Fotosearch/Getty Images; (e)* © *Stephen Durr*

18.4 Phylum Chlorophyta— the Green Algae

Learning Outcomes

a. Describe features of plankton.
b. List features found in a *Chlamydomonas* cell.
c. Identify the stages in the life cycle of *Ulothrix*.
d. Describe the unique features of *Spirogyra*.
e. Explain the meaning of the name *Oedogonium*.
f. Recognize that *Chlorella*, desmids, *Hydrodictyon*, *Acetabularia*, *Volvox*, and *Ulva* are green algae.

Phylum Chlorophyta includes species of organisms commonly known as the **green algae.** They occur in a rich variety of forms and in very diverse and widespread habitats; they possess some of the most beautiful chloroplasts of all photosynthetic organisms. Some are unicellular and microscopic; in fact, the green alga *Micromonas* is only 1 μm in diameter—the smallest eukaryotic cell known and similar in size to a bacterium. Most unicellular green algae are, however, considerably larger. Other green algae form threadlike filaments, platelike colonies, netlike tubes, or hollow spheres (Fig. 18.2).

Some green algae are seaweeds, resembling lettuce leaves or long, green ropes. Several unicellular species grow in greenish patches or streaks on the bark of trees, whereas others grow in large numbers on the fur of sloths and other jungle animals. The green fur camouflages the animals in their natural habitats. Still others thrive in snowbanks, live in flatworms and sponges or on rocks, or are found on the backs of turtles. They are the most common member in lichen "partnerships" (discussed in Chapter 19). The greatest variety, however, is found in freshwater ponds, lakes, and streams. Ocean forms are also varied; there they are an important part of the **plankton** (free-floating, mostly microscopic organisms) and, therefore, of food chains.

The chlorophylls (*a* and *b*) and other pigments of green algae are similar to those of higher plants. The green algae are believed to have been the ancestors of higher plants and, like the higher plants, they store their food in the form of starch within chloroplasts. Most green algae have a single nucleus in each cell and reproduce both asexually and sexually. How they do so illustrates the forms of reproduction of most of the organisms discussed in the chapters to follow, so we'll examine several different representative green algae in some detail.

Chlamydomonas

A small, actively moving little alga, *Chlamydomonas* (Fig. 18.3) is a common inhabitant of quiet freshwater pools. It has an ancient history among eukaryotic organisms, with fossil relatives occurring in rock formations reported to be nearly 1 billion years old. *Chlamydomonas* is unicellular, with a slightly oval cell surrounded by a complex, multilayered wall that is partially composed of glycoproteins. A pair of whiplike flagella at one end pull the cell very rapidly through the water. The flagella are, however, difficult to see with ordinary light microscopes. The cell itself is usually less than 25 micrometers (one ten-thousandth of an inch) long, which is, however, more

Figure 18.3 *Chlamydomonas* is a single-celled green alga that is typically found in freshwater pools. A pair of flagella pulls the alga through the water.

A dominant feature of each *Chlamydomonas* is a single, usually cup-shaped chloroplast that at least partially hides the centrally located nucleus. One or two roundish **pyrenoids** are located in each chloroplast. Pyrenoids are proteinaceous structures associated with the synthesis of starch. Most species also have a red *eyespot* on the chloroplast near the base of the flagella. The eyespot allows the alga to detect light; it is, however, merely part of an organelle within a single cell and is in no way complex and multicellular like an eye.

Reproduction

Chlamydomonas cells are haploid, carrying only one copy of each chromosome. However, these cells can undergo mitosis to produce genetically identical daughter cells. This is a form of asexual reproduction.

Under certain combinations of light, temperature, and additional unknown environmental forces, many cells in a *Chlamydomonas* population may congregate together in preparation for sexual reproduction. Pairs of cells in this mass fuse to form diploid zygotes. The collection of zygotes may remain dormant for several days, weeks, or even months, but under favorable conditions, a dramatic change occurs. The zygotes undergo meiosis, producing four genetically different haploid **zoospores** (motile spores). Each zoospore grows into a full-sized *Chlamydomonas* cell (Fig. 18.4).

Ulothrix

If you look carefully at dead twigs, rocks, and other debris in cold freshwater ponds, lakes, and streams (and a few marine habitats) you may find a threadlike alga called *Ulothrix* (Fig. 18.5), whose name is derived from the Greek words *oulos* (woolly) and *thrix* (hair). Each alga consists of a single row of cylindrical cells attached end to end to form a thread, or *filament*. The basal cell of each filament is slightly longer than the other cells and functions as an attachment cell, or **holdfast.** Each cell contains a single chloroplast with one or

than three times larger than a human red blood cell. Near the base of the flagella there are two or more vacuoles that contract and expand (contractile vacuoles). They may regulate the water content of the cell and remove waste products.

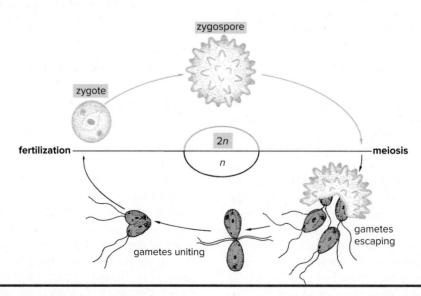

Figure 18.4 Sexual life cycle of *Chlamydomonas*.

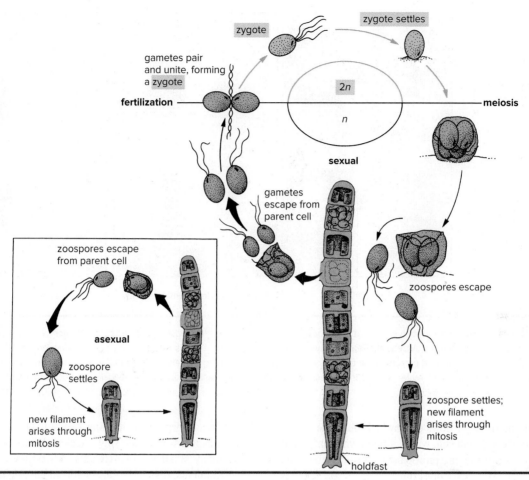

Figure 18.5 The life cycle of *Ulothrix* includes both sexual and asexual stages.

more pyrenoids. Any of the cells, except the holdfast, may divide, and as they do so, the filaments grow longer.

Reproduction

The contents of any cell except the holdfast appear to be able to condense inside the rigid cell wall, divide by mitosis, and develop into zoospores. The zoospores of *Ulothrix* are quite similar to *Chlamydomonas* cells. After swimming for a few hours to several days, they settle on submerged objects, the flagella are shed, and the cells divide. One of the first two daughter cells becomes a holdfast, while the other continues to divide, becoming a new algal body.

Asexual and sexual reproduction start out in the same way in *Ulothrix*. The contents of any cell except the holdfast condense and then divide, producing swimming cells that function as gametes. They unite with gametes from other filaments to form diploid zygotes, which become dormant. Eventually, the zygote cell contents undergo meiosis, giving rise to genetically variable zoospores that then can grow to become new filaments. Although the gametes of *Ulothrix* come from different filaments, they are identical in size and appearance. Sexual reproduction involving such gametes is called **isogamy** ("iso" means same). As in *Chlamydomonas*, the zygotes of *Ulothrix* are the only diploid cells in the life cycle. All the other cells are haploid.

Spirogyra

Watersilk, as *Spirogyra* is called, has watery sheaths, which make the alga feel slimy. Examination with a microscope reveals a beautiful alga with some of the most striking chloroplasts known.

These common freshwater algae, consisting of unbranched filaments of cylindrical cells, frequently float in masses at the surface of quiet waters. Each cell contains one or more long, frilly, ribbon-shaped chloroplasts that look as though they had been spirally wrapped around an invisible pole (vacuole) occupying most of the cell's interior. Most *Spirogyra* species have one or two chloroplasts in each cell. Every one of these elegant green ribbons has pyrenoids at regular intervals throughout its length.

Reproduction

Unlike *Chlamydomonas* and *Ulothrix*, *Spirogyra* produces no zoospores. Asexual reproduction occurs through the breakup, or *fragmentation,* of existing filaments. Fragmentation often occurs as a result of a storm or other disturbance.

In colonies of *Spirogyra,* the filaments usually are produced so close to each other that they touch. When sexual reproduction begins, the individual cells of adjacent filaments form small dome-shaped bumps that fuse to form cylindrical **conjugation tubes** between each pair of cells. Then, the

Figure 18.6 (a) A portion of a vegetative filament of *Spirogyra* shows ribbonlike chloroplasts spirally arranged in each cell. The centrally located darker object in each cell is a nucleus. (b) Papillae have grown out from opposite cells of two closely adjacent filaments and formed conjugation tubes. (c) The condensed protoplasts in the cells on the *left* are functioning as male gametes that are migrating through the conjugation tubes to the stationary female gametes in the cells on the *right*. (d) Zygotes have been produced in the cells on the *right* as a result of fusion of gametes. *(a–d)* © Kingsley Stern

protoplast of one filament flows or crawls through the conjugation tube to the adjacent cell, where it fuses with that cell to form a zygote. Each moving protoplast is considered a male gamete, while the stationary ones function as female gametes (Fig. 18.6).

Thick walls usually develop around the zygotes, which remain dormant for some time, often over the winter. Thick-walled zygotes are characteristic of most freshwater green algae. Eventually, each diploid zygote undergoes meiosis, producing four genetically unique haploid cells. Three disintegrate, and a single new *Spirogyra* filament grows from the interior of the old zygote shell. The type of sexual reproduction seen in *Spirogyra* is called **conjugation.**

Oedogonium

Oedogonium (pronounced ee-doh-góh-nee-um), a filamentous green alga, is often found attached to aquatic plants and algae. It is, however, strictly epiphytic and in no way parasitic. The basal cells of the unbranched filaments form holdfasts, and the terminal cell of each filament is rounded. The remaining cells are cylindrical and attached end to end. The name *Oedogonium* means "swollen reproductive cell" and is quite apt in describing the bulging female reproductive cells. Each cell contains a netlike chloroplast that forms a tube something like a loosely woven basket within the cell. There are pyrenoids throughout the net.

Reproduction

Oedogonium can reproduce asexually by either zoospore production or fragmentation. The tip of each filament may produce a zoospore (swimming haploid spore) (Fig. 18.7). The zoospores look like tiny, faceless heads because they have many small hairlike flagella. After they escape from their filaments, they eventually settle and divide to form new filaments in the same manner as *Ulothrix*.

Oedogonium has more sexual specialization than any of the other three green algae previously discussed. Short, boxlike cells in the filaments carry male gametes (sperm cells). The sperm cells resemble the zoospores but are smaller. Certain other cells become swollen and carry large female gametes (egg cells). Each specialized cell carries either two male gametes or one female gamete. Sperm cells swim to egg cells attracted by chemicals released by the eggs. One sperm unites with the egg, forming a zygote. After fertilization, the zygote wall thickens, and reddish oil accumulates in the protoplasm.

Zygotes may remain dormant for a year or more, but they eventually undergo meiosis, producing four haploid zoospores, each capable of developing into a new filament. The sexual reproduction exhibited by *Oedogonium*, in which one gamete is motile while the other gamete is larger and stationary, is called **oogamy.**

Other Green Algae

As indicated earlier, the green algae constitute a very diverse group, with an extraordinary variety of forms and chloroplast shapes. Obviously, each species has to reproduce in order to perpetuate itself. Most undergo both sexual and asexual reproduction, but a few do not. For example, the globally distributed algae that look like a splattering of green paint on tree trunks reproduce only asexually.

Figure 18.7 Life cycle of *Oedogonium*. © James E. Bidlack

Chlorella is a widespread green alga that is very easy to culture and is a favorite organism of research scientists. It has been used in many major investigations of photosynthesis and respiration, and in the future, it may become important in human nutrition. (See the section "Human and Ecological Relevance of the Algae" in this chapter.)

Chlorella could also play a key role in long-range space exploration. Because present exploration is severely limited by the weight of oxygen tanks and food supplies needed on a spacecraft, scientists have turned to *Chlorella* and similar algae as portable oxygen generators and food sources. Future spacecraft may be equipped with tanks of such algae. These would carry on photosynthesis, using available light and carbon dioxide given off by the astronauts, while furnishing them with oxygen. As the algae multiply, the excess could be either eaten or fed to freshwater shrimp that could, in turn, become food for the astronauts. Still other algae and bacteria could recycle other human wastes. Such a self-perpetuating *closed system,* as it is called, has already been successfully tested with mice and other animals. Many algae, however, are known to produce traces of carbon monoxide gas, which is deadly to most animal life, and until this problem is resolved, humans will not be subjected to such research.

The crescent, elliptical, and star-shaped cells of desmids (Fig. 18.8; see also Fig. 18.2b) are mostly free-floating and unicellular and reproduce by conjugation. Their chloroplasts are among the most beautiful in the green algae.

The striking *water nets* (*Hydrodictyon*) (Fig. 18.9) form netlike, tubular colonies with polygonal meshes. A daughter net may be asexually produced inside a parent cell and eventually released. Sexual reproduction is isogamous, with

Figure 18.8 A single cell of the common desmid, *Closterium*. ×400. © Kingsley Stern

Volvox (see Fig. 18.2a) is representative of a line of green algae that forms colonies similar to those of *Chlamydomonas*, held together in a secretion of gelatinous material. In some colonies, the cells are connected to one another by cytoplasmic strands. The flagella of individual cells beat separately but pull the whole colony along. A *Volvox* colony may consist of several hundred to many thousands of cells that resemble a hollow ball spinning as it moves. Reproduction may be either asexual or sexual, with smaller daughter colonies being formed inside the parent colony. The daughter colonies are released when the parent colony breaks apart.

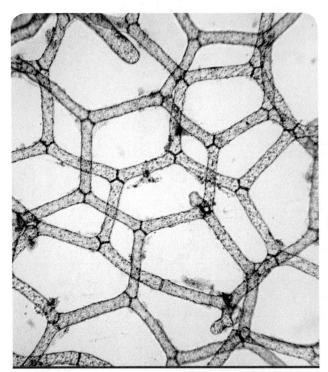

Figure 18.9 A portion of a water net (*Hydrodictyon*), a green alga, shows that cells form a network in the shape of a tube. ×400. © McGraw-Hill Education/Richard Gross

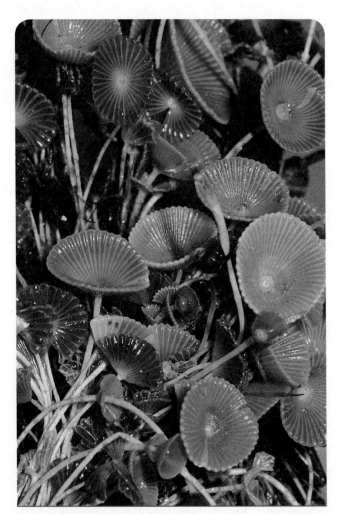

two flagellated gametes forming a zygote as they unite. The zygotes become zoospores that produce large, angular cells. Each new cell then becomes a new, netlike colony that is released from the parent cell as it breaks down.

Sexual reproduction is also isogamous in the *mermaid's wineglass* (*Acetabularia*), a marine alga consisting of a single, huge cell shaped like a delicate mushroom (Fig. 18.10). Each cell is up to 5 centimeters (2 inches) long. This alga has been used in classic experiments demonstrating the influence of the nucleus on the form of the cell. If the cap of an alga is removed and the nucleus is replaced with a nucleus taken from another species, the base regenerates a cap identical to the previous one. If this new cap is also removed, however, the next cap that develops shows characteristics of both species. If the intermediate cap is then removed, the next cap that develops is identical to that of the species from which the nucleus originally came. Clearly, the original nucleus directs development of cytoplasmic substances regulating cap form, and when these are gone, the replacement nucleus exerts its own influence.

Figure 18.10 Mermaid's wineglass (*Acetabularia*) is a large single-celled green alga. © Kingsley Stern

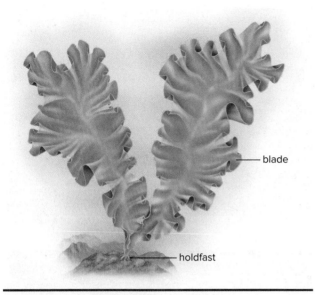

Figure 18.11 Sea lettuce (*Ulva*) is a multicellular green alga that looks like salad greens.

Sea lettuce (*Ulva*) is a multicellular seaweed with flattened, crinkly-edged green blades that may be 1 meter (3 feet) or more long (Fig. 18.11). A basal holdfast anchors the blades, which may be either haploid or diploid, to rocks. Except for the reproductive structures on diploid blades, the haploid and diploid blades of sea lettuce are indistinguishable from one another.

Cladophora is a branched, filamentous green alga whose species are represented in both fresh and marine waters. Unlike the cells of other green algae, those of *Cladophora* and its relatives are mostly multinucleate.

18.5 Phylum Chromophyta—the Yellow-Green Algae, Golden-Brown Algae, Diatoms, and Brown Algae

Learning Outcomes

a. List unifying features of members of Phylum Chromophyta.
b. Describe where most yellow-green algae are found.
c. Describe where you are most likely to find golden-brown algae.
d. List features found in all diatoms.
e. List the main components of the body of a brown alga.

Although many members of the Phylum Chromophyta are microscopic, kelps and other brown seaweeds are the best-known representatives. If the microscopic forms were larger, many surely would become collectors' items in the art and antique shops of the world because of their exquisite form and ornamentation (Fig. 18.12).

Figure 18.12 Some representatives of yellow-green algae include (a) *Vaucheria*, (b) *Stipitococcus*, and (c) *Dinobryon*.

The algae in this phylum can be grouped into several classes, including the *yellow-green algae* (Xanthophyceae), *golden-brown algae* (Chrysophyceae), *diatoms* (Bacillariophyceae), and *brown algae* (Phaeophyceae). Superficially, the organisms of each class may appear unrelated to each other, but they do share several features, including food reserves, specialized pigments, and other cell characteristics. Some members of the first three classes produce a unique resting cell called a *statospore* (Fig. 18.13). These cells resemble miniature corked bottles, complete with plugs that dissolve or "uncork," releasing the zoospore inside. Many statospores are striking in form, with finely sculptured ornamentations on the surface.

Yellow-Green Algae (Xanthophyceae)

Yellow-green algae are mostly freshwater organisms, with a few marine and terrestrial representatives. The two flagella of motile cells are oriented in opposite directions. The brownish pigment *fucoxanthin* (found in other members of this phylum) is not present except in *Vaucheria*, an oogamous, coenocytic, filamentous species. Nonmotile spores are commonly formed during asexual reproduction; less frequently, zoospores are produced. Sexual reproduction is relatively rare, but when it does occur (as in *Vaucheria*), it may exhibit specialization, with oogonia and antheridia being formed on special branches (see Fig. 18.12a).

Figure 18.13 Statospores, which often resemble corked bottles, are formed by many of the golden-brown algae.

Golden-Brown Algae (Chrysophyceae)

Most golden-brown algae are freshwater plankton and produce yellow-brown carotenoid pigments, giving them their distinctive color. The motile cells in this group have two flagella of unequal length inserted at right angles to each other, with a *photoreceptor* (light-sensitive area) on the short flagellum. The photoreceptor is usually shaded by an eyespot at the flagellar end of a large chloroplast. Silica scales are present in some species.

Diatoms (Bacillariophyceae)

Diatoms (Fig. 18.14) are among the best-known and economically most important members of the phylum. These mostly unicellular algae occur in astronomical numbers in both fresh and salt water, but they are particularly abundant in colder marine habitats. In fact, as a rule, the colder the water, the greater the number of diatoms present, and huge populations of diatoms are found on and within ice in both Antarctica and the Arctic. A major constituent of the foam that accumulates at the waveline on beaches is an oil produced by diatoms.

Diatoms usually also dominate the algal flora on damp cliffs, the bark of trees, bare soil, or the sides of buildings. Some can withstand extreme drought, and one species is known to have become active after lying dormant in dry soil for 48 years.

Diatoms look like tiny, ornate, glass boxes with lids. Half of the rigid, crystal-clear wall fits inside the other, overlapping half like a petri dish. Many marine diatoms are circular in outline when viewed from the top, while freshwater species viewed the same way tend to resemble the outline of a kayak. As much

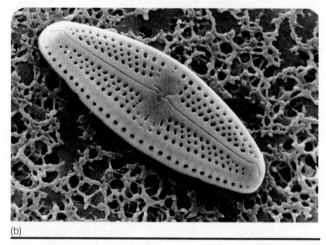

Figure 18.14 (a) A living diatom. ×200. (b) A diatom wall. ×1,500.
(a) © Melba Photo Agency/PunchStock; (b) © BSIP/Universal Images Group via Getty Images

as 95% of the wall content is silica, an ingredient of glass, deposited in an organic framework of pectin or other substances.

The diatom walls usually have exquisitely fine grooves and pores that are exceptionally minute passageways connecting the protoplasm with the watery environment outside the shell. These fine grooves and pores are so uniformly spaced they have been used to test the resolution of microscope lenses. When *phycologists* (scientists who specialize in the study of algae) want to examine the markings on diatom walls, they often obtain a better view by dissolving the cell contents with hot acid.

Each diatom may have one, two, or many chloroplasts per cell. In addition to chlorophyll *a*, the accessory pigments chlorophyll c_1 and chlorophyll c_2 are typically present. The chloroplasts usually are golden-brown in color because of the dominance of *fucoxanthin,* the brownish pigment also found in the brown algae. Food reserves are oils, fats, or the carbohydrate *laminarin.*

Freshwater diatoms that have a lengthwise groove, called a *raphe,* glide backward and forward with somewhat jerky motions, at a rate of up to three times their length in 5 seconds. It is believed that the movements occur in response to external stimuli such as light. Extremely tiny fibrils apparently take up water as they are discharged into the raphe or pores through which moving cytoplasm protrudes. When they come in contact with a surface, they stick and contract, moving the cell as the caterpillar treads on a tractor do and leaving a trail something like that of a snail.

Reproduction in diatoms is unique, with half of the cells becoming progressively smaller through several generations until, through sexual reproduction, the original cell size is restored. Before any cell division can occur, an adequate source of silicon must be present in the surrounding medium. In culture, the number of cells produced is directly proportional to the amount of silicon added to an otherwise nutrient-balanced medium. Division in culture is also rhythmic, with all the cells dividing at the same time. After a diatom's cell contents, which are diploid, have undergone mitosis and division, the two halves of the cell separate, with a daughter nucleus remaining in each half. Then a new half wall fitting *inside* the old half is formed. This occurs for a number of generations, with the result that half of the cells become progressively smaller. Eventually, however, when the ratio of cytoplasm to nucleoplasm reaches a critical point, a cell undergoes meiosis, producing four gametes, which then escape. These fuse with other gametes, becoming zygotes called *auxospores*. Auxospores are like any other zygotes except that the original size of the diatom is restored when the auxospores expand rapidly before rigid walls are formed (Fig. 18.15).

Brown Algae (Phaeophyceae)

Many species of seaweeds and other algae that are brown to olive green in color are assigned to the class Phaeophyceae in this phylum. Previously, the brown algae were in a phylum of their own (*Phaeophyta*), but cladistic and structural analyses suggest relatively close relationships with other members of Phylum Chromophyta. Many brown algae, such as kelp, are relatively large, and none are unicellular or colonial. Only a few of the known genera occur in fresh water; the vast majority grow in colder ocean waters, usually in shallower areas. The giant kelp (*Macrocystis pyrifera*) may be found in water 30 meters (100 feet) or more deep, and one brown alga (*Lobophora variegata*) was seen in clear tropical waters near the Bahamas growing at a depth of 220 meters (730 feet). One giant kelp measured 274 meters (900 feet) in length—a length believed to be a record for any single living organism.

Many of the brown algae have a **thallus** (plural: **thalli,** the term for multicellular bodies that are usually flattened and not organized into leaves, stems, and roots) that is differentiated into a holdfast, a **stipe,** and flattened, leaflike **blades** (Fig. 18.16). The holdfast is a tough, sinewy structure resembling a mass of intertwined roots. It holds the seaweed to rocks so tenaciously that even the heaviest pounding of surf will not break it loose.

The stalk of the stipe is often hollow, with a meristem either at its base or at the blade junctions. Since the meristem produces new tissue at the base, the oldest parts of the blades are at the tips.

The blades and most of the body are photosynthetic and may have gas-filled floats called *bladders* toward their bases (Fig. 18.17). These bladders help to hold the blades near the surface of the water, where they carry out photosynthesis. Sea otters even use the air bladders for support when they sleep.

The color of brown algae can vary from light yellow-brown to almost black, reflecting the presence of varying amounts of the brown pigment fucoxanthin, in addition to chlorophylls *a* and *c* and several other pigments in the chloroplasts. These pigments allow brown algae to grow in deeper water than most other algae.

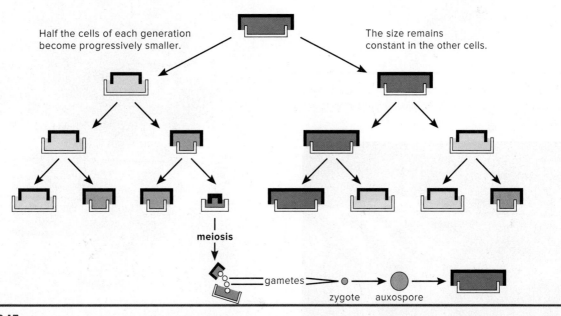

Figure 18.15 When diatoms reproduce, the two rigid halves of each petri dish-like cell separate and a new rigid half shell forms *inside* each original half. This results in half the new cells becoming smaller with each generation. At one point, however, the diploid nucleus of a reduced-size cell undergoes meiosis, and four gametes are formed. The zygote produced when two gametes unite becomes considerably enlarged. The enlarged cell, called an *auxospore,* develops into a diatom of the same size as the original diatom.

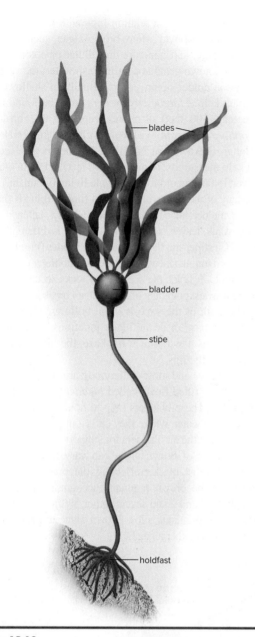

The main food reserve is the carbohydrate *laminarin*. Large kelp beds, such as that shown in the chapter opener, support a diverse array of life in cold ocean waters. **Algin,** or *alginic acid* (see Table 18.2 and the discussion under "Human and Ecological Relevance of the Algae"), occurs on or in the cell walls and can represent as much as 40% of the dry weight of some kelps. Reproductive cells are unusual in that the two flagella are inserted laterally (i.e., on the side) instead of at the ends. The only motile cells in the brown algae are the reproductive cells.

In some areas off the coast of British Columbia and Washington, herring deposit spawn (eggs) on both surfaces of giant kelp blades in late spring. In the past, and to a limited extent at present, native North Americans have harvested these spawn-covered blades, sun dried them, and used them for winter or feast food. Even today, some schoolchildren are given small pieces of the dried or preserved food for lunchbox snacks.

The Sargasso Sea, east of the United States, gets its name from a floating brown seaweed, *Sargassum,* that is washed on shore after tropical storms (Fig. 18.18). A species occurring

Figure 18.16 Features of the kelp *Nereocystis* include leaflike blades, a buoyant air bladder, a stemlike stipe, and a holdfast to anchor it to the ocean floor. $\times \frac{1}{10}$

Figure 18.17 Air bladders are located at the base of blades on this giant kelp (*Macrocystis pyritera*). © Corbis

Figure 18.18 *Sargassum* is a floating brown alga from which the Sargasso Sea got its name. It is also found in other marine waters. $\times \frac{1}{2}$.
© Kingsley Stern

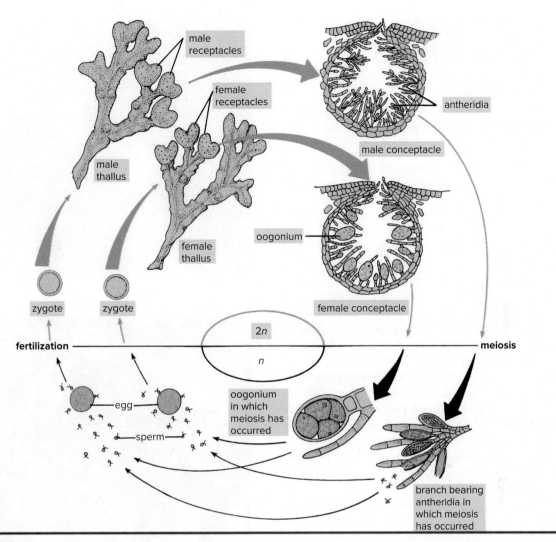

Figure 18.19 Life cycle of the common rockweed *Fucus*.

in the Pacific Ocean has been used, in chopped form, by native Hawaiians as a poultice on cuts from coral. This and several other brown algae reproduce asexually by fragmentation, while some produce spores. Sexual reproduction takes several different forms, depending on the species. The conspicuous phases of the life cycles are usually diploid.

In the common rockweed *Fucus* (Fig. 18.19), separate male and female thalli (bodies) are produced, or both sexes may develop on the same thallus. Swollen, fertile areas called *receptacles* develop at the tips of the branches of the thallus. The surface of each receptacle is dotted with pores (visible to the naked eye) that open into special, spherical hollow chambers called *conceptacles*. Within the conceptacles, **gametangia** (structures where gametes are produced) are formed. Eight eggs are produced in each *oogonium* (female gametangium) as a single diploid nucleus undergoes meiosis followed by mitosis (Fig. 18.20). Meiosis also occurs in each *antheridium* (male gametangium), but four mitotic divisions follow meiosis, so that 64 sperms are produced. Eventually, both eggs and sperm are released into the water, where fertilization takes place, and the zygotes develop into mature thalli, completing the life cycle.

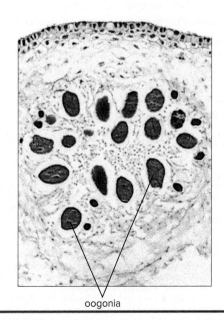

Figure 18.20 A sectioned female conceptacle of *Fucus* showing eggs in oogonia. ×100. © Kingsley Stern

18.6 Phylum Rhodophyta— the Red Algae

Learning Outcome

a. Describe the color, habitat, and life cycles that make members of Phylum Rhodophyta different from other algae.

Like many brown algae, most species of *red algae* are seaweeds (Fig. 18.21) that tend, however, to occur in warmer and deeper waters than their brown counterparts. Some grow attached to rocks in intertidal zones, where they may be exposed at low tide. Others grow at depths of up to 269 meters (884 feet), where the light is only 0.0005% of peak surface sunlight. A few are unicellular, but most are filamentous. However, the filaments frequently are so tightly packed that the plants appear to have flattened blades or to form branching segments. Some develop as beautiful, feathery structures that look like delicate works of art.[1] None of the red algae match the large kelps in size, with the largest species seldom exceeding a meter (3 feet) in length.

The red algae have relatively complex life cycles, often involving three different types of thallus structures. Meiosis usually occurs on a thallus called a *tetrasporophyte,* while gametes are produced on separate male and female thalli. All of the reproductive cells are nonmotile and are carried passively by water currents. Zygotes may migrate from one cell to another through special tubes, which form bizarre loops in some species.

In *Polysiphonia* (Fig. 18.22), a feathery red alga that is widespread in marine waters, the three types of thalli (male gametophyte, female gametophyte, and tetrasporophyte) all outwardly resemble one another. They are about 2 to 15 centimeters (1 to 6 inches) tall and are branched into many fine, threadlike segments. *Spermatangia,* the male sex structures, slightly resemble dense clusters of tiny grapes on slender branches of the male gametophyte thallus. Each spermatangium contains a single *spermatium* that functions as a nonmotile male gamete.

The female sex structures, called *carpogonia,* are produced on the female gametophyte thallus. Each carpogonium consists of a single cell that looks something like a microscopic bottle with a long neck called a *trichogyne.* A single nucleus at the base of the carpogonium functions as the

---

1. Most seaweeds produce their own "glue" and are easy to mount on paper for display. Fresh specimens can be laid directly on clean, high rag-content paper, covered with a layer or two of cheesecloth, and pressed between sheets of blotting paper for a day or two until they are dry. Feathery types will make better specimens if they are placed in a shallow pan of water, so that their delicate structures float out, as they do in the ocean. Paper should be slid under them and then carefully lifted, so that the seaweed spreads out naturally on the surface. Once the specimens are dry, the cheesecloth (which keeps them from sticking to blotting paper) is removed, and they remain glued to the paper. They can then be displayed or stored indefinitely. Green and brown seaweeds can be treated in similar fashion.

Certain marine algae, which form crusty growths or jointed-appearing, upright structures on rocks, accumulate calcium salts as they grow and often contribute to the development of coral reefs. These coralline algae need no special treatment to be displayed, although some may lose their natural pinkish or purplish color when they die.

(a)

(b)

(c)

Figure 18.21 Representative red algae. (a) *Gigartina canaliculata.* (b) *Rhodoglossum affine.* (c) *Chondrus crispus.*
(a-b) © Steven P. Lynch; *(c)* © Dan Ippolito

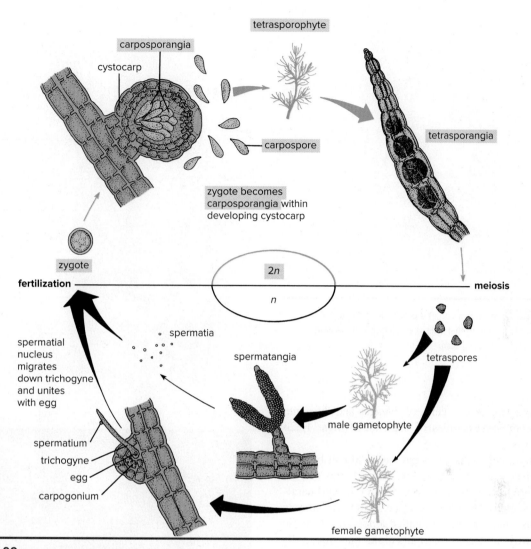

Figure 18.22 Life cycle of the red alga *Polysiphonia*.

female gamete, or *egg*. Since the spermatia have no flagella, they cannot move on their own, but currents may carry them considerable distances. If a spermatium should brush against a trichogyne, it may become attached. The walls between the spermatium and the trichogyne then break down, the nucleus of the spermatium migrates to the egg nucleus, and the two nuclei unite, forming a zygote.

Next, toward the base of a *pericarp* (an urn-shaped body, the outer part of which is formed by the female gametophyte thallus), the zygote begins to divide and eventually develops a cluster of clublike *carposporangia*. The pericarp and carposporangia combined constitute the *cystocarp*. Diploid asexual spores called *carpospores* are produced in the carposporangia, released, and then carried away by ocean currents.

When a carpospore lodges in a suitable location (e.g., a rock crevice or the hull of a ship), it usually germinates and grows into a *tetrasporophyte*, which closely resembles a gametophyte thallus. Tetrasporangia are formed along the branches of the tetrasporophytes. Each tetrasporangium undergoes meiosis, giving rise to four haploid *tetraspores*. When tetraspores germinate, they develop into male or female gametophytes, thereby completing the life cycle.

The red to purplish colors of most red algae are due to the presence of red and blue accessory pigments called *phycobilins,* similar to those found in the cyanobacteria. The similarity led to the belief of some phycologists that the red algae may have been derived from the cyanobacteria. Several other pigments, including chlorophyll *a* and sometimes chlorophyll *d,* are also present in the chloroplasts. The principal reserve food is a carbohydrate called *floridean starch.* A number of red algae also produce **agar** and other important gelatinous substances discussed in "Agar" under the section "Human and Ecological Relevance of the Algae."

18.7 Phylum Euglenophyta— the Euglenoids

Learning Outcome

a. Describe features of a *Euglena* cell.

Barnyard pools and sewage treatment ponds often develop a rich, green bloom of algae. If you were to look at the pond scum under a microscope, you would likely see large

Figure 18.23 *Phacus* is a euglenoid often found in polluted waters. The single flagellum that pulls the cell through the water is not visible. ×2,000. © *Kingsley Stern*

(a)

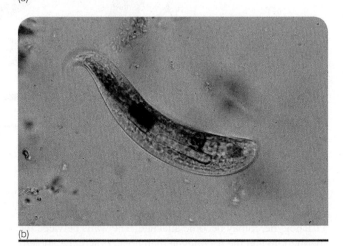

(b)

Figure 18.24 (a) Diagram of a single *Euglena*. (b) Photomicrograph of *Euglena*. ×400. *(b) © Melba Photo Agency/Alamy*

numbers of active, green cells of *Phacus* and *Euglena*, which are common examples of euglenoids (Figs. 18.23 and 18.24).

A *Euglena* cell is has no rigid wall and can be seen to change shape as it moves. Just beneath the plasma membrane are fine strips that spiral around the cell parallel to one another. The strips and the plasma membrane lack cellulose and together are called a *pellicle*. A single flagellum, with numerous tiny hairs along one side, pulls the cell through the water. A second, very short flagellum is present within a reservoir at the base of the long flagellum.

Other features of *Euglena* include the presence of a *gullet*, or groove, through which food can be ingested. The food of most *Euglena* species is ingested, with only about a third having chloroplasts that permit photosynthesis to take place. In the cytoplasm near the base of the flagella, red **eyespot** detects light. Small white bodies store a carbohydrate food reserve called *paramylon*.

Asexual reproduction is by cell division. The cell starts to divide at the flagellar end and eventually splits lengthwise, forming two complete cells. Sexual reproduction is suspected but has never been confirmed.

Some species of *Euglena* can live in the dark if appropriate food and vitamins are present. Others reproduce faster than their chloroplasts under certain circumstances, so that some chloroplast-free cells are formed. As long as a suitable environment is provided, these cells also can survive indefinitely. In the past, when only two kingdoms were recognized, *Euglena's* capacity to satisfy its energy needs through either photosynthesis or ingestion of food resulted in its being treated as a plant in botany texts and as an animal in zoology texts.

18.8 Phylum Dinophyta— the Dinoflagellates

Learning Outcome

a. Describe characteristic features of dinoflagellates.

Occasionally, visitors to an ocean beach in midsummer may notice a distinctly reddish tint to the water, usually as a result of a phenomenon known as a **red tide.** Red tides are caused by the sudden and not fully understood multiplication of unicellular organisms called *dinoflagellates*

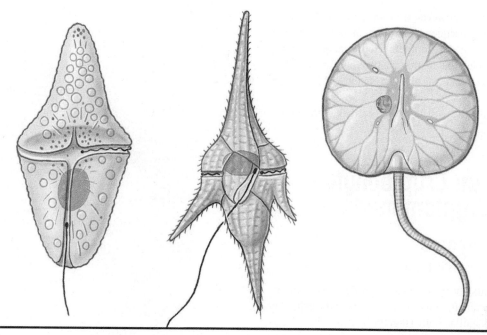

Figure 18.25 In this colorized micrograph of dinoflagellates, one species is colored blue. The brown species illustrates the grooves in which flagella beat.

(Fig. 18.25). When a red tide appears, marine biologists dip a cup of sea water and save it for examination with a microscope. (The material can be preserved indefinitely with the addition of a few drops of formaldehyde, vinegar, or other weak acid.)

This cup of sea water usually contains the shell-like remains of hundreds of dinoflagellates, the best-known representatives of Phylum Dinophyta. Some resemble armor-plated spaceships, while others may be smooth or have fine, lengthwise ribs. The "armor plates" are located just inside the plasma membrane and are composed mostly of cellulose.

Dinoflagellates have two flagella that are distinctively arranged, usually being attached near each other in two adjacent and often intersecting grooves. One flagellum, which acts as a rudder, trails behind the cell. The other, which encircles the cell at right angles to the first groove, gives the cell a spinning motion as it undulates in its groove like a tiny snake.

Most dinoflagellates have two or more disc-shaped chloroplasts, which contain distinctive brownish *xanthophyll* pigments in addition to various other pigments, including chlorophylls a and c_2. About 45% of the species are, however, nonphotosynthetic, and some ingest food particles, whether or not chlorophyll is present. Some have an *eyespot* (a pigmented organelle that is sensitive to light), and all have a unique nucleus in which the chromosomes remain condensed and clearly visible throughout the life of the cell. The chromosomes contain a disproportionately large amount of DNA—as much as 40 times that found in human cells. The food reserve is starch, which in dinoflagellates is stored outside the chloroplasts.

Many dinoflagellates have tiny projectiles that are fired out of the cells when they are irritated. The most common form of projectile is rod-shaped and is fired within milliseconds of irritation. The cells move rapidly in the opposite direction of the discharge, leading to speculation that they aid in escape, but there is also some evidence that they spear other organisms.

Dinoflagellates occur in most types of fresh and salt water, but those that cause red tides have received the most publicity because about 40 of the species also produce powerful neurotoxins that accumulate in shellfish such as oysters, mussels, scallops, and clams. The poisons apparently don't harm most shellfish, but about 2,000 persons become ill each year (15% fatally) from eating contaminated shellfish. Although fish in open waters can swim away from affected areas, large numbers still die and wash ashore after a red tide, and the devastation to caged fish can be enormous. Even pelicans, dolphins, whales, and manatees have been poisoned by dinoflagellate toxins in their food chains.

Three major types of dinoflagellate toxins have been studied for possible use in chemical warfare. They are so potent that as little as half a milligram (barely enough to cover two of the periods printed on this page) can be fatal. Symptoms in humans begin with a tingling feeling in the fingertips and lips and eventually progress to paralysis of the diaphragm, followed by death through suffocation. Other symptoms include nausea, abdominal cramps, muscular paralysis, amnesia, hallucinations, and diarrhea.

The havoc to the fishing industry caused by major red tides is so great that several laboratories have conducted research with dinoflagellates and their marine habitats to try to determine the causes of their blooms and to find a way of preventing such destruction in the future.

Until 1970, blooms of toxin-producing dinoflagellates were known only from temperate waters of the Northern

Hemisphere. By 1990, however, the blooms were also occurring with increasing frequency throughout both temperate and tropical waters of the Southern Hemisphere. As mentioned in the introduction to this chapter and discussed in Chapter 11, dinoflagellates also play a major role in oceanic bioluminescence.

Reproduction is by cell division. Sexual reproduction appears to be rare.

18.9 Phylum Cryptophyta— the Cryptomonads

Learning Outcome

a. Describe characteristic features of cryptomonads.

The **cryptomonads** constitute a small group of asymmetrical, somewhat flattened, unicellular, marine and freshwater algae with two flagella, both flagella bearing short hairs throughout their length. Each cryptomonad cell has one or more plates on the inside of the plasma membrane, with a few also having plates outside of the plasma membrane. The plates may be green, blue, yellowish green, red, or brown. There is a single, two-lobed chloroplast with starch granules surrounding a central pyrenoid. In addition to the nucleus, there is a distinctive *nucleomorph* that possibly represents a vestigial nucleus of a primitive organism that may have, in ancestral forms, lived symbiotically within the larger cell. In most species, the *gullet,* which is analagous to that of a euglenoid, is lined with two or more rows of *ejectosomes*. Ejectosomes are structurally different from the stinging, harpoonlike projectiles (trichocysts) of protozoans and some dinoflagellates but unreel as long threads similar to trichocysts. Asexual reproduction is by mitosis; sexual reproduction is unknown (Fig. 18.26).

18.10 Phylum Prymnesiophyta (Haptophyta)—the Haptophytes

Learning Outcome

a. Describe characteristic features of haptophytes.

Most *haptophytes* are unicellular, with two smooth flagella of similar length inserted at the apex. Their pigments and carbohydrate reserves are similar to those of members of the Chromophyta, from which the majority differ in having a third flagellum—the *haptonema*. Unlike true flagella, however, the haptonema, which is located between the true flagella, does not function in propulsion; instead, it has a sticky tip that aids in capturing food. Haptophyte cells may be naked, but more often they are covered with scales that are sloughed off into the water as they are replaced from

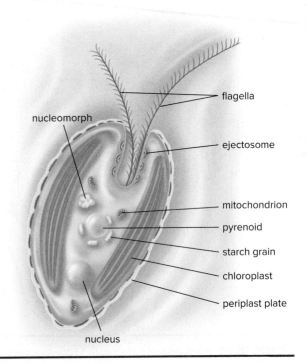

Figure 18.26 Cellular features of *Cryptomonas* sp., a cryptomonad. ×2,000.

beneath, where they are produced by dictyosomes. One group of haptophytes, the *coccolithophores,* has scales formed of calcite crystals. Two disc-shaped chloroplasts are present in each cell. Haptophytes, which occur in both salt and fresh water, are smaller than most diatoms and dinoflagellates but occur in huge numbers, especially in the tropics. Haptophytes are a major component of marine plankton.

18.11 Phylum Charophyta— the Stoneworts

Learning Outcome

a. Describe characteristic features of stoneworts.

Stoneworts (Fig. 18.27) are primarily aquatic organisms of shallow, freshwater lakes and ponds; they often carry calcium precipitate on their surfaces. A stonewort body consists of an axis with short, lateral branches in whorls, so it looks somewhat like a small horsetail plant (discussed in Chapter 21). Many scientists place stoneworts in their own phylum (Charophyta), but others consider them a class of green algae because of their pigmentation and a few other green algal features. Their sexual reproduction is oogamous (nonmotile female gamete and motile male gamete), and, unlike other algae discussed to this point, male gametes of stoneworts are produced by multicellular structures. Other features of both their vegetative growth and reproduction are more complex than those of any of the green algae. Some botanists consider them more closely related to mosses than to algae.

A comparative summary of the food reserves, chlorophylls, and flagella of the phyla of eukaryotic algae is given in Table 18.1.

18.12 Human and Ecological Relevance of the Algae

Learning Outcomes

a. List examples of human uses of diatoms.
b. Discuss the potential contribution of *Chlorella* to human nutrition.
c. Describe the uses of algin.
d. Discuss the uses of kelp by humans.
e. Explain how agar is used in plant propagation.

Diatoms

We have already noted that members of cyanobacteria are at the bottom of aquatic food chains, but members of all the protistan algal phyla play similar roles. Diatoms, for example, are consumed by fish that feed on plankton. Up to 40%

Figure 18.27 *Chara* is an example of a a stonewort. © *Kingsley Stern*

Table 18.1 Comparison of the Phyla of Algae

Phylum	Food Reserves	Chlorophylls[1]	Flagella
Chlorophyta			
7,500 species	Starch	Chlorophyll b	2, 4, or more at apex, or none
Chromophyta			
7,600 species	Oils, fats, mannitol, laminarin	Chlorophyll c_1 Chlorophyll c_2	2, 1, or none, usually unequal, lateral, or at or near apex
Rhodophyta			
More than 5,000 species	Floridean starch, mannitol	Chlorophyll d	None
Euglenophyta			
750 species	Paramylon	Chlorophyll b	2, 1, or 3, at or near apex
Dinophyta			
3,000 species	Starch	Chlorophyll c_2	2; 1 trailing, 1 girdling; few with none
Cryptophyta			
200 species	Starch	Chlorophyll c_2	2; apical or lateral
Prymnesiophyta (Haptophyta)[2]			
500 species	Laminarin	Chlorophyll c_1 Chlorophyll c_2	2; apical, equal
Charophyta			
350 species	Starch	Chlorophyll b	2; subapical

[1]All phyla of algae have chlorophyll a and yellow to orange pigments called carotenes and *xanthophylls*, although the combinations and specific types of xanthophylls vary with the phyla. The *fucoxanthin* of the golden-brown and brown algae is a xanthophyll.

[2]The name Haptophyta was changed to Prymnesiophyta because the international rules governing naming of organisms require that the name of a phylum be based on a generic name within the phylum, and the name Haptophyta was not based on a generic name.

KEY THEME 18.2 ecology

Biofuels from Algae

When we hear about biofuels, most of us think of ethanol from corn, sorghum, and switchgrass. Of course, there are many other sources of biofuels, including waste vegetable oil and methane from livestock operations. However, there is a growing interest in using algae to harvest and store the sun's energy for our consumption. Imagine a cornfield and the volume of air from the ground to 5 meters above those plants. Most of that volume is air, which of course is not fixing carbon. Now, imagine the one-acre surface of a pond and the volume of water 5 meters deep that can be occupied by algae. If they have abundant nutrients, the algae can grow very densely. It is estimated that biofuel production from algae can be up to 300 times greater than that from a crop on land. In addition to efficiently using space, algae are appealing because they offer a tremendous variety of species, with different types of metabolism and growth requirements, resulting in many types of fuels. Algae can be used to produce biobutanol, biodiesel, biogasoline, ethanol, methane, and even jet fuel. Research efforts are underway to identify optimal combinations of species and environments for efficient biofuel production. A new, novel approach is to create photobioreactors using algae growing in glass tubes (Box Figure 18.2A). These are highly controlled systems that maximize the production of biofuels, while minimizing space and resources. This is essentially an industrial system to cultivate algae and harvest the energy they have captured through photosynthesis.

Box Figure 18.2A Tubular bioreactors filled with green algae fixing CO_2, in Andalusia, Spain. © Santiago Urquijo/Getty Images

---

of a diatom's mass consists of oils that are converted into cod and other liver oils that are rich sources of vitamins for humans. In the past, the oils may also have contributed to petroleum oil deposits. The National Renewable Energy Laboratory in Golden, Colorado, is working on a project to convert oil from cultured diatoms into a "clean" diesel fuel substitute.

Diatoms also have other extensive and more direct industrial uses. In the past, they apparently have been at least as numerous as they are now. As billions upon billions of them reproduced and died, their microscopic, glassy shells accumulated on the ocean floor, forming deposits of *diatomaceous earth* (Fig. 18.28). These deposits have accumulated to depths of hundreds of meters in some parts of the world and are quarried in several areas where past geological activity has raised them above sea level. At Lompoc, California, beds of diatomaceous earth are more than 200 meters (650 feet) deep, while in the Santa Maria oil fields of California, deposits reach a depth of 1,000 meters (3,280 feet).

Diatomaceous earth is a light, porous, and powdery material that contains about 6 billion diatom shells per liter (1.057 quarts), yet a liter weighs only eight-tenths of a kilogram (1.76 pounds). It also has an exceptionally high melting point of 1,750°C (3,182°F) and is insoluble in most acids and other liquids. These properties make it ideal for a variety of industrial and domestic uses, including many types of filtration. The sugar industry uses diatomaceous earth in sugar refining, and its use for swimming-pool filters is widespread. It is also used in silver and other metal

Figure 18.28 A diatomaceous earth quarry. © Steven P. Lynch

polishes, in toothpaste, and in the manufacture of paint that reflects light. Reflectorized paint is used on highway markers and signs and on the license plates of cars in some states. Diatomaceous earth is also packed as insulation around blast furnaces and boilers and is manufactured into the construction panels used in prefabricated housing. Gardeners sometimes use diatomaceous earth for slug control. The tiny silica shells are like shards of glass that cut the soft bodies of the slugs.

Pacific Coast fish farmers have experienced losses of salmon and cod when dense concentrations of the diatom

Chaetoceros developed in the aquaculture pens. The diatoms have long, hollow spines that break off and penetrate the fish gills, disrupting gas exchange and causing bleeding. This damage, in turn, may permit secondary infections and excessive mucus production to occur.

Other Algae

A few green algae have occasionally been used for food, but members of this phylum have generally been used less by humans than algae of other phyla. With dwindling world food supplies, this could change. Sea lettuce has been used for food on a limited scale in Asian countries for some time, and several countries are experimenting with the suitability of plankton for human consumption. *Chlorella* is high in protein and contains most of the vitamins needed in human nutrition, except for vitamin C. Since *Chlorella* is so easy to culture, it may become an important protein source in many parts of the world. *Chlorella* has also been investigated as a potential oxygen source for atomic submarines, in addition to its possible use in space exploration.

Algin

Commercially produced ice cream, salad dressing, beer, jelly beans, latex paint, penicillin suspensions, paper, textiles, toothpaste, ceramics, and floor polish today all share a common ingredient, *algin,* produced by the giant kelps and other brown algae. It is now used in so many products that it is difficult to imagine how the world got along without it (Table 18.2).

Algin has the unique capacity to regulate water "behavior" in a wide variety of products. It can, for example,

Table 18.2 Some Uses of Algin

Food
1. As a thickening agent in toppings, pastry fillings, meringues, potato salad, canned foods, gravies, dry mixes, bakery jellies, icings, dietetic foods, flavored syrups, candies, puddings
2. As an emulsifier and a suspension agent in soft drinks and concentrates, salad dressings, barbecue sauces, frozen-food batters
3. As a stabilizer in chocolate drinks, eggnog, ice cream, sherbets, sour cream, coffee creamers, party dips, buttermilk, dairy toppings, milkshakes, marshmallows

Paper
1. Provides better ink and varnish holdout on paper surfaces; provides uniformity of ink acceptance, reduction in coating weight and improved holdout of oil, wax, and solvents in paperboard products; makes improved coating for frozen food cartons
2. Used for coating greaseproof papers

Textiles
1. Thickens print paste and improves dye dispersal; reduces weaving time and eliminates damage to printing rolls or screens

Pharmaceuticals and Cosmetics
1. As a thickening agent in weight-control products, cough syrups, suppositories, ointments, toothpastes, shampoos, eye makeup
2. As a smoothing agent in lotions, creams, lubricating jellies; binder in manufacture of pills; blood anticoagulant
3. As a suspension agent for liquid vitamins, mineral oil emulsions, antibiotics; gelling agent for facial beauty masks, dental impression compounds

Industrial Uses
1. Used in manufacture of acidic cleaners, films, seed coverings, welding rod flux, ceramic glazes, boiler compounds (prevents minerals from precipitating on tubes), leather finishes, sizing, various rubber compounds (e.g., automobile tires, electric insulation, foam cushions, baby pants), cleaners, polishes, latex paints, adhesives, tapes, patching plaster, crack fillers, wall joint cement, fiberglass battery plates, insecticides, resins, tungsten filaments for lightbulbs, digestible surgical gut (disappears by time incision is healed), oil well-drilling mud, clarification of beet sugar; mixed with alfalfa and grain meals in dairy and poultry feeds

Brewing
1. Helps create creamier beer foam with smaller, longer-lasting bubbles

Figure 18.29 Boats return to the Onnemoto Fish Port with kelp harvested near Nemuro, Hokkaido, Japan. Kelp harvesting in some areas of the world is a popular and profitable business. © *Asahi Shimbun via Getty Images*

control the development of ice crystals in frozen foods, regulate the penetration of water into a porous surface, and generally stabilize any kind of suspension, such as a milkshake or other thick fluid containing water. It is produced by several kinds of seaweeds, but a major source is the giant kelp found in the cooler ocean waters of the world, usually just offshore where there are strong currents. This large seaweed is known to grow to 92 meters (300 feet) or more long and sometimes grows at the rate of 3 to 6 decimeters (1 to 2 feet) per day. After cutting, the blades grow back from basal meristems, so cutting a kelp bed is like mowing a lawn. The blades will grow back, for the giant kelps are valuable renewable resources. Specially equipped ocean-going vessels harvest the kelp by mowing off the top meter (3 feet) of growth, taking the chopped material aboard, then transferring it (Fig. 18.29) to processing centers onshore, where algin is extracted and refined.

Minerals and Food

Brown algae also produce a number of other useful substances. Many seaweeds, but particularly kelps, build up concentrations of iodine to as much as 20,000 times that of the surrounding sea water. Although it is cheaper to obtain iodine from other sources in North America, dried kelp has been used in other parts of the world in the treatment of goiter, which results from iodine deficiency. Kelps are relatively high in nitrogen and potassium and have been used as fertilizer for many years. Before they are used, however, the seaweeds need to be rinsed to remove salt. Seaweeds also have been used as livestock feed in northern Europe and elsewhere. In Asia, many marine algae are used for food—in soups, confections, meat dishes, vegetable dishes, and beverages. In Japan, acetic acid is produced through fermentation of seaweeds.

During the Irish famine of 1845–1846, *dulse,* a red seaweed, became an important substitute for the potato crop that had been destroyed by blight. Dulse also occurs along both the Atlantic and the Pacific coasts of North America. In Maine and eastern Canada, where it is a popular snack food, dulse is referred to as "Nova Scotia popcorn." Another red seaweed, *purple laver* ("nori"), occurs in both American and Asian waters and is used extensively for food, particularly in Asia. In China and Japan, it is cultured on nets or bamboo stakes set out in shallow marine bays (Fig. 18.30). It is harvested when the thin, crinkly, gelatinous blades are several centimeters (2 to 3 inches) in diameter and is used in meat and macaroni dishes; soups; and dry, spiced delicacies.

Irish moss is another important edible red alga. It is also used in bulking laxatives, cosmetics, and pharmaceutical preparations. Blancmange is a dessert made from Irish moss and milk. *Carrageenan* is a mucilaginous substance extracted from Irish moss and used as a thickening agent in chocolate milk and other dairy products. *Funori,* obtained from yet another red alga, is used as a laundry starch, as an adhesive in hair dressings, and in some water-based paints.

Agar

One of the most important of all algal substances is agar, produced most abundantly by the red alga *Gelidium.* This substance, which has the consistency of gelatin, is used around the world in biology labs and medical institutions as a solidifier of nutrient culture media for the growth of bacteria. When other nutrients are added to it, it can also be used to solidify culture media for the growth of both plant and animal cells. Full-sized plants have been induced to develop from pollen grains sown on nutrient agar cultures. Orchid tissues are cultured commercially on such media and induced to grow into full-sized plants (as discussed in the section "Micropropagation" in Chapter 14), and agar is now used worldwide in making capsules containing drugs and vitamins. It is also used as an agent in bakery products to retain moistness, as a base for cosmetics, and as an agent in gelatin desserts to promote rapid setting.

Current research involving red algae and other seaweeds indicates they contain a number of substances of potential medicinal value. More than 20 seaweeds have been used in preparations designed for the expulsion of digestive-tract worms, the control of diarrhea, and the treatment of cancer. Some have shown considerable potential as antibiotics and insecticides. Chemical relatives of DDT have been found to be produced by certain red algae. The sea hare and other marine animals feed on such algae, and it is possible that such animals degrade (break down) the DDT-like compounds to simpler substances—a feat unknown among land animals.

Figure 18.30 Beds of purple laver ("nori") growing in the in the sea near Xiapu City, Fujian Province, China. © *Qiang Fu/Getty Images*

18.13 Other Members of Kingdom Protista

Learning Outcome

a. Explain why protozoans and sponges are not discussed in this textbook.

Protozoans (Phylum Protozoa) and *sponges* (Phylum Porifera) are included within Kingdom Protista; they have, however, traditionally been regarded as animals and are not covered in this book. The *slime molds,* which appear to be related to the protozoa, are discussed next.

18.14 Phylum Myxomycota— the Plasmodial Slime Molds

Learning Outcome

a. Describe characteristic features of plasmodial slime molds.

In the Ripley's "Believe It or Not" pavilion at the Chicago World's Fair in 1933, there was an exhibit of "hair growing on wood." The "hairs," while indeed superficially resembling short human hair, were actually the reproductive structures of a species of **slime mold** (Fig. 18.31). These curious organisms do not produce chlorophyll, so they cannot manufacture their own food. They are a bit of a puzzle to biologists because they are distinctly animal-like during much of their life cycle but also distinctly fungus-like when they reproduce.

The tiny, roundish spores of slime molds average only 10 to 12 micrometers in diameter and are individually invisible to the naked eye. Nevertheless, they are present nearly everywhere and are especially abundant in airborne dusts.

If you place almost any dead leaf or piece of bark in a dish and add a food source, such as a few dry oatmeal flakes to which a drop or two of water has been added, the odds are that slime mold spores will be present and will germinate. If the dish is covered, the spores will sometimes germinate in as little as 15 minutes, but germination usually takes several hours or longer. Within a few days, a glistening mass of active slime mold protoplasmic material, looking something like the netted veins of a leaf, may appear. This material, whose "veins" tend to merge into the shape of a fan at its leading edges, is a slime mold **plasmodium** (Fig. 18.32).

The plasmodium has no cell walls. This allows the protoplasm in the veins to flow very rapidly and rhythmically. The protoplasmic movement may stop momentarily at regular intervals and then resume, sometimes in the opposite direction.

Plasmodia are often white, but they also may be brilliantly colored in shades of yellow, orange, pink, blue, violet, or black. A few are colorless and more or less transparent. Plasmodia are found on damp forest debris, under logs, on

Figure 18.31 Reproductive bodies (sporangia) of a plasmodial slime mold. © Ed Reschke/Getty Images

Figure 18.32 A plasmodium of the slime mold *Physarum*. ×10. © Kingsley Stern

Figure 18.33 Common slime mold sporangia include *Lamproderma*. ×25. © Kingsley Stern

old shelf or bracket fungi, sometimes on older mushrooms, and in other moist places where there is dead organic matter. Plasmodia tend to flow forward at a rate of up to 2.5 centimeters (1 inch) or more per hour, often against slow moisture seepage, feeding on bacteria and other organic particles as they go. They contain many diploid nuclei, all of which divide often and simultaneously as growth occurs. With an adequate food supply, a plasmodium may increase to 25 times its original size in just 1 week.

Dramatic events take place when significant changes in food supplies, available moisture, light, and other environmental features occur. The plasmodium is converted, often quite rapidly, into many separate, small **sporangia** (Fig. 18.33), each containing thousands of tiny, one-celled **spores.** The sporangia often are globe-shaped, but in some species, they look like stacks of tubes or hairs. The sporangia may or may not have slender stalks, depending on the species. Others exhibit combinations of body forms. The spores are often distributed throughout a jumbled mass of threads called a *capillitium*.

When a spore is formed, a single nucleus and a small amount of cytoplasm become surrounded by a wall. Meiosis takes place in the spore, and three of the four resulting nuclei degenerate. When the spore germinates, one or more amoeba-like cells called *myxamoebae* emerge. Sometimes, these have flagella and are called *swarm cells*. Either form may become like the other through the development or loss of flagella. At first, myxamoebae or swarm cells feed on bacteria and other food particles. Sooner or later, however, they function as gametes, fusing in pairs and forming zygotes. A new plasmodium usually develops from the zygote, although occasionally zygotes or small plasmodia may fuse and form larger plasmodia (Fig. 18.34).

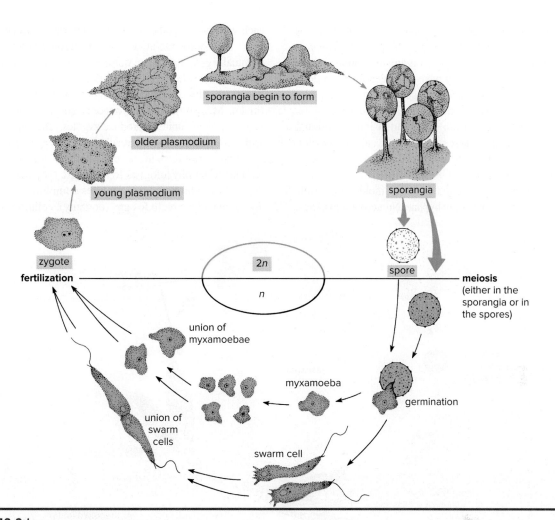

Figure 18.34 Life cycle of a plasmodial slime mold.

18.15 Phylum Dictyosteliomycota—the Cellular Slime Molds

Learning Outcomes

a. Distinguish between plasmodial and cellular slime molds.
b. Discuss the ecological significance of slime molds.

Two or more phyla of slime molds are recognized. Most species have typical plasmodia and follow the patterns just discussed. About two dozen species of *cellular slime molds* are, however, evidently not closely related to the other slime molds. Individual amoeba-like cells of cellular slime molds feed independently, dividing and producing separate new cells from time to time. When the population reaches a certain size, the cells stop feeding and clump together, forming a mass called a *pseudoplasmodium*. The pseudoplasmodium resembles and crawls like a garden slug. It eventually becomes stationary and transforms into a sporangium-like mass of spores.

Human and Ecological Relevance of the Slime Molds

The slime molds, like the bacteria, contribute to ecological balance in forests and woodlands by breaking down organic particles to simpler substances; they also reduce bacterial populations in their habitats. They can often be seen on bark mulch after a rain. One atypical species occasionally attacks cabbages and another causes powdery scab of potatoes. Yet another species causes a disease of watercress, but slime molds otherwise are of little economic significance.

18.16 Phylum Oomycota— the Water Molds

Learning Outcomes

a. Describe characteristic features of water molds.
b. Discuss the impact of water molds on humans.

If you have kept tropical fish aquaria or have seen salmon at the end of a spawning run, you may have noticed cottony growths on cuts and bruises or the eyes of a fish's body or have seen similar growths on the eyes of sick fish. The cottony growth is caused by aquatic organisms, called **water molds,** or *oomycetes*. You may have also seen them on dead insects such as houseflies. They range in form from single, spherical cells to branching, threadlike, coenocytic hyphae. Coenocytic hyphae, which are not divided into individual cells, may form large masses of threads called *mycelia*. Motile cells with two flagella are produced at various stages of their life cycles (Fig. 18.35).

During asexual reproduction in water molds, crosswalls form just below the tips of certain hyphae. Numerous zoospores, each with two flagella, are produced in these special tip chambers. The zoospores, after emerging through a terminal pore, eventually give rise to new water mold mycelia. The slime mold body produces hormones that stimulate the formation of the sexual reproduction structures **oogonia** (female) and **antheridia** (male). Meiosis within the oogonia and antheridia produces gametes, which unite to produce zygotes. The zygotes develop within the oogonia and eventually grow into new mycelia.

Water molds share several features with brown algae, leading some phycologists to speculate they may have been derived from brown algae. Features common to water molds and brown algae include eggs (oogamy), cellulose in the cell

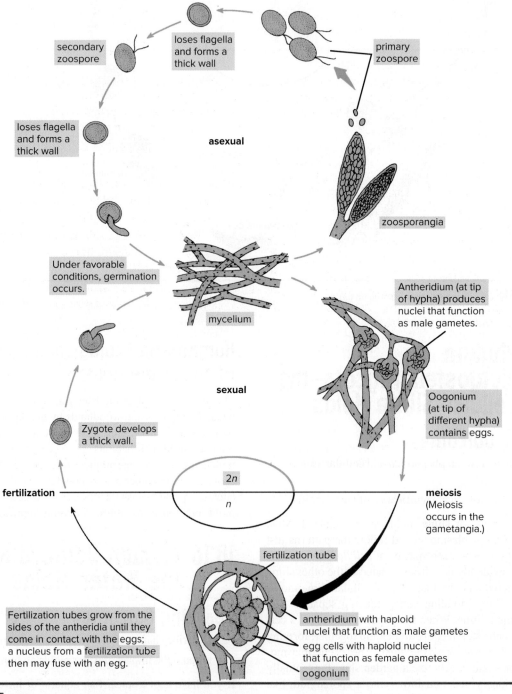

Figure 18.35 Life cycle of the water mold *Saprolegnia*.

walls, a predominantly diploid life cycle, and zoospores with two flagella.

Human and Ecological Relevance of the Water Molds

Two important members of Phylum Oomycota cause serious diseases of higher plants. Although dew or rain water is needed for their reproduction, neither grows under water. One, called *downy mildew of grapes,* completes its life cycle on grape leaves, usually killing the leaves and, if not controlled, the vine.

This disease seriously threatened the French wine industry toward the end of the 19th century, after it reached the vineyards on imported American cuttings. Within a few years, it was controlled when it was discovered that Bordeaux mixture, a combination of copper sulfate and lime (calcium oxide), inhibited the growth of downy mildew. This mixture, which makes the grapes look unappetizing, is the first substance known to have been used as a fungicide. It was originally sprayed on grape vines to discourage grape thieves.

Another important oomycete causes *late blight of potato.* In the past, the late blight pathogen plagued potato farmers so much that at times it altered the course of history. In the summer of 1846, the entire potato crop of Ireland was wiped out, causing a famine during which over a million people starved to death and many more emigrated to the United States. The organism produces its mycelium inside the leaves, so once its spores land on the surface and send hyphae into the interior tissues, it is relatively immune to spraying. Spraying leaves with Bordeaux mixture or other fungicides before the spores have germinated, however, checks development of the oomycete. Despite decades of research aimed at controlling the disease, late blight is still a serious problem in many parts of the world.

SUMMARY

18.1 Introduction to Kingdom Protista
- There is evidence that all living organisms were confined to the ocean less than 1 billion years ago. It is believed that green algae began making the transition from water to land about 400 million years ago and gave rise to land plants.

18.2 Features of Kingdom Protista
- Kingdom Protista includes many diverse organisms that all have eukaryotic cells. Members may be unicellular or multicellular and occur as either colonies or filaments. Modes of nutrition include photosynthesis, ingestion of food, a combination of both, or absorption of food in solution. Some members are nonmotile, but most are motile.

18.3 Algae
- There are many forms of algae. In general, algae have no true leaves or flowers.

18.4 Phylum Chlorophyta—the Green Algae
- Green algae (Chlorophyta) have cells with the same pigments and reserve food (starch) as those of higher plants. The chloroplasts of *Chlamydomonas* have pyrenoids; the cells have two or more vacuoles and often a red eyespot. Asexual reproduction is by mitosis; sexual reproduction is isogamous.
- *Ulothrix* is a filamentous green alga that may be attached to objects by means of a holdfast; each cell contains a curved plate chloroplast around the periphery. Asexual reproduction is by zoospores; sexual reproduction is isogamous.
- *Spirogyra* is a floating, filamentous green alga with spiral, ribbonlike chloroplasts. Asexual reproduction is by fragmentation. Sexual reproduction is by conjugation.
- *Oedogonium* is an epiphytic, filamentous green alga; it has cylindrical, netted chloroplasts. In asexual reproduction, zoospores are produced. Sexual reproduction is oogamous.
- Other green algae include *Chlorella,* desmids, *Acetabularia, Volvox, Ulva,* and *Cladophora.*

18.5 Phylum Chromophyta—the Yellow-Green Algae, Golden-Brown Algae, Diatoms, and Brown Algae
- The several classes of Chromophyta include the yellow-green algae, the true golden-brown algae, the diatoms, and the brown algae. Some members produce statospores.
- Diatoms are very abundant; they have a glassy shell that consists of two "halves" that fit together like a box with a lid.
- Fucoxanthin gives a golden-brown color to most chromophytes. Some diatoms move by contact of the cytoplasm with a surface as it protrudes through the pores.
- Diatoms reproduce asexually by mitosis and sexually through the fusion of gametes that form a zygote.
- Many brown algae (Phaeophyceae) are large seaweeds; their thalli often may be differentiated into a stipe, flattened blades, and a holdfast.
- The color of brown algae is largely due to fucoxanthin; the main carbohydrate food reserve is laminarin. Some brown algae produce algin, a useful gelatinous substance. The reproductive cells have lateral flagella. The common rockweed, *Fucus,* produces eggs and sperms that form zygotes in the water.

18.6 Phylum Rhodophyta—the Red Algae
- Most red algae (Rhodophyta), represented by *Polysiphonia,* are seaweeds with life cycles that involve three different types of thalli and nonmotile gametes.
- Red and blue phycobilins are partially responsible for the colors of red algae; chlorophyll *d* may be present in the chloroplasts. The main carbohydrate

food reserve is floridean starch. Some red algae produce agar, an economically important gelatinous substance.

18.7 Phylum Euglenophyta—the Euglenoids

- Euglenoids (Euglenophyta) have no rigid cell wall, one functional flagellum, a gullet, and paramylon as a carbohydrate food reserve. Reproduction is by cell division. Sexual reproduction has not been confirmed.

18.8 Phylum Dinophyta—the Dinoflagellates

- Dinoflagellates (Dinophyta) are unicellular organisms with two flagella inserted at right angles to each other. Some cause red tides that can kill fish and poison humans. Dinoflagellates in tropical waters exhibit bioluminescence (emission of light) when disturbed.

18.9 Phylum Cryptophyta—the Cryptomonads

- The cryptomonads (Cryptophyta) are biflagellated, unicellular algae with two chloroplasts and an extra vestigial nucleus (nucleomorph).

18.10 Phylum Prymnesiophyta (Haptophyta)—the Haptophytes

- Haptophytes (Prymnesiophyta) have biflagellated cells with a flagellum-like haptonema that aids in the capture of food.

18.11 Phylum Charophyta—the Stoneworts

- Stoneworts (Charophyta) are branched green organisms that superficially resemble horsetails.

18.12 Human and Ecological Relevance of the Algae

- Algae are ecologically and economically important. Diatomaceous earth is used for filtering, polishes, insulation, and reflectorized paint. *Chlorella* is a potential food and oxygen source. Algin is used as a stabilizer and thickening agent in hundreds of products.
- Some brown algae are a source of fertilizer, iodine, and food. Red algae are a source of agar and food and have potential medicinal value.

18.13 Other Members of Kingdom Protista

- Protozoans and sponges are protists but are regarded as animals, so are not discussed in this textbook.

18.14 Phylum Myxomycota—the Plasmodial Slime Molds

- The plasmodial slime molds, which are animal-like in their vegetative state, consist of a multinucleate mass of protoplasm called a plasmodium that flows over damp surfaces, ingesting food particles.
- Slime molds form stationary sporangia that contain spores from which myxamoebae or swarm spores emerge upon germination. Myxamoebae or swarm spores function as gametes, with new plasmodia developing from the zygotes.

18.15 Phylum Dictyosteliomycota—the Cellular Slime Molds

- Cellular slime molds produce a pseudoplasmodium that crawls like a slug and can convert to a stationary, sporangium-like mass of spores.

18.16 Phylum Oomycota—the Water Molds

- Water molds have coenocytic mycelia and include organisms that cause diseases of fish and other aquatic organisms. Asexual reproduction involves zoospores; gametes are produced in oogonia and antheridia.

REVIEW QUESTIONS

1. What forms of sexual and asexual reproduction occur in the green algae?
2. How would you distinguish *Chlamydomonas* from *Euglena*?
3. *Spirogyra, Ulothrix,* and *Oedogonium* all form filaments. How can you tell them apart?
4. In the green algae studied, where in the life cycles does the chromosome number change from haploid to diploid, and vice versa?
5. How do cells of diatoms differ from those of other organisms?
6. Which groups of algae produce the following important products: (1) agar, (2) algin, (3) nerve poisons, (4) abrasives for polishes?
7. Why are some green algae red and some red algae green?
8. Where and how is algin obtained?
9. Is there any difference in structure between the holdfasts of microscopic green algae and those of brown algae?
10. What is unique about the cells of cryptomonads and those of haptophytes?

DISCUSSION QUESTIONS

1. Some algae are attached to solid objects or other organisms, while others are free-floating. What are the advantages and disadvantages of each type of growth?
2. The variety and sizes of algae found in the oceans are considerably greater than those of freshwater forms. Can you suggest reasons?
3. Seaweeds that grow in intertidal zones, where they may be exposed between tides, are often more gelatinous than their continually submerged counterparts. Explain.
4. Why do some algae grow on one side of a tree and not all around the trunk?
5. Should the bladders (floats) of some of the kelps give these algae any advantage over those algae that do not possess such structures?

CHAPTER 19

Kingdom Fungi

19.1 Introduction to the Kingdom Fungi
19.2 Distinctions between Kingdoms Protista and Fungi
19.3 Kingdom Fungi
    Phylum Chytridiomycota—the Chytrids
    Phylum Zygomycota—the Coenocytic True Fungi
    Human and Ecological Relevance of the Coenocytic True Fungi
    Phylum Ascomycota—the Ascomycetes (Sac Fungi)
    Human and Ecological Relevance of the Sac Fungi

Key Theme 19.1: Bat White Nose Syndrome

    Phylum Basidiomycota—the Basidiomycetes (Club Fungi)
    Human and Ecological Relevance of the Club Fungi
    Phylum Deuteromycota—the Deuteromycetes (Imperfect Fungi)
    Human and Ecological Relevance of the Imperfect Fungi

19.4 Lichens
    Human and Ecological Relevance of Lichens

OVERVIEW

After a short introduction, the distinctions between Kingdoms Protista and Fungi are discussed. A summary of the features of Kingdom Fungi and a review of how the kingdom came to be recognized are given.

Kingdom Fungi includes five phyla of fungi. Life cycles and discussions of the economic importance of representative members of each phylum are presented. Nematode-trapping fungi, *Pilobolus,* truffles, morels, ergot, yeasts, stinkhorns, puffballs, bracket fungi, bird's-nest fungi, smuts, rusts, poisonous and hallucinogenic fungi, shiitake mushrooms, mushroom culture, antibiotics, industrial products obtained from fungi, and fungi in nature are discussed. The chapter concludes with an overview of lichens. Natural dyeing is briefly explored, and the economic importance of lichens is reviewed.

Lichens, such as these "British Soldier" lichens, are often found on decaying matter and help break down biomass to release nutrients, as well as fix nitrogen for use by plants. The British Soldier lichen exhibits a symbiotic relationship between the fungus *Cladonia cristatella* and the alga *Trebouxia erici.*
© *Steven P. Lynch*

19.1 Introduction to the Kingdom Fungi

Learning Outcome

a. Distinguish between hyphae and mycelia.

When it comes to breaking down organic materials and making them available for recycling, fungi and bacteria are the most important organisms on the planet. The British Petroleum oil spill that occurred in the Gulf of Mexico in 2010 intensified efforts to find better and faster ways of cleaning up such disasters. Although some bacteria break down oil molecules, very few survive the salt water of the ocean for any length of time. Attention now, however, is being given to fungi that break down many types of oils into harmless, simpler molecules, such as carbon dioxide.

The oyster mushroom (*Pleurotus ostreatus*) can tolerate exposure to seawater and is capable of degrading soil contaminated with diesel fuel. Studies are underway to determine whether a system can be developed to use this fungus or others to remove oil from spills in the open ocean.

The word *fungus* may evoke images of mushrooms or some sort of creeping growth. Although mushrooms are indeed fungi, and while many fungi do appear to be creeping along the ground, the forms of fungi seem almost infinite. There may be over 5 million species of mushrooms, rusts, smuts, mildews, molds, stinkhorns, puffballs, truffles, and other organisms in Kingdom Fungi. Many are yet undiscovered or undescribed, and our daily ravaging of rain forests and other fungal habitats undoubtedly is dooming many species to extinction before they can be found.

As they grow, most fungi produce an intertwined mass of delicate threads that branch and often fuse together (Fig. 19.1a). The individual threads are called **hyphae** (singular: **hypha**). Hyphae may or may not be partitioned into cells.

A mass of hyphae is collectively referred to as a **mycelium** (singular: **mycelia**) (Fig. 19.1b). When appropriate food is available, fungi grow very rapidly. In fact, despite the microscopic size of individual hyphae, all those produced by a single fungus in 1 day when laid end to end could extend more than a kilometer (0.6 mile). Some fungi thrive in freezers if the temperature is not lower than –5°C (23°F), while others grow best at temperatures of 55°C (131°F) and higher.

Besides being vital to the natural recycling of dead organic material, fungi, along with bacteria, also cause huge economic losses through food spoilage and disease; these topics are explored in the sections on human and ecological relevance of fungi later in the chapter. Scientists who study fungi are known as *mycologists* (from the Greek word *myketos*, a fungus). Consumers of mushroom-type fungi are called *mycophagists*.

(a)

(b)

Figure 19.1 (a) Fungal hyphae growing on the surface of wood. (b) Colonies of fungi growing on the surface of stale brewed coffee.
(a) © Chanus/Shutterstock. (b) © Kingsley Stern

19.2 Distinctions between Kingdoms Protista and Fungi

Learning Outcome

a. Distinguish between protists and fungi.

In the past, the true fungi, slime molds, and bacteria were all placed in a single division (phylum) of the Plant Kingdom. Once the fundamental differences between prokaryotic and eukaryotic cells became known, however, the bacteria were placed in the prokaryotic Kingdoms Archaea and Bacteria. Then it became increasingly apparent that the metabolism, reproduction, and general lines of diversity of fungi were different from those of members of the Plant Kingdom, with the fungi evidently having been independently derived from ancestral unicellular organisms. Accordingly, fungi were placed in their own kingdom (see Chapter 16).

All true fungi have chitin in their cell walls. Chitin is a large nitrogenous polysaccharide that is tough and flexible.

In addition, all fungi are filamentous or unicellular heterotrophs, most of which absorb their food in solution through their cell walls. Some are *saprobes* (organisms that live on dead organic matter); others are parasitic decomposers; still others (mycorrhizal fungi) have a mutualistic relationship with plants. Unlike other fungi, chytrids produce flagellated reproductive cells. However, chytrids have chitin in their cell walls, and recent molecular research suggests the chytrids are probably more closely related to the true fungi than they are to members of Kingdom Protista; therefore, they are discussed here under Kingdom Fungi. The water molds, however, which have cellulose but no chitin in their cell walls, continue to be more appropriately included in Kingdom Protista.

The members of Kingdom Fungi are placed in five phyla. With the exception of some chytrids and all yeasts, they are filamentous. Most fungi lack motile cells. Filamentous fungi produce hyphae that grow at their tips. Structures such as mushrooms are formed from hyphae tightly interwoven and packed together. Modes of sexual reproduction vary among the five phyla of fungi and have been important in taxonomic classification. The food substances that most fungi absorb through their cell walls are often broken down with the aid of enzymes secreted to the outside by the cells. Because of the great variety of form and reproduction throughout Kingdom Fungi, a neat pigeonholing of all the members into distinct groups is difficult. Broad groups can, however, be recognized, based mainly on vegetative and reproductive structures.

Recent DNA studies have revealed that fungi are more closely related to animals than they are to plants. It is still conventional, though, to include fungi in botany textbooks. As discussed in Chapter 16, the placement of organisms in taxonomic groupings is in a state of flux due to new insights from DNA sequence data. The old classification system for fungi focused on reproduction strategies. DNA sequence data often do not support those former groupings. This has led to disagreement about a common taxonomic grouping and frequent name changes.

19.3 Kingdom Fungi

Learning Outcomes

a. Describe characteristic features of chytrids.
b. Describe characteristic features of the coenocytic true fungi.
c. Discuss the impacts of the coenocytic true fungi on humans.
d. Describe characteristic features of the sac fungi.
e. Discuss the impacts of the sac fungi on humans.
f. Describe characteristic features of the club fungi.
g. Discuss the impacts of the club fungi on humans.
h. Describe characteristic features of the imperfect fungi.
i. Discuss the impacts of the imperfect fungi on humans.

Phylum Chytridiomycota— the Chytrids

If you were to immerse dead leaves or flowers, old onion bulb scales, dead beetle wing covers, or other organic material in water that has been mixed with a little soil, within a day or two, thousands of microscopic **chytrids** (pronounced kahy′trids) probably would appear on the surfaces of the immersed objects (Fig. 19.2). These simple, mostly one-celled organisms include

Figure 19.2 These chytrids are on a dead leaf submerged in water. The term *chytrid* comes from the Greek term *chytridion*, which means little pot, referring to the spore-containing vessels shown in this illustration.

many parasites of protists, aquatic fungi, aquatic flowering plants, and algae. One chytrid species, *Batrachochytrium dendrobatidis,* causes a disease in amphibians called chytridiomycosis. This disease is believed to be a major cause of amphibian population declines worldwide. Some chytrids parasitize pollen grains, and many other species are *saprobic* (feed on nonliving organic material). Some of the most common chytrids consist of a spherical cell with colorless, branching threads called **rhizoids** at one end. The rhizoids anchor the organism to its food source.

Other chytrids may develop short hyphae or even a complete mycelium whose hyphae contain many nuclei. Such multinucleate mycelia without crosswalls are said to be **coenocytic** (pronounced seè-no-sitt-ik). The cell walls of a few chytrids contain cellulose in addition to chitin.

Chytrids are the most primitive group of fungi. They are typically saprobic and aquatic. Many chytrid species reproduce only asexually through the production of zoospores within a spherical cell. The zoospores, each of which has a single flagellum, settle upon release and grow into new chytrids. Some species undergo sexual reproduction through the fusion of two motile gametes with haploid nuclei or by the union of two nonmotile cells whose diploid zygote nuclei undergo meiosis. The zygote typically becomes a resting spore. The origin of chytrids is unknown, but the presence of a flagellum on the motile cells has led some authorities to suggest they may have originated from the protozoa.

Phylum Zygomycota— the Coenocytic True Fungi

Although black bread molds are probably the best-known members of Phylum Zygomycota, they are not the only fungi that grow on bread. So many organisms can, in fact, contribute to bread spoilage that nearly all commercially baked goods have, in the past, had chemicals such as calcium propionate added to the dough to retard the growth of such organisms. There is now a trend toward eliminating preservatives from bread, pies, and other bakery items, because the chemicals that have made the goods a less suitable medium for the growth of fungi apparently are unhealthy for humans after prolonged use. Alternative ways of retarding fungal growth and spoilage are being sought.

Rhizopus (Fig. 19.3), a well-known representative black bread mold, has spores that are exceedingly common everywhere. They have been found in the air above the North

Figure 19.3 Sexual (top) and asexual (bottom) life cycles of the black bread mold *Rhizopus.*

Figure 19.4 Bread mold. Colored scanning electron micrograph (SEM) of a black bread mold (Rhizopus stolonifer) sporangium (orange, spherical). ca ×2,500. © Steve Gschmeissner/Science Source

Pole, on the inside of buildings, in soils, in clothing, in cars, and hundreds of kilometers out to sea, easily carried there by prevailing winds.

Asexual Reproduction

When a spore lands in a suitable growing area, it germinates into hyphae that may become an extensive mat, or mycelium, which is *coenocytic* (not partitioned into individual cells) and contains numerous haploid nuclei. After the mycelium has developed, certain hyphae called **sporangiophores** grow upright and produce globe-shaped **sporangia** at their tips (Fig. 19.4). Numerous black **spores** are formed within each sporangium. When these spores are released through the breakdown of the sporangium wall, they may blow away and repeat the cycle.

Sexual Reproduction

Black bread molds reproduce sexually by conjugation. Although there is no visible difference in form, black bread mold mycelia occur in two different mating strains, often indicated as "+" and "−." When a hypha of one strain encounters a hypha of the other, the chemicals they produce create an attraction, and swellings called *progametangia* develop opposite each other on the hyphae. The progametangia grow toward each other until they touch. A crosswall is formed just behind each tip, and the two *gametangia* merge, becoming a single, large, multinucleate *coenozygote* in which the nuclei of the two strains fuse in pairs. A thick, ornamented wall then develops around this coenozygote or zygospore with its numerous diploid nuclei. This structure, called a *zygosporangium*, is the characteristic sexual spore of members of this division. A zygosporangium may lie dormant for months, but eventually, it may crack open, and one or more sporangiophores with sporangia at their tips grow out. Meiosis apparently takes place just before this occurs, and thousands of haploid black spores are produced in the sporangia. In some species, the spores are produced externally on hyphae instead of being formed in sporangia.

One dung-inhabiting genus in this phylum has the scientific name of *Pilobolus*, derived from two Greek words meaning "cap thrower." The name is quite appropriate, as the mature sporangia are catapulted a distance of up to 8 meters (26 feet), where they adhere to grass or other vegetation (Fig. 19.5). When the vegetation is ingested by animals, the spores germinate in the digestive tract and are already growing in the dung when it is released.

The sporangia of *Pilobolus* fungi are forcibly released precisely in the direction of light, to which the sporangiophores are very sensitive. This action can be demonstrated by placing some horse dung (preferably at least 2 days old) in a glass dish that has a lid and then covering the dish with black paper. If an opening is cut in the paper and the dish is placed in light for a few days, *Pilobolus* sporangia that have been produced and forcibly discharged will form a black pattern on the glass closely corresponding to the cut-out area. In grassy fields, the sporangia are thrown far from the dung, frequently sticking to blades of grass, where they will be ingested again and ultimately end up in a new dung pile.

Human and Ecological Relevance of the Coenocytic True Fungi

One species of bread mold is used in Indonesia and adjacent areas to produce a food called *tempeh*. Tempeh basically consists of boiled, skinless soybeans that have been inoculated with a bread mold and set aside for 24 hours. The mycelium that develops holds the soybeans together, produces enzymes that increase the content of several B vitamins, and renders the soybean protein more digestible. The tempeh is fried, roasted, or diced for soup and is prepared fresh daily. Other bread mold species are used with soybeans to make a Chinese cheese called *sufu*.

At least two species of *Rhizopus* (*R. arrhizus, R. nigricans*) are used commercially to carry out important steps in the manufacture of birth control pills and anesthetics, while others are used in the production of industrial alcohols and as a meat tenderizer. Yet another species produces a yellow pigment used for coloring margarine.

Phylum Ascomycota— the Ascomycetes (Sac Fungi)

If you travel to the south of France in the summer, you might be surprised to see men and women with coils of rope slung

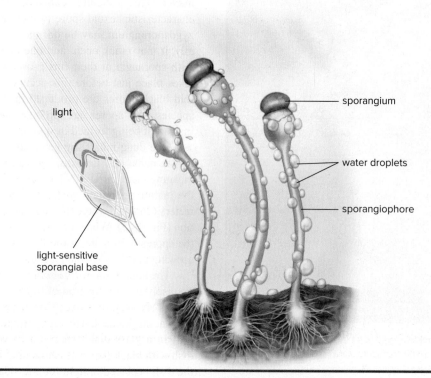

Figure 19.5 The dung fungus *Pilobolus* produces spores in a cap, which are released with force toward a light source.

around their shoulders pushing pigs in wheelbarrows. They happen to be wheeling the pigs into the woods to help them find *truffles*.

Truffles are gourmet "mushrooms," which grow mostly between 2.5 and 15 centimeters (1 and 6 inches) beneath the surface of the ground, usually near oak trees. They look somewhat like prunes and may be more than 10 centimeters (4 inches) in diameter, although most are smaller. They give off a tantalizing aroma that has been shown to contain a chemical similar to pig sex *pheromones*. Pigs can detect truffles a meter (3 feet) below the surface and more than 15 meters (50 feet) away. The animals, in an attempt to get to them, strain at the ropes that are tied around their necks. The owners dig up the truffles, which sell for hundreds of dollars per pound, and reward the pigs with acorns or other less expensive food. Trained dogs are now often substituted for pigs in truffle hunting, as they are less likely to eat the truffles and do less damage to the ground.

Truffles are the reproductive bodies of representatives of a large and varied phylum of true fungi called **ascomycetes (sac fungi).** More than 32,000 species, including yeasts, powdery mildews, brown fruit rots, ergot, morels, microscopic parasites of insects, fungi found in canned fruit, and many others, have thus far been described. Most produce mycelia superficially similar to those of the water molds and coenocytic true fungi, but the hyphae are partitioned into individual cylindrical cells. Each crosswall has one or more pores through which the tiny nuclei can pass. Little bodies that serve as plugs are located near the pores. The "plugs" seal off individual cells if they become damaged.

Asexual Reproduction

Asexual reproduction is by means of **conidia** (singular: **conidium**). Conidia are spores that are produced externally—outside of a sporangium—either singly or in chains at the tips of hyphae called *conidiophores* (see Fig. 19.34). Asexual reproduction in yeasts is by **budding** (Fig. 19.6). As a yeast cell buds, the nucleus divides, and a small protuberance appears to balloon out slowly from the cell. One daughter nucleus moves into the bud, which becomes pinched off as it grows to full size.

Sexual Reproduction

As a fungus body (mycelium) grows, it penetrates through its environment, consuming organic matter along the way. However, when the fungus reproduces, it needs to disperse its spores. Consequently, fungi often produce elaborate fruiting bodies (for example, mushrooms) to release spores into the air where they can be caught by a breeze. The sac fungi get their name from their sexual reproduction stage, in which they produce spores in tiny, fingerlike sacs called **asci** (singular: **ascus**). When hyphae of two different "sexes" become closely associated in the more complex sac fungi, male *antheridia* may be formed on one and female *ascogonia* on the other, although in many species, antheridia and ascogonia may be produced on the

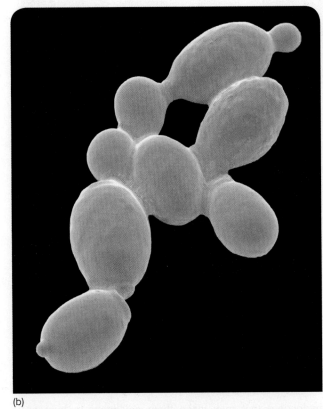

Figure 19.6 Asexual reproduction in yeast via budding, illustrated in (a) and as seen in a photomicrograph (b). *(b)* © Steve Gschmeissner/Science Photo Library/Getty Images

same mycelium. Hyphae grow and connect an antheridium and an ascogonium to each other; male nuclei then migrate into the ascogonium. There, the male nuclei pair with the female nuclei but do not unite. New hyphae (*ascogenous hyphae*), whose cells each contain one male and one female nucleus, grow from the ascogonium and the cells divide in a unique way so that each cell has one of each kind of nucleus. This special cell division begins when the cell at the tip of an ascogenous hypha forms a hook called a *crozier*. The two nuclei in the crozier divide simultaneously, and crosswalls then form, so that there are now three cells; the middle cell contains one male and one female nucleus. This middle cell becomes an ascus in which the two haploid nuclei unite and become a diploid zygote nucleus. The fingerlike, tubular asci develop in a layer (referred to as the *hymenium*) at the surface of a structure called an *ascoma* (previously called an *ascocarp*). The zygote nucleus undergoes meiosis, and the resulting four genetically unique haploid nuclei usually divide once more by mitosis, so that there is a row of eight nuclei in each ascus. Walls form to produce eight *ascospores*, each with a haploid nucleus and a small amount of cytoplasm (Fig. 19.7).

Thousands of asci may be packed together in a mushroom-like structure called an ascoma, which often is cup shaped (*apothecium*—Fig. 19.8). Ascoma may also be completely enclosed (*cleistothecium*—Fig 19.9) or flask shaped with a little opening at the top (*perithecium*). Truffles are an example of apothecia. Cup-shaped ascomata may be several centimeters (2 to 3 inches) in diameter and may be brilliantly colored on the inside. When ascospores are mature, they are often released with force from the asci. If an open ascoma is jarred at maturity, it may appear to release fine puffs of smoke consisting of thousands of ascospores that are dispersed to new locations by air currents.

When an ascospore lands in a suitable area, it may germinate, producing a new mycelium, and then repeat the process. In many instances, however, a number of asexual generations involving conidia are produced between the sexual cycles.

Sexual reproduction in yeasts is more streamlined because individual yeast cells function as ascogonia and antheridia. When two haploid cells unite, they become a diploid fusion cell that may serve as an ascus in which meiosis takes place and ascospores are formed—all within the original cell wall. In some yeasts, however (including the common baking and brewing yeast *Saccharomyces cerevisiae*), the fusion cell may multiply asexually into a diploid colony, and then all the cells will eventually undergo meiosis.

Human and Ecological Relevance of the Sac Fungi

Morels, which some people call the world's most delicious mushrooms, and *truffles* have been prized as food for centuries. Most mushrooms are, however, included in the club fungi discussed in the next section, "Phylum Basidiomycota—the Basidiomycetes (Club Fungi)." Wealthy Romans and Greeks insisted on preparing morels personally according to various recipes they had concocted, and they are still prized as gourmet food today. In the past, numerous unsuccessful attempts were made to cultivate them under controlled conditions, but morels now can be

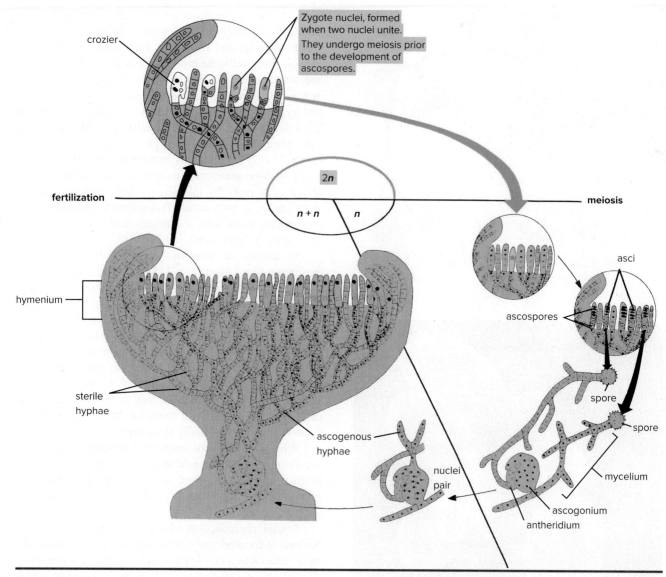

Figure 19.7 Life cycle of a sac fungus. When hyphae of two different strains of a sac fungus become closely associated, male antheridia may be formed on one and female ascogonia on the other. Male nuclei migrate into an ascogonium and pair but do not fuse with the female nuclei present. Then new hyphae (ascogenous hyphae), whose cells each contain a pair of nuclei, grow from the ascogonium. After fusion of the pairs of nuclei, followed by meiosis, fingerlike asci, each containing four or eight ascospores with haploid nuclei, are formed in a layer called a *hymenium*, which lines an ascoma. When mature, the ascospores are released into the air, where they can be carried to new locations to germinate and grow into new mycelia.

mass-produced commercially and are presently being marketed. Morels (Fig. 19.10) are tan in color in North America (and can be black in other parts of the world) and have a spongelike, somewhat cone-shaped top on a stalk that resembles a miniature tree trunk. The numerous depressions between the ridges each contain thousands of asci. Although well-cooked morels are perfectly edible by themselves, some persons have become ill after consuming undercooked specimens or consuming them with alcohol. Caution in this regard is advised.

A related mushroom called a *false morel* or *beefsteak morel* is considered a delicacy by many but has caused death in others. False morels contain a volatile toxin (monomethyl hydrazine) that may make them poisonous to humans if cooking has failed to eliminate all of the toxin.

When rye, and to a lesser extent other grains, comes into flower in a field, it may become infected with ergot fungus (Fig. 19.11). This fungus seldom causes serious damage to the crop, but as it develops in the maturing grain, it produces several powerful drugs. If the infected grain is harvested and milled, a disease called **ergotism** may occur in those who eat the contaminated bread. The disease can affect the central nervous system, often causing hysteria, convulsions, and sometimes death. Another form of ergotism causes gangrene of the limbs, which can result in loss of the affected limb. It can also be a serious problem for cattle grazing in infected fields.

Figure 19.8 An ascoma of a sac fungus. © Kingsley Stern

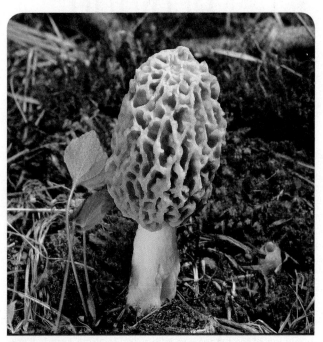

Figure 19.10 This morel is an example of a sac fungus.
© Robert Marien/Getty Images

Figure 19.9 Closed ascomata (*cleistothecia*) of a powdery mildew on an oak leaf. ×200. © Kingsley Stern

Figure 19.11 Ergot in wheat. The darker objects are kernels infected with ergot fungus. © Nigel Cattlin/Science Source

They frequently abort their fetal calves and may succumb themselves.

Ergotism was common in Europe in the Middle Ages. Known then as *St. Anthony's fire,* it killed 40,000 people in an epidemic in A.D. 994. In 1722, the cavalry of Czar Peter the Great was felled by ergotism just as the czar was about to conquer Turkey, and the conquest never took place. Some historians have implicated ergotism in violent social upheavals such as the French Revolution and the Salem Witchcraft Trials. In 1951, an outbreak hospitalized 150 victims in a French village. Five persons died and 30 became temporarily insane, imagining they were being chased by snakes and demons.

In small, controlled doses, ergot drugs have several medical applications. They stimulate contraction of the uterus to initiate childbirth and have been used in abortions and in the treatment of migraine headaches. Ergot is also an initial source for the manufacture of the hallucinogenic drug lysergic acid diethylamide, popularly known as LSD.

KEY THEME 19.1 ecology

Bat White Nose Syndrome

Bat populations across North America are being decimated by an emerging ascomycete fungal pathogen. *Pseudogymnoascus destructans* causes white-nose syndrome in hibernating bats (Box Figure 19.1A). Since its discovery in 2006, the pathogen has killed millions of bats. Four species have experienced declines of greater than 90%, and another is on the brink of extinction. The pathogen is present in Europe but is not nearly as destructive there. The fungus appears to be an invasive species from overseas, where the bats there have coevolved with the fungus for a long time. Spelunking tourists may have brought the fungus over the Atlantic Ocean. During hibernation, the fungus invades muzzle, wing, and ear tissues of bats and causes more frequent arousals during the winter. This depletes fat stores, leaving the animals emaciated by the end of the hibernation period. Bat population surveys in the United States point to a 75% decline in just two years. Genetic analysis has confirmed that a single genotype of the fungus was introduced to North America, specifically to a single cave in New York state. Since only a single clone is present here, the fungus cannot undergo sexual recombination in the wild. DNA sequencing of North American strains reveals that mutations are rare. White-nose syndrome is one of the most devastating wildlife epidemics ever recorded. It highlights a trend toward an increase in new fungal diseases in wildlife worldwide. Scientists are frantically working to devise strategies to control white-nose syndrome. A relatively new field of biology, the study of the microbiome, may offer a solution. We are beginning to understand that the collection of microbes on the skin can influence disease development. Scientists have discovered bacteria (*Pseudomonas* species) that occur naturally on the skin of some bats that can inhibit the growth of *P. destructans*. The presence or absence of these bacteria may help to explain why some populations are more vulnerable to the pathogen than others. This discovery may also lead to the development of a probiotic to protect bats from the disease. Another ray of hope comes from observations that resistance appears to be present in some bat populations that have persisted despite the presence of the pathogen. These colonies have stabilized over time, but they are less than a third of their original size.

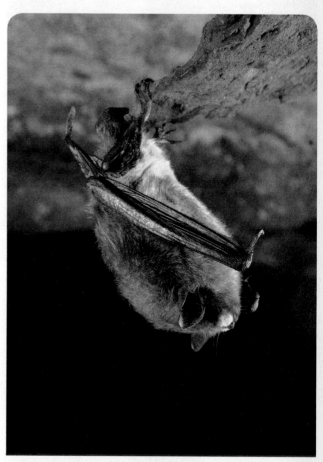

Figure 19.1A Little brown bat with white nose syndrome. *Provided by Al Hicks/New York Department of Environmental Conservation/U.S. Geological Survey*

Fungi in this phylum play a basic role in the preparation of baked goods and alcohol. Enzymes produced by yeast aid in fermentation, producing ethyl alcohol and carbon dioxide in the process. The carbon dioxide produced in bread dough forms bubbles of gas, which are trapped, causing the dough to rise and giving bread its porous texture. Part of the flavor of wines, beers, ciders, sake, and other alcoholic beverages is imparted by the species, or strain, of yeast used to ferment the fruits or grains.

Yeasts are at least indirectly involved in the manufacture of a number of other important products. Ephedrine, a drug also produced by mormon tea, a leafless western desert shrub, is obtained commercially from certain yeasts. The drug is used in nose drops and in the treatment of asthma. Yeasts are also a rich source of B vitamins and are used in the production of glycerol for explosives. Ethyl alcohol is used in industry as a solvent and in the manufacture of synthetic rubber, acetic acid, and vitamin D. Yeast contains about 50% protein and makes a nourishing feed for livestock. The production of the enzyme invertase by yeast catalyzes the conversion of sucrose into glucose and fructose and is used to soften the centers of chocolate candies after the chocolate coating has been applied.

Several very important plant diseases are caused by ascomycete pathogens. *Dutch elm disease* (caused by *Ophiostoma ulmi* or *O. nova-ulmi*), a disease originally described in Holland, has devastated the once numerous and stately American elm trees in many towns and cities in the United States. When Dutch elm disease was first discovered, limited control of the disease was achieved by spraying the trees with DDT to kill the elm bark beetles that spread the disease from tree to tree. DDT killed many useful organisms as well, however, and its use for such purposes was eventually banned.

Other sprays have since been used, again with limited success, and biological controls are now being sought. One such control involved injecting the vascular wilt fungus *Verticillium albo-atrum* into elm trees. The strain of fungus used in this work is pathogenic to potatoes, but it does not cause disease symptoms in elm trees. In fact, it does not move within the tree or survive for more than a few weeks. Apparently, it induces a disease resistance response without actually causing disease. In that sense, it acts as a vaccine.

Chestnut blight has decimated the once numerous populations of American chestnut trees in eastern deciduous forests. Wounding of oak trees in areas where houses are being built has facilitated the spread of the pathogen that causes another serious tree disease, *oak wilt*. Attempts to control these diseases have not been very successful. Much better success has been obtained in controlling *peach leaf curl*, caused by a pathogen that attacks the leaves of some stone fruit trees, especially peach trees (Fig. 19.12). Sprays that contain copper or zinc salts apparently inhibit the germination of spores of the peach leaf curl fungus and have been effective in preventing serious damage when applied to trees before the dormant buds swell and open in the spring.

Phylum Basidiomycota— the Basidiomycetes (Club Fungi)

If you spend much time in shaded garden areas, you may be able to use your nose to find a smelly fungus, which produces fruiting bodies about the size of human fingers. They have the consistency and appearance of a sponge, with the tips covered with a slimy and putrid-smelling substance. *Stinkhorn* fungi (Fig. 19.13) attract flies with their stinky odor; the flies disseminate the sticky spores that adhere to their bodies.

Stinkhorns are interesting but relatively unimportant representatives of another large phylum of true fungi, the **basidiomycetes (club fungi).** Other members of this phylum include mushrooms (Fig. 19.14a), puffballs, earth stars (Fig. 19.14b), shelf or bracket fungi (Fig. 19.14c), rusts, smuts, jelly fungi, and bird's-nest fungi. Common names of members of this group describe their fruiting bodies, which are easily recognized. Their vegetative bodies (mycelium) all look like a white fluffy mass and are hidden in the soil or in other places where they are decomposing organic matter. As mentioned earlier, the elaborate fruiting bodies are produced to get the sexual spores into the air so the fungus can spread to new locations. This group is called club fungi because within their fruiting bodies, they produce their spores at the tips of swollen hyphae that often resemble small clubs. The swollen hyphal tips are called **basidia**

Figure 19.12 A peach leaf infected with peach leaf curl.
© Kingsley Stern

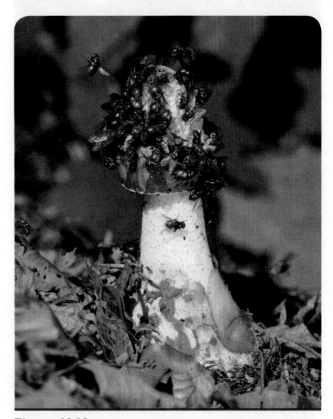

Figure 19.13 A common stinkhorn. The stench of the slimy mass of spores toward the tip attracts flies. © *Hans Lang/Getty Images RF*

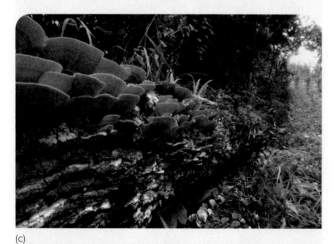

Figure 19.14 Common club fungi include (a) a *Russula* mushroom, (b) Earth star (*Astraeus* spp.), (c) and a shelf, or bracket, fungus. *(a)* © *Kingsley Stern; (b)* © *LASELVA2012/Getty Images RF; (c)* © *topten22photo/Getty Images RF*

(singular: **basidium**). The hyphae, like those of sac fungi, are divided into individual cells. These cells, however, have either a single nucleus or, in some stages, two nuclei. They are produced externally on the tips of the clubs, while sexual spores of ascomycetes are produced internally in sacs. The crosswalls have a central pore (*dolipore*) that is surrounded by a swelling, and both the pore and the swelling are covered by a cap. This cap, with some exceptions, blocks the passage of nuclei between cells but allows cytoplasm and small organelles to pass through.

Asexual Reproduction

Asexual reproduction is much less frequent in club fungi than in the other phyla of fungi. When it does occur, it is mainly through conidia, although a few species produce buds similar to those of yeasts, and others have hyphae that fragment into individual cells, each functioning as a spore and forming a new mycelium after germination.

Sexual Reproduction

Sexual reproduction in many club fungi mushrooms begins in the same way it does for members of the two fungal phyla previously discussed. When a spore lands in a suitable place—often an area with good organic material—it germinates and produces a mycelium just beneath the surface. The hyphae of the mycelium are divided into cells that each contain a single haploid nucleus. Such a mycelium is said to be **monokaryotic.** Monokaryotic mycelia of club fungi often occur in four mating types, usually designated simply as types 1, 2, 3, and 4. Only types 1 and 3, or types 2 and 4, can mate with each other.

If the growth of the hyphae of compatible mating types happens to bring them close together, cells of each mycelium may unite, initiating a new mycelium in which each cell has two nuclei. Such a mycelium is said to be **dikaryotic.** Dikaryotic mycelia sometimes have little, walled-off bypass loops called *clamp connections* between cells on the surface of the hyphae (Fig. 19.15). The clamp

Figure 19.15 The development of a clamp connection, *from left to right*. A protrusion appears in the wall of a cell, and one nucleus migrates into the loop. Both nuclei undergo mitosis, and the loop carrying the daughter nucleus turns toward the cell wall. A clamp connection is established, and the daughter nuclei pair in two cells.

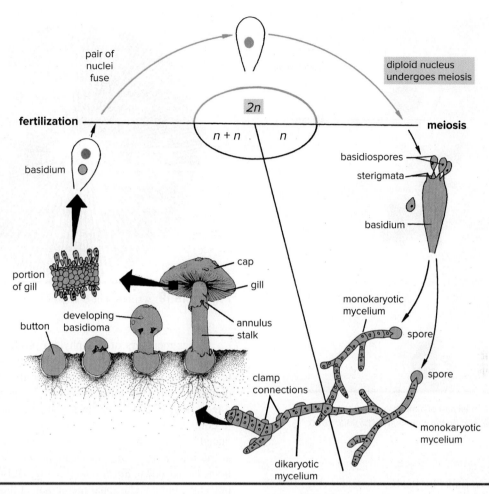

Figure 19.16 Life cycle of a typical club fungus.

connections develop as a result of a unique type of mitosis that ensures each cell will have one nucleus of each original mating type within it.

After developing for a while, the dikaryotic mycelium may become very dense and form a compact, solid-looking mass called a *button*. This pushes above the surface and expands into a *basidioma* (formerly known as a *basidiocarp;* commonly called a fruiting body). This is a spore-producing structure that is visible to the naked eye. A common example of a basidioma is **a mushroom** (Fig. 19.16). Most mushrooms have an expanded, umbrella-like cap (*pileus*) and a stalk (*stipe*). Some have a ring called an *annulus* on the stalk. It is the remnant of a membrane that extended from the cap to the stalk and tore as the cap expanded. Some mushrooms, such as the notorious *death angel* and *destroying angel* mushrooms (*Amanita* spp.), also have a swollen structure called a *volva* at the base (Fig. 19.17). Thin, fleshy-looking plates called **gills** radiate out from the stalk on the underside of the cap. Microscopic examination of a gill reveals it is composed of compacted hyphae, with large numbers of basidia oriented at right angles to the flat surfaces of the gill.

As each basidium matures, the two nuclei unite, and the diploid nucleus undergoes meiosis. The four nuclei that result from meiosis migrate through four (in a few species, two) tiny pegs at the tip of the basidium, with walls forming around the nuclei in the process. The resultant cells are now **basidiospores**. The tiny pegs, called *sterigmata* (singular:

Figure 19.17 In these death angel mushrooms (*Amanita virosa*), note the egglike volva at the base. *Courtesy of Clark Ovrebo*

Figure 19.18 A small part of a mushroom gill showing basidiospores produced on the club-shaped basidia. ×500. © Kingsley Stern

Figure 19.19 A spore print. © Kingsley Stern

Other members of this phylum produce spores within ball-like basidiomata called *puffballs*. Puffballs, which prior to developing their spores are generally edible, range in diameter from a few millimeters to 1.2 meters (0.125 inch to 4 feet) (Fig. 19.23). They have minimal or no stalks and rest in contact with the ground. Trillions of spores may be produced by a large puffball. These are released through a pore at the top or from random locations when the outer membrane breaks down. *Earth stars* (see Fig. 19.14b) are similar to puffballs, but an additional membrane at the base splits and resembles a set of woody flower petals.

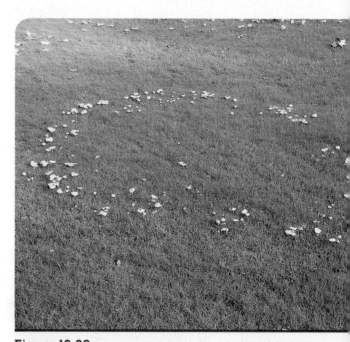

Figure 19.20 A fairy ring of mushrooms shows the typical growth pattern of the fungus vegetative body, radiating outward from a single spore. The mushrooms have emerged from the ground at the edges of the circular body. *Courtesy of Clark Ovrebo*

sterigma), serve as stalks for the basidiospores (Fig. 19.18). One large mushroom may produce several billion basidiospores within a few days. These are forcibly discharged into the air between the gills. They then drift downward and blow away with the slightest breeze.

If you remove a mushroom stalk and place the cap gill-side down on a piece of paper, covering it with a dish to eliminate air currents, the spores will fall and adhere to the paper in a pattern perfectly reflecting the arrangement of the gills. The dish and cap can be removed a day later, and the *spore print* (Fig. 19.19) can be fixed with the application of clear varnish or shellac. These spore prints can help to identify the fungus. White paper is used for dark-colored spores and black paper is used for white or light-colored spores.

In nature, a dikaryotic mycelium radiates out from its starting point, periodically producing basidiomata in so-called *fairy rings* (Fig. 19.20). If conditions are favorable, the mycelium continues to grow at the edges for many years while dying in the center as food resources are depleted. Some mycelia have been known to grow in this way for over 500 years.

Some mushrooms, such as *boletes*, produce their spores on the surfaces of thousands of tiny pores instead of on gills (Fig. 19.21).

Shelf, or *bracket, fungi* (Fig. 19.22) grow out horizontally from the bark or dead wood from which they have grown, some adding a new layer of growth each year. Perennial species can become large enough and so securely attached that they can support the weight of a human adult.

Figure 19.21 A bolete (*Suillus* sp.) mushroom. The basidiospores are produced at the margins of pores instead of along gills. © *Kingsley Stern*

Figure 19.22 A shelf, also called bracket, fungus growing out from the trunk of a tree. *Courtesy of Cliff Pelchat*

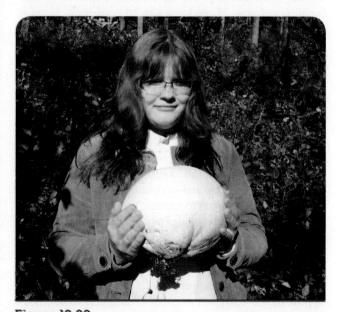

Figure 19.23 A giant puffball. © *Shelley H. Jansky*

Figure 19.24 Bird's-nest fungi. Each "egg" in a nest contains basidiospores that are dispersed when raindrops splash them out. © *Kingsley Stern*

Bird's-nest fungi (Fig. 19.24) grow on wood or manure and form nestlike cavities in which small, egglike bodies containing basidiospores are produced. In some species, each "egg" has a sticky thread attached to it. When raindrops fall in the nests, the eggs may be splashed out, and as they fly through the air, the sticky threads catch on nearby vegetation, whipping the eggs around it (Fig. 19.25). When animals graze on the vegetation, the spores pass unharmed through the intestinal tract.

Smuts are parasitic club fungi that don't form basidiomata. They do considerable damage to corn, wheat, and other grain crops. With corn smut (Fig. 19.26), the mycelium grows between the cells of the host. The hyphae absorb nourishment from host cells and secrete substances that stimulate them to divide and enlarge, forming tumors on the

Figure 19.25 How the "eggs" in a bird's-nest fungus are dispersed. *(From Harold J. Brodie. 1951. "The splash-cup dispersal mechanism in plants."* Canadian Journal of Botany *29:224–34. © 2008 NRC Canada or its licensors.)*

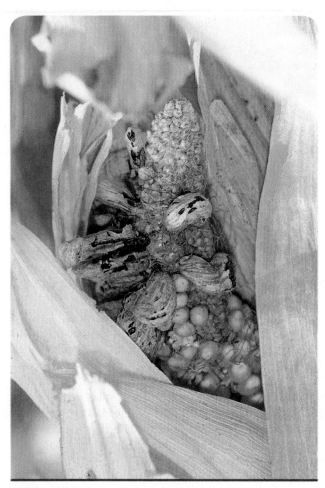

Figure 19.26 Corn smut fungus on an ear of corn.
© Kingsley Stern

Producing rust-resistant strains of wheat has helped, but even as new strains are developed, the rusts themselves hybridize or mutate, producing new races capable of attacking previously resistant varieties of cereals—a striking example of adaptation and natural selection in action.

Another serious rust with two hosts is the *white pine blister rust,* which has caused huge losses of valuable timber trees in both the eastern and western United States. Basidiospores infect the pine trees, and when the basidiospores germinate, other types of spores are produced. These different spores, in turn, infect currant and gooseberry bushes, and the spores formed on the currants and gooseberries eventually give rise to new basidiospores, completing the cycle. The U.S. Forest Service had a program of gooseberry bush eradication in operation for many years in an attempt to alleviate the problem, but the program has been only partially successful. Spraying programs have more recently been implemented with some success, and rust-resistant trees are being selected and bred as alternatives.

Other rusts with two hosts include *cedar-apple rust* (alternate host: cedar trees), *poplar leaf spot* (alternate host: larch or tamarack trees), and *corn rust* (alternate host: sorrel).

A relative of black stem rust is found in the Rocky Mountains of Colorado. This rust causes its rock cress host plants to produce fake flowers that look and smell so real that many insects are fooled by them. When bees and butterflies visit the fake flowers, they find a sugary, nectarlike secretion. While gathering the sticky fluid, they inadvertently also pick up fungal sex cells, which are spread to other rock cress plants.

Human and Ecological Relevance of the Club Fungi

Of the approximately 31,000 described species of club fungi, fewer than 75 are known to be poisonous. Many of the latter are, however, common and not readily distinguishable by amateurs from edible species. Also, some edible forms, such as the inky cap mushrooms, cause no problems by themselves but may make one very ill if consumed with alcohol. Few of the poisonous forms are fatal, but unfortunately, some—such as *Amanita* spp., which cause 90% of the fatalities attributed to mushroom poisoning—are relatively common.

Poisoning from death angels and similar species is due to *alpha-amanitin* (an alkaloid). The poison completely blocks RNA synthesis, and symptoms of poisoning usually take from 6 to 24 hours to appear. Treatment, which in the United States usually includes pumping of the stomach and ingestion of activated charcoal, is seldom successful by the time the intense stomachache, blurred vision, violent vomiting, and other symptoms occur.

In Europe, injections of *silibinin* and *silymarin* (extracts of milk thistle—*Silybum marianum*) do appear to be at least partially effective as antidotes. Oral forms of silymarin are

surfaces of the corn kernels. These eventually break open, revealing millions of sooty black spores, which are blown away by the wind. Some smuts affect only the flowering heads or grains, whereas others infect the whole plant.

Rusts are also parasites that don't form basidiomata. They attack a wide variety of plants. Some rusts grow and reproduce on only one species of flowering or cone-bearing plant. Others, however, require two or more different hosts to complete their life cycles. Rusts have had devastating effects on food supplies since grains were first cultivated nearly 10,000 years ago. Severe losses of grain crops due to stem rust disease in wheat probably contributed to the fall of the Roman Empire. *Black stem rust of wheat* requires both common barberry plants and wheat to complete its life cycle (Fig. 19.27).

Because two hosts are necessary for black stem rust of wheat to complete its life cycle, it was believed that control of the disease could be accomplished through eradication of common barberry bushes. In an attempt to eradicate the disease, an estimated 600 million such plants were destroyed in the United States between 1918 and 1990, but it has proved impossible to eliminate the species altogether, partly because the *urediniospores* (binucleate spores produced in the summer) can overwinter in southern fields and reinfect wheat seedlings in the spring.

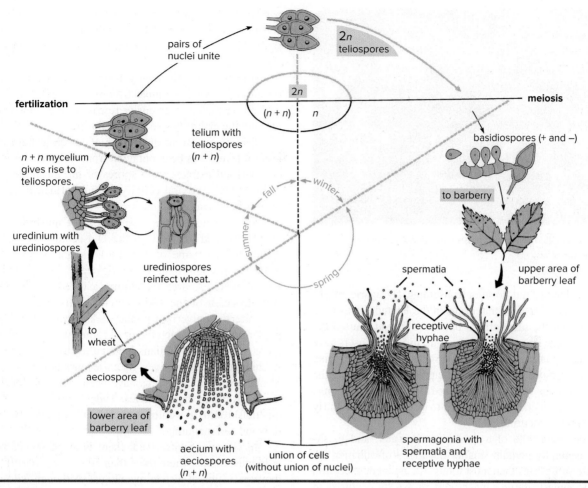

Figure 19.27 Life cycle of black stem rust of wheat.

available in North America, but they are much less effective than the injectable forms. Injectable silibinin and silymarin have not yet received FDA approval for treatment of mushroom poisoning in the United States. When a small, nonfatal ingestion has occurred, the mushroom poison usually leaves the patient with hypoglycemia (a blood sugar deficiency), which can be countered with the intravenous administration of glucose. Some wild mushroom lovers have fed a part of their collection to a dog or cat, and when nothing happened to the animal after an hour or two, they have eaten the mushrooms themselves. Both they and the animals later succumbed. Others have mistakenly believed that the toxic substances are destroyed by cooking. Records show, however, that before the discovery of milk thistle antidotes, death ensued in 50% to 90% of those who had eaten just one or two of the deadly mushrooms, cooked or raw, and that even as little as 1 cubic centimeter (less than 0.5 cubic inch) can be fatal.

There are many widespread but very unreliable beliefs about distinctions between edible and poisonous mushrooms. Some believe that a silver coin placed in the cooking pan will turn black while the mushrooms are cooking if any poison is present. However, there are both edible and poisonous species that will turn such a coin black, and others

of both types will not. Another superstition is that edible species can be peeled, while poisonous ones cannot. Again, this is a fallacy, for *Amanita* mushrooms peel quite easily. Still other erroneous beliefs maintain that poisonous mushrooms appear only in the fall or the early spring, or that all mushrooms eaten by snails and beetles are edible, or that all purplish-colored mushrooms are poisonous, or that all mushrooms growing in grassy areas are edible. Again, these notions are simply not supported by the facts. Wild mushrooms can be eaten with some confidence *only* if they have been identified by a knowledgeable authority. It is foolhardy for anyone to do otherwise.

Some poisonous mushrooms cause hallucinations in those who eat them. During the Mayan civilization in Central America, a number of *teonanacatl* ("God's flesh") sacred mushrooms (Fig. 19.28) were used in religious ceremonies. The consumption of these mushrooms has continued among native groups in Mexico and Central America. Ingestion of small amounts results in sharply focused, vividly colored visions. Similar use of the striking fly agaric mushroom (Fig. 19.29) in Russia, and for a while in the Indus valley of India, dates back to many centuries B.C.E. Users appear to go into a state of intoxication. It is believed that the ancient Norwegian berserkers, who occasionally exhibited fits of exceptionally

Figure 19.28 Teonanacatl mushrooms (*Psilocybe cyanescens*).
© neil hardwick/Alamy

savage behavior, did so after consuming fly agaric mushrooms. Related species that occur in the United States have not produced the same effects but have, instead, caused the user to become nauseated and to vomit. In Siberia, users have noted that the intoxicating principle is passed out in the urine, and some persons have adopted the practice of drinking the urine of persons who have consumed fly agaric mushrooms. Reindeer, incidentally, are reported to be attracted to both fly agaric and human urine.

More than 90% of a mushroom is water, and mushrooms generally contain smaller quantities of nutritionally valuable substances than most foods. An apparent exception is the legendary *shiitake* mushroom, grown for centuries in China and Japan on oak logs and now cultured in the United States. It has more than double the protein of ordinary, commercially grown mushrooms and is very rich in calcium, phosphorus, and iron. It has excellent flavor and contains significant amounts of B vitamins, vitamin D_2 (ergosterol), and vitamin C.

Ancient Chinese royalty believed that eating shiitake mushrooms would promote healthful vigor and retard the aging process, but no extensive research either supporting or refuting the claims has yet been conducted.

Lentinacin, an agent capable of lowering human cholesterol levels, has been obtained from shiitake mushrooms, and purified extracts from spores of the mushrooms have demonstrated antiviral activity against influenza and polio viruses in laboratory animals.

Since ancient times, many types of mushrooms have been cultivated for food. In the 2nd century B.C., a Greek doctor by the name of Nicandros taught people how to grow mushrooms underneath fig trees on soil fertilized with manure. Andrea Cesalpino, a noted Italian botanist and physician of the 16th century, cultivated mushrooms by scattering pulverized poplar bark on very rich soil near poplar trees with which the mushrooms were associated. In more recent times, Italians have cultivated mushrooms on waste material from olives, on coffee grounds, on remnants of oak leaves after tannins have been extracted for leather tanning, and on laurel berries. Today, jelly fungi and various mushrooms are cultivated in Asia on media composed of wood and manure.

In Geneva, Switzerland, there is a special market for wild mushrooms where more than 50 species are sold under the supervision of a state mycologist. Although wild mushrooms are consumed by many in North America, fewer such species are sold in markets. Only one species of mushroom (*Agaricus bisporus*—portabella mushroom) is extensively cultivated commercially, although several others, such as shiitake and oyster, are steadily gaining in popularity. Mushrooms have been grown in basements, caves, and abandoned mines, but contrary to widespread belief, light does not affect their growth.

Large-scale mushroom-growing operations generally use windowless warehouses with stacked rows of shallow planting beds because temperatures, humidity, and other climatic factors are easier to control in such buildings (Fig. 19.30).

Figure 19.29 Fly agaric mushroom. © *Gallo Images-Nigel Dennis/ Digital Vision/Getty Images*

Figure 19.30 Mushrooms growing in beds in a commercial mushroom farm in Petaluma, California. © *Kingsley Stern*

The mushrooms are grown on compost made from straw and horse or chicken manure. Prior to use, the compost is pasteurized for a week to destroy microorganisms, unwanted fungi, and insects and their eggs. Then it is inoculated with spawn, which is compact mycelium grown from germinated basidiospores sown on bran of wheat or other grains. After inoculation, the spawn is covered with a thin layer of soil. The moisture content of the compost is controlled with regular, light waterings, and a humidity of about 75% is maintained.

The mycelium grows in temperatures ranging from just above freezing to 33°C (91.4°F), but commercial growers try to keep the temperatures between 9°C and 13°C (48°F and 55°F) because the mushrooms are less subject to disease or insect attacks and are firmer when grown under cool conditions. The mushroom buttons appear within a week to 10 days after spawn is planted and continue appearing for 6 months or more. About 1 kilogram (2.2 pounds) of mushrooms is obtained from each 10 square decimeters (slightly over 1 square foot) of growing area. An estimated 416 million kilograms (917 million pounds) of mushrooms are produced annually in the United States, and lesser but significant amounts are produced in Canada.

Fungi are constantly breaking down dead wood and debris and returning the components to the soil, where they can be recycled. Sometimes, as we have seen, they can be very destructive from a human viewpoint, attacking everything from living plants and harvested or processed food to shoe leather, paper, cloth, construction timbers, paint, petroleum products, upholstery, and even glass, particularly in warmer, humid climates.

Phylum Deuteromycota—the Deuteromycetes (Imperfect Fungi)

Any fungus for which a sexual stage has not been observed is classified as an **imperfect fungus.** Many otherwise unrelated fungi are grouped together in this artificial phylum, which includes several well-known disease organisms as well as fungi important in disease control and food processing. If a member of this group is studied further and a sexual stage is discovered, the fungus is reassigned to its appropriate phylum. Because members are placed into this group because they lack a known sexual cycle and not because they are unified by common features, the Deuteromycota is considered an informal group, rather than a taxonomic unit. However, this "catch-all" phylum is important because there are many fungi for which a sexual cycle has not been found. Hence, there is no place for them except in this phylum.

Imperfect fungi most commonly reproduce by means of conidia (asexual spores). One such group grows on dead leaves and debris at the bottom of streams with rapidly moving water. The spores often have four long extensions on them. These arms are arranged in such a way that three may catch like a tripod on a flat surface, keeping the fungus from being washed downstream.

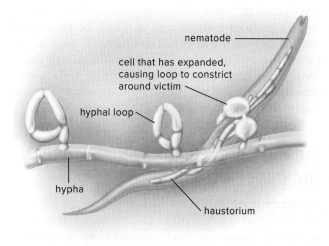

Figure 19.31 A nematode-trapping fungus with a victim. *Source: Drechsler, C. 1937. "Some hyphomycetes that prey on free-living terricolous nematodes."* Mycologia *29:447–552.*

One interesting group of imperfect fungi parasitizes protozoans and other small animals. Some develop an unbranched body inside their victim and slowly absorb nutrients until the host dies. They then produce a chain of spores that may stick to, or be eaten by, another victim. Others capture their prey on sticky hyphae to which passing amoebae adhere. One soil-dwelling group captures nematodes (eelworms) in hyphal rings or loops. In some, the rings are a little smaller than the circumference of a nematode, which tapers at both ends. When a nematode randomly happens to stick its head through such a loop, it frequently tries to struggle forward instead of backing out and becomes trapped (Fig. 19.31). The fungus then produces small, rhizoid-like outgrowths called *haustoria* that grow into the worm's body and digest it.

More specialized species of nematode-trapping fungi have loops that constrict around the animal less than a tenth of a second after contact. The loops are spaced so that sometimes the tail gets caught in a second loop while the nematode is thrashing around after being caught by the head. A number of these fungi can easily be grown on agar media, but they do not form loops unless a nematode is placed in the dish. The nematodes evidently produce a substance that promotes the development of the loops. Several species of imperfect fungi are found in ant and termite nests. The insects cultivate their fungus gardens by bringing in bits of leaves, other plant debris, caterpillar droppings, and their own feces, and in so doing, they form a rich growing medium. They also lick the hyphae and constantly probe them, using them for food as the fungi grow.

Some wood-boring beetles have pouches in their bodies that function as fungus spore containers. The spores germinate in the wood tunnels, their mycelia forming a lining that produces yeastlike cells on which the beetles and their larvae feed. As discussed in Chapter 5, most higher plants have mycorrhizal fungi associated with their roots. These fungi greatly increase the absorptive surface area around the roots and may be far more important than root hairs in this regard, particularly in mature roots.

Human and Ecological Relevance of the Imperfect Fungi

Among the best known of the medically important fungi are the *Penicillium* molds (Fig. 19.32), which secrete *penicillin,* the well-known and widely used **antibiotic** (a substance produced by a living organism that interferes with the normal metabolism of another living organism). In 1929, Sir Alexander Fleming of England noticed that certain bacteria would not grow near the mycelium of a *Penicillium* mold. He gave the name penicillin to the chemical in the mold that prevented the bacterial growth.

Fleming apparently did not grasp the significance of his findings, and neither did the medical profession until the outbreak of World War II some 10 years later. At that time, a team of British and American scientists at Northern Regional Laboratories in Peoria, Illinois, set out to see if they could coax *Penicillium* molds into producing more of this antibacterial substance. The war created a need for greater quantities of more effective medicines that could keep wounds from becoming infected and to prevent the spread of venereal diseases.

The scientists began with cultures from the original mold observed by Fleming, but the amounts of penicillin it produced were so small that it was very expensive to obtain significant quantities for human use. The scientific community then appealed to the general public, asking them to send in any material they found with a greenish or bluish mold on it. They received trainloads of moldy trash from all over the United States.

The breakthrough in the research came, however, when a different species of *Penicillium* mold—one that yielded 25 times the penicillin produced by the original culture—was found on a moldy cantaloupe from a local market. The scientists set to work germinating individual spores of this new mold on culture media, and by careful selection, they were eventually able to isolate a strain that produced more than 80 times the original quantity of penicillin. Later, when this strain was subjected to X-radiation, still other forms were produced that increased the penicillin output to 225 times that of Fleming's mold. Today, most of the penicillin produced around the world comes from descendants of that cantaloupe mold. Hundreds of other antibiotics effective in combating human and animal diseases have been discovered since the close of World War II, and the production of these drugs is a huge, worldwide industry.

Penicillium molds are also used in other ways. Some are introduced into the milk of cows, sheep, and goats during the production of gourmet cheeses, such as blue (Fig. 19.33), Camembert, Roquefort, Gorgonzola, and Stilton. The molds produce enzymes that break down proteins and fats in the milk, giving the cheeses their characteristic flavors.

Since the early 1980s, organ transplants have been aided by the discovery and production of a "wonder drug" from an imperfect fungus found in soil. Called *cyclosporine,* the drug suppresses immune reactions that cause rejection of transplanted organs without risking the development of leukemia and other undesirable side effects associated with other drugs.

Aspergillus is a genus of imperfect fungi that produces dark-brown to blackish or yellow spores (Fig. 19.34). It is closely related to the *Penicillium* molds and is extensively used in industry. One or more species are used commercially for the production of citric acid, a substance for flavoring foods and for the manufacture of effervescent salts. Citric acid is also used in the manufacture of inks and in medicines, and it is even used in some chewing gums.

Figure 19.32 A *Penicillium* colony. © *Kingsley Stern*

Figure 19.33 Blue cheese. The dark areas are conidia of a species of *Penicillium* that gives the cheese its unique flavor. © *Kingsley Stern*

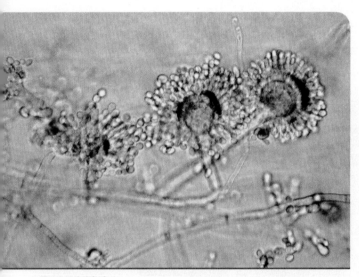

Figure 19.34 Conidiophores of *Aspergillus*. Rows of conidiospores (also called *conidia*) are radiating out from the conidiophores. © NNehring/E+/Getty Images

Aspergillus fungi also produce gallic acid, used in dyes and indelible black ink. Other species are used in the production of artificial flavoring and perfume substances, chlorine, alcohols, and several acids. Further uses are in the manufacture of plastics, toothpaste, and soap and in the silvering of mirrors.

One species of *Aspergillus* is used in Asia and elsewhere to make soy sauce, or *shoyu*, by fermenting soybeans with the fungus. A Japanese food called *miso* is made by fermenting soybeans, salt, and rice with the same fungus.

A number of diseases of both humans and animals are caused by *Aspergillus* species. The diseases, called *aspergilloses,* attack the respiratory tract after the spores have been inhaled. One type thrives on and in human ears. Other diseases caused by different genera of imperfect fungi include those responsible for the widespread problem of athlete's foot and ringworm, for white piedra (a mild disease of beards and mustaches), and for tropical diseases of the hands and feet that cause the limbs to swell grotesquely.

Aspergillus flavus, which grows on moist seeds, secretes *aflatoxin,* the most potent natural carcinogen known. The toxin causes liver cancer, and no more than 50 parts per billion is allowed in human food. In humid climates, such as those of the southeastern United States and adjacent Mexico, improperly stored grain can become moist enough to support the fungus, so foods such as peanuts, peanut butter, and peanut-based dairy feeds become carcinogenic. Dairy cattle feed is even more strictly controlled because aflatoxin can accumulate in milk.

One serious disease called *valley fever* (*coccidiomycosis*), found primarily in the drier regions of the southwestern United States, usually starts with the inhalation of dust-borne spores of an imperfect fungus (*Coccidioides immitis*) that produces lesions in the upper respiratory tract and lungs. The disease may spread in the body, with sometimes fatal results.

19.4 Lichens

Learning Outcomes

a. State what types of organisms are needed to form a lichen.
b. Describe the layers of a lichen.
c. List three different growth forms of lichens and how they may be identified.

Lichens traditionally have been referred to as prime examples of *symbiotic* relationships. Each consists of a fungus and an alga (or cyanobacterium) intimately associated in a spongy **thallus.** The thallus can range in diameter from less than 1 millimeter to more than 2 meters (0.04 inch to 6.5 feet). The photosynthetic component supplies the food for both organisms. The fungus protects the photosynthetic organism's partner from harmful light intensities, produces a substance that accelerates the rate of photosynthesis, and absorbs and retains water and minerals for both organisms. Neither the fungus nor the photosynthetic organisms grow independently where the lichen grows. This supports the belief that the components of a lichen are in symbiotic association. The physiological evidence suggests, however, that it would probably be more correct to say that the fungus parasitizes the photosynthetic component in a controlled fashion, actually destroying chlorophyll-containing cells in some instances.

There are about 14,500 known species of lichens. The photosynthetic component is either a green alga or a cyanobacterium, with a few lichens having two species of algae present. The photosynthetic partner in lichens may be any of about 100 species of algae or blue-green bacteria. Each lichen also has its own unique species of fungus. With the exception of about 20 tropical species of lichens that have a club fungus and one species (associated with bald cypress trees) that has a bacterial component, the fungal components of lichens are sac fungi. It is possible to isolate and culture the components separately. When this is done, however, the fungus takes on a very different, compact but indefinite shape, and the algae or cyanobacteria grow faster than they do when they are part of a lichen. The fungal component is very rarely found growing independently in nature, while the photosynthetic component is known to thrive independently of the lichen in some instances. Lichen species, therefore, are identified according to the fungus present.

Lichens grow very slowly, at a maximum rate of 1 centimeter (0.4 inch) and a minimum of as little as 0.1 millimeter (0.004 inch) per year. They are capable of living to an age of 4,500 or more years and are tolerant of environmental conditions that kill most other forms of life. They are found on bare rocks in the blazing sun or bitter cold in deserts, in both arctic and antarctic regions, on trees, and just below the permanent snow line of high mountains where nothing else will grow. One species grows completely submerged on ocean rocks. Lichens even attach themselves to manufactured substances, such as glass, concrete, and asbestos.

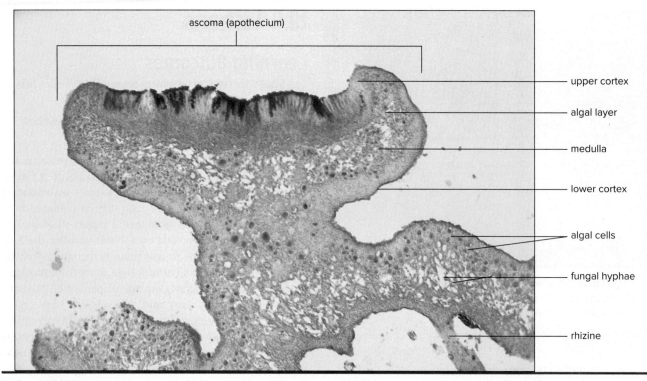

Figure 19.35 A section through a foliose lichen. *Courtesy of G.S. Ellmore*

Part of the reason for the wide range of tolerance of lichens is the presence in the lichens' thalli (bodies) of gelatinous substances, which enable them to withstand periods of rapid drying alternating with wet spells. While they are dry, their water content may drop to as low as 2% of their dry weight, and the upper part of the thallus becomes opaque enough to exclude much of the light that falls on them. In this state, most environmental extremes do not affect them at all, as they temporarily become dormant and do not carry on photosynthesis.

The lichen thallus usually consists of three or four layers of cells or hyphae (Fig. 19.35). At the surface is a protective layer constituting the *upper cortex*. The hyphae are so compressed they resemble parenchyma cells, and it is here that the gelatinous substances play a protective role. Below the upper cortex is an *algal layer* in which algal (or cyanobacterial) cells are scattered among strands of hyphae. Next is a *medulla* consisting of loosely packed hyphae, which occupies at least half the volume of the thallus. A number of substances produced by the lichen are stored here. A fourth layer, called the *lower cortex*, is frequently but not always present. It resembles the upper cortex but is usually thinner and is often covered with anchoring strands of hyphae called *rhizines*.

Lichens have been loosely grouped into three major growth forms, which have no basis in their natural relationships but are convenient as a first step in their identification (Fig. 19.36). *Crustose lichens* are attached to or embedded in their substrate over their entire lower surface. They often form brightly colored, crusty patches on bare rocks and tree bark. The hyphae of some that grow on sedimentary rocks penetrate as much as 1 centimeter (0.4 inch) into the rock.

Others grow just beneath the cuticle of the leaves of tropical hardwood trees, with no apparent harm to the leaves. *Foliose lichens* have somewhat leaflike thalli, which often overlap one another. They are weakly attached to the substrate. The edges are frequently crinkly or divided into lobes. *Fruticose lichens* may resemble miniature upright shrubs, or they may hang down from branches. Their thalli, which are usually branched, are basically cylindrical in form and are attached at one point.

It should be stressed that while lichens may be attached to trees or other plants, the majority do not parasitize them. There are, however, a very small number of species that do produce parasitic rhizines that penetrate the cortical parenchyma cells of their hosts.

Although the fungal component of a lichen usually reproduces sexually, lichens are naturally dispersed in nature through asexual structures. In about a third of the species, small, powdery clusters of hyphae and algae called *soredia* (singular: *soredium*) are formed and break away from the thallus in a set pattern as it grows. Rain, wind, running water, and animals act as agents of dispersal. In other lichens, specialized parts of the thallus, known as *isidia*, may break off or be separated by decay.

Sexual reproduction in lichens is similar to that of the sac fungi except that the ascomata produce spores continuously for many years. No one has yet observed the initiation of a new thallus in nature, but it is believed that thalli arise after ascospores carried by the wind come in contact with independently living algae, germinate, and parasitize them. Lichen algae reproduce by mitosis and simple cell division.

Human and Ecological Relevance of Lichens

Industrialization has had a significant negative impact on lichens. They are exceptionally sensitive to pollution, particularly sulfur dioxide. In fact, their sensitivity to this air pollutant is so great that it has been possible to calculate the amount of sulfur dioxide present in the air solely by mapping the occurrence or disappearance of certain lichens in a given area. These studies have shown that some species of lichens have disappeared entirely from industrial areas and are now facing extinction due to air pollution. They are also very sensitive to nuclear radiation and were used on one occasion to monitor radioactive contamination when a satellite fell to earth in a remote area. After the nuclear meltdown in Chernobyl, reindeer moss lichens in Lapland absorbed radioactive elements. The meat from reindeer that ate contaminated lichens was unsafe to eat.

Lichens contribute to the degradation of historic ruins and other exposed rock or stone materials. Archaeological sites such as those of Troy (in modern Turkey) and Machu Picchu in Peru have suffered damage from the acids produced by lichens, and sometimes older gravestones slowly become covered by various lichen species.

Lichens provide food for many lower animals as well as large mammals. Reindeer and caribou can survive exclusively on fruticose lichens in Lapland when other food is unavailable, while North African sheep graze on a crustose lichen. By themselves, lichens do not make good food for human consumption, but they have been used as a food supplement (e.g., in soups) in parts of Europe. Most have acids that make them unpalatable, and some (e.g., rock tripe) have had harsh laxative effects on those who have tried them.

More than half of the lichens investigated have antibiotic properties. One lichen substance has been used in Europe in combination with another antibiotic in the treatment of tuberculosis. Europeans have also used lichen antibiotics to produce salves that have been effective in treating cuts and skin diseases.

Lichens were used for dyes by the Greeks and Romans, and a lichen dye industry persisted for many centuries in Europe. Native Americans and others also used lichens for dyes. Coal tar dyes now have largely replaced those of lichens, but lichens are still used to manufacture Scottish tweeds and East Indian cotton fabrics. Lichens are used in the preparation of the litmus paper used in chemistry laboratories as an acid–alkaline indicator. The paper turns red under acid conditions and blue under alkaline ones.

Soaps are scented with extracts from lichens, and such extracts are still used in the manufacture of some European perfumes. Because of their resemblance to miniature trees and shrubs, some fruticose lichens are used by toy makers for the scenery of model railroads and car tracks. The importance of lichens in initiating soil formation is discussed in Chapter 25.

(a)

(b)

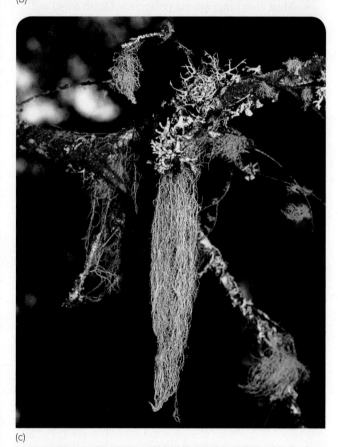
(c)

Figure 19.36 The three lichen growth forms are (a) crustose, (b) foliose, and (c) fruticose. *(a) © Kingsley Stern; (b) © McGraw-Hill Education/David Cox, photographer; (c) © McGraw-Hill Education/Richard Gross, photographer*

SUMMARY

19.1 Introduction to the Kingdom Fungi
- A mass of fungal threads collectively is called a mycelium.

19.2 Distinctions between Kingdoms Protista and Fungi
- True fungi, water molds, and slime molds apparently were derived independently of each other.

19.3 Kingdom Fungi
- Kingdom Fungi, whose members differ from those of other kingdoms in reproduction, metabolism, and lines of diversity, consists of five phyla of organisms.

- True fungi are all filamentous or unicellular with chitinous walls; they are mostly nonmotile, and most produce spores. Sexual reproduction is varied. Fungi may thrive at temperatures below freezing or well above human comfort levels. They are major natural decomposers and cause huge economic losses through food spoilage and diseases.

- Phylum Zygomycota (coenocytic fungi)—represented by *Rhizopus*—has coenocytic mycelia that produce sporangiophores with sporangia containing numerous spores at their tips. Sexual reproduction involves gametangia produced on hyphae of opposite strains that merge, creating a single, large cell, which becomes a zygosporangium and in which nuclei fuse in pairs. Meiosis occurs prior to the zygosporangium cracking open.

- In *Rhizopus*, spores are produced externally on hyphae and released through a breakdown of the sporangium wall. *Pilobolus* releases its sporangia with force toward a light source. Bread molds are used to make tempeh and sufu and can be used for various industrial purposes.

- The hyphae of most true fungi are partitioned into cells with pores in their crosswalls.

- Phylum Ascomycota (sac fungi) includes truffles, yeasts, powdery mildews, brown fruit rots, ergot, morels, and insect parasites. Asexual reproduction is by conidia or budding. Sexual reproduction involves the formation of asci following the union of male and female structures.

- Ergotism is a disease produced when rye infected with ergot is ingested. Yeasts produce carbon dioxide and alcohol used in baking and brewing, the drug ephedrine, and proteins and vitamins used for human and livestock consumption. Dutch elm disease, chestnut blight, and peach leaf curl are caused by ascomycetes.

- Phylum Basidiomycota (club fungi) includes mushrooms, bracket fungi, rusts, smuts, puffballs, stinkhorns, earth stars, jelly fungi, and bird's-nest fungi. The hyphae are partitioned into cells with either one nucleus (monokaryotic) or two nuclei (dikaryotic).

- Asexual reproduction in basidiomycetes is usually by conidia. Sexual reproduction involves the fusion of cells of two mating types of monokaryotic hyphae. The union initiates the development of a dikaryotic mycelium that usually has clamp connections between adjacent cells. The dikaryotic mycelium often develops basidiomata.

- The cap of a mushroom is usually on a stalk, which may have an annulus around it and sometimes a volva at the base.

- Basidia and basidiospores are produced on gills that radiate out from the stalk under the cap.

- Puffballs and earth stars are stalkless club fungi that may each produce trillions of spores. Shelf fungi may add a new layer of growth each year.

- Bird's-nest fungi produce their basidiospores in "eggs" that may become attached to vegetation ingested by animals. The eggs pass through the intestinal tract of the animals unharmed.

- Smuts infest cereals. Rusts, which may have more than one host, infest cereals and other important plants.

- Most club fungi are not poisonous. Death angel and destroying angel mushrooms cause about 90% of the fatalities attributed to mushroom poisoning. In Europe, milk thistle extracts appear to be at least partial antidotes to mushroom poisoning.

- For safety, poisonous and edible mushrooms should be distinguished from each other only by authorities.

- Some mushrooms have been used for intoxication and hallucinatory purposes. Most mushrooms are not high in nutritional value. *Agaricus bisporus* is the mushroom most under commercial cultivation in the United States.

- Fungi and bacteria play a major role in natural recycling processes. They are also destructive of many economically important substances.

- Members of Phylum Deuteromycota (imperfect fungi) all lack known sexual reproductive phases but otherwise resemble members of the other phyla of true fungi.

- Some imperfect soil fungi trap and digest nematodes. The *Penicillium* molds, which produce penicillin, and *Aspergillus*, some species of which are used in the production of shoyu, miso, and other commercial products, are primarily imperfect fungi. Other imperfect fungi also cause diseases, such as athlete's foot and ringworm.

- Some imperfect fungi are cultivated by ants and termites in their nests and form mycorrhizal associations with the roots of higher plants.

19.4 Lichens

- The bodies (thalli) of lichens consist of an alga or a cyanobacterium and a fungus in intimate symbiotic relationship. The photosynthetic component is usually a green alga or a cyanobacterium that may occur in many different species of lichens. Each lichen species has its own unique fungus (mostly sac fungus) component by which it is identified.
- Lichens grow very slowly and may live to be thousands of years old. They are exceptionally sensitive to industrial pollution.
- The lichen thallus usually consists of several layers of hyphae, one of which has algae or cyanobacteria between the threads.
- Lichens occur in three growth forms: crustose, foliose, and fruticose.
- Lichens reproduce sexually but are naturally dispersed primarily by soredia and other asexual means. Lichens provide food for animals, antibiotics,

REVIEW QUESTIONS

1. Why are fungi placed in a kingdom separate from protists and bacteria?
2. How do the cells of fungi differ from those of other organisms?
3. What does *coenocytic* mean? To which groups of organisms does it apply?
4. What means of asexual reproduction do the fungi exhibit?
5. If you were looking at hyphae of a bread mold, a sac fungus, and a club fungus through a microscope, could you distinguish among the three? How?
6. Which fungi produce zoospores?
7. Define conidia, haustoria, zygosporangium, sporangiophore, ascus, mycelium, ascoma, monokaryotic, dikaryotic, and clamp connection.
8. Is there any one best way to tell a poisonous mushroom from an edible one?
9. What is ergotism?
10. Discuss the human relevance of the sac fungi.
11. What is a spore print?
12. To which phylum of fungi do each of the following belong: truffles; puffballs; fungi involved with LSD; stinkhorns; peach leaf curl; late blight of potato; bird's-nest fungi; jelly fungi; downy mildew of grape; death angel; shelf, or bracket, fungi?
13. What is a fairy ring, and why does it form?
14. How do rusts and smuts attack their hosts?
15. Under what conditions do commercially produced mushrooms grow best?
16. Why are some fungi referred to as being "imperfect"?
17. Discuss the economic importance of imperfect fungi.
18. What is the relationship between an alga and the fungus in a lichen?
19. What makes up the basic structure of a typical lichen?
20. How is it possible for lichens to live to such great ages?

DISCUSSION QUESTIONS

1. Because fungi produce so many trillions of spores, why is it we are not overrun by them?
2. Why are fungal diseases much more common now in the United States than they were at the time of the Declaration of Independence?
3. If all rusts had only one host, would they (theoretically, at least) be easier to control?
4. Because most lichens are not good for food and are not very important economically, should we be as concerned about some of them becoming extinct as we are about whales disappearing?

CHAPTER

20

Introduction to the Plant Kingdom: Bryophytes

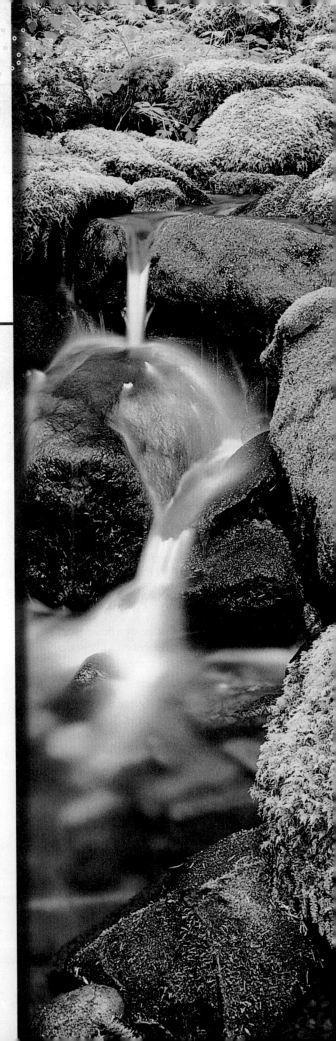

20.1 Introduction to the Plant Kingdom
20.2 General Characteristics of Bryophytes
20.3 Phylum Hepaticophyta—Liverworts
   Structure and Form
   Thalloid Liverworts
   "Leafy" Liverworts
20.4 Phylum Anthocerophyta—Hornworts
   Structure and Form
   Asexual Reproduction
   Sexual Reproduction
20.5 Phylum Bryophyta—Mosses
   Structure, Form, and Classes
   Sexual Reproduction

   Key Theme 20.1: Hibernating Mosses
   Asexual Reproduction
20.6 Human and Ecological Relevance of Bryophytes

OVERVIEW

This chapter opens with a discussion of the features of the Plant Kingdom and its members and then introduces bryophytes. Liverworts are explored after the habitats and general life history of bryophytes are discussed. Next, the chapter examines hornworts and their life cycles and then goes over mosses in greater detail. The chapter concludes with observations on the human and ecological relevance of bryophytes.

Moss-covered rocks in a small stream bed. © *Ron Crabtree/Digital Vision/Getty Images*

20.1 Introduction to the Plant Kingdom

Learning Outcome

a. Identify characteristic features of each of the three phyla of bryophytes.

Older botany texts tended to apply the term "plants" to many of the simpler organisms mentioned in previous chapters. Today, however, botanists confine this term to bryophytes, ferns, cone-bearing plants, flowering plants, and relatives of each of these groups. These groups are the subjects of the next several chapters.

Plants and green algae, discussed in Chapter 18, share several major pigments (e.g., chlorophyll *a,* chlorophyll *b,* carotenoids). Plants and green algae also have starch in common as the primary food reserve and cellulose in their cell walls. When plant and some algal cells divide, they develop *phragmoplasts* and *cell plates,* both of which are unknown in other organisms. These shared features suggest plants and green algae are each other's closest relatives and hence were derived from a common ancestor.

The invasion of land was a significant advance in plant evolution. The first land plants were comprised of simple filaments; organ and tissue systems evolved over time. Another important adaptation to land was the development of a life cycle comprised of an alternation of generations. In this life cycle, a haploid plant (gametophyte) produces gametes that unite at fertilization to form a zygote that grows into a diploid plant (sporophyte) that undergoes meiosis to produce haploid spores that grow into gametophytes.

The chapter focuses on the most primitive plants, the bryophytes. Bryophyte sporophytes are nutritionally dependent on gametophyte plants. These gametophytes are photosynthetic and produce food for the diploid sporophyte. When you see a moss plant growing by a stream, for example, you are seeing gametophytes. During the evolution of plants, this interdependence was broken, so the sporophyte no longer needed the gametophyte for its nutrition. This produced new opportunities for sporophytes in higher plants to develop into the tremendous array of forms we see in our world today. If you look out your window and see green plants, you are likely looking at sporophytes.

By the time plants became established on land, they had developed several features that allowed them to function in their new environment. (1) Plant surfaces developed fatty cuticle that retards water loss. (2) The **gametangia** (gamete-producing structures) and **sporangia** (spore-producing structures) of plants became multicellular and surrounded by a sterile cell jacket. (3) Plant zygotes developed into multicellular **embryos** (immature plantlets) within parental tissues that originally surrounded the egg.

Most members of the Plant Kingdom are more complex and varied in form than those of the kingdoms discussed in the preceding chapters, and their tissues are correspondingly more specialized for photosynthesis, conduction, support, anchorage, and protection. The sporophyte phases become larger and more independent in more advanced taxonomic groups; reproduction is primarily sexual. Plants produce multicellular embryos; organisms in the previous chapters do not.

This chapter provides an introduction to members of the Plant Kingdom typically referred to as *mosses.* Former taxonomic keys classified mosses together with similar organisms, including *liverworts* and *hornworts.* However, these organisms are now recognized in one of three distinct groups. In this text, three phyla of nonvascular plants, the *bryophytes,* and several phyla of **vascular plants** (plants with xylem and phloem) are included in the Plant Kingdom. The bryophytes, presented in this chapter, are currently classified into the three following phyla:

1. **Phylum Hepaticophyta** (*liverworts*). Gametophytes grow horizontally and have a flattened or leafy appearance, which may resemble lobes of animal liver. The gametophyte of some liverworts may form cuplike structures that produce propagules asexually, although sexual reproduction also takes place to produce the sporophyte.
2. **Phylum Anthocerophyta** (*hornworts*). Gametophytes may resemble those of liverworts, sometimes with a crumpled appearance and, after fertilization, the sporophyte resembles a hornlike structure that projects out of the gametophyte.
3. **Phylum Bryophyta** (*mosses*). Gametophytes vary but can have stemlike and leaflike structures that make them distinguishable from liverworts and hornworts. Sporophytes produced after fertilization often tower above the gametophyte generation and release spores from unique sporangia.

20.2 General Characteristics of Bryophytes

Learning Outcome

a. Describe the ecological significance of bryophytes.

In the midst of heavy battles in France during World War I, nurses ran out of bandages for the wounded soldiers. In desperation, they substituted some soft, green plant material they found growing in the water at the edge of a nearby lake. The material turned out to be a great substitute for the bandages; there were fewer infections in the wounds with the plant bandages than in those with the cotton bandages.

The material the nurses used was a species of Sphagnum moss (peat moss), which has since been experimentally demonstrated to have antiseptic properties. This moss has specialized water-absorbing "leaves" (see Fig. 20.10) and has been used as a packing material in the past. It is still widely used as a soil conditioner. The "bandage" Sphagnum is one of about 23,000 species of bryophytes that include mosses, liverworts, and hornworts, many

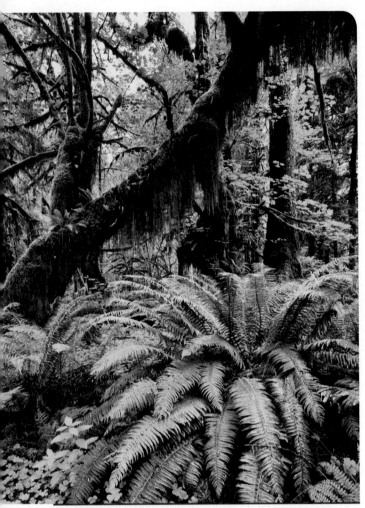

Figure 20.1 Moss on the trees and ferns are shown here at the Olympic National Park (Quinault Rain Forest). © Donald E. Hall/Getty Images

Figure 20.2 *Grimmia*, a rock moss that survives on bare rocks, often in scorching sun. © Kingsley Stern

of which frequently make a soft and cool-looking green covering on damp banks, trees, and logs that are shaded for at least a part of the day (Fig. 20.1).

In contrast, some bryophytes can withstand long periods of desiccation and are found on bare rocks in the scorching sun (Fig. 20.2), while others occur on frozen alpine slopes. The habitats of bryophytes range in elevation from sea level near ocean beaches up to 5,500 or more meters (18,000 or more feet) in mountains.

Some bryophytes are restricted to very specific habitats. For example, a few species are found only on the antlers and bones of dead reindeer. Others are confined to the dung of herbivorous animals, while still others grow only on the dung of carnivores. A few tropical bryophytes thrive only on large insect wing covers. The pygmy mosses, which appear annually on bare soil after rains, are only 1 to 2 millimeters (0.04 to 0.08 inch) tall and can complete their whole life cycle in a few weeks.

Bryophytes of all phyla often have mycorrhizal fungi associated with their rhizoids. In some instances, the fungi are apparently at least partially parasitic. One species of completely colorless liverwort that lives underneath mosses is totally dependent nutritionally on its fungal associate. In bryophytes, the sporophyte is nutritionally dependent on the gametophyte. As you will see in the next chapters, the sporophyte becomes nutritionally independent in higher plants.

The peat mosses are ecologically very important in bogs and in the transformation of bogs to dry land. Peat mosses sometimes form floating mats over water and keep conditions acid enough to inhibit the growth of bacteria and fungi. Organisms that die in such waters or bogs are often preserved for hundreds or even thousands of years. In the mid-twentieth century, two bodies were found near each other in a bog in Denmark. They are estimated to have died around 280 B.C.E. One body, the so-called Tollund Man, was so well preserved that it was initially believed that he had recently died. The combination of acid from the *Sphagnum* moss, the low oxygen in the bog, and cold climate allowed even soft tissues to be preserved. Several other "bog bodies" have been discovered. The oldest with intact flesh is the Cashel Man, who died around 2000 B.C.E.

The luminous mosses are found in caves near the entrances and in other dark, damp places. They are called luminous because they glow an eerie golden-green in reflected light. The upper surfaces of their cells are slightly curved, and each cell functions as a tiny magnifying glass, concentrating the dim light on the chloroplasts at the base. This allows photosynthesis to take place in light otherwise too faint for it to occur (Fig. 20.3).

None of the bryophytes have true xylem or phloem, and to be able to sexually reproduce, all bryophytes must have external water, usually in the form of dew or rain. Many mosses have special water-conducting cells called *hydroids* in the centers of their stems, and a few have food-conducting cells called *leptoids* surrounding the hydroids. Neither type of cell, however, conducts as efficiently as vessel elements or tracheids of xylem and sieve tube members of phloem, and most water is absorbed directly through the surface. The absence of xylem and phloem makes most bryophytes soft and pliable, and it is not surprising, therefore, that birds often use them to line their nests. In one study in the Appalachian Mountains of Virginia, for example, David Breil and Susan Moyle found birds native to the area used at least 65 species of bryophytes in the construction of their nests.

Figure 20.3 Lenslike cell of the protonema of the luminous moss. The cell concentrates light on the chloroplasts at the base.
Source: Adapted from Springer Science + Business Media: From Strasburger's Textbook of Botany, 30th German ed., 1976, D. von Denffer, W. Schumacher, K. Magdefrau, and F. Ehrendorfer, translated by P. Bell and D. Coombe.

Alternation of generations is more conspicuous in bryophytes and ferns than in most other organisms. In mosses, the "leafy" plant is the *gametophyte* generation that produces the *gametes*. The *sporophyte* generation, which grows from a "leafy" gametophyte, produces the *spores*.

It usually resembles a tiny can with a rimmed lid at the tip of a slender, upright stalk.

The relationship between vascular plants and bryophytes is still uncertain. Liverworts group with mosses, but additional studies are needed to confidently place hornworts and vascular plants in the phylogenetic tree. All bryophytes have similar life cycles and growth habits. However, based on their structure and reproduction, they are separated into three distinct phyla.

20.3 Phylum Hepaticophyta— Liverworts

Learning Outcomes

a. Discuss the relationship between a liverwort protonema and thallus.
b. Describe major features of sexual and asexual reproduction in *Marchantia*.
c. Distinguish between thalloid and leafy liverworts.

The word *wort* simply means plant or herb. In medieval times, when the *Doctrine of Signatures*[1] held sway, the herbalists of the day thought some of the bryophytes—specifically, those with flattened bodies and liver-shaped lobes (Fig. 20.4)—were useful for treating liver diseases. This belief proved to be without merit, but the name *liverwort* is still universally used today.

---

1. The *Doctrine of Signatures* held that any plant part that resembled a human body part would be helpful in healing that body part if it were diseased or hurting.

Figure 20.4 *Marchantia* (a thalloid liverwort). (a) Thalli with male gametophores (antheridiophores). (b) Thalli with female gametophores (archegoniophores). *(a) © Kingsley Stern; (b) © Steven P. Lynch*

Structure and Form

There are about 6,000 known species of liverworts. The most common and widespread liverworts have flattened, lobed, somewhat leaflike bodies called **thalli** (singular: **thallus**). The *thalloid liverworts,* however, constitute only about 20% of the species. The other 80% are "leafy" and superficially resemble mosses.

Liverworts differ from mosses in several details and are considered less complex. Their thalli or leafy stages (gametophytes) develop from spores. When the spores germinate, they may produce a *protonema,* which is an immature gametophyte consisting of a short filament of cells. Typically, the protonema develops further into a more extensive gametophyte, or in some instances, the protonemal stage may be absent, with the mature gametophyte then developing directly from the spore. Thalloid liverworts have smooth upper surfaces as well as various markings and pores, and the corners of cell walls of most liverworts are specially thickened. The lower surfaces have many one-celled *rhizoids.* Growth is prostrate instead of upright, and the rhizoids, which look like tiny roots, anchor the plants.

Thalloid Liverworts

The best-known species of thalloid liverworts are in the genus *Marchantia* (named in honor of French botanist N. Marchant) (Fig. 20.4). The most widespread *Marchantia* species is often found on damp soil after a fire. The thallus, which is about 30 cells thick in the center and 10 cells thick at the margin, forks dichotomously as it grows. Each branch has a notch at the apex and a central groove that extends back lengthwise behind the notch. The thalli grow in size as the meristematic cells at the notches continue to divide. Older tissues at the rear decay as the new growth is added. The upper surface of the thallus is divided into diamond-shaped or polygonal segments, the segment lines marking the limits of chambers below. Each segment has a small, bordered pore opening into the interior.

Seen through a microscope, a sectioned liverwort thallus looks like groups of short, erect, branching rows of cells with chloroplasts, sitting on a floor of colorless "bricks" (Fig. 20.5). The "brick wall," which may constitute most of the thallus, consists of parenchyma cells that have few, if any, chloroplasts. The tissue apparently stores substances produced in other cells. The bottom layer of cells is an epidermis from which rhizoids and scales arise. The individual groups of upright chlorenchyma cells are enclosed by vertical walls covered by a slightly dome-shaped layer of epidermal cells. A conspicuous pore is located in the center of each "roof" and remains open at all times. It resembles a tiny, short, suspended, open-ended barrel.

Marchantia–*Asexual Reproduction*

Marchantia reproduces asexually by means of **gemmae** (singular: **gemma**). Gemmae are tiny, lens-shaped pieces of tissue that become detached from the thallus. They are produced in small gemmae cups scattered over the upper surface of the liverwort gametophyte (Fig. 20.6). Raindrops may splash the gemmae as much as 1 meter (3 feet) away. While gemmae are in the cup, lunularic acid inhibits their further development, but each is capable of growing into a new thallus as soon as it leaves the cup. In addition, parts of an older thallus may die, isolating patches of active tissue, which may then continue to grow independently.

Marchantia–*Sexual Reproduction*

The gametangia of *Marchantia* are produced on separate male and female gametophytes and are more specialized than those of other liverworts. Both types of gametangia are formed on **gametophores** (umbrella-like structures borne on slender stalks rising from the central grooves of the thallus) (Fig. 20.6). The top of the male gametophore, or **antheridiophore,** is disclike with a scalloped margin, while that of the female gametophore, or **archegoniophore,** looks like the hub and spokes of a wagon wheel.

Figure 20.5 A longitudinal section through a portion of a *Marchantia* thallus.

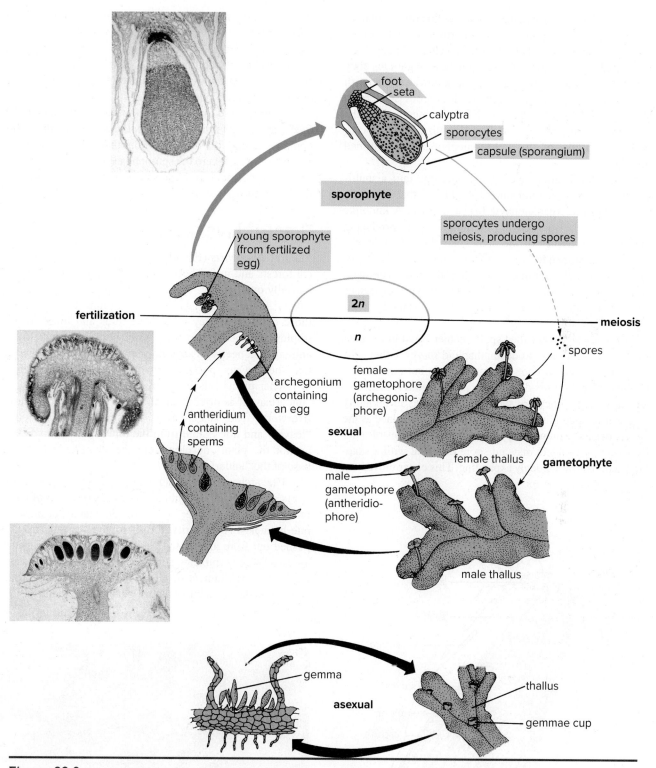

Figure 20.6 Life cycle of the thalloid liverwort *Marchantia*. © James E. Bidlack

Club-shaped male gametangia (**antheridia**) containing numerous *sperms* are produced in rows just beneath the upper surface of the antheridiophore. **Archegonia** are flask-like female gametangia, each containing a single *egg;* they are also produced in rows and hang neck downward beneath the spokes of the archegoniophore. Raindrops sometimes splash the released sperms, which have numerous flagella, more than 0.5 meter (1.5 feet) away. Fertilization may occur before the stalks of the archegoniophores have finished growing.

After fertilization, the zygote develops into a multicellular embryo (an immature *sporophyte*) that is totally dependent on the gametophyte for sustenance. A knoblike **foot** anchors the sporophyte (the diploid, spore-producing phase) in the tissues of the archegoniophore. The sporophyte hangs suspended from the underside of the archegoniophore by a short, thick stalk called the **seta**. The main part of the sporophyte, in which different types of tissues develop, is called a **capsule**. Liverwort sporophytes typically have no stomata.

Sporocytes (also called spore mother cells) in the capsule undergo meiosis, producing haploid *spores*. Other cells inside the capsule do not undergo meiosis but remain diploid and develop instead into long, pointed **elaters** with spiral thickenings. They are sensitive to changes in humidity (Fig. 20.7). Spore dispersal in *Marchantia* takes place as the elaters twist and untwist rapidly. In the sporophytes of other liverworts, the elaters may aid spore dispersal with a snapping action or by suddenly expanding. This elater movement helps to break up the spore mass.

Until the young sporophyte is mature, it is protected by the **calyptra**, a caplike tissue that grows out from the gametophyte, and by other membranes covering the capsule. The capsule splits at maturity, and air currents carry the spores away. Under favorable conditions, the spores germinate, producing new gametophytes.

Other thalloid forms, such as the floating or amphibious liverworts, do not produce gametophores. Instead, the archegonia and antheridia develop within the thallus beneath the central grooves, where the sporophytes also are formed. The spores are liberated from the submerged sporophytes as the thallus decays.

"Leafy" Liverworts

"Leafy" liverworts (Fig. 20.8) are often abundant in tropical forests and in fog belts. They always have two rows of partially overlapping "leaves" whose cells contain distinctive oil bodies. The "leaves" have no midribs, and unlike those of mosses, they often have folds and lobes. In the tropics, the lobes form little water pockets in which tiny animals are nearly always present. It has been suggested that these water pockets may function like the pitchers of pitcher plants (discussed in Chapter 7). A third row of "underleaves" is often present on the underside of "leafy" liverworts. The "underleaves" are smaller than the other "leaves" and not visible from the top. A few rhizoids that anchor the plants develop from the stemlike axis at the base of the "underleaves."

The archegonia and antheridia of the "leafy" liverworts are produced in cuplike structures composed of a few modified "leaves," either in the axils of "leaves" or on separate branches. At maturity, the sporophyte capsule may be pushed out from among the "leaves" as the seta elongates. When a spore germinates, it produces a **protonema** consisting of a short filament of photosynthetic cells. The protonema soon develops into a mature gametophyte plant.

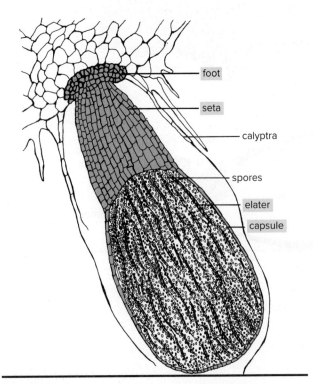

Figure 20.7 A longitudinal section through a sporophyte of *Marchantia*.

Figure 20.8 *Frullania*, a "leafy" liverwort on a tree in British Columbia. © Steven P. Lynch

20.4 Phylum Anthocerophyta—Hornworts

Learning Outcomes

a. List at least two features of hornworts that distinguish them from other bryophytes.
b. Explain how bryophytes reproduce asexually.
c. Describe how a hornwort sporophyte grows.

Structure and Form

Hornwort sporophytes look like miniature, greenish to blackish rods that may curve slightly; gametophytes resemble filmy versions of thalloid liverworts. They are usually less than 2 centimeters (0.8 inch) in diameter and thrive mostly on moist earth in shaded areas, although some occur on trees. There are only about 300 species worldwide. They differ from liverworts and mosses in several respects and appear to be only distantly related to them. An advanced feature of hornworts includes the presence of stomata, which are absent in liverworts and most mosses. Hornworts usually have only one large chloroplast in each cell (a few species have up to eight). Each chloroplast has pyrenoids similar to those of green algae. The thalli have pores and cavities filled with mucilage, in contrast to the air-filled pores and cavities of thalloid liverworts. Nitrogen-fixing cyanobacteria often grow in the mucilage. Rhizoids anchor the plants.

Asexual Reproduction

Hornworts reproduce asexually primarily by fragmentation, or as lobes separate from the main part of the thallus. A few hornworts form tiny tubers that are capable of becoming new gametophytes.

Sexual Reproduction

In sexual reproduction, archegonia and antheridia are produced in rows just beneath the upper surfaces of the gametophytes. Like both mosses and liverworts, some species of hornworts have *unisexual* plants, whereas other species are *bisexual*.

The distinctive sporophytes of hornworts have numerous stomata. They have no setae (stalks) and look like tiny, green broom handles or horns rising through a basal sheath from a foot beneath the surface of the thallus (Fig. 20.9). A meristem above the foot continually increases the length of the sporophyte from the base when conditions are favorable. As growth occurs, sporocytes surrounding a central rodlike axis in the sporophyte undergo meiosis, producing spores. Diploid elaters that function similarly to those of liverworts are intermingled with the spores. The tip of the sporophyte horn splits into two or three ribbonlike segments, releasing the spores, and the segments continue to peel back as long as the meristem is producing new tissue at the base.

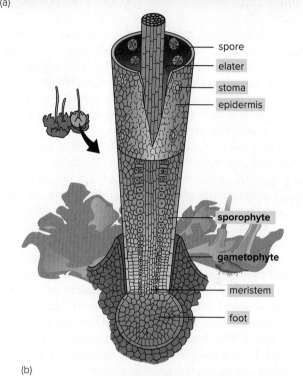

Figure 20.9 (a) Hornwort sporophytes. (b) Detail of a hornwort sporophyte. *(a) Kingsley Stern*

20.5 Phylum Bryophyta—Mosses

Learning Outcomes

a. Describe the main features of moss gametophytes.
b. Distinguish between a moss gametophyte and a sporophyte.
c. Explain how mosses reproduce asexually.

Structure, Form, and Classes

Many different organisms have been called *mosses*. In fact, almost any greenish covering or growth on tree trunks and forest floors has probably been called "moss." Examples

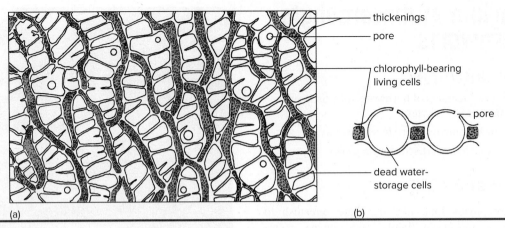

Figure 20.10 An enlargement of a portion of a peat moss (*Sphagnum*) leaf. (a) Surface view. (b) Enlarged cross section of living and dead cells.

include lichens (e.g., reindeer moss), red algae (e.g., Irish moss), flowering plants (e.g., Spanish moss), and club mosses. Club mosses look somewhat like large true mosses but are vascular plants with xylem and phloem. About 10,000 species of mosses are currently known. Mosses are distinct, both in form and reproduction, and possibly in origin, from any other group of organisms.

The "leaves" of moss gametophytes have no mesophyll tissue, stomata, or veins such as those of the leaves of vascular plants. The blades are nearly always only one cell thick, except at the *midrib,* which runs lengthwise down the middle, and they are never lobed or divided, nor do they have a *petiole* (leaf stalk). The midrib (absent in some genera) occasionally projects beyond the tip in the form of a hair or spine. The "leaf" cells usually contain numerous lens-shaped chloroplasts, except at the midrib. The "leaves" of peat mosses, however, have large, transparent cells (without chloroplasts) that absorb and store water. Small, green photosynthetic cells are sandwiched between the large cells (Fig. 20.10). The "leaves" are initially formed in three ranks and usually end up appearing to be arranged in a spiral or alternately on an axis that twists as it grows.

The axis is somewhat stemlike. It has no xylem or phloem, although there is often a distinctive central strand of hydroids. At the base, there are rootlike rhizoids consisting of several rows of colorless cells that anchor the plant. Some water absorbed by rhizoids rises up the central strand, but most water used by the plant apparently travels up the outside of the plant by means of capillarity. The closely packed habit of many mosses, and the fact that they rarely extend more than a few centimeters (2 to 3 inches) into the air, favors such outside movements of water. Water is absorbed directly through the plant surfaces.

Sexual Reproduction

Sexual reproduction in mosses begins with the formation of multicellular gametangia, usually at the apices of the "leafy" shoots of gametophytes (see Fig. 20.13), although they frequently form on special separate branches. Both male and female gametangia are often produced on the same plant, but in some species, they occur on separate plants. The archegonia (female gametangia) are somewhat cylindrical and project upward from the base of the expanded gametophyte tip (Fig. 20.11). A single *egg* cell is produced in a cavity that develops when cells break down in the swollen base of the archegonium. The part of the archegonium above the cavity (known as the **venter**) is called the *neck*. The neck may taper toward the tip and contains a narrow *canal*. The canal is at first plugged with cells, but these break down as the archegonium matures, leaving an opening to the outside at the top. Several archegonia usually are produced at the same time, with sterile, hairlike, multicellular filaments called *paraphyses* (singular: *paraphysis*) scattered among them.

Male gametangia also have paraphyses among them and are sausage-shaped to roundish, with walls that are one cell thick. These *antheridia* (Fig. 20.12) are borne on short stalks. A mass of tissue inside each antheridium develops into numerous coiled or comma-shaped *sperm* cells. This mass of sperms is forced out of the top of the antheridium when it absorbs water and swells. After release, the sperm mass breaks up into individual cells, each with a pair of flagella. It is believed the breakup of the sperm mass is aided by fats produced by the moss or by rain splash.

Archegonia release sugars, proteins, acids, or other substances that attract the sperm, and eventually a sperm, after swimming down the neck of an archegonium, unites with the egg, forming a diploid *zygote* (Fig. 20.13). The zygote usually grows rapidly into a spindle-shaped *embryo*. The embryo breaks down the cells at the base of the archegonium and becomes firmly established in the tissues of the stem by means of a swollen knob called a *foot*. As the embryo grows, cells around the venter divide to accomodate its increasing size. The length of the embryo soon exceeds the length of the cavity in the venter. The top of the venter is split off and is left sitting like a cap on top of the embryo. By this time, the embryo is a developing *sporophyte*. The cap, called a *calyptra*, remains until the sporophyte is mature. In one genus with the common name of "extinguisher mosses," the calyptra looks just like a little candlesnuffer, and in the hairy cap mosses, it resembles a miniature, pointed, goatskin cap such as might be worn by a Shakespearean actor.

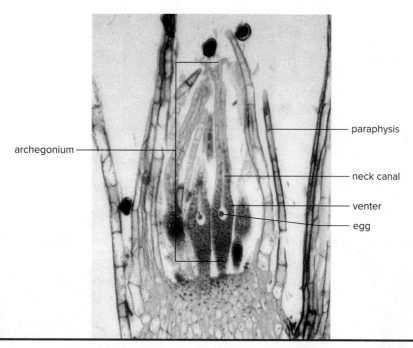

Figure 20.11 A longitudinal section through the tip of a female gametophyte of the moss *Mnium*. © *James E. Bidlack*

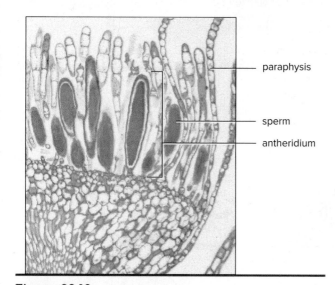

Figure 20.12 A longitudinal section through the tip of a male gametophyte of the moss *Mnium*. © *Kingsley Stern*

the stomata normally remain open until the capsule begins to age, and then they close permanently. The free end of the capsule is usually protected by a little rimmed lid, the **operculum,** which falls off at maturity.

As the capsule matures, *sporocytes* inside it undergo meiosis, producing haploid *spores*. These spores, often numbering in the millions, are released from the capsule, usually through a structure called a **peristome,** after the operculum falls off. Most peristomes consist of a circular row or two of narrowly triangular and membranous teeth arranged around the rim of the capsule, each row having 16 teeth. The teeth are frequently colored orange or red and are often beautifully sculptured with bars and fringes. They open or close in response to changes in humidity. In a few species of mosses, the peristome is a cone-shaped structure with pores through which the spores are released.

Some mosses have neither a peristome nor an operculum. The spores in these mosses are released when the capsule splits lengthwise along four lines. In the dung mosses, a putrid odor is given off when the spores are ready for release. Some of the spores adhere to the legs and bodies of flies, which are attracted by the odor, and are disseminated as the insects clean themselves. Most moss spores are, however, simply blown away by the wind, and if they fall in a suitable damp location, they usually germinate relatively quickly.

In most mosses, fine, green, tubular threads, consisting of single rows of cells with chloroplasts, first emerge from the spores. These soon branch and grow, forming an alga-like *protonema*. The protonema can be distinguished from a filamentous green alga by the oblique crosswalls of its cells and by the lens-shaped chloroplasts. If light and other conditions are favorable, tiny, "leafy" *buds* appear at intervals along the protonemal filaments after about 2 to 4 weeks of growth. These "leafy" buds develop rhizoids at the base and grow into new "leafy" gametophytes, completing the cycle.

The cells of the sporophyte become photosynthetic as it develops, remaining so until maturity. The sporophyte, however, depends on the gametophyte for some of its carbohydrate needs as well as for at least a part of its water and minerals, which are absorbed through the foot.

The mature sporophyte is at first green and photosynthetic; it consists of a *capsule* located at the tip of a slender stalk called the *seta*. Depending on the species, the seta may be less than 1 millimeter (0.04 inch) long, or it may be up to 15 centimeters (6 inches) long. Most, however, are less than 5 centimeters (2 inches) long. The capsule may resemble a tiny apple, a pear, an urn, a box, or a wingtip fuel tank of an airplane and usually has from three to over 200 stomata at or near its surface. Unless extremely dry conditions prevail,

Figure 20.13 Life cycle of a moss. © *James E. Bidlack*

KEY THEME 20.1 ecology

Hibernating Mosses

Mosses are the "amphibians" of the plant world, so called because they are at home in either semiaquatic environments, such as moist stream banks, or drier habitats, such as rock surfaces or the arctic tundra. Unlike their towering cousins, the flowering plants, mosses depend on a watery landscape to reproduce. Because moss sperms are flagellated, a film of water is needed for the sperm to swim to the egg. But life on land is not easy, especially when water becomes limited or nonexistent for months. The ecological diversity of mosses is exceptional, and they inhabit some of the most inhospitable habitats on earth. Dry heaths, rock faces, tree trunks, and even deserts are home to these remarkable plants. Distinctive adaptations allow moss species this wide range of habitats, including the ability to tolerate drying out, a process known as **desiccation**.

Vascular plants have structural mechanisms that maintain an adequate water supply within plant tissues. These include an impermeable, waxy cuticle on leaf surfaces, an internal water transport system (xylem), water-absorbing organs (roots), and leaf pores (stomata) that can close, conserving water. Parts of the life cycle, such as seeds, are especially tolerant of desiccation. Most mosses do not possess these adaptations to living in a dry land environment. This means the internal water balance in mosses is in equilibrium with the atmosphere. When the air is dry, mosses are dry. When it rains, mosses quickly absorb water and become rehydrated. They are "opportunists" in this regard.

Some species of mosses are incapable of withstanding this desiccation-rehydration cycle, but those that have this capacity can "return from the dead" with each cycle. It is as if they have been hibernating. There are several mechanisms that slow water loss from mosses that make the cycle less drastic. Many mosses form dense mats or tufts that create a moist micro atmosphere within the tufts and over the plants. *Polytrichum commune* is a common moss that has moderate desiccation tolerance. It lives in habitats such as bogs or temperate, moist forests and can grow up to 40 centimeters tall. It grows luxuriantly during rainy periods (Box Figure 20.1A) but then twists into rusty red mats upon drying in the sun. These mosses have a thin, waxy cuticle covering their tiny "leaves" (8 to 10 millimeters in length) that retards water loss. They also have primitive water-conducting cells (*hydroids*) that move water through the plant when water is lost to the atmosphere by evaporation. Instead of root hairs, they have a rhizoid system that absorbs water from the soil. However, if dry conditions persist, all available water is lost to the atmosphere.

Additional mechanisms by which mosses are capable of surviving these desiccation-rehydration cycles are being discovered. *Tortula ruralis* is one of the most desiccation-tolerant mosses known, and research has centered on its ability to recover after prolonged or repeated desiccation events. Hydrated *T. ruralis* cells are similar to mesophyll cells in the leaves of vascular plants. There are chloroplasts with

Box Figure 20.1A *Polytrichum commune* (hairy cap moss) growing in a forest after heavy rainfall. © *Tainas/yaymicro.com*

stacks of grana and a prominent nucleus. Mitochondria are numerous and similar in size and shape to those of higher plants, with internal membranes folded into cristae.

During desiccation, *T. ruralis* dries out, and its "leaves" fold up around the stem. Cells of desiccated plants are damaged. There is extensive plasmolysis as water is lost from the protoplast. Chloroplasts become smaller and more spherical, and starch is not present. Internal thylakoid membranes are collapsed and disorganized. Mitochondria in the hydrated cell are elongated with numerous cristae membranes. In the desiccated state, mitochondria are smaller and rounded, with few internal cristae membranes. The nucleoplasm of nuclei becomes dense, and chromatin is condensed. Compact-appearing nucleoli are prominent. The plasma membrane is damaged, and electrolytes leak out. In all desiccated mosses, photosynthesis stops, and respiration slows or ceases.

When the plants are rehydrated, dried "leaves" unfold and return to the normal hydrated state within 2 minutes. Most of the internal damage is repaired within a few more minutes. Activity of a drought-repair gene increases in the minutes after rehydration, and repair proteins are quickly manufactured and mobilized for action. Respiration resumes after a few minutes. Photosynthesis resumes up to 24 hours later.

Scientists at the USDA's Agricultural Research Service are attempting to locate the genes responsible for drought repair in *Tortula* species in the hope of transferring them to crop plants. With crops having more drought tolerance, rain shortage would be less of a problem than it is now. Arid lands currently unsuited for agriculture could blossom with crop plants genetically engineered with drought-tolerant genes. In addition, even in less arid regions, crops with drought tolerance would require less irrigation. World population increases of 1.6% yearly mean that an additional 78,000 metric tons of grain per day are required just to maintain current consumption levels. A lowly moss might someday be responsible for a revolution in food production.

D. C. Scheirer

Asexual Reproduction

Mosses can reproduce asexually through fragmentation. It has been demonstrated under laboratory conditions that cells of archegonia and antheridia, paraphyses, "leaves," stems, and rhizoids can develop protonemata. Two American biologists once collected bryophyte gametophyte fragments from a snowbed in the Canadian high arctic and found that 12% of their samples resumed growth in various ways when cultured in the laboratory. They calculated that for each cubic meter (1.3 cubic yards) of snow at their study site, there were over 4,000 bryophyte fragments capable of becoming new plants. They suggested that wind dispersal of fragments may be routine in arctic regions. Such dispersal and vegetative reproduction also occur widely in more temperate areas.

20.6 Human and Ecological Relevance of Bryophytes

Learning Outcome

a. Discuss the significance of bryophytes to humans.

Some bryophytes and lichens are pioneers on bare rock after volcanic eruptions or other geological upheavals and after the retreat of glaciers. They slowly accumulate mineral and organic matter, which can then be inhabited or utilized by other organisms. This process, called *succession*, is discussed in Chapter 25. Mosses, in particular, retain moisture, slowly releasing it into the soil. They reduce flooding and erosion and contribute to humus formation. Some mosses grow only in soils that are rich in calcium; the presence of others indicates higher than usual soil salinity or acidity.

When certain mosses are present in a dry area, it is a good indication that running water occurs there sometime during the year. A few mosses are occasionally a problem in water reservoirs, where they may plug entrances to pipes. A few bryophytes are reported to be grazed, along with lichens, by foraging mammals in arctic regions, but bryophytes are not generally edible. Some mosses have been used for packing dishes and stuffing furniture, and Native Americans are reported to have used mosses for diapers and under splints when setting broken limbs.

By far, the most important bryophytes to humans are the peat mosses. When allowed to absorb water, 1 kilogram (2.2 pounds) of dry peat moss will take up 25 kilograms (55 pounds) of water. Its extraordinary absorptive capacity has made it very useful as a soil conditioner in nurseries and as a component of potting mixtures (Fig. 20.14). Live shellfish and other organisms are shipped in it. The natural acidity produced inhibits bacterial and fungal growth and gives it antiseptic properties. The absorbency, which is greater than that of cotton, combined with the antiseptic properties, has made it a useful poultice material for application to wounds.

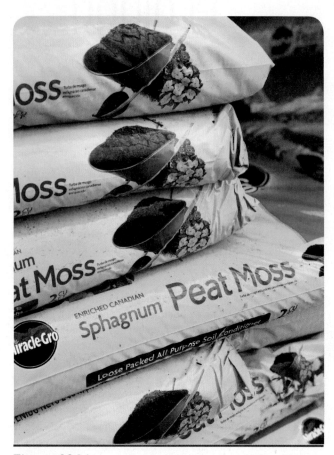

Figure 20.14 Peat moss can be found at many greenhouse stores and nurseries for commercial or home use. © *jadimages/Shutterstock*

It was used for this purpose during the Crimean War of 1854 to 1856 and, as indicated in the chapter introduction, on an emergency basis during World War I. Extensive peat deposits have been formed from the remains of peat mosses that flourished in past eras. Peat, like the undecomposed peat mosses, is used around the world as a soil conditioner and as a fuel. In the manufacture of Scotch whiskey, sprouted barley is dried on a screen over a peat fire. The peat smoke permeates the barley and imparts a smoky flavor to the beverage. See Appendix 1 for the scientific names of all the bryophytes discussed.

SUMMARY

20.1 Introduction to the Plant Kingdom

- Members of the Plant Kingdom usually have a cuticle and produce their gametes and spores in multicellular organs surrounded by a sterile jacket of protective cells. Their zygotes develop into embryos; and tissues specialized for photosynthesis, conduction, support, anchorage, protection, and reproduction are produced.

- Cell plates and phragmoplasts appear when plant cells divide. Outside of the Plant Kingdom, these occur only in certain green algae. The similarity in pigments, food reserve (starch), and occurrence of cell plates suggests a common ancestor for the green algae and plants. The Plant Kingdom includes three phyla of bryophytes and several phyla of vascular plants.

20.2 General Characteristics of Bryophytes

- Bryophytes (liverworts, hornworts, mosses) occur in highly varied and very specific habitats.
- Water is essential to bryophyte reproduction. Most water is absorbed directly through the plant surfaces.

20.3 Phylum Hepaticophyta—Liverworts

- Liverwort gametophytes with flattened, dichotomously forking thalli are common, but about 80% of the liverwort species are "leafy." Liverworts have distinct upper and lower surfaces, with one-celled rhizoids that function in anchorage on the lower surface.
- A Marchantia thallus has a central lengthwise groove along its upper surface and is chambered, each chamber containing chlorenchyma cells and having a surface pore. Rhizoids and scales arise from the thallus base.
- Marchantia reproduces asexually by means of gemmae produced in surface cups and by thallus fragmentation.
- Marchantia reproduces sexually by means of eggs and sperms produced in archegonia and antheridia on archegoniophores and antheridiophores that arise from the thallus.
- The zygote develops into a sporophyte that is anchored to the archegoniophore by a foot, from which is suspended a capsule connected to the foot by a seta. Sporocytes in the capsule undergo meiosis, producing spores. Diploid elaters that aid in spore dispersal do not undergo meiosis.
- The calyptra and other membranes protect the spores until they are released as the capsule splits; the spores may then develop into new gametophytes.
- "Leafy" liverworts have two rows of overlapping "leaves" and frequently a third row of "underleaves" not visible from above. The "leaves" often have lobes that retain rain water.

20.4 Phylum Anthocerophyta—Hornworts

- Hornworts have one chloroplast with pyrenoids in each cell, and they resemble thalloid liverworts in their gametophytes. Their sporophytes are hornlike and have a meristem above the foot. Hornwort thalli have pores and cavities filled with mucilage, where cyanobacteria often grow.
- Asexual reproduction in hornworts is by fragmentation. Sexual reproduction involves archegonia and antheridia produced in rows beneath the upper surface of a thallus. The tip of the hornlike sporophyte splits vertically, releasing the spores.

20.5 Phylum Bryophyta—Mosses

- A moss gametophyte consists of an axis to which "leaves" are attached, with rhizoids at the base. The "leaves" have no mesophyll, stomata, or veins. Water is absorbed primarily directly through the plant surfaces.
- Multicellular archegonia and antheridia are produced at the tips of "leafy" shoots. Each archegonium has a cavity, the venter, containing a single egg and a neck through which a sperm gains access to the egg. Sperms are produced in antheridia.
- After fertilization, the zygote grows into an embryo that is attached to the gametophyte by an embedded foot. The sporophyte developing from the embryo consists of a capsule, a seta, and a foot. A calyptra derived from the gametophyte partially covers the capsule. Sporocytes in the capsule undergo meiosis, producing spores that are released through the teeth of the peristome, a structure at the tip of the capsule.
- An operculum that falls off when the spores mature initially covers the peristome and protects sporocytes and spores. When moss spores germinate, protonemata with "leafy" buds develop. The buds grow into new gametophytes.

20.6 Human and Ecological Relevance of Bryophytes

- Mosses may be pioneers, along with lichens, on bare rocks. They are indicators of soil calcium, salinity, and acidity. Mosses are not generally edible, although a few are grazed in arctic regions. Some mosses are used for packing material, but the most significant use is that of peat mosses for soil conditioners. Peat mosses can absorb and retain large amounts of water, and their natural acidity gives them antiseptic properties. Peat deposits, from peat mosses that flourished in past eras, are used for fuel and as a soil conditioner.

REVIEW QUESTIONS

1. What basic features distinguish members of the Plant Kingdom from those of other kingdoms?
2. What features distinguish the bryophytes discussed in this chapter from other plants?
3. How could you tell a hornwort thallus from that of a thalloid liverwort?
4. Contrast the sporophytes of mosses, liverworts, and hornworts.

5. What is a protonema? Do all bryophytes have them? How would you tell a protonema from a green alga?
6. What adaptations do bryophytes have for their particular habitats?
7. Which parts of the life cycles of bryophytes have haploid cells? Which parts have diploid cells?
8. Why is a bryophyte "leaf" technically not the same as a flowering plant leaf?
9. Define calyptra, operculum, capsule, peristome, paraphysis, foot, seta, archegoniophore, thallus, underleaves, and elaters.
10. If you were to single out one bryophyte as being the most important member of its phylum from a human viewpoint, which would you choose? Why?

DISCUSSION QUESTIONS

1. Very few fossils of bryophytes have been found. Suggest reasons for this.
2. Do the multicellular sex structures of plants give them any advantages over other organisms with unicellular sex structures?
3. After reading about the characteristics and uses of peat mosses, can you suggest some possible new uses for these plants?
4. Some bryophytes produce unisexual gametophytes, while others produce bisexual gametophytes. Should one type have any survival or adaptive advantage over the other? Explain.

CHAPTER 21

The Seedless Vascular Plants: Ferns and Their Relatives

21.1 Introduction to the Seedless Vascular Plants
21.2 Phylum Psilotophyta—the Whisk Ferns
    Structure and Form
    Reproduction
    Fossil Whisk Fern Look-Alikes
21.3 Phylum Lycophyta—the Ground Pines, Spike Mosses, and Quillworts
    Lycopodium—Ground Pines
    Selaginella—Spike Mosses
    Isoetes—Quillworts
    Ancient Relatives of Club Mosses and Quillworts
    Human and Ecological Relevance of Club Mosses and Quillworts
21.4 Phylum Equisetophyta—the Horsetails and Scouring Rushes
    Structure and Form
    Reproduction
    Ancient Relatives of Horsetails
    Human and Ecological Relevance of Horsetails and Scouring Rushes
21.5 Phylum Polypodiophyta—the Ferns
    Structure and Form
    Reproduction
    Fossil Relatives of Ferns
    Human and Ecological Relevance of Ferns
21.6 Fossils
    Introduction
    Molds, Casts, Compressions, and Imprints
    Key Theme 21.1: Ferns and Fossil Fuels
    Petrifactions
    Coprolites
    Unaltered Fossils

OVERVIEW

This chapter opens with a brief review of the features that distinguish the major groups of vascular plants without seeds from one another and from the bryophytes, and then discusses representatives of each phylum. Included in the discussion are whisk ferns (*Psilotum*), club mosses (*Lycopodium, Selaginella*), quillworts (*Isoetes*), horsetails (*Equisetum*), and ferns. A digest of the human and ecological relevance of each group is given, and life cycles of representatives are illustrated. The chapter concludes with an examination of fossils, and each type is briefly described. A table showing the geologic time scale is provided.

The underside of a fern leaf. The numerous brown structures are sori, each of which consists of many sporangia. Mother cells within the sporangia undergo meiosis to produce spores which are released when the sporangia dry out and split open.
© George Bernard/Science Source

21.1 Introduction to the Seedless Vascular Plants

Learning Outcomes

a. Explain the basic structural differences between bryophytes and vascular plants.
b. List and describe the distinguishing characteristics of the four phyla of seedless vascular plants.

Thus far, our survey has taken us from simple, one-celled prokaryotic organisms, through more specialized eukaryotic protists and fungi, and on to the bryophytes. We now take up several phyla of plants whose members have internal conducting tissues but do not produce seeds.

In the introduction to the Plant Kingdom in Chapter 20, we mentioned that the ancestors of the bryophytes and the vascular plants probably were multicellular green algae that became established on the land over 420 million years ago. As green organisms moved ashore, they developed several new features, including sterile jackets of cells around gametangia, embryos within protective tissues, stomata for gas exchange, and a cuticle. Most of these features at least partly protected vital parts from drying out. Bryophytes have no significant internal tissues for conducting water, and most water used internally is absorbed directly through the aboveground parts. External water is required for fertilization.

During the early stages of vascular plant evolution, internal conducting tissues (xylem and phloem) began to develop, true leaves appeared, and roots that function in absorption and anchorage developed. At the same time, gametophytes became progressively smaller and more dependent on sporophytes that became progressively larger. Initially the sporophyte also afforded gametophyte protection.

Unlike conifers and flowering plants, the primitive vascular plants discussed in this chapter do not produce seeds. These seedless vascular plants are sporophyte dominant, and the spores produced germinate into a free-living gametophyte generation. They include the ferns and a number of their relatives, often referred to as "fern allies." Four phyla of seedless vascular plants are recognized:

1. **Phylum Psilotophyta** (*whisk ferns*). Features unique to members of this phylum of vascular plants include sporophytes that have neither true leaves nor roots, and stems and rhizomes that fork evenly.
2. **Phylum Lycophyta** (*club mosses* and *quillworts*). The stems of these plants are covered with **microphylls** (leaves with a single vein whose trace is not associated with a leaf gap—see Chapter 6). Most microphylls are photosynthetic.
3. **Phylum Equisetophyta** (*horsetails* and *scouring rushes*). The sporophytes of these plants have ribbed stems containing silica deposits and whorled, scalelike microphylls that lack chlorophyll.
4. **Phylum Polypodiophyta** (*ferns*). The sporophytes of ferns have **megaphylls** (leaves with more than one vein and a leaf trace that is associated with a leaf gap—see Chapter 6) that are often large and much divided.

Scientists generally agree that the seedless vascular plants share a single ancestor. However, these plants did not necessarily evolve from a single branch on the tree of life because seed plants also descended from this group. *Fossils,* discussed at the end of the chapter, give us clues to the ancestry of some of these plants and to the origins of seed plants, discussed in the chapters to follow. Molecular studies are currently being conducted to determine the relationships among the four phyla of seedless vascular plants.

21.2 Phylum Psilotophyta— the Whisk Ferns

Learning Outcome

a. Draw a whisk fern and label the enations, sporangia, and rhizomes.

There is nothing very fernlike in the appearance of *whisk ferns,* commonly referred to by their scientific name of *Psilotum,* but they do loosely resemble small, green whisk brooms. Exactly where these plants fit in the Plant Kingdom is not clear. Traditionally, they have been associated with a

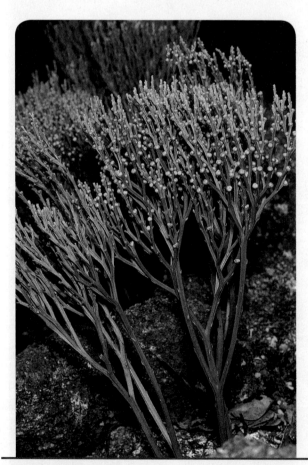

Figure 21.1 *Psilotum* sporophytes. The small, yellowish structures are three-lobed sporangia. © *Kingsley Stern*

number of extinct plants called *psilophytes* that flourished perhaps 400 million years ago, but there is not much fossil evidence to substantiate the link. Because whisk fern and true fern gametophytes share several features, some botanists classify them with the true ferns. They will be discussed first because, by most present criteria, they are among the simplest of all living seedless vascular plants.

Structure and Form

Psilotum sporophytes consist almost entirely of dichotomously forking (evenly forking) aerial stems and are unique among living vascular plants in having neither leaves nor roots. They usually grow up to 30 centimeters (1 foot) tall but occasionally grow as much as 1 meter (3 feet) or more tall (Fig. 21.1). The aboveground stems arise from short, branching rhizomes just beneath the surface of the ground.

Enations, which are tiny, green, superficially leaflike, veinless, photosynthetic flaps of tissue, are spirally arranged along the stems. Photosynthesis takes place in the outer cells of the stem (epidermis and cortex). A central cylinder of xylem is surrounded by phloem. The xylem is star-shaped in cross section. Rhizoids, which act as root hairs, aided by mycorrhizal fungi, are scattered along the surfaces of rhizomes. Some of the mycorrhizal fungi become established in cortical cells just beneath the epidermis.

Reproduction

Small *sporangia* that are fused together in threes and resemble miniature yellow pumpkin-like structures are produced at the tips of very short, stubby branches in the upper parts of the angular stems. Spores released from the sporangia germinate slowly in the soil, in tree bark or tree fern "bark" crevices, or in other similar habitats. In Hawaii, they also grow in old lava flows.

The gametophytes, which lack pigmentation, develop from the spores beneath the soil surface and are easily overlooked. Sometimes they resemble tiny, transparent dog bones that are only about 2 millimeters (0.08 inch) wide and seldom more than 6 millimeters (0.25 inch) long. They are cylindrical and, like the aerial stems, may branch dichotomously. They have no chlorophyll and absorb their nutrients via one-celled, rootlike *rhizoids* aided by mycorrhizal fungi. Archegonia and antheridia develop randomly over the surface of the same gametophyte. After a sperm unites with an egg in an archegonium, the zygote develops a foot and a rhizome. As soon as the rhizome becomes established, upright stems are produced, and the rhizome separates from the foot (Fig. 21.2).

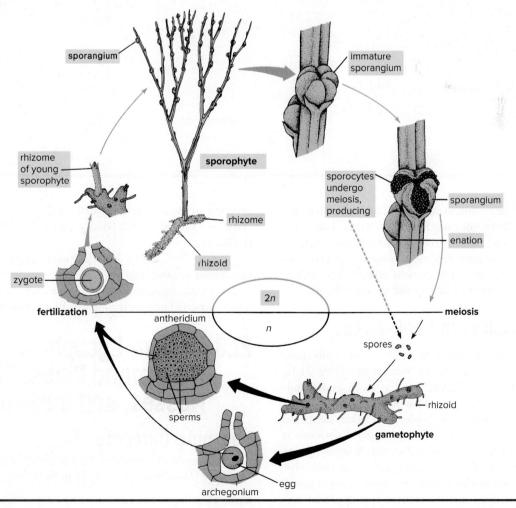

Figure 21.2 Life cycle of *Psilotum*.

Table 21.1 Geologic Time Scale

Era	Period	Age	Earliest Evidence of Taxonomic Group	Duration (Millions of Years)	Million Years Before Present
CENOZOIC	Quaternary	Age of modern seed plants		63	
	Tertiary				65
MESOZOIC	Cretaceous			65	
					130
	Jurassic		Grasses Flowering plants	45	
					175
	Triassic	Age of ancient seed plants		45	
					220
PALEOZOIC	Permian		Ginkgoes Conifers	50	
					270
	Carboniferous			80	
			Seed ferns		
					350
	Devonian	Age of spore-bearing plants	Lycopods Horsetails Bryophytes	50	
					400
	Silurian			40	
					440
	Ordovician	Age of bacteria and algae		60	
					500
	Cambrian		Marine algae	100	
					600
PRECAMBRIAN				1,700?	2,300?

Whisk ferns are native to tropical and subtropical regions. In the United States, they are found in Florida, Louisiana, Texas, Arizona, and Hawaii. They are extensively cultivated in Japan and sometimes become a weed in greenhouses around the world. *Tmesipteris,* a close relative, has leaflike appendages; it is native to Australia and the South Pacific.

Fossil Whisk Fern Look-Alikes

Fossil plants that somewhat resemble whisk ferns have been found in Silurian geological formations (Table 21.1). These formations are estimated to be as much as 400 million years old. One group of these fossil plants, of which *Cooksonia* and *Rhynia* are examples, had naked stems (no photosynthetic flaps) and terminal sporangia. *Cooksonia* is the oldest plant known to have had xylem. A second group of fossils, represented by *Zosterophyllum,* had somewhat rounded sporangia produced along the upper parts of naked stems. *Zosterophyllum* and its relatives first appeared during the Devonian period (Table 21.1). They are thought to be ancestral to the club mosses, discussed in the next section, "Phylum Lycophyta—the Ground Pines, Spike Mosses, and Quillworts."

Whisk ferns are of little economic importance. Their spores have a slightly oily feel and were once used by Hawaiian men to reduce loincloth irritations of the skin. Hawaiians also made a laxative liquid by boiling whisk ferns in water.

21.3 Phylum Lycophyta— the Ground Pines, Spike Mosses, and Quillworts

Learning Outcome

a. Explain the differences in the life cycles of ground pines (*Lycopodium*) and spike mosses (*Selaginella*).

Some of the most treasured moments in our history include photographs that were taken with flash powder, which illuminated the area and was followed by a large billow of smoke. That particular phenomenon was provided by the spores of **club mosses**, which are flammable and burn very quickly. It is interesting that this was one of the most popular uses of club mosses during the early years of photography.

The 950 or so known species of club mosses (referred to in this text as *ground pines* and *spike mosses*) mostly look enough like large true mosses that the Swedish naturalist Linnaeus lumped both together in a single class. Once details of the structure and the form of club mosses were known, however, it became obvious that they are quite unrelated to true mosses.

Today, there are living representatives of two major genera and two minor genera of club mosses. Several others became extinct about 270 million years ago. The sporophytes of all species have microphylls that are usually quite small; they also have true stems and true roots.

The two major genera of club mosses with living members are *Lycopodium,* with about 50 species, and *Selaginella,* with over 700 species. They are distributed throughout the world, but they are more abundant in the tropics and wetter temperate areas.

Lycopodium—Ground Pines
Structure and Form

Lycopodium plants often grow on forest floors. They are sometimes called *ground pines,* partly because they resemble little Christmas trees, complete with "cones" that are usually upright or, in a few species, hang down (Fig. 21.3). The stems of ground pine sporophytes are either simple or branched. The plants are mostly less than 30 centimeters (1 foot) tall, although some tropical species grow to heights of 1.5 meters (5 feet) or more. The upright, or sometimes pendent, stems develop from branching rhizomes. The leaves may be whorled or in a tight spiral and are rarely more than 1 centimeter (0.4 inch) long. Adventitious roots, whose epidermal cells often produce root hairs, develop along the rhizomes.

Reproduction

At maturity, some species of ground pines produce kidney bean–shaped sporangia on short stalks in the axils of specialized leaves. Such sporangium-bearing leaves are called **sporophylls.** In other species, the sporophylls have no chlorophyll, are smaller than the other leaves, and are in terminal conelike clusters called **strobili** (singular: **strobilus**). In the sporangia, *sporocytes* undergo meiosis, producing spores that are released and carried away by air currents. The spores of some species germinate in a few days if they land in a suitable location, but spores of other species may not germinate for up to several years.

After germination, independent gametophytes develop from the spores. The gametophytes vary in shape, some

(a)

(b)

Figure 21.3 Club mosses. (a) *Lycopodium cernum,* a native to the tropics. (b) *Lycopodium obscurum,* native to northern and eastern North America. *(a-b)* © *Kingsley Stern*

Figure 21.4 Gametophyte of *Lycopodium*. © *Steven P. Lynch*

resembling tiny carrots (Fig. 21.4). The gametophyte body usually develops in the ground in association with mycorrhizal fungi. Some gametophytes, however, develop primarily on the surface, where the exposed parts turn green. All types produce both antheridia and archegonia on the same gametophyte that, in some species, may live for several years. Because the sperm are flagellated, water is essential for fertilization to occur.

Zygotes first become embryos, each with a foot, stem, and leaves, which then develop into mature sporophytes (Fig. 21.5). If the gametophyte is underground, chlorophyll does not develop in the young sporophyte until it emerges into the light. Several sporophytes may be produced from a single gametophyte. A number of ground pines also reproduce asexually by means of small bulbils (bulbs produced in the axils of leaves), each of which is capable of developing into a new sporophyte.

Selaginella—Spike Mosses
Structure and Form

The sporophytes of *Selaginella*, the larger of the two major genera of living club mosses, are sometimes called

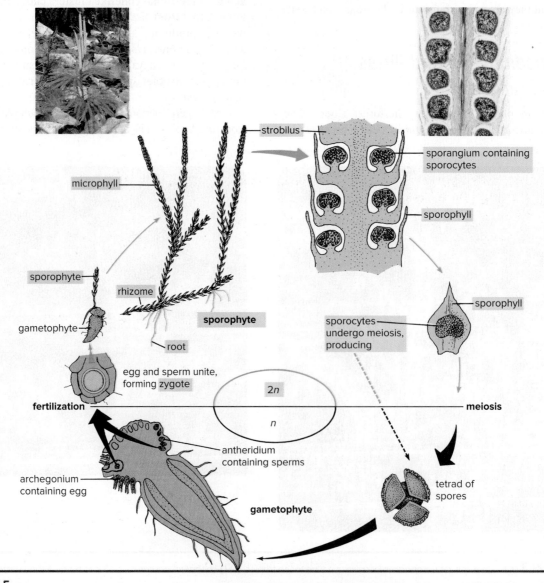

Figure 21.5 Life cycle of the ground pine *Lycopodium*, a homosporous lycopod. © *James E. Bidlack and Shelley H. Jansky*

The Seedless Vascular Plants: Ferns and Their Relatives   391

Figure 21.6 *Selaginella,* a spike moss. © *Kingsley Stern*

spike mosses (Fig. 21.6). The approximately 700 species are widely scattered around the world in wetter areas, but they are especially abundant in the tropics. A few are common weeds in greenhouses. They tend to branch more freely than ground pines, from which they differ in several respects. The two most obvious differences are these: (1) their leaves each have a tiny extra appendage, or tongue, called a **ligule,** on the upper surface near the base, and (2) they produce two different kinds of spores and gametophytes—an advanced feature referred to as **heterospory.** The seed-bearing coniferous and flowering plants discussed in chapters to follow are all heterosporous.

Reproduction

Sexual reproduction of both spike mosses and ground pines involves the production of sporangia, but spike moss sporangia develop on either *microsporophylls* or *megasporophylls* (Fig. 21.7). **Microsporophylls** bear *microsporangia* containing numerous **microsporocytes** that undergo meiosis, producing tiny **microspores.** The *megasporangia* of **megasporophylls** usually contain a **megasporocyte** that, after meiosis, becomes four comparatively large **megaspores.**

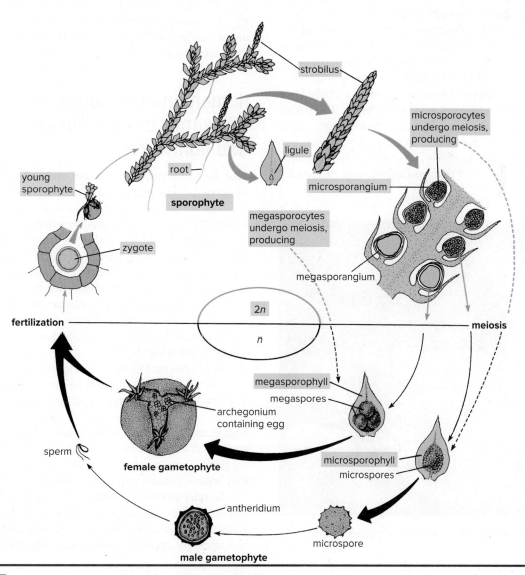

Figure 21.7 Life cycle of the spike moss *Selaginella,* a heterosporous lycopod.

Each microspore may become a male gametophyte consisting simply of a somewhat spherical antheridium within the microspore wall. Either 128 or 256 sperm cells with flagella are produced in each antheridium. A megaspore develops into a female gametophyte that is also relatively simple in structure. By the time this gametophyte is mature, however, it consists of many cells that have been produced inside the megaspore. As it increases in size, it eventually ruptures its thickened spore wall and produces several archegonia in the exposed seams. The development of both male and female gametophytes often begins before the spores are released from their sporangia. Fertilization and development of new sporophytes are similar to those of ground pines.

Isoetes–Quillworts
Structure and Form

There are about 60 species of *quillworts* (all in the genus *Isoetes*) (Fig. 21.8). Most are found in areas where they are at least partially submerged in water for part of the year. Their leaves (microphylls) are slightly spoon-shaped at the base and look like green porcupine quills, although they are not rigid. They are arranged in a tight spiral on a stubby stem, which resembles the corm of a gladiola or a crocus. *Ligules* occur toward the leaf bases. The corms have a vascular cambium and may live for many years. Wading birds and muskrats often eat the corms. The plants are generally less than 10 centimeters (4 inches) tall, but the leaves of one species become 0.6 meter (2 feet) long.

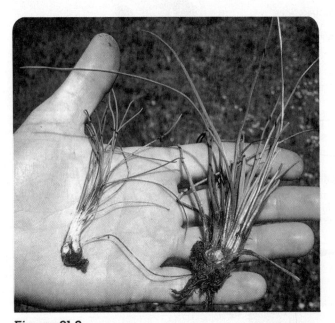

Figure 21.8 Quillwort (*Isoetes*) sporophytes. *Courtesy of Robert A. Schlising*

Reproduction

Reproduction is similar to that of spike mosses, except that no strobili are formed. Both types of sporangia are produced at the bases of the leaves (Fig. 21.9). Up to 1 million microspores may occur in a single microsporangium.

Ancient Relatives of Club Mosses and Quillworts

Some of the ancient and extinct relatives of the club mosses were large and treelike and grew up to 30 meters (100 feet) tall; their trunks were up to 1 meter (3 feet) in diameter. They were dominant members of the forests and swamps of the Carboniferous period, which reached its peak some 325 million years ago (Table 21.1). Quillworts first appeared in the Cretaceous period, about 130 million years ago. Their fossils reveal that even the oldest were remarkably similar to their present-day relatives (Fig. 21.10). A knowledge of the life histories and ancestry of various members of this phylum enhances our understanding of the development and diversity of the Plant Kingdom as a whole.

Human and Ecological Relevance of Club Mosses and Quillworts

Like the whisk ferns, club mosses and quillworts are of little economic importance today. As mentioned earlier, large numbers of club moss spores produce a flash of light when ignited. This characteristic was exploited at one time in the manufacture of theatrical explosives and photographic flashlight powders. In the past, druggists mixed spore powder with pills and tablets to prevent them from sticking to one another. The spore powder itself has been used for centuries in folk medicine, particularly for the treatment of urinary disorders and stomach upsets. Some Native Americans used it as a talcum powder for babies, snuffed it to arrest nosebleeds, and applied it following childbirth to stanch hemorrhaging. It was also sometimes used to stop bleeding from wounds.

Club moss extracts have been used in several countries in the past to reduce fevers, but partly because of undesirable side effects, medicinal use of club mosses has now largely been abandoned. Native Americans of Washington, Oregon, and British Columbia became mildly intoxicated after chewing parts of one local species of club moss. It is reported that they became unconscious if they chewed too much.

Novelty stores sometimes sell a species of spike moss native to Mexico and the southwestern United States. Known as the *resurrection plant,* it shrivels and rolls up in a ball when dry, appearing to be completely dead, but quickly

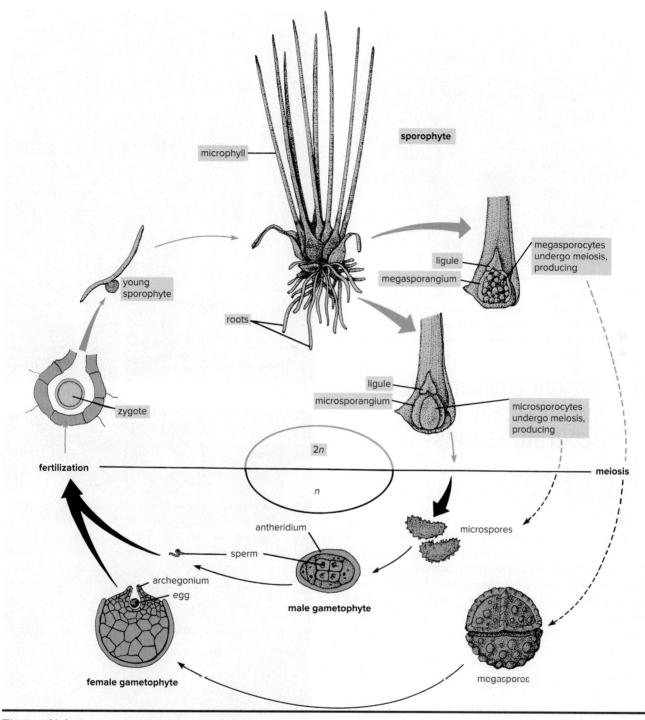

Figure 21.9 Life cycle of a quillwort (*Isoetes*), a heterosporous member of the phylum.

unfolds and turns green when sprinkled with water. Other spike mosses have been placed on shelves indoors without water for nearly 3 years and then have resumed growth when given water. Ground pines and spike mosses have been used ornamentally indoors and outdoors as ground covers. Some ground pines are spray painted and used as Christmas ornaments or in floral wreaths. Several species of *Lycopodium* have been exploited to the extent that they are now on rare and endangered or threatened species lists; they should no longer be collected.

Quillwort corms have been eaten by domestic and wild animals, waterfowl, and humans.

Figure 21.10 Fossilized scale-like bark of a tree-sized lycopod or club moss, which grew in hot and humid swampland. The fossil shows leaf (microphyll) bases similar to those seen in modern lycopods.
© Colin Keates/Dorling Kindersley/Natural History Museum, London/Science Source

21.4 Phylum Equisetophyta— the Horsetails and Scouring Rushes

Learning Outcome

a. Summarize the structural features of horsetail (*Equisetum*) sporophytes.

Many backpackers and campers have become aware of a unique use of a relatively common and widespread genus of plant (*Equisetum*) known as *horsetails* and *scouring rushes*. Significant deposits of silica accumulate on the inner walls of the epidermal cells of the stems, which make excellent scouring material for dirty metal pots and pans. Native Americans were aware of this scouring property of the plants and used them extensively.

Structure and Form

About 25 species of horsetails (the name usually applied to branching forms that look like a horse's tail) and scouring rushes (unbranched forms) (Fig. 21.11) are scattered throughout all continents, including Australia, where they are weeds. They usually grow less than 1.3 meters (4 feet) tall, but some in the tropics and coastal redwood forests of California exceed 4.6 meters (15 feet) in height. Where branches occur, they are normally in whorls at regular intervals along the jointed stems. Both branched and unbranched species have tiny, scalelike leaves (microphylls) in whorls at the nodes. These leaves are fused together at their bases, forming a collar. They are green when they first appear, but they soon wither and bleach, and virtually all photosynthesis occurs in the stems.

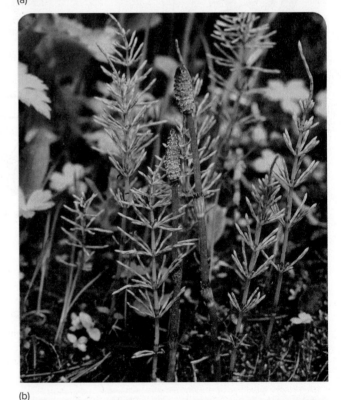

Figure 21.11 Horsetails (*Equisetum*). (a) An unbranched species. (b) A branched species. *(a)* © Kingsley Stern; *(b)* © Steven P. Lynch

The stems are distinctly ribbed and have obvious nodes and internodes; there are numerous stomata in the grooves between the ribs. A cross section of a stem (Fig. 21.12) reveals that the pith breaks down at maturity, leaving a hollow central canal.

Figure 21.12 A horsetail (*Equisetum*) stem in cross section. ×20. © *James E. Bidlack*

There are two cylinders of smaller canals outside of the pith. The inner cylinder consists of water-conducting *carinal canals* that are aligned opposite the ribs of the stem. Each canal has a patch of xylem and phloem to the outside. The canals of the outer cylinder, called *vallecular canals,* contain air. They are larger than the carinal canals and are aligned opposite the "valleys" between the ribs.

The aerial stems develop from horizontal rhizomes, which also have regular nodes, internodes, and ribs. In some species, the rhizomes have adventitious roots and may form extensive branching systems as much as 2 meters (6.5 feet) below the surface. Both internal stem structure and external features help distinguish the various species of horsetails from one another.

Reproduction

If stems or rhizomes are broken up by a disturbance, such as a storm or foraging animals, the fragments can grow into new sporophyte plants. In most species, however, reproduction usually involves a sexual process. In the spring, some species produce special cream- to buff-colored non-photosynthetic stems from rhizomes (Fig. 21.13). Small, conelike strobili develop at the tips of these special stems or, in other species, at the tips of regular photosynthetic stems.

The strobili are usually about 2 to 4 centimeters (0.75 to 1.5 inches) long. Hexagonal, dovetailing plates at the surface of the strobilus give it the appearance of an ellipsoidal honeycomb.

Each hexagon marks the top of a *sporangiophore* that has 5 to 10 elongate sporangia connected to the rim. The stalks of the sporangiophores are attached to the central axis of the strobilus. The sporangia surround the sporangiophore stalks and point inward. These hidden sporangia are not visible until maturity, when the sporangiophores separate slightly. The spores are then released.

When the sporocytes in the sporangia undergo meiosis, distinctive-appearing green spores are produced. At one pole, the spores have four ribbonlike appendages that are slightly expanded at the tips (Fig. 21.14). The appendages are called **elaters** (not structurally or otherwise related to the elaters of liverworts and hornworts). The elaters are very sensitive to changes in humidity and aid spores in their dispersal. While spores are being carried by an air current, the elaters are more or less extended like wings. If a spore enters a humid air pocket above a damp area below, the elaters coil, causing the spore to drop in an area that is more likely to support germination and growth.

Germination of spores usually occurs within a week of their release. Lobed, cushionlike, green gametophytes develop and seldom grow to more than 8 millimeters (0.36 inch) in diameter. Rhizoids anchor them to the surface. At first, about half of the gametophytes are male with antheridia, and the other half are female with archegonia. After a month or two, however, the female gametophytes of most species become bisexual, producing only antheridia

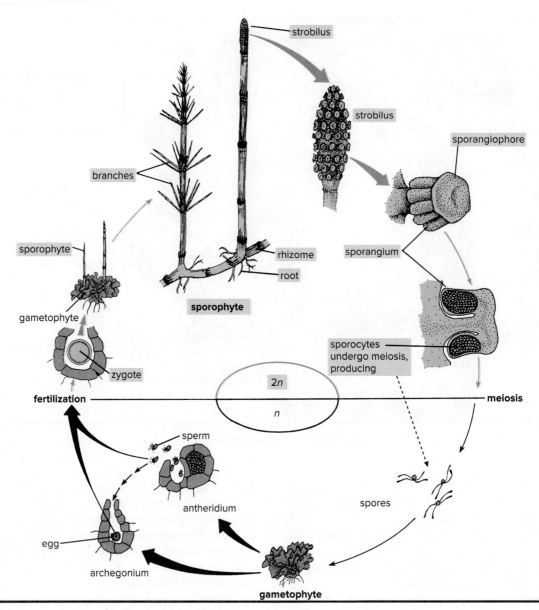

Figure 21.13 Life cycle of a horsetail (*Equisetum*) that has separate vegetative and reproductive shoots.

from then on. When water contacts mature antheridia, sudden changes in water pressure cause the sperms produced within to be explosively ejected. Several eggs on a female or bisexual gametophyte may be fertilized, and the development of more than one sporophyte is common.

Ancient Relatives of Horsetails

Horsetails and scouring rushes belong to the single remaining genus (*Equisetum*) of a phylum of several different orders that flourished in the Carboniferous period about 300 million years ago. At that time, some members of this phylum, like the ancient club mosses, were treelike (Fig. 21.15) and grew over 15 meters (50 feet) tall, while others were vinelike. Some were similar to present-day horsetails in having leaves in whorls (Fig. 21.16), jointed stems with internal canals, and sporangiophores with sporangia hanging down from the rims.

Human and Ecological Relevance of Horsetails and Scouring Rushes

The 7th-century Romans ate the boiled young strobili of field horsetails or fried them after mixing them with flour. Native Americans peeled off the tough epidermis of young stems and ate the inner parts either raw or boiled, while others cooked the rhizomes of the giant horsetail for food. Cows, goats, muskrats, bears, and geese also have been known to eat horsetails. Hopi Native Americans ground dried horsetail stems to flour, which they mixed with cornmeal to make bread or mush. Horsetails have, however, occasionally been reported to be poisonous to horses, and they are not recommended for human consumption.

Native Americans and Asians had several other uses for these plants. One tribe drank the water from the carinal canals of the stem, and another thought the shoots were "good for the blood." It is not known if there was any basis for this latter

Figure 21.14 Horsetail spores. (a) With elaters coiled. (b) With elaters spread.

idea, but there is an old, unconfirmed report that field horsetail consumption "produces a decided increase in blood corpuscles." At least one or two species are known to have a mild diuretic effect (a *diuretic* is a substance that increases the flow of urine), and they have been used in the past in folk medicinal treatment of urinary and bladder disorders. Some have also been used as an antacid or an astringent (an *astringent* is a substance that arrests discharges, particularly of blood). One species was used in the treatment of gonorrhea, and others were used for tuberculosis.

At least two Native American tribes burned the stems and used the ashes to alleviate sore mouths or applied the ashes to severe burns. Members of another tribe ate the strobili of a widespread scouring rush to cure diarrhea, and still others boiled stems in water to make a hair wash for the control of lice, fleas, and mites.

During pioneer times when the covered wagons moved westward, the use of scouring rush stems for scouring and sharpening was widespread. They were used not only for cleaning pots and pans but also for polishing brass and hardwood furniture and flooring and for honing mussel shells to a fine edge. Scouring rushes are still in limited use for these purposes today. Some species of horsetails accumulate certain minerals in addition to silica. Veins of such minerals have been located beneath populations of horsetails by analyzing the plants' mineral contents.

In the geological past, the giant horsetails and club mosses were a significant part of the vegetation growing in vast swampy areas. In some instances, the swamps were stagnant and slowly sinking, permitting the gradual accumulation of plant remains, which, because of the lack of oxygen in the water, were not readily attacked by decay bacteria. Such circumstances, over aeons of time, were ideal for the formation of coal. Today, if you section a lump of coal thinly enough and examine it under the microscope, you can still see bits of tissue and spores of plants that were living hundreds of millions of years ago. One soft coal known as *cannel* consists primarily of the spores of giant horsetails and club mosses that, through the ages, were reduced to carbon.

Figure 21.15 Reconstruction of a coal age (Carboniferous) forest. © *Ludek Pesek/Science Source*

Figure 21.16 Reconstruction of the fossil giant horsetail, *Calamites*.

21.5 Phylum Polypodiophyta—the Ferns

Learning Outcome

a. Draw the life cycle of a fern, label all structures, and explain why ferns demonstrate alternation of generations.

Ferns include some of the most popular indoor (and outdoor patio) plants throughout the world. They are seen in business offices, restaurants, residential homes, shopping malls, universities, and virtually any place where patrons appreciate plants. Some of the reasons ferns are used as ornamentals include their easy care and resilience under many conditions. Some ferns have even been reported to reduce the amount of pollutants, such as benzene, formaldehyde, and trichloroethylene, in the air. Their leaves are so infinitely varied in form and aesthetically pleasing that just about any dwelling can be accessorized with a fern matched to the setting.

Structure and Form

The approximately 11,000 known species of ferns vary in size from tiny floating forms less than 1 centimeter (0.4 inch) in diameter to giant tropical tree ferns up to 25 meters (82 feet) tall. Fern leaves are *megaphylls* (leaves associated with leaf gaps and having branching veins—see Chapter 6) that are commonly referred to as **fronds.** They are typically divided into smaller segments and are feathery in appearance (Fig. 21.17), but some are undivided, pleated, or tonguelike, and others resemble a four-leaf clover or grow in such a way as to form "nests." In the tropics, the "nest" ferns often accumulate enough humus to provide food and shelter for huge earthworms that are up to 0.6 meter (2 feet) long. Because ferns require external water for their sexual reproduction, they are most abundant in wetter tropical and temperate habitats, but a few are adapted to drier areas.

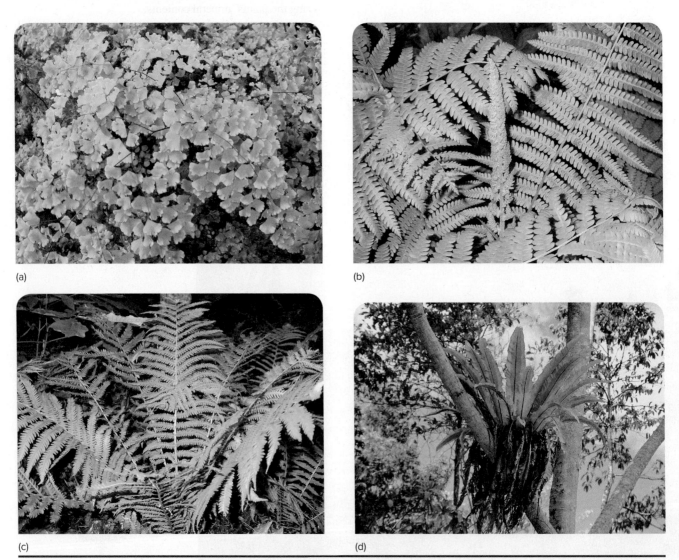

Figure 21.17 Common ferns with typically dissected fronds. (a) A maidenhair fern. (b) Cinnamon fern. The cinnamon-colored fertile frond in the center is non-photosynthetic and produces large numbers of sporangia. (c) Ostrich fern. (d) Bird's-nest fern growing in a forest tree of the Andaman islands. *(a-c) © James E. Bidlack; (d) © ephotocorp/Alamy*

Reproduction

The sporophyte is the conspicuous phase of the life cycle in all the plants discussed in this chapter (Fig. 21.18). A fern sporophyte consists of the fronds, a stem in the form of a rhizome, and adventitious roots that arise along the rhizome. The fronds, regardless of their ultimate form, usually first appear tightly coiled at their tips. These *croziers* (unrelated to the croziers of sac fungi), or "fiddleheads" (Fig. 21.19), then unroll and expand, revealing the blades. At maturity, the blades are often divided into segments called **pinnae** (singular: **pinna**) that are attached to a midrib, or rachis. A stalk, or petiole, is usually present at the base.

When the fronds have expanded and matured, small, often circular, rust-colored patches of powdery-looking material may appear on the lower surfaces of some or all of the blades. Because these patches appear similar to fungal rusts, some fern owners have thought their plants were diseased and have turned to sprays to deal with the "problem" or have even carefully scraped the patches off. The development of the patches is normal and healthy, however, and examination with a hand lens or dissecting microscope

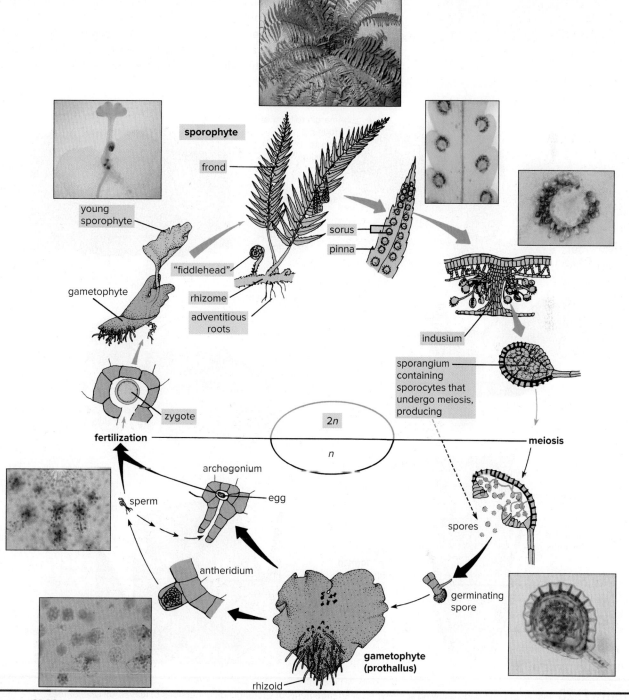

Figure 21.18 Life cycle of a fern. © *James E. Bidlack and Kingsley Stern.*

will reveal that they are actually clusters of *sporangia* (Fig. 21.20). The sporangia may be scattered evenly over the lower surfaces of the fronds or confined to the margins. The sporangia are mostly found in numerous, discrete clusters called **sori** (singular: **sorus**). In many ferns, the sori, while they are developing, are protected by thin, individual flaps of colorless tissue called **indusia** (singular: **indusium**). As the sporangia mature, the indusium, which often resembles a tiny, semi-transparent umbrella attached by its base to the frond surface, shrivels and exposes the sporangia beneath.

Most of the sporangia are microscopic and stalked, and they look something like tiny, transparent baby rattles with a conspicuous row of heavy-walled, brownish cells along the edge. This row of cells, which looks like a tiny millipede, is called an **annulus**. It functions in catapulting spores out of the sporangium with a distinct snapping action influenced by moisture changes in the cells (Fig. 21.21).

Figure 21.19 A crozier (fiddlehead) of a tropical fern. © *Kingsley Stern*

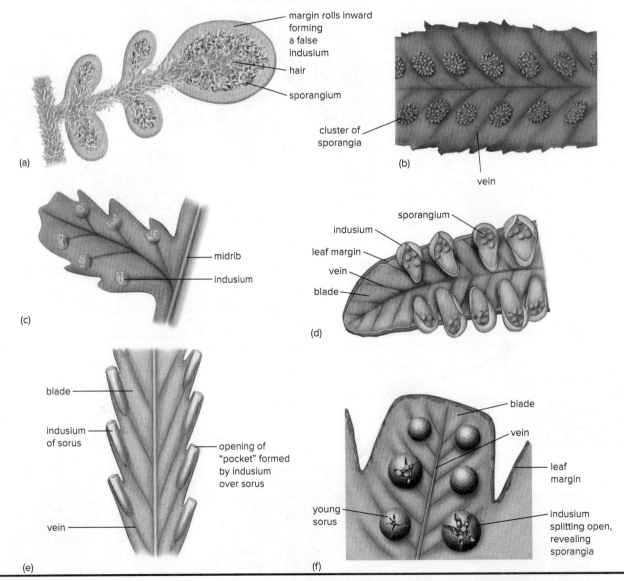

Figure 21.20 Pinnae of fern fronds, showing some types of arrangements of sporangia on undersides of leaves. (a) *Cheilanthes.* (b) *Polypodium.* (c) *Cystopteris.* (d) *Cibotium.* (e) *Davallia.* (f) *Cyathea.*

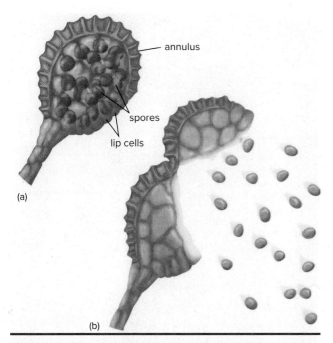

Figure 21.21 Release of spores from a fern sporangium. (a) An intact sporangium. (b) Spores being ejected as the sporangium splits; the annulus first draws back and then snaps forward.

Sporocytes undergo meiosis in the sporangia, usually producing either 48 or 64 spores per sporangium. Sporangia of some of the primitive adder's tongue ferns may have up to 15,000 spores, however, and the number of sporangia is often so great that it has been estimated that a single beech fern plant will produce a total of 50 million spores.

Two kinds of spores are produced in certain aquatic or amphibious ferns, such as the clover fern (*Marsilea*) and *Salvinia*. Single, large *megaspores* are produced in some sporangia, while numerous, tiny *microspores* are produced in others. The vast majority of fern species, however, produce only one kind of spore.

After the spores have been flung out of their sporangia, they are dispersed by wind; relatively few end up in habitats suitable for their survival. Such habitats include shady, wet ledges and rock crevices or moist soil. The spores can also easily be germinated on damp clay flower pots in the home or greenhouse. Those that germinate in favorable locations produce little "Irish valentines," or **prothalli** (singular: **prothallus**), as the green, heart-shaped gametophytes of these and other seedless vascular plants are more properly called (Fig. 21.22). These free-living structures often curl slightly at their edges and may be 5 to 6 millimeters (0.25 inch) in diameter; they are visible without a microscope.

Prothalli are only one cell thick, except toward the middle, where they are slightly thicker. Antheridia are

Figure 21.22 Fern prothalli. (a) Surface view. ×10. (b) A prothallus as seen with the aid of a light microscope. ×20. (c) Archegonia. ×100. (d) Antheridia. ×100. *(a-d) © Kingsley Stern*

interspersed among the rhizoids produced on the lower surface of the central area of most prothalli, with archegonia also being produced, usually closer to the notch of the heart-shaped gametophyte. The archegonia are somewhat flask-shaped, with curving necks that protrude slightly above the surface, whereas the antheridia are more spherical and often elevated above the surface on short stalks. In a few species, antheridia and archegonia are produced on separate prothalli.

A single antheridium may produce from 32 to several hundred sperms, each with few to many flagella. Fertilization of an egg takes place within an archegonium and is chemically influenced so that it usually occurs with a sperm from a different prothallus. Only one zygote develops into a young sporophyte on any prothallus, regardless of the number of eggs that may be fertilized. Hence, what was the gametophyte (or portion of the gametophyte) dies and leaves the sporophyte growing independently. This sporophyte usually has smaller, simpler fronds during its first season of growth, but typical full-sized fronds grow from the persisting rhizomes in succeeding years.

Fossil Relatives of Ferns

Fossil remains of ferns and plants thought to be ancestors of ferns abound in ancient deposits. Possible ancestors of ferns (Fig. 21.23) are found in Devonian formations (Table 21.1) estimated to be 375 million years old. Most resemble ferns in growth habit, but they have no broad fronds and otherwise are more like whisk ferns than true ferns. Well-preserved fossils of tree ferns related to large, tropical ferns of the present day are found in geological strata dating back to between 250 and 320 million years ago (Fig. 21.24).

Ferns (especially tree ferns) became so abundant during the latter part of the Carboniferous period that this era in the

Figure 21.24 A fossil fern. © *Kingsley Stern*

past was referred to as the *Age of Ferns*. The discovery of seeds on some of them, however, raised questions, and the term *Age of Ferns* was dropped.

Human and Ecological Relevance of Ferns

In some parts of North America, it is unusual to find a house without a Boston fern growing in it. Ferns make ideal houseplants because many of them are adapted to low light, and most are not as susceptible as other plants to aphids, mites, mealybugs, and similar pests. Outdoors, ferns are equally popular as ornamentals. Some are subjects for artists and photographers.

Apart from the pleasing aesthetic aspects of ferns, they function well as air filters. For example, during a single hour, one average-sized Boston fern can remove about 1,800 micrograms of formaldehyde (a common pollutant from carpets) from the air in a typical room measuring 3 meters by 3 meters (10 feet by 10 feet). According to the Environmental Protection Agency, if properly maintained, the fern can almost completely rid the air of pollutants on a daily basis.

Ferns also have other practical value. Commercial growers of the brilliantly colored anthuriums in Hawaii and elsewhere have found that native tree ferns provide the ideal amount of shade and other environmental conditions needed

Figure 21.23 Possible ancestors of ancient ferns. (a) *Aglaophyton*. (b) *Psilophyton* (?). The 19th-century origin of this drawing apparently was a composite of three distantly related genera.

for bringing their flowers to perfection. Tree fern rhizome or root bark[1] and rhizome bark of certain other species, such as the royal fern (which produces osmunda bark), are a favorite medium of orchid, bromeliad, and staghorn fern growers. Its texture is well suited to the growth of the orchids' aerial roots, and as the bark slowly decomposes, rain water trickling over its surface picks up nutrients that are particularly appropriate for these plants. The demand for fern bark for orchids has exceeded the supply for a number of years, and it has become very expensive (Fig. 21.25).

As the young "fiddleheads" of many species of ferns unroll, a dense covering of hairs is visible on the petiole and rachis. In the past, the silky hairs of some of the larger tropical tree ferns (Fig. 21.26) were stripped and used for upholstery, pillow, and mattress stuffing. During the late 1800s, over 1,900 metric tons (2,094 tons) of this material were shipped from Hawaii to the mainland, and if it were not for

Figure 21.25 An orchid plant growing on fern bark.
© Kingsley Stern

the eventual substitution of alternative materials, these magnificent plants might have been totally destroyed. Some tropical hummingbirds use these hairs along with scales of other ferns to line their nests, and at one time, Polynesians used them in a form of embalming for their dead. The trunks of tree ferns have been used in the construction of small houses in the tropics. Parts of one Hawaiian species of small tree fern and the fronds of an Asian fern yield red pigments used for dyeing cloth.

The bracken fern, which is distributed worldwide and is a weed in parts of Europe, has been used and even cultivated

---

1. Ferns do not have a cork cambium and do not, therefore, produce bark that is typical of woody seed plants. The cells of the outer layers of fern roots and rhizomes, however, do become impregnated with suberin and consequently become barklike in appearance and function.

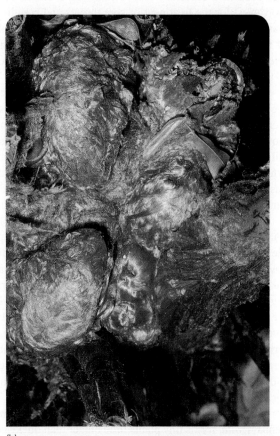

(a) (b)

Figure 21.26 (a) A Brazilian tree fern. (b) The growing tip of a tree fern, showing the protective, rust-colored hairs. *(a-b)* © Kingsley Stern for human food for many years, particularly in Japan and New Zealand. It has, however, also long been known to be mildly poisonous to livestock. Research in both Europe and Japan has shown conclusively that bracken fronds fed to experimental animals produce intestinal tumors, and because of this, the consumption of these fronds for food is now discouraged.

Indigenous peoples of many areas where ferns occur have eaten the cooked rhizomes and young fronds of various ferns. Native Americans often baked the rhizomes of sword ferns, lady ferns, and others in stone-lined pits, removed the outer layers, and ate the starchy inner material. Similarly, native Hawaiians ate the starchy core of their tree ferns as emergency food. In Asia, the oriental water fern is still sometimes grown for food in rice paddies and used as a raw or cooked vegetable. In Malaysia, a relative of the lady fern is frequently used as a vegetable.

Uses of ferns in folk medicine abound. They have been used in the treatment of diarrhea, dysentery, rickets, diabetes, fevers, eye diseases, burns, wounds, eczema and other skin problems, leprosy, coughs, stings and insect bites; as a poison antidote; and for labor pains, constipation, dandruff, and a host of other maladies. The male fern, which is more common in Europe than in North America, contains a drug that is effective in expelling intestinal worms (e.g., tapeworms). Its use for this purpose dates back to ancient times, and it is still occasionally used, although synthetic medicines have now largely replaced it. The licorice fern was used in the past by Native Americans of the Pacific Northwest for the treatment of sore throats and coughs and was used as a flavoring agent and sugar substitute. Members of a tribe in California chewed stalks of goldback ferns to quell toothaches and snuffed a liquid made from the fronds of the bird's-foot fern to arrest nosebleeds.

The fronds of bracken and other ferns have been used in the past for thatching houses. Anyone who has placed such fronds in compost piles knows they break down much more slowly than do the leaves of other plants. Bracken fronds are still occasionally used as an overnight bedding base by fishermen and hunters. A substance extracted from these fronds has been used in the preparation of chamois leather, and the rhizome is used in northern Europe in the brewing of ale.

The chain fern has large fronds up to 2 meters (6.5 feet) long, which have two flexible, leathery strands in the petiole and rachis of each frond. Native Americans and others have gathered the fronds for many years to strip these strands for use in basketry and weaving. They do so by gently cracking the long axis with stones to expose the strands, which are then easily removed. The glossy, black petioles of the five-finger fern have also been used in intricate basketry patterns by Native Americans. In Southeast Asia, the climbing fern has fronds with a rachis that may grow up to 12 meters (40 feet) long; it is still a favorite material (when available) for the weaving of baskets.

The mosquito fern, which is a floating water fern in the genus *Azolla,* forms tiny plants little bigger than duckweeds. It is found over wide areas where the climate is relatively mild. It sometimes forms such dense floating mats that it is believed to suffocate mosquito larvae, which periodically need to reach the surface for air. This same fern frequently has cyanobacteria living symbiotically in cavities between cells. The cyanobacteria fix nitrogen, and it has been shown experimentally that plants without these organisms do not grow as well as those with them. In some parts of the world, mosquito ferns are thrown into rice paddies, so that the cyanobacteria associated with them will fix nitrogen and reduce the need for added fertilizers.

Many ferns are now commercially propagated through tissue culture (discussed in Chapter 14). The scientific names of ferns mentioned in this chapter are given in Appendix 1.

21.6 Fossils

Learning Outcome

a. Provide an accurate definition of a fossil, describe different types of fossils, and explain how they are formed.

Introduction

Several references to fossils have been made in this and preceding chapters, and other references occur in the chapters that follow. We have become increasingly aware of worldwide limitations of energy sources, and we now frequently hear references to fossil fuels as nonrenewable resources. Exactly what is a fossil, and how did it come into existence?

Originally, the word *fossil* was applied to anything unusual found in rocks, but its use is now confined to any recognizable prehistoric organic object (or its impression) that has been preserved from past geological ages in the earth's crust (Table 21.1). Such a definition includes teeth marks, borings, impressions of footprints and tracks, dung deposits, and deposits of chemicals that are evidence of the activities of algae and bacteria. Age itself does not make a fossil. Human remains thousands of years old that have been found in caves and tombs are not regarded as fossils, but it is conceivable that other remains of the same age could be preserved in rocks and regarded as fossils.

Fossils are formed in a number of different ways. The conditions for their formation almost always include the accumulation of sediments in an area where plants, animals, or other organisms are present or to which they have been transported. Such areas are found in swamps, oceans, lakes, or other bodies of water. Hard parts, such as wood or bones, are more likely to be preserved than soft parts. Other environmental factors, such as the presence of salt or antiseptic chemicals or stagnation resulting in low oxygen content of the water, all favor fossil formation.

Quick burial is another factor enhancing the chances of organic material becoming fossilized. Such burial might result if an organism was unable to avoid a rain of ash from a volcanic eruption or was trapped in a cave or sudden sandstorm. Other forms of quick burial occur when an organism falls into quicksand, a body of water, or asphalt. Descriptions of several common types of fossils follow.

KEY THEME 21.1 ecology

Ferns and Fossil Fuels

Ferns and their relatives were the first land plants with significant internal tissues for conducting water and roots for absorbing water and nutrients, morphological innovations that expanded the environments in which these ancient plants could live. In the ancient geological period known as the Carboniferous, giant horsetails and club mosses grew in great abundance in swampy areas. The biomass produced by these plants accumulated in huge deposits. The lack of oxygen inhibited bacteria-mediated decay of the plant biomass, which was eventually transformed into coal. These coal deposits represent large quantities of carbon dioxide removed from the ancient atmosphere by ferns and their relatives. The reintroduction of this carbon dioxide into the atmosphere in modern times, as we burn coal (Box Figure 21.1A), may produce rapid climate change.

Box Figure 21.1A Closeup view of burning coal. © *Brand X Pictures/PunchStock*

Figure 21.27 A compression fossil. © *Kingsley Stern*

original objects. When objects such as leaves are buried by layers of sediment, the sheer weight of the overlying material may compress them to as little as 5% of their original thickness. When this occurs, all that is left is a thin film of organic material and an outline showing some surface details. Virtually no preservation of cells or other internal structure takes place. Such fossils are very common and are called *compressions* (Fig. 21.27). The image of a compression, like the details of a foot or a hand pressed into wet cement, is called an *imprint*. Coal is a special type of compression, often involving different plant parts that are thought to have been subjected to enormous pressures after the fallen plants slowly accumulated in a swamp.

Molds, Casts, Compressions, and Imprints

After silt or other sediment has buried an object and hardened into rock, the organic material may be slowly washed away by water percolating through the pores of the rock. This leaves a space, which may then be filled in with silica deposits. If only a space is left, it is called a *mold*; if it is filled in with silica, it is called a *cast*. Artificial casts of plaster or wax compounds may be made from molds of the

Petrifactions

Petrifactions (Fig. 21.28) are uncompressed, rocklike materials in which the original cell structure has been preserved. About 20 different mineral substances, including silica and the salts of several metals, are known to bring about petrifaction. At one time, it was believed the process occurred through the replacement of the plant parts by minerals in solution, one molecule at a time. It is now believed that chemicals in solution infiltrate the cells and cell walls, where they crystallize and harden, permanently preserving the original material.

Petrifactions can be studied by cutting thin sections with a diamond or carborundum saw and polishing the material with extremely fine grit powder until it is thin enough for

Figure 21.28 Petrified wood. © *Sharon Stern*

light to pass through. Then it can be examined with a compound microscope. Another, simpler method of studying petrifactions involves etching the cut surface with a dilute acid and then applying a plastic or similar film. As soon as the film hardens, it can be peeled off. Such peels display lifelike microscopic details of the surface with which they were in contact. They are commonly used by paleobotanists (botanists who study fossil plant materials) in their research.

Coprolites

Coprolites are the fossilized dung of prehistoric animals and humans. They may contain pollen grains and other plant and animal parts that provide clues to the food and feeding habits of past organisms and cultures.

Unaltered Fossils

Some plants or animals fell into bodies of oil or water that, because of substances present or a nearly total lack of oxygen, did not permit decay to occur. Some animals died in snowfields, and their bodies were permanently frozen. In such instances, preservation in the unaltered state occurred. Unaltered fossils, particularly frozen ones, are very rare.

SUMMARY

21.1 Introduction to the Seedless Vascular Plants
- Four phyla of seedless vascular plants are recognized: Psilotophyta (whisk ferns); Lycophyta (club mosses and quillworts); Equisetophyta (horsetails and scouring rushes); and Polypodiophyta (ferns).

21.2 Phylum Psilotophyta—the Whisk Ferns
- Whisk ferns (*Psilotum*) are the simplest of all living vascular plants, consisting of evenly forking, green stems that have small protuberances called enations, but no leaves; roots are also lacking. The stem contains a central cylinder of xylem and phloem.
- Whisk fern spores germinate into tiny gametophytes, with antheridia and archegonia scattered over their surfaces. The zygote develops into a foot and a rhizome. Upright stems are produced when the rhizome separates from the foot.
- *Tmesipteris,* an Australian relative of whisk ferns, has leaflike appendages. Fossil plants resembling whisk ferns have been found in Silurian geological formations.

21.3 Phylum Lycophyta—the Ground Pines, Spike Mosses, and Quillworts
- There are two genera of club mosses with living members. Ground pines (*Lycopodium*) develop sporangia in the axils of sporophylls. Several sporophytes may be produced from one gametophyte. Spike mosses (*Selaginella*) are heterosporous and have a ligule on each microphyll; microspores develop into male gametophytes with antheridia, and the megaspores develop into female gametophytes with archegonia.
- Quillworts (*Isoetes*) are heterosporous and have quill-like microphylls that arise from a cormlike base. The corms have a cambium that remains active for many years.
- Some fossil relatives of club mosses were large, dominant members of the forests and swamps of the Carboniferous period.
- Club moss spores have been used for flash powder, for medicinal purposes, as talcum powder, and to stanch bleeding. The plants themselves have been used as ornamentals, novelty items, and Christmas ornaments, and for intoxicating purposes.

21.4 Phylum Equisetophyta—the Horsetails and Scouring Rushes
- Horsetails and scouring rushes (*Equisetum*) accumulate deposits of silica in their epidermal cells and have made

good scouring material. They occur in both unbranched and branched forms. The stems are jointed and ribbed and have tiny, scalelike leaves in whorls at each joint.

- The stems of *Equisetum* are centrally hollow and contain cylindrically arranged carinal and vallecular canals. The stems arise from rhizomes that branch extensively below the surface of the ground.
- Some species of *Equisetum* produce nonphotosynthetic stems. Conelike strobili are produced in the spring in all species. *Equisetum* spores have ribbonlike elaters that are sensitive to changes in humidity.
- *Equisetum* plants produce equal numbers of male and female gametophytes, but female gametophytes may become bisexual. The development of more than one sporophyte from a gametophyte is common.
- Ancient relatives of horsetails were the size of trees when they flourished in the Carboniferous period of 300 million years ago.
- Horsetails have been used for food after removal of the parts containing silica, but they are not recommended for human consumption. They have also been used medicinally as a diuretic, as an astringent, and in the treatment of venereal disease and tuberculosis. Other uses include a hair wash, a mineral indicator, and a metal polish. Cannel coal consists primarily of spores of giant horsetails that were reduced to carbon.

21.5 Phylum Polypodiophyta—the Ferns

- Fern leaves (fronds) are typically divided and feathery in appearance but vary greatly in form. They usually first appear as croziers that unroll and expand.
- Patches of sporangia appear on the lower surfaces of fern fronds. The sporangia commonly occur in sori, which may be protected by indusia. Each sporangium has an annulus that catapults mature spores out of the sporangium.
- Fern gametophytes (prothalli) develop from spores. Most prothalli contain both archegonia and antheridia; only one zygote develops into a sporophyte.
- Possible ancestors of ferns are found in Devonian deposits estimated to be 375 million years old.
- Ferns are used as ornamentals, air filters, a source of "bark" for growing orchids and other plants, and a source of stuffing materials for bedding; in tropical construction; as food; and in numerous folk medicinal applications. Some ferns provide basketry and weaving material, an ingredient in brewing ale, and an ingredient in the preparation of chamois leather. One floating fern forms dense mats and is believed to suffocate mosquito larvae.

21.6 Fossils

- Fossils are recognizable prehistoric organic objects that are formed in different ways. Molds, casts, compressions, and imprints are formed when material buried by silt or other sediment has hardened into rock and the organic material has slowly been washed away by water.
- Petrifactions are uncompressed, rocklike materials in which the original cell structure has been preserved. Coprolites are fossilized dung that may contain pollen grains and other plant and animal parts. Unaltered fossils are those of plants or animals that may have fallen into bodies of oil or water or snowfields and were not subjected to decay.

REVIEW QUESTIONS

1. What basic features of the ferns and their relatives distinguish them from any organisms studied thus far?
2. How does the gametophyte of a whisk fern differ from that of a true fern?
3. Which of the fern relatives have significantly functional leaves? In those without conspicuous leaves, how are the carbohydrate needs of the plants met?
4. How does one distinguish among ground pines, spike mosses, and quillworts?
5. How did the ancient members of the Lycophyta phylum differ from those of the present day?
6. How do the spores and the female gametophytes of horsetails differ from those of any other plants studied thus far?
7. What is the location and function of carinal canals?
8. In your opinion, which have the most human relevance today: club mosses and horsetails of the present or those of the geological past? Why?
9. Define crozier, rachis, pinna, indusium, and prothallus.
10. Diagram the life cycle of a typical fern.
11. Summarize present and past human uses of ferns.
12. How are fossils formed, and what different types are recognized?

DISCUSSION QUESTIONS

1. Would you assume there is any significance to the fact that both the sporophytes and the gametophytes of whisk ferns branch in the same manner?
2. Do spike mosses, which produce two kinds of spores and gametophytes, have any advantage over ground pines, which produce only one kind of spore and gametophyte?
3. Some gametophytes of fern relatives develop underground, while others develop at the surface. If you were to be a gametophyte, which would you prefer? Why?
4. After looking at the internal structure of a horsetail stem, can you suggest a function for the silica in the ribs?
5. How would we be affected if all ferns were to become extinct in a few years?

CHAPTER

22

Introduction to Seed Plants: Gymnosperms

22.1 Introduction to Gymnosperms
22.2 Phylum Pinophyta—the Conifers
  Pines
  Key Theme 22.1: Resilient and Useful Gymnosperms
  Other Conifers
22.3 Other Gymnosperms
  Phylum Ginkgophyta—*Ginkgo*
  Phylum Cycadophyta—the Cycads
  Phylum Gnetophyta—the Gnetophytes
22.4 Human Relevance of Gymnosperms
  Conifers
  Other Gymnosperms
  Key Theme 22.2: A Living Fossil?

OVERVIEW

The chapter begins by exploring the differences between ferns and seed plants and then gives a brief geological history of gymnosperms. The leaves, roots, and stems of pine trees are discussed, and the life cycle of a typical gymnosperm (pine) is given. Other conifers, such as yews, podocarps, junipers, and redwoods, are mentioned, and a brief discussion of *Ginkgo,* cycads, *Ephedra, Gnetum,* and *Welwitschia* follows. The chapter concludes with a digest of the human relevance of gymnosperms, with particular emphasis on the conifers.

A female cycad cone (*Cycas* sp.). Cycads are slow-growing, superficially palmlike gymnosperms. *Gymnosperm* means naked-seeded, and the developing seeds can be seen here out in the open at the edges of the cone scales.
© Steven P. Lynch

22.1 Introduction to Gymnosperms

Learning Outcomes

a. Explain how seeds differ from spores, and explain the evolutionary significance of the seed coat.
b. List the four phyla of gymnosperms, and list distinguishing features and examples of each phylum.

Some giant tree ferns have large, graceful leaves held above an unbranched trunk (see Fig. 21.26) and superficially resemble coconut palms. Anyone who has sat at the base of a coconut palm when a coconut has fallen, however, has been reminded of a basic difference between ferns and coconut palms. The palms are flowering plants that produce seeds such as those of pine trees and other cone-bearing plants, whereas the ferns produce no seeds at all.

The oldest known seeds were produced by plants that appeared late in the Devonian period, more than 350 million years ago. Seeds provided a significant adaptation for plants that had invaded the land. Unlike spores, seeds have a protective seed coat and a supply of food (usually endosperm) for the embryo. The embryo may be capable of lying dormant through long periods of freezing weather and of drought, and may, in some instances, even survive fire. This survival value of seeds undoubtedly played a major role in seed plants becoming the dominant vegetation on earth today.

The first seed plants were so fernlike in appearance that they were originally classified as ferns. When fossils with obvious seeds on the fronds were discovered, however, these *pteridosperms* ("seed ferns") were reclassified as *gymnosperms*.

The name **gymnosperm** is derived from two Greek words: *gymnos,* meaning naked, and *sperma,* a seed. The name refers to the exposed nature of the seeds, which are produced on the surface of sporophylls or similar structures instead of being enclosed within a fruit as they generally are in the flowering plants (Fig. 22.1). The seed-bearing sporophylls of the sporophyte are often spirally arranged in strobili (seed cones) that develop at the same time as smaller pollen-bearing strobili (pollen cones). The pollen cones produce *pollen grains.*

The female gametophyte is produced inside an **ovule** that contains a fleshy tissue called the **nucellus** (see Fig. 22.7). The nucellus is itself enclosed within one or more outer layers of tissue. These outer layers of tissue constitute an **integument** that becomes a **seed coat** after the fertilization and development of an embryo takes place (see Fig. 22.8). All of these tissues are derived from the mother plant.

The sporophytes of gymnosperms are mostly trees and shrubs, with a few species being vines. The gametophytes are even more reduced in cell number than they are in ferns and their relatives. Unlike the gametophytes discussed so far, they don't grow independently but develop within sporophyte structures.

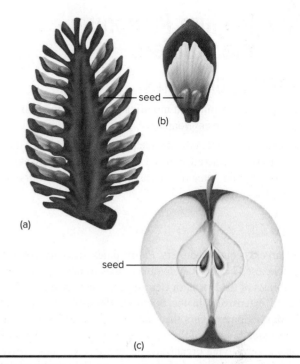

Figure 22.1 A comparison between exposed gymnosperm seeds and enclosed angiosperm seeds. (a) Exposed seeds on a woody seed cone of a pine tree. (b) A single seed cone scale with two seeds. (c) A section through an apple (angiosperm fruit), showing the enclosed seeds.

Molecular studies currently indicate that gymnosperms may share a common ancestor with another plant group, but these exciting investigations are still being pursued. Fossil records suggest that flowering plants may have evolved from within gymnosperm ancestors, and scientists are actively researching how this may have occurred. At this time, there are four phyla of living gymnosperms:

1. **Phylum Pinophyta** (*Conifers*). These members bear distinct strobili ("pine cones") and have leaves that are needlelike and arranged in clusters of up to five leaves each. There are about 600 species of pines, firs, spruces, hemlocks, cedars, redwoods, and other coniferous woody plants living today.
2. **Phylum Ginkgophyta** (*Ginkgo*). Notched, broad, fan-shaped leaves with veins that fork evenly are distinct among *Ginkgos*. The mature seeds resemble small plums and are enclosed in a fleshy covering. Only one species, *Ginkgo biloba,* exists within this phylum.
3. **Phylum Cycadophyta** (*Cycads*). These slow-growing plants have the appearance of a cross between a tree fern and a palm but are not directly related to either group. The strobili of cycads are located in the center of the plant. About 300 species of cycads are known to exist, many of which have been discovered during the past two decades.
4. **Phylum Gnetophyta** (*Gnetophytes*). A unique characteristic of gnetophytes is the presence of vessels and tracheids in the xylem. Many members superficially resemble flowering plants, although they still produce strobili. About 70 species of gnetophytes exist.

22.2 Phylum Pinophyta— the Conifers

Learning Outcomes

a. Explain how pine leaf modifications enable them to adapt to harsh environments.

b. Draw a generalized pine tree life cycle and identify the following structures: male and female gametophytes, pollen grain, integuments, archegonium, egg, embryo, and sporophyte.

Pines

The largest genus of conifers, *Pinus* (pines) has over 100 living species. They are the dominant trees in the vast coniferous forests of the Northern Hemisphere. They have also been planted extensively in the Southern Hemisphere, but only the Merkus pine occurs there naturally, and its distribution barely extends south of the equator. They include the world's oldest known living organisms, the bristlecone pines, native to the White Mountains of eastern central California and the Snake range on the central Nevada-Utah border. Some trees still standing are nearly 4,600 years old, and one that was, unfortunately, cut down in 1964 was found to have been about 4,900 years old. The oldest living tree is believed to be a spruce in Sweden that is about 9,550 years old. It grew from a seed that germinated during the last Ice Age.

Structure and Form

Pine leaves are needlelike and are arranged in clusters of two to five leaves each (a handful of species have as many as eight or as few as one to a cluster). Regardless of the number of leaves, each cluster (*fascicle*) forms a cylindrical rod if the leaves are held together (Fig. 22.2).

Figure 22.2 A cluster (fascicle) of pine needles in cross section, showing how the needles form a loose, cylindrical rod. © *Steven P. Lynch*

The fascicles are short shoots with restricted growth, a feature of some gymnosperms not found in flowering plants. Pines often live in areas where the topsoil is frozen for a part of the year, making it difficult for the roots to obtain water. In addition, the leaves may be exposed to high winds and bitterly cold temperatures. Accordingly, they have several modifications that enable them to survive in harsh environments.

Just beneath the epidermis is a **hypodermis**, consisting of one to several layers of thick-walled cells. The epidermis itself is coated with a thick cuticle. The stomata, instead of being at the surface, are recessed in small cavities. The veins and their associated tissues are surrounded by an endodermis, and the mesophyll cells do not have the obvious air spaces typical of the spongy mesophyll of the leaves of flowering plants (see Figs. 7.6 and 7.11).

Conspicuous **resin canals** develop in the mesophyll. These resin canals, which are found throughout other parts of the plant as well, consist of tubes lined with special cells that secrete *resin*. Resin is aromatic and antiseptic and prevents the development of fungi; it also deters insect attacks. Other conifers apparently produce resin canals in response to injury. Pine fascicles usually *abcise* (i.e., fall off—see Chapter 7) within 2 to 5 years of their maturing, but those of bristlecone pines persist for up to 30 years. The fascicles are lost a few at a time, so that some functional leaves are always present on a healthy tree.

Wood varies considerably in hardness. Most gymnosperm wood, including that of pines, consists primarily of tracheids and differs from the wood of dicots in having no vessel members or fibers. Conifer wood, because of the absence of thick-walled cells, is said to be *soft*, while the wood of broadleaf trees is described as *hard*.

In many conifers, the annual rings of the xylem are often fairly wide as a result of a comparatively rapid growth rate during the growing season. Resin canals are formed both vertically and horizontally throughout various tissues (Fig. 22.3). The bark includes the secondary phloem and may be relatively thick. It often becomes 7.5 centimeters (3 inches) or more wide, and in the giant redwood, it may become as much as 60 centimeters (2 feet) wide. Companion cells are absent from the phloem, but similar *albuminous cells* apparently perform the same function.

Mycorrhizal fungi are associated with the roots of most conifers. In fact, pine seedlings that germinate in sterilized soil do not grow well at all until the fungi are introduced or allowed to develop. The roots of adjacent pines often interweave. New England pioneers made use of this characteristic in eastern white pines when clearing land. After trees were felled, they would tip over the stumps with the roots still attached and use them for fences. The fences survived for many years.

Reproduction

Pines, like spike mosses, quillworts, and a few of the ferns, produce two kinds of spores (see Fig. 22.8). **Pollen cones**

KEY THEME 22.1 ecology — Resilient and Useful Gymnosperms

The evolution of seeds by the seed plants, including the gymnosperms, facilitated their invasion of land. Encased in a seed, the embryo of seed plants can withstand the extremes of the terrestrial environment, such as freezing weather, drought, and even fire. In addition to protection of embryos within the seed, gymnosperms have other characteristics that enable them to survive during adverse conditions. These anatomical features include a thick cuticle on the epidermis of leaves, sunken stomata, a resilient hypodermis, and expandable mesophyll cells, all of which are particularly advantageous for survival in freezing winters as well as withstanding the heat in dry desert areas. Resins produced by gymnosperms provide defense against fungal and insect attacks, while mycorrhizal fungi increase availability of soil nutrients and water to gymnosperms. Mutualistic associations with squirrels and birds, such as nutcrackers, promote gymnosperm seed dispersal. Gymnosperms provide humans with a wealth of "nature's services." Next time you see a gymnosperm, give the plant a hug (Box Figure 22.1A). Among the many things gymnosperms provide, you can thank the cedars for pencils, hemlocks for dyes, firs for Christmas trees, pines for wood and turpentine, spruce for paper, and yews for cancer treatment.

Box Figure 22.1A Give a tree a hug and say thanks.
© Ferencz Teglas/123RF

Figure 22.3 A portion of a cross section of a pine stem, showing annual rings. © Kingsley Stern

(male strobili) consist of papery or membranous scales, called **microsporophylls,** arranged in a spiral or in whorls around an axis; they are usually produced in the spring. The pollen cones usually develop toward the tips of the lower branches in clusters of up to 50 or more (Fig. 22.4) and are mostly less than 4 centimeters (about 1.5 inches) long. **Microsporangia** develop in pairs toward the bases of the scales.

Each of the **microsporocytes** in the microsporangia undergoes meiosis, producing four haploid **microspores.** These then develop into **pollen grains;** each grain consists

411

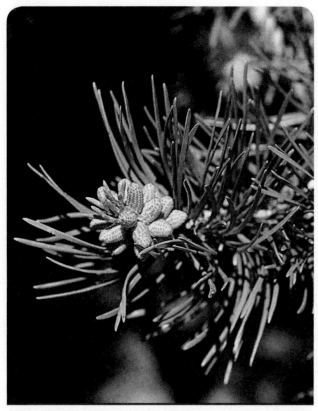

Figure 22.4 A cluster of pine pollen cones. © *Kingsley Stern*

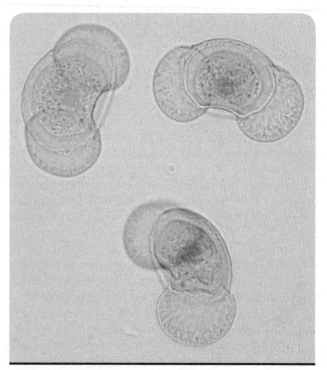

Figure 22.5 Pollen grains of a pine, as seen with the aid of a light microscope. Each pollen grain has a pair of air sacs that provide added buoyancy. ×300. © *Kingsley Stern*

of four cells and a pair of external air sacs. The air sacs look something like tiny water wings (Fig. 22.5) and give the pollen grains added buoyancy that may result in the grains being carried great distances by the wind.

Pines, like all gymnosperms, are wind pollinated and produce pollen grains in astronomical numbers. For example, it has been estimated that each of the 50 or more pollen cones commonly found in a single cluster may produce more than 1 million grains, and there may be hundreds of such clusters on one tree. The grains accumulate as a fine yellow dust on cars, shrubbery, or anything else in the vicinity, and they often form an obvious scum on pools and puddles. Within a few weeks after the pollen has been released, the now shriveled pollen cones fall from the trees.

Megaspores are produced in **megasporangia** located within ovules at the bases of the seed cone scales, called **megasporophylls.** The **seed cones** (female strobili) are much larger than the pollen cones, becoming as much as 60 centimeters (2 feet) long in sugar pines and weighing as much as 2.3 kilograms (5 pounds) in Coulter pines. When mature, they have woody scales, with inconspicuous bracts between them, arranged in a spiral around an axis. They are mostly produced on the upper branches of the same tree on which the pollen cones appear (Fig. 22.6).

The ovules (Fig. 22.7) occur in pairs toward the base of each scale of the immature seed cones and are larger and more complex than the microsporangia of pollen cones.

Each ovule has within it a *megasporangium* containing the nucellus and a single **megasporocyte.** This, in turn, is surrounded and enclosed by a thick, layered integument. The integument has a somewhat tubular channel or pore called a **micropyle** that is pointed toward the cone's central axis. One of the integument layers later becomes the seed coat of the seed.

A single megasporocyte within the megasporangium of each ovule undergoes meiosis, producing a row of four relatively large megaspores. Three of the megaspores soon degenerate. Over a period of months, the remaining one slowly develops into a *female gametophyte* that ultimately may consist of several thousand cells. The nucellus is used as the food source for the growing gametophyte.

As gametophyte development nears completion, two to six *archegonia* differentiate at the end facing the micropyle. Each archegonium contains a single, large egg. When a stained, thin, lengthwise section of a pine ovule is examined with a microscope, only one archegonium or egg may be seen, or the micropyle may appear to be missing. This is due to the section not having been sliced precisely through the middle of all the structures.

Seed cones are at first usually reddish or purplish in color and commonly take two seasons to mature into the green and finally the brownish woody structures with which we are all familiar (Fig. 22.8). During the first spring, the immature cone scales spread apart, and pollen grains carried by the wind sift down between the scales. There they

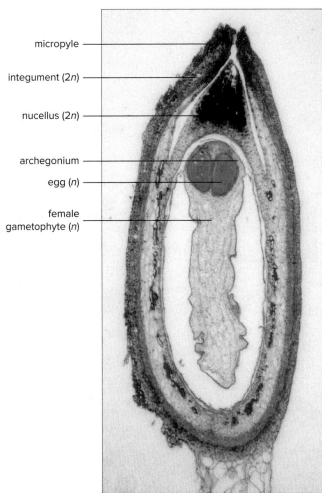

Figure 22.7 A longitudinal section through a pine ovule. ×40.
© James E. Bidlack

Figure 22.6 Seed cones of pines. (a) Immature cones shortly after being produced. (b) A mature cone. *(a-b)* © *Kingsley Stern*

catch in sticky drops of fluid (*pollen drops*) oozing out of the micropyles. As the fluid evaporates, the pollen is drawn down through the micropyle to the top of the nucellus.

After pollen transfer, the scales grow together and close, protecting the developing ovule. Meiosis and megaspore development don't occur until about a month after pollination. After a functional megaspore is produced, the female gametophyte and its archegonia don't mature until more than a year later.

Meanwhile, the pollen grain (immature *male gametophyte*) produces a **pollen tube** that slowly grows and digests its way through the nucellus to the area where the archegonia develop. While the pollen tube is growing, two of the original four cells in the pollen grain enter it. One of these, called the **generative cell,** divides and forms two more cells, called the sterile cell and the **spermatogenous cell.** The spermatogenous cell divides again, producing two male gametes, or **sperms.** The sperms have no flagella (unlike the sperms of other organisms discussed in the preceding chapters) and are confined to the pollen tube until just before fertilization occurs. The germinated pollen grain, with its pollen tube and two sperms, constitutes the mature *male gametophyte*. Notice that no antheridium has been formed.

About 15 months after pollination, the tip of the pollen tube arrives at an archegonium, unites with it, and discharges the contents. One sperm unites with the egg, forming a zygote.

Figure 22.8 Life cycle of a pine. © James E. Bidlack

The other sperm and remaining cells of the pollen grain degenerate. The sperms of other pollen grains present may unite with the eggs of other archegonia, and each zygote begins to develop into an **embryo** that is nourished by the female gametophyte. This is similar to the development of fraternal twins or triplets in animals. At a later stage, an embryo may divide in such a way as to produce the equivalent of identical twins or quadruplets in animals. Embryos develop at the ends of long suspensor cells, which, in turn, are connected to rosette cells. Normally, however, only one embryo completes development. While this development is occurring, one of the layers of the integument hardens, becoming a *seed coat*. A thin, membranous layer of the cone scale becomes a "wing" on each seed. The wing may aid in the seed's dispersal. Squirrels and other animals may also help dispersal by breaking open the cones.

In other species, such as the lodgepole, jack, and knobcone pines, the cones remain on the tree with the scales closed until they are seared by fire or open with old age. Sometimes, these cones are slowly buried as the cambium adds tissues that increase the girth of the branch. Seeds of such engulfed cones have been reported to germinate after they have been dug out of the stem.

Other Conifers

Some conifers don't produce woody seed cones with conspicuous scales, nor do all produce both pollen and seed cones on the same plant. For example, yew (*Taxus*) and California nutmeg (*Torreya*) produce ovules singly at the tips of short axillary shoots. Each ovule is at least partially surrounded by a fleshy, cuplike covering called an **aril**. In yews, this is bright red and open at one end, giving the fleshy seed the appearance of a small, red hors d'oeuvre olive with its stuffing removed (Fig. 22.9). The seeds with their fleshy arils are produced only on female plants, while the pollen cones are produced only on male plants.

Podocarps are conifers of the Southern Hemisphere and are widely planted as ornamentals in regions with milder climates. Their fleshy-coated seeds are produced singly and are similar to those of yews. They are not, however, open at one end, and they have an additional, larger appendage at the base (Fig. 22.10). The origin of these fleshy seeds is not clear and has led to speculation that yews and podocarps may have diverged from other conifers very early in the evolution of gymnosperms.

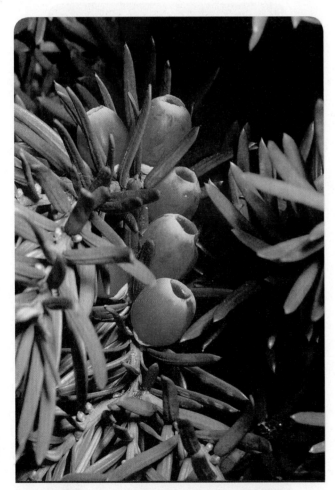

Figure 22.9 The seeds of yew (*Taxus*) are not produced in cones but are surrounded at maturity by a red, fleshy, cuplike structure called an *aril*. © Kingsley Stern

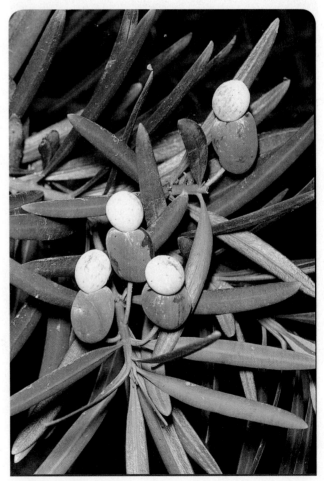

Figure 22.10 Fleshy seeds of a podocarp (*Podocarpus*). Note the large appendage at the base of each seed. © Kingsley Stern

The scales of the seed cones of junipers tend to be fleshy at maturity, and they look more like berries than cones. Juniper pollen grains, as well as those of a number of other conifers, do not have air sacs. Cypress and redwood seed cone scales are flattened at the tips and narrow at the base, where they do not overlap one another as do pine cone scales.

The two California species of redwoods are both renowned for their size, height, and longevity. Coastal redwoods (Sequoia sempervirens) occasionally grow to a height of 90 meters (295 feet). At 115.6 meters (379 feet), a coastal tree called "Hyperion" is believed to be the world's tallest conifer. It grows in Northern California. The other species (Sequoiadendron giganteum), usually referred to as *Big Tree* or *Giant Redwood,* is confined to the western slopes of California's Sierra Nevada range. It does not grow quite as tall as the coastal redwood but exceeds it in total mass. The General Sherman tree in Sequoia National Park, for example, is 31 meters (101.5 feet) in circumference at the base and over 24 meters (79 feet) in circumference 1.5 meters (5 feet) above the ground. It weighs an estimated 5,594 metric tons (6,167 tons). There are 600,000 board feet of timber in this single tree—enough to build more than 75 five-room houses (although the wood is generally not suitable for construction purposes) or to make 20 billion toothpicks. It is over 3,500 years old.

22.3 Other Gymnosperms

Learning Outcome

a. State the name and describe one example for each of the following phyla: Ginkgophyta, Cycadophyta, and Gnetophyta.

Living representatives of other gymnosperms are not as numerous or as well known as the conifers and outwardly don't resemble them at all. In fact, some of them look more like leftover props from a science-fiction movie set. They include the **ginkgos,** the **cycads,** and the **gnetophytes** (pronounced neet'oh-fite).

Phylum Ginkgophyta—*Ginkgo*

There is only one living species of *Ginkgo* (Fig. 22.11), whose name is derived from Chinese words meaning "silver apricot."

(a)

(b)

(c)

Figure 22.11 Ginkgo. (a) A mature tree in the fall. (b) Seeds and leaves. (c) Male strobili. *(a-c)* © Kingsley Stern

The fossil record indicates *Ginkgo* and other members of its Family (Ginkgoaceae) were once widely distributed, especially in the Northern Hemisphere. Despite isolated reports to the contrary, there are doubts that ginkgos now exist anywhere they have not been cultivated, and the plant has often been called a living fossil.

Ginkgos are often referred to as *maidenhair trees* because their notched, broad, fan-shaped leaves look like larger versions of the individual pinnae of maidenhair ferns. They are widely cultivated in the United States and are popular street trees in some areas. The leaves are mostly produced in a spiral on short, slow-growing spurs and have no midrib or prominent veins. Instead, hairlike veins branch *dichotomously* (fork evenly) and are relatively uniform in their width. They are deciduous and turn a bright golden yellow before abscission in the fall.

Although its life cycle is similar to that of pines, *Ginkgo* is **dioecious**, with male and female reproductive structures produced on separate trees. Fertilization after pollination is assisted by sperm that have flagella. The mature seeds resemble small plums and are enclosed in fleshy seed coats. The flesh is, however, unrelated to that of true fruits. In North America, male trees (propagated from cuttings) are preferred for ornamental purposes because the seed flesh has a nauseating odor and is irritating to the skin of some individuals. In China and Korea, however, the seeds are considered a delicacy, and female trees predominate; a minimal number of males are propagated to ensure pollination.

Phylum Cycadophyta—the Cycads

Cycads have the appearance of a cross between a tree fern and a palm but are related to neither. These slow-growing plants of the tropics and subtropics have unbranched trunks that grow more than 15 meters (50 feet) tall in a few species and have a crown of large, pinnately divided leaves. Several of the approximately 300 known living species of cycads are facing extinction. During the Mesozoic era, now extinct gymnosperms known as *cycadeoids* were abundant. Cycadeoids bore a superficial resemblance to cycads but had very different reproductive structures and are not related.

Cycad life cycles are similar to those of conifers, except that pollination of cycads is sometimes brought about by beetles instead of wind. In addition, each sperm of cycads has from 10,000 to 20,000 spirally arranged flagella. Cycads are dioecious, with either the pollen strobili or the seed strobili typically situated in the center of different plants. In some species, the strobili are huge (e.g., more than a meter [3 feet 3 inches] long) with a weight of over 220 kilograms (100 pounds) (Fig. 22.12). The scales of seed strobili of some species are covered with feltlike or woolly hairs.

(a)

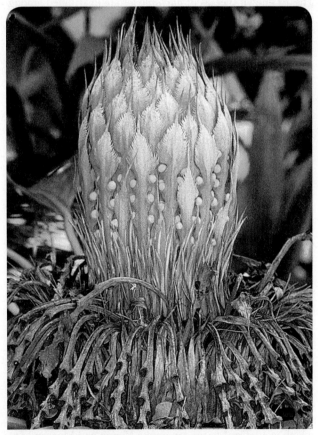
(b)

Figure 22.12 (a) A male cycad with a strobilus. (b) A female cycad with a strobilus. *(a)* © Kingsley Stern; *(b)* © Steven P. Lynch

Figure 22.13 Joint fir (*Ephedra*). (a) Part of a single plant. (b) Male strobili. ×20. (c) A female strobilus. ×20. *(a-c)* © *Kingsley Stern*

Phylum Gnetophyta—the Gnetophytes

The 70 known species of *gnetophytes* are distributed among three distinctive genera. They are unique in having vessels in the xylem, which are an evolutionary advance compared with the more primitive tracheids found in other gymnosperms. More than half of the gnetophytes are species of *joint firs* in the genus *Ephedra.* These shrubby plants inhabit drier regions of southwestern North America, as well as South America, Asia, and Europe. Their tiny leaves are produced in twos and threes at a node and turn brown soon after they appear. The stems and branches, which are often whorled, are slightly ribbed; they are photosynthetic when they are young (Fig. 22.13).

Before pollination, the ovules of *Ephedra* produce a small, tubular extension resembling the neck of a miniature bottle extending into the air. Sticky fluid oozes out of this extension, which constitutes the micropyle, and airborne pollen catches in the fluid. Male and female strobili may be produced on the same plant or on different ones, depending on the species.

Most of the remaining species in this division are in the genus *Gnetum* (Fig. 22.14), which has not been given an English common name. Its members occur in the tropics of South America, Africa, and Southeast Asia. Most are vinelike, with broad leaves similar to those of flowering plants. The best-known species of *Gnetum,* however, is a tree that grows up to 10 meters (33 feet) tall.

The third genus, *Welwitschia,* has only one species, which is confined to the temperate Namib and Mossamedes deserts of southwestern Africa. There the average annual rainfall is only 2.5 centimeters (1 inch), and in some years,

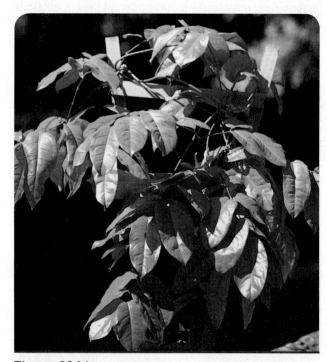

Figure 22.14 A climbing *Gnetum.* © *Kingsley Stern*

it does not rain at all. The plants carry on CAM photosynthesis, and their stomata are open at night. The plants apparently obtain most of their water from dew and condensate from fog that rolls in off the ocean at night.

Welwitschia plants are also truly extraordinary in appearance. The stem rises only a short distance above the ground's surface and is in the form of a large, shallow cup that tapers at the base into a long taproot. The plants may live to be 100 years old and at maturity have a crusty,

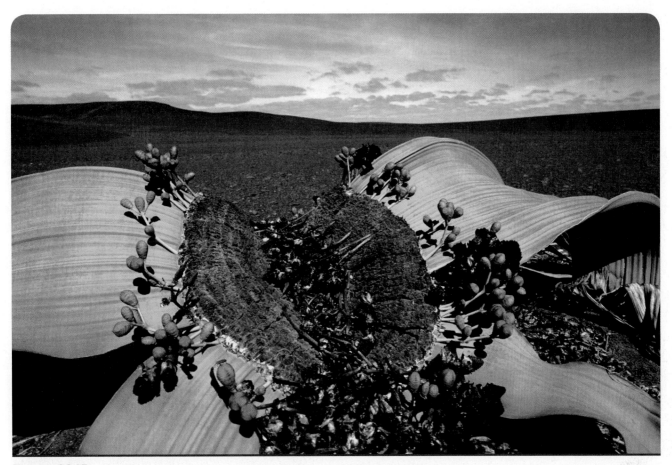

Figure 22.15 A female *Welwitschia* plant in the Namibian Desert. © FRIEDRICHSMEIER/Alamy

barklike covering on the surface of the stem cup. The stems may be more than 1 meter (3 feet 3 inches) in diameter (Fig. 22.15).

Throughout their life span, *Welwitschia* plants usually produce only two leaves. The leaves are wide and straplike, each with a meristem at the base. The meristems constantly add to the length of the leaves, but as the leaves flap about in the wind, they become tattered and split, wearing off at the tips, so that they are seldom more than 2 meters (6.5 feet) long. *Welwitschia* is dioecious, with both male and female strobili being produced on axes that emerge from the axils of the leaves. This makes the strobili appear to be growing around the rim of the stem cup.

22.4 Human Relevance of Gymnosperms

Learning Outcome

a. List 10 uses of gymnosperms, and provide a specific source for each.

As a group, the gymnosperms are second only to the flowering plants in their impact on our daily lives. Space does not permit a detailed account of all they contribute, but the following sections provide an overview of some of their uses, past and present.

Conifers

In the early 1970s, the late author-naturalist Euell Gibbons filmed a series of television commercials in which he mentioned uses of several wild plants for food. In one of the commercials, he rhetorically inquired if his audience had ever eaten a pine tree and added, "Many parts are edible."

The edibility of parts of many conifers was known to Native Americans long before Europeans set foot on the North American continent. In fact, early explorers found large numbers of pines stripped of their bark. For centuries, these inner parts (phloem, cambium) had been used for emergency food. The Adirondack Mountains of New York are believed to have received their name from a Mohawk word meaning "tree eater," in reference to Native American use of the inner bark of eastern white pines. This material (specifically, the phloem) contains sugars that make it taste sweet. Some tribes ate the material raw, some dried it and ground it to flour, and others boiled it or stored dried strips for winter food. Early settlers in New England candied strips of eastern white pine inner bark. To prevent scurvy, both they and local Native Americans drank a tea made of the needles, which are rich in vitamin C.

The seeds of nearly all pines are edible, but those of western North America include the larger and better-tasting species. The protein content of those analyzed generally ranges between 15% and 30%, with much of the remainder consisting of oils.

California Native Americans relished gray pine seeds in particular, but even the small seeds of ponderosa pine were eaten raw or made into meal for soups and bread. Cones of pinyons were collected by tribes of the Southwest and thrown on a fire to loosen the seeds. These were then pounded and made into cakes or soup. The soup was often fed to infants. In Siberia, local residents crush the seeds of Siberian white pine to obtain a nutritious oil, but its use has declined since corn and cottonseed oils became available.

Italians and other Europeans cook "pignolias," the seeds of the stone pine, in stews and soups. The seeds are also used in cakes and cookies, and some are exported to the United States for this purpose. Many of the so-called nuts used by commercial American bakers in cakes and confectionery, however, are really seeds from the east Himalayan chilghoza pine. Other sources include the Mexican stone pine and a few pinyons.

Eastern white pines were often used as masts in sailing vessels. In colonial days, the royal surveyors marked certain trees for the use of the Crown, and severe penalties were imposed on colonists who ignored the ban on the use of any white pine not growing on private land. It was, however, legal for colonists to use white pines that had blown down, which gave rise to the term *windfall.* Eastern white pine wood contains less resin than that of other species and was extensively used for crates, boxes, matchsticks, furniture, flooring, and paneling. By the end of the 19th century, eastern white pines, which originally occurred over vast tracts of the northeastern United States and Canada, had been decimated by wholesale logging done with no thought to conservation. Bald cypress trees in the southeastern United States met a similar fate. White pine blister rust also took its toll. Although new growth is now being promoted, most white pine lumber used today comes from large stands of western white pine in the Pacific Northwest.

The trunks of lodgepole pines are used in both the United States and Canada for telephone poles; the straight-grained wood is also used for railroad ties, mine timbers, and pulp.

Smog has severely damaged ponderosa and other native pines in California. For a number of years, the U.S. Forest Service has experimented with Afghanistan pine, a smog- and drought-resistant pine from Russia and adjacent areas, as a replacement for native trees. Growth rates in tests have been very rapid. Rapid growth is a desirable commercial feature, because considerably more timber can be produced in the same time needed to obtain it from slower-growing species, but the wisdom of introducing non-native plants into natural communities is in question, because there are many examples of such activities thoroughly disrupting delicate ecological balances.

The resin produced in the resin canals of conifers is a combination of a liquid solvent called *turpentine* and a waxy substance called *rosin.* When a conifer tree is wounded or damaged by insects, resin usually covers the area, sometimes trapping the insects. Out in the air, the turpentine evaporates quickly, leaving a protective layer of rosin, which deters water loss and fungal attacks. Both turpentine and rosin are very useful products, and a large industry centered in the southern United States and in the south of France is devoted to their extraction and refinement. They are often referred to as *naval stores,* a term that originated when the British Royal Navy used large quantities for caulking and sealing their sailing ships. Today, most naval stores and a third or more of the lumber used in the United States come from a group of southern yellow pines, particularly slash pine.

Pitch pine, also a source of naval stores before slash and other yellow pines became more profitable, was used in the past for the waterwheels of mills. Pitch pine wood was also used as fuel for steam engines, as it produces considerable heat when it burns. Turpentine is considered a premier paint and varnish solvent; rosin is used by musicians on violin bows and by baseball pitchers to improve their grip on the ball. Batters apply pine tar to the handles of bats to minimize slippage.

In the past, pine pitch was used by Native Americans for patching canoes, and it has been suggested that Noah's ark was sealed with pitch from aleppo pine. Pine resin was used for purifying wine in the 1st century A.D., and today, Greeks still add it to certain wines. *Colofonia* is the Spanish word for a type of resin Monterey pines produced abundantly around the old Spanish capital of Monterey. The early California priest Padre Arroyo suggested during the first half of the 19th century that California received its name from this Spanish word. California was, however, the name of a mythical paradise in a Spanish novel published in the early 16th century, and no more than an interesting similarity between two Spanish words may be involved.

The huge kauri pines (*Agathis*) of New Zealand are in a family separate from that of true pines. They are the source of a mixture of resins called *dammar,* which is used in high-quality, colorless varnishes. Dammar was also the resin originally used in the manufacture of linoleum.

In New Zealand, dammar, also called *amber,* is obtained primarily in fossil form from former or present kauri pine forest areas. Most amber, however, has come from extinct conifers that flourished 60 to 70 million years ago in the Baltic area of the former Soviet Union and from other extinct conifers in what is now the Dominican Republic. It occurs as lumps of translucent material with a deep orange-yellow tint. According to Greek mythology, amber was the congealed tears of Phaëthon's sisters who, while weeping

Figure 22.16 A Dominican amber fossil of an insect (order Diptera). The amber is estimated to be between 20 and 30 million years old. © *Philippe Gouveia/Getty Images*

ways of reducing paper consumption in the United States, tried trimming 2.5 centimeters (1 inch) from the width of all rolls of toilet paper in their building facilities. They found that the employees still used the same number of rolls per month as they had previously. From this, it was calculated that if all rolls of toilet paper were similarly trimmed in width, 1 million trees would be saved each year in the United States alone.

Split roots of the white spruce are quite pliable and were used by Native Americans for lashing canoes and for basketry. Spruce beer, brewed from young twigs and leaves with an added sugar source, such as honey or molasses, was once used as a remedy for scurvy. Resin of white, red, black, and Sitka spruces was used as a type of chewing gum by Native Americans, who sometimes hardened it slightly in cold water. Europeans who have tried it report that it has to be of the right consistency to be enjoyable, because it behaves like unhardened taffy if it is too soft and is bitter if it is too hard. In southeast Alaska, Sitka spruce buds are boiled in sugar water to make spruce bud syrup.

The tracheids of spruces have spiral thickenings on the inner walls. These apparently are responsible for giving the wood a resonance that makes it ideal for use as soundboards for musical instruments (Fig. 22.18). Sitka spruce of the Pacific Northwest produces a strong, resilient wood that is a favorite material of manufacturers of light aircraft.

Larches, which along with the dawn redwood and bald cypress are exceptions to the rule that conifers are evergreens,

Figure 22.17 Large paper rolls waiting to be put on the printing press of a modern newsprint factory. *Kenneth Sponsler/Shutterstock*

over his death, were turned into trees. Some of the lumps weigh up to 45 kilograms (100 pounds). The supply was at its peak at the beginning of the 20th century but is now nearing exhaustion. Remarkably lifelike preservations of prehistoric insects millions of years old have occurred in amber (Fig. 22.16).

Other products refined from resin are used in the manufacture of menthol for cigarettes (menthol also occurs naturally in members of the Mint Family), floor waxes, printer's ink, paper coatings, varnishes, and perfumes.

White spruce is the chief source of pulpwood for newsprint (Fig. 22.17) and other paper in North America. Enormous quantities of paper are used every day. A single midweek issue of a large metropolitan newspaper may use an entire year's growth of 50 hectares (123 acres) of these trees, and that amount may double for weekend editions. A large American publishing company, in an attempt to find

Figure 22.18 A violin with a soundboard made from red spruce. © *Kingsley Stern*

have some of the most durable of all conifer woods. Fence posts of larch are known to last 20 years. In the southern and southwestern United States, posts of juniper wood and bald cypress last even longer, some remaining usable for 40 to 50 years or more. The resin of the western larch has been used in the manufacture of baking powder, and the European larch is the source of a type of turpentine.

There are about 40 species of true firs that are widely used in the construction, plastic, and paper industries, as ornamentals, and as Christmas trees. The balsam fir produces blisters containing a clear resin on its bark. This resin, known as *Canada balsam,* was used in the past for cementing optical lenses and is still occasionally used for making permanent mounts on microscope slides. It has medicinal properties, too, and was used by New England colonists in sore throat medications.

Douglas fir, found in the mountains of the West, is not a true fir. In the Pacific Northwest, it grows into giant trees that are second only to the redwoods in size. It is probably the most desired timber tree in the world today. The wood is strong and relatively free of knots as a result of rapid growth, with less branching than most other conifers. It is heavily used in plywoods and is a major source of large beams. A useful wax is extracted from the bark of Douglas firs. Exploitation has nearly eliminated old-growth stands, but large numbers of new trees are being grown in managed forests.

Coastal redwoods are also prized for their wood, which contains substances that inhibit the growth of fungi and bacteria. The wood is light in weight, strong, and soft, but it splits easily. It is used for some types of construction, furniture, posts, greenhouse benches, and many other purposes. California wineries use it extensively for wine barrels. The Giant Redwoods (Big Trees) are so huge that a double bowling alley was built on the log of the first specimen cut down. They are no longer logged and now serve almost exclusively as tourist attractions in California. They are, however, being planted as timber trees in Romania and Yugoslavia.

Wood of the bald cypress, found in southern swamps of the United States, is like that of the redwood in being very resistant to decay. In the past, it was used for railroad ties, coffins, general construction, guttering, and shingles. The trees are well known for their "knees," which rise above the water as tapering growths from the roots (Fig. 22.19). At one time, it was widely believed that these were a means of admitting oxygen to the roots, but this now appears doubtful. They are favored for making knickknacks, such as wall ornaments and lamp bases. The leaves of the bald cypress yield a red dye.

Figure 22.19 Bald cypress trees in a southern swamp. Note the "knees" protruding above the surface. © *Kingsley Stern*

KEY THEME 22.2 evolution

A Living Fossil?

Just imagine a real-life Jurassic Park, where prehistoric dinosaurs long thought extinct actually are alive. Well, something close to this happened in Australia in 1994. David Noble, while hiking on his Christmas holiday in the Wollemi National Park near Sydney, stepped into a stand of trees that, until that moment, were thought to have been extinct for over 150 million years. Known only as fossils, there they stood—about 40 pine trees growing in an isolated, rugged, rain-forest gully that had protected them all these years. This is one of the most significant botanical finds of the 20th century—and the trees are among the rarest plants alive.

Now called the *Wollemi pine* (*Wollemia nobilis*) (Box Figures 22.2A and 22.2B), these conifer (cone-bearing) trees are so distinct that they constitute a new genus. While most conifers have dark green leaves, Wollemi pine leaves are bright and lighter green, almost the color of green apples. The leaves are complex and unusual. The trees are bisexual, with bright green seed cones and brown, cylindrical pollen cones on the tips of upper branches. Also distinctive is the corklike, knobby bark, which is an unusual chocolate brown. The tallest tree is estimated to be 150 years old, towering 130 feet from the ground, with a trunk about 3 feet wide. The new trees belong to the plant family Araucariaceae, which, while found only in the Southern Hemisphere today (Norfolk Island pine is an example), had a worldwide distribution during the Jurassic and Cretaceous periods (208–66 million years ago). This suggests that the Wollemi pines' closest relatives lived when Australia, New Zealand, Africa, South America, and India were all parts of the supercontinent Gondwana. During this period, the eastern coast of Australia lay close to the South Pole, when worldwide climates were relatively warm to hot and wet.

Sydney's Royal Botanic Gardens and the New South Wales National Parks and Wildlife Service are jointly studying the Wollemi pine with methods ranging from scanning electron microscopy of the leaf and pollen to DNA extraction and gene-sequencing studies. One priority is to study its propagation methods, so that the plant can be established in cultivation. Although the Wollemi site is within 200 kilometers of Sydney, Australia's largest city, the exact location remains a secret to protect the plant from seed collectors and poachers.

The Wollemi pines, protected in their sheltered spot, not only escaped when their closest relatives died some 50 million years ago but remained hidden until this exciting, remarkable find. *Wollemi* is an aboriginal word meaning "look around you." What an appropriate name for this real living fossil—it reminds all of us that there remains much to understand and explore in our world.

D. C. Scheirer

(A)

(B)

Box Figures 22.2A and 22.2B Wollemi pine (A) Top: Pollen cone (male) at end of branch; Bottom: Seed cone (female). (B) Wollemi pine showing leaves and cones. *(A)* © *The Sydney Morning Herald/Contributor/Getty Images (B)* © *Claver Carroll/Getty Images*

A dull red dye can be obtained from the younger bark of the eastern hemlock, which is also a source of tannins for shoe leather. The tanning industry so depleted the native eastern hemlocks that it now has to use tropical substitutes. The wood of these small trees contains exceptionally hard knots that can chip an axe blade. It sputters and throws out sparks freely when placed on a fire. Native Americans made a poultice for scrapes and cuts by pounding the inner bark. British Columbia Native Americans used to scrape out the cambium and phloem of both the western and mountain hemlocks for food, killing the trees in the process.

Northern white cedar, also known as *arborvitae,* is a favorite ornamental in temperate areas. The wood is pliable, and several Native American tribes used it for canoes. The Atlantic or southern white cedar was the first tree to be used for the construction of pipes for pipe organs in North America. During World War II, old logs of this species found in a swamp in New Jersey were milled and used in the construction of patrol torpedo boats.

The fleshy red aril surrounding the seeds of yews is sweet and edible, but the seeds themselves and other parts of the plants are very poisonous. The wood is tough and resilient and is favored for making bows.

English yew, the wood of longbows, changed history in 1415 and brought an end to the Middle Ages at the Battle of Agincourt, when the English longbow proved its superiority over heavily armored cavalry. A red or purple dye can be obtained from the bark and roots of yew. Podocarps, two species of which are valuable timber trees in New Zealand, have edible seeds.

In 1989, researchers at the Johns Hopkins Oncology Center reported that 12 women with advanced ovarian cancers that had not responded to traditional therapies (including, in some cases, surgery) had a decrease of 50% or more in the size of their tumors and one woman's tumor disappeared altogether after treatment with paclitaxel (also known by its trade name Taxol), a drug obtained from the bark of the Pacific yew tree. In 2001, the National Institute for Health and Clinical Excellence recommended that taxol can be used for certain types of lung cancer and ovarian cancers, as well as first-line and second-line treatment of ovarian cancer. Pacific yew trees are small and do not occur in extensive stands. They also grow so slowly it takes more than 70 years to attain their full size, and removal of their bark kills them. Scientists have developed methods to produce paclitaxel though chemical manipulation of a precursor found in more common yew trees. In addition, cell lines established through tissue culture are being used to synthesize paclitaxel. Annual worldwide sales of paclitaxel exceed one billion dollars.

The wood of incense cedars has been used in the manufacture of venetian blinds and pencils. Red cedar wood is also used for pencils, cedar chests, closet lining, fence posts, and cigar boxes. It was used at one time in Germany for smoking hams. An aromatic oil used in floor-sweeping compounds is extracted from red cedar wood, and the "berries" of this and related junipers are widely eaten by birds. Many Native American tribes used the berries and inner bark as survival food during bleak winters. Some roasted the dried berries and brewed a beverage from them. Western red cedar was the most important single plant of Native Americans of the Northwest, who used it for housing, clothing, nets, canoes, totem poles, medicines, and other purposes. Berries of the dwarf juniper are used to flavor gin. Some authorities indicate that the word *gin* may have been derived from *genievre,* the French word for "juniper berry."

Other Gymnosperms

Despite the foul odor of the fleshy seed coat of seeds of *Ginkgo,* the starchy food reserves of the seeds themselves have a nutlike flavor punctuated with a hint of shrimp. In Asia, *Ginkgo* seeds are widely used for food, either boiled or roasted, and they are found imported in canned form in Chinese food stores of large metropolitan areas in the United States and Canada. Extracts of *Ginkgo* plants have been clinically demonstrated to improve blood circulation in humans. Clinical studies have shown that *Ginkgo* leaf extracts improve blood supply to the brain and lungs. The effects on memory enhancement, though, have not been conclusively demonstrated.

Florida arrowroot starch was once obtained from the extensive cortex and pith of a species of cycad whose northernmost distribution occurs in Florida. Before it could be used for human consumption, however, a poisonous substance had to be leached out. Because cycads grow too slowly to make continued preparation of the arrowroot starch profitable, the practice has been abandoned. Today, cycads seen outside of their natural habitats are being grown primarily for ornamental or educational purposes. In Louisiana and elsewhere, however, the large compound leaves are used in Palm Sunday religious services. Some species are nearing extinction in the wild and may soon be known only in botanical gardens and conservatories.

In the southwestern United States, joint firs (*Ephedra*) are grazed by livestock, and the stems are still brewed into "Mormon tea." To offset a slight bitterness, a teaspoon of sugar, honey, or jam is added to each cup of tea. Native Americans and pioneers used a concentrated version of the tea in the treatment of venereal diseases. They also ground the seeds into flour, from which a bitter bread was made. The drug *ephedrine,* widely used in the treatment of asthma and other respiratory problems, still is extracted from a Chinese species, but most of the drug now in use is synthetically produced. The Chinese used ephedrine medicinally more than 4,700 years ago. Continued medicinal use of ephedrine is now questionable because of potentially serious side effects.

One species of *Gnetum* is cultivated in Java for the shoots, which are eaten after being cooked in coconut milk. Fibers from the bark are made into a rope.

See Appendix 1 for scientific names of species discussed in this chapter.

SUMMARY

22.1 Introduction to Gymnosperms

- The exposed seeds of gymnosperms are produced on sporophylls, in a strobilus, or cone. The seed (female) cones and smaller pollen (male) cones are produced on sporophytes.
- A female gametophyte develops within an ovary and is nourished by a nucellus, which is enclosed by a protective integument that will form the seed coat after fertilization.
- Four phyla of living gymnosperms are recognized. Phylum Pinophyta includes the conifers. *Ginkgo* is the sole living representative of Phylum Ginkgophyta. The cycads are in Phylum Cycadophyta. Phylum Gnetophyta includes the gnetophytes *Ephedra, Gnetum,* and *Welwitschia.*

22.2 Phylum Pinophyta—the Conifers

- Pines are the most numerous conifers. The needlelike leaves are arranged in clusters of two to five and have modifications adapting them to harsh environments.
- Resin canals, occurring throughout the plant body, secrete resin that inhibits fungi and certain insect pests.
- Pine xylem lacks vessel members and fibers and is relatively soft. The phloem lacks companion cells but has albuminous cells that apparently perform the same function. Pine roots are always associated with mycorrhizal fungi, which are essential to normal plant development.
- Two kinds of spores are produced. Microspores are produced in papery pollen cones that develop in clusters toward the tips of lower branches. The microsporangia develop in pairs toward the base of the pollen cone scales and give rise to four-celled pollen grains that occur in huge numbers.
- Megaspores are formed in ovules at the bases of seed cone scales. The integument of the ovule has a pore called the micropyle. One megaspore develops into a female gametophyte. The mature female gametophyte contains archegonia.
- Before the archegonia mature, pollen grains are caught in sticky pollination drops between the cone scales. Each pollen grain produces a pollen tube that digests its way down to the developing archegonia, and two of the original four cells in the pollen grain migrate into the tube as it grows. The generative cell divides and produces a sterile cell and a spermatogenous cell that itself divides, producing two sperms.
- After pollination, one sperm unites with the egg, forming a zygote. The zygote develops into an embryo of a seed that has a membranous wing formed from a layer of the cone scale.
- Some conifers produce seeds enclosed in fleshy or berry-like coverings. Their evolutionary origin is not clear.

22.3 Other Gymnosperms

- *Ginkgo* has small, fan-shaped leaves with evenly forking veins. The life cycle of *Ginkgo* also is similar to that of cycads. The edible seeds are enclosed in a fleshy covering that has a rank odor at maturity.
- Cycads superficially resemble palm trees with unbranched trunks and crowns of large, pinnately divided leaves. Typically, the strobilus of a cycad is found in the center of the plant. Life cycles of cycads are similar to those of conifers, but the sperms of cycads include sperms with numerous flagella.
- Gnetophytes are the only gymnosperms with vessels in their xylem. Half the species are in the genus *Ephedra,* whose members have jointed stems and leaves reduced to scales. *Gnetum* species have broad leaves and occur in the tropics, primarily as vines. *Welwitschia* is confined to southwest African deserts. Its stem is in the form of a shallow cup with straplike leaves that extend from the rim; basal meristems on the leaves constantly add to their length.

22.4 Human Relevance of Gymnosperms

- The seeds and inner bark of pines are edible, and a tea has been made from the leaves. Eastern white pine stems were used as masts for sailing vessels and for crates, furniture, flooring, paneling, and matchsticks. Western white pine is the source of most such lumber today.
- Resin from pines consists of turpentine and rosin. Turpentine is used as a solvent, and rosin is used by musicians and by baseball players. Dammar from kauri pines is used in colorless varnishes. Amber is fossilized resin. Resin is also used in floor waxes, printer's ink, paper coatings, perfumes, and the manufacture of menthol.
- White spruce is the chief source of newsprint. It was also used for basketry and canoe lashing by Native Americans, with molasses or honey for treating scurvy, and in brewing a type of beer. Spruce resin was used for a type of chewing gum. The wood is used as soundboards for musical instruments and in the construction of aircraft.
- Larch and juniper woods are used for fence posts. Firs are used in the construction, paper, ornament, and Christmas-tree industries. Douglas fir is probably the most desired timber tree in the world today.
- Coastal redwoods are also prized for their wood, which is resistant to fungi and insects. Bald cypress wood, used in the past for coffins and shingles, is also resistant to decay.
- A dye and tannins are obtained from the eastern hemlock. Native Americans used parts of hemlocks for poultices and for food.

- Eastern white cedar's wood was used for canoes, and that of the Atlantic cedar was used for construction of pipes for pipe organs. Yew wood is used for making bows, and an extract from yew bark has potential for the treatment of human ovarian cancer.
- Podocarps of New Zealand have edible seeds. Incense cedar wood is used for cedar chests, cigar boxes, pencils, and fence posts. Juniper berries are used to flavor gin and were used by Native Americans for food and a beverage.
- *Ginkgo* seeds are edible, and *Ginkgo* plant extracts are used to improve blood circulation. Arrowroot starch was once obtained from a cycad. Mormon tea is brewed from the leaves and stems of joint firs (*Ephedra*), which in the past were also a source of the drug ephedrine and a venereal disease treatment. One *Gnetum* species is cultivated in Java for food.

REVIEW QUESTIONS

1. What is a gymnosperm? How is it distinguished from other major groups of plants?
2. What is the difference between a seed and a spore?
3. How are the leaves of pines different from those of broadleaf flowering plants?
4. What is a resin canal, and what is its function? Where are resin canals found?
5. How do pines differ in their reproduction from ground pines and ferns?
6. How do pollen grains differ from spores or sperms?
7. Which conifers discussed in this chapter do not have woody seed cones?
8. If you had samples of leaves of a pine, *Ginkgo*, a cycad, a joint fir, and *Welwitschia* all together, indicate how you could tell them apart by constructing a key.
9. What parts of a pine are considered edible?
10. What is resin? Discuss some of its uses, past and present.

DISCUSSION QUESTIONS

1. *Ginkgo* and cycads have broad leaves, whereas those of pines are needlelike. Can you suggest any significance of this difference in terms of the climates and habitats involved?
2. If no distinction were made at the level of kingdom between plants and animals, what would be the equivalent, if any, of sporophyte and gametophyte in humans?
3. Both bristlecone pines and redwoods can live to be thousands of years old. What do you suppose makes this possible?
4. Most of the old-growth stands of conifers in North America are now gone, and others will be gone soon. Much of what has been harvested is being replaced with new growth, sometimes with hybrids and non-native plants. Our forests are essential to our economy as we know it. If you had the power to change the way we manage and exploit our natural forest resources, what would you do differently, assuming you did not want to damage the economy?
5. If money were no object and you wished to landscape your yard primarily with gymnosperms, realistically, what would you include, taking into account your geographical area? Why?

CHAPTER

23

Seed Plants: Angiosperms

23.1 Introduction to Angiosperms
23.2 Phylum Magnoliophyta—the Flowering Plants
   Development of Gametophytes
   Pollination
   Fertilization and Development of the Seed
   Apomixis and Parthenocarpy
   Trends of Specialization and Classification in Flowering Plants

   Key Theme 23.1: The Difference between "n" and "x" in Plant Life Cycles

23.3 Pollination Ecology
23.4 Herbaria and Plant Preservation
   Methods
   A Word of Caution to Collectors

OVERVIEW

This chapter begins with a probe of the differences between angiosperms and gymnosperms and then discusses the origin of flowering plants. The exceptional diversity of the flowering plants is reviewed next, and the chapter continues with a description of the parallel development of the gametophytes in the anthers and ovules. This leads up to pollination, followed by some details of fertilization and the development of a seed. A discussion of trends of specialization and classification in flowering plants is followed by a section on pollination ecology. Plant preservation, including simple herbarium techniques and practice, is next. The chapter closes with a brief survey on the uses of herbaria and a word of caution to plant collectors concerning unnecessary depletion of native floras.

Lady's slipper (*Cypripedium reginae*) orchid, which can be found in swampy areas. In some states, it is a threatened or endangered species.
© Doug Sherman/Geofile

23.1 Introduction to Angiosperms

Learning Outcomes

a. State the basic differences between angiosperms and gymnosperms.
b. Hypothesize how flowering plants originated based upon knowledge of other plant phyla and explain how DNA evidence might support this hypothesis.

The Pacific bleeding heart (Fig. 23.1) is one of more than 250,000 known species of flowering plants that constitute, by far, the largest and most diverse of the phyla of the Plant Kingdom. These flowering plants are called **angiosperms.**

The term *angiosperm* is derived from two Greek words: *angeion,* meaning "vessel," and *sperma,* meaning "seed." The "vessel" is the **carpel,** which is like an inrolled leaf with seeds along its margins. A green pea pod, for example, is a carpel that resembles a leaf that has folded over and fused at the margins, enclosing the attached seeds. Many flowers have *pistils* (see Chapter 8) composed of either a single carpel or two more united carpels. A seed develops from an *ovule* within a carpel and is part of an *ovary* that becomes a *fruit*. Although the angiosperms generally have organs and tissues similar to those of the gymnosperms, the enclosed ovules and seeds of the angiosperms distinguish them from gymnosperms, which have exposed ovules and seeds.

All angiosperms are presently considered to be in the Phylum Magnoliophyta (previously known as the Anthophyta). Other classifications of flowering plants are discussed in Chapter 16. Phylum Magnoliophyta has been divided into two large classes: the Magnoliopsida (dicots) and the Liliopsida (monocots). As indicated in Chapter 16, DNA and cladistic evidence now suggests that two groups of dicots should be recognized.

Angiosperms are the most diverse of land plants, whose ancestors evolved from now-extinct gymnosperms. Fossil evidence and DNA analyses are currently being used to learn more about events that led to the development of flowers. Most scientists agree that flowers evolved to take advantage of motile animals in the reproduction process. Numerous strategies have developed, over millions of years, for pollination and seed dispersal and, as such, have resulted in the great assortment of flowering plants in today's world.

Since Darwin's *Origin of Species* appeared in 1859, there have been two major theories concerning the origin of angiosperms. The older and now more or less disregarded theory was held by the German botanist Adolph Engler and his followers. It suggested that flowering plants evolved from conifers and that primitive flowers are similar in structure to the strobili of conifers. Such flowers are inconspicuous and in clusters, such as those of the grasses, oaks, willows, and cattails.

Most contemporary botanists, however, point to anatomical, chemical, and fossil evidence, along with cladistic interpretations, that indicate angiosperms evolved independently from the *pteridosperms* (seed ferns—discussed in Chapter 22). They also hypothesize that a flower is really a modified stem bearing modified leaves. The most primitive flower is thought to be one with a long receptacle and many spirally arranged flower parts that are separate and not differentiated into sepals and petals. In addition, the stamens and carpels are flattened and numerous. Such flowers are found among relatives of magnolias and buttercups, leading many modern botanists to postulate that all present-day flowering plants are derived from a primitive stock with such characteristics. A further discussion of primitive and advanced characteristics follows later in the chapter.

23.2 Phylum Magnoliophyta— the Flowering Plants

Learning Outcomes

a. Diagram the life cycle of a flowering plant, indicating shifts from haploid to diploid cells, and vice versa.
b. Explain what happens during male and female gametophyte development.
c. State what is meant by double fertilization.
d. Define *monoecious* and *dioecious.*

The plants of members of this phylum vary greatly in size, shape, texture, form, and longevity. The phylum includes, for example, the tiny duckweeds that may be less than 1 millimeter long, all the grasses and palms, many aquatic

Figure 23.1 Pacific bleeding hearts (*Dicentra formosa*).
© Kingsley Stern

and epiphytic plants, and most shrubs and trees, including the huge *Eucalyptus regnans* trees of Tasmania that rival the redwoods in total volume.

About 4,000 flowering plant species are considered parasitic. Dodders, for example, occasionally cause serious crop losses as they twine about their hosts and, by means of haustoria (shown in Fig. 5.15a), intercept food and water in the host xylem and phloem. Broomrapes also parasitize a variety of plants, as do mistletoes. Mistletoes produce some chlorophyll and depend only partially on their hosts for food. Still others, such as the beautiful snowplant (Fig. 23.2) and some of the orchids, are *saprophytes* (i.e., their nutrition comes mostly from the absorption in solution of dead organic matter). The vast majority of flowering plants, however, produce their food independently through photosynthesis.

Like the gymnosperms, the angiosperms are *heterosporous* (produce two kinds of spores), and the sporophytes are even more dominant than in the gymnosperms. The female gametophytes are wholly enclosed within sporophyte tissue and reduced to only a few cells. At maturity, the male gametophytes consist of a pollen grain with three nuclei.

Development of Gametophytes

While the flower is developing in the bud, a diploid **megasporocyte** cell differentiates from all the other cells in the ovule (Fig. 23.3). This megasporocyte undergoes meiosis, producing four haploid *megaspores.* Soon after they are produced, in most flowering plants, three of these megaspores degenerate and disappear, but the nucleus of the fourth undergoes mitosis, and the cell enlarges. While the cell is growing larger, its two haploid nuclei divide once more. The resulting four nuclei then divide yet another time. Consequently, eight haploid nuclei in all are produced (without walls being formed between them). By the time these three successive mitotic divisions have been completed, the cell has grown to many times its original volume. At the same time, two outer layers of cells of the ovule differentiate from maternal tissue. These layers, called **integuments,** later become the **seed coat.** As the integuments develop, they leave a pore, called the **micropyle,** at one end of the ovule.

At this stage, there are eight haploid nuclei in two groups, four nuclei near each end of the large, oblong cell. One nucleus from each group then migrates toward the middle of the cell. These two **central cell nuclei** may fuse to form a diploid nucleus or they may remain separate, so that the central cell is binucleate. Plasma membranes and cell walls form around each of the remaining six haploid nuclei. In the group closest to the micropyle, one of the cells functions as the female gamete, or *egg.* The other two cells, called *synergids,* which assist in guiding a *pollen tube* toward the egg, are later destroyed or degenerate. At the other end, there are three remaining cells, called *antipodals,* which may help with embryo nutrition, and later degenerate. The large sac, usually containing eight nuclei in seven cells, constitutes the *female gametophyte* **(megagametophyte),** formerly known as the *embryo sac* (Fig. 23.4).

Usually while the megagametophyte is developing, a parallel process that leads to the formation of male gametophytes takes place in the **anthers.** As an anther develops, four patches of tissue differentiate from the main mass of cells. These patches of tissue contain many diploid **microsporocyte** cells, each of which undergoes meiosis, producing a *quartet* (also referred to as a *tetrad*) of **microspores.** Four chambers (pollen sacs) lined with nutritive *tapetal* cells are visible in an anther cross section by the time the microspores have been produced. As the anther matures, the walls between adjacent pairs of chambers break down, so that only two larger sacs remain.

After meiosis, the haploid microspores in the pollen sacs undergo several changes more or less simultaneously, the processes usually taking from a day or two to

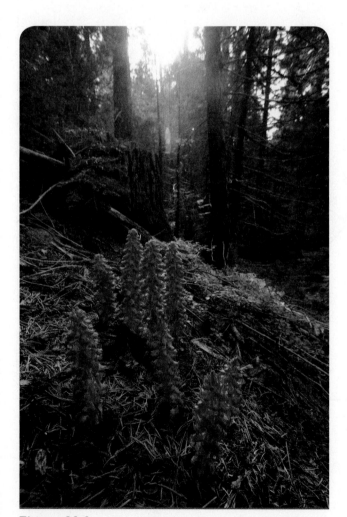

Figure 23.2 Red snowplant (*Sarcodes sanguinea*) flowers growing in the Stanislaus National Forest, east of Sonora, California.
© Phil Schermeister/Getty Images

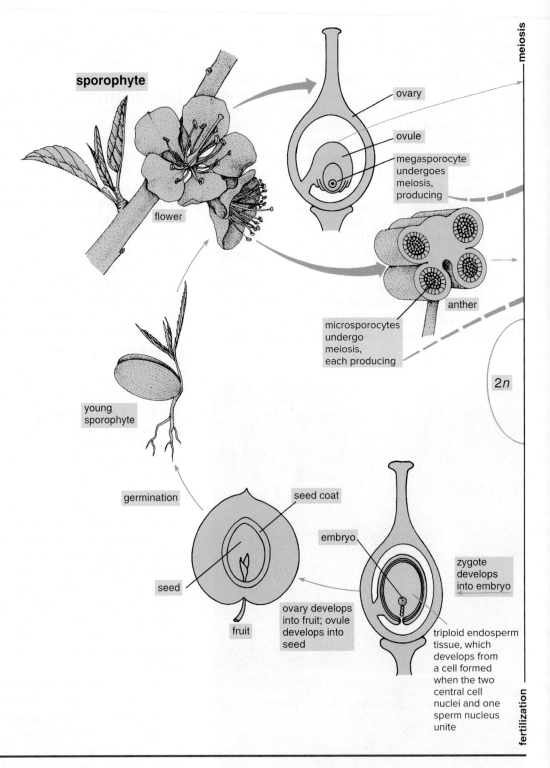

Figure 23.3 Life cycle of a typical flowering plant.

a couple of weeks. The following three changes are the most important:

1. The microspore divides once by mitosis to produce two cells. One of the cells (the generative cell) is small and is formed inside the larger cell (the tube cell).
2. The members of each quartet of microspores separate from one another.
3. A two-layered wall, whose outer layer is often finely sculptured, develops around each microspore.

When these events are complete, the microspores have become **pollen grains** (Fig. 23.5). The outer layer of the pollen grain wall, called the **exine,** which is often sculptured, contains chemicals that may later react with other chemicals in the stigma of a flower. As a result of these reactions, the

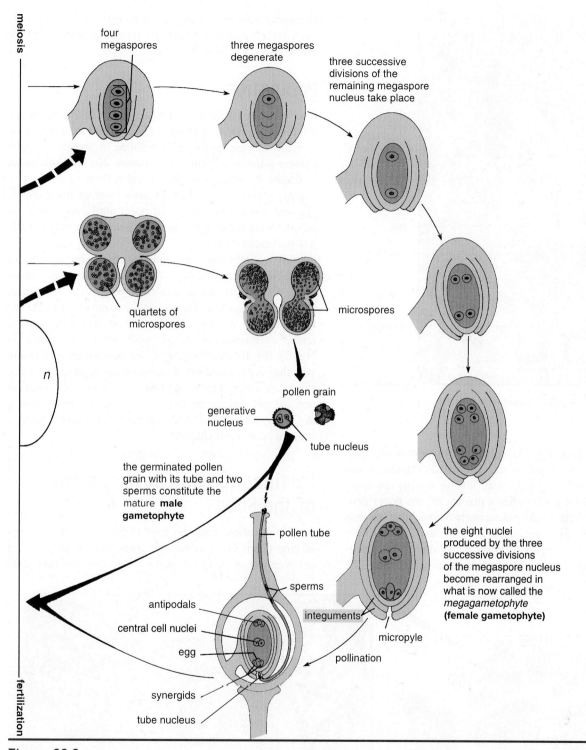

Figure 23.3 Continued.

pollen grain may germinate or its further development may be blocked, depending on whether or not it originated from the same plant, another plant of the same species, or a plant of a different species.

The cytoplasm of pollen grains is often rich in vitamins, and pollen is often collected and sold in health-food stores as a food supplement. The exine of many pollen grains is virtually indestructible, and pollen has been found preserved in deposits thousands of years old. This has proved invaluable in archaeological investigations. The sculpturing of pollen grain exine of various species is also distinctive enough to allow identification to family, genus, and even species in some instances. This feature has resulted in its use in forensic investigations of criminal cases.

Many pollen grains have three porelike thin areas (*apertures*) in the wall, but the number of apertures may range

Figure 23.4 A mature megagametophyte of a lily (*Lilium* sp.).
© Kingsley Stern

remaining **vegetative nucleus** (often referred to as the *tube nucleus*) is involved in events that take place after the pollen grain has left the anther.

Pollination

One often reads popular magazines or learns from the Internet about flowers being pollinated, and the implication is made that the dusting of the pollen of one flower upon another is the equivalent of fertilization. Not so! **Pollination** is simply the transfer of pollen grains from an anther to a stigma, nothing more. **Fertilization** involves the union of egg and sperm, and it may not occur until days, weeks, or months after pollination has taken place, or it may not follow pollination at all.

Most pollination is brought about by insects or wind, but in many species, water, birds, bats, other mammals, and gravity act as agents or pollinators. The adaptations between the flower and its pollinators can be intricate and precise and in certain instances involve force, chemicals, or sexual enticement (as shown in Fig. 23.18). In some instances, self-pollination may occur, with the pollen grains germinating on the stigma of the same flower in which they were produced. The genetic implications of self-pollination and cross-pollination are discussed in Chapters 13 and 14. A discussion of pollination ecology is given later in this chapter.

from one to many (Fig. 23.6). The development of the male gametophyte, discussed in the section "Fertilization and Development of the Seed," may involve any of the apertures. One of the pollen grain's two nuclei, the **generative nucleus,** will later divide, producing two nuclei that become surrounded by a plasma membrane and function as *sperm* cells. Unlike the sperms of many plants discussed in preceding chapters, angiosperm sperms have no flagella. The

Fertilization and Development of the Seed

After pollination, with the notable exception of a number of crop plants (e.g., wheat, rice, oats, peas, beans), further development of the male gametophyte may not take place

Figure 23.5 (a) Six anthers, each containing microspores, surrounding a central ovary in *Lillium* sp. ×10. (b). Enlargement of one anther to show the details of microspores, which will eventually become pollen grains. ×40. *(a-b)* © *James E. Bidlack*

Figure 23.6 Scanning electron micrograph of a poppy pollen grain. The furrows are apertures through which a pollen tube may later emerge. © Kingsley Stern

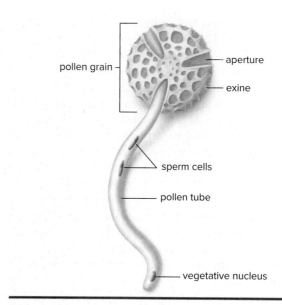

Figure 23.7 A mature male gametophyte of a flowering plant.

unless the pollen grain is (1) from a different plant of the same species, or (2) from a variety different from that of the flower receiving it.

Under suitable conditions, the dense cytoplasm of the pollen grain absorbs fluids from the stigma and bulges out through one of the apertures in the form of a tube. This **pollen tube** then grows down between the cells of the stigma and style until it reaches the micropyle of the ovule in the ovary. In corn, it may have to grow more than 50 centimeters (20 inches) through the silks before it arrives at its destination, but in most plants, the distance is considerably less.

The pollen tube's journey may last less than 24 hours and usually does not take more than 2 days, although there are a few plants in which growth takes over a year. As the tube grows, most of the contents of the pollen grain are discharged into it. The vegetative nucleus of the tube cell stays at the tip, while the generative cell lags behind and divides by mitosis, usually in the tube, producing two sperm cells; no flagella develop. Sometimes, the generative cell divides before the pollen tube has formed. The germinated pollen grain with its vegetative nucleus and two sperms within the tube cell constitutes the *mature male gametophyte* (*microgametophyte*) (Fig. 23.7).

When the pollen tube reaches the micropyle, it continues on to the *female gametophyte* (*megagametophyte*), which it enters, destroying a now degenerating synergid in the process; it then either bursts or forms a pore, discharging its contents. Next, an event unique to angiosperms (and a few gnetophytes), called **double fertilization** (or **double fusion**), takes place. One sperm migrates from the synergid to the egg, losing most of its protoplasm along the way. The sperm cell nucleus then unites with the egg nucleus, forming a **zygote.**

The other sperm cell also migrates from the synergid into the central cell, and, upon reaching the central cell nuclei, unites with them, producing a $3x$, triploid, *endosperm nucleus*. The endosperm nucleus becomes exceptionally active and divides repeatedly by mitosis, with cell cycles being completed every few hours, usually without formation of cell walls. This nutritive $3x$ tissue, called **endosperm,** may eventually have hundreds of thousands of nuclei; it surrounds the embryo that develops from the zygote. The formation of endosperm is considered to be a major development in the evolution of angiosperms. It supplies a superior form of nutrition for developing embryos. The endosperm is also a major source of nutrition for human beings. The three major world food crops (wheat, rice, and maize) are consumed mainly for the nutritional quality of the endosperm in their seeds.

In some monocots, such as corn and other grasses, the endosperm tissue becomes an extensive part of the seed, but in most dicots, such as members of the Legume Family (Fabaceae; e.g., peas, beans), the endosperm provides nutriment for the embryo that develops from the zygote but is absorbed into the cotyledons by the time the seed is mature. The integuments harden, becoming a seed coat, and the remaining haploid nuclei or cells (antipodals, synergids, and tube nucleus) degenerate. At the conclusion of these events, the ovule has become a seed, and at the same time, the ovary matures into a fruit. Seed dispersal and fruits are discussed in Chapter 8.

Other Types of (Female) Gametophyte Development

The process of female gametophyte development just described in the section "Fertilization and Development of the

Seed" occurs in about 70% of the known flowering plants. The remaining 30% exhibit variations in which the female gametophyte has from four to 16 nuclei or cells at maturity, and the endosperm may be $5x$, $9x$, or even $15x$.

One such variation occurs in lilies, a favorite of microscope slide manufacturers for showing female gametophyte development. When a megasporocyte undergoes meiosis in a lily, all four of the haploid megaspore nuclei produced remain functional. Three of the megaspore nuclei unite, forming one $3x$ nucleus; the fourth megaspore nucleus remains x. Both the $3x$ nucleus and the x nucleus undergo mitosis twice, resulting in a large cell with four $3x$ nuclei plus four x nuclei. The large cell with eight nuclei becomes the female gametophyte; one of the central cell nuclei is $3x$, and the other central cell nucleus is x. During fertilization when one sperm (x) unites with the two central cell nuclei, a $5x$ endosperm nucleus is produced (Fig. 23.8).

Apomixis and Parthenocarpy

Some embryos can develop *apomictically*—that is, without fusion of gametes (sex cells) but with the normal structures (e.g., ovaries) otherwise being involved.

An embryo may develop, for example, from a $2x$ nutritive cell or other diploid cell of an ovule, instead of from a zygote. After germination, this makes the plant that develops from such a seed genetically equivalent to a vegetatively propagated plant. *Apomixis* was discussed in Chapter 15.

Fruits that develop from ovaries having unfertilized eggs are said to be **parthenocarpic.** In contrast to fruits containing apomictic seeds, fruits formed by parthenocarpy are seedless. Navel oranges, supermarket bananas, and certain varieties of figs and grapes are examples of parthenocarpic fruit.

Figure 23.8 Development of a lily megasporocyte. (a) The process begins with a diploid ($2x$) megasporocyte. (b) and (c) The megasporocyte undergoes meiosis, producing four haploid (x) megaspore nuclei. (d) and (e) The megaspore nuclei unite, forming a triploid ($3x$) nucleus; the other haploid nucleus remains separate. (f) and (g) Both the $3x$ and x nuclei undergo two consecutive mitotic divisions, resulting in four $3x$ and four x nuclei. (h) Three of the $3x$ nuclei function as antipodals; the fourth $3x$ nucleus and one x nucleus function as central cell nuclei. The remaining haploid nuclei function as an egg and two synergids. © *James E. Bidlack*

KEY THEME 23.1 molecular — The Difference between "n" and "x" in Plant Life Cycles

In both plants and animals, the product of fertilization (the zygote) undergoes mitosis to produce a multicellular structure. However, in plants, the product of meiosis also undergoes mitosis to produce a multicellular structure. Consequently, plants, but not animals, exhibit an alternation of generations. That is, the life cycle alternates between the gametophyte generation (the product of meiosis) and the sporophyte generation (the product of fertilization). Because of this complex life, plant biologists have developed a set of terms specific to plants. The gametophyte generation is abbreviated "n," while the sporophyte generation is abbreviated "$2n$." These terms do not apply to animals. In addition, it is irrelevant to refer to anything other than n or $2n$. The life cycle contains only the gametophyte (n) generation and the sporophyte ($2n$) generation. There are no values of n other than 1 or 2. The abbreviation for the number of sets of chromosomes in a cell of an organism is "x." This term can be used in both the plant and the animal worlds, and the number preceding the "x" can be anything from 1 to over 100. A plant or animal with two sets of chromosomes is diploid; this is abbreviated $2x$. Triploid endosperm or a triploid plant is $3x$ and contains three sets of chromosomes. A wheat plant, with six sets of chromosomes, would be appropriately described using the abbreviation $2n = 6x$ (every cell in the plant—the sporophyte generation—has six sets of chromosomes). Pollen from that wheat plant would be referred to as $n = 3x$ (each cell in the pollen grain—the gametophyte generation—contains three sets of chromosomes).

To further complicate matters, some seedless fruits aren't parthenocarpic. For example, when Thompson seedless grapes are fertilized, the ovules don't develop with the fruit. Also, applying dilute hormone sprays to flowers can bring about artificial parthenocarpy. Seedless tomatoes are often produced in this way. Crossing watermelon varieties with different chromosome numbers results in hybrid watermelons that are seedless because the chromosomes can't pair properly during meiosis, and seed formation does not occur.

Trends of Specialization and Classification in Flowering Plants

Various classifications of plants have been proposed ever since the 4th century B.C., when Theophrastus grouped plants into trees, shrubs, and herbs. The first classifications were merely for convenience and did not necessarily reflect natural relationships. Even today, plants may be lumped together in unnatural groupings in order to make them easier to identify. For example, some wildflower books arrange together all white-flowered species or all yellow-flowered species. There is nothing wrong with such identification arrangements, but because such schemes do not reflect natural relationships (i.e., relationships based on evolution), it is often difficult to recognize family characteristics or lineage. Modern botanists, therefore, try to group plants according to their natural relationships. These are based on evidence gleaned from breeding experiments, form and structure, chemical components, fossil records, and other features such as DNA sequence homology. There are, however, many interpretations of trends in the specialization of flowering and other plants that are based primarily on inference from evidence presently available.

Although the information is very incomplete, the fossil record suggests that the flowering plants first appeared about 160 million years ago during the late Jurassic period (see Table 21.1), although the oldest known fossil of a flowering plant (discovered in China in 1998) has been dated at 142 million years. The flowering plants then developed during the Cretaceous period and the ensuing periods of the Cenozoic era into the dominant elements of the flora they are today. Most botanists hypothesize that primitive flowering plants had their origins in the Jurassic period and had the following features: their leaves were simple; the flowers had numerous, spirally arranged parts that were not fused to each other and were variable in number; the flowers were also *radially symmetrical,* or *regular* (i.e., the flowers could be divided into two equal halves along more than one lengthwise plane); and the flowers had both stamens and pistils.

Although today there still are many plants whose flowers have primitive features, various specializations and modifications have occurred since flowers first appeared. Flower parts have become fewer in number. Some parts have fused, and spiral arrangements have been compressed to whorls (Fig. 23.9).

The first pistil is believed to have been formed from a leaflike structure with ovules along its margins. The edges of the blade, called a *carpel,* apparently rolled inward and fused together. In due course, the separate carpels of primitive flowers fused together, forming the common *compound pistil* consisting of several carpels, which is found in many of today's angiosperms (Fig. 23.10). Each segment of an

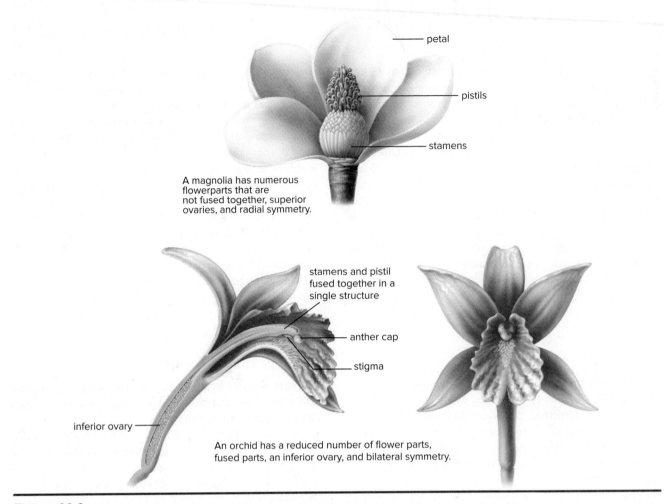

Figure 23.9 Comparison between a primitive flower, magnolia (*top*), and an advanced flower, orchid (*bottom*).

orange, for example, represents a single carpel, and three carpels are easily distinguishable in a cross section of a cucumber.

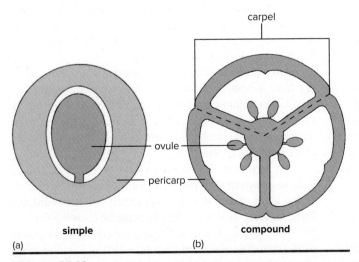

Figure 23.10 Ovaries in cross section. (a) The ovary of a simple pistil as found in a peach. (b) The ovary of a compound pistil as found in a lily.

In advanced flowers, the *receptacle* or other flower parts have fused to the ovary and grown up around it, so that the calyx and the corolla appear to be attached at the top of the ovary. When the ovary is embedded in the receptacle and other parts, it is said to be an **inferior ovary,** and the flower parts attached at the top of the ovary are said to be **epigynous.** A more primitive **superior ovary** is produced on top of the receptacle with the other flower parts attached around its base; such flower parts are said to be **hypogynous.** Some flowers appear to have ovaries in an intermediate position with **perigynous** flower parts, which are usually attached to a corolla tube of fused petals; the floral tube itself is generally not fused to the ovary (Fig. 23.11). Flowers have also tended to become *bilaterally symmetrical,* or *irregular* (i.e., capable of being divided into two symmetrical halves only by a single lengthwise plane passing through the axis), as in sweet peas and orchids.

If a flower has a calyx, a corolla, stamens, and a pistil, it is *complete*. If, however, the corolla or other flower parts are missing, the flower is *incomplete*. If, as is the case with most flowers, both stamens and a pistil

Figure 23.11 Ovary positions in flowers: *superior* (parts hypogynous; e.g., geranium); *superior* (parts perigynous; e.g., cherry); and *inferior* (parts epigynous; e.g., orchid).

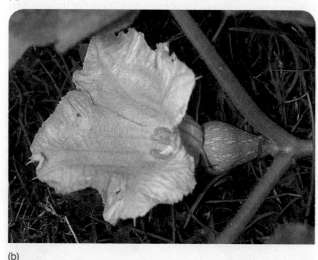

Figure 23.12 Unisexual flowers of a squash. (a) Male. (b) Female. Note the inferior ovary beneath the corolla. *(a-b)* © Kingsley Stern

(or pistils) are present, it is said to be *perfect,* regardless of whether or not it also happens to be complete or incomplete. In some families, however, the flowers are not only incomplete but also have become *imperfect* (*unisexual*). Each imperfect flower has either stamens or a pistil, but not both. The Pumpkin Family (Cucurbitaceae) includes pumpkins, squashes, watermelons, cantaloupes, cucumbers, and other species with imperfect flowers (Fig. 23.12). When both male and female imperfect flowers occur on the same plant (as is the case with many members of the Pumpkin Family), the species is **monoecious.** If a plant bears only male flowers and other plants of the same species bear only female flowers, the species is said to be **dioecious.**

In both the dicots and the monocots, evolutionary specialization has involved reduction and fusion of parts and a shifting of the ovary from a superior to an inferior position, as hypothetical progression is made from families with primitive characteristics through intermediates to families with advanced features.

Observations on the human and ecological relevance of the flowering plants occupy literally thousands of volumes. A brief overview of several aspects of this subject is given in Chapters 24 and 25.

23.3 Pollination Ecology

Learning Outcome

a. Describe five different pollination strategies and the floral characteristics used to promote pollination.

When certain *consumers,* such as insects, forage for food among photosynthetic *producers,* such as plants, they often come in contact with flowers. Many insects and other animals become dusted with pollen, and as they feed or collect pollen and/or nectar, they unknowingly but effectively bring about pollination of the plants they visit. Throughout the evolutionary history of the flowering plants, the pollinators have evidently coevolved with plants. In some instances, the relationship between the two has become highly specialized.

Twenty thousand different species of bees are included among the pollinators of present-day flowering plants. By far, the best known of these are honey bees. Their chief source of food is nectar, but they also gather pollen for their larvae. The flowers that bees visit are generally brightly colored and mostly blue or yellow—rarely pure red. Pure red appears black to bees, and they generally overlook red flowers. Flowers often have lines or other distinctive markings, which may function as honey guides that lead the bees to the nectar. Bees can see ultraviolet light (a part of the spectrum not visible to humans), and some flower markings are visible only in ultraviolet light, making patterns seen by bees sometimes different from those seen by humans (Fig. 23.13).

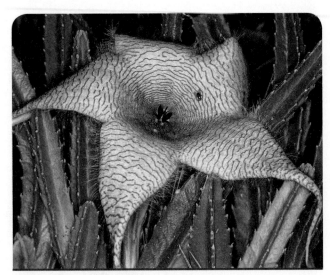

Figure 23.14 A *Stapelia gigantea* (carrion flower) plant. © *Steven P. Lynch*

(a)

(b)

Figure 23.13 Flower markings on coneflowers (*Rudbeckia*). (a) In ordinary light. (b) In ultraviolet light. *(a-b)* © *Kingsley Stern*

Many bee-pollinated flowers are delicately sweet and fragrant. In contrast, flowers pollinated by beetles tend to have stronger, yeasty, spicy, or fruity odors. Beetles don't have keen visual senses, and flowers pollinated by them are usually white or dull in color. Some beetle-pollinated flowers don't secrete nectar but either furnish the insects with pollen or have food available on the petals in special storage cells, which the beetles consume.

Some flowers, including the *stapelias* of Africa (Fig. 23.14), smell like rotten meat. Short-tongued flies pollinate such flowers, which tend to be dull red or brown. These plants are related to our milkweeds, although superficially they don't resemble milkweeds at all. They are often called *carrion flowers* because of their foul odor and appearance. Flies with longer tongues may also pollinate bee-pollinated flowers.

Moth- and butterfly-pollinated flowers (Fig. 23.15), like bee-pollinated flowers, often have sweet fragrances. Night flying moths visit flowers that tend to be white or yellow—colors that stand out against dark backgrounds in starlight or moonlight.

Red flowers are sometimes pollinated by butterflies, some of which can detect red colors, but these insects are more often found visiting bright blue, yellow, or orange flowers. The nectaries of these flowers are at the bases of corolla tubes or spurs, where only moths and butterflies with longer tongues can reach. However, an enterprising bumblebee will occasionally bypass convention and chew through the base of a spur to get at the nectar.

Birds—particularly the hummingbirds of the Americas and the sunbirds of Africa—and the flowers they pollinate are also adapted to one another (Fig. 23.16). The birds do not have a keen sense of smell, but they have excellent vision. The flowers they visit are often bright red or yellow

Figure 23.15 Butterfly pollination of the brightly colored *Lantana* inflorescence. © *James E. Bidlack*

Figure 23.16 A ruby-throated hummingbird (*Archilochus colubris*) feeding from a *Salvia* flower. © *Mike Bons/Getty Images*

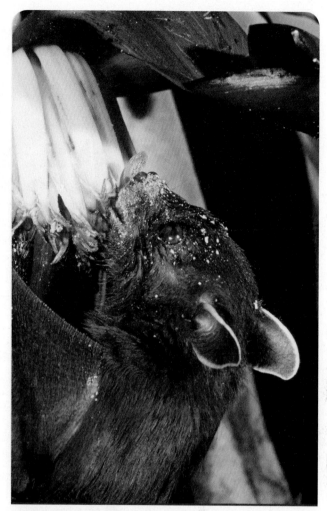

Figure 23.17 A greater short-nosed fruit bat (*Cynopterus sphinx*) on a banana flower in southeast Asia. The banana is a prime example of a bat-dependent plant. Pollen will be carried on the bat's fur to the next plant, thus achieving cross-pollination. © *Merlin D. Tuttle/Science Source*

and usually have little if any odor. Bird-pollinated flowers also are typically large; are part of a large, sturdy inflorescence (flower cluster); or, in some cases, are produced on trunks of trees.

Birds are highly active pollinators and tend to burn energy rapidly. They must feed frequently to sustain themselves. Many of the flowers birds prefer produce copious amounts of nectar, thereby assuring repeated visits. The nectar is often produced in long floral tubes that keep most insects out. Some native California fuchsias and their relatives have long threads extending from each pollen grain. When a hummingbird inserts its long bill into such a flower, the pollen grain threads catch on short, stiff hairs located toward the base of its bill, and in this way, the bird unknowingly transfers pollen from one flower to another.

Bat-pollinated flowers, found primarily in the tropics, tend to open only at night when the bats are foraging (Fig. 23.17). These flowers are dull in color, and like flowers pollinated by birds, they either are large enough for the animal to insert part of its head or consist of ball-like inflorescences containing large numbers of small flowers that dust the visitor with pollen.

The very large Orchid Family, with approximately 35,000 species, has pollinators among all the types mentioned. Some of the adaptations between orchid flowers and their pollinators are extraordinary (Fig. 23.18). The pollen grains of most orchids are produced in little sacs called **pollinia** (singular: **pollinium**) that typically have sticky pads at the bases. When a bee visits such a flower, the pollinia are usually deposited on its head. The "glue" of the sticky pads dries almost instantly, causing the pollinia to adhere tightly. In some orchids, the pollinia are forcibly slapped on the pollinator through a trigger mechanism within the flower.

Figure 23.18 Some pollinating mechanisms in orchids.

Members of *Ophrys*, a genus of orchids found in Europe and in North Africa, have a modified petal that resembles a female bumblebee or wasp. Male bees or wasps emerge from their pupal stage a week or two before the females and apparently mistake the flowers for potential mates. They try to copulate with the flowers, and while they are doing so, pollinia are deposited on their heads. When they visit other flowers, the pollinia catch in sticky stigma cavities. During a single visit by the pollinators, the pollinia removed from one flower are replaced with fresh pollinia from the next flower visited.

Seed Plants: Angiosperms   441

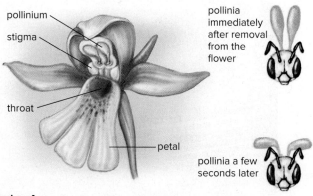

In the showy orchids, there are two separate stigma patches. When an insect visits a flower, the pollinia are attached to its head; then they twist outward and forward so that when the insect visits another flower, the pollinia are simultaneously deposited in the separate stigmas while new pollinia are attached to the insect's head.

mechanism 4

One genus of orchids native to Europe and North Africa evolved a petal that resembles a female wasp. Male wasps try to copulate with the "female" and, in doing so, bump a sensitive area of the flower containing pollinia. The pollinia are "glued" to a wasp's head in the process and carried to another flower. The wasps thus effectively bring about cross-pollination.

mechanism 5

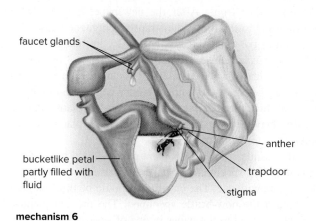

The bucket orchids of South America have faucet glands that partially fill a bucketlike petal with fluid. In visiting the flower, the insects may fall into the fluid. They escape by a trapdoor. The pathway of an insect assures contact with both the stigma and anthers containing pollinia, because they are located adjacent to the trapdoor.

mechanism 6

Figure 23.18 Continued.

The pollinia of orchids pollinated by moths and butterflies are attached to their long tongues with sticky clamps instead of pads. The pollinia of some bog orchids become attached to the eyes of their pollinators, which happen to be female mosquitoes. After a few visits, the mosquitoes are blinded and unable to continue their normal activities.

Among the most bizarre of the orchid pollination mechanisms are those in which the pollinator is dunked in a pool of watery fluid secreted by the orchid itself; the pollinator escapes under water through a trapdoor. The route of the insect ensures contact with pollinia and stigma surfaces. In other orchids with powerful narcotic fragrances, pollinia are slowly attached to the drugged pollinator. When the transfer of pollinia has been completed, the fragrance abruptly disappears. The temporarily stupefied insect then recovers and flies away.

23.4 Herbaria and Plant Preservation

Learning Outcome

a. Describe the functions of a herbarium and the techniques of preparing herbarium specimens.

The botanical resources of many universities and other institutions include **herbaria** (singular: **herbarium**), which are often part of larger museums. Some of the larger herbaria contain literally millions of specimens. Unfortunately, today, many institutional herbaria are understaffed, underfunded, and underutilized.

Herbaria are essentially libraries of dried, pressed (and/or liquid-preserved) plants, algae, and fungi, arranged and labeled so that specific specimens can easily be retrieved. Properly prepared and maintained specimens should remain in excellent condition for 300 or more years. In fact, DNA from herbarium specimens made in 1600 were recently used to determine the origin of the European white potato. Formal training and experience are not needed to make one's own herbarium. The materials, some simple equipment (that may be homemade), and the ability to follow a few relatively elementary procedures are all that is necessary.

Methods

Fungi and bryophytes are usually dried and stored in small boxes or packets. Larger algae tend to have unique properties that lend themselves to special preservation, which may require spreading out specimens and allowing them to adhere to blotting paper. The specimens that make up the bulk of the holdings of most herbaria, however, consist of vascular plants, and what follows pertains primarily to such plants.

The moisture content of flowers and other plant parts to be preserved should be reduced as quickly as possible, with a minimum of distortion. This is usually done with the aid of a *plant press* (Fig. 23.19). This simple device consists of two pieces of plywood (or other wood materials or thin metal plates) with dimensions of approximately 30 × 46 centimeters (12 × 18 inches) and a pair of webbing or leather straps to go around the boards. A number of felts (sheets of heavy blotting paper) of similar dimensions are placed between the boards. A folded page of newspaper is placed between each pair of blotters and a few sheets of stiff cardboard are interspersed between the blotters.

Any soil clinging to the roots of a specimen to be pressed is washed off, and the plant is laid out on one of the newspaper sheets. Leaves are carefully straightened out, as are petals and other plant parts, so that they are not folded during pressing. Notes on where, when, and by whom the specimen

Figure 23.19 A plant press. © *Kingsley Stern*

was collected are penciled on the newspaper, or a number corresponding to such notes in a separate field notebook is used. The newspaper is then folded over the specimen and placed between two blotters (felts).

Many specimens may be placed in a press at one time or, if there is space, between the same sheets of newspaper. Only one species should be placed in a single fold of newspaper, however, as mixing species invariably leads to some being pressed better than others, due mostly to the varying amounts of woody tissues within the stems. If the plant is small, press the entire plant, and try to have both upper and lower leaf or flower surfaces visible. If, however, the plant is large (e.g., a tree), press representative parts, especially flowers and/or fruits. After the newspaper and specimen are returned to the press, the straps are tightened around it as much as possible, and it is left to dry in the sun or near a heater (but not close enough to scorch!) for 3 or 4 days. Unless the leaves were succulent or wet at the time they were placed in the press, they should be dry enough to mount on paper at this time. If a press is not available, plants may be pressed between newspapers and blotters by placing heavy weights on top of them.

If possible, 100% rag-content paper should be used for mounting the specimen, as papers made from wood pulp deteriorate with age. The paper also should be thick enough to support the now-brittle specimen. The pressed plant(s) may be attached to the paper in one of several ways. Some workers spread a good white library glue on a glass plate and then place the specimen on the wet glue, so that one side is covered with it. They then transfer the specimen to herbarium paper that normally measures 29 × 42 centimeters (11.5 × 16.5 inches). Others place the specimen on the paper first and add glue or liquid plastic at strategic points. The bottom right-hand corner of the paper should be kept clear for a label indicating the scientific name of the plant, collection information, the collector's name, and the collection date (Fig. 23.20).

Professional botanists also number their individual collections, giving a collection number after the collector's name. The label should measure about 7.5 × 12.5 centimeters

Figure 23.20 A mounted herbarium specimen. © *Kingsley Stern*

(3 × 5 inches), and its paper should also be of 100% rag content. Specimens are then placed in manila folders and stored in some systematic fashion so retrieval of individual specimens is easily accomplished. Major institutional herbaria usually arrange the plants by families in a presumed evolutionary sequence, and then the genera and species are alphabetically arranged within the families. Paradichlorobenzene, mothballs, or similar material should be added to the cabinet or storage area to prevent insect damage during storage.

Although other forms of flower preservation are not associated with herbaria or professional botanical activities, a few will be mentioned here simply for reference. For example, some people are interested in pressing flowers for use in dry arrangements on cards, place mats, or other decorative items. In such cases, the pressing is done as though they were herbarium specimens, but the material is then further manipulated in one of several ways. For wall art, the specimens can be mounted on a piece of smooth cardboard, covered with clear plastic or glass, matted, and framed (Fig. 23.21). Decorative notepaper can be made by placing pressed flowers on the paper and covering them with rice paper or facial tissue. A mixture of one part white glue to three parts water is then brushed over the tissue, causing it to become a permanent mount, with the specimen clearly visible through the thin paper film. Clear contact paper cut to the appropriate size makes a good mount for pressed flowers placed on cards or place mats. Pressed flowers can also be embedded in clear plastic poured into molds.

Flowers can also be dried without pressing. With a little patience, relatively lifelike, three-dimensional preservations can be made, although flowers with petals that are not easily detached lend themselves more readily to three-dimensional drying than others. Most such drying is done in a shoe box, but almost any type of container can be used. The bottom of the box is covered with about 2 centimeters (0.8 inch) of sand, silica gel, or borax mixture. The fresh flower, with a little of the peduncle or stem still attached, is then gently laid on the surface. After this, more sand, silica gel, or borax mixture is slowly drizzled by hand into the box until the entire flower is

Figure 23.21 Pressed flowers framed as wall art.
© Kingsley Stern

Figure 23.22 Three-dimensional drying. A flower placed in a box is embedded in sand, silica gel, or a borax mixture. © Kingsley Stern

buried, with care being taken not to create air pockets around any parts (Fig. 23.22).

Each of the three drying agents mentioned has certain advantages and drawbacks. Sand must be thoroughly washed several times to be certain it is perfectly clean before use. The sand also needs to be completely dry, be uniformly fine, and if possible, have individual grains that are relatively rounded rather than angular. Such sand with rounded grains is found around the Great Salt Lake in Utah and is available in arts and crafts stores. It takes about 2 weeks to dry most flowers with sand. Silica gel is also available commercially and is the quickest drying agent of the three mentioned, usually completing the job in 4 to 5 days (if the box is placed in a microwave oven, the drying process may be completed in as little as 5 minutes). Its granules tend to be of different sizes, however, giving the flower surface a slightly irregular texture. It also tends to darken certain colors, and it is expensive. When borax is used, two parts of it should be mixed with one part sand or cornmeal. It dries flowers in about 3 weeks but is sometimes difficult to remove completely from the dried flower surfaces.

When drying has been completed, the container is tilted, and the sand or other drying material is slowly and gently poured out in an uninterrupted motion. If any petals have come loose, they may be glued in place later. Wire may be inserted in the stem to add rigidity, and after any clinging granules have been carefully removed, soft, powdered colored chalk may be dusted on to restore any fading of color. It is recommended that beginners use fairly large flowers whose petals are not easily detached (e.g., zinnias, rose buds) for their first attempts at three-dimensional flower drying.

A Word of Caution to Collectors

Dried plant collections of herbaria, in particular, have proved invaluable in botanical research in the past and will continue to do so in the future. They have facilitated quick identifications of plants in emergency situations where children have eaten plants or plant parts suspected of being poisonous, or they have helped pin down specific plants that have caused allergic reactions. Herbaria also have been involved in archaeological research where the uses of plants by past cultures have been determined, and they have been used for teaching purposes at various educational levels. Herbarium specimens have been useful in criminal litigation for both the prosecution and the defense, and have been the primary source of information on the distribution of plants with potential for new agricultural and horticultural crops or those with possible medicinal values. Most of the unraveling of problems pertaining to natural relationships of plants begins in a herbarium, and without these plant libraries, increasing our knowledge along many practical and theoretical lines would be severely restricted.

Hundreds of plants native to North America are now on rare and endangered species lists, and thousands more are in similar predicaments on other continents. *The day has come when both professional and amateur persons interested in plants must discipline themselves to exercise extreme caution in collecting native plants. Collectors should first know what they are collecting or otherwise refrain, and collecting for private collections without serious purpose should be strictly limited. Except for certain types of research, a good photograph of a plant may actually be preferable to a dried specimen and aesthetically more pleasing. It is sincerely hoped that as much as possible, each reader will confine a collection to photographs of native plants.* In any case where you need to remove a plant specimen for research, you must obtain collection permits from the private or public agency that manages the area from which the specimen is collected.

SUMMARY

23.1 Introduction to Angiosperms

- Flowering plants (angiosperms) have ovules and seeds completely enclosed within carpels; the carpels comprise ovaries that become fruits. There is one phylum of flowering plants (Magnoliophyta); it is divided into two classes (dicots and monocots).
- Flowering plants are believed to have evolved from a gymnosperm ancestor. Most contemporary botanists hypothesize that flowering plants evolved separately from pteridosperms and that the first flowers had many separate, flattened parts spirally arranged on an elongated receptacle.

23.2 Phylum Magnoliophyta—the Flowering Plants

- Flowering plants are heterosporous.
- Flowering plant gametophytes develop in separate structures. The female gametophyte (megagametophyte) develops in the ovule. Integuments, which later become a seed coat, surround the megagametophyte. Pollen grains developed in anthers become male gametophytes.
- Pollination is the transfer of pollen grains from an anther to a stigma; it is brought about by insects, wind, and other agents.
- After pollination, a pollen tube may grow from a pollen grain to the female gametophyte; the tube nucleus remains at its tip, and the generative cell divides, producing two sperm nuclei.
- Following the discharge of the contents of the pollen tube into the female gametophyte, a sperm unites with the egg, forming a zygote; the other sperm simultaneously unites with the two central cell nuclei, forming $3x$ endosperm.
- The nutritive endosperm tissue may become part of the seed or be absorbed by the seed's embryo.
- Due to variations in how a female gametophyte develops, some flowering plants produce $5x$, $9x$, or $15x$ endosperm tissue.
- Some artificial groupings of flowers to aid identification do not reflect natural relationships. Sources of evidence used to try to group plants naturally include fossils, which suggest the flowering plants first appeared about 160 million years ago.
- Primitive flowering plants had simple leaves and numerous, spirally arranged flower parts that were not fused to each other; they possessed both stamens and pistils and were radially symmetrical (regular).
- Specializations include a reduction in the number of parts, fusion of parts, appearance of compound pistils composed of several individual carpels, inferior ovaries, bilateral symmetry (irregular flowers), and unisexual flowers.
- Monoecious species have both male and female flowers on the same plants; dioecious species have male and female flowers on separate plants.

23.3 Pollination Ecology

- Bee-pollinated flowers are delicately sweet and fragrant and tend to be blue or yellow in color.
- Beetle-pollinated flowers tend to have stronger odors and are usually white or dull in color.
- Some fly-pollinated flowers emit foul odors.
- Moth-pollinated flowers tend to be white or yellow.
- Bird-pollinated flowers are usually bright red or yellow and have much nectar but little odor.
- Most orchids produce pollen grains in pollinia that adhere or clamp onto parts of visiting insects. The flowers of some orchid species have developed bizarre pollination mechanisms.

23.4 Herbaria and Plant Preservation

- Herbaria are essentially libraries of dried, pressed, or otherwise preserved plants, fungi, and algae arranged so that specific specimens may be readily located. Properly preserved plants may last for hundreds of years.
- A plant to be pressed is placed in a plant press between sheets of newspaper and absorbent material. Dry specimens are affixed to sheets of high-quality paper, with a label giving collection information.
- Plant parts may be pressed for artwork, place mats, and so on. Flowers may be preserved three-dimensionally.
- Because so many plants are now on rare and endangered species lists, collectors should try to confine their future collecting to photographs as much as possible.

REVIEW QUESTIONS

1. What are the basic differences between gymnosperms and angiosperms?
2. What is the function of the integuments?
3. How are the pollen grains of flowering plants different from those of pine trees?
4. What is the functional equivalent of an archegonium in a flower?
5. What is the function of endosperm, and how does it originate?
6. What differences are there between the female gametophytes of peaches and those of lilies?
7. Distinguish between radial and bilateral symmetry.
8. What distinguishes flowers with primitive features from those that are advanced?
9. In general, how do flowers pollinated by bees differ from those pollinated by moths, flies, beetles, and birds?
10. What is a herbarium?

DISCUSSION QUESTIONS

1. The world's largest flowers are on inconspicuous parasitic plants that resemble a fungus mycelium. If the flowers were not present, would it be possible to tell that the plant is not a fungus? How?
2. In most female gametophytes with eight nuclei, only the egg and the central cell nuclei do not degenerate. What would happen if the megaspore itself functioned as an egg and did not go on to produce other nuclei in a female gametophyte?
3. It takes only one pollen grain to initiate the development of an ovule into a seed, yet a single flower may produce many thousands of pollen grains. Do you suppose such huge numbers of pollen grains are really necessary? Why do you suppose wind-pollinated plants produce much more pollen than insect-pollinated plants?
4. Are such items as the name of the collector and the date significant on a herbarium specimen?
5. Is it really important that plants be classified? What would be the consequences if they were not?

CHAPTER 24

Flowering Plants and Civilization

24.1 Introduction to Flowering Plants and Civilization
24.2 Origin of Cultivated Plants
  Near-Eastern Region
  Chinese Region
  African Continent
  South Asia and the Islands of the Pacific
  North America
  South and Central America
24.3 Selected Families of Flowering Plants
24.4 Dicots
  The Buttercup Family (Ranunculaceae)
  The Laurel Family (Lauraceae)
  The Poppy Family (Papaveraceae)
  The Mustard Family (Brassicaceae)
  The Rose Family (Rosaceae)
  The Legume Family (Fabaceae)
  The Spurge Family (Euphorbiaceae)
  The Cactus Family (Cactaceae)
  The Mint Family (Lamiaceae)
  The Nightshade Family (Solanaceae)
  The Carrot Family (Apiaceae)
  The Pumpkin Family (Cucurbitaceae)
  The Sunflower Family (Asteraceae)
24.5 Monocots
  The Grass Family (Poaceae)
  The Lily Family (Liliaceae)
  The Orchid Family (Orchidaceae)

Key Theme 24.1: Wild Rice—More Than Just Food
Key Theme 24.2: Coffee and Caffeine

OVERVIEW

This chapter begins with comments on some of the problems in distinguishing between fact and fiction in reports of past uses of plants. It continues with a brief discussion of Vavilov's centers of origin of cultivated plants and subsequent modifications of Vavilov's theories. A survey of 16 well-known flowering plant families is presented and brief comments are made on family characteristics. Some past, present, and possible future uses of family members are explored.

Pods of a tropical cacao tree (*Theobroma cacao*). The seeds inside the pods are the source of chocolate. © *Kingsley Stern*

24.1 Introduction to Flowering Plants and Civilization

Learning Outcome

a. Design an experiment to determine whether the consumption of seeds of a certain native plant can decrease the growth rate of cancer cells growing in petri dishes.

Reminder: In this chapter, there are references to a large number of plants. As in all chapters throughout the book, a few scientific names are given within the text, but the scientific names of all the plants mentioned are listed in Appendix 1.

Puccoons (*Lithospermum* spp.) are herbaceous plants that grow on dry plains and slopes throughout the western United States and British Columbia. One puccoon with greenish yellow flowers (*L. ruderale*) has seeds that are so hard, it was given the common name of *stoneseed* in Nevada and California.

Native American women of the Shoshoni tribe of Nevada reportedly drank a cold water infusion of the roots of stoneseed every day for 6 months to ensure permanent sterility. Biologist Clellan Ford became curious about these reports and gave extracts of stoneseed plants to mice. He found the extracts effectively eliminated the estrous cycle of the mice and decreased the weights of the ovaries, the thymus, and the pituitary glands.

Although this reported Native American use of a plant was demonstrated experimentally to have a basis in fact, distinguishing between fact and fiction in recorded past uses of plants is often difficult, particularly if the plants have become rare or extinct. However, this type of research is essential today if we are going to save potential sources of medicinal drugs and other useful plant products before they are eliminated by clearing of land and other practices in the name of progress.

Earlier generations, despite occasional misguided superstition and folklore, did cure some of the diseases they treated with plants, even though they did not know the scientific reasons for the results. As indicated earlier, botanists who have recognized this have teamed with anthropologists, medical doctors, and interpreters to interview tribal and other medical practitioners all over the world. They are sifting through and gleaning as much of this information as possible before it is too late. Their work is already leading to useful new medicinal and other discoveries.

24.2 Origin of Cultivated Plants

Learning Outcomes

a. Describe features of regions where plants were first cultivated.
b. List plants that originated in the Near-Eastern region.
c. List plants that originated in the Chinese region.
d. List plants that originated on the African continent.
e. List plants that originated in South Asia and the islands of the Pacific.
f. List plants that originated in North America.
g. List plants that originated in South and Central America.

In the 1880s, Alphonse de Candolle, a Swiss botanist, published a book entitled *Origin of Cultivated Plants*, based on data he had gathered from many sources. He deduced that cultivated plants probably originated in areas where their wild relatives grow.

In 1916, Nikolai Vavilov, a Russian botanist and geneticist, began a follow-up of de Candolle's work. During the next 20 years, he expanded on de Candolle's work and focused on geographic regions at a smaller scale. Vavilov believed that plants were domesticated in geographic regions where high genetic diversity occurs in wild relatives. He showed that the majority of domesticated plants originated from mountain regions, rather than large river valleys, as previously thought.

Vavilov recognized eight centers of diversity of cultivated plants, with some plants originating in more than one center. The centers, some of which were subdivided, are shown in Figure 24.1. The centers of origin of cultivated plants are regions of abundant genetic diversity, but they were also home to advanced civilizations at the time of domestication.

A significant number of additional studies were undertaken by agricultural geneticist Jack R. Harlan and his students at the Crop Evolution Laboratory of the University of Illinois. Harlan concluded that some crops do not have centers of origin, and he revised Vavilov's concept, preferring instead to associate crop origins with regions. A sampling of cultivated plants that appear to have originated in six major regions follows.

Near-Eastern Region

The Near-Eastern region encompasses the Mediterranean area, northern Europe (including the former Soviet Union), Turkey, the Balkans, Pakistan, and Iran. Among the many cultivated plants that appear to have originated in this region are barley, wheat and other cereals, garden pea, lentil, chickpea, lupine, asparagus, beet, carrot, turnip, olive, stone fruits (cherry, plum, apricot), apple, pear, onion, garlic, cabbage, broccoli, lettuce, parsley, flax, clover, pistachio, several range grasses, digitalis, belladonna, opium poppy, and psyllium.

Chinese Region

Many of the cultivated plants believed to have originated in this region apparently came from the temperate and southern parts of China. Examples of widely known plants from this part of the world include bamboo, peach, litchi, walnut, persimmon, ginger, ginseng, gourds, tea, soybean, buckwheat, horseradish, Chinese cabbage, and cucumber. While China is known for its centuries-old cultivation and consumption of rice, this cereal probably originated elsewhere in Asia.

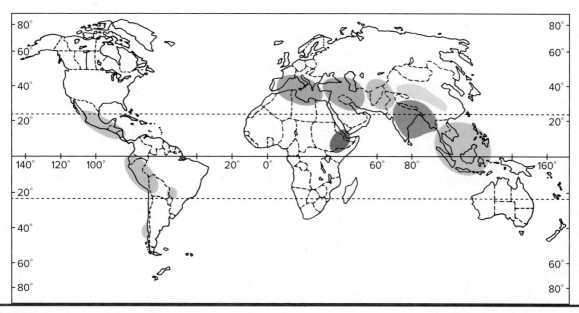

Figure 24.1 Major centers of diversity of cultivated plants according to N. I. Vavilov. *After Vavilov, N. I. 1951.* The origin, variation, immunity and breeding of cultivated plants. *Source: K. Starr Chester, Trans. New York: The Ronald Press Company*

African Continent

Although there is evidence that some cultivated plants originated in South and East Africa, most African plants appear to have come from Ethiopia, Eritrea, West Africa, and other areas between the Sahara and the Congo. Better-known plants include yam, sorghum, okra, oil palm, baobab, the sweet melons, coffee, some cotton, and castor bean.

South Asia and the Islands of the Pacific

This region encompasses South Asia, Malaysia, Indonesia, Melanesia, the Philippines, and thousands of islands extending across the southern Pacific Ocean. Among the plants from this region are sugarcane, eggplant, mango, banana, citrus fruits, safflower, nutmeg, clove, cardamom, turmeric, black pepper, jute, hemp fiber, coconut, taro, rice, mung bean, sesame, betel nut, onion, and jackfruit.

North America

Few cultivated plants are believed to have originated in North America. Most are small fruits and include strawberry, cranberry, blueberry, grape, raspberry, pumpkin, and sunflower.

South and Central America

There is no distinct line between South America and areas to the north with respect to some of the cultivated plants that originated in the New World. The wild relatives of common garden beans, for example, are distributed throughout parts of both North and South America. Cultivated plants from the region include corn, peanut, white potato, lima bean, common bean, cashew, pineapple, papaya, avocado, red pepper, tomato, cotton, cocaine, amaranth, quinoa, cacao (chocolate), guava, sweet potato, pumpkin, squash, rubber, prickly pear, peyote, chayote, vanilla, and cassava.

24.3 Selected Families of Flowering Plants

Learning Outcome

a. Describe five characteristics that can be used to identify families of flowering plants.

The following is a survey of a few well-known plant families, indicating not only past uses of some of their members but also possible future uses, along with biological notes and miscellaneous observations. While DNA sequence data are leading to a reorganization of plant relationships, the morphological key presented here continues to be useful for identifying members of plant families.

The reader is cautioned against assuming that past medicinal uses were always effective or without harmful side effects. Do not experiment with these plants.

More than 300 families of flowering plants have been recognized but are currently undergoing extensive reorganization on the basis of DNA, chemical, and cladistical information. Flowers and fruits are considerably more reliable and stable indicators of heredity than leaves or other vegetative plant parts, so flowering plant families have been distinguished from one another mostly on the basis of flower and fruit parts and structure. Some families include a mere handful of species, while others are very large, with thousands of members. We will focus on only a few of the larger, better-known families here.

The families are presented in a phylogenetic (evolutionary) sequence, beginning with those that have a more generalized flower structure and progressing to those whose flowers are considered specialized. Generalized flowers tend to have numerous parts that are not fused together, and the ovaries are superior. In specialized flowers, the parts are reduced in number and often fused together; the ovary is inferior (see Chapter 23).

The following is an abbreviated key to the families discussed. (See Chapter 16 for a discussion of keys and their uses.)

24.4 Dicots

The Buttercup Family (Ranunculaceae)

Learning Outcome

a. List 10 dicot families and provide an example of each.

Nearly all the 1,500 members of the Buttercup Family are herbaceous. The flowers, whose petals often vary in number, have numerous stamens and several to many pistils with superior

1. Flowers with parts in fours or fives or multiples thereof; seeds with two cotyledons (DICOTS)
    2. Petals separate from one another, or lacking
        3. Petals present
            4. Stamens more than twice as many as the petals
                5. Stamens, petals, and sepals attached to the rim of a cup surrounding the one to many pistils .... Rose Family (Rosaceae)
                5. Stamens, petals, and sepals not attached to the rim of a cup
                    6. Pistils several to many in each flower .... Buttercup Family (Ranunculaceae)
                    6. Pistil one
                        7. Ovary superior .... Poppy Family (Papaveraceae)
                        7. Ovary inferior .... Cactus Family (Cactaceae)
            4. Stamens not more than twice as many as the petals
                8. Herbaceous vines; fruit a pepo .... Pumpkin Family (Cucurbitaceae)
                8. Primarily herbs, shrubs, and trees; fruit not a pepo
                    9. Fruit a legume .... Legume Family (Fabaceae)
                    9. Fruit not a legume
                        10. Fruit a silique or silicle..Mustard Family (Brassicaceae)
                        10. Fruit not a silique or silicle
                            11. Ovary superior; stems square in cross section; leaves opposite; fruit of four nutlets... Mint Family (Lamiaceae)
                            11. Ovary inferior; stems rounded in cross section; leaves alternate; fruit a schizocarp...Carrot Family (Apiaceae)
        3. Petals lacking; calyx sometimes petal-like
            12. Ovary of three carpels and usually elevated on a gynophore; anthers splitting lengthwise............................Spurge Family (Euphorbiaceae)
            12. Ovary of one carpel; gynophore lacking; anthers splitting by raised flaps .... Laurel Family (Lauraceae)
    2. Petals fused together
        13. Flowers not in a head; each flower with its own receptacle; ovary superior .... Nightshade Family (Solanaceae)
        13. Flowers in a head, several to many florets on a common receptacle, comprising a "flowerlike" inflorescence; ovary inferior .... Sunflower Family (Asteraceae)
1. Flowers with parts in threes or multiples thereof; seed with one cotyledon (MONOCOTS)
    14. Flowers inconspicuous; without petals or sepals .... Grass Family (Poaceae)
    14. Flowers conspicuous; petals and sepals mostly similar in coloration
        15. Ovary superior; petals all alike .... Lily Family (Liliaceae)
        15. Ovary inferior; one petal different in form from the other two............................... Orchid Family (Orchidaceae)

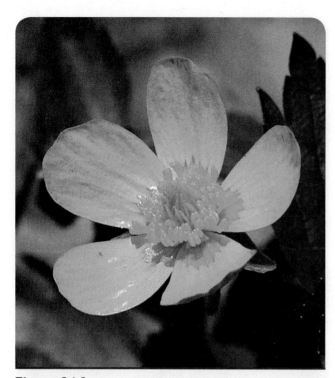

Figure 24.2 A buttercup flower. © *Kingsley Stern*

ovaries (Fig. 24.2). Most have dissected leaves with no stipules and with petioles that are slightly expanded at the base. Well-known representatives include ornamental plants such as buttercup, columbine, larkspur, anemone, monkshood, and *Clematis* (Fig. 24.3). Most members of the Buttercup Family are concentrated in north temperate and arctic regions.

Columbine flowers have five spurred petals that resemble a circle of doves. The name comes from *columba,* the Latin word for "dove." A blue and white species of columbine is the state flower of Colorado. Native Americans controlled diarrhea with a tea made from boiled columbine roots, and members of at least two tribes believed columbine seeds to have aphrodisiac properties. A man would pulverize the seeds, and after rubbing them in the palms of his hands, he would try to shake hands with the woman of his choice, believing the woman would then succumb to his advances. Others crushed and moistened the seeds and applied them to the scalp to repel lice.

Most members of the family are at least slightly poisonous, but the cooked leaves of cowslips have been used for food, and the well-cooked roots of the European bulbous buttercup are considered edible. The European buttercup causes blistering on the skin of sensitive individuals. East Indian fakirs are reported to deliberately blister their skin with buttercup juice in order to appear more pitiful when begging. Native Americans of the West gathered buttercup achenes, which they parched and ground into meal for bread. Others made a yellow dye from buttercup flower petals. Karok Indians made a blue stain for the shafts of their arrows from blue larkspurs and Oregon grape berries.

Goldenseal, sold in health-food stores, is a plant that was once abundant in the woods of temperate eastern North America. It has become virtually extinct in the wild because of relentless collecting by herb dealers. They sell the root for medicinal uses, including remedies for inflamed throats, skin diseases, and sore eyes. At least one Native American tribe mixed the pounded root in animal fat and smeared it on the skin as an insect repellent.

Monkshood yields a drug complex called *aconite* that was once used in the treatment of rheumatism and neuralgia. Although popular as garden flowers, monkshoods are very poisonous. Death may follow within a few hours of ingestion of any part of the plant. Most species have purplish to bluish or greenish flowers, but one Asian monkshood, called *wolfsbane,* has yellow flowers. Wolf hunters in the past poisoned the animals with a juice obtained from wolfsbane roots.

The Laurel Family (Lauraceae)

The Laurel Family's primitive flowers have no petals but have six sepals that are sometimes petal-like. The stamens, which occur in three or four whorls of three each, are a curiosity because the anthers open by flaps that lift up. The ovary is superior. Most of the approximately 1,000 species in this family are tropical evergreen shrubs and trees, many with aromatic leaves. The family received its name from the famous laurel cultivated for centuries in Europe. Its foliage was used by the ancient Greeks to crown victors in athletic events and later was used in the conferring of academic honors. The phrase "to rest on one's laurels" refers to victory and status associated with this Greek symbol. The title Nobel laureate also references the laurel plant.

Several important spices come from members of this family. Cinnamon is the pulverized bark of a small tree native to India and Sri Lanka, although it is also grown commercially elsewhere. Cassia is very similar and is often sold interchangeably with cinnamon. Cinnamon oil is distilled from young leaves of the trees. Use of cinnamon and cassia dates back thousands of years. They were used in perfumes and anointing oils at the time of Moses, and other records reveal their use in Egypt at least 3,500 years ago.

Camphor has been used since ancient times. This evergreen tree, native to China, Japan, and Taiwan, is the main source of camphor essence still used in cold remedies and inhalants, insecticides, and perfumes. The essence is distilled from wood chips. Camphor trees are smog resistant, and some American cities and towns with milder climates have been using them as street trees.

Sassafras trees, native to the eastern United States and eastern Asia, also have spicy-aromatic wood. A flavoring widely used in toothpaste, gum, mouthwashes, and soft drinks was obtained by distillation of wood chips and bark. Sassafras tea is still considered a refreshing beverage. Sassafras is also an ingredient of some homemade root beers, and in the southern states, an alcoholic beer has been made by adding molasses to boiled sassafras shoots and

Figure 24.3 Representatives of the Buttercup Family include the following: (a) Columbine. (b) *Hepatica*. (c) Monkshood. (d) *Isopyrum*.
(a-b) © Kingsley Stern; (c) © Chris Burrows/Getty Images; (d) © Kingsley Stern

allowing the mixture to ferment. In Louisiana, powdered sassafras leaves (called filè) have been used as a thickening and flavoring agent for gumbo. In the past, it has been used by country physicians for treating hypertension and for inducing a sweat in those with respiratory infections. Reports indicate that in large doses, it has a narcotic-stimulant effect, and it is also reported to be carcinogenic. Most sassafras flavorings now in use are artificial.

The sweet bay, used as a flavoring agent in gravies, sauces, soups, and meat dishes, comes from the leaves of the laurel. Leaves of the related California bay (Fig. 24.4) are sometimes used as a substitute for sweet bay and for making Christmas wreaths. This tree, which is native to California, also occurs in southwestern Oregon, where it is known as myrtle (true myrtles, however, belong to a different family). California bay wood is hard and can be polished to a high luster; it is used for making a variety of bowls, ornaments, and other smaller wooden articles.

Early settlers in the West and Native Americans of the region used California bay to relieve rheumatism by bathing

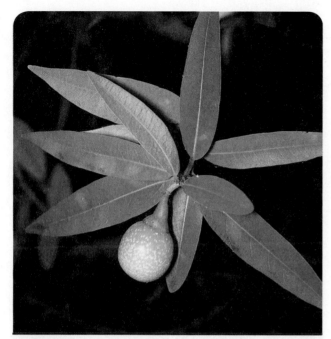

Figure 24.4 A fruit and leaves of a California bay tree (also known as Oregon myrtle). © *Kingsley Stern*

Figure 24.5 A prickly poppy flower. © *Kingsley Stern*

in hot water to which leaves had been added. The nutlike fruits (drupes) were roasted and used for winter food. A leaf was placed under a hat on the head to "cure" a headache (but even a small piece of leaf placed near the nostrils can produce an almost instant headache!). A few leaves placed on top of flour or grains in a canister will keep weevils away, and small branches have been used as chicken roosts to repel bird lice and fleas. A leaf or two rubbed on exposed skin functions as a mosquito repellent.

Avocados are also members of the Laurel Family. The fruits have more energy value by weight than red meats and are rich in vitamins and iron.

The Poppy Family (Papaveraceae)

Most members of the Poppy Family are herbs distributed throughout temperate and subtropical regions north of the equator, but several poppies occur in the Southern Hemisphere, and a number are widely planted as ornamentals. Poppies, like buttercups, tend to have numerous stamens, but most have a single pistil (Fig. 24.5). Most also have milky or colored sap, and their sepals usually fall off as the flowers open. All members produce alkaloidal drugs.

Bloodroot (*Sanguinaria canadensis*) is a pretty, early-flowering spring plant of eastern North American deciduous forests. A bright reddish sap that is produced in its rhizomes was used by some Native Americans as a facial dye, an insect repellent, and a cure for ringworm; children today paint their nails with it. It has a very bitter taste, except when ingested in minute amounts. The bitterness made it effective in inducing vomiting, but members of one tribe treated sore throats with it after compensating for the bitterness by squeezing a few drops on a lump of maple sugar so that it could be held in the mouth.

Figure 24.6 Opium poppy (*Papaver somniferum*) capsule. Opium poppy is the commercial source of the narcotic analgesics morphine and codeine. An immature capsule is cut to release opium, seen as a whitish fluid oozing from within the structure. © *Michael Freeman/Getty Images*

Opium poppies have had a significant impact on societies of both the past and the present. Opium itself was described by Dioscorides in the 1st century A.D., and ancient Assyrian medical texts refer to both opium and opium poppies. Opium smoking, which does not extend back nearly as far as the use of the drug in other ways, became a major problem in China in the 1600s. Smoking opium has given way to other forms of use in recent years. The substance is obtained primarily by making small gashes in the green capsules of the poppies (Fig. 24.6). The crude opium appears as a thick, whitish fluid oozing out of the gashes and is scraped off. It contains two groups of drugs. One group contains the narcotic and addicting drugs morphine and codeine, best known for their widespread medicinal use as painkillers and cough suppressants. Members of the other group are neither narcotic nor addictive. They include papaverine, used in the treatment of circulatory diseases, and noscapine, used as a codeine substitute because it functions like codeine in suppressing coughs but does not have its side effects.

Heroin, a scourge of modern societies, is a derivative of morphine. It is four to eight times more powerful than morphine as a painkiller, and less than 100 years ago, it was advertised and marketed in the United States as a cough suppressant. The loss to society in terms of its economic, physical, and moral impact is enormous, because addicts frequently commit violent crimes to obtain the funds needed to support their habits.

The seeds of opium poppies contain virtually no opium and are widely used in the baking industry as a garnish. However, there are reports that even the mere traces of opium in some poppy seeds have produced positive results in urine drug tests. They also contain up to 50% edible oils that are used to manufacture margarines and shortenings. Another type of oil obtained from the seeds after the edible oils have been extracted is used in soaps and paints.

The Mustard Family (Brassicaceae)

The original Latin name for the Mustard Family was *Cruciferae*. The name describes the four petals of the flowers that are arranged in the form of a cross. The flowers also have four sepals, usually four nectar glands, and six stamens, two of which are shorter than the other four. All members produce siliques or silicles that are unique to the family (see Fig. 8.14). All 2,500 species of the family produce a pungent, watery juice, and nearly all are herbs distributed primarily throughout the temperate and cooler regions of the Northern Hemisphere.

Among the widely cultivated edible plants of the Mustard Family are cabbage, Chinese cabbage, cauliflower, brussels sprouts, broccoli, radish, kohlrabi, turnip, horseradish, watercress, and rutabaga. The widely used condiment mustard is a mixture of the ground, dried seeds of two species of *Brassica*. Some edible members are also widespread weeds. The leaves of shepherd's purse (Fig. 24.7), for example, can be cooked and eaten, and the seeds can be used for bread meal. Other wild edible members include several cresses, peppergrass, sea rocket, toothwort, and wild mustard. Wild mustards are often weeds in row crops. Their leaves are sometimes sold as vegetable greens in markets.

The seeds of wild mustard, shepherd's purse, and several other members of this family produce a sticky mucilage when wet. Biologists at the University of California at Riverside discovered a potential new use for these seeds. They fed pelleted alfalfa rabbit food to mosquito larvae in water tanks they were using for experiments on mosquito control. They noticed the larvae, which had to come to the surface at frequent intervals for air, often stuck to the pellets and suffocated. Curious, the workers examined the pellets under a microscope and found that they contained mustard seeds.

Figure 24.7 Shepherd's purse. © *Kingsley Stern*

Evidently, the field where the alfalfa had been harvested had also contained mustard plants. The scientists then tried heating the mustard seeds to kill them and found this did not affect production of mucilage by wet seeds. It was calculated that 0.45 kilogram (1 pound) of such seeds could kill about 25,000 mosquito larvae. A few mosquito abatement districts have used the seeds effectively, but experiments are needed to determine if there is a practical way to harvest many more seeds and control mosquitoes on a much larger scale by such non-polluting means.

Native Americans mixed the tiny seeds of several members of this family with other seeds and grains for bread meal and gruel. To prevent or reduce sunburn, Zuni Indians applied a water mixture of ground western wallflower plants to the skin. Watercress is widely known as a salad plant and has had many medicinal uses ascribed to it. During the 1st century A.D., for example, Pliny listed more than 40 medicinal uses. Native Americans of the west coast of the United States treated liver ailments with a diet consisting exclusively of large quantities of watercress for breakfast, abstinence from any further food until noon, and then resumption of an alcohol-free but otherwise normal diet for the remainder of the day. This was repeated until the disease, if curable, disappeared. Dyer's woad, a European plant that has become naturalized and established in parts of North America, is the source of a blue dye that was used for body markings by the ancient Anglo-Saxons. The seeds of other members of the family produce camelina and canola oils. Camelina oil has been used in soaps and was once used as an illuminant for lamps. Canola oil (obtained from rape seed) is a source of heart-healthy (HDL) fats and is widely used in cooking.

The Rose Family (Rosaceae)

The Rose Family includes more than 3,000 species of trees, shrubs, and herbs distributed throughout much of the world. The flowers characteristically have the basal parts fused into a cup, with petals, sepals, and numerous stamens attached to the cup's rim (Fig. 24.8). The family is divided into subfamilies on the basis of flower structure and fruits. The flowers of one group have inferior ovaries and produce pomes for fruits. Flowers of other groups have ovaries that are superior or partly inferior and produce follicles, achenes, drupes, or clusters of drupelets.

The economic impact of members of the Rose Family is enormous, with large tonnages of stone fruits (e.g., cherries, apricots, peaches, plums), pome fruits (e.g., apples, pears), and aggregate fruits (e.g., strawberries, blackberries, loganberries, raspberries) being grown annually in temperate regions of the world (Fig. 24.9).

Members of this family have been and continue to be relevant to humans in many other ways. Roses, for example, have for centuries been favorite ornamentals for gardeners and are especially prized for their fragrance. In Bulgaria and neighboring countries, a major perfume industry has grown up around the production from damask roses of a perfume oil known as attar (or otto) of roses. The oil is often blended with less expensive substances in the perfume industry. Perfume workers are rarely reported to develop respiratory disorders, suggesting that the plant extracts may have medicinal properties.

Figure 24.9 A raspberry. © *Kingsley Stern*

Figure 24.10 Mature rose hips. © *Kingsley Stern*

The fruits of wild roses, called *hips* (Fig. 24.10), are exceptionally rich in vitamin C. In fact, they may contain as much as 60 times the vitamin C of a comparable quantity of citrus fruit. Native Americans from coast to coast included rose hips in their diet (except for members of a British Columbia tribe, who believed they gave one an "itchy seat"), and it is believed that this practice contributed to scurvy being unknown among Native Americans. During World War II when food supplies became scarce in some European countries, children in particular were kept healthy on diets that included wild rose hips. In addition to vitamin C, the hips contain significant amounts of iron, calcium, and phosphorus. Today many Europeans eat *Nyppon Sopa,* a sweet, thick purèe of rose hips, whenever they have a cold or influenza.

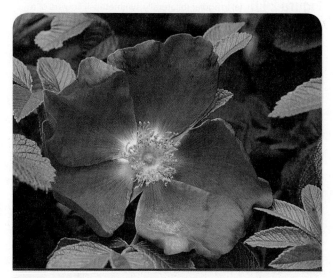

Figure 24.8 A Sitka rose. © *Kingsley Stern*

Figure 24.11 (a) Irregular (bilaterally symmetrical) flowers of a coral tree. (b) Parts of a sweet pea flower. (c) An inflorescence of a powderpuff "flower" consisting of regular, radially symmetrical legume flowers. *(a)* © Kingsley Stern; *(c)* © Kingsley Stern

After giving birth, the women of one western Native American tribe drank western black chokecherry juice to stanch the bleeding. Other tribes frequently made a tea from blackberry roots to control diarrhea. Five hundred Oneida Native Americans once cured themselves of dysentery with blackberry root tea, while many nearby white settlers, who refused to use "Indian cures," died from the disease. Men of certain tribes used older canes of roses for arrow shafts (presumably after removing the prickles!). Wild blackberries, raspberries, salmonberries, thimbleberries, dewberries, juneberries, and strawberries all provided food for Native Americans and early settlers, and they are still eaten today, either fresh or in pies, jams, and jellies. A spiced blackberry cordial is still a favorite for "summer complaints" (diarrhea) in southern Louisiana. Wild strawberries are considered by many to be superior in flavor to cultivated varieties.

The Legume Family (Fabaceae)

The Legume Family, originally referred to as the *Leguminosae*, is the third largest of the approximately 300 families of flowering plants, with only the Sunflower and Orchid Families having more species. The 13,000 family members of the Fabaceae are cosmopolitan in distribution and include many important plants. The flowers range in symmetry from radial (regular) to bilateral (irregular). The irregular flowers have a boat-shaped *keel* composed of two fused petals that enclose the pistil, two *wing petals,* and a larger *banner* petal (Fig. 24.11). The stamens in such flowers are generally fused in the form of a tube around the ovary. The legume fruit is the common feature shared by all members of the family (see Fig. 8.13).

Important crop plants include peas, many kinds of beans (e.g., kidney, lima, garbanzo, broad, mung, tepary), soybeans, lentils, peanuts, alfalfa, sweet clover, jicama, licorice, and wattle. Wattle is an Australian tree that is grown commercially as a source of tannins for leather tanning. Carob, another member of this family, is widely used as a chocolate substitute. Several copals (hard resins used in varnishes and lacquers) are obtained from certain legume plants, as are gum arabic and gum tragacanth used in mucilages, pastes, paints, and cloth printing. The hard wood of several tropical leguminous trees (e.g., rosewood) is prized for furniture.

Important dyes, such as indigo, logwood (used in staining tissues for microscope slides and now scarce), and woadwaxen (a yellow dye), come from legume plants. Locoweeds, belonging to *Astragalus*—a genus of about

1,600 species—have killed many horses, cattle, and sheep, particularly in the southwestern United States. The poisonous compound in those species affecting livestock seems to vary in concentration according to the soil type in which the plants are growing. Other poisonous legumes include lupines, jequirity beans, black locusts, and mescal beans.

About 90% of the members of the Legume Family exhibit leaf movements, but few are as rapid as those of the sensitive plant (*Mimosa pudica*), whose leaves fold within seconds in response to a disturbance. Sensitive plants grow as weeds in the tropics and the Deep South of the United States; they are discussed in Chapter 11. Many of the movements of other legume plant leaves are correlated primarily with day length.

Clovers have been widely used in the past by gatherers of wild food plants. The leaves are difficult to digest in quantity, but the rhizomes were gathered and usually roasted or steamed in salt water, then dipped in grease before being eaten. The seeds of both clovers and vetches also were gathered and either ground for meal or cooked in a little water and eaten as a vegetable. Today, seeds of several legumes, including alfalfa and mung beans, are popular for their sprouts, which are widely used in salads and Asian dishes. A tropical bean called *winged bean* (Fig. 24.12) has unusually high levels of protein, and all parts of the plant are edible. It is presently being grown in several widely scattered tropical and subtropical regions and is being marketed on a limited scale in some temperate zones. It is believed to have great potential for improving the diet of undernourished peoples throughout the tropics.

The Spurge Family (Euphorbiaceae)

Although many of the members of the Spurge Family are tropical, they are widespread in temperate regions both north and south of the equator. The stamens and pistils are produced in separate flowers that often lack a corolla and are inconspicuous. In true spurges (*Euphorbia*), the female flower is elevated on a stalk called a *gynophore* and is surrounded by several male flowers that each consist of little more than an anther. Both the female and the male flowers are inserted on a cup composed of fused bracts, the cup usually having distinctive glands on the rim. This type of inflorescence is called a *cyathium* (Fig. 24.13). Sometimes, the inconspicuous flowers are surrounded by brightly colored bracts (e.g., poinsettia) that give the inflorescence the appearance of a single, large flower. Most members of this large family produce a milky latex, and a number of species are poisonous.

Sooner or later, many gardeners experience "urges to purge spurges," as some members of the family (e.g., spotted spurge) are exceptionally aggressive weeds that reproduce very rapidly. Other large tropical spurges closely resemble columnar cacti.

Several economically important plants are cultivated, particularly in frost-free regions. For example, an estimated 278 million metric tons (306 million tons) of cassava are harvested annually from plants cultivated in South America, Africa, and eastern Asia. The roots develop thickened storage areas that resemble large sweet potatoes and are a diet staple of the tropics, much as white potatoes and cereals are of temperate areas. Poisonous compounds are removed by boiling, fermenting, or squeezing out the juice. In dried and powdered form, cassava is known as farina. In Western countries, tapioca is prepared by forcing heated cassava pellets through a mesh while it is being agitated. Cassava starch is also used as a base for the production of alcohol, acetone, and other industrial chemicals.

Another cultivated spurge of the tropics is the Pará rubber tree, the source of the crude rubber from which most rubber products are made today. Although wild South American trees were the original commercial source of rubber, rubber trees have been widely planted in Indonesia, Africa, and adjacent areas. The trees vary in height from less than 5 to over 50 meters (16 to 164 feet) and produce most of the latex in the inner bark. The laticifers in which latex is secreted spiral around inside the trunk at an angle of about 30 degrees. So, cuts are made at the same angle into the inner bark to obtain maximum yields of latex, which trickles down into collecting cups that are attached to the tree.

Figure 24.12 A winged bean. Winged beans are highly nutritious, are easy to grow in the tropics, and hold promise for improving diets in developing countries. © *Kingsley Stern*

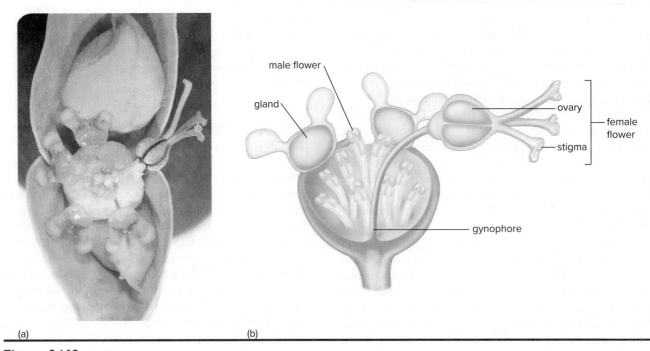

Figure 24.13 A spurge. (a) Inflorescence. (b) An individual cyathium. *(a) © Kingsley Stern*

After collection, the latex is coagulated by chemicals or smoke and then shipped in sheet or crumbled form to processing plants. Sometimes, an anticoagulant is mixed with it, and the liquid is transferred in tankers. Much of the world's rubber is used to manufacture car and airplane tires, but other products made from rubber are numerous.

The Pará rubber tree should not be confused with a broad-leaved ornamental known as the rubber plant. The rubber plant is popular as a houseplant and produces latex; it is, however, a member of the Fig Family (Moraceae).

The latex of other spurges may hold a key to future sources of fuel and lubricating oils. In 1976, Melvin Calvin, a University of California Nobel Prize winner, proposed the use of latex of gopher plants as a source of materials for oil. He estimated that such plants, which can grow in semidesert areas, would produce 10 to 50 barrels of oil per year on 0.4 hectare (1 acre) of land at a cost of $3 to $10 per barrel.[1]

A spurge called *candelilla* occurs in remote areas of Mexico. It produces a wax on its stems that is used in the making of candles and other wax products. Still another spurge produces seeds with a special oil used in plasticizers, and castor oil (from castor beans) is used to manufacture nylon, plastics, and soaps. Castor beans themselves are very poisonous—as few as one to three are sufficient to kill a child. The plants grow very rapidly and are popular as ornamentals despite their being one of the leading natural causes of poisoning among American children.

A Mexican jumping bean is the seed of a certain spurge in which a small moth has laid an egg. When the egg hatches, the grub periodically changes position with a jerk, causing the seed to jump. Crown-of-thorns is an ornamental plant with somewhat flexible twisting stems bearing vicious-looking spines. Some believe it to have been the plant from which the crown of thorns for the head of Christ was made. Poinsettias, or Christmas flowers, are favorite yuletide plants in various parts of the world. Tung oil, used in oil paints and varnishes, and Chinese vegetable tallow, a substance used in the manufacture of soap and candles, are two more commercially important products obtained from the seeds of cultivated members of the Spurge Family.

The Cactus Family (Cactaceae)

Cacti are native only to the Americas but include many highly regarded ornamentals that have been exported around the world. The flowers are usually showy (Fig. 24.14), with numerous stamens, petals, and sepals. The sepals are often colored like the petals, and the inferior ovary develops into a berry. There are about 2,000 species, most occurring in drier subtropical regions. The leaves of many are reduced in size or more often in the form of spines, with the fleshy, flattened or cylindrical, often fluted

---

1. Jojoba, a member of the Box Family (Buxaceae), is a desert shrub with an acorn-sized capsule containing a large, oily seed with about 50% liquid wax content. This high-grade wax and another found in the seeds of meadowfoams, which are members of the Meadowfoam Family (Limnanthaceae), are the only known natural substitutes for sperm whale oil, a vital ingredient in engine lubricants. The importing of sperm whale oil into the United States was banned in 1970 to protect the nearly extinct large ocean mammals, and an expensive synthetic substitute is being used. Experiments and research are in progress to find improved strains of jojoba and to test the feasibility of its being grown on a large scale.

Most cacti have edible fruits, and only three cacti (peyote, living rock, and hedgehog cactus) are known to be poisonous. Prickly pear fruits are occasionally sold in American grocery stores and taste somewhat like pears. Prickly pear fruits also have seeds that Native Americans of the Southwest dried and ground for flour that they mixed with water and used as an *atole* (thickened, souplike) staple food. Syrup can be produced by boiling the fruits of prickly pears and those of the giant saguaros. In the past, cactus candy was made by partly drying strips of barrel cactus and boiling them in saguaro fruit syrup, but the cactus is now usually boiled in cane sugar syrup.

Native Americans of the Southwest scooped out barrel cacti, dried them, and used them for pots. They also mixed the sticky juice of prickly pear cacti in the mortar used to build their adobe huts. In Texas, a poultice of prickly pear stem was applied to spider bites. Hopi Native Americans chewed raw cholla cactus as a treatment for diarrhea, and the skeletons of these cacti were used for flower arrangements.

In the middle of the 19th century, Australians planted a few imported prickly pear cacti in the dry interior. These cacti found no natural enemies in their new environment and multiplied rapidly, infesting more than 24 million hectares (60 million acres) within 75 years. In 1925, in an effort to control them, Australia introduced an Argentine moth among the cacti. The moth's caterpillars feed on prickly pear cacti and have gradually brought the plants under control, making the land usable again.

Another cactus parasite, the cochineal insect (related to the mealybug, a common houseplant pest), feeds on prickly pear cacti in Mexico. At one time, the insects were collected for a crimson dye they produce; the dye was used in lipstick and rouge before aniline dyes derived from coal tar were introduced in the 1930s.

Peyote cacti are small, buttonlike plants that have no spines, with roots resembling those of carrots. They contain several drugs, the best known of which is *mescaline*, a powerful hallucinogen. Dried slices of peyote have been used in native religious ceremonies in Mexico for centuries and more recently by at least 30 additional tribes of Native Americans.

(a)

(b)

Figure 24.14 Flowers of two different cacti. *(a-b)* © Kingsley Stern

stems carrying on photosynthesis (Fig. 24.15). Many cacti can tolerate high temperatures, and some can withstand up to several years without moisture. They vary in size from pinheadlike forms to the giant saguaro that can attain heights of 15 meters (50 feet) and weigh more than 4.5 metric tons (5 tons). Cacti are generally exceptionally slow-growing and, because they need so little care, make good houseplants for sunny windows.

In 1944, a marine pilot was forced to bail out of his aircraft over the desert near Yuma, Arizona. Until he was rescued 5 days later, he survived the intense heat and low humidity of the area by chewing the juicy pulp of barrel cacti in the vicinity. Since then, the use of cacti for emergency fluids and food has been recommended in most survival manuals.

The Mint Family (Lamiaceae)

The 3,000 members of the Mint Family are relatively easy to distinguish because they have a unique combination of angular stems that are square in cross section, opposite leaves, and bilaterally symmetrical (irregular) flowers (Fig. 24.16). Most also produce aromatic oils in the leaves and stems. The superior ovary is four-parted, with each of the four divisions developing into a nutlet. Included in the family are such well-known plants as rosemary, thyme, sage (not to be confused with sagebrush of the Sunflower Family), oregano, marjoram, basil, lavender, catnip, peppermint, and spearmint.

Mint oils can be distilled at home with ordinary canning equipment. Whole plants (or at least the leaves) are loosely

Figure 24.15 Cacti. (a) A barrel cactus. (b) Prickly pear cacti. (c) Peyote. (d) An organ-pipe cactus. *(a-d) © Kingsley Stern*

packed to a depth of about 10 centimeters (4 inches) or more in the bottom of a large canning pot. Then a wire rack or other support is also put in the pot, and a bowl is placed in the middle on the rack. Enough water is added to cover the vegetation, the pot is placed on a range, and the lid is inverted over it. The water is brought to boil, and as it does so, ice is placed on the inverted lid. The oils vaporize and condense when they contact the cold lid, dripping from the low point into the bowl (Fig. 24.17). Of course, some moisture also condenses, but since the oil is lighter, it, floats on top. Peppermint oil is easy to collect this way and will keep for a year or two in a refrigerator.

Mint oils have been used medicinally and as an antiseptic in different parts of the world. Mohegan Native Americans used catnip tea for colds, and dairy farmers in parts of the midwestern United States used local mint oils

A leaf extract is still used in horehound candy and cough medicines. In England, it is a basic ingredient of horehound beer. Vinegar weed, also known as blue curls, is a common fall-flowering plant of western North America. Native Americans used it in cold remedies, for the relief of toothaches, and in a bath for the treatment of smallpox. It was also used to stupefy fish.

Menthol, the most abundant ingredient of peppermint oil (Fig. 24.18), is widely used today in toothpaste, candies, gum, liqueurs, and cigarettes. Most American mint is grown commercially in the Columbia River basin of Oregon and Washington. Geese are sometimes used in mint fields to control both insects and weeds, because they do not interfere with the growth of the mint plants themselves.

Ornamental mints include salvias and the popular variegated-leaf *Coleus* plants, neither of which has typical mint oils in the foliage. *Chia* (Fig. 24.19), another relatively odorless mint, is confined to the drier areas of western North America. Native Americans parched chia

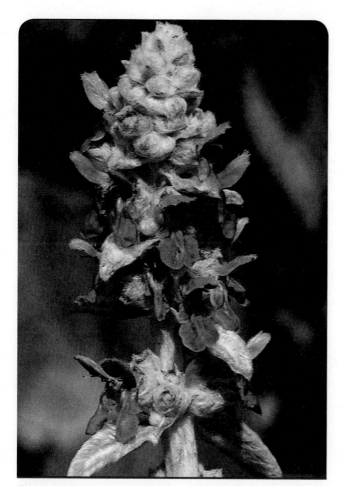

Figure 24.16 Flowers of lamb's ear mint. © *Kingsley Stern*

Figure 24.17 A simple apparatus for distilling mint oil at home. See the text for an explanation of the process.

to wash their milking equipment before antibiotics became popular for this purpose. As a result, mastitis, a common disease of dairy cattle, was seldom encountered in their herds. Horehound, a common mint weed of Europe, has become naturalized on other continents and is cultivated in France.

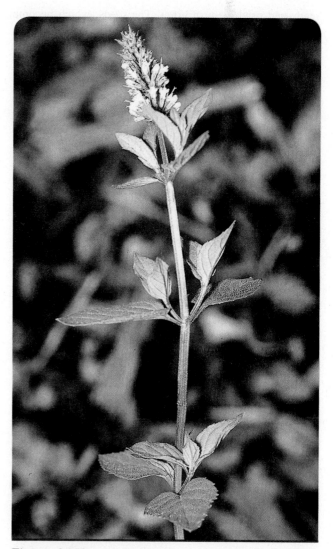

Figure 24.18 A peppermint plant in flower. Oil from the leaves is a source of menthol. © *Kingsley Stern*

Figure 24.19 A flowering head of chia. © Kingsley Stern

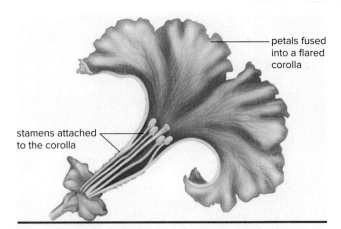

Figure 24.20 A sectioned petunia flower.

seeds and used them in gruel. The seeds, which become mucilaginous when wet, were also ground into a paste that was placed in the eye to aid in the removal of dirt particles. The paste was also used as a poultice for gunshot wounds, and Spanish Californians made a refreshing drink from ground chia seeds, lemon juice, and sugar. Chia seeds reportedly contain an unidentified substance that has effects similar to those of caffeine. Over 100 years ago, one physician reported that a tablespoon of chia seeds was sufficient to sustain a man on a 24-hour endurance hike. Since that time, backpackers have experimented with the seeds, and results tend to support that claim. A thorough scientific investigation of the matter is needed. Chia seeds are sold commercially for making into a paste that is spread on clay models and then watered. The seeds resemble green hair when they sprout.

The Nightshade Family (Solanaceae)

The Nightshade Family is concentrated in the tropics of Central and South America. The flowers have fused petals, with the stamen filaments fused to the corolla so that they appear to be arising from it (Fig. 24.20). The superior ovary develops into a berry or a capsule. The more than 3,000 species of the family have alternate leaves and occur as herbs, shrubs, trees, or vines. Well-known representatives include tomato, white potato, eggplant, pepper, tobacco, and petunia.

Many nightshades produce poisonous drugs, some of which have medicinal uses. One of the best-known medicinal drug producers is the deadly nightshade of Europe. A drug complex called *belladonna* is extracted from its leaves. Belladonna was used in the "magic potions" of the past and for dilating human pupils for cosmetic purposes. It is now the source of several widely used drugs, including atropine, scopolamine, and hyoscyamine. Atropine is used in shock treatment, for relief of pain, to dilate eyes, and to counteract muscle spasms. Scopolamine is used as a tranquilizer, and hyoscyamine has effects similar to those of atropine. Capsicum, obtained from a pepper, is used as a gastric stimulant and is a principal ingredient of mace, which is used to repel human or animal assailants. Capsaicin, derived from peppers, is used in ointments for the relief of arthritic and neuropathic pain.

Jimson weed (see Fig. 8.6) is also a source of medicinal drugs that have been used to treat asthma and other ailments. The drugs can be fatal if ingested in high quantities but were used in controlled amounts in Native American rituals of the past. Records indicate that users became temporarily insane and had no recollection of their activities when the effects of the drug wore off. The drug solanine is present in most, if not all, members of the family. Many arthritis sufferers apparently are sensitive to solanine, and a number of arthritics have reported partial relief through total abstinence from members of this family (including potatoes, tomatoes, peppers, and eggplant). In its dried form, tobacco contains 1% to 3% of the drug nicotine. Nicotine is used in certain insecticides, and it is used for killing intestinal worms in farm livestock. It is, however, an addictive drug. The evidence of human tobacco use as a primary cause of heart and respiratory diseases, including lung cancer and other cancers such as those of the mouth and throat, is indisputable. The only "benefit" it may have to humans appears to be as a killer of leeches. It is said that leeches attaching themselves to heavy smokers will drop off dead within 5 minutes from nicotine poisoning—a very dubious justification for continued human use!

Tomatoes are among the most popular of all "vegetables." The plants are day-neutral (day lengths are discussed in Chapter 11), and even though they require warm night temperatures (16°C, or 60°F) to set fruit well, they are easily cultivated in greenhouses when natural conditions are unfavorable. Most commercially grown American tomatoes are processed into juice, tomato paste, and catsup. In Italy, a small amount of edible oil is extracted from the seeds after

Figure 24.21 Harvesting a field with a mechanical tomato harvester. © *Kingsley Stern*

the pulp has been removed. Most American tomatoes are grown in California and Florida, where they are harvested with machinery developed during the 1960s when inexpensive labor became unavailable (Fig. 24.21).

The white potato is the third most important food crop in the world. The leading producers are China, India, and the Russian Federation. It is believed that white potatoes from Chile were sent to Europe by Spanish invaders of South America in the 16th century. In the 1840s, late blight infested and destroyed the potato crop of Ireland, causing severe famine. Irish settlers subsequently emigrated to the United States, Canada, Australia, and other parts of the world.

When potato tubers are exposed to light, they turn green at the surface. Poisonous drugs (e.g., solanine) are produced in the green areas. They can cause intestinal distress in animals and humans and should not be eaten.

The Carrot Family (Apiaceae)

Members of the Carrot Family are widely distributed in the Northern Hemisphere, and many have savory-aromatic foliage. The flowers tend to be small and numerous and are arranged in umbels. The ovary is inferior, and the stigma is two-lobed. The leaves are generally dissected, and the bases of their petioles usually form sheaths around the stem. Included in the 2,000 members of the family are dill, celery, carrot, parsley, caraway, coriander, fennel, anise, and parsnip.

Anise is one of the earliest aromatics mentioned in literature. It is used for flavoring cakes, curries, pastries, and candy. Pocket gophers apparently are attracted by its aroma, and some poison baits are enhanced with anise. A liqueur known as *anisette* is flavored with it.

Caraway seeds are used to flavor a Danish liqueur, called *kümmel*. The seeds are well known for their use in rye and pumpernickel breads.

Some members of the Carrot Family are poisonous. Water hemlock (Fig. 24.22) and poison hemlock are common weeds in ditches and along streams. All parts of the plants are deadly, often having been fatal to unwary wild food lovers. Socrates is believed to have died as a result of ingesting poison hemlock, which should not be confused with cone-bearing hemlock trees.

Several members of the Carrot Family, such as cow parsnip, squawroot, and hog fennel, have edible roots and were used for food by Native Americans. Be advised, however, to be absolutely certain of the identity of such plants before experimenting with them.

The Pumpkin Family (Cucurbitaceae)

Although most species in the Pumpkin Family are tropical or subtropical, many occur in temperate areas of both the Northern and the Southern Hemispheres. Plants are

Figure 24.22 A water hemlock in flower. © *Kingsley Stern*

(a)

(b)

(c)

Figure 24.23 Fruits and items associated with members of the Pumpkin Family (Cucurbitaceae). (a) A Hawaiian ceremonial gourd with feathers attached and gourds used in South America for drinking maté. A metal straw that strains out the maté leaves is resting in one gourd. (b) Cantaloupes. (c) A luffa (vegetable sponge). *(a-c) © Kingsley Stern*

prostrate or climbing herbaceous vines with tendrils. The flowers have fused petals, and female flowers have an inferior ovary with three carpels. All are unisexual. Some species have both male and female flowers on the same plant, while others have only male or only female flowers on one plant. In male flowers, the stamens cohere to varying degrees, depending on the species. The family has over 900 members, several of which have many horticultural varieties.

This family includes many important edible plants, and some have been cultivated for so long that they are unknown in the wild state. Well-known members of the family include pumpkins, squashes, cucumbers, cantaloupes (Fig. 24.23b), and watermelons.

The vegetable sponge (Fig. 24.23c), when it is growing, resembles a large cucumber and has a highly netted, fibrous skeleton that can serve as a bath sponge after the soft tissues have been removed.

Gourds found in Mexican caves and subjected to radiocarbon dating have proved to be many thousands of years old. Various types of gourds (Fig. 24.23a), serving many purposes, are still grown today. Some are scooped out and used for carrying liquids or for storing food, particularly grains. South Americans drink maté (a tea) from gourds, which are also used for several types of musical instruments. In parts of Africa, gourds are used to catch monkeys. A type of gourd with a narrow neck is scooped out and partly filled with corn or other grains. One end of a rope is then tied to the gourd and the other to a stake driven into the ground. When a monkey tries to grab a fistful of grain, it finds that the neck will not allow its bulging hand to be removed. Most do not realize that letting go of the grain would allow them to escape, and they stubbornly hang on until they are captured.

Melonette, a small cucumber-like vine of the southeastern United States, has seeds that can be *purgative* (drastically laxative). Other cucumber-like plants of the western states, manroots, produce huge water-storage roots, some weighing as much as 90 kilograms (200 pounds). These roots were crushed by Native Americans and thrown into dammed streams to stupefy fish. An oil from the seeds was applied to the scalp as a remedy for infections that caused hair loss.

The Sunflower Family (Asteraceae)

The Sunflower Family is the second largest of the flowering plant families in terms of number of species (over 23,000). The individual flowers are called **florets.** They are usually tiny and numerous but are arranged in a compact inflorescence, so that they resemble a single flower. A sunflower or daisy, for example, consists of dozens if not hundreds of tiny flowers crowded together, with those around the margin having greatly developed corollas that extend out like straps, forming what appear to be the "petals" of the inflorescence (Fig. 24.24 gives details of a sunflower inflorescence). In dandelions, all the individual florets of the inflorescence have narrow, straplike extensions.

Well-known members of this family include lettuce, endive, chicory, Jerusalem artichoke, globe artichoke, dahlia, chrysanthemum, marigold, sunflower, and thistle.

Santonin, obtained from flower buds of a relative of sagebrush that is native to the Middle East, is used as an intestinal worm remedy. Tarragon, used as a spice in meat dishes and pickles, comes from another relative of sagebrush. Pyrethrum is a natural insecticide obtained from certain chrysanthemum flowers. Fructose, a sugar, is obtained from the tubers of Jerusalem artichokes and dahlias. Dahlias are also renowned for their huge, showy flowers, while Jerusalem artichokes are often eaten as a vegetable.

Marigolds are favorite plants of organic gardeners. Their roots are said to release a substance that repels nematodes, and the odor of the leaves repels white flies and other insects. Unfortunately, snails and slugs seem to be immune and voraciously consume the foliage.

Many members of this family were used widely by Native Americans. The dry fruits of balsamroot and mule ears were used for food. Balsamroot plants were eaten raw or cooked, and in the West, extracts of both mule ears and tarweeds were used to treat poison oak inflammations. Salsify and thistle roots were also used for food.

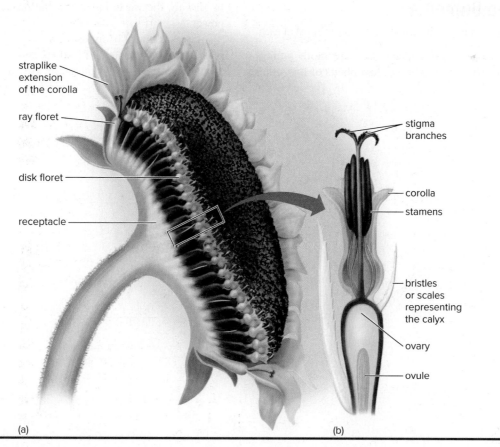

Figure 24.24 Parts of a sunflower. (a) A section through an inflorescence. (b) A section through a single floret.

Young leaves and roots of dandelions have been eaten for centuries, and the flowers have been used to make wine. Roasted dandelion and chicory roots have been used as a coffee substitute. During World War II, chicory was grown as a crop specifically for use as a coffee adulterant. Even today, coffee is flavored with chicory as a New Orleans specialty.

The dried and boiled leaves of American yarrow are said to make a nourishing broth. European yarrow has become naturalized in North America. Its rhizomes contain an anesthetic that, when chewed, numbs the tongue and gums, and it has been used to ease the discomfort of teething toddlers; the substance has also been used in suppression of menstruation. This plant is believed to have been used by Achilles in treating the wounds of his soldiers.

Sunflowers themselves were widely used by Native Americans for their dry fruits, which were ground into a meal for bread. They are grown commercially today primarily for the edible oil extracted from the seeds (after removal from the husks), but the seeds are increasingly being eaten by modern Americans.

24.5 Monocots

The Grass Family (Poaceae)

Learning Outcome

a. List three monocot families and provide an example of each.

Individual plants of the grass family are more numerous and more widely distributed than plants of any other flowering plant family. There are, however, possibly only one-fourth as many species of grasses as there are members of the Sunflower Family. The flowers of grasses are typically wind-pollinated. As such, they are highly specialized in structure and have a terminology all their own (Fig. 24.25). The calyx and corolla are represented by tiny, inconspicuous scales, and the flowers are protected by boat-shaped bracts. The stigmas, when they are exposed, are feathery, and the leaves sheathe the stem at their bases.

Nearly all of the cereals, including wheat, barley, rye, oats, rice, and corn, belong in this family. Indeed, civilization as we know it would be vastly different without them. More than 2.8 billion metric tons (3.1 billion tons) of cereals, feeding more than half of the world's population, are harvested each year, primarily in Asia, North America, and Europe.

Sugarcane (Fig. 24.26), from which 80% of the world's sugar supply is extracted annually, is grown at lower elevations throughout humid tropical areas. It is a large grass, often growing to heights of 6 meters (20 feet). After the cane is harvested, raw liquid sugar is squeezed out by a milling process. The juice then is centrifuged, with the solids being crystalized into table sugar. The dark remnant (molasses) may be used to produce rum or alcohol. The solid waste is sometimes made into paper or particle board and has been converted to gasoline in South Africa. In Hawaii, the waste is used in the production of electric power.

Grasses have been used widely by primitive peoples for making mats and baskets and for thatching huts. Some varieties of sorghum are grown for their fibers,

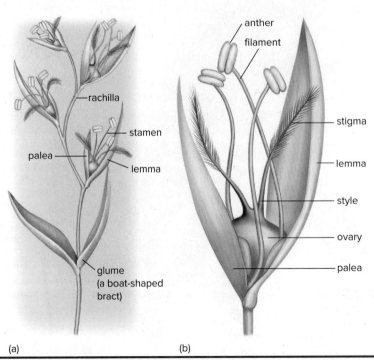

Figure 24.25 Grass flowers. (a) An expanded grass spikelet. (b) An enlargement of a single flower (floret).

Figure 24.26 Sugarcane plants. © *Kingsley Stern*

Figure 24.27 A tiger lily flower. © *Kingsley Stern*

which are made into brooms, although natural broom fiber now has been replaced to a considerable extent by synthetic materials. Some varieties are grown for the seeds, which are processed into cereal flours. Others are a source of silage and a carnauba-like wax. Citronella oil, a common ingredient of mosquito repellents, is obtained from a grass grown in Indonesia. It is now used in cheaper soaps, cosmetics, and perfumes. Related grasses are the source of lemon grass oil, used for the same purposes as citronella oil.

Juice squeezed from fresh, young grass mowings has a high protein content and is being investigated as a source of protein for future human consumption.

The Lily Family (Liliaceae)

In recent classifications, the traditional members of this family have been augmented by members of several other families now included within the Liliaceae. Lily family members are particularly abundant in the tropics and subtropics, but they occur in almost any area that supports vegetation. The flowers are often large, and their parts are all in multiples of three, with the sepals and petals frequently resembling each other in color and form (Fig. 24.27).

In addition to many types of lilies and daffodils used widely as ornamentals, the family includes asparagus, sarsaparilla, squill, meadow saffron, bowstring hemp, and *Aloe*. Sarsaparilla was at one time widely used for flavoring soft drinks and medicines and is obtained from the roots of a genus of woody vines (*Smilax*) whose stems are often covered with prickles. The bulbs of squills are the source of a rodent poison and of a drug used as a heart stimulant.

Meadow saffron is the source of *colchicine,* a drug once used to treat rheumatism and gout but now much more widely used in experimental agriculture to interfere with spindle formation in cells so that the chromosome number of plants may be artificially increased. This increasing of the chromosome number can result in larger and more vigorous varieties of plants. (Meadow saffron should not be confused with true saffron, a member of the Iris Family and the source of the world's most expensive spice and a powerful yellow dye.)

Bowstring hemps are related to the familiar, seemingly indestructible houseplants called *sansevierias* (Fig. 24.28), which have long, narrow, stiff, upright leaves. The plants are cultivated in tropical Africa for their long fibers, which are used for string, rope, bowstrings, mats, and cloth. New Zealand flax, a larger plant, is grown in South America and New Zealand for similar purposes but is also widely used in ornamental plantings.

Figure 24.28 *Sansevieria* plants. © Kingsley Stern

Figure 24.29 An uprooted California soaproot plant. © Kingsley Stern

Several *Aloe* species produce juices used in shampoos, cosmetics, and sunburn lotions and in the treatment of burns. African *Aloe* species are prized as ornamentals in areas with milder climates. Their thick, fleshy leaves have short spines along the margins.

Many lily bulbs (e.g., onions and garlic) are edible, and wild lily bulbs were used for food extensively by Native Americans. Wild bulbs should no longer be eaten, however, as doing so may endanger the survival of native species. *(Caution: Lily bulbs should not be confused with those of daffodils and their relatives. Daffodil and related bulbs are highly poisonous.)*

The California soaproot (Fig. 24.29) is confined to California and southern Oregon and was important for food to Native Americans of the region. It has also had several other uses. The large bulbs are covered with coarse fibers that were removed and tied to sticks to make small brooms. The bulbs themselves produce a lather in water and were used for soap. Sometimes, numbers of bulbs were crushed and thrown in a small stream that had been dammed. Fish would be stupefied and float to the surface. The bulbs were generally eaten after being roasted in a stone-lined pit in which a fire had been made. While they were roasting, a sticky juice would ooze out. This was used for gluing feathers to arrow shafts.

A resin used in stains and varnishes exudes from the stem of dragon's blood plants. Grass trees of Australia yield resins used in sealing waxes and varnishes.

The Orchid Family (Orchidaceae)

This very large family has, according to some authorities, more than 35,000 species that are widely distributed. Like members of the Lily Family, they are especially abundant in the tropics. In many genera, the number of individual plants at any one location may be quite small—sometimes limited to a single plant.

The flowers are exceptionally varied in size and form, and the habitats of the plants are equally diverse. The flowers of one Venezuelan species have a diameter of less than 1 millimeter (1/25 of an inch), while those of a species native to Madagascar may be more than 45 centimeters (about 18 inches) long. One species of *Dendrobium* orchid from Java has flowers that are so delicate they wither within 5 or 6 minutes after opening. Many orchids are epiphytic on the bark of trees. During its 5-month

KEY THEME 24.1 ecology

Wild Rice—More Than Just Food

Anyone who travels in the northern Great Lakes region is likely to encounter wild rice for sale in local markets. Wild rice (*Zizania palustris*) is not directly related to Asian rice (*Oryza sativa*), although both are in the Family Poaceae, Oryzeae tribe, of flowering plants. Wild rice grows in shallow water in the abundant lakes and rivers scattered throughout northern Wisconsin and Minnesota and along the south shore of Lake Superior. It is harvested by canoeing through a patch of plants in the fall, bending the ripe grain heads with wooden sticks, and brushing them into the canoe (**Box Figure 24.1A**). Some seeds fall to the bottom of the lake and germinate to provide the next year's crop. Several Native American cultures, such as the Ojibwe (also Ojibwa or Ojibway), consider wild rice to be sacred. Wild rice is so central to the Ojibwe that it is part of the Nation's migration story. In the earliest teachings of the tribe's history, a prophet told the Anishinaabe people in the eastern United States to travel west until they returned to the place where food grows on water. When the people reached the Great Lakes region, they discovered wild rice and, due to its abundance, storability, and nutritional quality, quickly adopted it as a major source of sustenance. Wild rice became the cultural centerpiece for the Ojibwe Nation. The wild rice harvest continues to be an important cultural activity today.

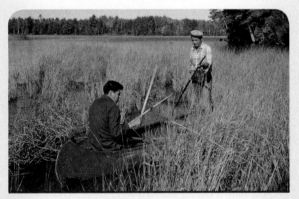

Box Figure 24.1A Wild rice harvest.
© Volkmar K. Wentzel/Contributor/Getty Images

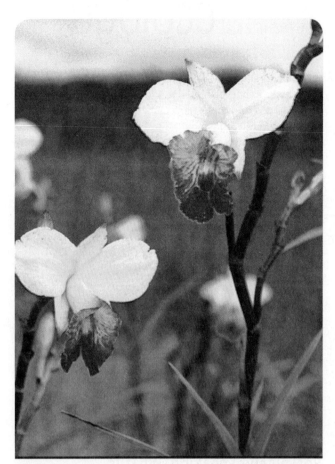

Figure 24.30 Bamboo orchids growing wild in Hawaii.
© Kingsley Stern

Orchids have three sepals and three petals, with one of the petals (the *lip petal*) differing in form from the other two (Fig. 24.30). The stamens and pistil are united in a unique, single structure, the *column*. The stigma usually consists of a sticky depression on the column. The anthers contain sacs of pollen called *pollinia* and are covered with a cap until they are removed by an insect or other pollinator. The specific adaptations between orchid flowers and their pollinators are extraordinary and sometimes bizarre.

Orchids have tiny seeds that are often produced in large numbers (e.g., a single fruit of certain orchid species may contain up to 1 million seeds). Each seed consists of only a few cells, and in order for a seed to germinate, it must become associated with a specific mycorrhizal fungus that produces substances necessary for its development. Once a seed has germinated, it may take from 6 to 12 or more years before the first flower appears.

Contrary to popular belief, some orchids can be grown relatively easily on a windowsill that has bright light, but not direct sunlight. Because orchids are among the most beautiful and prized of flowers, a large industry has grown up around their culture and propagation, discussed in the section on tissue culture in Chapter 14. One species, the vanilla orchid, is grown commercially in the tropics for its fruits, which are the source of true vanilla flavoring.

flowering season, one epiphytic species of Malaysia and the Philippine Islands produces 10,000 flowers on plants that weigh more than 1 metric ton. Others are aquatic or terrestrial, and a saprophytic species native to western Australia grows and flowers entirely under ground.

KEY THEME 24.2 ecology

Coffee and Caffeine

Among all the foods, beverages, and products provided by flowering plants, coffee (*Coffea arabica*) is one that millions of people, particularly Americans, have contact with every day. We take our coffee break seriously. We drink some 400 million cups of coffee daily, which is enough coffee to turn the Rose Bowl Stadium into a deep caffeine pool. Here are some frequently asked questions concerning coffee and caffeine. So grab a cup of coffee, pull up your favorite chair, sit down, and relax while you read about one of the world's favorite beverages.

Where Does Coffee Grow?

Coffee is a small tree or shrub native to the mountains of Ethiopia. Because of its popularity as a drink, coffee was spread by traders to many tropical regions of the world. Today, Brazil and Colombia are the world's leading coffee producers. Coffee is derived from the seeds (beans) of *Coffea arabica* or *Coffea robusta*, two species of 70 or more recognized varieties. The trees require tropical or subtropical climate and need about 60 to 100 inches of rain per year. They grow best at mountain elevations of 3,000 to 6,000 feet where the temperature is cool and relatively stable (68°F) because they are unable to tolerate frost. Trees start producing fruit at around 3 to 5 years of age.

Botanically Speaking, What Is a Coffee "Bean"?

In Costa Rica, the crop season begins with the first "flowering rains" that come in March or April. Flowering buds that were produced at the end of the dry season are now conditioned to flower by the gentle rains. The white, delicate flowers have little aroma and bloom for 3 days before falling off the trees. Each pollinated flower will produce one fruit, which is called a cherry because of its red color. The coffee cherry contains two seeds that are tightly packed against each other, resulting in each seed having a flat side. The coffee "beans" are actually the two seeds of each fruit. Coffee fruits change color from dark green to deep red and are handpicked when ripe (red color). Each tree is usually harvested three or four times because of the uneven maturation of fruits. A single tree can yield up to 5 to 6 pounds of fruit per year.

How Much Caffeine Does a Cup of Coffee Have?

A 7-ounce cup of coffee prepared by the drip method contains 115 to 175 milligrams of caffeine, while instant coffee contains 65 to 100 milligrams. Decaffeinated instant coffee contains 2 to 3 milligrams of caffeine.

How Much Caffeine Is in Soft Drinks?

The following table compares the caffeine content of several brands of soft drinks.

Soft Drink (12-oz can)	Caffeine Content (mg)
Jolt	72.0
Mountain Dew	54.0
Coca-Cola	45.6

SUMMARY

24.1 Introduction to Flowering Plants and Civilization
- Distinguishing between fact and fantasy in reported past uses of plants is often difficult. Teams of specialists are interviewing tribal medicine men and women of the tropics to try to save potentially useful plants from extinction.

24.2 Origin of Cultivated Plants
- Nikolai Vavilov proposed eight major centers of distribution of cultivated plants. More recent studies by Harlan and others have modified the concept of centers of distribution.

24.3 Selected Families of Flowering Plants
- Flowering plant families are surveyed in evolutionary sequence, discussing first those considered primitive.

24.4 Dicots
- The Buttercup Family (Ranunculaceae) includes buttercup, columbine, larkspur, anemone, monkshood, *Clematis*, goldenseal, and wolfsbane.
- The Laurel Family (Lauraceae) includes cinnamon, camphor, sassafras, sweet bay, California bay (or myrtle), avocado, and laurel.
- The Poppy Family (Papaveraceae) includes bloodroot and opium poppy (the source of medicinal drugs and heroin).
- The Mustard Family (Brassicaceae) includes cabbage, cauliflower, brussels sprouts, broccoli, radish, kohlrabi, turnip, horseradish, watercress, rutabaga, shepherd's purse, western wallflower, dyer's woad, and camelina.
- The Rose Family (Rosaceae) includes stone fruits (e.g., cherry, apricot, peach, plum), strawberry, raspberry, rose, and related wild species.
- The Legume Family (Fabaceae) includes pea, bean, lentil, peanut, soybean, alfalfa, clover, licorice, wattle, indigo, logwood, locoweed, jicama, sensitive plant (*Mimosa pudica*), and winged bean.
- The Spurge Family (Euphorbiaceae) includes poinsettia, spurge, cassava, Pará rubber, gopher plant,

Soft Drink (12-oz can)	Caffeine Content (mg)
Dr. Pepper	39.6
Pepsi-Cola	38.4
RC Cola	36.0

Chemically Speaking, What Is Caffeine?

Caffeine is a plant alkaloid found not only in coffee beans but also in tea leaves, cola nuts, and cacao seeds. It works as a stimulant of the central nervous system and the respiratory system, as well as increasing heart rate and constricting blood vessels. As a stimulant, caffeine makes us more aware and alert and able to work without drowsiness. It is reported that coffee was discovered in the Near East (perhaps in Yemen) by goats, which became frisky and sleepless after eating the wild fruit. A shepherd is said to have located the shrub the goats had chewed and described it as "a shrub with leaves like laurel, flowers like jasmine, and little dark berries the chief object of the animals' nibbling."

What Are the Effects of Caffeine on the Body?

Caffeine stimulates or increases heartbeat rate, respiration, basal metabolic rate, and the production of stomach acid and urine. These effects translate into a stimulating "lift" that a person may feel. Generally, one feels less drowsy, less fatigued, and more alert and focused after consuming caffeine.

Where Does the Word "Coffee" Come From?

The word *coffee* is probably derived from the Arabic word *kahveh*, which means "stimulating," and is thought to have been used to describe the effects of eating coffee seeds (beans).

What Does Caffeine Taste Like?

Caffeine is very bitter to the taste, as are other plant alkaloids. Caffeine is added as a flavoring agent to beverages such as root beer because it produces the sharp bitterness characteristic of the drink.

How Is Coffee Decaffeinated?

If green (unroasted) coffee beans are rinsed with hot water above 175°F, caffeine is extracted from the beans. However, in the process of removing caffeine, hot water alone is not used because it strips away too many of the essential flavors and aromatic elements contained in the beans. Along with water, a decaffeinating chemical such as methylene chloride or ethyl acetate is generally used. One common method involves placing the green beans in a rotating drum and softening them with steam for 30 minutes. They are then rinsed for approximately 10 hours with methylene chloride, which removes the caffeine from the beans. Methylene chloride containing the caffeine is drained off, and the beans are steamed a second time for 8 to 12 hours to remove any remaining methylene chloride solvent. Finally, the beans are air- or vacuum-dried to remove excess moisture before the roasting process. Most decaffeinated coffee contains less than 0.1 part per million (ppm) of residual methylene chloride, 100 times less than the maximum level of 10 ppm allowed by the U.S. Food and Drug Administration.

D. C. Scheirer

---

candelilla, castor bean, crown-of-thorns, Mexican jumping bean, and tung oil tree.
- The Cactus Family (Cactaceae) includes all cacti (e.g., prickly pear, cholla, barrel cactus, peyote).
- The Mint Family (Lamiaceae) includes rosemary, thyme, sage, oregano, marjoram, basil, lavender, catnip, peppermint, spearmint, horehound, salvia, *Coleus,* and chia.
- The Nightshade Family (Solanaceae) includes tomato, white potato, eggplant, pepper, tobacco, belladonna, petunia, and jimson weed (a source of hallucinogenic drugs).
- The Carrot Family (Apiaceae) includes dill, caraway, celery, carrot, parsley, coriander, fennel, anise, parsnip, water hemlock, poison hemlock, cow parsnip, squawroot, and hog fennel.
- The Pumpkin Family (Cucurbitaceae) includes pumpkin, squash, cucumber, cantaloupe, watermelon, vegetable sponge, gourd, melonette, and manroot.
- The Sunflower Family (Asteraceae) includes sunflower, dandelion, lettuce, endive, chicory, Jerusalem and globe artichokes, dahlia, chrysanthemum, marigold, thistle, sagebrush, pyrethrum, balsamroot, tarweed, and yarrow.

24.5 Monocots

- The Grass Family (Poaceae) includes nearly all cereals (e.g., wheat, barley, rye, oats, rice, corn), sugarcane, sorghum, citronella, and lemon grass.
- The Lily Family (Liliaceae) includes lilies, asparagus, sarsaparilla, squill, meadow saffron, bowstring hemp, *Aloe,* and New Zealand flax.
- The Orchid Family (Orchidaceae) has highly specialized flowers. It includes a species that is the source of vanilla flavoring.

REVIEW QUESTIONS

1. To which flowering plant family does each of the following belong: poinsettia, lupine, columbine, peach, pear, cinnamon, sarsaparilla, belladonna, peyote, horehound, rubber, gourd, jimson weed, parsley, sorghum, asparagus, broccoli, lettuce, tomato, opium?
2. Make a list of the poisonous plants in the families discussed.
3. Which plants mentioned are or have been used for medicines?
4. Which plants mentioned have been used for tools or utensils?
5. Native Americans made extensive use of plants for a wide variety of purposes in the past. List such uses for as many plants as possible.

DISCUSSION QUESTIONS

1. Scientific investigations often take a great deal of time and money. Is it worth the effort to check out scientifically the past uses of plants?
2. A return to exclusively herbal medicines is being advocated in some quarters. Is this a good idea? What are the pros and cons?
3. Would you expect drugs produced naturally by plants to be more effective or better than drugs produced synthetically? Why?
4. If you were asked to single out the three most important families of the 16 discussed, which would you choose? Why?
5. A number of wild edible plants were mentioned in this chapter. What would happen if a large portion of the population were to gather these wild plants as a major source of food?

CHAPTER 25

Ecology

25.1 Introduction to Ecology
25.2 Plants and the Environment
  Populations, Communities, and Ecosystems
25.3 Life Histories
25.4 Natural Cycles
  The Water Cycle
  The Carbon Cycle
  The Nitrogen Cycle
25.5 Succession
  Key Theme 25.1: Plant Population Ecology
  Primary Succession
  Secondary Succession
25.6 Impact of Humans on Plant Communities
  Global Warming
  Carbon Dioxide
  Methane
  Ozone Depletion
  Erosion
  Aquifer Depletion
25.7 Loss of Biodiversity
  Acid Deposition
  Water Contamination
  Wetland Destruction
  Hazardous Waste
  Invasion of Foreign Species
25.8 Restoration of the Land
  Key Theme 25.2: John Muir, Father of America's National Park System

Overview

This chapter explores some of the ecological topics not already discussed in the text. An introduction to populations, communities, and ecosystems is followed by a brief look at producers, primary and secondary consumers, decomposers, and food. Then energy flow through an ecosystem is considered, and the nitrogen, carbon, and water cycles are explored. Discussions of succession and fire ecology follow. The chapter concludes with a discussion of human impact on plant vegetation and ecosystem cycles at global and regional levels.

An alpine ecosystem near Mount Rainier, Washington. © *Craig Tuttle/Getty Images*

25.1 Introduction to Ecology

Learning Outcome

a. Define *ecology*.

Ecology is a broad, integrated discipline that examines the relationships of organisms to each other and to their environment. The term *ecology*, which has become a household word, was proposed by Ernst Haeckel in 1869, but the discipline traces its origins back to early civilizations when humans learned to modify their surroundings through the use of tools and fire.

Various aspects of ecology have been mentioned in earlier chapters. For example, soils and organisms associated with them were discussed in Chapter 5. Some of the adaptations and modifications of leaves associated with specific environments were discussed in Chapter 7. The effects of light and temperature on plants were covered in Chapters 10 and 11.

In recent years, scientists, and increasingly the general public, have become alarmed about the effects of human carelessness and greed on our environment. In the 1970s, damage to forests and lakes caused by acid deposition (acid rain); contamination of ground water by nitrates and pesticides; soil erosion; reduction of the stratospheric ozone shield; major global climatic changes; loss of biodiversity in general; loss of tropical rain forests in particular; and other impacts gained widespread publicity. The growing, and now nearly universal, understanding that humans are the cause of these problems has led many people to adopt environmentalism as a guiding philosophy.

Because it is possible to review only a few of the major aspects of ecological study and environmental considerations in a text of this nature, this chapter will briefly explore some of the issues at several broad levels. These include (1) individual plants, plant populations, and plant communities; (2) global matters; and (3) regional issues.

25.2 Plants and the Environment

Learning Outcomes

a. Give examples of biotic and abiotic components of an urban park.
b. Distinguish among populations, communities, and ecosystems.

The interactions between organisms and their environment determine whether or not a species, or an individual member of a species, can survive and reproduce in a particular habitat. The environment of each habitat is determined by both living and nonliving factors. Living, or *biotic,* factors include all of the other organisms in the habitat with which the organism interacts. If we consider a maple tree in the forest, the biotic factors include the other maple trees that might provide pollen for reproduction, or the animals that might eat its leaves, or the fungi and other microorganisms that are associated with its roots. The nonliving, or *abiotic,* factors in the environment include the wind, rain, sunlight, soil, and temperature to which the maple tree is exposed during its lifetime. Each habitat in the world has a different combination of biotic and abiotic factors, with different habitats favoring different kinds of plants and animals. Thus, maples might grow well in places where there are freezing winters, while palm trees would not. Conversely, not all species that could grow in a place are actually found there, even though habitats in different places support similar species. Humans have transported many species between regions, and alien species now thrive in their new habitats.

The fact that a particular plant species can flourish in one habitat but not another suggests that plants reflect and respond to their environment. For example, when a seed germinates, it must be capable of growing into a mature plant and reproducing in the same environment. When the plant reproduces, its genes are passed along to the next generation; these genes allow it to grow in its environment. In other words, the plant is *adapted* to its environment (see Chapter 15). If the plant were not adapted to its environment, its seeds might germinate but the plants would die before reproduction could occur. By observing the plants growing in a particular habitat, one can tell the general nature of their environment. For example, cacti are indicative of desert conditions that normally include little precipitation and warm daytime temperatures. On the other hand, the presence of natural grasslands indicates that the environment provides higher levels of precipitation, but at insufficient levels to sustain the growth of trees.

Populations, Communities, and Ecosystems

We recognize that plants, animals, and other organisms tend to be associated in various ways with one another and with their physical environment. For example, forests consist of **populations** (groups of individuals of the same species) of trees or other plants that form a **plant community** (unit composed of all the species of plants occurring in a given area). The lichen and moss flora on a rock also constitute a community, as do the various seaweeds in a tidepool (Fig. 25.1). However, these communities also invariably have animals and other living organisms associated with them. It is preferable, therefore, to refer to the unit composed of all the populations of living organisms in a given area as a **biotic community.** Considered together, the communities and their physical environments, which interact and are interconnected by physical, chemical, and biological processes, constitute **ecosystems.** Some populations, communities, and ecosystems may be microscopic, while others can be much larger, or even global.

Figure 25.1 A tidepool community at low tide. © *Kingsley Stern*

Populations

Populations vary in numbers, in density, in genetic diversity, and in the total mass of individuals. Depending on circumstances, a field biologist may investigate a population in various ways. If, for example, a conservation organization is concerned about the preservation of a rare or threatened species, the organization may simply count the *number* of individuals, although this may not always be feasible. If such a count is not feasible, the organization may estimate population *density* (number of individuals per unit volume—e.g., five blueberry bushes per square meter). If the individuals in a population vary greatly in size or are unevenly scattered, a better estimate of the population's importance to the ecosystem may be calculated by determining the **biomass** (total mass of the living individuals present).

Communities

Communities are composed of populations of different species living together in the same place. Within a region, similar environments have similar species composition. However, each species is distributed on the landscape according to its own responses to changes in the physical and biotic environment. As a result, the identification of communities is subjective because there is no sharp discontinuity between them. The vegetation on a particular landscape is not the inevitable result of predictable processes because the species composition of a community is also determined by the local availability of species, by unique historical events, and by chance.

There are practical reasons to analyze and classify plant communities. Vegetation maps are useful in land-use planning, forest and natural resource management, biological conservation, landscape restoration, and even military maneuvers. Changes in vegetation over time can be readily visualized when the vegetation is classified. The type of vegetation can be used to infer qualities of the environment.

Ecosystems

Living organisms interacting with one another and with factors of the nonliving environment constitute an ecosystem. The nonliving factors of the environment (abiotic factors) include light, temperature, concentrations of oxygen and other gases, air circulation, precipitation, rocks, and soil type. Disturbances such as fire, strong wind, and erosion are additional abiotic factors. The distribution of a plant species in an ecosystem is controlled mostly by temperature, precipitation, soil type, and the effects of other living organisms (biotic factors). For example, in Mediterranean climates, such as those that occur in parts of California and Chile, nearly all precipitation occurs during the winter months, and the summers are dry. This type of climate favors spring annuals that complete their life cycles by summer and evergreen shrubs that can tolerate long periods of drought and can withstand fire. Forests may occur in areas where melting heavy winter snowfall soaks deeply into the soil, compensating somewhat for the lack of summer moisture.

The distribution of plant species is strongly determined by the mineral content of soil. Heavy metals, such as arsenic, copper, lead, nickel, and cadmium, occur in many mine tailings in concentrations that prevent plant growth. Serpentine soils, derived from metamorphic magnesium silicate rocks that are common along the Pacific Coast, in Scotland, in New Zealand, and elsewhere, naturally contain heavy metals. These occur as small "islands" surrounded by more normal soils and are rich in magnesium, iron, and chromium, the effects of which can mimic those found on mine wastes. They are also low in calcium, nitrogen, and phosphorus. Most plants cannot distinguish between calcium and magnesium, so they absorb toxic levels of the latter. Only species and ecotypes that have evolved adaptations to exclude magnesium can grow on serpentine. A trade-off for this ability is relatively low competitive ability, so that serpentine plants cannot grow on nonserpentine soils in association with other species. Serpentine habitats are rich in species and sustain many endemic species of plants. Studies of serpentine vegetation have helped develop methods to restore soils polluted by heavy metals. Plant distributions are also strongly influenced by competition for resources, by herbivory and dispersal by animal members of the biotic community, and by the pattern of available water.

The leaves and other parts of plant species that occur naturally in areas of low precipitation and high temperatures (*xerophytes*) generally are adapted to their particular environment through modifications that reduce transpiration. These modifications are discussed under "Specialized Leaves" in Chapter 7 and "Regulation of Transpiration" in Chapter 9. Plants of arid areas may also have specialized forms of photosynthesis, such as CAM photosynthesis. Similarly, plants that grow in water (*hydrophytes*) are modified for aquatic environments.

Ecosystems may be sustained entirely through photosynthetic activity, energy flow through **food chains** (discussed in the next paragraph), and the recycling of nutrients. Organisms called **producers** are capable of carrying on photosynthesis (e.g., plants, algae) and storing energy that may be released by other organisms. Animals such as cows, caribou, caterpillars, and other organisms that feed directly on producers are called **primary consumers. Secondary consumers,** such as tigers, toads, and tsetse flies, feed on primary consumers. These categories are arbitrary, because many consumers eat plants and herbivores. Grizzly bears, for example, eat berries and salmon. **Decomposers** break down organic materials to forms that can be reassimilated by the producers. The foremost decomposers in most ecosystems are bacteria and fungi. Note that there is not necessarily a sharp distinction between consumers and decomposers.

Decomposers, for example, depend just as much as consumers on organic matter for their nutrition.

In any ecosystem, the producers and consumers interact, forming food chains or interlocking *food webs* that determine the flow of energy through the different levels (Fig. 25.2). In such chains or webs, each link feeds on the level below and is consumed by one at a higher level, with photosynthetic organisms at the bottom and the largest non-herbivorous animals at the top. Because most non-plant organisms have more than a single source of food and are themselves often consumed by a variety of consumers, there are considerable differences in the length and intricacy of

Figure 25.2 A simple food web connecting aquatic and terrestrial habitats. *Source: K. Starr Chester, Trans. New York: The Ronald Press Company*

food chains or webs, but there are rarely more than four levels in any given terrestrial system.

Light energy itself, which enters at the producer level, can't be recycled in an ecosystem. Only about 1% of the light energy falling on a temperate zone community contributes to the production of organic material. As the organisms at each level respire, energy dissipates as heat into the atmosphere. Some parts of organisms may not be consumed at each level of the food chain. For example, leaves that are grazed by an herbivore are broken down and much of their energy is released within the animal. However, the energy stored within the uneaten leaves of the same plant isn't released until the leaves fall from the plant and decomposers (bacteria and fungi) break them down.

Trophic efficiency—the percentage of available energy actually transferred from one trophic level to the next—varies greatly among systems and between different trophic levels. Aquatic systems have relatively high efficiencies, ranging from 2% to 24%, with an average of 10.1%. Terrestrial systems are less efficient. In a deciduous forest, 96% of the primary production goes to decomposers and to respiration, leaving at most only 4% for all consumers. To illustrate how quickly energy available to secondary consumers drops off, consider this example. Assume that corn stores 100 calories of sun energy, converting photons at 1% efficiency. When the corn is fed to a cow, she converts 10% of this energy into her body. Disregarding the fact that not all of the corn is eaten by the cow, this leaves at most 10 calories of energy. When we, as secondary consumers, consume this beef, we get only about 10% of the energy in the beef. Thus, for every 100 calories stored by the corn, we get 1 calorie. When we consume tuna and swordfish, we are eating even higher in the trophic chain. Were we to consume the corn directly, cutting out the cow from the food chain, we would increase the efficiency tenfold and reduce several forms of environmental damage.

From this, it follows that a vegetarian diet makes much more efficient use of solar energy than one that relies heavily on meats, and where food is scarce or humans are very abundant (as in India or Ethiopia), humans may be forced to favor a vegetarian diet. It also follows that in terms of the numbers of individuals and the total mass, there is a sharp reduction at each level of the food chain. In a given portion of ocean, for example, there may be billions of microscopic algal producers supporting millions of tiny crustacean consumers, which, in turn, support thousands of small fish, which meet the food needs of scores of medium-sized fish, which are finally consumed by one or two large fish (Fig. 25.3). In other words, one large fish may very well depend on a billion tiny algae to meet its energy needs every day.

The interrelationships and interactions among the components of an ecosystem can be quite complex, but many function together in a somewhat regulatory fashion. An increase in food made available by producers can result in an increase in consumers, but the increased number of consumers and competition among them reduce the available food,

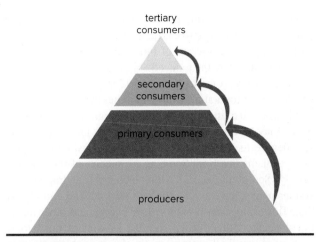

Figure 25.3 An energy pyramid of an ecosystem. There is much more energy at the bottom than at the top.

which then inevitably leads to a reduction of consumers to earlier levels. While cyclical in nature, the net result is sustained self-maintenance of the ecosystem. This is the basis for the so-called *Balance of Nature*.

Interactions among Plants, Herbivores, and Other Organisms

While it is easy to see that the total mass of consumers is largely determined by the total mass of food made available by the producers, the interactions among different producers themselves, between predators and prey, and between the decomposers and the other members of the ecosystem are usually more subtle. Many flowering plants produce chemical substances that can inhibit or promote the growth of other flowering plants. The negative effects of black walnut hulls on understory plants are called *allelopathy*. Many plants, including sagebrush, eucalyptus, and an array of crop weeds, seem to inhibit other species using chemicals leached from their litter, root exudates, or even volatiles released from leaves. However, allelopathic effects are subtle and can depend on soil properties, microorganisms, plant stress, evolutionary relationships, and climate. A careful experimental study of invasive knapweed (*Centaurea*) suggested that the success of this weed was due to allelopathy. Knapweeds do not appear allelopathic in their native habitats, suggesting that their associates have evolved adaptations to tolerate these chemicals. Many other plants produce *phytoalexins* (chemicals that kill or inhibit disease fungi or bacteria), making them resistant to various diseases (see "Use of Resistant Varieties" in Appendix 2).

Plants sometimes provide crucial protection to the seedlings of other species, thus enhancing the biodiversity of a community. Facilitation by nurse plants is common in deserts and in alpine habitats, where stresses make seedling establishment difficult. In central Mexico, a tall columnar cactus can only establish beneath a nitrogen-fixing shrub. Ironically, the cactus eventually suppresses and kills the shrub through competition for water and light. The nurse plant phenomenon rarely benefits the nurse, so facilitation is unlikely to be an adaptive trait. Rather, the beneficiary can establish by being tolerant of the competitive environment and benefiting from the reduced abiotic stress.

Conversely, some bacteria and fungi limit higher plant growth by producing inhibitory chemical compounds. Population size in other bacteria, fungi, and flowering plants may be limited by nutritional needs and availability. The degree to which this occurs varies considerably with the organisms involved. Some of the species of the Figwort Family (Scrophulariaceae), which includes snapdragons and similar plants, have no chlorophyll; they depend entirely on their flowering plant hosts for their energy and other nutritional needs. Other related species, as well as mistletoes, do have chlorophyll but apparently also require supplemental food from their hosts. Still other species often parasitize the roots of certain plants but are capable of existing independently.

Mycorrhizal fungi are intimately associated with the roots of most woody and many other plants in such a way that both organisms derive benefit (such associations, called *mutualisms*, are a major part of life in general; they are briefly discussed in Chapter 5). The fungi greatly increase the absorptive surface of the root, usually playing a major role in the absorption of phosphorus and other nutrients, while obtaining energy from root cells.

Thomas Belt, a naturalist who lived more than 100 years ago, first called attention to an association between tropical ants and thorny, rapidly growing species of *Acacia* (Fig. 25.4). The *Acacia* has large, hollow thorns at the base of each leaf and is host to ants that feed on sugars, fats, and proteins produced by petiolar nectaries and special bodies at the tip of each leaflet. The ants live within the hollow thorns and vigorously attack any other organism, from insects to large animals, that come in contact with the plant. They also kill, by *girdling,* any plant that touches the *Acacia* (girdling is discussed under "bridge grafting" in Appendix 4). Experiments by Daniel Janzen have shown that when ants are removed from these *Acacia* species, the plants grow very slowly and usually soon die from insect attacks or from shading by other plants.

Large herbivorous animals, such as deer and moose, feed on a wide variety of plants, each differing in nutritional value. Each plant species also produces different combinations, types, and amounts of chemical compounds in addition to proteins, fats, and carbohydrates. Many of the chemical compounds may be toxic, but some animals grazing on the plants that produce these chemicals do not display symptoms of poisoning because their digestive systems are capable of breaking down the compounds and eliminating or excreting them to a limited degree. The limitations imposed by such compounds result in the consumers varying their diet, seeking familiar foods, and being wary of new ones.

Some plants are dependent for their existence on predators controlling populations of grazing animals. Populations of others such as oaks may become limited in extent when squirrels and acorn-consuming birds remove

(a)

(b)

Figure 25.4 A species of *Acacia* that is host to ants that live in its hollow stipules. The ants attack any organisms, large or small, that come in contact with the plant. The plant provides food for the ants through nectaries at the bases of the stipules (a) and nitrogenous bodies at the tips of the leaflets (b). *(a-b)* © *Kingsley Stern*

acorns that would potentially become trees. All plant species have some natural defense, such as chemical compounds or structural modifications (for example, spines). If this were not so, primary consumers of all kinds from insects to elephants could soon render a species extinct. In an ecosystem, the defenses that both producers and consumers have against each other have been developed through a process of coevolution resulting from natural selection and are more or less balanced.

25.3 Life Histories

Learning Outcome

a. Explain the differences among annuals, biennials, and perennials.

In order for a species to maintain itself, it must be able to both survive and reproduce. It is critical to establish a proper balance between the use of resources for survival and those for reproduction. As you might expect, different strategies for allocating resources have developed. A species' *life history* is composed of the traits that control its survival and reproduction. Species adapted to regions with erratic weather conditions often use the *big bang reproduction* strategy. These species devote all their resources toward growth for most of their lives. Then, when conditions are favorable, most of their energy goes into a final single reproductive burst. For example, desert agave plants grow vegetatively for many years and then, during a moist year, all the energy goes into producing huge flower stalks with many seeds. The plant then dies. In contrast, some species have *repeated reproduction*, producing seeds throughout their lifetimes. Most trees require several years to reach reproductive maturity but produce seeds every year for the remainder of their lives.

Plants can be divided into three groups based on their reproductive strategies. As discussed in Chapter 8, annuals, biennials, and perennials differ in their seasonal growth cycles. During the growing season, annuals grow vegetatively, with reproductive growth occurring toward the end of the season. Once seeds have been produced, the plants die. Many weeds, wildflowers, and garden plants exhibit this type of growth. Biennials typically put all their energy into vegetative growth for one year. Then, during the second year, most energy is put into the production of reproductive structures, the plants dying after seeds are produced. Examples of biennial plants include parsley, carrots, celery, foxglove, and sweet William. Perennial plants produce vegetative structures that survive for many years. Herbaceous perennials grow actively during periods of adequate temperature and moisture but die back during unfavorable growing conditions such as a cold winter or a dry season. They survive through underground structures such as roots, rhizomes, bulbs, or tubers, which remain dormant until conditions are favorable for growth again. Examples of herbaceous perennials include coneflowers, tulips, and most grasses. Woody perennials do not die back during unfavorable periods in the growing season, but they do become dormant. Examples include trees, shrubs, and vines.

Life histories are also described in functional terms. One useful method starts with this premise: plants must adapt to physiological stress and to disturbance. Habitats that impose low stress and have little disturbance select for traits that confer competitive advantage. Competitive species are large, persistent, and fast growing but often are slow to reproduce. Habitats with high stress, but little disturbance, select for stress tolerance. Stress-tolerant species are small, are slow growing, have limited reproductive ability, and do not respond to nutrients. Finally, habitats with low stress, but substantial disturbance, select for weedy traits. These species are fast growing, are small (many are annuals), and reproduce quickly. The traits of any species can be determined by observation and experiment, so by characterizing the plants of a given habitat by these traits, it is possible to determine important characteristics of the habitat (an example of using indicator species). This functional model of plant responses demonstrates the fundamental evolutionary principle of trade-offs: a species cannot simultaneously be good at everything.

The timing of crucial events in the life of a plant, for example, germination, bud burst, flowering, and seed production, is called its *phenology*. Many species use photoperiod or light quality to trigger germination and flowering, but growth rates are controlled by local habitat factors, such as fertility, available moisture, and temperature. As global temperatures continue to increase, many temperate zone researchers are reporting that phenological stages (life cycle events) in plants continue to occur earlier and earlier. In a study of herbs in Connecticut, it was found that temperatures have increased about 4°C (7.2°F) since 1850, and that the typical species flowers about 4 days earlier for every one-degree increase in temperature. Herbarium collections confirm that flowering now occurs more than 2 weeks sooner than when Thoreau was at Walden Pond.

25.4 Natural Cycles

Learning Outcomes

a. Describe some ways in which humans have disrupted the water cycle.
b. Explain why the carbon cycle is vulnerable to alteration.
c. Describe ways in which soil nitrogen is replenished.

Water and elements such as carbon, nitrogen, and phosphorus are constantly cycling throughout nature. Such cycling involves transformation between organic and inorganic forms. We'll use the water, carbon, and nitrogen cycles as examples.

Nutrients constantly cycle and are recycled in the web of life. Carbon, nitrogen, water, phosphorus, and other molecules have been passing through cycles for eons. Some molecules that were a part of a primeval forest that became compressed and turned to coal may have become part of another plant after the coal burned. Then the new plant may have been eaten by an animal that, in turn, contributed molecules to a part of yet another living organism. Just think of where the molecules in your own body may have been in the past few billion years. They may well have been a part of some prehistoric seaweed, a saber-toothed tiger, a mighty dinosaur, or even all three!

The Water Cycle

Most of a living cell consists of water, and it is in this medium that the myriad of chemical reactions of living organisms takes place. In fact, life as we know it couldn't exist without water. The earth's water is constantly being recycled, and the total amount remains stable. However, as we will see later in this chapter, the distribution of water across the globe can change over time. Ninety-eight percent of the water is in oceans, rivers, and lakes that make up about two-thirds of the earth's surface. Most of the remaining water is in living organisms, glaciers, polar ice, water vapor, and the soil.

Some water that falls on land penetrates until it reaches an area of saturation known as the *water table*. Water in the water table may emerge from beneath the surface in the form of springs and artesian wells. Below the water table, porous rocks collect water and, when these rocks transmit water to wells and springs, they are called **aquifers.**

Water evaporates from bodies of water and is transpired from plants, which make up 80% of the earth's biomass. Water also evaporates from the surfaces of animals and damp areas of the earth as the sun shines on them. The water vapor rises into the atmosphere, condenses, and falls back to earth in the form of rain, snow, and hail in a constant cycle—the *water cycle* (Fig. 25.5).

The water cycle has been disrupted by humans in many ways beyond aquifer depletion. The creation of reservoirs, large-scale irrigation projects, and even the use of air conditioning have resulted in major redistribution of fresh water. High global temperatures are increasing evaporation and transpiration rates, so that many parts of the world are becoming more arid.

The Carbon Cycle

In the process of "feeding" (acquiring the energy they need to maintain themselves and grow; i.e., respire), bacteria process carbon (and nitrogen) in a way that allows carbon to be recycled (Fig. 25.6). Bacteria "eat" organic matter (carbohydrates), obtaining the energy present in the molecules. As a result of this activity (respiration), bacteria break down carbohydrates, contributing to the atmosphere's current carbon dioxide concentration of 0.040%. Every year, photosynthetic organisms have the amazing ability to recombine billions of tons of carbon dioxide back into carbohydrates, thereby contributing toward the fixed carbon that already exists in aquatic and terrestrial ecosystems. As the cycle continues, respiration by all living organisms constantly converts these carbohydrates right back into carbon dioxide, with perhaps as much as 90% or more of it being produced by the incredible numbers of decay bacteria and fungi as they decompose tissues.

The burning of fossil fuels by the internal combustion engines of industry and transportation releases lesser but

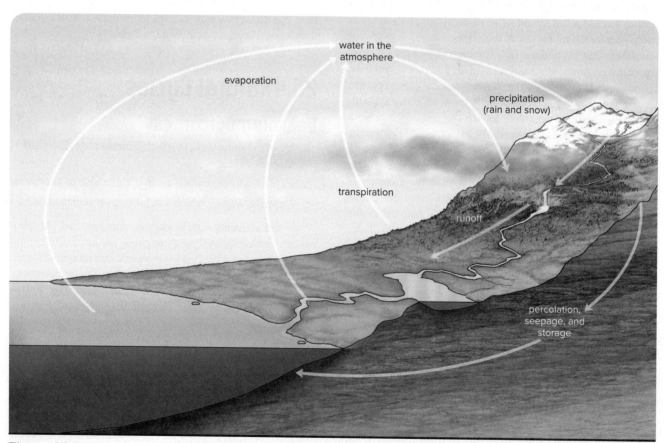

Figure 25.5 The water cycle.

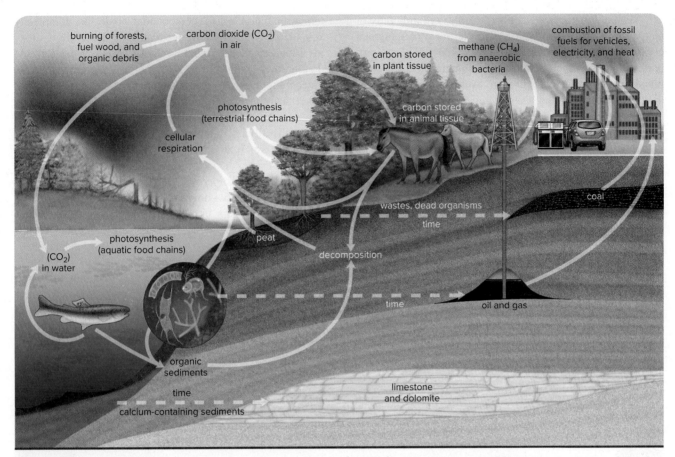

Figure 25.6 The carbon cycle.

significantly increasing amounts of carbon dioxide into the air, and a small amount originates with fires and volcanic activity. The carbon cycle is delicate and has been unbalanced by human actions.

Scientists have found that plant growth initially can be accelerated by increasing levels of carbon dioxide in the air, and some commercial nurseries pump carbon dioxide along rows of bedding plants as a "fertilizer." However, the accelerated growth soon becomes limited by other factors in the environment and does not necessarily result in crop yield increases. Recent studies indicate that while C_3 plants may respond to increased carbon dioxide levels, C_4 plants may not, suggesting that in nature, the response of C_3 plants to higher levels of carbon dioxide may give them a competitive edge over C_4 plants. This could cause major redistributions of prairie species and affect production of maize (corn), sugarcane, and other C_4 crops.

The carbon cycle is extremely vulnerable to alteration. The amount of atmospheric carbon is relatively small, and the rates of transfer between components of the cycle are large. Carbon moves through the ecosystem via photosynthesis, respiration, and burning at a rapid rate, and because the reservoirs of coal, oil, and methane are large, the potential to increase levels of atmospheric and oceanic carbon is huge. In addition, soil erosion releases soil organic matter, while land clearing and fires release additional carbon stored in vegetation. Carbon dioxide increases in the atmosphere because absorption in the oceans has not kept up and global photosynthesis may be reduced due to habitat conversion and desertification. It has, however, caused the oceans to become more acid, making it more difficult for any organism that makes cell coverings, shells, or skeletons to function. Corals, echinoderms, crustaceans, mollusks, and many unicellular organisms such as foraminifera are vulnerable to reduced dissolved carbonate that results when the pH declines.

The Nitrogen Cycle

Most of the nitrogen in living organisms is in the protoplasmic proteins of their cells. Nitrogen gas makes up about 78% of our atmosphere but constitutes only about 18% of the protein in living cells. There are nearly 69,000 metric tons of nitrogen in the air over each hectare of land (35,000 tons per acre), but the total amount of nitrogen in the soil seldom exceeds 3.9 metric tons per hectare (2 tons per acre) and is usually considerably less. This discrepancy results from the nitrogen of the atmosphere being chemically inert, which is another way of saying that it will not combine readily with other molecules. It is, therefore, largely unavailable to plants and animals for their use in building proteins and other substances containing nitrogen.

Most of the nitrogen supply of plants (and indirectly of animals) is derived from the soil in the form of inorganic compounds and ions taken in by the roots. These compounds and ions include those that contain nitrogen combined with oxygen or hydrogen. Animals, through their digestive processes, and bacteria and fungi break down the more complex molecules of dead plant and animal tissues to simpler ones. Some nitrogen from the air is also *fixed*—that is, converted to ammonia or other nitrogenous compounds by various *nitrogen-fixing bacteria*. Some of these organisms gain access to various species, particularly legumes (e.g., peas, beans, clover, alfalfa), through the root hairs, with the plant producing root nodules in which the bacteria multiply (root nodules are shown in Chapter 5). Others live free in the soil.

Figure 25.7 shows there is a constant flow of nitrogen from dead plant and animal tissues into the soil and from the soil back to the plants. Decay bacteria and fungi can break down enormous quantities of dead leaves and other tissues to tiny fractions of their original volumes within a few days to a few months. If they were abruptly to cease their activities, the available nitrogen compounds would be completely exhausted within a few decades and the carbon dioxide supply needed for photosynthesis seriously depleted. At present, even with the various bacteria in the nitrogen cycle functioning normally, the total amount of nitrogen in the soil is not being increased by their activities but is merely being recycled. Note, however, that addition of nitrogenous fertilizers will artificially increase soil nitrogen content.

Large amounts of nitrogen are continually being lost as water leaches it out or carries it away through erosion of topsoil. More is lost with each harvest, the average crop removing about 25 kilograms per hectare (25 pounds per acre) per year. This nitrogen loss from the harvesting of crops can be sharply reduced if vegetable and animal wastes are recycled and returned to the soil each year. While bacteria are decomposing tissues, they use nitrogen, and little is available until they die and release their accumulations into the soil. Accordingly, until bacteria have completed their breakdown of organic matter, crops should not be planted in soils to which only partially decomposed materials have been added. Likewise, when sawdust, straw, or other organic

Figure 25.7 The nitrogen cycle.

mulches are spread around plants in a garden to control weeds and conserve soil moisture, the soil nitrogen will be less available to the growing plants until the mulches have been decomposed.

Weeds and stubble are often controlled by burning. Fire, however, causes serious loss of nitrogen, which has to be replaced. It has been estimated that the annual combined loss of nitrogen from the soil in the United States from fire, harvesting, and other causes exceeds 21 million metric tons (23 million tons), and only 15.5 million metric tons (17 million tons) are replaced by natural means. To offset the net loss, some 32 million metric tons (35 million tons) of inorganic fertilizers are applied to the soils each year. This inorganic fertilizer is produced by expending large quantities of electricity produced by fossil fuel. Much of the nitrogen from this fertilizer is lost by leaching because nitrates remain in the water and do not become attached to soil particles. If organic matter is not added at the same time, however, this application of inorganic fertilizers, combined with the annual burning of stubble, may eventually result in the creation of a *hardpan* soil.

Hardpan develops through the gradual accumulation of salt residues, which dissolve humus and disrupt the structure of the soil, causing the clay particles to clump and producing colloids that are impervious to moisture. In hardpan soils and others low in oxygen (e.g., flooded areas), denitrifying bacteria use nitrates instead of oxygen in their respiration, depleting the remaining soil nitrogen.

Precipitation returns some nitrogen to the soil from the atmosphere, where it has accumulated as a result of the action of light on industrial pollutants, fixation by flashes of lightning, and diffusion of ammonia released through decay. The activities of nitrogen-fixing bacteria and volcanoes also contribute to the natural replenishment of nitrogen by converting it to forms that plants can use.

The nitrogen cycle is vulnerable to human disruption because most of it depends on the activities of microorganisms that are sensitive to changes in pH and pollution. Greater efficiency in crop production can be achieved by fostering soil organic matter, reducing erosion, and breeding varieties with higher nitrogen use efficiency.

25.5 Succession

Learning Outcomes

a. Describe the general events that happen during succession.

b. Define *climax community*.

c. Explain how secondary succession differs from primary succession.

After a volcano spews lava over a landscape or after an earthquake or a landslide exposes rocks for the first time, there is initially no sign of life on the lava, rock, or other surfaces. Soon, living organisms begin to appear, and a sequence of events known as **succession** takes place. During succession, the plants and other organisms that first appear gradually alter their environment as they carry on metabolism and reproduction. In time, the accumulation of wastes, dead organic material, and inorganic debris and other changes (such as of shade and water content in the habitat) favor different species that may replace the original ones. These, in turn, modify the environment until no further changes in species composition occur. The process of succession may take centuries to unfold.

Succession occurs whenever and wherever there has been a disturbance of natural areas on land or in water. It proceeds at varying rates, depending on the climate, the soils, and the animals or other organisms in the vicinity. Prior to the 1970s, ecologists believed that each community could establish a state of balance that would remain stable unless disturbed by humans. We now realize that communities are constantly changing in response to an array of disturbances. Humans may cause disturbances, but so does nature, in the form of hurricanes, floods, landslides, and fire.

Disturbances may appear to have a negative impact on communities, but their effects may not be as devastating as they first appear. In fact, they can help to enhance the species diversity. For example, in 1988, forest fires burned 45% of Yellowstone National Park, which covers about 1 million hectares (2.5 million acres). While the scenes of destruction

---

KEY THEME 25.1 ecology

Plant Population Ecology

Unlike the growth of any other single species, explosive human population growth has disrupted ecosystems across the entire planet. These disruptions have included changes in atmospheric composition, overenrichment of ecosystems with nutrients, and large-scale land clearing, which threatens to produce a massive extinction of species. Knowledge of the basic ecology of populations (groups of individuals of the same species), communities (all the populations occurring in a given area), and ecosystems (communities and the physical environments with which they interact) is necessary to understand the magnitude and significance of the ecological changes produced by the growing human population. Plant population ecology includes processes such as pollination, seed dispersal, germination, and seedling survival and establishment. Plant community classification and analysis is an important aspect of land-use planning, natural resource management, and conservation of biodiversity. Knowledge of succession, energy flow, and nutrient cycling is critical to assessing ecosystems and managing them for health and sustainability.

seemed horrific, long-term studies of the aftermath confirm that fires can rejuvenate forest ecosystems. In fact, for several reasons, the park has recovered more quickly than expected. Although the landscape looked uniformly barren after the fire, some areas were less severely damaged than others. The pockets of relatively unscathed ground even near the largest burn areas allowed the forest to regenerate more quickly than expected. In addition, roots and rhizomes of many wildflowers survived the fire because the soil was charred for only a few centimeters in depth. These plants were able to grow rapidly and disperse seeds throughout the burned areas. This allowed the charred areas to recover quickly because seeds did not need to be dispersed from untouched regions. As a result of the fires, uniform stands of old lodgepole pine trees have been replaced by a rich diversity of species undergoing a new round of succession. However, ecologists studying the recovery of Yellowstone are concerned that global warming will dramatically alter the course of succession in the park. Based on computer models, wetter winters caused by global warming could allow for the establishment of larch, scrub oak, and other trees not currently in this vegetation. And, of course, with additional warming, more fires can be expected, allowing new rounds of succession to take place and leading to a deteriorating ecosystem.

Ecologists recognize several variations of two basic types of succession. *Primary succession* involves the formation of soil in the beginning stages, whereas *secondary succession* takes place in areas that had previously developed vegetation and had experienced a drastic environmental change such as conversion to farmland.

Primary Succession

One of the most universal types of primary succession begins with bare rocks and lava that have been exposed through glacial or volcanic activity or through landslides. Initially, the rocks are subjected to alternate thawing and freezing, at least in temperate to colder areas. Tiny cracks or flaking may occur on the surface as a result, and these cracks permit vascular plants to invade rapidly (Fig. 25.8). Lichens and mosses often become established on such surfaces. They produce acids that very slowly etch the rocks, and as they die and contribute organic matter, they are replaced by other species. Certain rock mosses adapted to long periods of desiccation also may become established, and a small amount of soil begins to build up. This is augmented by dust, debris blown in by the wind, seeds, and even insects and spiders. Eventually, a mat is formed to permit plants to become established on the surface, and the pace of soil buildup and rock breakdown accelerates.

If deep cracks appear in the rocks, the larger seeds may widen them further as they germinate and the roots expand in girth. It has been calculated that germinating seedlings can exert a force of up to 31 kilograms per square centimeter (450 pounds per square inch). Indeed, seedlings can split rocks that weigh several tons (Fig. 25.9).

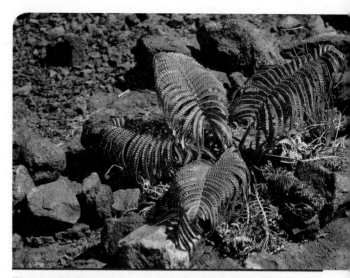

Figure 25.8 Early stages of primary succession in cracks in lava on the Hawaiian volcano Kilauea. The spores of the fern *Sadleria cyantheoides* ('ama'u) blow into cracks in young lava and can establish easily. © *Roger del Moral*

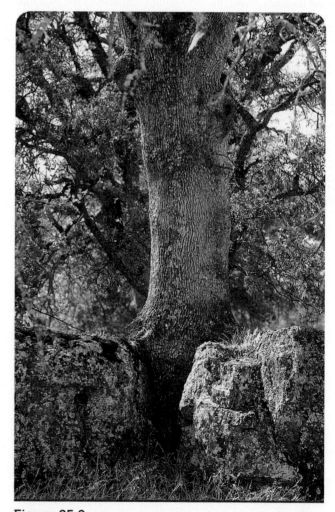

Figure 25.9 A blue oak that grew from an acorn lodged in a small crack of the rock. Over the years, the oak has split the rock apart. © *Kingsley Stern*

As soil buildup continues, larger plants take over, and eventually the vegetation reaches an equilibrium in which the associations of plants and other organisms remain the same until another disturbance takes place or climatic changes occur. These relatively stable plant communities may be referred to as **climax communities,** which differ from one part of the world to another, depending on climate and soil or major disruptions such as fires or floods. However, even within a region, these communities can differ significantly. So-called alternative stable states develop in response to the available species and chance events. Increasingly, primary succession in landscapes dominated by crops or weeds occurs in unique ways. For example, the flora on mudflows spawned by Mount Pinatubo (1993) are dominated by exotic species.

Often, when a volcano produces ash (an airborne deposit also called tephra) that buries existing vegetation, pioneering stages involving lichens or mosses are bypassed. Vascular plants are the first colonists. This occurred after the blast and a series of ash eruptions and mudflows on Mount St. Helens (Washington State) during the early 1980s. Most life in the path of destruction was destroyed (Fig. 25.10a), but within weeks organisms began to recolonize the devastated zones. However, it took nearly two decades for vegetation to become prominent. In the early 1990s, vegetation was limited to wetlands, a few areas that escaped severe damage on the northeastern flank of the volcano, and areas where lupines became established (Fig. 25.10b). Lupines, in turn, promoted the establishment of other species because, as legumes, they could fix nitrogen. Several different community types have developed since, but they do not appear related to landscape features, nor have they previously been recorded in the region. Instead, random factors associated with seed dispersal have determined community composition in the early stages of succession. Factors that act at the local scale, such as competition and soil moisture status, have yet to influence the array of species in communities. It is expected that, as succession continues and the vegetation becomes more abundant, competitive interactions will limit species to those sites in which they are well adapted.

Mount St. Helens created many disturbances, not just one, and these affected plants in different ways. Where ash deposits were shallow, burrowing pocket gophers survived, and they subsequently mixed up the soil, ash, seeds, plant parts, and microbes. This allowed mycorrhizal fungi to associate with plant roots, allowing plants to thrive in an infertile habitat. The variability in patterns and rates of succession across the Mount St. Helens landscape demonstrates that succession is a complex phenomenon that cannot be interpreted as a single, linear, predictable event.

In 1883, the island of Krakatau, Indonesia, underwent one of the most powerful volcanic eruptions in human history, almost eliminating the island and covering the surroundings with very thick ash. Virtually all plant and animal life was destroyed on Krakatau and two surrounding islands. The recolonization is of great interest to biogeographers

(a)

(b)

Figure 25.10 (a) The slopes of Smith Creek valley, east of Mount St. Helens, show trees blown down by a lateral blast in 1980. The direction of the blast, shown here from left to right, is apparent in the alignment of the downed trees. Over 4 billion board feet of usable timber, enough to build 150,000 homes, was damaged or destroyed. (b) Dense, lupine-dominated vegetation on the slope of Mount St. Helens in 2007. *(a) Source: Lyn Topinka/U.S. Geological Survey (USGS); (b) © Roger del Moral*

and ecologists who study colonization. New plants reached Krakatau from 40 kilometers (25 miles) away. Seeds and spores wafted in by air, while others floated across in seawater or rode logs or swimming animals. Others were dispersed by seabirds that also colonized Krakatau. These islands now host over 400 species of vascular plants. However, dispersal from the very diverse pool of species on the surrounding islands was selective, and the composition of the forests is now very different than it was before the volcanoes erupted.

Succession may also take place in wet habitats. For example, ponds and lakes abound in the northern parts of midwestern states such as Michigan, Wisconsin, and Minnesota. Many of these bodies of water were left behind by retreating glaciers and often have no streams draining them.

Figure 25.11 (a) Succession in a pond in North Cascades, Washington. (b) Duckweeds floating on a pond in early stages of succession. *(a) © Roger del Moral; (b) © Kingsley Stern*

The water that evaporates from them is replaced annually by precipitation runoff. They also grow a bit smaller each year as a result of succession (Fig. 25.11a).

This succession often begins with algae either carried in by the wind or transported on the muddy feet of waterfowl and wading birds. Although algae multiply throughout the upper, sunlit levels of the entire pond, they tend to become concentrated in shallow water near the margin, and with each reproductive cycle, the dead parts sink to the bottom. Floating plants, such as duckweeds, may then appear, often forming a band around the body of water just offshore (Fig. 25.11b). When nutrients, oxygen, pH, and temperatures are low, peat mosses encroach from the sides and become the dominant floating plants. There are presently about 4 million square kilometers (1.5 million square miles) of peat bogs throughout the world, and peat remains an important energy source.

Water lilies and other rooted aquatic plants with floating leaves often become established, each group of plants contributing to the organic material on the bottom, which slowly turns to muck. Cattails and other flowering plants that produce their inflorescences above the water often take root in the muck around the edges, and the accumulation of organic material accelerates.

Meanwhile, the algae, duckweeds, peat mosses, and other plants move farther out, and the surface area of exposed water gradually diminishes. Sedges become established along the damp margins and sometimes form floating mats as their roots interweave with one another. Dead organic material accumulates and fills in the area under the sedge mats, and herbaceous and shrubby plants then move in. As the margins become less marshy, coniferous trees whose roots can tolerate considerable moisture (e.g., tamaracks or eastern white cedars) become established, eventually growing across the entire site as the pond or lake disappears. The trees continue to contribute to the formation of new soil and, in due course, a stable climax community becomes established. At this point, no visible trace of the pond or lake remains, and the only evidence of its having been there lies beneath the surface, where fossil pollen grains, bits of wood, fossilized fish skeletons, and other material reveal the history. Such succession can take thousands of years before it is complete. The evidence that it does occur, however, is extensive and compelling.

Under natural conditions, some stream-fed lakes and ponds eventually become filled with silt and debris, although this, too, may take hundreds of years to occur. The streams that feed these lakes bring in silt, and the nutrient content of the water rises as dissolved organic and inorganic materials (particularly nitrogen and phosphorus) are brought in. This gradual to relatively rapid enrichment, called **eutrophication,** facilitates the growth of algae and other organisms that add their debris to the bottom of the lake. When sewage and other pollutants enter the lake, the process of eutrophication may be greatly accelerated through stimulation of the growth of aquatic organisms. Eutrophication may also accelerate when trees are cleared from land surrounding lakes prior to the construction of summer homes and resorts. The cleared land erodes more readily, with precipitation runoff carrying soil into the water. Regardless of size, all bodies of water, including rivers, are subjected to these processes.

The reestablishment of plants on barren surfaces depends on dispersal from the immediate surroundings. Therefore, establishment is a highly variable process. It is aided by the presence of special microsites (e.g., cracks) and by the introduction of nutrients from outside the system. Many plants also facilitate succession, but others, as well as herbivores, retard the process. Eventually, stable communities are formed. As ecologists have come to understand primary succession in detail, the knowledge has been applied to restoring habitats damaged by human activities.

Secondary Succession

Secondary succession, which occurs more rapidly than primary succession, may take place if soil is already present and there are surviving species in the vicinity. In fact, survivors strongly condition subsequent succession. Many secondary successions follow human disturbances (e.g., land that was cleared when timber was harvested or land converted to farmland). Other secondary successions follow fires. Grasses and other herbaceous plants become established on burned or logged land. These usually are followed by trees and shrubs that have widely dispersed seeds (e.g., aspen and sumac in the Midwest and East and chaparral plants, such as chamise and gooseberries, in the West). After going through fewer stages than are typical of primary succession, a climax community becomes established, often in less than 100 years.

Fire Ecology

Fires, started primarily by lightning and the activities of prehistoric humans, have occurred for thousands of years in North America and other continents (Fig. 25.12a). Trees, such as the giant redwoods and ponderosa and lodgepole pines, although scarred by certain types of fires, often survive; the dates of fires can be determined by the proximity of the scars to specific annual rings. In the West, growth rings of ponderosa pines show that in the past, forests of such pines used to burn on an average of every 6 to 7 years. These fires and the climate, topography, and soil combined to have a profound effect on various biomes.

As human cultures developed, major and largely successful efforts to control fires were made, and this, in turn, significantly altered vegetational patterns. As knowledge of the role of fires in the maintenance of ecosystems has accumulated, it has become apparent that trying to eliminate fires, at least in certain areas, disrupts natural habitats more in the long run than allowing them to occur, and agencies such as the U.S. National Park Service may allow fires at higher elevations to run their natural course.

Fires also play a role in the composition of forests. In the mountains of east-central California, gooseberry and deerbrush appear in abundance after a fire, but their numbers stabilize within 15 to 30 years when larger trees return to the area. Ponderosa, jack and southern longleaf pines, and Douglas firs (which do not tolerate shade) are among the species that repeatedly replace themselves after fires; seeds of some species rarely germinate until they have been exposed to fire. Most chaparral species, both woody and herbaceous, are so adapted to fire that their seeds also will not germinate, as a rule, until fires remove accumulated litter and toxic wastes produced by the plants during growth.

Fires actually benefit grasslands, chaparral, and forests by converting accumulated dead organic material to mineral-rich ash, whose nutrients are recycled within the ecosystem. If the soil has been subjected to fire, some of its nutrients and organic matter may have been lost, and the composition of microorganisms originally present is likely to have been altered.

(a)

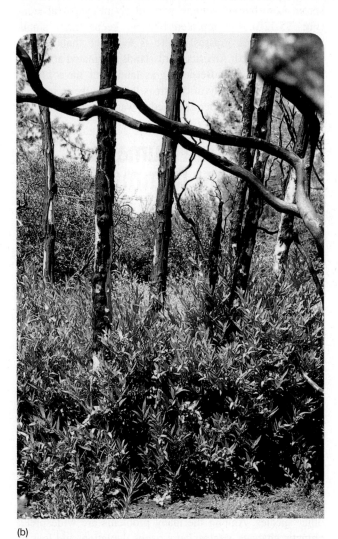
(b)

Figure 25.12 (a) Fires appear to be devastating, but they can be beneficial to forest ecosystems. (b) Secondary succession on a burned area. California bay trees are resprouting the first season after a burn.
(a) Source: John McColgan, Alaska Fire Service/Bureau of Land Management;
(b) © Kingsley Stern

Losses are offset, however, by the fact that soil bacteria, including cyanobacteria, which are capable of fixing nitrogen from the air, increase in numbers after a fire, and there is a decrease in fungi that cause plant diseases.

In some areas, such as the prairie states of the Midwest, grasses are better adapted to fire than are woody plants, producing seeds within a year or two after germination. Perennial grass buds that are at the tips of rhizomes close to or beneath the surface, where they escape the most intense heat of fires, usually survive, producing new growth the first season after a fire. Accordingly, a fire destroys only one season's growth of grass, often after growth has been completed. Shrubs, however, have much of their living tissue above ground, and a fire may destroy several years' growth. In addition, woody plants often do not produce seeds until several years after a seed germinates. Many shrubs do sprout from burned stumps, particularly in chaparral areas (Fig. 25.12b), but repeated burning keeps them small, thereby favoring grasses. There is evidence that at least some of the North American grasslands originated and were maintained by fire. Because grassland fires have largely been controlled, many of the areas have now been invaded by shrubs that were once confined to watercourses.

25.6 Impact of Humans on Plant Communities

Learning Outcomes

a. Discuss human impacts on ecosystems.
b. Explain why increases in atmospheric carbon dioxide are a concern.
c. Explain why levels of methane in the atmosphere are increasing.
d. Describe how the ozone problem has been addressed on a global scale.
e. Describe the effects of erosion.
f. Explain what is meant by aquifer depletion.

In the modern world, humans are found in almost every ecosystem on earth. A classification of the planet into anthropocentric biomes (landscapes divided into categories of human activity) estimates that about 60% of the surface is dominated by humans. Because of their large numbers and the huge amounts of resources they use on a daily basis, humans have a greater worldwide impact on ecosystems than any other species. Among the many problems are widespread climatic changes, stratospheric ozone depletion, and loss of biodiversity, all of which are global in scope and long lasting in impact. The origins of most of these problems have been traced to human activities. Strategies for attacking the problems emphasize the need for multinational dialogue and cooperation among specialists in fields as diverse as biology, chemistry, economics, and political science and a need to conserve the resources used.

Global Warming

Although there is evidence that the earth has gone through cycles of warming and cooling in the prehistoric past, there is little question that human activities are accelerating the rate at which global warming is now occurring. The five years from 2014 to 2019 were the five warmest years on record, and the 20 warmest years on record have occurred in the past 22 years. In 2003, the first-ever 37.8°C (100°F) temperatures in England were recorded, and thousands of human deaths were attributed to a summer heat wave that baked the area. Many scientists predict that such heat-related natural disasters will repeat themselves around the world as global warming pushes temperatures higher. In 2018, there were 14 natural disasters in the United States whose costs exceeded $1 billion. These climate-related catastrophes resulted in 247 deaths and $91 billion in damages. The most destructive were Hurricanes Florence and Michael and western wildfires. Globally, glaciers are shrinking, permafrost is disappearing, and sea levels continue to rise as warm water expands and glaciers melt. This *greenhouse effect* is the accumulation in the atmosphere of gases that permit radiation from the sun to reach the earth's surface but prevent the heat from escaping back into space, thus adding to the global rise in temperature. Carbon dioxide and methane, the two most common gases involved in potential global warming, have been part of our atmosphere for millions of years, but others, such as chlorofluorocarbons, are relatively recently produced by-products of the manufacture of refrigerants, plastics, and aerosol cans. In addition to an overall increase in global mean temperatures, climate scientists are predicting that weather patterns will become more erratic, with an increase in the frequency of severe weather events.

Carbon Dioxide

D. L. Lindstrom of the University of Illinois at Chicago and D. R. MacAyeal of the University of Chicago examined records of cores of ancient ice from Siberia, Scandinavia, and the Arctic Ocean. Using computers to simulate the status of ice and atmosphere going back 30,000 years, they found the earth's temperature had increased as the levels of carbon dioxide increased. At the end of the most recent ice age, the rise in temperature was sufficient to melt the ice. Their findings suggest that cycles of ice ages followed by shorter warm periods may have been caused solely by rising and falling levels of carbon dioxide.

Atmospheric carbon dioxide emissions from transportation and burning fossil fuels continue to increase dramatically, leading to universal concern about global warming and climate change. Since 1950, the level of carbon dioxide in the atmosphere has exceeded that of all of recorded history. It continues to climb dramatically (Fig. 25.13a). The rise in atmospheric carbon dioxide was associated with, and almost certainly responsible for, a rise in global temperatures, especially in recent years (Fig. 25.13b). The global temperature

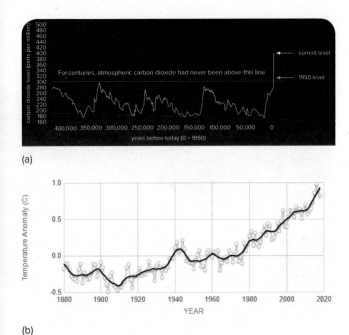

Figure 25.13 (a) Air bubbles trapped in ice have allowed scientists to measure atmospheric carbon dioxide levels for the past 400,000 years. This graph reveals that levels in the past 70 years have been higher than those in the past 400,000 years. (b) This graph shows the temperature anomaly (change compared with the baseline average), which is particularly noticeable from 1960 to 2020. Temperatures continue to climb steadily in the 21st century.
(a) Source: climate.nasa.gov (b) Source: National Oceanic and Atmospheric Administration

is now 0.8°C higher than that in the mid-twentieth century. While this may seem like a trivial change, slight differences in the Earth's thermal budget can make huge differences. At the height of the last Ice Age (20,000 years ago), when glaciers covered Canada and extended to the Ohio River, temperatures were only 4°C (7.2°F) colder than today. The best climate models cannot predict the details of global warming, but they agree that potentially catastrophic temperature increases will occur if trends in greenhouse gases continue. Average global surface temperature increases of 1.5°C (2.7°F) by 2050 and as much as 6°C (11°F) by 2100 are of tremendous concern.

A short list of unwelcome events includes sea levels rising, which would flood such places as the Gulf Coast and much of Bangladesh; more frequent and more extreme storms; huge swings between wet and drought years in places such as southern California; rapidly expanding deserts; dramatic drops in crop yields; massive extinctions due to habitat changes; and the expansion of diseases borne by tropical vectors (e.g., mosquitoes). Higher sea levels occur from both melting ice caps and the fact that water expands as it warms. The U.S. Environmental Protection Agency estimates that for each 30 centimeters (11.8 inches) of sea level rise, the sea advances 30 meters (100 feet) inland. If seas rise as predicted, the coast from Boston to Galveston will be severely impacted. A 50-centimeter rise in sea level could flood many of the world's cities and could create millions of "climate refugees" around the world. Small coral island nations such as Tuvalu and Vanuatu are already threatened with obliteration from rising seas and storm surges.

Changing climate will certainly result in major human population shifts. As deserts expand, agriculture may shift. The midwestern United States grain belt may shift into Ontario, while the fertile crescent of the Middle East may shift into Russia. Wheat production in Australia may cease. Major changes in weather patterns that will cause droughts, intensify desertification in some places, or cause torrential rains in others are likely to occur. Early warnings of such phenomena are noted with increasing frequency. Many species of plants and animals will need to migrate or disperse to new locations if they are to survive. Many species will be unsuccessful due to dispersal barriers and resistance to invasion by the species surviving in new areas.

Increasing temperatures are not spread evenly over the planet; rather, in 2008, the Arctic was experiencing summer temperatures at least 5°C (9°F) above average and an icepack extent 38% below average. This loss is wreaking havoc with indigenous people and mammals like the polar bear. More rapid Arctic and Antarctic warming is occurring because of positive feedback. Ice masses have a high *albedo*; they reflect most incoming solar radiation. As ice melts, the uncovered land or water has a much lower albedo; it absorbs most solar energy. Thus, warming is amplified.

This gloomy scenario of adverse effects of global climate change leaves room for optimism. It is possible, through concerted global action, to reduce the emission of carbon and to improve the earth's ability to absorb it. First there must be a rapid conversion from coal, oil, and natural gas to renewable energy sources such as wind, solar, and tidal energy. Devastated lands must be rehabilitated into productive ecosystems, and erosion must be mitigated. As Nobel Peace Prize laureate Al Gore has suggested, doing these things is necessary, is possible, and makes economic sense.

Methane

Swamps and wetlands have long been known to be sources of methane produced by anaerobic bacteria. Many animals produce methane during digestive processes, and large amounts of this gas are produced by the wood-digesting organisms within the guts of termites. Anaerobic bacteria in rice paddies produce significant quantities of methane. The total annual production of methane in the atmosphere has been slowly increasing in recent years. A small part of this increase may stem from the prodigious mushrooming in numbers of termites in cleared tropical rain-forest areas, which, in 1998, were being destroyed at the rate of more than 100 acres per minute, 24 hours a day.

Methane is a greenhouse gas 23 times as potent as carbon dioxide. As Arctic ice melts, the sea warms and permafrost on the seafloor thaws. Methane once trapped beneath bubbles to the surface. Thus, another positive feedback loop has been revealed. Similar processes are occurring on land as the permafrost melts. The historical record indicates that past sudden releases of methane have caused rapid global temperature increases and possible mass species extinctions. Methane is produced by decomposition under anaerobic conditions. The amount locked up in the Arctic may rival that of carbon found in global coal reserves. There is a potential for swift methane releases to cause a rapid temperature increase that would overwhelm efforts to curb it.

Ozone Depletion

Methane gas and chlorofluorocarbons (CFCs), the inert chemicals used for refrigeration and other industrial purposes, are broken down into active compounds by sunlight at high altitudes. The breakdown products destroy ozone, a form of oxygen that in the stratosphere provides a natural shield for living organisms against ultraviolet radiation. Increased ultraviolet radiation correspondingly increases skin cancers. There is evidence the ozone shield has gone through lengthy cycles of expansion and contraction over millions of years, but there appears to be little question that ozone breakdown has been accelerated by industrial pollutants. The weakening of the ozone shield has been recognized as a serious global problem by North American countries and the European Economic Community. In 1987, the Montreal Protocol on Substances that Deplete the Ozone Layer was finalized. This is a global agreement to protect the ozone layer in the stratosphere by phasing out ozone-depleting substances. Emissions decreased as a result of this protocol. However, they have been rising in recent years. It is difficult to determine the source of these emissions.

Chlorofluorocarbons apparently are not the only force actively destroying the protective ozone layer. Bromine-based compounds called *halons,* which are commonly found in electronic equipment, such as computer protection systems, and in portable fire extinguishers, are reported to be as much as 10 times more destructive of ozone than chlorofluorocarbons. Halon concentrations in the atmosphere increased about 20% a year between 1980 and 1986, according to the Environmental Protection Agency. Some scientists believe the actual concentrations to be as much as 50% higher than that, and as a result of their strong recommendations, powders and other inert gases are being substituted for halons in fire extinguishers.

Erosion

Erosion is a serious global problem. Wind and water attack the land surface, removing productive soil and degrading the land. Coastal erosion, often a dramatic result of a hurricane or storm surges, can reshape the coastal landscape. Dunes that protect property and productive landscapes can be removed in an hour. Cliffs, such as those in Pacifica, California, upon which development has occurred were undercut and collapsed after a major storm. In recent years, there has been a net loss of beach area as a result of intensification of storms and increases in sea level.

Figure 25.14 Trees help to absorb the impact of a tsunami.
Source: Henry Helbush/NOAA/NGDC

Trees are critical for the prevention of coastal erosion. The significance of trees for the protection of shorelines was apparent during the December 26, 2004, Indian Ocean tsunami, which killed 276,000 people. Dense growths of mangrove trees were especially effective at minimizing damage on land. It is estimated that 96% of the land occupied by dense tree growth was undamaged by the tsunami, while only 38% of the land without trees escaped serious damage. The trees were able to break the waves and absorb their energy (Fig. 25.14). Prior to the tsunami, about one-quarter of the mangroves in the Indian Ocean region had been destroyed to make room for agriculture, aquaculture, and development. Land managers in the area are now reconsidering the wisdom of that trend.

In August 2005, Hurricane Katrina hit the Gulf Coast, causing more devastation than any hurricane in U.S. history. Over 1,800 people died and tens of thousands were displaced as a result of the hurricane and its aftermath. Although the devastation is shocking, it did not come as a surprise to scientists. A number of ecological factors contributed to the events that transpired in the late summer of 2005. First, much of the damage occurred in New Orleans, the only major metropolitan area in the United States that is below sea level. Before the 1800s, the Mississippi Delta on which New Orleans was built was regularly replenished by floods that brought sediment from the Mississippi River. However, in the past 200 years, a series of dams were constructed on the Mississippi River, preventing most sediment from reaching the delta. In addition, the levees that were built to protect New Orleans channel any sediment that does reach the South directly into the Gulf of Mexico. The levees also direct sediment away from barrier islands between New Orleans and the Gulf of Mexico. These islands are important because they provide a buffer against hurricane winds and storm surges. Finally, the sea level continues to rise and is expected to rise another 50 to 100 centimeters (20 to 40 inches) in the next century, compounding the effects of storm surges.

Certainly, the hurricane was a natural disaster, but the effects of human intervention compounded the devastation.

The world human population now exceeds 7 billion. Amazingly, it has doubled in less than 50 years. The United Nations has developed high, medium, and low population growth estimates. Based on the medium estimate of 9 billion people by 2050, food production will need to increase by 50% in the next 40 years. With current farming methods, each person needs a half hectare (1.24 acres) to produce a food supply equivalent to that eaten by Western Europeans and North Americans. However, there is only half that amount of arable land in the world. It is obvious that there is a critical need to continue to increase agricultural production to keep up with the demand for food.

In addition, more land must be brought into production by reclaiming derelict and underused lands. For example, degraded pastures can be improved to increase production. Further, crops with greater efficiency must be developed, as described in Chapter 14. The use of genetically modified crops will have to be expanded. Human diets will certainly change in response to food limitations (see the discussion of energy pyramids and ecological efficiencies later in this chapter) and the human population will become increasingly vegetarian. Finally, every effort must be made to curtail human population growth.

The most significant limitation to sustainable productivity is soil erosion. The Midwest Dustbowl of the 1930s provides an illustration of the severe effects of soil erosion. Farmers moving west plowed the grasslands in the Great Plains and planted wheat. The crops grew well until a drought set in during the early 1930s. When the crops dried up, there was nothing to hold the soil, and it blew into the air in huge billows of dust. Settlers were forced westward in an attempt to find another region with productive soil. While the Dustbowl is an extreme example, soil erosion continues to be a major issue.

Topsoil is difficult to replace, yet erosion is removing it faster than ever before. At least 40% of the world's farmland is seriously degraded, and fertile soil in an area equal to the Ukraine is lost annually. Wind blows across semi-arid steppes, removing soil and changing the land to desert. In Turkey, 69% of the landscape suffers from serious soil erosion, and the map of arable land continues to shrink. Countries as different as Iceland and India suffer serious erosion problems. Soil is washed or blown away, taking with it the valuable organic matter that makes the soil fertile. Without organic matter, the ability to soak up water is lost, so water runs off the land, increasing erosion. In addition, runoff carries fertilizers and pesticides into streams and lakes, causing pollution.

Soil erosion is a direct result of overgrazing, clearing land for urbanization and roads, and plowing and is the leading cause of water pollution in the United States. Erosion may be reduced by practices that are aimed at increasing soil organic matter. Effective practices include following crop rotation and using green manure, where a crop is planted to protect the soil, then plowed back into the ground. Reduced tillage and the use of companion crops are tactics that reduce erosion and have other benefits. Increasing soil organic matter will help to offset carbon dioxide emissions.

Aquifer Depletion

While soil erosion is visible as gullies in fields or billowing clouds of dust, another serious resource issue often goes undetected until it is too late to address. The overpumping of aquifers is probably the most underestimated ecological problem worldwide. Roughly 70% of water pumped from underground is used for irrigation, 20% is used by industry, and 10% is used for homes. However, the demand for water for all three uses is increasing rapidly. The United Nations has predicted that by the year 2050, 1.7 billion people will not have enough water to satisfy their basic needs. Most of the world water deficit is in China, India, the Middle East, North Africa, and North America. This problem is relatively recent, because the use of water has tripled since 1950. The demand for water is growing while the sustainable yield of aquifers is fixed, so water deficits grow larger every year. For example, in the Chenaran Plain of northeastern Iran, the water table has been dropping by 2.8 meters (9 feet) per year. However, in 2001, the aquifer plummeted 8 meters (26 feet) due to a drought and additional development.

25.7 Loss of Biodiversity

Learning Outcomes

a. Explain how the loss of biodiversity impacts ecosystems.
b. Describe the effects of acid deposition.
c. Describe ways in which water can become contaminated.
d. Describe how attitudes toward wetlands have changed in recent decades.
e. Discuss ways in which the management of hazardous waste disposal has changed.
f. Discuss the impacts of foreign species on ecosystems.

Most plants and animals are adapted, sometimes in subtle ways, to the habitats in which they occur. When their natural habitats are destroyed, a few species may be able to adapt to new habitats, but most are not capable of doing so and ultimately perish. Ever since living organisms first appeared on this planet, various species have become extinct as climates and other environmental factors changed. However, the rates of extinction have accelerated enormously in just the last half-century as many types of habitats, from tropical rain forests and swamps to deserts, have been damaged or destroyed. Thousands of species are now facing extinction, many within the next 10 years.

If food and fiber plants are not particularly endangered by the loss of natural habitats, why should we be concerned if seemingly unimportant wild species disappear? One of the many consequences of such losses pertains to the origins of our crop plants. Virtually all food, fiber, medicinal, and other useful plants have been developed from wild species, and keeping crops from succumbing to diseases often depends on our ability to breed new disease-resistant strains by tapping the gene pools of the plants' wild cousins. The gene pools usually have developed over thousands or even millions of years, and if they disappear, the most skillful plant breeders may be no match for rapidly mutating fungal and other diseases that attack our crops.

The loss of biodiversity in an ecosystem has been shown to reduce the efficiency of production and nutrient use and to make the ecosystem less resistant to disturbances.

Although funding for solving many types of ecological problems is considered by citizens interested in the environment to be inadequate, the allocations to regional issues usually gain better public support because they are highly visible, and progress in dealing with them can be followed.

In the past three decades, concern about the effects of pollution on lands, waters, and peoples has been widely reported in the media and in scientific journals. During this time, attempts to correct or stem environmental damage have resulted in requirements that many construction projects file environmental impact reports before proceeding. These reports provide information that helps various agencies evaluate the effects a proposed project may have on the flora, fauna, and physical environment and determine what *mitigations* (alleviations of the problem) may be necessary before the project can or should be approved. The process has sometimes resulted in a great deal of controversy and emotional debate between landowners who believe their rights are being usurped and those who are convinced that preservation of the environment should supersede material considerations. Brief discussions of some regional issues follow.

Acid Deposition

Acid deposition occurs after fossil fuels are burned. The burned fuels release sulfur and nitrogen compounds into the atmosphere. There, chemical reactions brought about by sunlight and rain convert the compounds to nitric acid (HNO_3). Burning may result in the conversion of sulfur dioxide (SO_2) to sulfuric acid (H_2SO_4). These may combine with moisture droplets and become a liquid acid deposition (*acid rain*), which has a pH of less than 5.5, as compared with a pH of about 5.7 for unpolluted rain water. Some of the acid deposition, however, may be in the form of solid particulates. The acid deposition, in turn, adversely affects or even kills many living organisms on which it falls and increases the concentration of other toxic elements. Mycorrhizal fungi are beneficially associated with the roots of many trees and are susceptible to acidified soil, which may be linked to acid deposition. Higher acidity of soil water also reduces the capacity of plant roots to obtain needed mineral nutrients.

Trees have been dying prematurely in parts of the world's forests. Some have died because of a lack of water during successive years of low rainfall or from other causes, such as insect infestations. Salt scattered to melt ice and snow on roads has affected trees in the immediate vicinity. There seems to be little question that acid deposition has had a major role in stunting or destroying trees downwind from industrial sites. In recent years, there have been increasing international attention and cooperation in reducing industrial and vehicle emissions that play a role in the production of acid deposition.

Acid deposition also alters soil fertility because large amounts of nitrogen accumulate (eutrophication). Soils that are well buffered to resist pH changes can become more fertile. While greater fertility may appear beneficial, the result is often the loss of plant species in the face of competition. Boreal forests with dense moss layers and a preponderance of stress-tolerant plants are particularly susceptible. Fast-growing species can overwhelm both. In many cases, invasive grasses can alter the composition of pastures.

Water Contamination

Water becomes contaminated in different ways. Much of the pollution in our lakes and streams comes from the dumping of toxic industrial wastes and from runoff over polluted land. Other sources include the spraying of pesticides, the exhaust and other emissions of aircraft and ships, and airborne pollutants originating with the combustion of fossil fuels. Ground water supplies become polluted from pesticides and wastes trickling through the soil, from septic tanks, and from garden and farm fertilizers. Even deep wells, which seldom became polluted in the past, are increasingly becoming contaminated with unacceptably high levels of nitrates and other substances that harm the health of water-dependent humans and animals. Blue-baby syndrome, or cyanosis in newborns, is one example of the effects of high nitrate levels in drinking water. In response to the problem, many communities are improving their water treatment plants, and many families are installing water filtration systems in their homes.

Long-range goals for curbing water pollution include a tightening of restrictions on the dumping of wastes and greater improvements in municipal water purification plants and systems. Genetic engineering and bacteria also will play a major role in the future. For example, a bacterium that can remove more than 99% of 2,4,5-T (a major component of Agent Orange—discussed in Chapter 11) from a contaminated environment has already been bred, and several other bacteria are being genetically engineered to enhance their capacities for breaking down other toxic wastes.

Wetland Destruction

Swamps, marshes, bogs, lagoons, river margins and estuaries, floodplains, and other wetland areas have historically been regarded as wastelands and have routinely been drained and converted to drier, more "productive" agricultural and industrial sites. For example, only 6% of California's original wetlands, 2% of Iowa's original wetlands, and less than 10% of the wetlands once along the shores of Lake Ontario remain today. Wetlands are, however, far from the wastelands they were once thought to be. Consider that the plants, algae, and other organisms in tidal wetland can perform the same recycling functions that wastewater treatment equipment is capable of—at no cost to taxpayers. The wetlands

Ecology

Figure 25.15 A view of Florida wetlands as seen from the air.
© Jupiterimages/Getty Images

(Fig. 25.15), at the same time, provide habitat, shelter, and breeding grounds for a wide variety of wildlife, many species of which are now threatened with extinction. They also provide many other ecosystem functions. Wetlands purify streams and lakes and reduce erosion. By trapping and gradually releasing water, they smooth out stream flows and reduce flooding. Finally, wetlands offer intangible aesthetic and educational value and direct economic benefits related to many recreational activities, such as fishing and canoeing.

Some attitudes toward wetlands have begun to change as the public has slowly become aware of the folly of eliminating them. City and regional planners in many areas now either ban new developments on wetlands or require a proposed development that would encroach on a wetland to provide mitigation measures, such as the creation of new wet areas equal in extent and quality to those that may be lost. In addition, projects now in progress in many areas involve the restoration of damaged or lost wetlands.

Hazardous Waste

Members of earlier generations routinely drained the oil from their vehicle crankcases directly onto the ground or into storm drains, or they took the used oil and other hazardous wastes to the dump. Highly toxic industrial wastes were also disposed of both within and at the outskirts of cities and towns. In 1996, it was estimated that some 12 million children in the United States were living less than 4 miles from a hazardous waste site. Disasters have occurred when living organisms (including humans) have not been properly isolated from radioactive wastes produced by atomic energy plants. Even when hazardous wastes aren't unceremoniously dumped, serious accidents and spills take place, with the effects sometimes lingering indefinitely, as, for example, in and around the former Soviet Union's Chernobyl atomic meltdown site.

Today, there are concerted efforts to curb the disposal of hazardous wastes and to greatly reduce the probabilities of accidents and spills. At most solid-waste dumps, it is illegal to dispose of even empty latex paint cans, let alone more toxic materials, and heavy fines are levied on those found disposing of industrial wastes in an improper manner. Monies from a U.S. government "superfund" are being used, with some success, to clean up selected old hazardous waste sites. The process thus far is slow and inadequate, but the increased restrictions on disposal methods and, as previously noted, the genetic engineering of bacteria that can dismantle and render harmless many types of wastes hold promise for the future.

Invasion of Foreign Species

Each year, alien plants, animals, and pathogens cause billions of dollars in damage in the United States alone, entering the country as stowaways in airplanes, trucks, or ships and sometimes escaping from the gardens, fields, and yards of people who unwittingly contribute to the problem. These invasive species are often aggressive weeds, reproducing quickly and crowding out native plants and crops. Hawaii, which is home to about 10,000 endemic species, is vulnerable to invasive species, because it is one of the most isolated groups of islands in the world. As a trade and tourist hub, Hawaii is visited by countless ships and planes each year, each potentially carrying foreign plants. Indeed, Hawaii is now home to 47 of the 100 most invasive species in the world. When these alien plants are introduced in areas where there are no natural pests or herbivores, they will experience selection for individuals that devote more energy to reproduction and less to defense, allowing them to outcompete native plants. Most established foreign species are successful because they exhibit more phenotypic plasticity (physiological tolerance) or genetic differentiation (rapid evolution) than do their natural counterparts.

Kudzu, a legume from Japan, was brought into the United States in the late 19th century to help control soil erosion. Today, the climbing vine affects millions of acres of forests and farmland all over the southern United States, killing other plants with shade and weight, and covering houses, light poles, and electric lines. The tamarisk (salt cedar) was brought to the United States from its native western Asia habitats in the 1830s as an ornamental plant that could also be used as a windbreak. In the United States, the tamarisk has no natural enemies, can withstand heat and drought, and can grow in extremely salty soils. Each shrub produces 500,000 to 1 million seeds each year, which has helped the tamarisk to spread throughout the dry areas of many western states. Because each plant can use up to 300 gallons of water per day—much more than the native plants that it is replacing—the tamarisk is threatening the water resources of the areas it has invaded.

In view of the tremendous economic damage caused by exotic species, foresters, agronomists, and conservation biologists are striving diligently to control these species.

KEY THEME 25.1 ecology

John Muir, Father of America's National Park System

Today, America's national parks are overcrowded as some 270 million people enjoy their beauty each year, relaxing in the great outdoors, hiking in fields and mountains, and navigating their rivers. Whether it is Yosemite, the Grand Canyon, Acadia, or one of the many others, each park has a unique appeal. These magnificent parks are the result of the vision and foresight of many people, perhaps none more so than John Muir, who lived from 1838 to 1914. Often called the "Father of the National Park System," he influenced presidents, members of Congress, and "just plain folks" by his love of nature and his writings about it. Always traveling and exploring various ecosystems, he visited Alaska five times and walked 1,000 miles from Indianapolis to the Gulf of Mexico, as well as across California's San Joaquin Valley to explore the Sierra Nevada. Later he wrote of viewing the Sierra Nevada range for the first time: "Then it seemed to me the Sierra should be called not the Nevada, or 'Snowy Range,' but the Range of Light, the most divinely beautiful of all the mountain chains I have ever seen."

Wherever he roamed in the wilderness, he noticed increasing waste and destruction from various sources, whether overgrazing by cattle and sheep or overcutting by loggers. It was his love for the American forest that drove him to work tirelessly for its preservation. He wrote some 300 articles and 10 books that told of his travels and contained his naturalist philosophy. He called to everyone to "climb the mountains and get their good tidings." He wrote that the wilderness should be preserved, viewed as a treasure rather than a resource to be exploited. Through his writing, he drew attention to the destruction forests were facing and worked to find remedies.

One solution was the formation of national parks. Largely through Muir's efforts, in 1890 Congress created Yosemite National Park. He was also personally involved in the creation of Sequoia, Grand Canyon, Mount Rainier, and Petrified Forest national parks. He went one step further by founding the Sierra Club in 1892 and served as its first president until his death in 1914. He and his friends conceived the Sierra Club with the mission to "do something for wilderness and make the mountains glad." His idea was to form an association that would work to protect Yosemite National Park.

This brought him to the attention of President Theodore Roosevelt, who visited Muir in Yosemite in 1903. Together with Muir, under the trees of the park, Roosevelt's innovative conservation programs began to take form.

John Muir's words and life inspire environmental activism today; they also teach the importance of enjoying nature and protecting it. Muir wrote a warning that is just as timely today as it was 100 years ago:

> The axe and saw are insanely busy, chips are flying thick as snowflakes and every summer thousands of acres of priceless forests, with their underbrush, soil, springs, climate, scenery and religion, are vanishing away in clouds of smoke, while, except in the national parks, not one forest guard is employed. Any fool can destroy trees. They cannot run away; and if they could, they would still be destroyed—chased and hunted down as long as fun or a dollar could be got out of their bark hides, branching horns or magnificent bole backbones. Few that fell trees plant them; or would planting avail much towards getting back anything like the noble primeval forest.[1]

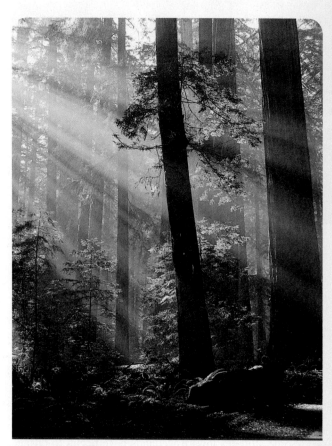

Box Figure 25.1A Shafts of light penetrate the coastal redwoods of Muir Woods. © *Kingsley Stern*

A living memorial to John Muir is located 12 miles north of San Francisco, Muir's home in his later years. Muir Woods (Box Figure 25.1A) is a 560-acre national monument consisting of a grove of majestic coastal redwoods along a canyon floor. The towering redwoods and lush canyon ferns growing along Redwood Creek make this a forest of tranquility. Only a 30-minute drive from San Francisco, Muir Woods is visited by over 1.5 million people each year. He declared that this grove of coastal redwoods was the "best tree-lovers' monument that could possibly be found in all the forests of the world."

D. C. Scheirer

1. Atlantic Monthly, *No. 80*, 1897.

It is a daunting challenge, but many tactics can be applied. Where a single plant species invades a landscape, it may be controlled by carefully introducing a specific herbivore. A cactus (*Opuntia tomentosa*) was imported into Australia from South America in the mid-19th century to use as fence posts and to establish a dye industry based on the cochineal bug that eats *Opuntia*. Neither plan worked, but the cactus soon rendered 40,000 square kilometers of farming land next to useless. The inspired introduction of a moth (*Cactoblastis cactorum*) from the native range of the cactus in 1925 brought the cactus under control. In other cases, manual or chemical control can be used, but in most cases, invasive species persist. They thrive because they are reasonably well adapted to the climate of the invaded area and no longer are subjected to their old pests and diseases.

25.8 Restoration of the Land

Learning Outcome

a. Describe the goals of restoration ecology.

The recognition that damaged lands and devastated ecosystems must be returned to productive use led to the relatively recent development of the science of restoration ecology. While conservation biology attempts to save natural systems and species, restoration ecology assumes that much environmental damage can be mitigated. The focus of this science is distinctly applied, defined as taking actions that initiate or accelerate the return of ecosystem health, integrity, and sustainability.

Earlier in this chapter, succession was discussed in terms of the natural recovery of communities. Left unsaid was that many forms of damage inflicted by human activities are so severe that natural recovery cannot occur. R. J. Hobbs describes severely damaged systems as having been driven beyond a threshold from which they cannot recover. Restoration is the art and science of applying successional concepts to assist and accelerate the recovery process. Restoration ecology is developing into a force to help repair the many forms of damage reflected in many urban, industrial, and agricultural landscapes. E. O. Wilson emphasized that restoration ecology "is the means to end the great extinction spasm. The next century will, I believe, be the era of restoration in ecology." Without concerted efforts to return the planet to health, the future well-being of our species, and all the others, is in doubt.

SUMMARY

25.1 Introduction to Ecology
- Ecology is the study of the relationships of organisms and their environment. It is a vast field with many facets.

25.2 Plants and the Environment
- The environment of a plant's habitat is determined by both living and nonliving factors. Plants reflect and respond to their environment.
- Populations vary in numbers, density, and the total mass of individuals. Populations of one community may also occur in another community, and ecotypes may be specifically adapted to a single community.
- Precipitation, temperature, soils, and biotic factors play roles in determining the distributions of plant species in an ecosystem.
- Xerophytes and hydrophytes have modifications of leaves and other organs.
- Species distribution is influenced by soil mineral content and biotic factors, such as competition for light, nutrients, and water.
- Ecosystems sustain themselves through photosynthetic activity, energy flow through food chains, and the recycling of nutrients. Producers carry on photosynthesis; consumers feed on producers. Primary consumers feed directly on producers, and secondary consumers feed on primary consumers.
- Decomposers break down organic materials to forms that can be reassimilated by the producers. In any ecosystem, the producers and consumers compose food chains that determine the flow of energy through the different levels. Food chains vary in length and intricacy and, because of their interconnections, form food webs.
- Energy itself is not recycled in an ecosystem; it escapes in the form of heat as it passes from one level to another. In a long food chain, the final consumer gains only a tiny fraction of the energy originally captured by the producer at the bottom of the chain.
- When producers increase the amount of food available, consumers may increase correspondingly and compete for the food; this reduces what is available, resulting in a self-maintaining ecosystem. The composition of an ecosystem may be influenced by its living components through the secretion of growth-inhibiting and growth-promoting substances.

25.3 Life Histories
- Annual plants complete their life cycle in 1 year, while biennials require 2 years. Perennial plants live for more than 2 years.

25.4 Natural Cycles
- Some nitrogen from the air is fixed by the nitrogen-fixing bacteria found in legumes and other plants. In the nitrogen cycle, there is a constant flow of nitrogen from dead plant and animal tissues into the soil and from the soil back to the plants.

- Water leaches nitrogen from the soil and carries it away when erosion occurs. Other nitrogen is lost from harvesting crops, but the loss can be reduced if wastes are decomposed and annually returned to the soil. Fire also causes nitrogen loss.
- Replacement of nitrogen loss by the application of chemical fertilizers may eventually create hardpan by altering the soil structure.
- Bacteria also recycle carbon and other substances, such as water and phosphorus.

25.5 Succession

- Succession occurs whenever there has been a disturbance of natural areas on land or in water.
- Primary succession involves the formation of soil in the beginning stages, while secondary succession takes place in areas previously covered with vegetation. When a primary succession begins with bare rock or lava, the activities of lichens, plants, and physical forces convert the rock or lava to soil in a progression of events over a period of time.
- A stable community becomes established at the conclusion of succession and remains until or unless a disturbance disrupts it.
- A primary succession can be initiated in a wet habitat and culminates in a climax community. As a lake or other body of water is filled in with silt and debris, eutrophication facilitating the growth of algae and other organisms occurs.
- Secondary succession, which proceeds more rapidly than primary succession, may take place if soil is present. It may occur after fires.

25.6 Impact of Humans on Plant Communities

- Impacts of humans on plant communities occur at both global and regional levels.
- A global rise in temperature due to carbon dioxide, methane, and other gases preventing the sun's radiant heat from escaping back into space is referred to as the greenhouse effect.
- The earth's temperature and the carbon dioxide and methane levels in the earth's atmosphere have been rising. It is predicted that polar ice will melt as a result, and flooding of low-lying coastal areas will occur.
- In the stratosphere, sunlight converts methane, chlorofluorocarbons, and halons into active compounds that destroy the ozone shield that protects us from intense ultraviolet radiation.
- Beach and soil erosion are increasingly serious environmental problems with no simple solutions.
- The depletion of aquifers may lead to serious water shortages in the near future.

25.7 Loss of Biodiversity

- Loss of biodiversity has serious consequences, including the potential loss of means with which to combat crop diseases.
- Regional ecological issues include acid deposition, water contamination, wetland destruction, hazardous waste, and the invasion of foreign species.
- Acid deposition damages or kills living organisms; it is produced when sulfur and nitrogen compounds released by the burning of fossil fuels are converted to nitric acid and sulfur dioxide by sunlight and rain. It can also cause eutrophication in both water and land habitats.
- Water contamination occurs when toxic wastes, pesticides, septic tanks, and fertilizers wash or leach into surface and ground water. Bacteria may in the future be used to break down various contaminants.

25.8 Restoration of the Land

- Wetlands were once considered wastelands. They have been drained but now increasingly are being protected because they provide many valuable services.
- Hazardous waste sites are gradually being detoxified, but progress is slow.
- Foreign invasive species may crowd out or destroy native vegetation.
- Restoration ecology is a branch of science devoted to the mitigation of environmental damage.

REVIEW QUESTIONS

1. How does acid deposition occur?
2. What are the principal threats to global drinking-water sources?
3. Distinguish among a plant community, a biotic community, an ecosystem, and a biome.
4. What is the difference between a primary consumer and a secondary consumer?
5. How are short food chains more efficient in the use of solar energy than long ones?
6. Why is the diet of larger herbivorous animals varied?
7. What types of checks and balances exist in an ecosystem?
8. How are nitrogen and carbon recycled in nature?
9. What is primary succession as opposed to secondary succession?
10. What are general characteristics of climax vegetation as compared with nonclimax vegetation?
11. What is the greenhouse effect?

12. What is the ozone shield, and what significant role does it play? What threats are there to its existence?

DISCUSSION QUESTIONS

1. Humans have disrupted ecosystems almost everywhere they have established themselves, at least in industrialized countries. Do you believe that humans could also improve an ecosystem?
2. If a vegetarian diet makes more efficient use of solar energy, should we all strive to become vegetarians for this reason?
3. Besides the hypothetical example given (in which it was observed that eating corn directly was a more efficient use of solar energy than feeding it to cattle and then eating the beef), can you think of other ways in which food chains might be shortened?
4. Peanut butter has many of the nutrients needed in human nutrition. On the basis of what you have learned about the diet of animals in an ecosystem, do you think it would be a good idea to live on peanut butter and water as a means of saving money?
5. Could succession take place in an abandoned swimming pool?

CHAPTER

26

Biomes

26.1 Introduction to Biomes
26.2 Major Biomes of the World
　　Tundra
　　Taiga
　　Temperate Deciduous Forest
　　Grassland
　　Key Theme 26.1: Alpine Flora as an Indication of Climate Change: The GLORIA Project
　　Desert
　　Tropical Rain Forest
　　Savanna
　　Chaparral

OVERVIEW

This chapter examines and discusses the major biomes of North America, as well as other biomes of the world. Characteristic climatic and other factors associated with each biome are mentioned. Descriptions and listings of principal plant and animal life are included. A discussion of the future of the tropical rain forest and introductory information about other biomes are presented at the end of the chapter.

During the brief growing season, plants of the arctic tundra in central Alaska produce a stunning array of flowers. © *Digital Vision/PunchStock RF*

26.1 Introduction to Biomes

Learning Outcome

a. Describe how the composition of a biome is determined.

Terrestrial biotic communities considered on a global or at least on a continental scale are referred to as **biomes.** Several important biomes occupy the North American continent, and most are represented on other continents (Fig. 26.1). Each biome has a characteristic mix of plants and animals adapted to that region's environment. Climate is the major influence determining the composition of a biome. While each biome is considered as a separate entity in this chapter, it is important to remember that biomes blend together and interact with each other.

26.2 Major Biomes of the World

Learning Outcome

a. List major biomes of the world, and describe the characteristics and principal living members of each.

Tundra

Tundra occupies vast areas of the earth's land surfaces (about 20% in all), primarily above the Arctic Circle (Fig. 26.2). It looks treeless, although miniature willow and birch only 2.5 to 50 centimeters (1 inch to 2 feet) tall do survive in some areas. Another characteristic of arctic tundra is the presence of *permafrost* (permanently frozen soil) at a depth of from 10 centimeters (4 inches) to about 1 meter (3 feet 3 inches) below the surface. The level of the permafrost determines the depth to which plant roots can penetrate. Precipitation averages less than 25 centimeters (10 inches) a year, but because of the permafrost, it is largely held at or near the surface of the land. There is a very short growing season of only 2 to 3 months, with frost possible on any day of the year. However, temperatures can soar to 27°C (81°F) or higher during a long midsummer day. The vegetation of arctic tundra is dominated by dwarf shrubs, sedges, grasses, lichens, and mosses. During the brief growing season, tiny perennial plants produce brightly colored flowers and form brilliant mats over the topsoil, which is largely organic and generally only 5.0 to 7.5 centimeters (2 to 3 inches) deep.

Patches of alpine tundra occur above timberline in mountains at latitudes below the Arctic Circle. Permafrost is frequently absent from alpine tundra, and grasses, sedges, and herbs tend to predominate.

The tundra biome is exceptionally fragile. A car driven across the tundra compresses the soil sufficiently to kill plant roots, and the tracks are evident many years later. Occasionally, sheep grazing tundra have been observed to pull up patches of the matted vegetation, leaving exposed edges. High winds then can catch the exposed edges and rip away larger segments of mat, leaving barren patches called *blowouts*.

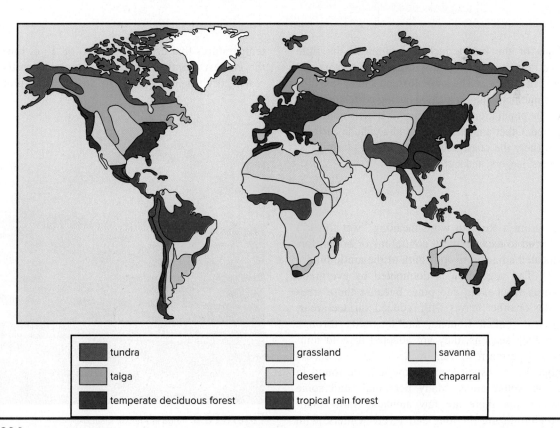

Figure 26.1 Major biomes of the world.

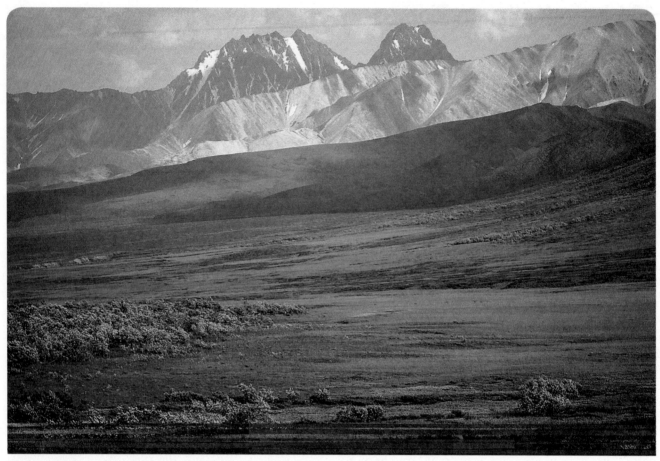

Figure 26.2 Tundra in Alaska. © *Kingsley Stern*

Animals of the tundra include lemmings, the little rodents that reproduce prodigiously every 3 or 4 years and then decline in numbers. The arctic fox and snowy owl also fluctuate in numbers, the fluctuations being correlated with the size of the populations of lemmings that constitute their principal food. Other animals of the tundra include caribou, polar bears (near the coast), shrews, marmots, ptarmigans, loons, plovers, jaegers, and arctic terns.

Taiga

The *taiga* (from a Russian word meaning "wet forest") is also referred to as a northern coniferous or boreal forest. It is located adjacent to and south of the arctic tundra (Fig. 26.3). The vegetation is dominated by evergreen trees, such as spruce, fir, and pine. Because these trees have needles or other leaves with reduced surface areas, thick cuticles, sunken stomata, and longevity through several growing seasons, they are adapted to cold temperatures and permafrost. Tamarack (larch) also occurs in the taiga. Birch, aspen, and willow may be found in some of the wetter areas. Many perennials and some shrubs occur, but there are few annuals. Snow blankets the region during the long and severe winters, with temperatures that drop to −50°C (−58°F) or lower during the coldest months. In summer, the temperatures often reach 20°C to 30°C (68°F to 86°F). Most precipitation

Figure 26.3 Taiga in Alaska. © *Kingsley Stern*

occurs in the summer, ranging from about 25 centimeters (10 inches) to more than 100 centimeters (39 inches) in parts of western North America.

A variety of birds, including jays, warblers, and nuthatches, inhabit the region, which is dotted with many lakes, ponds, and marshes. Rodents, such as shrews and jumping mice, and larger mammals, notably moose and deer, are found in the taiga. Ermine and wolverines also make their home in the taiga, and caribou overwinter in this biome.

Temperate Deciduous Forest

Most deciduous trees are broad-leaved species that shed their leaves annually during the fall and remain dormant during the shorter days and colder temperatures of winter. Most *temperate deciduous forests* (Fig. 26.4) occur, like the taiga, on large continental masses in the Northern Hemisphere. In North America, an example of this type of forest occurs from the Great Lakes region south to the Gulf of Mexico and extends roughly from the Mississippi River to the eastern seaboard, although some do not consider the eastern coastal plain to be part of the temperate deciduous forest biome. Temperatures within the area vary a great deal but can range from −30°C (−22°F) in midwinter to 30°C (86°F) in summer. The climax trees

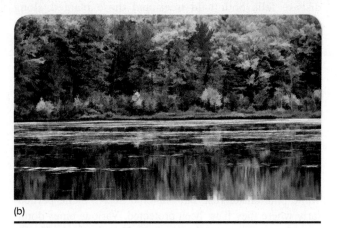

Figure 26.4 Splendor of the forest. (a) Scene within the eastern deciduous forest. (b) Morning mist over the colorful autumn forest landscape. *(a) © Robert A. Schlising; (b) © Larry Gerbrandt/Getty Images*

of the forest are well adapted to subfreezing temperatures as long as the cold is accompanied by precipitation or snow cover that protects the roots from temperature fluctuations. Most of the annual precipitation, which totals between 50 and 165 centimeters (20 to 65 inches), occurs in the summer.

Some of the most beautiful of all the broad-leaved trees are found in a variety of associations in this biome.

In the upper Midwest, sugar maple and American basswood predominate. Sugar maple, birch, and oak are also found to the northeast, where they tend to be associated with the stately American beech. In the west-central part of the forest, oak and hickory abound. Oaks also are abundant along the eastern slopes of the Appalachian Mountains, where the American chestnut was once a conspicuous part of the flora. The chestnut has now virtually disappeared, having succumbed to chestnut blight disease, which was introduced and began killing the trees during the early 1900s. Oak and hickory extend into the southeastern United States, where they become associated with pine, bald cypress, and other species of trees.

Before the arrival of European immigrants, a mixture of large deciduous trees, which included maple, ash, basswood, beech, buckeye, hickory, oak, tulip tree, and magnolia, was found on the eastern slopes and valleys of the Appalachian Mountains. Some of the trees grew over 30 meters (100 feet) tall and had trunks up to 3 meters (10 feet) in diameter. Except for a few protected pockets in the Great Smoky Mountains National Park, the largest trees of this rich forest have been almost eliminated through logging. An extension of this forest, but with smaller trees, is found in an area northeast of Baton Rouge, Louisiana, in western Tennessee and Kentucky, and in southern Illinois. The American elm, also once a part of the forest, is rapidly disappearing as Dutch elm disease fells both wild trees and those planted along city streets. One small midwestern town that had some 600 elm trees planted along its streets found itself left with only six live trees within a year or two after Dutch elm disease struck.

A mixture of deciduous trees and evergreens occurs on the northern and southeastern borders of the forest. Hemlock and eastern white pine are found from New England west to Minnesota and south to the Appalachians. The once vast stands of eastern white pine are now largely gone, their valuable lumber having been used for construction and other purposes. Some have been lost to still another tree disease, white pine blister rust, but scattered trees remain, particularly where they have been protected. Various pines dominate the eastern coastal plain from New Jersey to Florida and westward to east Texas, with pitch pine being common in New Jersey. Some of the southern pines (e.g., longleaf, slash, loblolly) are now cultivated in the southeastern United States for wood pulp, turpentine, lumber, and other commercially valuable products.

During the summer, the trees of the deciduous forests form a relatively solid canopy that keeps most direct sunlight from reaching the floor. Many of the showiest spring flowers of the region (such as bloodroot, hepatica, trillium, and violet) flower before the trees have leafed out fully. These wildflowers complete most of their growth within a few weeks. Other plants that can tolerate more shade, principally members of the Sunflower Family (e.g., aster and goldenrod), flower in succession in forest openings from midsummer through fall.

Animal life in the eastern deciduous forest includes red and gray foxes; raccoons; opossums; many rodents, such as gray squirrels; snakes, such as copperheads and black rat snakes; salamanders; and a wide variety of birds, including hawks, flickers, and mourning doves.

Included in the temperate deciduous forest biome on many maps showing major biomes of the world is the mountain and coastal forest biome. This biome deserves special treatment as a significant type of biome with particularly interesting characteristics.

Mountain and Coastal Forest

In the geologic past, deciduous forests were found throughout western North America. As the climate changed and summer rainfall was reduced, conifers largely replaced the deciduous trees, although some (e.g., maple, birch, aspen, oak) still remain, particularly at lower elevations. Today, coniferous forests occupy vast areas of the Pacific Northwest and extend south along the Rocky Mountains and the Sierra Nevada and California coast ranges. Isolated pockets of the mountain and coastal biome also occur in other parts of the West, particularly toward the southern limits of the mountains. The trees tend to be very large, particularly in and to the west of the Cascade Mountains of Oregon and Washington and on the western slopes of the Sierra Nevada. Part of the reason for the huge size of trees such as Douglas fir is the high annual rainfall, which exceeds 250 centimeters (100 inches) in some areas. The world's tallest conifers, the coastal redwoods of California (Fig. 26.5), grow at low elevations between the Pacific Ocean and the California outer coast ranges. Here their size and longevity apparently depend more on fog, which reduces transpiration rates, than on large amounts of rain. One of the characteristics of mountain forests is a fairly conspicuous altitudinal zonation of species. In other words, one encounters different associations of plants as one proceeds from sea level up the mountain sides. At lower elevations in both the Rocky Mountains and the Sierra Nevada, the predominant conifer is ponderosa pine. At lower elevations in the northern part of the Cascades, Douglas fir, western red cedar, and western hemlock are more common. At intermediate elevations in the Sierra Nevada, sugar pine, white fir, and Jeffrey pine take over, while at higher elevations, other species of pine, fir, and hemlock predominate.

Most of the mountain forest biome has comparatively dry summers. This has led to frequent forest fires, even before human carelessness became a factor. Several tree species are well adapted to fires. The Douglas fir, for example, has a very thick, protective bark that can be charred without transmitting sufficient heat to the interior to damage more delicate tissues, and its seedlings thrive in open areas after a fire. When the bark of the giant redwoods of the Sierra Nevada is burned, the trees are rarely killed. This undoubtedly has contributed to the great age and size of many of the trees. The cones of some of the pine trees (e.g., knobcone pine) remain closed and do not release their seeds until a fire opens them, while seeds of several other species germinate best after they have been exposed to fire. These attributes of members of the mountain forest biome, as mentioned in the section on fire ecology in Chapter 25, have led to the practice in some of our national parks of occasionally allowing fires at higher elevations to run their natural course. The higher-than-normal incidence of fires occurring since humans arrived in large numbers to the forest has made it necessary, however, to control fires in most instances, even though in doing so we may be interfering with natural cycles that would otherwise occur.

Animal life in the mountain forests includes many different rodents (especially chipmunks and voles), bears, mountain lions, bobcats, mountain beavers (not related to true beavers), mule deer, and elk. Large birds, such as golden eagles, and many small birds, including mountain chickadees, warblers, and juncos, are an integral part of this biome.

Grassland

Naturally occurring *grasslands* are found toward the interiors of continental masses (Fig. 26.6). They tend to intergrade with forests, woodlands, or deserts at their margins, depending on precipitation patterns and amounts. A grassland may receive as little as 25 centimeters (10 inches) of rainfall or as much as 100 centimeters (39 inches) annually. Temperatures can range from 45°C (113°F) in midsummer to –45°C (–50°F) in midwinter. North American grasslands known as *prairies* grew on fertile soils and provided grazing for huge herds of buffalo. The buffalo disappeared, however, as the settlers cultivated more and more of the land and hunters increasingly slaughtered the large animals. By 1889, there were only 551 left, but breeding and conservation subsequently increased their numbers. Large areas are now used for growing cereal crops (particularly corn and wheat) and for grazing cattle.

Before they were destroyed, American prairies were a remarkable sight. In Illinois and Iowa, the grasses grew over 2 meters (6 feet 6 inches) tall during an average season and another meter taller during a wet one. A dazzling display of wildflowers began before the young perennial grasses emerged in the spring and continued throughout the growing season. Over 50 species of flowering plants can be seen in bloom at the same time on 1 hectare (2.47 acres) of protected grassland in the middle of spring.

Areas with a Mediterranean climate (e.g., the Great Central Valley of California), where most of the precipitation takes place in the winters, generally support grasses that are shorter than those of the midwestern prairies. Such grasslands usually include *vernal pools*, some of which may be more than 50,000 years old. The vernal pools are temporary accumulations of rain water that evaporate after the rains come to an end. Their

(a)

(b)

Figure 26.5 The world's largest conifers. (a) Coastal redwoods in the mountain forests of Jedediah Smith State Park, California. (b) Relative size of a redwood tree; the photo shows that its base can exceed the width of a car. *(a) © Jim Zuckerman/Getty Images; (b) © Kingsley Stern*

Figure 26.6 Naturally occurring grasslands. (a) Little bluestem (*Schizachyrium scoparium*) among other species from a prairie setting in Illinois. (b) View of the Tallgrass Prairie Preserve (Oklahoma and Kansas), which is the largest remnant of tallgrass prairie on earth. *(a) © Doug Sherman/Geofile; (b) © Clint Farlinger/Alamy*

Figure 26.7 Zonation in a vernal pool in northern California. *© Robert A. Schlising*

unique flora include an orderly sequence of flowering plants, some appearing initially at the pool margins, with each species forming a distinct zone or band until the water is gone (Fig. 26.7). Some species flower only in the damp soil and drying mud that remain. The seeds of many species germinate under water.

Grassland animals include cottontails, jackrabbits, gophers, mice, and pronghorns. As indicated earlier, buffalo were once abundant. Many of these large animals now are protected in national parks, game preserves, and private ranches. Various sparrows (e.g., vesper and savannah sparrows) and other birds still find homes in uncultivated tracts of this biome.

KEY THEME 26.1 ecology
Alpine Flora as an Indication of Climate Change: The GLORIA Project

During the past century, scientists have observed changes in alpine flora and in the elevation of tree lines. In Scandinavia, for example, tree lines have risen by 160 meters (525 feet). When conditions become less harsh, a greater number of species can survive. Species richness of many alpine peaks has increased, and on one summit in the European Alps it increased from one to 10 species over 100 years. That would be expected if the cold alpine elevations were warming. But until the last 10 years, there was not a standard, universal, and replicable process for assessing such changes.

In response to the need for better scientific data and more valid comparisons, the Global Observation Research Initiative in Alpine Environments (GLORIA, http://www.gloria.ac.at/) was conceived at the University of Vienna and established at field sites in Europe in 2001. GLORIA defines a protocol, with resurveys every 5 years, that can be applied to alpine summits anywhere in the world. The alpine zone is ideal for studying the biological effects of global climate change (Box Figure 26.1A) because human disturbance is often minimal, and alpine regions span nearly all latitudes and elevations. Alpine studies also help scientists learn about many of the major climate zones of the world, including maritime and continental, tropical, mid-latitude, and polar.

The alpine zone is defined as that above the upper tree line—a remarkably consistent elevation in a given region, with the trees dropping out completely over elevation increases of just meters. Tree foliage is effectively cooled by the airflow. Tree root zones are shaded by the crowns and remain close to the average air temperature. In effect, a tree experiences a cold temperature limit dictated by average air temperature. Average temperature in the lower atmosphere declines steadily with altitude. Therefore, above tree line elevation, the growing season is simply too cool to sustain tree growth.

Alpine plants survive in the cooler zone because their small stature reduces airflow cooling and allows them to be sun-warmed well above air temperature. Unshaded alpine soils are also warmed above average air temperature.

As highly adapted as they are, alpine plants still live at the edge of adequate warmth and benefit from the lack of competition from the cold-limited trees, so they are very vulnerable to temperature changes. Alpine plant ecosystems are also dependent on snow distribution. Snow shelters many of them from extremes of winter cold and desiccation. Late-season snowfields retard growth until they melt and, upon melting, release a flush of nutrients and water. Alpine vegetation patterns are strongly controlled by the distribution of snowfields.

It is expected that at least two variables related to climate change could significantly affect alpine ecosystems: warmer temperatures and the amounts and patterns of precipitation and snowmelt. Warmer temperatures can raise the tree line and will raise the elevations at which a variety of alpine and subalpine plants can be more competitive. Changes in patterns of snow deposition and melt schedule can also alter alpine plant distributions.

Preliminary results from a 1994–2004 study of the European Alps have shown reductions in cover of the highest elevation species (those in the zone of permanent snow, or the "nival" zone), with lesser reductions and some increases within the alpine subzones below that. These changes suggest that the highest plants are losing ground to plants from lower elevations, at least over a recent decade.

Even if the changes observed in the GLORIA resurveys are significant, it is not possible to attribute change over the short term to long-term climate trends. Interseasonal variation can easily affect the flora represented in a given year. The value of the current comparisons is that they demonstrate the potential of the protocols to reveal change. However, changes in a direction consistent with a warming world, sustained over two or more decades, will indeed be compelling evidence.

Box Figure 26.1A Volunteer botanists make a detailed survey of all plant species contained in a gridded square-meter plot on a peak in the Carson Range, Northern Sierra Nevada, California.
© Jim Bishop

Jim Bishop

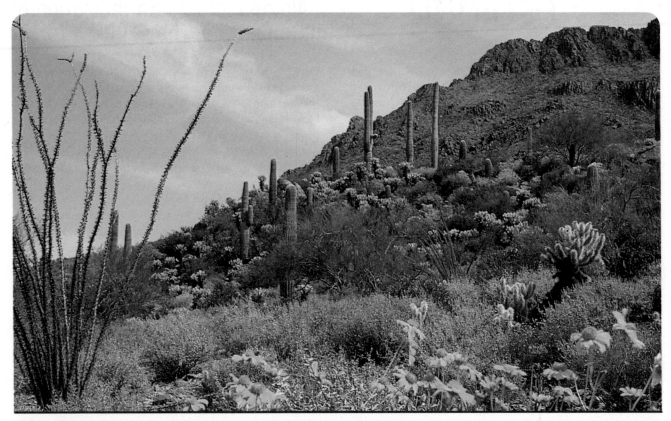

Figure 26.8 A desert community in the southwestern United States. © Kingsley Stern

Desert

If you ask the average person to describe a *desert*, the response will probably include words such as *sand, heat, mirage, oasis,* and *camels* (Fig. 26.8). These are indeed found in some of the world's large deserts that are located in the vicinity of 30° latitude both north and south of the equator, but deserts occur wherever precipitation is consistently low or the soil is too porous to retain water. Most receive less than 12.5 centimeters (5 inches) of rain per year. The low humidity results in wide daily temperature ranges. On a summer day when the temperature has reached over 35°C (95°F), it will generally fall below 15°C (59°F) the same night. The light intensities reach higher peaks in the dry air than they do in areas where atmospheric water vapor filters out some of the sun's rays. Some desert plants have become adapted to these higher light intensities through the evolution of crassulacean acid metabolism (CAM) photosynthesis (discussed in Chapter 10), a process by which certain organic acids accumulate in the chlorophyll-containing parts of the plants during the night and are converted to carbon dioxide during the day. This permits much more photosynthesis to take place than would otherwise be possible, because most of the carbon dioxide of the atmosphere is excluded from the plants during the day by the stomata, which remain closed (thereby also retarding water loss).

Other adaptations of desert plants include thick cuticles, water-storage tissues in stems and leaves, leaves with a leathery texture and/or reduced size, or even the absence of leaves. In cacti and other succulents without functional leaves, the stems take over the photosynthetic activities. Desert perennials are adapted to the biome in various ways. Cacti and similar succulents have widespread, shallow root systems that can rapidly absorb water from the infrequent rains. The water can then be stored for long periods in the interior of the stems, which are modified for such storage. Other perennials grow from bulbs that are dormant for much of the year. Annuals provide a spectacular display of color and variety, particularly during an occasional season when above-average precipitation has occurred. The seeds of the annuals often germinate after a fall or winter rain, and the plants then grow slowly or remain in a basal circular cluster of leaves (rosette) for several months before producing flowers in the spring. Hundreds of different species of desert annuals may occur within a few square kilometers (1 or 2 square miles) of typical desert in the southwestern United States.

Desert animals are adapted, in many instances, to foraging at night, when it is cooler. These include various mice and kangaroo rats, snakes (notably rattlesnakes and king snakes), chuckwallas, and lizards (e.g., gila monsters). Various thrashers, doves, and flycatchers are included in the bird life.

Figure 26.9 A waterfall in the tropical rain forest of Akaka Falls State Park, Hawaii. © *Digital Vision/Getty Images*

Tropical Rain Forest

About 5% of the earth's surface, representing nearly half of the forested areas of the earth, is included in the *tropical rain forest* biome (Fig. 26.9). The biome is distributed throughout those areas of the tropics where annual rainfall amounts normally range between 200 and 400 centimeters (79 and 157 inches) and where temperatures range between 25°C and 35°C (77°F and 95°F). Although monthly rainfall amounts vary, there is no dry season, and some precipitation occurs throughout all 12 months of the year, frequently in the form of afternoon cloudbursts. The humidity seldom drops below 80%.

Such climatic conditions support a diversity of flora and fauna so great that the number of species exceeds those of all the other biomes combined. The forests are dominated by broad-leaved evergreen trees, whose trunks are often unbranched for as much as 40 or more meters (160 feet), with luxuriant crowns that form a beautiful dark green and several-layered canopy. The root systems are shallow, and the tree trunks often buttressed (buttress roots are shown in Fig. 5.14). There are hundreds of species of such trees, each usually represented by widely scattered specimens.

Most of the plants of the rain forests are woody, although not all of them are evergreen. Several of the deciduous tree species shed their leaves from some branches, retain the leaves on others, and flower on yet other branches all at the same time, while branches of adjacent trees of the same species are losing their leaves or flowering at different times. Many hanging woody vines and even more numerous *epiphytes*—especially orchids and bromeliads—can be seen on or attached to tree branches. The epiphyte roots are not parasitic and have no contact with the ground. The plants are sustained entirely by rain water that accumulates in their leaf bases and by their own photosynthesis. Traces of minerals, also necessary to the growth of the epiphytes, accumulate in the rain water as it trickles over decaying bark and dust.

The multilayered canopy is so dense that very little light penetrates to the floor, and the few herbaceous plants that survive are generally confined to openings in the forest. Despite the lush growth, there is little accumulation of litter or humus, and the soil is relatively poor. Decomposers rapidly break down any leaves or other organic material on the forest floor, and the nutrients released by decomposition are quickly recycled or leached by the heavy rains.

A few larger animals, with adaptations for moving through the mesh of branches in the rain forest, are found on the forest floor. Such animals include peccaries, tapirs, and anteaters. Most of the great numbers of animals, however, live out their lives in the canopies. Tree frogs, with adhesive pads on their toes, are common in tropical rain forests, as are various monkeys, bats (especially fruit bats), sloths, opossums, tree snakes, and lizards. Ants abound, and many of the extraordinary variety of other insects are adapted to their environment through excellent camouflages. Large flocks of parrots and other birds feed on the abundance of insects as well as the available fruits.

The Future of the Tropical Rain Forest Biome

In the 1960s, major plans were developed to convert the Amazon rain forest into large farms, hydroelectric plants, and mines. During the 1990s and early 2000s, gold-mining activities were filling the rivers with silt, and, as indicated in the discussion of the greenhouse effect in Chapter 25, the tropical rain forests were being destroyed or damaged for commercial purposes at the rate of more than 100 acres per minute. In one study of Amazon rain-forest birds conducted in 1989, it was found that a population of low-flying species in a 10-hectare (25-acre) section of the forest fell by 75% within just 6 weeks after adjacent land had been cleared, and 10 of the 48 species apparently disappeared completely.

When a rain forest is cleared for development, the area changes from a climax community to one of the earlier stages of succession. It is dominated by annuals that put energy into seed production instead of extensive root growth. Consequently, the plant growth cannot utilize many of the nutrients in the soil, and they are lost as runoff. In addition, many of the nutrients in tropical soils are bound up in plant biomass instead of the soil. When plants are harvested, so are the mineral resources for the ecosystem.

Soon after trees are harvested, the land is so nutrient-poor that it is useless even for farming.

This situation has been repeated on numerous occasions and confirmed by a project cosponsored by the Smithsonian Institution and Brazil's Institute for Research in the Amazon. During 10 years of study, this project demonstrated that, when large expanses of forest are cut into smaller pieces separated by as little as 10 meters (33 feet) of cleared land, it can have disastrous effects on the ecology of the entire forest, with most of the bird and other animal species originally present disappearing permanently. Very little of this biome, which is the home of more than 50% of all living species of organisms, is now protected from commercial development. Many of the organisms are doomed to extinction, often before they have been described for the first time, and the biome will essentially have vanished within 20 years if governments and individuals do not take definitive action to halt the large-scale destruction.

Savanna

Savannas are not found in the United States but occupy about 20% of the earth's land area. The largest savanna is in Africa, and there are large geographic regions of savanna in Australia, India, and South America. Typically a transitional area between forest and desert or prairie, the savanna vegetation can be described as rolling grassland with isolated shrubs and trees. Limited rainfall, particularly during the dry season, prevents the growth of large trees. Rainfall can be as little as 15 centimeters (6 inches) but averages about 50 to 125 centimeters (20 to 50 inches) per year. The temperature of the savanna varies little and ranges from 20° to 30°C (68 to 86°F). Most savannas are home to many grass species, such as elephant grass, love grass, star grass, and several species of *Themeda*, and, when present, trees and shrubs, including many *Acacia* species, jackel berrytree, and sickle brush. Animals of the savanna include elephants, lions, giraffes, and zebras; reptiles, such as cobras, crocodiles, monitor lizards, and tortoises; and birds, such as flycatchers, ostriches, and vultures.

Chaparral

The chaparral biome is a geographically small but important biome found in California and the Baja California peninsula of Mexico, as well as other areas scattered around the world. It is typically characterized as a shrubland with infrequent fires. Chaparral temperatures vary annually from −2° to 38°C (30° to 100°F), and rainfall is limited from 25 to 50 centimeters (10 to 20 inches) per year. Shrubs of the chaparral are shorter than trees of forest biomes and have a woody stem with no dominant trunk. Many shrubs have extensive underground root systems that help them survive during dry weather and after fires. This allows them to be adapted to chaparral biomes, which are characterized by cool, moist winters and hot, dry summers. Fire is an important and regular disturbance in shrubland ecosystems.

Some shrubs are actually dependent on fires because their seeds require heat for germination. Other shrubs escape fires by resprouting from roots. Often, the shrubs are evergreen, allowing them to hold on to nutrients during the short growing season. In California, chaparral is used to define regions of dense, spiny shrubs. Plants found in the chaparral include buckwheat, chamise, lupine, mountain mahogany, sages, scrub oak, sumac, and yucca. Animals that may be seen in the chaparral include the coyote, gray fox, mountain lion, and spotted skunk, as well as an assortment of birds, including the acorn woodpecker, brushtit, canyon wren, greater roadrunner, and sage sparrow. The chaparrel of California looks similar to shrublands in Chile and the Mediterranean, although the species of plants in each area are different. Plant species in the chaparral biome are locally adapted to their unique environment and are not found elsewhere. Because many of these areas are popular for development, many of the species are in danger of extinction.

SUMMARY

26.1 Introduction to Biomes

- Biomes are biotic communities considered on a global or at least a continental scale. Major biomes of North America include tundra, taiga, temperate deciduous forests, grasslands, deserts, mountain forests, and tropical rain forests.

26.2 Major Biomes of the World

- Tundra is found primarily above the Arctic Circle. It includes few trees and many lichens and grasses, and is characterized by the presence of permafrost below the surface. Alpine tundra occurs above timberline in mountains below the Arctic Circle.

- Taiga is dominated by coniferous trees, with birch, aspen, and willow in the wetter areas. Many perennials and few annuals occur in taiga.

- Temperate deciduous forests are dominated by trees such as sugar maple, American basswood, beech, oak, and hickory, with evergreens, such as hemlock and eastern white pine, toward the northern and southeastern borders.

- Mountain and coastal forests occupy much of the Pacific Northwest and extend south along the Rocky Mountains and California mountain ranges. The primarily coniferous tree species tend to be in zones determined by altitude. These forests have mostly dry summers, and some of the species associated with them have thick bark, which protects the trees from fires that occur frequently in this type of climate. Other species (e.g., knobcone pine) depend on fires for normal distribution and germination of seeds.

- Grasslands are found toward the interiors of continental masses. Those located in Mediterranean climatic zones usually include vernal pools with unique annual flora. Many grasslands have been converted to agricultural use.
- Deserts are characterized by low annual precipitation and wide daily temperature ranges, with plants adapted in both form and metabolism to the environment.
- The tropical rain forests constitute nearly half of all forest land and contain more species of plants and animals than all the other biomes combined. Numerous woody plants and vines form multilayered canopies that permit very little light to reach the floor. The soils are poor, with nutrients released during decomposition being rapidly recycled.
- The savanna biome has limited rainfall and fairly constant temperatures, and it can be described as rolling grassland with isolated shrubs and trees.
- A chaparral is typically characterized as a shrubland with infrequent fires. Plant species in the chaparral are locally adapted to their unique environment and are not found elsewhere.

REVIEW QUESTIONS

1. Characterize desert and tundra biomes.
2. What is the function of fire in grassland and mountain forest biomes?
3. What kinds of plants predominate in tropical rain forests?
4. Why does the loss of the tropical rain forest biome have very serious implications for the future?

DISCUSSION QUESTIONS

1. Fire has been a natural phenomenon in several biomes for thousands of years, and most plant species are adapted to it in various ways in the ecosystem where it occurs regularly. Should we then not extinguish forest and grassland fires when they occur?
2. Much of the tropical rain forest biome is located in developing countries, which have precarious economies. Bearing that in mind, what could be done to halt the destruction of this priceless and vital part of our ecosystem?

APPENDIX 1
Scientific Names of Organisms Mentioned in the Text

This is an alphabetical list of the organisms whose scientific names may or may not be mentioned in the text. The common names are listed alphabetically along with the scientific names. Common and scientific names of organisms mentioned in Appendices 2 through 4 are provided within the respective appendices.

Common Names and Scientific Names of Organisms

Common Name	Scientific Name	Common Name	Scientific Name
Aardvark	*Orycteropus* spp.	Algae, bark	*Pleurococcus* spp. and others (see footnote under Bark, green algae that inhabit)
Abrasives, horsetail source of	*Equisetum* spp.	Algae, brown	members of Phylum Chromophyta, Kingdom Protista
Absinthe liqueur, source of ingredients	*Pimpinella anisum, Artemisia absinthium,* and others	Algae, carrageenan-producing	*Chondrus crispus, Eucheuma* spp., and others
Acacia	*Acacia* spp.	Algae, coralline	*Bossiella* spp., *Corallina* spp., *Lithothamnion* spp., and others
Aconite, source of	*Aconitum* spp.		
Actinomycetes	*Actinomyces* spp. and others	Algae/cyanobacteria, edible[1]	*Chlorella,* Irish moss (*Chondrus crispus*), kelp (*Laminaria* spp.), laver or nori (*Porphyra* spp.), spirulina (*Spirulina* spp.), wakame (*Undaria* spp.), and others
Adder's tongue fern, reticulate	*Ophioglossum reticulatum* (has highest known diploid chromosome number—1,260)		
Adder's tongue ferns	*Ophioglossum* spp.		
Afghanistan pine	*Pinus eldarica*		
Aflatotoxin, source of	*Aspergillus flavus*	Algae/cyanobacteria, toxic	*Anabaena* sp., *Caulerpa* sp., *Chlorella* sp., *Chondria armata, Gambierdiscus toxicus, Hizikia* sp., *Lyngbya majusculis, Oscillatoria nigroviridis, Protogonyaulax (Gonyaulax)* sp., *Prototheca* sp., *Prymnesium parvum, Ptychodiscus (Gymnodinium brevis), Schizothrix calcicola,* and others
African sausage tree	*Kigelia pinnata*		
Agar, source of	*Chondrus crispus, Eucheuma* spp., *Gelidium* spp., *Gracilaria* spp., and other red algae		
Agave	*Agave angustifolia, A. palmeri, A. tequilana,* and other *Agave* spp.		
Air plant—see also Bromeliad	*Kalanchoë* spp.	Algae/cyanobacteria used as fertilizers or soil conditioners	*Anabaena azollae, Chlamydomonas mexicana,* kelps, and others
Alder	*Alnus* spp.		
Alfalfa	*Medicago sativa*	Algae, flatworm	*Platymonas* spp.
Alfalfa caterpillar	*Colias philodice*	Algae, golden brown	members of Phylum Chromophyta, Kingdom Protista
Algae	members of Kingdom Protista—all phyla	Algae, green	members of Phylum Chlorophyta, Kingdom Protista
Algae, agar-producing	*Acanthopeltis* spp., *Ahnfeltia* spp., *Gelidium* spp. (principal source), *Gracilaria* spp., *Pterocladia* spp., and others	Algae, green colonial	*Chaetopeltis* spp., *Eudorina* spp., *Pandorina* spp., *Pediastrum* spp., *Scenedesmus* spp., *Volvox* spp., and others
Algae, alginate-producing	*Ascophyllum* spp., *Durvillea* spp., *Ecklonia* spp., *Laminaria* spp., *Macrocystis* spp., and others		

[1]More than 150 species of algae and cyanobacteria are known to be edible, but most of the approximately half million tons of dried algae consumed annually consists of species of *Laminaria, Porphyra,* and *Undaria.* Most commercially grown nori consists of the fronds of *Porphyra tenera,* but other species of *Porphyra* are edible. During their life cycles, *Porphyra* spp. alternate between the familiar frond (bladed) form and a relatively inconspicuous filamentous form that was discovered after a British phycologist germinated spores of *Porphyra umbilicaulis* in a culture dish in her laboratory. The filamentous form previously had been considered a distinct species that had been named *Conchocelis rosea.* Species of *Ascophyllum, Fucus, Laminaria,* and *Macrocystis* are harvested for animal and poultry feeds. *Dunaliella bardawil* is cultured commercially as a source of beta-carotene and glycerol. *Spirulina* spp. (cyanobacteria) have a protein content of up to 70%; they are commercially cultivated for human consumption, particularly in Mexico and Israel, and have been a staple food of natives of the Lake Chad region in Africa for centuries.

Common Names and Scientific Names of Organisms

Common Name	Scientific Name	Common Name	Scientific Name
Algae, green filamentous	*Oedogonium* spp., *Spirogyra* spp., *Ulothrix* spp., *Zygnema* spp., and others	Arrowroot	*Maranta arundinacea*, *Tacca leontopetaloides*
Algae, medicinal	*Laminaria* spp., *Digenia* spp., and many others	Arrowroot, Florida, source of	*Zamia floridana*
Algae, metal-removing	*Chlamydomonas reinhardtii*	Artichoke, Chinese (Crosne)	*Stachys affinis*
Algae, red	members of Phylum Rhodophyta, Kingdom Protista	Artichoke, globe	*Cynaria scolymus*
		Artichoke, Jerusalem	*Helianthus tuberosus*
Algae, snowbank	*Chlamydomonas nivale* and others	Arum Lily (Arum) Family	Araceae
Algae, sponge	*Chlorella* spp., *Zoochlorella* spp.	Ascomycete	member of Phylum Ascomycota, Kingdom Fungi
Algae, yellow-green	members of Phylum Chromophyta, Kingdom Protista	Ash, blue	*Fraxinus quadrangulata*
		Ash, Oregon	*Fraxinus latifolia*
Almond	*Prunus amygdalus*	Ash, white	*Fraxinus americana*
Aloe juice, source of	*Aloe barbadensis*, *A. ferox*, *A. vera*, and others	Asparagus	*Asparagus officinalis*
Amaranth	*Amaranthus* spp.	Aspen, quaking	*Populus tremuloides*
Amaryllis	*Amaryllis* spp.	Aspergillosis, causal agent(s) of	*Aspergillus fumigatus* and other *Aspergillus* spp.
Ama'uma'u	*Sadleria cyatheoides*		
American chestnut	*Castanea dentata*	Aster	*Aster* spp.
American elm	*Ulmus americana*	Astringent, horsetail source of	*Equisetum arvense*, *E. debile*, and others
Amoeba	*Amoeba proteus* and others		
Amoeba, fungal internal parasites of	*Cochlonema verrucosum* and others	Athlete's foot, fungal causal agent of	*Trichophyton* spp.
Amoeba, fungal trappers of	*Dactylella* spp. and others	Autograph tree	*Clusia rosea*
Anabaena	*Anabaena* spp. (including nitrogen-fixing spp. such as *A. azollae*)	Avocado	*Persea americana* and others
		Azalea	*Rhododendron* spp.
Anemone	*Anemone* spp.	Baby blue eyes	*Nemophila menziesii*
Angelica	*Angelica archangelica*	Baby powder, ground pine source of	*Lycopodium clavatum*
Anise	*Pimpinella anisum*		
Anise swallowtail butterfly	*Papilio zelicaon*	Bacteria, acetone-producing	*Clostridium acetobutylicum* and others
Annatto	*Bixa orellana*		
Ant	*Formica* spp. and many others	Bacteria, acidophilus	*Lactobacillus acidophilus*
Anteater	*Myrmecophaga jubata*	Bacteria, ammonifying	*Clostridium* spp., *Micrococcus* spp., *Proteus* spp., *Pseudomonas* spp., and others
Ants, bullhorn Acacia	*Pseudomyrmex ferruginea*		
Aphid	*Anuraphis* spp., *Aphis* spp., and others		
		Bacteria, anthrax	*Bacillus anthracis*
		Bacteria, blue-green—see Cyanobacteria	
Aphid, root (pest of grape vines)	*Phylloxera* spp.	Bacteria, botulism	*Clostridium botulinum*
Apple[2]	*Malus domestica* (= *Malus pumila*)[2]	Bacteria, brucellosis	*Brucella abortus*, *B. suis*, *B. melitensis*
Apple brown rot, causal agent	*Monolinia fructigena*	Bacteria, Bt	*Bacillus thuringiensis*
Apple scab, causal agent	*Venturia inaequalis*	Bacteria, bubonic plague	*Yersinia pestis*
Apricot	*Prunus armeniaca*	Bacteria, buttermilk	*Streptococcus lactis*, *S. cremoris*, *Leuconostoc citrovorum*, and others
Apricot brown rot	*Sclerotinia fructicola*		
Arabidopsis (Mouse-ear cress)	*Arabidopsis thaliana*	Bacteria, butyl alcohol	*Clostridium acetobutylicum* and others
		Bacteria, cholera	*Vibrio cholerae*
Arborvitae (American/Northern)	*Thuja occidentalis*	Bacteria, common	*Escherichia coli*
		Bacteria, decay/decomposer	*Clostridium* spp., *Micrococcus* spp., *Proteus* spp., *Pseudomonas* spp., and others
Archaebacteria	members of Phylum Archaebacteria, Kingdom Archaea		
Archaefructus, extinct plant believed to be the earliest flowering plant		Bacteria, denitrifying	*Micrococcus denitrificans*, *Thiobacillus denitrificans*, and others

[2]There are more than 1,000 varieties of apples, mostly of hybrid origin. The principal ancestors of *Malus pumila* probably include *M. sylvestris*, *M. dasyphylla*, and *M. praecox*. Some authorities include *Malus* within the genus *Pyrus* and refer to most cultivated apples as *Pyrus malus*. Others distinguish between the two genera on the basis of leaf pubescence and stone cells within the fruit, referring those cultivars with sparse leaf pubescence and sclereids as *Pyrus* and those with the opposite features as *Malus*.

Common Names and Scientific Names of Organisms

Common Name	Scientific Name	Common Name	Scientific Name
Bacteria, dextran	*Leuconostoc mesenteroides*	Bacteria, pneumonia (some forms of pneumonia are viral)	*Streptococcus pneumoniae* and others
Bacteria, diphtheria	*Corynebacterium diphtheriae*		
Bacteria, ensilage	*Lactobacillus delbrueckii, L. plantarum,* and others	Bacteria, PPLO (*mycoplasmas*)	*Mycoplasma pneumoniae*
Bacteria, ethanol-producing	*Bacillus stearothermophilus* (mutant form)	Bacteria, pseudomonad	*Pseudomonas* spp.
Bacteria, frost-damage preventing	*Pseudomonas syringiae*	Bacteria, purple nonsulfur	*Rhodomicrobium* spp., *Rhodopseudomonas* spp., *Rhodospirillum* spp.
Bacteria, gas gangrene	*Clostridium novyi, C. perfringens, C. septicum*	Bacteria, purple sulfur	*Amoebobacter* spp., *Lamprocystis* spp., *Rhodothece* spp., and others
Bacteria, giant	*Epulopiscium fishelsonii*		
Bacteria, glutamic acid-producing	*Arthrobacter* spp., *Brevibacterium* spp., *Micrococcus* spp.	Bacteria, salmonella (food-poisoning bacteria)	*Salmonella* spp.
Bacteria, gonorrhea	*Neisseria gonorrhoeae*	Bacteria, salt	*Halococcus* spp., *Halobacterium* spp.
Bacteria, grease- and oil-dissolving	*Pseudomonas aeruginosa*	Bacteria, sauerkraut	*Leuconostoc* spp. and others
Bacteria, green sulfur	*Chlorobium* spp., *Chloropseudomonas* spp., *Prosthecochloris* spp., and others	Bacteria, sorbose	*Acetobacter suboxydans*
		Bacteria, spotted fever	*Rickettsia rickettsii*
		Bacteria, strep throat	*Streptococcus* spp.
Bacteria, hot water (Sulfolobus)	*Pyrodictium* spp.	Bacteria, sulfolobus	*Sulfolobus* spp., *Thermoplasma* spp., *Thermoproteus* spp.
Bacteria, human ulcer-causing	*Helicobacter pylori*	Bacteria, sulfur	*Desulfovibrio* spp., *Thiobacillus* spp., and others
Bacteria, hydrogen	*Hydrogenomonas* spp.	Bacteria, syphilis	*Treponema pallidum*
Bacteria, ice-minus	*Pseudomonas syringiae*	Bacteria, tetanus	*Clostridium tetani*
Bacteria, iron	*Gallionella* spp., *Sphaerotilus* spp.	Bacteria, tularemia	*Francisella tularensis*
Bacteria, kefir	*Lactobacillus bulgaricus, Streptococcus lactis*	Bacteria, typhoid fever	*Salmonella typhi*
		Bacteria, typhus fever	*Rickettsia prowazekii* and others
Bacteria, lactic acid	*Lactobacillus delbrueckii* and others	Bacteria, vinegar	*Acetobacter* spp.
Bacteria, Legionnaire's disease	*Legionella pneumophilia*	Bacteria, whooping cough	*Bordetella pertussis*
		Bacteria, yogurt	*Streptococcus thermophilus*
Bacteria, luminescent	*Achromobacter* spp., *Flavobacterium* spp., *Photobacterium* spp., *Pseudomonas* spp., *Vibrio* spp., and others	Bald cypress	*Taxodium distichum*
		Balsa	*Ochroma lagopus*
		Balsam fir	*Abies balsamea*
		Bamboo	*Bambusa* spp., *Phyllosytachys* spp.
Bacteria, meningitis	*Neisseria meningitidis* and others	Banana	*Musa paradisiaca* and others[3]
Bacteria, methane	*Methanobacterium* spp., *Methanococcus* spp., *Methanosarcina* spp., and others	Banana Family	Musaceae
		Banyan tree	*Ficus* spp.
		Baobab, African	*Adansonia digitata*
Bacteria, milky spore disease	*Bacillus popilliae*	Baobab, Australian	*Adansonia gregorii*
		Barbasco	*Lonchocarpus nicou* var. *utilis, Derris elliptica,* and others
Bacteria, mosquito-killing	*Bacillus thuringiensis* var. *israelensis*		
Bacteria, nitrate (nitrifying)	*Nitrobacter* spp.	Barberry	*Berberis verruculosa* and other *Berberis* spp.
Bacteria, nitrite (nitrosifying)	*Nitrosomonas* spp.	Barberry, Common/European	*Berberis vulgaris*
Bacteria, nitrogen-fixing	*Azorhizobium* spp., *Azotobacter* spp., *Brachyrhizobium* spp., *Clostridium pasteurinum, Rhizobium* spp., *Sinorhizobium* spp., and others	Bark, green algae that inhabit	*Protococcus* spp.[4]
		Barley	*Hordeum vulgare*
		Barn swallow	*Hirundo rustica erythrogaster*
Bacteria, paratyphoid fever	*Salmonella paratyphi*		

[3]The domestic banana was developed from hybrids between *Musa acuminata* and *M. balbisiana*, and its genetic history is complex. N. W. Simmonds, a recognized authority on the genus *Musa*, believes that only cultivars and not species of domestic banana should be recognized; others prefer to retain Linnaeus's species name of *Musa paradisiaca*.

[4]These algae are known under several names (*Desmococcus, Phytoconis, Pleurococcus, Protococcus*), and uncertainty exists as to which name has priority. The green algal component of certain lichens, *Trebouxia*, also occurs independently on bark.

Common Names and Scientific Names of Organisms

Common Name	Scientific Name	Common Name	Scientific Name
Barrel cactus	*Ferocactus* spp., *Mammillaria* spp., and others	Big tree	*Sequoiadendron giganteum*
		Birch	*Betula papyrifera* and others
Barrel cactus, Coville's	*Ferrocactus covillei*	Bird's-nest fungus	*Crucibulum levis*
Basil	*Ocimum basilicum*	Birth control pills, fungi used in manufacture of	*Rhizopus nigricans, R. arrhizus*
Basswood	*Tilia* spp.		
Basswood, American	*Tilia americana*	Bison	*Bison bison*
Bat	*Eidolon* spp., *Epomophorus* spp., *Leptonycteris sanbornii*, and others	Bittersweet	*Celastrus scandens*
		Blackberry	*Rubus argutus, R. laciniatus, R. procerus, R. ursinus,* and others
Bay, California (also known as Oregon myrtle)	*Umbellularia californica*	Blackbird	*Euphagus* spp. and others
		Black bread mold	*Rhizopus stolonifer* and others
Bay laurel	*Lauris nobilis*	Black locust	*Robinia pseudo-acacia*
Bay, sweet	*Laurus nobilis*	Black stem rust of wheat	*Puccinia graminis*
Beach strawberry	*Fragaria chinensis*	Bladderwort	*Utricularia minor* and other *Utricularia* spp.
Bean, broad	*Vicia faba*		
Bean, castor	*Ricinus communis*	Blazing star	*Liatris ligulistylis*
Bean Family	Fabaceae (formerly Leguminosae)	Bleeding, ground pine used to arrest	*Lycopodium clavatum*
Bean, garbanzo	*Cicer arietinum*		
Bean, garden	*Phaseolus vulgaris*	Bleeding heart	*Dicentra* spp.
Bean, green	*Phaseolus vulgaris*	Bleeding heart, eastern	*Dicentra eximia*
Bean, jequirity	*Abrus precatorius*	Bleeding heart, Pacific	*Dicentra formosa*
Bean, kidney	*Phaseolus vulgaris*	Bloodroot	*Sanguinaria canadensis, S. isabellinus*
Bean, lima	*Phaseolus lunatus*		
Bean, mescal—*see* Mescal bean		Blueberry	*Vaccinium* spp.
Bean, Mexican jumping	*Sebastiana* spp. and others	Blue curls	*Trichostema* spp.
Bean, mung	*Phaseolus aureus* (= *Vigna radiata*)	Blue-green algae—*see* Cyanobacteria	
		Blue-green bacteria—*see* Cyanobacteria	
Bean, navy	*Phaseolus vulgaris*	Blue jay	*Cyanocitta cristata*
Bean, pinto	*Phaseolus vulgaris*	Bobcat	*Felis rufus*
Bean, scarlet runner	*Phaseolus coccineus*	Bolete	*Boletus* spp., *Suillus* spp., and others
Bean, tepary	*Phaseolus acutifolius* var. *latifolius*	Bollworm	*Pectinophora gossypiella*
Bean, winged	*Psophocarpus tetragonolobus*	Bowstring fibers, source of	*Sansevieria metalaea*
Bear	*Ursus* spp. and others	Bowstring hemp, source of	*Sansevieria* spp.
Bear, polar	*Thalarctos maritimus*	Box elder	*Acer negundo*
Bearberry (Kinnikinnick)	*Arctostaphylos uva-ursi*	Boysenberry	*Rubus* hybrids, with *R. ursinus* as one parent
Beaver, mountain	*Aplodontia rufa*		
Bedstraw	*Galium* spp.	Bracken fern	*Pteridium aquilinum*
Bee, honey	*Apis mellifera*	Brazil nut	*Bertholettia excelsa*
Beech, American	*Fagus grandifolia*	Breadfruit	*Artocarpus altilis*
Beefsteak morel	*Gyromitra* sp.	Bridalwreath	*Spiraea vanhouttei* hybrids and others
Beet, garden	*Beta vulgaris*		
Beet, sugar	*Beta vulgaris* (horticulturally selected strains)	Broccoli	*Brassica oleracea* var. *botrytis*
		Bromeliad ("Air plant")	Member of the Bromeliad Family (Bromeliaceae)[5]
Beetle	member of Order Coleoptera, Class Insecta, Phylum Arthropoda, Kingdom Animalia		
		Broomrape	*Orobanche* spp.
		Brown algae	Member of Phylum Chromophyta, Kingdom Protista. Representative genera include *Ascophyllum, Durvillea, Ecklonia, Ectocarpus, Hizikia, Laminaria, Undaria,* and others
Beetle, scarab	member of Family Scarabaeidae—*see* Beetle		
Begonia	*Begonia* spp.		
Belladonna, source of	*Atropa belladonna*		
Bermuda grass	*Cynodon dactylon*		
Betel nut	*Areca catechu*	Brussels sprouts	*Brassica oleracea* var. *gemmifera*
Betony, wood	*Pedicularis canadensis*		

[5]There are more than 2,000 species of bromeliads, which include pineapple (*Ananus comosus*), Spanish moss (*Tillandsia usneoides*), and many popular houseplants in genera such as *Aechmea, Bilbergia, Cryptanthus* (not to be confused with *Cryptantha*, which is in the Boraginaceae), *Neoregelia, Nidularium, Quesnelia,* and *Vriesia*.

Common Names and Scientific Names of Organisms

Common Name	Scientific Name	Common Name	Scientific Name
Bryophyte (see also individual listings)	member of Phyla Anthocerotophyta, Hepaticophyta, or Bryophyta, Kingdom Plantae	California poppy	*Eschscholzia californica*[6]
		Camel	*Camelus* spp.
		Camelina	*Camelina sativa*
Bryopsid	member of Phylum Chlorophyta, Kingdom Protista	Camellia[7]	*Camellia* spp.[7]
		Camphor, source of	*Cinnamomum camphora*
Bt	*Bacillus thuringiensis*	Candelilla	*Euphorbia antisyphilitica*
Buckeye	*Aesculus* spp.	Candlenut	*Aleurites moluccana*
Buckwheat	*Fagopyrum esculentum*	Cankerworm	*Alsophila pometaria* and others
Buffalo	*Bison bison*	Canna	*Canna edulis* and other *Canna* spp. and hybrids
Bullhorn acacia	*Acacia cornigera*		
Bunchberry	*Cornus canadensis*	Cantaloupe	*Cucumis melo*
Burn treatment, horsetail source of ashes for	*Equisetum hyemale* and others	Caraway	*Carum carvi*
		Cardamon/Cardamom	*Elettaria cardamomum*
Butcher's broom	*Ruscus aculeata*	Caribou	*Rangifer tarandus*
Buttercup	*Ranunculus* spp.	Carnation	*Dianthus caryophyllus*
Buttercup, European bulbous	*Ranunculus bulbosa*	Carnaubalike wax, source of	*Stipa tenacissima*
Buttercup Family	Ranunculaceae	Carnauba wax, source of	*Copernicia cerifera*
Butterfly	member of Superfamily Papilionoidea, Order Lepidoptera, Phylum Arthropoda, Kingdom Animalia	Carob	*Ceratonia siliqua*
		Carpetweed Family	Molluginaceae
		Carrot	*Daucus carota*
		Carrot Family	Apiaceae (formerly Umbelliferae)
Butterwort	*Pinguicula grandiflora, P. vulgaris,* and other *Pinguicula* spp.	Cashew	*Anacardium occidentale*
		Cassava	*Manihot esculenta*
Button snakeroot	*Eryngium* spp.	Cassia[8]	*Cinnamomum cassia*[8]
Cabbage (green or red)	*Brassica oleracea* var. *capitata*	Catalpa	*Catalpa* spp.
Cabbage, Chinese	*Brassica chinensis*	Caterpillar	larval stage of member of Order Lepidoptera, Phylum Arthropoda, Kingdom Animalia
Cabbage Family	Brassicaceae (formerly Cruciferae)		
Cabbage looper	*Trichoplusia ni*		
Cabbage worm	*Pieris rapae*	Catnip	*Nepeta cataria*
Cacao	*Theobroma cacao*	Cattail	*Typha* spp.
Cactus	*Hamatocactus setispinus*	Cattle—see Cow	
Cactus, barrel	*Mamillaria* spp., *Ferocactus* spp., and others	Cauliflower	*Brassica oleracea* var. *botrytis* (= *B. oleracea* var. *cauliflora*)[9]
Cactus, cholla	*Opuntia* spp. (cylindrical forms)	Caussu wax, source of	*Calathea lutea*
Cactus family	Cactaceae	Cedar, Atlantic white	*Chamaecyparis thyoides*
Cactus, giant saguaro	*Carnegia gigantea*	Cedar, eastern red	*Juniperus virginiana*
Cactus, hedgehog	*Echinocereus* spp. and others	Cedar, incense	*Calocedrus decurrens*
Cactus, living rock	*Ariocarpus fissuratus* and others	Cedar, northern white	*Thuja occidentalis*
Cactus, organ-pipe	*Lemaireocereus* spp.	Cedar, southern white	*Chamaecyparis thyoides*
Cactus, prickly pear	*Opuntia* spp.	Cedar, western red	*Thuja plicata*
Cajuput, source of	*Melaleuca cajuputi*	Celery, Celeriac	*Apium graveolens*
Calabash	*Lagenaria siceraria*	Cell-from-hell (dinoflagellate)	*Pfiestera piscicida*
Calabazilla	*Cucurbita foetidissima*	Cellular slime mold	member of Phylum Dictyosteliomycota, Kingdom Protista
California bay (also known as Oregon myrtle)	*Umbellularia californica*		
		Century plant	*Agave americana* and others

[6]Although the generic name was given in honor of Johann Friedrich Eschscholtz, an early 19th-century German naturalist and surgeon, the name was first published as *Eschscholzia*, making the spelling *Eschscholtzia* an orthographic variant.

[7]More than 80 species of *Camellia* and 2,000 horticultural varieties are recognized, with most of the ornamental varieties having been derived from *C. japonica* and *C. sasanqua*. The late George Petersen of Chico, California, produced 700 of the horticultural varieties. Other important members of the genus include *C. sinensis* (tea), and *C. oleifera*, whose seeds yield tea tree oil.

[8]This should not be confused with the genus *Cassia*, the source of senna in the Legume Family, or cassie, a perfume oil whose source is *Acacia farnesiana*, another member of the Legume Family.

[9]Cauliflower and broccoli are two different forms of the same variety.

Common Names and Scientific Names of Organisms

Common Name	Scientific Name	Common Name	Scientific Name
Chamise	*Adenostoma fasciculatum*	Coastal redwood	*Sequoia sempervirens*
Chara	*Chara* spp.	Cobra plant	*Darlingtonia californica*
Chard	*Beta vulgaris* var. *cicla*	Coca/Cocaine, source of	*Erythroxylum* (often misspelled *Erythroxylon*) *coca*. *E. novogravatense* is a lesser source.
Cheese bacteria—see Bacteria, buttermilk			
Cheese fungi	*Penicillium camembertii* (for Camembert cheese), *P. roquefortii* (for blue, Gorgonzola, Roquefort, and Stilton cheeses)	Cochineal insect	*Dactylopius coccus*
		Cocklebur	*Xanthium strumarium*
		Cockroach	*Blatta orientalis*, *Blatella germanica*, and others
Cherry, sour	*Prunus cerasus*		
Cherry, sweet	*Prunus avium*	Cockroach plant	*Haplophyton cimicidum*
Chestnut, American	*Castanea dentata*	Cockscomb	*Celosia* spp.
Chia	*Salvia columbariae*	Coffee, Arabian	*Coffea arabica*
Chickadee, mountain	*Parus gambeli*	Coffee, Liberian	*Coffea liberica*
Chickpea	*Cicer arietinum*	Coffee, robusta	*Coffea canephora*
Chickweed (Himalayan)	*Stellaria decumbens*	Coffee Family (= Madder Family)	Rubiaceae
Chicle, source of	*Manilkara zapota*		
Chicory	*Cichorium intybus*	Coleus	*Coleus blumei*, *C. x hybrida*, and others
Chimpanzee	*Pan troglodytes* and others		
China grass	*Boehmeria nivea*	Columbine	*Aquilegia formosa*, *Aquilegia* spp.
Chinese vegetable tallow	*Sapium sebiferum*	Compass plant	*Lactuca serriola*; (*Silphium laciniatum* is also known as Compass plant)
Chipmunk	*Eutamias* spp., *Tamias* spp., and others		
		Coneflower	*Rudbeckia* sp.
Chlamydomonas	*Chlamydomonas* spp.	Coneflower, Asian	*Strobilanthes* spp.
Chloroxybacteria	member of Chloroxybacteriae, Phylum Eubacteria, Kingdom Bacteria	Copal, sources of	*Agathis alba*, *Copaifera demeussei*, *Hymenea coubaril*, *Trachylobium verrucosum*, and others
Chocolate, source of	*Theobroma cacao*		
Chokecherry	*Prunus virginiana* var. *melanocarpa*	Copperhead	*Ancistrodon contortrix*
Cholla (cactus)	*Opuntia* spp. (cylindrical forms)	Coral tree	*Erythrina crista-galli*
Christmas flower	*Euphorbia pulcherrima*	Cordage fibers, source of	*Agave sisalina*, *A. heterocantha*, *A. lophantha*, *Phormium tenax*, and others
Chrysanthemum	About 160 spp.; many garden cultivars are hybrids of *Chrysanthemum frutescens* and *C. morifolium*		
		Coriander	*Coriandrum sativum*
Chuckwalla	*Sauromalus obesus*	Corn (Maize)	*Zea mays*
Chufa	*Cyperus esculentus*	Corn borer, European	*Pyrausta nubialis*
Chytrid	*Allomyces arbusculus* and many other members of Phylum Chytridiomycota, Kingdom Fungi	Corpse flower	*Amorphophallus titanum*
		Cotton	*Gossypium arboreum*, *G. barbadense*, *G. herbaceum*, *G. hirsutum*, *G. raimondii*
Cilantro	*Coriandrum* sp.		
Cinnamon, cassia	*Cinnamomum cassia*, *C. burmannii*, *C. loureii*	Cottonwood	*Populus deltoides*, *P. fremontii*, and others
Cinnamon, true	*Cinnamomum zeylanicum*	Cow	*Bos* sp.
Citric acid, fungal producers of	*Aspergillus niger* and others	Cow parsnip	*Heracleum lanatum*
		Cowslip	*Caltha palustris*
Citronella oil, source of	*Cymbopogon nardus*	Crab apple	*Crataegus* spp., *Malus cortonaria* (= *Malus sylvestris?*)
Citrus	*Citrus* spp.		
Citrus Family	Rutaceae	Crabgrass	*Digitaria sanguinalis*
Cladophora	*Cladophora* spp.	Cranberry, American	*Vaccinium macrocarpon*
Clematis	*Clematis* spp.	Cress, garden	*Lepidium sativum*, *Barbarea verna*, and others
Clover	*Trifolium* spp.		
Clover, bur	*Medicago polymorpha*	Cress, rock	*Arabis* spp.
Cloves	*Syzygium aromaticum* (formerly *Eugenia caryophyllus*)	Crocus, autumn/fall	*Colchicum autumnale* and other *Crocus* spp.
Club fungus	member of Phylum Basidiomycota, Kingdom Fungi	Crown of thorns	*Euphorbia milii* var. *splendens* and others
Club moss	member of Phylum Lycophyta, Kingdom Plantae		

Common Names and Scientific Names of Organisms

Common Name	Scientific Name	Common Name	Scientific Name
Crozier, tropical tree fern	*Sadleria cyatheoides*	Dinoflagellate	member of Phylum Dinophyta, Kingdom Protista. Representative genera include *Gambierdiscus*, *Gonyaulax*, and *Gymnodinium*
Crustacean	member of Class Crustacea, Phylum Arthropoda, Kingdom Animalia		
Cryptomonad	member of Phylum Cryptophyta, Kingdom Protista	Dinoflagellate, midnight-bioluminescent	*Gonyaulax polyedra*
Cucumber	*Cucumis sativus*	Dischidia	*Dischidia rafflesiana*
Cucumber, squirting	*Ecballium elaterium*	Divi-divi	*Caesalpinia coriaria*
Cyanobacteria	member of Kingdom Bacteria; common genera include *Anabaena*, *Lyngbya*, *Oscillatoria*, *Phormidium*, *Schizothrix*, and *Spirulina*	Dodder	*Cuscuta* spp.
		Dogbane	*Apocynum* spp.
		Dogwood	*Cornus* spp.
		Douglas fir	*Pseudotsuga menziesii*
Cyanobacteria, Lake Chad edible	*Spirulina* sp.	Dove	member of Family Columbidae, Class Aves, Phylum Vertebrata, Kingdom Animalia
Cyanobacteria, Red Sea	*Trichodesmium erythraeum*		
Cyanobacteria, thermal	*Bacillosiphon induratus*, *Synechococcus* spp., and others	Dove, mourning	*Zenaidura macroura*
		Downy mildew of grape	*Plasmopora viticola*
Cycad	*Cycas* sp., *Dioon edule*, *Encephalartos altensteinii*	Dragon's blood	*Dracaena* spp., *Daemonorops* spp.
		Drimys	*Drimys winteri* and other *Drimys* spp.
Cycadeoid (extinct gymnosperm with palmlike leaves)	*Cycadeoidea* and other genera	Duckweed	*Lemna* spp., *Wolffia* spp., and others
Cyclamen	*Cyclamen* spp.	Dulse	*Rhodymenia* spp.
Cypress	*Cupressus* spp.	Dung mosses (on dung of carnivores)	*Tayloria* spp.
Cypress, bald	*Taxodium distichum*		
Daffodil	*Narcissus* spp. (see note under *Narcissus*)	Dung mosses (on dung of herbivores)	*Splachnum* spp.
Dahlia	*Dahlia* spp.	Dutch elm disease, causal agent of	*Ophiostoma ulmi/O. nova-ulmi*
Daisy	*Dimorphotheca* spp., *Layia* spp., and others	Dutchman's breeches	*Dicentra cucullaria*
Daisy fleabane	*Erigeron* spp.	Dyer's woad	*Isatis tinctoria*
Dandelion	*Taraxacum officinale* (Scandinavia only), elsewhere, *Taraxacum* sp. aff.	Dyes, sources of—see listing in Appendix 3	
		Eagle, golden	*Aguila chrysautos*
Dandruff, fern(s) used in treatment of	*Adiantum capillus-veneris*, *Polystichum munitum*	Earth star	*Astraeus* spp., *Geaster* spp., and others
Date	*Phoenix dactylifera*	Earthworm	*Lumbricus* spp. and others
DDT-like compound, algal producers of	*Laurencia* spp. and others	Ebony	*Diospyros ebenum*
		Eelworm (nematode)	member of Class Nematoda, Phylum Aschelminthes, Kingdom Animalia
Death angel (Death cap)	*Amanita* spp.		
Deer	*Odocoileus* spp. and others		
Deer, mule	*Odocoileus hemionus*	Eelworm (nematode) fungi, those that trap with constricting rings	*Dactylaria* spp., *Arthrobotrys actyloides*
Dendrobium (orchid)	*Dendrobium* spp. and hybrids		
Desmids	*Closterium* spp., *Cosmarium* spp., and others	Eelworm (nematode) fungi, those that trap with passive rings	*Dactylella* spp.
Destroying angel	*Amanita virosa*		
Dewberry	*Rubus* hybrids with *R. ursinus* as one parent	Eggplant	*Solanum melongena*
		Elderberry, common	*Sambucus canadensis*
Diatom	*Biddulphia* spp., *Cymbella* spp., *Navicula* spp., *Cymatopleura solea Thalassiosira elsayedii*, *Delphineis karstenii*, *Pseudonitzchia australis*, and many others	Elephant	*Elephas* spp., *Loxodonta* spp.
		Elephant ears	*Colocasia* spp.
		Elk	*Cervus canadensis*
		Elm, American	*Ulmus americana*
		Elm bark beetle	*Hylurgopinus rufipes*, *Scolytus multistriatus*
Dicot	member of Class Magnoliopsida, Phylum Magnoliophyta, Kingdom Plantae	Endive	*Cichorium endivia* spp. *divaricatum*
		Endive, Belgian	*Cichorium intybus*
Digitalis, source of	*Digitalis purpurea, D. lanata*	Ergot	*Claviceps purpurea*
Dill	*Anethum graveolens*	Ermine	*Mustela erminea*
		E. coli	*Escherichia coli*
		Eucalyptus, source of bark/wood for tannins	*Eucalyptus wandoo*

Common Names and Scientific Names of Organisms

Common Name	Scientific Name	Common Name	Scientific Name
Eucalyptus, Tasmanian giant	*Eucalyptus regnans*	Fern(s) used as laxative	*Asplenium trichomanes, Polypodium vulgare*
Eucalyptus oil, source of	*Eucalyptus* spp.; there are more than 250 spp. of *Eucalyptus*	Fern used in treating leprosy	*Marsilea quadrifolia*
Euglenoid	member of Phylum Euglenophyta, Kingdom Protista	Fern, licorice	*Polypodium glycyrrhiza*
Fennel	*Foeniculum vulgare*	Fern(s) poisonous to livestock	*Onoclea sensibilis, Pteridium aquilinum*
Fenugreek	*Trigonella foenum-graecum*	Fern, edible Malaysian (relative of Lady fern)	*Athyrium esculentum*
Fern(s), adder's tongue	*Ophioglossum* spp.		
Fern, a´ma´u ma´u	*Sadleria cyantheodies*	Fern, male	*Dryopteris filix-mas*
Fern(s), amphibious	*Marsilea* spp. and others	Fern, marsh	*Thelypteris* sp.
Fern(s), aquatic (floating)	*Azolla* spp., *Salvinia* spp.	Fern, mosquito	*Azolla caroliniana*
Fern(s), source of astringent	*Actiniopteris radiata, Drynaria quercifolia, Pteridium aquilinum,* and others	Fern, nest	*Asplenium nidus*
Fern, bird's-foot	*Pellaea mucronata*	Fern used to arrest nosebleeds	*Pellaea mucronata*
Fern, bird's-nest	*Asplenium nidus*	Fern(s) used for orchid bark	*Cibotium* spp., *Osmunda* spp.
Fern, Boston	*Nephrolepis exaltata*	Fern, Oriental water	*Ceratopteris thalictroides*
Fern, bracken	*Pteridium aquilinum*	Fern, ostrich	*Matteuccia struthiopteris*
Fern, Brazilian tree	*Cyathea* sp.	Fern used as poison antidote	*Polystichum squarrosum*
Fern used in treating burns	*Polystichum munitum*	Fern(s) used in treating rickets	*Asplenium ruta-muraria, Osmunda regalis*
Fern, chain	*Woodwardia fimbriata*		
Fern, cinnamon	*Osmunda cinnamomea*	Fern(s) used for stuffing mattresses, pillows, upholstery	*Cibotium* spp., *Sadleria* spp.
Fern, climbing (Asian)	*Lygodium salicifolium*		
Fern(s) used in treating coughs	*Adiantum aethiopicum, A. lunulatum, Polypodium glycyrrhiza*	Fern, sword	*Polystichum munitum*
Fern(s) used in treating dandruff	*Adiantum capillus-veneris, Polystichum munitum*	Fern used in treating toothache	*Pentagramma triangularis*
Fern used in treating diabetes	*Adiantum caudatum*	Fern, tree	*Cyathea* spp., *Ctenitis* spp., *Dicksonia* spp., *Marattia* spp., *Sphaeropteris* spp., and others
Fern(s) used in treating diarrhea	*Botrychium lunaria, B. ternatum, Pteridium aquilinum,* and others	Fern, tropical	*Dicranopteris linearis*
Fern(s) used as diuretic	*Adiantum venustum, Lygodium japonicum*	Fern, tropical tree	*Cibotium* sp.
Fern(s) source of dyes	*Sadleria cyatheoides* (trunk), *Sphenomeris chusana* (fronds)	Fern used for expelling worms	*Dryopteris filix-mas*
Fern used in treating eczema	*Lygodium flexuosum*	Fern(s) used for treating wounds	*Lygodium circinatum, Ophioglossum vulgatum*
Fern used in treating eye diseases	*Asplenium adiantum-nigrum*	Fevers, fern used to reduce	*Marsilea quadrifolia*
Fern used to reduce fevers	*Marsilea quadrifolia*	Fevers, ground pine used to reduce	*Lycopodium clavatum*
Fern, five-finger	*Adiantum pedatum*	Fig, common	*Ficus carica*
Fern(s) used as food	*Athyrium filix-femina, Dryopteris austriaca, D. filix-mas, Polystichum munitum,* and others	Fig, tropical	*Ficus macrophyllus, Ficus* spp.
		Figwort Family	Scrophulariaceae
		Filaree	*Erodium* spp.
Fern(s), fossil	*Psaronius* spp., *Thamnopteris* spp., and others	Fir, balsam	*Abies balsamea*
		Fir, Douglas	*Pseudotsuga menziesii*
Fern, gold-back	*Pentagramma triangularis*	Fir, white	*Abies concolor*
Fern(s), Hawaiian tree	*Cibotium* spp., *Sadleria* spp.	Fireweed	*Epilobium angustifolium*
Fern, holly	*Polystichum lonchitis*	Fish	member of Class Pisces, Phylum Vertebrata, Kingdom Animalia
Fern(s) used by hummingbirds	*Cyathea arborea, Lophosoria quadripinnata, Nephelea mexicana*		
Fern used for treating insect stings and bites	*Adiantum capillus-veneris*	Fish, flashlight	*Anomalops katoptron, Photoblepharon palpebratus*
Fern used for easing labor pains	*Athyrium filix-femina*	Fish molds	*Saprolegnia* spp. and others
Fern, lady	*Athyrium filix-femina*		

Common Names and Scientific Names of Organisms

Common Name	Scientific Name	Common Name	Scientific Name
Flashlight powder, ground pine source of	*Lycopodium* spp.	Fungi, hallucinogenic	*Amanita muscaria*, *Conocybe* spp., *Panaeolus* spp., *Psilocybe* spp., and others
Flatworm	*Convoluta roscoffensis*		
Flax	*Linum* spp.	Fungi, horse dung	*Pilobolus* spp.
Flax, New Zealand	*Phormium tenax*	Fungi, industrial alcohol-producing	*Aspergillus* spp.
Flea	member of Order Siphonaptera, Phylum Arthropoda, Kingdom Animalia	Fungi, insect-parasitizing	members of Order Laboulbeniales, Phylum Ascomycota, Kingdom Fungi, and others
Flicker	*Colaptes* spp.		
Florida arrowroot	*Zamia integrifolia*		
Flour, Hopi Indian horsetail source of	*Equisetum laevigatum*	Fungi, meat-tenderizing	*Thamnidium* spp.
		Fungi, ringworm	*Epidermophyton* spp., *Microsporium* spp., *Trichophyton* spp.
Flowerpot leaf plant	*Dischidia rafflesiana*		
Fly	member of Order Diptera, Phylum Arthropoda, Kingdom Animalia	Fungi, shoyu	*Aspergillus oryzae*, *A. soyae*
Fly agaric	*Amanita muscaria*	Fungi used in silvering of mirrors	*Aspergillus* spp.
Flycatcher	*Empidonax* spp., *Myiarchus* spp., and others	Fungi used in manufacturing soap	*Penicillium* spp.
Fly, tsetse	*Gossinia morsitans*, *G. palpalis*	Fungi, soil	*Fusarium* spp., and others
Fly, white	*Aleurocanthus woglumi* and others	Fungi, soy sauce	*Aspergillus oryzae*, *A. soyae*
Fossil, compression	*Annularia radiata*	Fungi, sufu	*Actinomucor elegans*, *Mucor* spp.
Fossil, ground pine (*Lycopodium*)	*Lepidodendron*	Fungi, teonanacatl (sacred)	*Conocybe* spp., *Panaeolus* spp., *Psilocybe* spp., and others
Four-o'clock Family	Nyctaginaceae		
Fox, arctic	*Alopex lagopus*	Fungus, bolete	*Serillus pungens*
Fox, gray	*Urocyon cinereoargentus*	Fungus, bracket/shelf	*Grifola sulphurea*
Fox, red	*Vulpes fulva*	Fungus, downy mildew of grape	*Plasmopora viticola*
Foxglove	*Digitalis purpurea*		
Frangipanni	*Plumeria rubra* and other *Plumeria* spp.	Fungus, "foolish seedling" (of rice)	*Gibberella fujikuroi*
Frog	*Rana* spp., and others	Fungus used in Beadle & Tatum genetic experiments	*Neurospora crassa*
Fruit fly, common	*Drosophila melanogaster* (there are many other species of fruit fly)	Fungus, jelly	*Auricularia* spp., *Exidia* spp., *Tremella* spp., and others
Fuchsia	*Fuchsia* spp.		
Fuchsia, California	*Epilobium canum*	Fungus, kidney bean leaf (production of fungal inhibitors stimulator)	*Colletotrichum lindemuthianum*
Fumitory, Himalayan	*Corydalis gerdae*		
Fungi that produce antibiotics	*Penicillium* spp., *Cephalosporium* spp., and others		
Fungi that cause aspergilloses	*Aspergillus fumigatus*, *Candida albicans*, *Coccidiodes immitis*, and others	Fungus, miso	*Aspergillus oryzae*
		Fungus, causal agent of Panama disease (of bananas)	*Fusarium oxysporum*
Fungi that cause athlete's foot	*Trichophyton* spp.	Fungus used in producing plastics	*Aspergillus terreus*
Fungi used by beetles for food	*Ambrosiella* spp., *Monilia* spp.	Fungus, tempeh	*Rhizopus oligosporus*
Fungi, bird's-nest	*Nidularia* spp., *Crucibulum levis*	Fungus used in manufacturing toothpaste	*Aspergillus niger*
Fungi used in manufacturing birth control pills	*Rhizopus nigricans*, *R. arrhizus*	Fungus, white piedra	*Trichosporon beigeli*
Fungi, cap-thrower	*Pilobolus* spp.	Fungus used in manufacturing yellow food-coloring agent	*Blakeslea trispora*
Fungi, cheese	*Penicillium camembertii* (for Camembert cheese), *P. roquefortii* (for blue, Gorgonzola, Roquefort, and Stilton cheeses)	Funori, source of	*Gloiopeltis* spp.
		Fur, green algae that inhabit animals	*Trentepohlia* spp.
Fungi, flavor-producing	*Aspergillus* spp.	Gentian, source of	*Gentiana* spp.

Common Names and Scientific Names of Organisms

Common Name	Scientific Name	Common Name	Scientific Name
Geranium	*Geranium* spp., *Pelargonium* spp.	Ground pine used as intoxicant	*Lycopodium selago*
Geranium Family	Geraniaceae		
Gila monster	*Heloderma suspectum*	Ground pine used for ornaments	*Lycopodium clavatum*, *L. complanatum*, *L. obscurum*, and other *Lycopodium* spp.
Ginger	*Zingiber officinale* and others		
Ginseng, source of	*Panax quinquefolium* and others		
Giraffe	*Giraffa camelopardalis*	Ground pine used to reduce fevers	*Lycopodium clavatum*
Gladiolia/Gladiolus	*Gladiolus* spp.		
Gloeocapsa	*Gloeocapsa* spp.	Ground pine, fossil relatives of	*Baragwanathia* spp., *Drephanophycus* spp., *Protolepidodendron* spp., and others
Goat	*Capra* spp.		
Goldback fern	*Pentagramma triangularis*		
Golden brown algae	members of Phylum Chromophyta, Kingdom Protista	Guava	*Psidium guajava*
		Gum arabic, source of	*Acacia senegal*
Golden chain tree	*Laburnum anagyroides*	Gum tragacanth, source of	*Astragalus echidenaeformis*, *A. gossypinus*, *A. gummifer*, and others
Goldenrod	*Solidago* spp.		
Goldenseal	*Hydrastis canadensis*		
Goldenweed	*Haplopappus gracilis*[10]	Guppy	*Lebistes reticulatus*
Goose	*Branta* spp. and others	Hairy cap moss	*Polytrichum commune*
Gooseberry	*Ribes* spp.	Haptophyte	member of Phylum Prymnesiophyta, Kingdom Protista
Goosefoot Family	Chenopodiaceae		
Gopher plant	*Euphorbia lathyrus* (*E. lathyris* = *E. lathyrus*)	Hawk	*Buteo* spp., *Falco* spp., and others
		Hazelnut	*Corylus* spp.
Gopher, pocket	*Geomys bursarius, Thomomys* spp., and others	Hazelnut, European	*Corylus avellana*
		Heath	*Erica* spp. and others
Gourd	*Lagenaria siceraria* and others	Heath Family	Ericaceae
Grape	*Vitis* spp.	Hemlock, eastern	*Tsuga canadensis*
Grape, wine/table	*Vitis vinifera*	Hemlock, mountain	*Tsuga mertensiana*
Grapefruit	*Citrus paradisi*	Hemlock, poison	*Conium maculatum*
Grass (including lawn grasses)	*Bromus* spp. and others[11]	Hemlock, water	*Cicuta* spp.
		Hemlock, western	*Tsuga heterophylla*
Grass, Bermuda	*Cynodon dactylon*	Hemp	*Cannabis sativa*
Grass, crested wheat	*Agropyron cristatum*	Hemp, Manila	*Musa textilis*
Grass, Indian	*Sorghastrum nutans*	Hemp, Mauritius	*Furcraea gigantea*
Grass, pampas	*Cortaderia selloana*	Henbit	*Lamium amplexicaule*
Grass Family	Poaceae (formerly Gramineae)	Henna	*Lawsonia inermis*
Grass tree (Australian)	*Xanthorrhea* spp.	Hepatica	*Hepatica americana, Hepatica* spp.
Gray pine[12]	*Pinus sabiniana*	Hickory	*Carya* spp.
Green algae	member of Phylum Chlorophyta, Kingdom Protista; representative genera include *Caulerpa, Chlorella, Codium, Dunaliella, Enteromorpha, Hydrodictyon, Microcystis, Pandorina, Pithophora, Scenedesmus, Spirogyra, Tetraselmis*; there are more than 200 genera and about 7,500 species	Hog	*Sus scrofa*, and others
		Hog fennel	*Lomatium* spp.
		Holly, American	*Ilex opaca*
		Honey bee	*Apis mellifera*
		Hop hornbeam	*Ostrya virginiana*
		Hops	*Humulus lupulus*
		Horehound	*Marrubium vulgare*
		Hornwort	*Anthoceros* spp.
Greenbrier	*Smilax* spp.	Horse	*Equus caballus*
Ground pine	*Lycopodium* spp.	Horse chestnut	*Aesculus hippocastanum*
Ground pine used for baby powder	*Lycopodium clavatum*	Horsetail	*Equisetum hyemale, Equisetum telmateia, Equisetum* spp.
Ground pine used to arrest bleeding	*Lycopodium clavatum*	Horsetail used as abrasive	*Equisetum* (all spp.)

[10]This species has a diploid number of $2x = 4$; i.e., each body cell has four chromosomes.

[11]The Grass Family (Poaceae) comprises about 4,500 species of grasses. Some plants with *grass* in their name are in other families and are not true grasses, e.g., Grass of parnassus (*Parnassia californica*); Grass pink (*Petrorhagia dubia*).

[12]Gray pines were formerly known as Digger pines. The common name was changed in deference to Native Americans who consider *digger* a derogatory term.

Common Names and Scientific Names of Organisms

Common Name	Scientific Name	Common Name	Scientific Name
Horsetail used as astringent	*Equisetum arvense*, *E. debile*, and other *Equisetum* spp.	Jacaranda	*Jacaranda* spp.
		Jaeger	*Stercorarius* spp.
Horsetail used for treating burns	*Equisetum hyemale* and others	Japanese yew	*Taxus cuspidata*
		Jicama	*Pachyrhizus erosus*
Horsetail used for treating diarrhea	*Equisetum hyemale*	Jimson weed	*Datura stramonium*
		Jojoba	*Simmondsia californica*, *S. chinensis*
Horsetail used as diuretic	*Equisetum arvense*, *E. debile*, and others	Joshua tree	*Yucca brevifolia*
		Jumping mouse	*Zapus hudsonius*, *Napaeozapus insignis*
Horsetail used for treating dysentery	*Equisetum hyemale*		
Horsetail, field	*Equisetum arvense*	Junco	*Junco* spp.
Horsetail, fossil	*Equisetites* spp., *Hyenia* spp., *Sphenophyllum* spp., and others	Junco, slate-colored	*Junco hyemalis*
		Juneberry	*Amelanchier* spp.
		Juniper	*Juniperus* spp.
Horsetail, giant	*Equisetum telmateia*	Juniper, dwarf	*Juniperus communis* and others
Horsetail, Hopi Indian flour source	*Equisetum laevigatum*	Jute	*Corchorus capsularis* and others
		Kauri pine/resin	*Agathis australis*, *A. robusta*
Horsetail, treelike fossil	*Calamites* spp.	Kelp	*Alaria* spp., *Dictyoneurum* spp., *Ectocarpus* spp., *Egregia* spp., *Laminaria* spp., *Lessoniopsis* spp., *Nereocystis* spp., and others
Horsetail used as hair wash	*Equisetum hyemale*		
Horsetail used as water source	*Equisetum telmateia*		
Hot springs, blue-green bacteria (cyanobacteria) of	*Bacillosiphon induratus*, *Synechococcus* spp., and others	Kelp, giant	*Macrocystis pyrifera*
		Knotweed	*Polygonum aviculare*, *P. arenastrum*
"Human hair" slime mold	*Stemonitis* spp.	Kohlrabi	*Brassica oleracea* var. *caulorapa* (= *B. oleracea* var. *gongyloides*)
Hummingbird	*Archilocus* spp. and others		
Hummingbird, Oasis	*Rhodopis vesper*		
Hummingbirds, ferns used by (for nest material)	*Cyathea arborea*, *Lophosoria quadripinnata*, *Nephelea mexicana*	Koonwarra angiosperm (extinct angiosperm whose fossil was discovered in Australia) (appears to be similar to members of the pepper family—Piperaceae)	
Hummingbirds, tropical	*Chlorostilbon maugaeus* and others	Kudzu	*Pueraria thunbergiana* (= *P. lobata*)
Hyacinth	*Hyacinthus orientalis* and other *Hyacinthus* spp.		
		Kumquat	*Fortunella japonica*
Hyacinth, bean	*Lablab purpureus*	Lamb's ears	*Stachys byzantina*
Hyacinth, grape	*Muscari* spp.	Lantana	*Lantana camara* and others
Hyacinth, water	*Eichhornia crassipes*	Larch, eastern	*Larix laricina*
Hyssop	*Hyssopus officinalis*	Larch, European	*Larix decidua*
Ice plant	*Carpobrotus* spp. (esp. *C. edulis*), *Mesembryanthemum crystallinum*, and others	Larch, western	*Larix occidentalis*
		Larkspur, blue	*Delphinium* spp.
		Larkspur, red	*Delphinium nudicaule*
India, toxic blue-green bacteria (cyanobacteria) of	*Lyngbya majuscula*	Late blight of potato	*Phytophthora infestans*
		Laurel	*Laurus nobilis*
Indian paintbrush	*Castilleja indivisa*	Laurel Family	Lauraceae
Indian pipe	*Monotropa uniflora*	Lavender	*Lavandula officinalis*, *L. angustifolia*
Indian warrior	*Pedicularis densiflora*	Leaf hopper	member of Order Homoptera, Phylum Arthropoda, Kingdom Animalia
Indigo	*Indigofera tinctoria*		
Insects—see individual entries		Leaf miner	*Agromyza* spp. and others
Insects, fern used for treating stings and bites of	*Adiantum capillus-veneris*	Leaf roller	*Archips argyrospila* and others
		Leafy liverwort—see Liverwort, leafy	
Ipecac, source of	*Cephaelis ipecacuanha*	Legume Family	Fabaceae (formerly Leguminosae)
Iris	*Iris* spp.	Lemming	*Lemmus* spp., *Dicrostonyx groenlandicus*
Iris, butterfly	*Moraea* spp.		
Iris Family	Iridaceae	Lemon	*Citrus limon*
Ironwood, South American	*Krugiodendron ferreum*	Lemongrass oil, source of	*Cymbopogon citratus*, *C. flexuosus*
Isopyrum	*Isopyrum occidentale*		
Ivy, Boston	*Parthenocissus tricuspidata*	Lentil	*Lens esculenta* (= *Lens culinaris*)
Ivy, English	*Hedera helix*	Lettuce	*Lactuca sativa*
Ivy, poison	*Toxicodendron radicans* (formerly *Rhus toxicodendron*)	Lichen (symbiotic association of an alga and a fungus)	member of Phylum Ascomycota, Kingdom Fungi[13]

[13]The lichens are arbitrarily treated under Phylum Ascomycota within Kingdom Fungi because (1) the vast majority of fungal components of each species are ascomycetes, and (2) the fungal component of each species of lichen is unique to the species, while the algal component is often common to more than one species of lichen.

Common Names and Scientific Names of Organisms

Common Name	Scientific Name	Common Name	Scientific Name
Lichen, British Soldier	*Cladonia cristatella*	Logwood	*Haematoxylon campechianum*
Lichen, crustose		Loon	*Gavia* spp.
black	*Rinodina* sp.	Lotus, Oriental sacred	*Nelumbo nucifera*
chartreuse	*Acarospora citrina*	Louse	Orders Mallophaga and Anaplura, Class Insecta, Phylum Arthropoda, Kingdom Animalia
gray	*Psora* sp.		
orange-red	*Caloplaca elegans*		
yellow	*Candelariella vitellina*	Love-lies-bleeding	*Amaranthus caudatus*
Lichen, foliose	*Physcia* sp., *Parmelia* sp.	Lucerne—see Alfalfa	
Lichen, fruticose	*Usnea* sp.	Luffa	*Luffa cylindrica, L. acutangula*
Lichen, grazed by North African sheep	*Lecanora* spp.	Lupine	*Lupinus* spp.
		Lupine, tree with seed valves	*Lupinus arboreus*
Lichen, litmus	*Rocella* spp.	Madder Family	Rubiaceae
Lichen used as miniature trees and shrubs	*Cladonia* spp. and others	Magnolia	*Magnolia* spp.
Lichen, natural dye	*Parmelia* spp., *Usnea* spp., and others	Magnolia Family	Magnoliaceae
Lichen, perfume stabilizer	*Evernia* spp.	Mallow	*Malva* spp.
Lichen, reindeer (reindeer moss)	*Cladonia* spp., *Cetraria islandica*	Mallow Family	Malvaceae
		Mango	*Mangifera indica*
		Mangrove	*Rhizophora mangle, R. candelaria,* and others
Licorice, source of	*Glycyrrhiza glabra*		
Lignum vitae	*Guaiacum officinale*		
Lilac, common	*Syringa vulgaris*	Mangrove, black	*Avicennia germinans, A. nitida*
Lily	*Lilium regale, L. auratum, L. martagon, Lilium* spp., and others	Manila hemp	*Musa textilis*
		Manioc—see Cassava	
Lily, giant water	*Victoria amazonica*	Manroot	*Marah* spp.
Lily, kaffir	*Clivia* sp.	Maple	*Acer* spp.
Lily, tiger	*Lilium pardalinum*	Maple, big leaf	*Acer macrophyllum*
Lily, wood	*Lilium superbum*	Maple, hard	*Acer saccharum*
Lily Family	Liliaceae	Maple, red	*Acer rubrum*
Lime	*Citrus aurantifolia*	Maple, silver	*Acer saccharinum*
Litchi	*Litchi sinensis*	Maple, sugar	*Acer saccharum*
Litmus indicator dye, source of	*Rocella* spp.	Marigold	*Tagetes erecta* and other *Tagetes* spp.
Live oak	*Quercus chrysolepis, Q. virginiana, Q. wislizenii*, and others	Marijuana	*Cannabis sativa*
		Marjoram	*Majorana hortensis* (= *Origanum majorana*), pot marjoram = *Origanum onites*
Liverwort	member of Phylum Hepaticophyta, Kingdom Plantae		
Liverwort, leafy	*Calopogeia* sp., *Bazzania trilobata, Frullania* spp., *Jungermannia* spp., *Porella* spp., and others[14]	Maté	*Ilex paraguariensis*
		Meadow foam	*Limnanthes* spp.
		Mealy bugs	*Pseudococcus* spp.
		Melon	*Cucumis melo*
Liverwort, thalloid	*Conocephalum* spp., *Lunularia* spp., *Marchantia* spp., and others	Melon, honeydew	*Cucumis melo* (variety)
		Melonette	*Melothria pendula*
Lizard	*Sceloporus* spp. and others	Mermaid's wineglass	*Acetabularia* spp.
Lobeline sulfate, source of (used in formulas to assist in stopping smoking)	*Lobelia inflata*	Mescal bean	*Sophora secundiflora*
		Mesquite	*Prosopis glandulosa*
		Milkweed	*Asclepias syriaca* and other *Asclepias* spp.
Locoweed	*Astragalus mollisimus* and other *Astragalus* spp.	Milkweed, swamp	*Asclepias incarnata*
		Millet[15]	
Locust, black	*Robinia pseudo-acacia*	Millipede	member of Class Diplopoda, Phylum Arthropoda, Kingdom Animalia
Loganberry	*Rubus* hybrids, with *R. ursinus* as one parent; *Rubus vitifolius*		

[14]There are thousands of species of leafy liverworts assigned to about 200 genera.

[15]Several species of grain are called *millet*, but the most extensively cultivated taxa are *Pennisetum glaucum* (pearl millet) and *Eleusine coracana* (finger millet). Other millets, some of which are used as pasturage, include *Panicum milliaceum* (broomcorn millet); *P. maximum; P. obtusum; P. purpurascens; P. ramosum; P. texanum; P. virgatum; Echinochloa colona; E. crus-galli; Paspalum* sp.; *Eragrostis* sp.; *Setaria italica*; and others.

Common Names and Scientific Names of Organisms

Common Name	Scientific Name	Common Name	Scientific Name
Mint—see Peppermint, Spearmint, etc.		Moss, pygmy—see Moss, annual	
Mint Family	Lamiaceae (formerly Labiatae)	Moss, rock	*Andreaea* spp., *Grimmia* spp., and others
Mistletoe	*Phoradendron* spp.	Moss, saline (salty) soil indicator	*Pottia* spp.
Mistletoe, dwarf	*Arceuthobium* spp.	Moss, seasonal running-water indicator	*Fontinalis* spp.
Mite	member of Order Acarina, Phylum Arthropoda, Kingdom Animalia	Moss, sphagnum	*Sphagnum* spp.
Mock orange	*Philadelphus x virginalis* and other *Philadelphus* spp. and hybrids	Moss rose	*Portulaca grandiflora*
Mollusc	member of Phylum Mollusca, Kingdom Animalia	Moth	member of Order Lepidoptera, Class Insecta, Phylum Arthropoda, Kingdom Animalia
Monkey	*Ateles dariensis* and many others		
Monkey flower	*Mimulus* spp.	Moth, Argentine, used to control cactus in Australia	*Cactoblastus cactorum*
Monkey plant	*Lunaria annua*	Moth, codling	*Carpocapsa pomonella*
Monkshood	*Aconitum columbianum*	Moth, gypsy	*Porthetria dispar*
Monocot	member of Class Liliopsida, Phylum Magnoliophyta, Kingdom Plantae	Moth, Mexican jumping bean	*Carpocaps asaltitans*
Moose	*Alces americana, A. alces*	Moth, Yucca	*Pronuba* spp., *Tegeticula* spp.
Morel	*Morchella esculenta* and other *Morchella* spp.	Moth mullein	*Verbascum blattaria*
Morel, false	*Gyromitra* sp.	Mountain beaver	*Aplodontia rufa*
Morning glory	*Ipomoea violacea* and others	Mouse	*Mus musculus, Peromyscus* spp., and others
Mosquito	*Anopheles* spp., *Culex* spp., and others	Mouse, jumping	*Zapus hudsonius, Napaeozapus insignis*
Moss	member of Phylum Bryophyta, Kingdom Plantae	Mulberry	*Morus* spp.
Moss, annual (bare soil)	*Acaulon* spp., *Ephemerum* spp., and others	Mulberry, red	*Morus rubra*
Moss, antler and bone	*Tetraplodon* spp.	Mulberry, white	*Morus alba*
Moss, used as burn treatment by Native Americans	*Bryum* Spp., *Minium* Spp.	Mule ears	*Wyethia* spp.
		Mullein	*Verbascum thapsus*
		Mullein, moth	*Verbascum blattaria*
Moss, copper-rich substrate-inhabiting	*Mielichhoferia* spp., *Scopelophila* spp.	Mushroom[16]	*Agaricus* spp. and others
Moss, carnivore dung-inhabiting	*Tayloria* sp.	Mushroom, common cultivated edible	*Agaricus bisporus*
Moss, calcium absence indicator	*Andreaea* spp., *Rhacomitrium lanuginosum*	Mushroom, common red	*Russula*
		Mushroom, fairy ring	*Marasmius oreades*
		Mushroom, fly agaric	*Amanita muscaria*
Moss, calcium presence indicator	*Didymodon* spp., *Desmatotodon* spp., and others	Mushroom, inky cap	*Coprinus* spp.
		Mushroom, oyster	*Pleurotus ostreatus*
Moss, exceptionally desiccation-resistant	*Tortura ruralis*	Mushroom, pore	*Serillus pungens*
		Mushroom, portabella	*Agaricus bisporus* (variety)
Moss, European roof-waterproofing	*Dicranoweisia* sp.	Mushroom, shaggy mane	*Coprinus comatus*
Moss, extinguisher	*Encalypta* spp.	Mushroom, shiitake	*Lentinus edodes*
Moss, hair(y) cap	*Polytrichum* spp.	Muskrat	*Ondatra zibesthicus*
Moss, herbivore dung-inhabiting	*Splachnum* spp.	Mustard	*Brassica campestris, B. nigra,* and others
Moss, luminous	*Schisostega pennata, Mittenia* sp.	Mustard, cultivated	*Brassica alba, B. juncea,* and others
Moss, mammal dung-inhabiting	*Splachnum luteum*	Mustard Family	Brassicaceae (formerly Cruciferae)
Moss, peat	*Sphagnum* spp.	Myrrh, source of	*Commiphora abyssinica*
Moss, pollution-sensitive	*Hypnum* spp. and others	Myrtle[17]	*Umbellularia californica*[17]

[16]*Mushroom* is a term generally applied to the fruiting bodies with stalked, caplike structures produced by members of Phylum Basidiomycota, Kingdom Fungi. The term is also loosely applied to some of the fruiting bodies of members of other classes of true fungi. There are thousands of known species.

[17]This plant, also known as the California bay, is in the Laurel Family (Lauraceae). True myrtles are in the Myrtle Family (Myrtaceae).

Common Names and Scientific Names of Organisms

Common Name	Scientific Name	Common Name	Scientific Name
Narcissus	*Narcissus* spp. and hybrids[18]	Orchid, vanilla	*Vanilla planifolia* and others
Nasturtium (garden)	*Tropaeolum majus*	Orchid with cladophylls	*Epidendrum* spp.
Nectarine	*Prunus persica*	Orchid Family	Orchidaceae
Neem tree	*Azadirachta indica*	Oregano	*Origanum vulgare* and others
Nematode	member of Class Nematoda, Phylum Aschelminthes, Kingdom Animalia	Oregon grape	*Berberis aquifolium* (= *Mahonia aquifolium*) and other *Mahonia* spp.
Nettle	*Urtica* spp.		
Nicotine relative (nornicotine), source of	*Duboisia hopwoodii*, *Nicotiana tabacum*	Organ-pipe cactus	*Lemaireocereus* spp.
		Osage orange	*Maclura pomifera*
Nightshade, deadly	*Atropa belladonna*	Oscillatoria	*Oscillatoria* spp.
Nightshade Family	Solanaceae	Our Lord's Candle	*Yucca whipplei*
Nori—see Purple laver		Owl, snowy	*Nyctea scandiaca*
Nostoc	*Nostoc* spp.	Painted lady	*Echeveria derenbergii*
Nutmeg	*Myristica fragrans*	Palm, carnauba wax	*Copernicia cerifera*
Nutmeg, California	*Torreya californica*	Palm, coconut	*Cocos nucifera*
Nutmeg Family	Myristicaceae	Palm, date	*Phoenix dactylifera*
Oak	*Quercus* spp.	Palm, oil	*Elaeis guineensis*
Oak, black	*Quercus velutina*	Palm, panama hat	*Carludovica palmata*
Oak, blue	*Quercus douglasii*	Palm, Seychelles Island	*Lodoicea maldivica*
Oak, cork	*Quercus suber*	Palm Family	Arecaceae (formerly Palmae)
Oak, Hooker	*Quercus lobata*	Pansy	*Viola tricolor*
Oak, live	*Quercus wislizenii* (other live oaks include *Quercus chrysolepis* and *Q. virginiana*)	Papaya	*Carica papaya*
		Pará rubber tree	*Hevea brasiliensis*
		Parsley	*Petroselinum crispum*
Oak, poison	*Toxicodendron diversilobum*	Parsley Family	Apiaceae (formerly Umbelliferae)
Oak, red	*Quercus borealis*	Parsnip	*Pastinaca sativa*
Oak, white	*Quercus alba*	Passion fruit	*Passiflora edulis*, *P. mollissima*, and other *Passiflora* spp.
Oak wilt	*Ceratocystis fagacearum*		
Olibanum tree	*Boswellia* spp.	Patchouli oil, source of	*Pogostemon cablin* and others
Olive	*Olea europaea*	Pea (garden)	*Pisum sativum*
Onion	*Allium cepa*	Pea, sweet	*Lathyrus odoratus*
Oomycete	member of Phylum Oomycota, Kingdom Fungi	Peach	*Prunus persica*
		Peach leaf curl	*Taphrina deformans*
Opuntia—see Prickly pear		Peanut	*Arachis hypogaea*
Orchid	*Cattleya* spp., and many others[19]	Pear	*Pyrus communis*
Orchid, bamboo	*Arundina graminifolia*	Peat moss	*Sphagnum* spp.
Orchid "bark," fern sources of	*Cibotium* spp., *Osmunda* spp.	Pecan	*Carya illinoensis*
		Peccary	*Pecari angulatus*, *Tayassus pecari*
Orchid, Bletilla	*Bletilla* sp.	Penicillin mold[20]	*Penicillium* spp.[20]
Orchid, bucket	*Coryanthes* spp.	Pennyroyal	*Hedeoma pulegioides*
Orchid, showy	*Orchis* spp.	Peony	*Paeonia* spp.
Orchid, underground-flowering	*Rhizanthella gardneri*	Peperomia	*Peperomia* spp.
		Pepper	*Capsicum annuum, C. frutescens*[21]

[18]The 27 known species of *Narcissus* (native to Europe and the Mediterranean regions) have been extensively cultivated and hybridized. There is a botanical classification for wild forms and a horticultural classification based primarily on the extent of the corona, flower color, and fragrance. The horticultural forms are known by common names such as narcissi, daffodils, jonquils, pheasant's eye, angel's tears, etc.

[19]Depending on which authorities are followed, the number of known orchid species (all in the family Orchidaceae) may exceed 30,000. Popularly cultivated orchids include species of *Cattleya, Cymbidium, Dendrobium, Odontoglossum, Oncidium, Paphiopedilum, Phalaenopsis, Vanda,* and both interspecific and intergeneric hybrids.

[20]The original producer of penicillin discovered by Sir Alexander Fleming was *Penicillium notatum;* current commercially used producers of penicillin are strains of *Penicillium chrysogenum*. Other commercially cultivated *Penicillium* species include *P. roquefortii* (used to make roquefort cheese), *P. camembertii* (used to make blue cheese), and *P. griseofulvum* (used for the production of a ringworm and athlete's foot antibiotic known as *griseofulvin*).

[21]The drug *capsicum*, whose active ingredient is the oleoresin *capsaicin*, is derived from these species, and garden peppers include these and other species of *Capsicum*.

Common Names and Scientific Names of Organisms

Common Name	Scientific Name	Common Name	Scientific Name
Pepper, black	Piper nigrum	Pine, stone—see Pine, European stone; and Pine, Mexican stone	
Pepper, red	Capsicum annuum, C. baccatum, C. chinense, C. frutescens, C. pubescens	Pine, sugar	Pinus lambertiana
		Pine, western white	Pinus monticola
		Pine, western yellow	Pinus ponderosa
Peppergrass	Lepidium spp.	Pine, Wollemi	Wollemia nobilis
Peppermint	Mentha piperita	Pineapple	Ananas comosus
Persimmon	Diospyros spp.	Pinedrops	Pterospora spp.
Petit grain oil, source of	Citrus aurantium var. amara	Pistachio	Pistacia vera
Petunia	Petunia spp. and hybrids	Pitcher plant	Sarracenia spp. and others
Peyote	Lophophora williamsii	Pitcher plant, Asian	Nepenthes spp. and others
Phoebe	Sayornis phoebe	Plantain	Plantago spp. (cooking bananas, also called plantains, are mostly Musa x paradisiaca)
Pigweed[22]	Amaranthus spp., Chenopodium spp.		
Pigweed Family[22]	Amaranthaceae, Chenopodiaceae		
Pill bug	Cylisticus convexus and others	Plastic, fungus used in production of	Aspergillus terreus
Pine	Pinus spp.		
Pine, Afghanistan	Pinus eldarica	Plasticizers, source of oil for	Euphorbia agascae
Pine, Aleppo	Pinus halepensis	Plover	Charadrius spp. and others
Pine, bristlecone	Pinus longaeva	Plum, cherry	Prunus cerasifera
Pine, Chilghoza	Pinus gerardiana	Plum, European[23]	Prunus domestica[23]
Pine, Colorado bristlecone	Pinus aristata	Podocarps, New Zealand timber	Podocarpus dacrydoides, P. totara
Pine, Coulter	Pinus coulteri	Podocarps, ornamental	Podocarpus macrophylla, P. nagi, and others
Pine, eastern white	Pinus strobus		
Pine, European stone	Pinus pinea	Poinsettia	Euphorbia pulcherrima
Pine, gray (formerly Pine, digger)	Pinus sabiniana	Poison ivy	Toxicodendron radicans
		Poison oak	Toxicodendron diversilobum
Pine, jack	Pinus banksiana	Poison sumac	Toxicodendron vernix
Pine, Jeffrey	Pinus jeffreyi	Polyanthus	Primula polyanthus and hybrids
Pine, kauri	Agathis australis, A. robusta	Pomegranate	Punica granatum
Pine, knobcone	Pinus attenuata	Poor man's pepper	Lepidium virginicum
Pine, loblolly	Pinus taeda	Popcorn	Zea mays (horticultural variety)
Pine, lodgepole	Pinus contorta	Poplar	Populus spp.
Pine, longleaf	Pinus palustris	Poppy—see also California poppy	Papaver spp. and others
Pine, Merkus	Pinus merkusii		
Pine, Mexican pinyon	Pinus cembroides	Poppy, bush	Dendromecon rigida
Pine, Mexican stone	Pinus cembroides	Poppy, Mexican	Hunnemannia spp.
Pine, Monterey	Pinus radiata	Poppy, opium	Papaver somniferum
Pine, pinyon	Pinus edulis, P. monophylla, P. quadrifolia	Poppy, Oriental	Papaver orientale
		Poppy, prickly	Argemone glauca
Pine, pitch	Pinus rigida	Poppy Family	Papaveraceae
Pine, ponderosa	Pinus ponderosa	Porcupine	Erethizon spp., Hystrix spp.[24]
Pine, red	Pinus resinosa	Portulaca Family	Portulacaceae
Pine, Scotch	Pinus sylvestris	Potato, Irish	Solanum tuberosum
Pine, shortleaf	Pinus echinata	Potato, sweet	Ipomea batatas
Pine, Siberian white	Pinus sibirica	Potato vine	Solanum jasminoides
Pine, slash	Pinus caribaea, P. elliottii	Powder-puff flower	Calliandra inaequilatera
Pine, southern yellow—see Pine, loblolly; Pine, longleaf; Pine, shortleaf; and Pine, slash			

[22]Species of *Amaranthus* in the Amaranth family (Amaranthaceae) and *Chenopodium* in the Goosefoot family (Chenopodiaceae) have been called *pigweeds,* and both families have also been referred to as the Pigweed family. Because of this, many botanists prefer to refer to the Amaranthaceae as the Amaranth family and the Chenopodiaceae as the Goosefoot family. Most *Amaranthus* spp. are called amaranths and most *Chenopodium* spp. are called *goosefoot,* although the widespread weed *Chenopodium album* is popularly called lamb's quarters. Pig's weed (*Oryza* sp.) is completely unrelated to either of the two families just mentioned; it is a grass related to rice.

[23]Although the European plum was involved in the development of the majority of older plum varieties, more than 600 varieties of plum have been developed from American plums such as *Prunus subcordata* and *P. americana,* along with various hybrids involving at least one American parent.

[24]*Hystrix* is also a name for a genus of grasses.

Common Names and Scientific Names of Organisms

Common Name	Scientific Name	Common Name	Scientific Name
Powdery mildew	*Erysiphe* spp. and others	Redbud, eastern	*Cercis canadensis*
Prayer plant	*Maranta* spp.	Redbud, western	*Cercis occidentalis*
Preferns	*Cladoxylon* spp., *Protopteridium* spp. and others	Redwood, coastal	*Sequoia sempervirens*
		Redwood, dawn	*Metasequoia glyptostroboides*
Prickly pear	*Opuntia* spp.	Redwood, giant	*Sequoiadendron giganteum*
Primrose	*Primula vulgaris* and about 400 other *Primula* spp.	Reindeer	*Rangifer* spp.
		Reserpine, source of	*Rauvolfia serpentina*
Pronghorn	*Antilocarpa americana*	Resurrection plant	*Selaginella lepidophylla*
Psyllium	*Plantago ovata*	Rhododendron	*Rhododendron* spp.
Ptarmigan	*Lagopus* spp.	Rhubarb	*Rheum rhaponticum*
Pteridosperms	*Lyginopteris* spp., *Medullosa* spp., and others	Rice[25]	*Oryza sativa*
		Rice-paper plant	*Tetrapanax papyriferus* (some authors refer to *Fatsia japonica* [*F. papyrifera*] as rice-paper plant)
Puffball	*Calvatia cyathiformis* and other *Calvatia* spp., *Lycoperdon* spp.		
Pulque, source of	*Agave* spp.	Rice, wild	*Zizania aquatica*, *Zizania palustris*
Pumpkin	*Cucurbita pepo*	Robin	*Turdus migratorius*
Pumpkin Family	Cucurbitaceae	Rock cress	*Arabis* sp.
Puncture vine	*Tribulus terrestris*	Rock-rose, European	*Helianthemum vulgare*
Purple laver	*Porphyra tenera*	Rock tripe	*Umbilicaria* spp.
Puya (rare)	*Puya raimondii*	Rockweeds	*Fucus* spp., *Pelvetia* spp., and others
Pyrethrum	*Chrysanthemum cinerariifolium*, *C. coccineum*, *C. marschallii*	Rose	*Rosa odorata* and other *Rosa* spp. (there are more than 25,000 rose cultivars)
Quillwort	*Isoetes* spp.		
Quillwort, fossil relatives of	*Cinchona ledgeriana* and other *Cinchona* spp.	Rose, damask	*Rosa damascena*
		Rose Family	Rosaceae
Quince	*Cydonia oblonga*	Rose, Sitka	*Rosa rugosa*
Quinine, source of	*Cinchona officinalis*, *C. ledgeriana*	Rosemary	*Rosmarinus officinalis*
Quinoa	*Chenopodium quinoa*	Rotenone, source of	*Derris elliptica*, *Lonchocarpus nicou*
Rabbit	*Oryctolagus cuniculus*	Rotenone, relative of	*Tephrosia vogelii*
Rabbit, cottontail	*Sylvilagus* spp.	Rubber, Pará	*Hevea brasiliensis*
Rabbit, jack	*Lepus* spp.	Rubber plant	*Ficus elastica*
Raccoon	*Procyon lotor*	Ruellia	*Ruellia portellae* and others
Radish	*Raphanus sativus*	Rust, black stem of wheat	*Puccinia graminis*
Rafflesia	*Rafflesia micropylora*	Rust, cedar-apple	*Gymnosporangium juniperi-virginianum*
Ragweed	*Ambrosia* spp.		
Rape/rapeseed	*Brassica napus*	Rust, corn	*Puccinia sorghi*
Raspberry, red	*Rubus idaeus*, *R. strigosus*, and their hybrids	Rust, poplar leaf spot	*Melampsora medusae*
		Rust, rock cress	*Puccinia monoica*
Rat	*Rattus norvegicus*, *R. rattus*, and others	Rust, white pine blister	*Cronartium ribicola*
		Rutabaga	*Brassica campestris* var. *napobrassica* (= *Brassica napus*)
Rat, kangaroo	*Dipodomys* spp.		
Rat snake, black	*Elaphe obsoleta*	Rye	*Secale cereale*
Rattlesnake	*Crotalus* spp.	Ryegrass	*Lolium* spp.
Red algae	Member of Phylum Rhodophyta, Kingdom Protista; representative genera include *Chondrus*, *Eucheuma*, *Gelidium*, *Gigartina*, *Gracilaria*, *Polysiphonia*, *Porphyra*, and *Pterocladia*; there are about 3,900 spp. of red algae	Safflower	*Carthamus tinctorius*
		Saffron (true)	*Crocus sativus*
		Saffron, meadow	*Colchicum autumnale*
		Sage[26]	*Salvia officinalis*[26]
		Sagebrush	*Artemisia tridentata*
		Sage, Jerusalem	*Phlomis fruticosa*

[25] At the beginning of the year 2000, the International Rice Research Institute in the Philippines had in storage the seeds of 20 species and more than 81,000 different varieties of rice. The staple food of nearly 2 billion people, rice cultivation presently occupies 11% of agricultural land. Rice has been cultivated in Asian countries for more than 7,000 years. Although the great majority of rice cultivated is *Oryza sativa*, some forms of *Oryza glaberrima* are also cultivated.

[26] This sage, which is in the Mint Family (Lamiaceae), should not be confused with *sagebrush*, which is in the Sunflower Family (Asteraceae).

Common Names and Scientific Names of Organisms

Common Name	Scientific Name	Common Name	Scientific Name
Saguaro	*Carnegia gigantea*	Soaproot, California	*Chlorogalum pomeridianum*
Salmon	*Oncorhynchus* spp., *Salmo salar*, and others	Sorghum	*Sorghum bicolor* and other *Sorghum* spp.
Salmonberry	*Rubus spectabilis*	Sorrel	*Oxalis* spp.
Salsify	*Tragopogon* spp.	Southern yellow pine—see Pine, loblolly; Pine, longleaf; Pine, shortleaf; and Pine, slash	
Saltbush/Saltscale	*Atriplex* spp.		
Salvia	*Salvia* spp.	Soybean	*Glycine max*
Sansevieria	*Sansevieria trifasciata* and others *Sansevieria* spp.	Spanish moss	*Tillandsia usneoides*
		Sparrow, savannah	*Passerculus sandwichensis*
Santonin, source of	*Artemisia cina*	Sparrow, song	*Melospiza melodia*
Sargassum	*Sargassum* sp.	Sparrow, vesper	*Pooecetes gramineus*
Sarsaparilla, source of	*Smilax* spp.	Spearmint	*Mentha spicata*
Sassafras	*Sassafras albidum*	Spiderwort	*Tradescantia virginiana* and other *Tradescantia* spp.
Sausage tree, African	*Kigelia pinnata*		
Savory	*Satureia hortensis*	Spiderwort, European	*Tradescantia paludosa*
Saxifrage	*Saxifraga* spp.	Spike moss	*Selaginella* spp.
Screw pine	*Pandanus veitchii* and others *Pandanus* spp.	Spike moss, fossil relatives of	*Lepidodendron* spp., *Sigillaria* spp., and others
Sea anemone	*Stephanauge* spp. and others	Spinach	*Spinacia oleracea*
Sea hare	*Aplysia californica*	Spirogyra	*Spirogyra* spp.
Sea lettuce	*Ulva* spp.	Sponge	*Spongilla* spp. and others
Sea palm	*Postelsia palmaeformis*	Sponge, vegetable	*Luffa cylindrica*
Sea rocket	*Cakile edentula*	Spring beauty	*Claytonia virginica*
Sedge	*Carex* spp. and others	Spruce, black	*Picea mariana*
Seed ferns (Pteridosperms)	*Lyginopteris* spp., *Medullosa* spp., and others	Spruce, Norway	*Picea abies*
		Spruce, red	*Picea rubens*
Senna	*Cassia senna* and others	Spruce, Sitka	*Picea sitchensis*
Sensitive plant	*Mimosa pudica*	Spruce, white	*Picea glauca*
Sequoia	*Sequoiadendron giganteum*	Spurge	*Euphorbia peplus*, *Euphorbia* spp.
Sesame	*Sesamum indicum*	Spurge Family	Euphorbiaceae
Shallot[27]	*Allium cepa*	Squash	*Cucurbita maxima, C. mixta, C. moschata, C. pepo*, and others
Sheep	*Ovis* spp.		
Shepherd's purse	*Capsella bursa-pastoris*	Squawroot	*Perideridia* spp.
Shrimp	*Crago* spp. and others	Squill	*Scilla* spp.
Sisal	*Agave sisalina*	Squills	*Urginea maritima*
Skunk	*Mephitis* spp.	Squirrel	*Citellus* spp., *Sciuris* spp., and others
Slime mold	member of Phyla Myxomycota and Dictyosteliomycota, Subkingdom Myxobionta, Kingdom Protista *Lamproderma* sp., *Lycogala epidendrum*	Squirrel corn	*Dicentra canadensis*
		Squirrel, gray	*Sciurus carolinensis*
		Squirting cucumber	*Ecballium elaterium*
		Stapelia	*Stapelia* sp.
		Stinkhorn	*Mutinus* spp., *Phallus impudicus*, and others
Slime mold, cellular	member of Phylum Acrasiomycota, Kingdom Protista	Stinkhorn, common	*Mutinus caninus*
Slime mold, human-hair	*Stemonitis* spp.	Stonecrop	*Sedum* spp., *Crassula* spp., and others
Sloth	*Bradypus* spp., *Choleopus* spp.	Stoneseed	*Lithospermum ruderale*
Smut	*Ustilago* spp. and others	Stonewort	*Chara* spp., *Nitella* spp.
Smut, corn	*Ustilago maydis*	Strawberry	*Fragaria ananassa* and other *Fragaria* spp. and hybrids
Snail	*Haplotrema concava* and others		
Snapdragon	*Antirrhinum majus*	String-of-pearls	*Senecio rowellianus*
Snowplant	*Sarcodes sanguinea*	Strychnine, source of	*Strychnos nux vomica* and other *Strychnos* spp.
Snowy owl	*Nyctea scandiaca*		

[27]Linnaeus applied the name *Allium ascalonicum* to what was probably an *Allium* cultigen believed to have originated in Asia Minor, and the name *Allium ascalonium* since has generally been applied to shallots. Shallots seldom set seed, however, and seeds sold under the name *Allium ascalonicum* have sometimes proved to be those of other *Allium* spp. The current practice of considering shallots to be a form of *Allium cepa* should lessen the confusion.

Common Names and Scientific Names of Organisms

Common Name	Scientific Name	Common Name	Scientific Name
Sugar cane	*Saccharum officinarum*	Tree fern, small Hawaiian	*Sadleria cyatheoides*
Sumac	*Rhus* spp.		
Sunbird	*Anthodiaeta* spp., *Notiocinnyris* spp., and others	Tree-of-heaven	*Ailanthus altissima*
		Trillium	*Trillium* spp.
Sundew	*Drosera* spp.	Truffles	*Tuber* spp.
Sundew relative used for flypaper	*Drosophyllum lusitanicum*	Tulip	*Tulipa* spp.
		Tulip tree	*Liriodendron tulipifera*
Sunflower	*Helianthus annuus, H. debilis*	Tumbleweeds	*Amaranthus albus, Salsola pestifera*, and others
Sunflower Family	Asteraceae (formerly Compositae)		
Sweet pea	*Lathyrus odoratus*	Tung oil, source of	*Aleurites fordii*
Sword fern	*Polystichum munitum*	Turmeric, source of	*Curcuma longa, C. domestica*
Sycamore	*Platanus occidentalis* and others	Turnip	*Brassica rapa*
Tamarack	*Larix* spp.	Turtle	*Chelydra* spp., *Chrysemys* spp., and others
Tamarisk	*Tamarix* spp.		
Tangerine	*Citrus reticulata*	Twinflower	*Linnaea borealis*
Tapir	*Tapirus* spp.	Ulothrix	*Ulothrix* spp.
Taro	*Colocasia esculenta*	Ultraviolet light, flowers seen in	*Rudbeckia* sp.
Tarragon	*Artemisia dracunculus*		
Tarweed	*Grindelia* spp.	Unicorn plant	*Proboscidea* spp.
Tarweed, western	*Calycadenia* sp.	Venus's flytrap	*Dionaea muscipula*
Tea	*Camellia sinensis*	Vetch	*Vicia* spp.
Tent caterpillar	*Malacosoma americanum* and others	Vetchling, yellow	*Lathyrus aphaca*
		Vinegar weed	*Trichostema* spp.
Teosinte (annual)	*Zea mexicana*	Violet	*Viola odorata* and other *Viola* spp.
Teosinte (perennial)	*Zea diploperennis*		
Tequila, source of	*Agave* spp.	Violet, African	*Saintpaulia ionantha* and other *Saintpaulia* spp.
Termite	*Odontotermes* spp., *Reticulitermes* spp., and others		
		Violet, gold	*Viola douglasii*
Texas paintbrush	*Castilleja indivisa*	Virginia creeper	*Parthenocissus quinquefolia*
Thalloid liverworts	*Marchantia* spp., *Conocephalum* spp., and others	Virus[29]	
		Vole	*Microtus* spp. and others
Thimbleberry	*Rubus parviflorus*	Wahoo	*Euonymus alata* and others
Thistle	*Cirsium* spp. and others	Wake-robin	*Trillium* spp.
Thistle, Canada	*Cirsium arvense*	Wallflower, western	*Erysimum capitatum*
Thrasher	*Toxostoma* spp.	Walnut	*Juglans* spp.
Thyme	*Thymus vulgaris* and others	Walnut, black	*Juglans nigra*
Ti (Ki) plant	*Cordyline fruticosa*	Warbler	*Dendroica* spp. and others
Tiger	*Panthera tigris*	Watercress	*Nasturtium officinale*
Toad	*Bufo americanus*	Water fern, oriental	*Ceratopteris thalictroides*
Tobacco	*Nicotiana tabacum, N. rustica*	Watermelon	*Citrullus lanatus*
Tomato[28]	*Solanum esculentum*	Water mold	member of Phylum Oomycota, Subkingdom Mastigobionta, Kingdom Protista
Tomato, Galápagos	*Solanum esculentum* var. *minor, S. pimpinellifolium, S. cheesmanii* (salt tolerant sp.)		
		Water net	*Hydrodictyon* spp.
Tomato fruitworm	*Heliothis armigera*	Watersilk	*Spirogyra* spp.
	Heliothis armigera	Water weed	*Elodea* spp.
Tomato hornworm	*Protoparce quinquemaculata*	Water weed, yellow	*Ludwigia repens*
Toothwort	*Dentaria* spp.	Wattle	*Acacia decurrens, A. mearnsii*, and others
Tortoise, giant Galápagos	*Testudo elephantopus porteri*		
Touch-me-not	*Impatiens glandulifera* and others	Weaver birds	*Anaplectes* spp., *Hyphantoris* spp., and others
Tree fern	*Cibotium glaucum*		

[28]DNA evidence indicates the tomato, long known as *Lycopersicum esculentum*, belongs in the genus *Solanum* and should be transferred to that genus, making the correct name *Solanum esculentum*.

[29]Depending on the classification used, viruses may not have a scientific name. Many are named after the disease they cause; e.g., tobacco mosaic virus causes tobacco mosaic disease. One classification attempts to give them at least a Latin prefix, so that the virus for warts is *Papavovirus;* for smallpox, *Poxvirus;* for polio, *Picornavirus;* for measles and mumps, *Paramyxovirus*.

Common Names and Scientific Names of Organisms

Common Name	Scientific Name	Common Name	Scientific Name
Webworm, fall	*Hyphantria cunea*	Witch hazel	*Hamamelis virginiana*
Welwitschia	*Welwitschia mirabilis*	Woad, dyer's	*Isatis tinctoria*
Whale, sperm	*Physeter catodon*	Woadwaxen	*Genista tinctoria*
Wheat[30]		Wolfsbane	*Aconitum vulparia*
Wheel tree	*Trochodendron aralioides*	Wolverine	*Gulo luscus, G. gulo*
Whisk fern	*Psilotum* spp.	Woodpecker	*Dendrocopus* spp. and others
Whisk fern, fossil relatives of	*Asteroxylon* spp., *Psilophyton* spp., *Rhynia* spp., and others	Wormwood	*Artemisia annua, A. absinthium*
		Yam	*Dioscorea alata, D. cayensis, D. composita, D. esculenta, D. floribunda, D. rotundata, D. trifida*
Whisk fern, living relatives of	*Tmesipteris* spp.		
White pine blister rust	*Cronartium ribicola*	Yareta	*Azorella yareta*
Willow	*Salix* spp.	Yarrow, American	*Achillea lanulosum*
Willow Family	Salicaceae	Yarrow, European	*Achillea millefolium*
Window leaves, plants with	*Fenestraria* spp. and others	Yeast, baking/brewing	*Saccharomyces cerevisiae*
		Yellow-green algae	member of Phylum Chromophyta, Kingdom Protista
Wintergreen oil, sources of	*Gaultheria procumbens* and others	Yew	*Taxus* spp.
Wisteria	*Wisteria sinensis* and other *Wisteria* spp.	Yew, Japanese	*Taxus cuspidata*
		Zebra	*Equus zebra* and others
		Zinnia	*Zinnia elegans* and others

[30]More than 20,000 varieties of cultivated bread wheat, which has a history dating back thousands of years, are presently recognized. The ancestry and cytology are complex and still not fully understood. The principal ancestors appear to have been *Triticum monococcum* (which, after mutant forms were incorporated, became known as *einkorn* wheat) and species of *Aegilops,* especially *A. speltoides,* with several other mutations and natural hybridizations having occurred throughout the past several thousand years. Emmer wheat has been recognized as *Triticum dicoccom* or *T. turgidum* var. *dicoccum*; durum wheat as *T. durum* or *T. turgidum* var. *durum;* Polish wheat (also known as Jerusalem rye) as *T. polonicum*; and common bread wheat as *T. aestivum* (which is believed to be have been derived from *T. turgidum* and a genome from *Aegilops tauschii*). Other taxa believed to have played a role in the development of cultivated wheat include *T. longissima* and *T. searsii*. Uncertainty as to the precise evolutionary history of wheat persists, however, and awaits further investigation.

APPENDIX 2
Biological Controls

General Controls
    Establishment of Beneficial Organisms
        Ladybugs (Family Coccinellidae)
        Lacewings (Families Chrysopidae
            and Hemerobiidae)
        Praying Mantis (Family Mantidae)
        Trichogramma Wasps (Family
            Trichogrammatidae)
        Ichneumon Wasps (Family
            Ichneumonidae)
        Tachinid Flies (Family Tachinidae)
        Mites (Family Phytoseiidae and
            Others)
        Mosquito Fish (*Gambusia affinis*)
    Use of Pathogenic Bacteria
    Establishment of Toads and Frogs
    Use of Beneficial Nematodes
    Use of Limonoid Sprays
    Use of Liquefied Pest Sprays
    Use of Resistant Varieties
    Interplanting with Plants that Produce
        Natural Insecticides
        or Substances Offensive to Pests
Specific Controls
    Weeds
    Insects
Companion Planting
Additional Reading

If you were to ask a farmer or backyard gardener how to control a plant pest, you might be given the name of a chemical spray. If not used properly, though, chemical pesticides can be harmful to the environment and may become ineffective if the insects develop resistance (see Chapter 15). **Biological controls** provide an alternative to synthetic chemical sprays.

Chemical sprays may promote pest invasions because they may kill beneficial insects along with the undesirable ones. In addition, the pests, through mutations, may become resistant to the sprays. In undisturbed natural areas, weeds are never a problem, and even though pests may be present, they seldom destroy the community. Why is this so? You may recall from your reading that all members of a community are in ecological balance with one another. The plants produce a variety of substances that may either repel or attract insects, inhibit or promote the growth of other plants, and generally contribute to the health of the community as a whole.

Virtually all insects have their own pests and pathogens, as do most other living organisms. Each pest ensures, at least indirectly, that the various species of a community are perpetuated. This principle of nature can be applied, to a certain extent, to farming and gardening. The following are some general and specific biological controls now in widespread use. Others showing promise for the future are included.

General Controls
Establishment of Beneficial Organisms

The living organisms discussed here as biological control agents are typically available for purchase from commercial vendors. An Internet search will identify sources.

Ladybugs (Family Coccinellidae)

The small and often colorful beetles called *ladybugs,* and particularly their larval stages, consume large numbers of aphids, thrips, insect eggs, weevils, and other pests. While ladybugs may be purchased, if given a chance, they probably will establish themselves without being imported. When obtained from outside of the local area, they should be placed in groups at the bases of plants on which pests are present, preferably in the early evening after watering.

Lacewings (Families Chrysopidae and Hemerobiidae)

Lacewings are slow-flying, delicate-winged insects that consume large numbers of aphids, mealybugs, and other pests. They lay their eggs on the undersides of leaves, each egg being borne at the tip of a slender stalk. The larvae consume the immature stages of leafhoppers, bollworms, caterpillar eggs, mites, scale insects, thrips, aphids, and other destructive pests.

Praying Mantis (Family Mantidae)

About 20 species of *praying mantis* are now established in the United States. These are voracious feeders that prey somewhat indiscriminately on flying insects and sometimes even on other mantises. They can be established by tying their egg cases to tree branches or at other locations above the ground. The egg cases, which form compact masses about 2.5 to 5.0 centimeters (1 to 2 inches) long, are obtainable from commercial sources.

Trichogramma Wasps (Family Trichogrammatidae)

Trichogramma wasps are minute insects, mostly less than 1 millimeter (1/25 inch) long; they parasitize insect eggs and are known to have significantly reduced populations of well over 100 different insect pests, including alfalfa caterpillars, armyworms, cabbage loopers, cutworms, hornworms, tent caterpillars, and the larvae of many species of moths. As with other insects used as biological controls, trichogramma wasps should not be released unless there are pest eggs in the vicinity, as the wasps may otherwise parasitize eggs of beneficial butterflies and other useful insects.

Ichneumon Wasps (Family Ichneumonidae)

The *ichneumon wasps* belong to a very large family of wasps that are mostly stingless. These tiny wasps tend to be slender and have long ovipositors that are sometimes longer than the body. Most insects are parasitized by at least one species of ichneumon; many species parasitize the larval stages of insects, consuming the host internally after hatching from eggs deposited on the body; alternatively, they may complete development in a later stage. Ichneumons will usually appear naturally in a backyard or farm population of pests if toxic sprays and other unnatural conditions have not interfered with their normal activities.

Tachinid Flies (Family Tachinidae)

Many members of the large family of *tachinid flies* resemble houseflies or bumblebees. All parasitize other insects, including a large variety of caterpillars, Japanese beetles, European earwigs, grasshoppers, gypsy moths, tomato worms, sawflies, and various beetles.

Mites (Family Phytoseiidae and others)

Species of several genera of mites prey on red spider mites and have been used successfully to control other mites and scale insects.

Mosquito Fish (Gambusia affinis)

Mosquito fish have been added to bodies of fresh water all over the world to control mosquitoes. The fish feed on mosquito larvae, particularly as the larvae rise to the surface for air. One mosquito fish can consume thousands of larvae per day.

Use of Pathogenic Bacteria

Bacillus thuringiensis (Bt) is one of several pathogenic bacteria registered for use on edible plants in the United States. It reproduces only in the digestive tracts of caterpillars and is harmless to humans and all other wildlife, including earthworms, birds, and mammals. It is exceptionally effective against a wide range of caterpillars, such as tomato hornworms and fruitworms, cabbage worms and loopers, grape leaf rollers, corn borers, cutworms, fall webworms, and tent caterpillars. It is mass-produced and sold in a powdered spore form at nurseries and garden supply stores under the trade names of *Dipel, Biotrol,* and *Thuricide.* The powder is mixed with water and applied as a spray.

Establishment of Toads and Frogs

It has been estimated that a single adult toad will consume about 10,000 insects and slugs in one growing season. Toads and frogs feed at night when snails, slugs, sowbugs, earwigs, and other common pests are active.

Use of Beneficial Nematodes

Several species of these abundant microscopic roundworms are notorious for damaging economically important crops when they invade plant roots and other underground organs. Most species, however, are either harmless or beneficial to plants. They have been used successfully in parasitizing cabbage worm caterpillars, codling moth larvae, Japanese beetle grubs, and tobacco budworms and have shown considerable potential against other pests. One species that has been particularly effective in controlling ants, beetles, bugs, flies, wasps, and many other insects is the caterpillar nematode (*Neoaplectana carpocapsae*). It carries a symbiotic bacterium (*Xenorhabdus nematophilus*), which multiplies rapidly in the host, killing most insects within 24 hours after initial contact.

Use of Limonoid Sprays

Limonoids are bitter substances found in the rinds, seeds, and juice of citrus fruits (especially grapefruit). If the rinds and seeds of two or three fruits are ground up and soaked overnight in a pint of water, and the solid material is strained out, the liquid may then be sprayed on plants. The bitter principle apparently stops or reduces the feeding of larvae on the foliage. In experiments, limonoid sprays have proved effective against corn earworm, fall armyworm, tobacco budworm, and pink bollworm, but undoubtedly will deter many other pests as well.

Use of Liquefied Pest Sprays

Jeff Cox, an editor of Rodale's *Organic Gardening* magazine, called attention to this method of pest control in the magazine in October 1976, and again in May 1977. Insect pests or slugs are gathered in small quantities and liquefied with a little water in a blender. The material is then further diluted with water and sprayed throughout the infested area. It is not known why spraying with "bug juice" is effective against pests. It is known, however, that virtually all organisms harbor viruses. It has been theorized that even the inactive viruses carried by healthy insects and slugs may somehow be activated in the process of liquefaction. The viruses would be spread throughout an entire yard or farm if all parts of the area were sprayed. Most viruses are highly specific, generally attacking a single species of organism. M. Sipe, a Florida entomologist who recommended the "bug juice" technique, also suggested that the odor of the liquefied insects possibly attracts their predators and parasites or that the insects' distress **pheromones** (naturally produced insect chemicals that influence sexual or other behavior) are released by the blender, with the pheromones acting as an insect repellent. Possibly the observed effects of spraying "bug juice" are the result of a combination of viruses, predator attraction, and repellent pheromones. Sipe warns that if one tries this method of pest control, care should be taken to use only pest species and only those that are doing significant damage. Failure to heed this warning could disrupt the activities of natural predators and other natural controls present. This approach needs extended testing and investigation of its safety for use by humans, but test results over the past 20 years in various areas of North America have yielded impressive results with no evidence of harm to humans or beneficial organisms.

Use of Resistant Varieties

Many plants have genetic mechanisms to protect themselves against pests and pathogens. In some cases, resistance is controlled by many genes and is difficult to breed into improved cultivars. However, other forms of resistance are controlled by major genes, which breeders have introduced into cultivars. Several tomato varieties, for example, are listed as being VFN. The letters V and F indicate a resistance to *Verticillium* and *Fusarium* (common pathogenic fungi), while the letter N denotes a resistance to *root-knot nematodes*.

Interplanting with Plants That Produce Natural Insecticides or Substances Offensive to Pests

Many plant species produce substances that repel a significant number of pests, but none produce anything that repels all pests. Among the best-known plant producers of insect repellents are marigolds; garlic; and members of the Mint Family, such as pennyroyal, peppermint, basil, and lavender.

Specific Controls

Weeds

Table A2.1 provides information from the Weed Science Society of America about biological control of weeds through use of insects and plant pathogens known to decrease weed populations. The best-known example of this strategy is the introduction of a moth to control the prickly pear cactus in Australia. This has effectively controlled the cactus for over 50 years.

Insects

The maintenance of ecological balance in nature includes a vast array of predator-prey relationships between animals, birds, insects, and other organisms. Specific biological controls for several types of insect pests, in addition to the general controls previously discussed, are given in Table A2.2.

Companion Planting

Despite the scientific evidence on the subject to date, a significant amount of the "backyard biological control" that is practiced today is based primarily on empirical information. Such information has been obtained from thousands of gardeners and farmers who have tried various techniques with their plantings and pest controls. As a result, they have come to conclusions that certain things work, while others do not, but they have not deliberately set up controlled experiments, nor have they necessarily understood the scientific basis for what they have observed. This does not mean that their observations are not useful or that they are invalid. In fact, such empirical observations have often been the inspiration for investigations and experiments by scientists. The scientific investigations have sometimes revealed that the empirical observations were biased or not carefully made or that erroneous conclusions had been drawn, but frequently, sound scientific bases for these observations have been uncovered.

Further insights into how plants inhibit or enhance the growth of others and into the nature of their resistance to disease- or insect-repelling mechanisms continue to be discovered. Observations of such phenomena in the past have led organic gardeners and others to the practice of *companion planting*, which involves the interplanting of various crops and certain other plants in such a way that each species derives some benefit from the arrangement. The companion planting list in Table A2.3, based primarily on empirical information, appeared in the February 1977 issue of *Organic Gardening and Farming* magazine. It is included here with the permission of Rodale Press, Inc.

Additional Reading

Ananthakrishnan, T. N. 1998. *Technology in biological control.* Enfield, NH: Science Publishers, Inc.

Barboso, P. 1998. *Conservation biological control.* San Diego, CA: Academic Press.

Brar, S. T. 2012. *Biocontrol: Management, processes and challenges.* New York: Nova Science Publishers.

Carson, R. 1999. *Silent spring.* Boston: Houghton Mifflin.

Cook, R. J., and K. F. Baker. 1983. *Nature and practice of biological control of plant pathogens.* St. Paul, MN: American Phytopathological Society.

Dixon, A. F. G. 2005. *Insect predator-prey dynamics: Ladybird beetles and biological control.* New York: Cambridge University Press.

Flint, M. L., J. K. Clark, and S. H. Dreistadt. 1999. *Natural enemies handbook: The illustrated guide to biological pest control.* Berkeley, CA: University of California Press.

Gilbert, L. I., and S. S. Gill (Eds.). 2010. *Insect control: Biological and synthetic agents.* San Diego, CA: Academic Press.

Goeden, R. D., et al. 1981. Natural and applied control of insects by protozoa. *Annual Review of Entomology* 26: 49–73.

Hawkins, B. A., and H. V. Cornell (Eds.). 1999. *Theoretical approaches to biological control.* New York: Cambridge University Press.

Huffaker, C. B., and A. P. Gutierrez (Eds.). 1998. *Ecological entomology,* 2d ed. New York: John Wiley & Sons.

Jutsum, A. R., and R. F. S. Gordon (Eds.). 1989. *Insect pheromones in plant protection.* New York: John Wiley & Sons.

Pickett, C. H. 1998. *Enhancing biological control: Habitat management to promote natural enemies of agricultural pests.* Berkeley, CA: University of California Press.

Rechcigl, J. E., and N. A. Rechcigl. 1999. *Biological and biotechnological control of insect pests.* Los Angeles, CA: Lewis.

Rice, E. L. 1995. *Biological control of weeds and plant diseases: Advances in applied allelopathy.* Norman, OK: University of Oklahoma Press.

Table A2.1 Weeds and Agents Involved in Their Biological Control

Weed	Agent(s) of Biological Control
Alligator weed (*Alternanthera philoxeroides*)	Flea beetles (*Agasicles hygrophila*)
Bladder campion (*Silene cucubalus*)	Tortoise beetle (*Cassida hemisphaerica*)
Brazil peppertree (*Schinus terebinthifolius*)	Weevil (*Bruchus atronotatus*) and others
Brushweed (*Cassia surattensis*)	Imperfect fungus (*Cephalosporium* sp.)
Curly dock (*Rumex crispus*)	Rust (*Uromyces rumicis*)
Curse (*Clidemia hirta*)	Thrip (*Liothrips urichi*) and others
Cypress spurge (*Euphorbia cyparissias*)	Sphinx moth (*Hyles euphorbiae*)
Dalmatian toadflax (*Linaria dalmatica*)	Leaf miner (*Stagmatophora serratella*) and others
Emex (*Emex australis*)	Seed weevils (*Apion antiquum*) and others
Gorse (*Ulex europaeus*)	Seed weevils (*Apion ulicis*) and others
Halogeton (*Halogeton glomeratus*)	Casebearer (*Coleophora parthenica*) and others
Hawaiian blackberry (*Rubus penetrans*)	Sawflies (*Pamphilius sitkensis, Priophorus morio*) and others
Jamaica feverplant (puncture vine) (*Tribulus terrestris*)	Weevils (*Microlarinus* spp.)
Joint vetch (*Aeschynomene virginica*)	Imperfect fungus (*Colletotrichum gloeosporioides*)
Klamath weed (*Hypericum perforatum*)	Leaf beetles (*Chrysolina* spp.), buprestid beetle (*Agrilus hyperici*)
Lantana (*Lantana camara*)	Seed weevil (*Apion* sp.), ghost moth (*Hepialus* sp.), plume moth (*Platyptilia pusillidactyla*), hairstreaks (*Strymon* spp.), and others
Leafy spurge (*Euphorbia esula*)	Wood-boring beetle (*Oberea* sp.) and others
Mediterranean sage (*Salvia aethiopis*)	Snout beetles (*Phrydiuchus* spp.)
Milkweed vine (*Morrenia odorata*)	Oomycete fungus (*Phytophthora citrophthora*), rust (*Aecidium asclepiadinum*)
Prickly pear (*Opuntia* spp.)	Moth (*Cactoblastis cactorum*), cochineal insects (*Dactylopius* spp.), and others
Puncture vine (see Jamaica feverplant)	Weevils (*Microlarinus* spp.)
Scotch broom (*Cytisus scoparius*)	Seed weevil (*Apion fuscirostre*) and others
Skeleton weed (*Chondrilla juncea*)	Gall mite (*Aceria chondrillae*), root moth (*Bradyrrhoa gilveolella*), rust (*Puccinia chondrillina*), powdery mildews (*Erysiphe cichoracearum, Leveillula taurica*)
Spiny emex (*Emex spinosa*)	Seed weevil (*Apion antiquum*)
Tansy ragwort (*Senecio jacobaea*)	Seed fly (*Hylemya seneciella*), cinnabar moth (*Tyria jacobaeae*), leaf beetle (*Longitarsus jacobaeae*)
Thistles:	
Bull thistle (*Cirsium vulgare*)	Weevil (*Ceuthorrhynchidius horridus*), tortoise beetle (*Cassida rubiginosa*)
Canada thistle (*Cirsium arvense*)	Weevil (*Ceutorhynchus litura*), flea beetle (*Altica carduorum*), stem gall fly (*Urophora cardui*)
Diffuse knapweed (*Centaurea diffusa*)	Seed fly (*Urophora affinis*)
Italian thistle (*Carduus pycnocephalus*)	Flea beetles (*Rhinocyllus conicus, Psylliodes chalcomera*), weevil (*Ceutorhynchus trimaculatus*)
Milk thistle (*Silybum marianum*)	Flea beetle (*Rhinocyllus conicus*)
Musk thistle (*Carduus nutans*)	Weevils (*Ceutorhynchus trimaculatus, Ceuthorrhynchidius horridus, Rhinocyllus conicus*), flea beetle (*Psylliodes chalcomera*)
Perennial sowthistle (*Sonchus arvensis*)	Peacock fly (*Tephritis dilacerata*)
Plumeless thistle (*Carduus acanthoides*)	Tortoise beetle (*Cassida rubiginosa*), seed weevil (*Rhinocyllus conicus*), weevil (*Ceuthorrhynchidius horridus*)
Russian thistle (*Salsola kali* var. *tenuifolia*)	Casebearer (*Coleophora parthenica*) and others
Slenderflower thistle (*Carduus tenuiflorus*)	Weevil (*Ceutorhynchus trimaculatus*)
Spotted knapweed (*Centaurea maculosa*)	Seed fly (*Urophora affinis*) and others
Star thistle (*Centaurea nigrescens*)	Weevil (*Ceuthorrhynchidius horridus*)
Yellow star thistle (*Centaurea solstitialis*)	Seed fly (*Urophora siruna-seva*)
Water hyacinth (*Eichhornia crassipes*)	Weevils (*Neochetina bruchi, N. eichhorniae*), moth (*Sameodes albiguttalis*)
Water purslane (*Ludwigia palustris*)	Snout beetle (*Nanophyes* sp.)

Table A2.2 — Biological Controls for Several Types of Insect Pests

Insect	Control
Ants (about 8,000 spp. within the Superfamily Formicoidea)	Ants that carry aphids into trees and consume ripening fruits can be prevented from getting farther than the trunk by applying a band of sticky material around the trunk. A commercial preparation sold under the trade name of *Tanglefoot* is particularly effective. A water suspension of ground hot peppers (*Capsicum* spp.) used as a spray can act as an ant deterrent. *Caution:* Many ants are beneficial to a balanced ecology; they should not be decimated indiscriminately.
Grasshoppers (there are several families of grasshoppers, but the insects that usually constitute the most serious pests are species of *Melanoplus*, Family Acrididae)	In 1980, the Environmental Protection Agency permitted private companies to begin the mass culture of a protozoan, *Nosema locustae,* for use in controlling rangeland grasshoppers. Tests have shown that properly timed applications of spores mixed with wheat bran can reduce grasshopper populations by up to 50%.
Gypsy moths (*Porthetria dispar*)	Parasitic wasps (*Apanteles flavicoxis, A. indiensis*) imported from India lay their eggs in gypsy moth caterpillars and kill large numbers.
Japanese beetles (*Popillia japonica*)	The pathogenic bacterium *Bacillus popillae,* which is sold commercially, is specific for Japanese beetle larvae. It causes what is known as "milky spore disease" in the grubs while they are still in the soil, and it is very destructive.
Mealybugs (*Pseudococcus* spp.)	The small brown beetles called *crypts* (*Cryptolaemus montrouzieri*) effectively control mealybugs in greenhouses and also outdoors on apple, pear, peach, and citrus trees.
Mosquitoes (*Culex* spp., *Anopheles* spp., and others)	The bacterium *Bacillus thuringiensis* var. *israelensis* has proved to be very effective in destroying mosquito larvae. A fungus (*Lagenidium giganteum*) has also proved highly effective against mosquito larvae if the temperature is above 20°C (68°F).
Red spider mites (*Tetranychus telarius*)	Predatory mites (*Phytolesius persimilis,* which works best when weather is not hot, and *Amblyseius californicus,* which is more effective in hot weather) effectively control populations of red spider mites.
White flies (*Trialeurodes vaporariorum*)	A minute wasp, *Encarsia formosa,* parasitizes white flies exclusively. The wasps have been known to be very effective in greenhouses. Alternatively, white flies are attracted to the color yellow. Large numbers of white flies are trapped when a yellow board is sprayed or painted with any sticky substance and placed in the vicinity of the pests.

Table A2.3 — Companion Plants

Plant	Companions and Effects
Asparagus	Tomatoes, parsley, basil
Basil	Tomatoes (improves growth and flavor); said to dislike rue; repels flies and mosquitoes
Beans	Potatoes, carrots, cucumbers, cauliflower, cabbage, summer savory, most other vegetables and herbs; around houseplants when set outside
Beans (bush)	Sunflowers (beans like partial shade, sunflowers attract birds and bees), cucumbers (combination of heavy and light feeders), potatoes, corn, celery, summer savory
Beets	Onions, kohlrabi
Borage	Tomatoes (attracts bees, deters tomato worm, improves growth and flavor), squash, strawberries
Cabbage Family	Potatoes, celery, dill, chamomile, sage, thyme, mint, pennyroyal, rosemary, lavender, beets, onions. Aromatic plants deter cabbage worms
Carrots	Peas, lettuce, chives, onions, leeks, rosemary, sage, tomatoes
Catnip	Plant in borders; protects against flea beetles
Celery	Leeks, tomatoes, bush beans, cauliflower, cabbage
Chamomile	Cabbage, onions

Table A3.1 Wild Edible Plants and Fungi

Organism	Scientific Name	Uses
Amaranth	*Amaranthus* spp.	Young leaves used like spinach; seeds ground with others for flour
Arrow grass	*Triglochin maritima*	Seeds parched or roasted (**Caution: All other plant parts are poisonous**)
Arrowhead	*Sagittaria latifolia*	Tubers used like potatoes
Balsamroot	*Balsamorhiza* spp.	Whole plant edible, especially when young, either raw or cooked
Basswood	*Tilia* spp.	Fruits and flowers ground together to make a paste that can serve as a chocolate substitute; winter buds edible raw; dried flowers used for tea
Bearberry (Kinnikinnick)	*Arctostaphylos uva-ursi*	Berries are edible but much more palatable when cooked
Bedstraw (Cleavers)	*Galium aparine*	Roasted and ground seeds make good coffee substitute
Beechnuts	*Fagus grandifolia*	Seeds used as nuts; oil extracted from seeds for table use
Biscuit root	*Lomatium* spp.	Roots eaten raw or dried and ground into flour; seeds edible raw or roasted
Bitterroot	*Lewisia rediviva*	Outer coat of the bulbs should be removed to eliminate the bitter principle; bulbs are then boiled or roasted
Blackberry (wild)	*Rubus* spp.	Fruits edible raw, in pies, jams, and jellies
Black walnut	*Juglans nigra*	Nut meats edible
Bladder campion	*Silene cucubalus*	Young shoots (less than 5 cm tall) cooked as a vegetable
Blueberry	*Vaccinium* spp.	Fruits edible raw, frozen, and in pies, jams, and jellies
Bracken fern	*Pteridium aquilinum*	Young uncoiling leaves ("fiddleheads") cooked like asparagus; rhizomes also edible but usually tough (**Caution: Evidence indicates that frequent consumption of bracken fern can cause cancer of the intestinal tract**)
Broomrape	*Orobanche* spp.	Entire plant eaten raw or roasted
Bulrush (Tule)	*Scirpus* spp.	Roots and young shoot tips edible raw or cooked; pollen and seeds also edible
Butternut	*Juglans cinerea*	Nut meats edible
Camas	*Camassia quamash*	Roasted bulbs considered a delicacy
Caraway	*Carum carvi*	Young leaves in salads; seeds for flavoring baked goods and cheeses
Cattail	*Typha* spp.	Copious pollen produced by flowers in early summer is rich in vitamins and can be gathered and mixed with flour for baking; rhizomes can be cooked and eaten like potatoes
Chicory	*Cichorium intybus*	Leaves eaten raw or cooked; dried, ground roots (roasted) make good coffee substitute
Chokecherry	*Prunus virginiana*	Fruits make excellent jelly or can be cooked with sugar for pies and cobblers
Clover	*Trifolium* spp.	Roots edible
"Coffee" (wild)	*Triosteum* spp.	Berries dried and roasted make good coffee substitute
Common chickweed	*Stellaria media*	Plants cooked as a vegetable
Corn lily	*Clintonia borealis*	Youngest leaves can be used as a cooked vegetable
Cow parsnip	*Heracleum lanatum*	Roots and young stems cooked (**Caution: Be certain of identity; some other members of the family that are similar in appearance to cow parsnip are highly toxic**)
Cowpea	*Vigna sinensis*	"Peas" and young pods cooked as a vegetable (plant naturalized in southern U.S.)
Crab apple	*Pyrus* spp.	Jelly made from fruits
Cranberry (wild, bog)	*Vaccinium* spp.	Berries edible cooked, preserved, or in drinks; adding a small amount of salt while cooking significantly reduces amount of sugar needed to counteract acidity
Crowberry	*Empetrum nigrum*	Fruits should first be frozen, then cooked with sugar
Dandelion	*Taraxacum* sp. aff.	Leaves rich in vitamin A; dried roots make good coffee substitute; wine made from young flowers
Dock	*Rumex* spp.	Leaves cooked like spinach; tartness of leaves varies from species to species and sometimes from plant to plant—tart forms should be cooked in two or three changes of water
Douglas fir	*Pseudotsuga menziesii*	Cambium and young phloem edible; tea made from fresh leaves
Elderberry	*Sambucus* spp.	Fresh flowers used to flavor batters; fruits used in pies, jellies, wine (**Caution: Other parts of the plant are poisonous**)
Evening primrose	*Oenothera hookeri, O. biennis,* and others	Young roots cooked
Fairy bells	*Disporum trachycarpum*	Berries can be eaten raw
Fennel	*Foeniculum vulgare*	Leaf petioles eaten raw or cooked
Ferns	Most (but not all) spp.	Young coiled fronds (fiddleheads) may be cooked as a vegetable
Fireweed	*Epilobium angustifolium*	Young shoots and leaves boiled as a vegetable

Table A2.2 — Biological Controls for Several Types of Insect Pests

Insect	Control
Ants (about 8,000 spp. within the Superfamily Formicoidea)	Ants that carry aphids into trees and consume ripening fruits can be prevented from getting farther than the trunk by applying a band of sticky material around the trunk. A commercial preparation sold under the trade name of *Tanglefoot* is particularly effective. A water suspension of ground hot peppers (*Capsicum* spp.) used as a spray can act as an ant deterrent. *Caution:* Many ants are beneficial to a balanced ecology; they should not be decimated indiscriminately.
Grasshoppers (there are several families of grasshoppers, but the insects that usually constitute the most serious pests are species of *Melanoplus*, Family Acrididae)	In 1980, the Environmental Protection Agency permitted private companies to begin the mass culture of a protozoan, *Nosema locustae,* for use in controlling rangeland grasshoppers. Tests have shown that properly timed applications of spores mixed with wheat bran can reduce grasshopper populations by up to 50%.
Gypsy moths (*Porthetria dispar*)	Parasitic wasps (*Apanteles flavicoxis, A. indiensis*) imported from India lay their eggs in gypsy moth caterpillars and kill large numbers.
Japanese beetles (*Popillia japonica*)	The pathogenic bacterium *Bacillus popillae,* which is sold commercially, is specific for Japanese beetle larvae. It causes what is known as "milky spore disease" in the grubs while they are still in the soil, and it is very destructive.
Mealybugs (*Pseudococcus* spp.)	The small brown beetles called *crypts* (*Cryptolaemus montrouzieri*) effectively control mealybugs in greenhouses and also outdoors on apple, pear, peach, and citrus trees.
Mosquitoes (*Culex* spp., *Anopheles* spp., and others)	The bacterium *Bacillus thuringiensis* var. *israelensis* has proved to be very effective in destroying mosquito larvae. A fungus (*Lagenidium giganteum*) has also proved highly effective against mosquito larvae if the temperature is above 20°C (68°F).
Red spider mites (*Tetranychus telarius*)	Predatory mites (*Phytolesius persimilis,* which works best when weather is not hot, and *Amblyseius californicus,* which is more effective in hot weather) effectively control populations of red spider mites.
White flies (*Trialeurodes vaporariorum*)	A minute wasp, *Encarsia formosa,* parasitizes white flies exclusively. The wasps have been known to be very effective in greenhouses. Alternatively, white flies are attracted to the color yellow. Large numbers of white flies are trapped when a yellow board is sprayed or painted with any sticky substance and placed in the vicinity of the pests.

Table A2.3 — Companion Plants

Plant	Companions and Effects
Asparagus	Tomatoes, parsley, basil
Basil	Tomatoes (improves growth and flavor); said to dislike rue; repels flies and mosquitoes
Beans	Potatoes, carrots, cucumbers, cauliflower, cabbage, summer savory, most other vegetables and herbs; around houseplants when set outside
Beans (bush)	Sunflowers (beans like partial shade, sunflowers attract birds and bees), cucumbers (combination of heavy and light feeders), potatoes, corn, celery, summer savory
Beets	Onions, kohlrabi
Borage	Tomatoes (attracts bees, deters tomato worm, improves growth and flavor), squash, strawberries
Cabbage Family	Potatoes, celery, dill, chamomile, sage, thyme, mint, pennyroyal, rosemary, lavender, beets, onions. Aromatic plants deter cabbage worms
Carrots	Peas, lettuce, chives, onions, leeks, rosemary, sage, tomatoes
Catnip	Plant in borders; protects against flea beetles
Celery	Leeks, tomatoes, bush beans, cauliflower, cabbage
Chamomile	Cabbage, onions

Table A2.3 Companion Plants

Plant	Companions and Effects
Chervil	Radishes (improves growth and flavor)
Chives	Carrots; plant around base of fruit trees to discourage insects from climbing trunk
Corn	Potatoes, peas, beans, cucumbers, pumpkin, squash
Cucumbers	Beans, corn, peas, radishes, sunflowers
Dill	Cabbage (improves growth and health), carrots
Eggplant	Beans
Fennel	Most plants are supposed to dislike it
Flax	Carrots, potatoes
Garlic	Roses and raspberries (deters Japanese beetle); with herbs to enhance their production of essential oils; plant liberally throughout garden to deter pests
Horseradish	Potatoes (deters potato beetles); around plum trees to discourage curculios
Lamb's quarters	Nutritious edible weed; allow to grow in modest amounts around corn plants
Leek	Onions, celery, carrots
Lettuce	Carrots and radishes (lettuce, carrots, and radishes make a strong companion team), strawberries, cucumbers
Lovage	Plant here and there in garden
Marigolds	The workhorse of pest deterrents; keeps soil free of nematodes; discourages many insects; plant freely throughout garden
Marjoram	Here and there in garden
Mint	Cabbage family; tomatoes; deters cabbage moth
Mole plant	Deters moles and mice if planted here and there throughout the garden
Nasturtium	Tomatoes, radishes, cabbage, cucumbers; plant under fruit trees. Deters aphids and pests of cucurbits
Onion	Beets, strawberries, tomato, lettuce (protects against slugs), beans (protects against ants), summer savory
Parsley	Tomato, asparagus
Peas	Squash (when squash follows peas up trellis), plus grows well with almost any vegetable; adds nitrogen to the soil
Petunia	Protects beans; beneficial throughout garden
Pigweed	Brings nutrients to topsoil; beneficial growing with potatoes, onions, and corn; keep well thinned
Potato	Horseradish, beans, corn, cabbage, marigold, limas, eggplant (as trap crop for potato beetle)
Pot marigold	Helps tomato, but plant throughout garden as deterrent to asparagus beetle, tomato worm, and many other garden pests
Pumpkin	Corn
Radish	Peas, nasturtium, lettuce, cucumbers; a general aid in repelling insects
Rosemary	Carrots, beans, cabbage, sage; deters cabbage moth, bean beetles, and carrot fly
Rue	Roses and raspberries; deters Japanese beetle; keep it away from basil
Sage	Rosemary, carrots, cabbage, peas, beans; deters some insects
Southernwood	Cabbage; plant here and there in garden
Soybeans	Grows with anything, helps everything
Spinach	Strawberries
Squash	Nasturtium, corn
Strawberries	Bush beans, spinach, borage, lettuce (as a border)
Summer savory	Beans, onions; deters bean beetles
Sunflower	Cucumbers
Tansy	Plant under fruit trees; deters pests of roses and raspberries; deters flying insects; also Japanese beetles, striped cucumber beetles, squash bugs, deters ants
Tarragon	Good throughout garden
Thyme	Here and there in garden; deters cabbage worm
Tomato	Chives, onion, parsley, asparagus, marigold, nasturtium, carrot, limas
Turnip	Peas
Valerian	Good anywhere in garden
Wormwood	As a border, keeps animals from the garden
Yarrow	Plant along borders, near paths, near aromatic herbs; enhances essential oil production of herbs

APPENDIX 3
Useful and Poisonous Plants, Fungi, and Algae

Wild Edible Plants, Fungi, and Algae
    Words of Caution
Poisonous Plants and Fungi

Medicinal Plants, Fungi, and Algae
    Memory Problems or Enhancement
    Cholesterol-Lowering Plants or Plant Derivatives

Hallucinogenic Plants
Spice Plants
Dye Plants
Additional Reading

Wild Edible Plants, Fungi, and Algae

Words of Caution

At least some parts of literally thousands of native and naturalized plants, fungi, and algae have been used for food and other purposes by Native Americans. Many were also used by the immigrants who came later from other areas of the world. A representative compilation of wild edible plants and fungi is shown in Table A3.1.

The list of plants and fungi in Table A3.1 has been compiled from a variety of sources; the authors have had opportunities to sample only a fraction of these themselves, and cannot confirm the edibility of all of the organisms listed. *The reader is cautioned to be certain of the identity of a plant or fungus before consuming any part of it.* Cow parsnip (*Heracleum lanatum*) and water hemlocks (*Cicuta* spp.), for example, resemble each other in general appearance, but although cooked roots of cow parsnip have been used for food for perhaps many centuries, those of water hemlocks are very poisonous and have caused many human fatalities.

Many species of organisms are now on rare and endangered species lists, and a number of them will become extinct within the next few years. Although the wild edible plants and fungi discussed here may not presently be included in such lists, it might not take much indiscriminate gathering to endanger their existence as well. Because of this, one should exercise the following rule of thumb: *Never reduce a population of wild plants by more than 10% when collecting them for any purpose! If the population consists of less than 10 plants, do not disturb it.*

Poisonous Plants and Fungi

Literally thousands of plants and fungi contain varying amounts of poisonous substances. In many instances, the poisons are not present in sufficient quantities to cause adverse effects in humans when only moderate contact or consumption is involved, and cooking may destroy or dissipate the substance. Some plants and fungi have substances that produce toxic effects in some organisms but not in others. Ordinary onions (*Allium cepa*), for example, occasionally poison horses or cattle yet are widely used for human food, and poison ivy (*Toxicodendron radicans*) or poison oak (*Toxicodendron diversilobum*) produce dermatitis in some individuals but not in others. Table A3.2 and Table A3.3 include plants and fungi that are native to, or cultivated in, the United States and Canada.

Medicinal Plants, Fungi, and Algae

Today's worldwide multibillion-dollar pharmaceutical industry is producing many different drugs that are saving lives—lives that otherwise would have been lost to disease or other ailments. Because of the discovery and development of these drugs, human life expectancies more than doubled during the 20th century, and continue to increase even more during the present century. Unfortunately, however, thousands also die each year from the side effects and interactions of some of the prescription drugs. This, in part, has led to an increased interest in medicinal plants and fungi, which, with few exceptions when used in appropriate amounts, very rarely produce undesirable side effects. Like prescription drugs, however, medicines derived directly from plants need to have been tested in scientific studies that can readily be duplicated, despite the fact that they may have been in use since before the dawn of recorded history. A few purveyors of natural plant medicines have been unscrupulous in making unsubstantiated claims about their efficacy, and the Food and Drug Administration generally has prohibited putting claims of efficacy on the containers of such products. The FDA also spends a proportionately small amount of its research budget on testing natural plant medicines.

Natural plant medicines usually can't be patented, making financial incentives for needed research much less likely to be forthcoming than they are for the development of pharmaceutical drugs, which can be exceptionally profitable to investors.

Table A3.1 — Wild Edible Plants and Fungi

Organism	Scientific Name	Uses
Amaranth	*Amaranthus* spp.	Young leaves used like spinach; seeds ground with others for flour
Arrow grass	*Triglochin maritima*	Seeds parched or roasted (*Caution: All other plant parts are poisonous*)
Arrowhead	*Sagittaria latifolia*	Tubers used like potatoes
Balsamroot	*Balsamorhiza* spp.	Whole plant edible, especially when young, either raw or cooked
Basswood	*Tilia* spp.	Fruits and flowers ground together to make a paste that can serve as a chocolate substitute; winter buds edible raw; dried flowers used for tea
Bearberry (Kinnikinnick)	*Arctostaphylos uva-ursi*	Berries are edible but much more palatable when cooked
Bedstraw (Cleavers)	*Galium aparine*	Roasted and ground seeds make good coffee substitute
Beechnuts	*Fagus grandifolia*	Seeds used as nuts; oil extracted from seeds for table use
Biscuit root	*Lomatium* spp.	Roots eaten raw or dried and ground into flour; seeds edible raw or roasted
Bitterroot	*Lewisia rediviva*	Outer coat of the bulbs should be removed to eliminate the bitter principle; bulbs are then boiled or roasted
Blackberry (wild)	*Rubus* spp.	Fruits edible raw, in pies, jams, and jellies
Black walnut	*Juglans nigra*	Nut meats edible
Bladder campion	*Silene cucubalus*	Young shoots (less than 5 cm tall) cooked as a vegetable
Blueberry	*Vaccinium* spp.	Fruits edible raw, frozen, and in pies, jams, and jellies
Bracken fern	*Pteridium aquilinum*	Young uncoiling leaves ("fiddleheads") cooked like asparagus; rhizomes also edible but usually tough (*Caution: Evidence indicates that frequent consumption of bracken fern can cause cancer of the intestinal tract*)
Broomrape	*Orobanche* spp.	Entire plant eaten raw or roasted
Bulrush (Tule)	*Scirpus* spp.	Roots and young shoot tips edible raw or cooked; pollen and seeds also edible
Butternut	*Juglans cinerea*	Nut meats edible
Camas	*Camassia quamash*	Roasted bulbs considered a delicacy
Caraway	*Carum carvi*	Young leaves in salads; seeds for flavoring baked goods and cheeses
Cattail	*Typha* spp.	Copious pollen produced by flowers in early summer is rich in vitamins and can be gathered and mixed with flour for baking; rhizomes can be cooked and eaten like potatoes
Chicory	*Cichorium intybus*	Leaves eaten raw or cooked; dried, ground roots (roasted) make good coffee substitute
Chokecherry	*Prunus virginiana*	Fruits make excellent jelly or can be cooked with sugar for pies and cobblers
Clover	*Trifolium* spp.	Roots edible
"Coffee" (wild)	*Triosteum* spp.	Berries dried and roasted make good coffee substitute
Common chickweed	*Stellaria media*	Plants cooked as a vegetable
Corn lily	*Clintonia borealis*	Youngest leaves can be used as a cooked vegetable
Cow parsnip	*Heracleum lanatum*	Roots and young stems cooked (*Caution: Be certain of identity; some other members of the family that are similar in appearance to cow parsnip are highly toxic*)
Cowpea	*Vigna sinensis*	"Peas" and young pods cooked as a vegetable (plant naturalized in southern U.S.)
Crab apple	*Pyrus* spp.	Jelly made from fruits
Cranberry (wild, bog)	*Vaccinium* spp.	Berries edible cooked, preserved, or in drinks; adding a small amount of salt while cooking significantly reduces amount of sugar needed to counteract acidity
Crowberry	*Empetrum nigrum*	Fruits should first be frozen, then cooked with sugar
Dandelion	*Taraxacum* sp. aff.	Leaves rich in vitamin A; dried roots make good coffee substitute; wine made from young flowers
Dock	*Rumex* spp.	Leaves cooked like spinach; tartness of leaves varies from species to species and sometimes from plant to plant—tart forms should be cooked in two or three changes of water
Douglas fir	*Pseudotsuga menziesii*	Cambium and young phloem edible; tea made from fresh leaves
Elderberry	*Sambucus* spp.	Fresh flowers used to flavor batters; fruits used in pies, jellies, wine (*Caution: Other parts of the plant are poisonous*)
Evening primrose	*Oenothera hookeri*, *O. biennis*, and others	Young roots cooked
Fairy bells	*Disporum trachycarpum*	Berries can be eaten raw
Fennel	*Foeniculum vulgare*	Leaf petioles eaten raw or cooked
Ferns	Most (but not all) spp.	Young coiled fronds (fiddleheads) may be cooked as a vegetable
Fireweed	*Epilobium angustifolium*	Young shoots and leaves boiled as a vegetable

Table A3.1 Wild Edible Plants and Fungi

Organism	Scientific Name	Uses
Fritillary	*Fritillaria* spp.	Cooked bulbs are edible
Ginger (wild)	*Asarum* spp.	Rhizomes can be used as substitute for true ginger
Gooseberry	*Ribes* spp.	Berries eaten cooked, dried, or raw; make excellent jelly
Grape (wild)	*Vitis* spp.	Berries usually tart but can be eaten raw; make good jams and jellies
Grass	Many genera and species	Seeds of most can be made into flour; rhizomes of many perennial species can be dried and ground for flour
Greenbrier	*Smilax* spp.	Roots dried and ground; refreshing drink made with ground roots, sugar, and water
Groundnut	*Apios americana*	Tubers cooked like potatoes
Hawthorn	*Crataegus* spp.	Fruits edible raw and in jams and jellies
Hazelnut	*Corylus* spp.	Nuts eaten raw or roasted
Hickory	*Carya* spp.	Nuts edible
Highbush cranberry	*Viburnum trilobum*	Fruits make excellent jellies and jams
Huckleberry	*Vaccinium* spp.	Berries eaten raw or in jams and jellies
Indian paintbrush	*Castilleja* spp.	Flowers of many species edible (*Caution: On certain soils, plants absorb toxic quantities of selenium*)
Indian pipe	*Monotropa* spp.	Whole plant edible raw or cooked
June berries	*Amelanchier* spp.	Fruit edible fresh, dried, or preserved
Juniper	*Juniperus* spp.	"Berries" dried, ground, and made into cakes
Labrador tea	*Ledum* spp.	Tea made from young leaves
Lamb's quarters	*Chenopodium album*	Leaves and young stems used as cooked vegetable
Licorice	*Glycyrrhiza lepidota, G. glabra*	Roots edible raw or cooked
Mallow	*Malva* spp.	Leaves and young stems used as vegetable (use only small amounts at one time)
Manzanita	*Arctostaphylos* spp.	Berries eaten raw, in jellies or pies, or made into "cider" (*Caution: Raw berries can be somewhat indigestible*)
Maple	*Acer* spp.	Sugar maples (Acer saccharum) well known for the sugar content of the early spring sap; other species (e.g., box elder—A. negundo, bigleaf maple—A. macrophyllum) also contain usable sugars in their early spring sap
Mariposa lily	*Calochortus* spp.	Bulbs edible raw or cooked
Mayapple	*Podophyllum peltatum*	Fruit good raw or cooked (*Caution: Other parts of the plant are poisonous*)
Maypops	*Passiflora incarnata*	Fruits edible raw or cooked
Miner's lettuce	*Claytonia perfoliata*	Leaves eaten raw as a salad green
Mint	*Mentha arvensis* and others	Leaves of several mints used for teas
Mormon tea	*Ephedra* spp.	Tea from fresh or dried leaves
Mulberry	*Morus* spp.	Fruits of the red mulberry (M. rubra) are used raw and in pies and jellies; fruits of white mulberry (M. alba) edible but insipid
Mushrooms	Many genera and species	*Utmost caution should be exercised in identifying mushrooms before consuming them;* although poisonous species are in the minority, they are common enough; edible forms that are relatively easy to identify include morels (Morchella esculenta), most puffballs (Lycoperdon spp.), and inky cap mushrooms (Coprinus spp.)
Mustard	*Brassica* spp.	Leaves used as vegetable; condiment made from ground seeds
Nettles	*Urtica* spp.	Leaves and young stems cooked like spinach
New Jersey tea	*Ceanothus americanus*	Tea from leaves
Nutgrass	*Cyperus esculentus* and others	Tubers can be eaten raw
Oak	*Quercus* spp.	Acorns were ground for flour and widely used by native North Americans; all contain bitter tannins that must be leached out before use
Onion (wild)	*Allium* spp.	Bulbs edible raw or cooked
Orach	*Atriplex patula* and others	Leaves and young stems cooked as a vegetable
Oregon grape	*Berberis aquifolia, B. nervosa*	Berries edible raw or preserved
Ostrich fern	*Matteuccia struthiopteris*	Young coiled fronds cooked as a vegetable
Pawpaw	*Asimina triloba*	Fruit edible raw or cooked
Pennycress	*Thlaspi arvense*	Young leaves are edible raw
Peppergrass	*Lepidium* spp.	Immature fruits add zest to salads; seeds spice up meat dressings
Persimmon	*Diospyros virginiana*	Fully ripened fruits can be eaten raw or cooked

Table A3.1 — Wild Edible Plants and Fungi

Organism	Scientific Name	Uses
Pickerel weed	*Pontederia cordata*	Fruits edible raw or dried
Pigweed (see Amaranth)		
Pines	*Pinus* spp.	Cambium, young phloem, and seeds edible; tea from fresh needles rich in vitamin C
Pipsissewa	*Chimaphila umbellata*	Drink made from boiled roots and leaves (cool after boiling)
Plantain	*Plantago* spp.	Young leaves eaten in salads or as cooked vegetable
Poke	*Phytolacca americana*	Fresh young shoots boiled like asparagus (*Caution: Older parts of plant are poisonous*)
Prairie turnip	*Psoralea esculenta*	Turnip-like roots cooked like potatoes
Prickly pear	*Opuntia* spp.	Fruits and young stems peeled and eaten raw or cooked
Psyllium	*Plantago ovata*	Seed husks widely used as a bulking laxative
Purple avens	*Geum rivale*	Liquid from boiled root has chocolate-like flavor
Purslane	*Portulaca oleracea*	Leaves and stems cooked like spinach
Quackgrass	*Elytrigia repens*	Noxious weed whose rhizomes can be used as emergency food
Raspberry (wild)	*Rubus* spp.	Fruits edible raw or in pies, jams, and jellies
Redbud	*Cercis* spp.	Flowers used in salads; cooked young pods edible
River-beauty	*Epilobium latifolium*	Young shoots and fleshy leaves can be cooked as a vegetable
Rose (wild)	*Rosa* spp.	Fruits (hips) exceptionally rich in vitamin C; hips can be eaten raw, pureed, or candied
Salal	*Gaultheria procumbens, G. shallon*	Ripe berries edible raw, dried, or preserved
Salmonberry	*Rubus spectabilis*	Fruits edible raw, dried, or cooked
Salsify	*Tragopogon* spp.	Roots edible raw or cooked
Saltbush	*Atriplex* spp.	Seeds nutritious (*Caution: On certain soils, plants can absorb toxic amounts of selenium*)
Sassafras	*Sassafras albidum*	Tea from roots (*Caution: Large quantities have narcotic effect*); leaves and pith used for Louisiana filé
Serviceberry	*Amelanchier* spp.	All fruits edible (mostly bland)
Sheep sorrel	*Rumex acetosella*	Raw leaves have a pleasant sour taste; leaves can be used as seasoning in other dishes
Shepherd's purse	*Capsella bursa-pastoris*	Leaves cooked as vegetable; seeds eaten parched or ground for flour
Showy milkweed	*Asclepias speciosa*	Flowers eaten raw or cooked; young shoots cooked
Silverweed	*Potentilla anserina*	Cooked roots edible
Soap plant	*Chlorogalum pomeridianum*	Bulbs slow-baked and eaten like potatoes after fibrous outer coats are removed
Solomon's seal	*Polygonatum* spp.	Rootstocks dried and ground for bread flour
Sorrel	*Oxalis* spp.	Leaves mixed in salads
Spatterdock	*Nuphar polysepalum*	Seeds placed on hot stove burst like popcorn and are edible as such; peeled tubers eaten boiled or roasted
Speedwell	*Veronica americana* and others	Leaves and stems used in salads
Spring beauty	*Claytonia* spp.	Bulbs edible raw or roasted
Strawberry (wild)	*Fragaria* spp.	Fruits superior in flavor to cultivated varieties
Sunflower	*Helianthus annuus*	Seeds eaten raw or roasted; seeds yield cooking oil
Sweet cicely	*Osmorhiza* spp.	Roots have aniselike flavor
Sweet flag	*Acorus calamus*	Young shoots used in salads; roots candied
Thimbleberry	*Rubus parviflorus*	Fruits edible raw, cooked, dried, or preserved; dried leaves used for tea
Thistle	*Cirsium* spp.	Peeled stems edible; roots edible raw or roasted
Vetch	*Vicia* spp.	Tender green pods edible baked or boiled
Watercress	*Nasturtium officinale*	Leaves edible raw in salads or cooked as a vegetable
Waterleaf	*Hydrophyllum* spp.	Young shoots raw in salads; shoots and roots cooked as vegetable
Water plantain	*Alisma* spp.	The bulblike base of the plant is dried and then cooked
Water shield	*Brasenia schreberi*	Tuberlike roots are peeled and then dried to be ground for flour or boiled
Winter cress	*Barbarea* spp.	Leaves and young stem edible as cooked vegetable
Yarrow	*Achillea lanulosa*	Plant dried and made into broth (*Caution: The closely related and widespread European yarrow—A. millefolium—is somewhat poisonous*)
Yellow pond lily (see Spatterdock)		
Yew	*Taxus* spp.	Bright red pulpy part of berries edible (*Caution: Seeds and leaves are poisonous*)

Table A3.2 — Plants and Fungi Known to Have Caused Human Fatalities

Organism	Scientific Name	Poisonous Parts
Angel's trumpet	*Datura suaveolens*	All parts, especially seeds and leaves
Azalea	*Rhododendron* spp.	Leaves and flowers (however, poisoning is rare)
Baneberry	*Actaea* spp.	Berries and roots
Belladonna	*Atropa belladonna*	All parts, especially fruits and roots
Black cherry	*Prunus serotina*	Bark, seeds, leaves (*Caution: Seeds of most cherries, plums, and peaches contain a poisonous principle*)
Black locust	*Robinia pseudo-acacia*	Seeds, leaves, inner bark
Black snakeroot	*Zigadenus* spp.	Bulbs
Buckeye	*Aesculus* spp.	Seeds, shoots, flowers, leaves, roots; even the honey that bees make from buckeye flowers is poisonous
Caladium	*Caladium* spp.	All parts
Carolina jessamine	*Gelsemium sempervirens*	All parts; even visiting honeybees can be poisoned
Castor bean	*Ricinus communis*	Seeds
Chinaberry	*Melia azedarach*	Fruits and leaves
Daphne	*Daphne mezereum*	All parts
Death angel (Fly agaric)	*Amantia muscaria*	All parts (as little as one bite can be fatal)
Death camas (see Black snakeroot)		
Destroying angel	*Amanita verna*	All parts (as little as one bite can be fatal)
Dieffenbachia (Dumb cane)	*Dieffenbachia* spp.	All parts
Duranta	*Duranta repens*	Berries
Dutchman's breeches	*Dicentra cucullaria*	All parts
English ivy	*Hedera helix*	Berries and leaves
False hellebore	*Veratrum* spp.	All parts
Foxglove	*Digitalis purpurea*	All parts
Gloriosa lily	*Gloriosa superba* and other *Gloriosa* spp.	All parts (especially tubers)
Golden chain	*Laburnum anagyroides*	Seeds and flowers
Jequirity bean	*Abrus precatorius*	Seeds
Jimson weed	*Datura stramonium* and other *Datura* spp.	All parts, especially seeds
Lantana	*Lantana camara*	Unripe fruits
Lily of the valley	*Convallaria majalis*	All parts
Lobelia	*Lobelia* spp.	All parts
Mistletoe	*Phoradendron* spp.	Berries
Monkshood	*Aconitum* spp.	All parts
Moonseed	*Menispermum canadense*	Fruits
Mountain laurel	*Kalmia latifolia*	Leaves, shoots, flowers
Mushrooms	Many genera and species, especially *Amanita* spp.	All parts
Nightshade	*Solanum* spp.	Unripened fruits (*Caution: A poisonous principle is produced in common potato* [*Solanum tuberosum*] *tubers exposed to light long enough for the skins to turn green or greenish*)
Oleander	*Nerium oleander*	All parts
Poison hemlock	*Conium maculatum*	All parts
Poke	*Phytolacca americana*	Roots and mature stems
Rhododendron (see Azalea)		
Rhubarb	*Rheum rhaponticum*	Leaf blades (*Caution: Although young petioles are widely eaten, dangerous accumulations of a poisonous substance can occur in leaf blades*)
Rubber vine	*Cryptostegia grandiflora*	All parts
Sandbox tree	*Hura crepitans*	Milky sap and seeds
Tansy	*Tanacetum vulgare*	Leaves, flowers
Tung tree	*Aleurites fordii*	All parts, especially seeds
Water hemlock	*Cicuta* spp.	Roots
White snakeroot	*Eupatorium rugosum*	All parts
Yellow oleander	*Thevetia peruviana*	All parts, especially fruits
Yew	*Taxus* spp.	All parts except "berry" pulp

Table A3.3 — Other Organisms Producing Significant Quantities of Poisonous Substances

Organism	Scientific Name	Poisonous Parts
Amaryllis	*Amaryllis* spp.	Bulbs
Autumn crocus	*Colchicum autumnale*	All parts
Bittersweet	*Celastrus scandens*	Seeds
Bleeding hearts	*Dicentra* spp.	All parts
Bloodroot	*Sanguinaria canadensis*	All parts
Blue cohosh	*Caulophyllum thalictroides*	Fruits, leaves
Boxwood	*Buxus sempervirens*	Leaves
Buckthorn	*Rhamnus* spp.	Fruits
Bushman's poison	*Acokanthera* spp.	All parts
Buttercup	*Ranunculus* spp.	All parts; toxicity varies from species to species; mostly cause blistering
Buttonbush	*Cephalanthus occidentalis*	Leaves
Caladium	*Caladium* spp.	All parts
Carolina jessamine	*Gelsemium sempervirens*	All parts (even visiting honeybees can be poisoned)
Chincherinchee	*Ornithogalum thyrsoides*	All parts
Crown of thorns	*Euphorbia milii*	Milky latex
Culver's root	*Veronicastrum virginicum*	Root
Daffodil	*Narcissus* spp.	Bulbs
Desert marigold	*Baileya radiata*	All parts
Fly poison	*Amianthemum muscaetoxicum*	Leaves, roots
Four-o'clock	*Mirabilis jalapa*	Seeds, roots
Goldenseal	*Hydrastis canadensis*	Rhizomes, leaves
Holly	*Ilex aquifolium*	Berries
Horse chestnut	*Aesculus hippocastanum*	Seeds, flowers, leaves
Hyacinth	*Hyacinthus* spp.	Bulbs
Hydrangea	*Hydrangea* spp.	Buds, leaves
Jack-in-the-pulpit	*Arisaema triphyllum*	Roots, leaves
Jessamine	*Cestrum* spp.	Leaves, young stems
Jonquil (see Daffodil)		
Karaka nut	*Corynocarpus laevigata*	Seeds
Kentucky coffee tree	*Gymnocladus dioica*	Fruits
Larkspur	*Delphinium* spp.	Young plants, seeds
Lignum vitae	*Guaiacum officinale*	Fruits
Locoweed	*Astragalus* spp.	Location of poisonous principles varies from species to species; plants more of a problem for livestock than for humans
Lupine	*Lupinus* spp.	Location of poisonous principles varies from species to species; primarily in pods and seeds
Marijuana	*Cannabis sativa*	Resins secreted by glandular hairs among flowers
Mayapple	*Podophyllum peltatum*	All parts except ripe fruits
Mescal bean	*Sophora secundiflora*	Seeds
Narcissus (see Daffodil)		
Ngaio	*Myoporum laetum*	Leaves
Opium poppy	*Papaver somniferum*	Unripe fruits
Philodendron	*Philodendron* spp.	Leaves, stems
Pittosporum	*Pittosporum* spp.	Fruits, leaves, stems
Poinsettia	*Euphorbia pulcherrima*	Milky latex
Poison hemlock	*Conium maculatum*	All parts
Poison ivy	*Toxicodendron radicans*	Leaves
Poison oak	*Toxicodendron diversilobum*	Leaves
Poison sumac	*Toxicodendron vernix*	Leaves
Poke	*Phytolacca americana*	Roots, leaves, stems (uncooked fruits may be slightly poisonous)
Prickly poppy	*Argemone* spp.	Seeds, leaves
Privet	*Ligustrum vulgare*	Fruits
Rhododendron	*Rhododendron* spp.	All parts
Sneezeweed	*Helenium* spp.	All parts
Snow-on-the-mountain	*Euphorbia marginata*	Milky latex
Squirrel corn	*Dicentra canadensis*	All parts
Star-of-Bethlehem	*Ornithogalum umbellatum*	All parts
Sweet pea	*Lathyrus* spp.	Seeds
Tobacco	*Nicotiana tabacum*	Leaves (when eaten)
Water hemlock	*Cicuta* spp.	All parts

Also, plants can vary widely in the amounts of medicines they produce. Responsible persons involved in the growing, harvesting, and marketing of medicinal plants recognize not only the necessity for scientific testing of their products, but the additional need to standardize medicinal plant products sold for human use. A number of medicinal plant products are already standardized before they are sold, and many others will be in the future.

Positive responses to medicinal plant substances and extracts for specific ailments are often much slower than they are to prescription medicines, and one might ask why, then, one might want to use them alternatively in some instances. The expense, and sometimes dangerous side effects, of prescription medicines, are obvious answers, but there are other less obvious reasons. Prescription medicines invariably are specific drugs produced in pure, isolated form. Natural plant medicines, however, almost always are a combination of various chemicals, with many often having a synergistic effect (i.e., the combination works better than a single isolated component by itself). In fact, in Chinese and East Indian (Ayurvedic) medicine, which have been practiced for thousands of years, various combinations of medicinal plants usually are prescribed, primarily because of the synergistic effects.

Table A3.4 includes a sampling of plants, algae, fungi, and bacteria associated with past and some present medicinal uses. Some of the drugs concerned are prescribed for specific ailments by modern medical practitioners, whereas others are a part of folk medicine still practiced in mostly nonurban areas. ***Caution:*** *Do not use any of the plants, algae, fungi, or bacteria listed here for medicinal purposes without consulting a reputable, qualified health practitioner.*

The following groups of plants or plant derivatives have been used to treat memory and cholesterol problems. The reader should be aware that the human tendency to produce bad cholesterol varies considerably with the individual's heredity, as well as the types of food consumed. If an individual consumes a food known to contain relatively high amounts of cholesterol (e.g., eggs) it does not necessarily follow that the individual's bad cholesterol level will then increase, nor does removal of the cholesterol-containing food necessarily result in the lowering of bad cholesterol levels. Nevertheless, there are some plants or plant derivatives that have been demonstrated to aid in reduction of human bad cholesterol levels.

Table A3.4 — Plants, Algae, Fungi, and Bacteria Associated with Medicinal Uses

Organism	Scientific Name	Uses
Agrimony	*Agrimonia* spp.	High silica content makes aerial parts of plant useful as an astringent to stop bleeding
Alfalfa	*Medicago sativa*	Leaf concentrates shown to promote desirable balance between LDL and HDL cholesterol levels
Aloe	*Aloe* spp. (esp. *Aloe vera*)	Juice from leaves contains chrysophanic acid, which promotes healing of burns; used for relief of constipation
American mountain ash	*Pyrus americana*	Liquid made from steeping inner bark in water used as an astringent; tea of berries used as a wash for hemorrhoids; berries eaten to prevent or cure scurvy
Anemarrhena asphodelioides (see Chinese lily)		
Anemone	*Anemone canadensis*	Pounded boiled root applied to wounds as an antiseptic
Angelica	*Angelica archangelica*	Fruits used in treating colds and fevers; leaves used to stimulate appetite
Angelica dahurica (see Dahurian angelica)		
Angelica polymorpha (see Chinese angelica)		
Anise	*Pimpinella anisum*	Seed oil used to relieve indigestion, colds, and respiratory problems such as sinusitis
Apocynum androsaemifolium (see Dogbane, bitter)		
Apocynum venetium (see Dogbane, venetian)		
Apple	*Malus domestica*	Source of polyphenols and enzyme inhibitors that exhibit antioxidant and bactericidal activity (e.g., against gingivitis bacteria), especially when working in concert with bioflavonoids
Apricot	*Prunus armeniaca*	Seed extract said to function as a bronchodilator
Arnica	*Arnica* spp.	Plants applied as a poultice to bruises and sprains
Ashwagandha	*Withania somnifera*	Reported to have multiple immune system-boosting effects as well as alleviating menopausal symptoms
Asian epimedium	*Epimedium grandiflorum*	Plant extracts are said to have a stimulatory, hormone-like effect on the prostate gland and testes
Asian skullcap	*Scutellaria baicalensis*	Plant extracts contain flavonoids and antioxidants; some components function together as a bronchodilator and bactericide; they also reduce blood pressure and LDL cholesterol levels

Table A3.4 — Plants, Algae, Fungi, and Bacteria Associated with Medicinal Uses

Organism	Scientific Name	Uses
Asparagus, wild	*Asparagus racemosus*	Root flavonoid extracts used to relieve menopausal problems
Astragalus	*Astragalus* spp.	Root extracts used to boost the immune system; said to be good for colds, flu, and immune-deficiency disorders; also lowers blood pressure. (**Caution: Some *Astragalus* spp. *sequester toxic amounts of selenium; should not be taken if a fever is present*)**
Astragalus membranaceus (see Membranous milk vetch)		
Atractylodes lancea (see Southern tsangshu)		
Atractylodes macrocephala (see Southern tsangshu)		
Balm of Gilead	*Populus x gileadensis*	Buds used as an ingredient in cough syrup
Balsam poplar	*Populus balsamifera*	Buds made into ointment, which Native Americans placed in nostrils for relief of congestion
Barberry	*Berberis vulgaris*	Slows heartbeat rate
Beefsteak plant (see Perilla)		
Bifidobacteria	*Bifidobacterium bifidum* (*Helicobacter pylori*, *B. breve*, *B. infantis*, *B. longum*, and others)	Bifidobacteria destroy the bacteria that cause ulcers in humans
Bilberry/Blueberry	*Vaccinium* spp.	Evidence that regular consumption of fruit, which contains more than a dozen anthocyanosides, increases oxygen flow to eyes, slowing progression of cataracts, glaucoma, and macular degeneration; helps to balance insulin levels
Bitter melon	*Momordica charantia*	Plant extracts promote increased insulin production and are believed to reduce sugar damage to pancreatic cells
Bittersweet nightshade	*Solanum dulcamara*	Plant extracts used to treat skin problems such as acne, eczema, and boils. (**Caution: The fruits and other plant parts are poisonous**)
Blackberry	*Rubus* spp.	Tea of roots used by northern California Native Americans to cure dysentery
Black cohosh	*Cimicifuga racemosa*	Dried rhizomes used in cough medicines and for rheumatism; counters effects of declining estrogen levels in women (e.g., hot flashes, sleep disturbances); alleviates urinary tract problems
Black currant	*Ribes nigrum*	Oil from seeds used to improve suppleness of skin and to reduce skin dryness
Black haw	*Viburnum prunifolium*	Bark used in treatment of asthma and for relieving menstrual irregularities
Bloodroot	*Sanguinaria canadensis*	Native Americans used rhizome for ringworm, as an insect repellent, and for sore throat
Blue cohosh	*Caulophyllum thalictroides*	Tea of root drunk by Native Americans and early settlers a week or two before giving birth to promote rapid parturition
Boneset	*Eupatorium perfoliatum*	Water infusion of dried plant tops widely used to treat fevers and colds
Borage	*Borago officinalis*	Oil from seeds contains gamma linoleic acid (GLA) and other oils beneficial in human nutrition
Boswellia	*Boswellia serrata*	Extract of resin from this East Indian tree inhibits substances that cause joint swelling
Broom snakeweed	*Gutierrezia sarothrae*	Navajo Indians applied chewed plant to insect stings and bites of all kinds
Buckthorn	*Rhamnus catharticus*	Fruits used as a laxative
Bupleurum chinense (see Chinese thoroughwax)		
Burdock	*Arctium lappa*	Used as an insulin substitute in folklore; root extract used in 17th century for venereal diseases
Butcher's broom	*Ruscus aculeatus*	Plant extracts shown in clinical trials to strengthen capillaries and to relieve hemorrhoid symptoms; improve flow of blood to the hands and feet
Button snakeroot	*Eryngium* spp.	Natchez Indians inserted chewed stem in nostrils to arrest nosebleed
Cajuput	*Melaleuca cajuputi*	Oil obtained from leaves and twigs is used in treatment of muscular pain and as an antiseptic
California bay	*Umbellularia californica*	Yuki Indians put leaves in bath of hot water and bathed for relief of rheumatism; leaves used as an insect repellent
Camphor	*Cinnamomum camphora*	Oil from leaves and wood used in cold remedies and liniments

Table A3.4 — Plants, Algae, Fungi, and Bacteria Associated with Medicinal Uses

Organism	Scientific Name	Uses
Camptotheca	*Camptotheca acuminata*	Extracts from flowers and immature fruits yield camptothecin, which has given evidence of being effective against certain forms of cancer
Cardamon	*Elettaria cardamomum*	Seed oil has antibiotic properties and is used in treatment of colds, coughs, and other respiratory problems
Cascara	*Rhamnus purshiana*	Bark extract widely used as a laxative
Catnip	*Nepeta cataria*	Leaf tea used for treatment of colds and to relieve infant colic
Cat's claw	*Uncaria tomentosa*	Root bark extracts used in treatment of intestinal problems, including diverticulosis and Crohn's disease; extracts also shown to have anti-inflammatory properties; hirsutin component lowers blood pressure; rhynchophylline (alkaloid) inhibits clumping of blood platelets and has been shown in animals to increase brain serotonin levels
Cayenne pepper	*Capsicum frutescens*	Used to reduce mucous drainage (recent evidence, however, suggests it may be carcinogenic); capsaicin extracts used in ointments to relieve pains of arthritis and neuropathy; aids digestion
Celery	*Apium graveolens*	Seed contains an essential oil that acts like an antioxidant that fights free radicals that attack joints; oil believed to have sedative properties
Chamomile	*Chamaemelum nobile*	Tea used as a digestive aid
Chaste tree	*Vitex agnus-castus*	Extract of berries reduces symptoms of premenstrual syndrome
Chaulmoogra	*Hydnocarpus* spp.	Seed oil used in the treatment of skin diseases such as eczema, psoriasis, and leprosy
Cherry (wild)	*Prunus serotina*	Tea brewed from bark used for coughs and colds
Chia	*Salvia columbariae*	Mucilaginous seeds used by Spanish Californians to make a refreshing drink; seeds contain a caffeine-like principle that enabled Native Americans to perform unusual feats of endurance; seed paste used in eye irritated by foreign matter
Chinese angelica	*Angelica polymorpha*	Root extracts used to suppress or relieve asthma
Chinese cinnamon	*Cinnamomum cassia*	Pulverized bark ingested to improve urinary flow and reduce more than normal frequencies of urination
Chinese club moss	*Lycopodium serratum*	Source of an alkaloid (huperzine A) that inhibits destruction of acetylcholine involved in nerve transmissions, and thereby enhances memory
Chinese lily	*Anemarrhena asphodelioides*	Plant extracts apparently can aid in controlling blood glucose levels, hay fever, dermatitis, and other allergic symptoms; rhizome extracts used to quench thirst caused by fevers
Chinese magnolia	*Magnolia officinalis, Magnolia quinquepeta*	Bark extracts used to reduce nasal stuffiness and discharge, to drain sinuses, and to alleviate asthma and sinus headaches
Chinese rubber tree	*Eucommia ulmoides*	Bark extract improves circulation to the hands and feet; reduces high blood pressure; alleviates frequent urnination problems
Chinese thoroughwax	*Bupleurum chinense*	Root extracts found to have general calming effect and to promote sound sleep; usually used in combination with other herbal extracts
Chlorella	*Chlorella* spp.	Green algae that boost the immune system, relieve constipation, and can remove heavy metals from food
Chocolate	*Theobroma cacao*	Seed extracts are good source of L-arginine and magnesium and are believed (when combined with other chocolate constituents) to elevate serotonin levels; chocolate also contains theobromin (somewhat similar to caffeine in action) and phenylethylene, which are believed to produce sustained elevation of mood (Note: These attributes pertain primarily to chocolate that does not have milk, sugar, or other products added to it)
Chrysanthemum indicum (see Indian chrysanthemum)		
Cinchona	*Cinchona* spp.	Bark yields quinine drugs used in treating malaria
Cinquefoil (Eurasian)	*Potentilla erecta*	Dried rhizome used to control diarrhea
Club moss	*Lycopodium clavatum*	Spores dusted on wounds or inhaled by Native Americans to arrest nosebleeds
Coca	*Erythroxylon coca*	Cocaine from leaves used as a local anesthetic; South American laborers use it as a stimulant
Cola	*Cola nitida, C. acuminata*	Seeds contain up to 3.5% caffeine and 1% theobromine, which may lessen fatigue

Table A3.4 — Plants, Algae, Fungi, and Bacteria Associated with Medicinal Uses

Organism	Scientific Name	Uses
Coleus	Coleus for skolii	Plant extracts used in treatment of hypertension, allergies, glaucoma, and psoriasis
Cordyceps	Cordyceps sinensis	This fungus, which parasitizes caterpillars, apparently alleviates respiratory problems and elevates phagocyte action
Cornsilk	Zea mays	Cornsilk extracts used for centuries as a diuretic
Corydalis	Corydalis turtschaninovii	One isolate of several produced by the plant has multiple anti-inflammatory and calming effects; it has been used to relieve joint pain (Note: This plant also produces poisonous alkaloids)
Cotton	Gossypium spp.	Cotton root "bark" used by black slaves and Native Americans to induce abortions
Cranberry	Vaccinium oxycoccum	Fruit juice drunk to treat female yeast infections
Creosote bush	Larrea divaricata	Decoction from leaves used as a cure-all by Native Americans but especially for respiratory problems
Cubebs	Piper cubeba	Dried fruit best known as a condiment but is also used in treatment of asthma
Cynanchum	Cynanchum spp.	Plant extracts of these relatives of milkweeds are said to reduce mucous congestion of the lungs
Dahurian angelica	Angelica dahurica	Plant extracts used to treat allergy symptoms
Damiana	Turnera diffusa	Dried leaves used for minor pain, as a laxative, as a flavoring for a liqueur, and as a sexual stimulant; also said to improve blood circulation
Dandelion	Taraxacum sp. aff.	Root extracts said to stimulate the liver and facilitate its natural functioning in detoxification
Deadly nightshade	Atropa belladonna	Belladonna, a drug complex extracted from leaves, contains the drugs atropine, hyoscyamine, and scopolamine; these are used as an opium antidote, for shock treatments, and for dilation of pupils; scopolamine is also used as a tranquilizer and for "twilight sleep" in childbirth
Devil's claw	Harpagophytum procumbens	Tuber extracts reported to have anti-inflammatory and pain-relieving properties
Di-huang (see Rehmannia)		
Dogbane, bitter	Apocynum androsaemifolium	Roots boiled in water and resulting liquid used as a heart medication (contains a drug similar in action to digitalis)
Dogbane, venetian	Apocynum venetium	Leaves smoked for bronchitis relief; extracts lower blood pressure
Dogwood	Cornus spp.	Inner bark boiled in water and resulting liquid drunk to reduce fevers
Dong quai	Angelica sinensis	Root extracts (which contain flavonoids) used in the alleviation of hot flashes and other menopausal symptoms; also used to treat premenstrual syndrome
Echinacea	Echinacea purpurea	Leaves and roots have antiviral and anti-inflammatory properties; used to boost the immune system
Elderberry	Sambucus spp.	Source of Sambucol, which is reported to have antiviral properties, especially in controlling those viruses involved in the common cold
Ephedra	Ephedra spp.	Drug ephedrine, widely used to relieve nasal congestion and low blood pressure; obtained from inner bark, berries, flowers, leaves (most ephedrine now in use is synthetic); also known as ma huang (Caution: Stems contain toxic amounts of cyanide)
Epimedium grandiflorum (see Asian epimedium)		
Ergot	Source: Claviceps purpurea on cereal grains	Used to treat migraine headaches and to control bleeding after childbirth
Eucalyptus	Eucalyptus spp.	Oil extracted from leaves used to alleviate bronchitis and coughs
Eucommia ulmoides (see Chinese rubber tree)		
European birch	Betula pendula	Oil distilled from bark and leaves used in treatment of kidney stones and urinary tract infections
Evening primrose	Oenothera spp.	Seeds are source of GLA oils beneficial in human nutrition
Eyebright	Euphrasia officinalis	Plant extracts used as an eyewash and in the relief of allergic itching of the eyes
Fennel	Foeniculum vulgare	Extracts of roots, stems, and fruits used as an appetite suppressant and as an eyewash

Table A3.4 Plants, Algae, Fungi, and Bacteria Associated with Medicinal Uses

Organism	Scientific Name	Uses
Fenugreek	*Trigonella foenum-graecum*	Seeds used in bulking laxatives; reduces mucus resulting from asthma and sinus problems; reduces skin inflammation
Feverfew	*Chrysanthemum parthenium*	Dried flowers used in treatment of migraine headaches, to induce abortion and menstruation, and as an insecticide; dilutes bronchial mucus; keeps body from producing histamines
Field mint	*Mentha arvensis*	Oil distilled from aerial parts of plants used to alleviate symptoms of colds, fevers, bronchitis; also used as an antiseptic
Flax	*Linum usitatissimum*	Cold-processed seed oils are rich source of gamma linoleic acid (GLA), beneficial in human nutrition, and in suppressing or reversing atherosclerosis; crushed seeds used as a laxative and for treating bronchial problems
Flowering ash	*Chionanthus virginicus*	Bark used as a laxative and in treatment of liver ailments
Forsythia	*Forsythia suspensa*	Plant extract has anti-inflammatory properties; is used to relieve sinus congestion and headaches
Foxglove	*Digitalis purpurea*	Drug digitalis, widely used as a heart stimulant, obtained from leaves
Frankincense	*Boswellia serrata*	Used to reduce joint pain and stiffness
Gambir	*Uncaria rhyncophylla*	Experimentally shown to relax blood vessels
Garlic	*Allium sativum*	Evidence that allicin and other sulfur-containing compounds extracted from bulbs inhibit common cold and other viruses; decreases artery-plugging fibrin levels, and regular consumption appears to enhance general cardiovascular health, including the lowering of LDL cholesterol levels and the inhibition of blood platelet clumping; the risk of stomach cancer also appears to be lowered
Gastrodia orchid	*Gastrodia elata*	Used in treating epilepsy and blood circulation problems; glucosides lower blood pressure
Gentian	*Gentiana catesbaei*	Catawba Indians applied hot-water extract of roots to sore backs; liquid drunk as a remedy for stomach aches; aids digestion and boosts circulation
Geranium (wild)	*Geranium maculatum*	Dried roots used for dysentery, diarrhea, and hemorrhoids
Ginger	*Zingiber officinale*	Powerful antioxidant; aids digestion and reduces nausea (including that of motion sickness); helps promote normal bladder control
Ginger (wild)	*Asarum* spp.	Extract of rhizome used as a broad-spectrum antibiotic
Ginkgo	*Ginkgo biloba*	Concentrated leaf extract improves oxygen-carrying capacity of capillaries, especially those of the brain, and may improve memory; used for treating vertigo and tinnitus
Ginseng (see also Siberian ginseng)	*Panax* spp.	Considered a general panacea, especially in Asia; strengthens the reproductive and adrenal glands; said to increase stamina
Globe artichoke	*Cynara scolymus*	Leaf and root extract used to inhibit gallstone formation and to alleviate digestion problems
Goldenrod	*Solidago canadensis*	Inflorescences used in the treatment of kidney stones and urinary tract infections
Goldenseal	*Hydrastis canadensis*	Rhizome source of alkaloidal drugs used in treatment of inflamed mucous membranes; also used as a tonic (*Note:* Pregnant women should not use goldenseal)
Goldthread	*Coptis groenlandica*	Native Americans boiled plant and gargled the liquid for sore or ulcerated mouths
Gotu kola	*Centella asiatica*	Shown in clinical trials to reduce swelling of ankles, generally improve blood circulation, and accelerate healing of wounds
Grape	*Vitis vinifera*	Seed extract source of powerful antioxidants (including quercetin) that also improve blood flow to the retina, thereby retarding macular degeneration; red grapes in particular produce significant amounts of reservatrol, which has been demonstrated to enhance enzyme activity associated with the regeneration and stimulation of nerve cells
Grapefruit	*Citrus × paradisi*	Seed extract used to combat bacterial or fungal infections
Green hellebore	*Helleborus viridis*	Plant extract used to treat hypertension (drug now synthesized); Thompson Indians used plant in small amounts to treat syphilis

Table A3.4 — Plants, Algae, Fungi, and Bacteria Associated with Medicinal Uses

Organism	Scientific Name	Uses
Green tea	Camellia sinensis	Unfermented leaves source of polyphenols, which appear to reduce incidence of cancers in regular users through neutralization of free radicals; epigallocatechin gallate (EGCG) ingredient of green tea demonstrated by the Mayo Clinic to be particularly effective in control of prostate cancer
Gum plant	Grindelia camporum	Liquid from freshly and briefly boiled plants effective in treating poison oak and poison ivy rashes; used in treatment of coughs
Gymnema	Gymnema sylvestre	Extracts used to stabilize insulin levels in diabetics
Gynostemma	Gynostemma pentaphylla	Chinese plant related to melons; extracts believed to stimulate the immune system and aid in metabolism of fats that contribute to strokes
Hawthorn	Crataegus oxycantha	Anti-inflammatory that lowers LDL cholesterol levels; dilates coronary blood vessels
Hemlock	Tsuga spp.	Native Americans made tea of inner bark to treat colds and fevers (*Note:* Do not confuse this tree with poison hemlock [*Cicuta* spp.])
Hops	Humulus lupulus	Extracts are sedative and used in treating insomnia and nervous tension
Horehound	Marrubium vulgare	Extract from dried tops of plants used in lozenges for relief of sore throats and colds; dilutes mucus in bronchial tubes
Horse chestnut	Aesculus hippocastanum	Seed and leaf extracts used to improve blood flow; night cramps of legs; reduce varicose veins and leg swelling (*Caution: Plant is poisonous and only standardized extracts of demonstrated therapeutic value should be used; a coumarin component of horse chestnut leaves can interact adversely with aspirin and other anticoagulants*)
Horseradish	Armoracia rusticana	Roots used to treat infections of the urinary tract
Horsetail	Equisetum spp.	Plants boiled in water; liquid used as a delousing hairwash or as a gargle for mouth ulcers
Huperzia serrata = *Lycopodium serratum* (see Chinese club moss)		
Hydrangea	Hydrangea paniculata	Essential oil from roots acts as diuretic (*Caution: Leaves contain toxic amounts of cyanide*)
Hypericum perforatum (see St. John's wort)		
Indian chrysanthemum	Chrysanthemum indicum	Glucoside extract said to lower blood pressure
Indigo (wild)	Baptisia tinctoria	Native Americans boiled plant and used liquid as an antiseptic for skin sores
Ipecac	Cephaelis ipecacuana	Drug from roots and rhizome used to treat amoebic dysentery; also used as an emetic
Java plum	Syzygium cumini	Powdered seeds used to counter excessive thirst and excretion of sugar in the urine, characteristic of diabetics
Jimson weed	Datura spp.	Drugs atropine, hyoscyamine, and scopolamine obtained from seeds, flowers, and leaves; drug stramonium used for knockout drops and in treatment of asthma (*Caution: Jimson weeds are highly poisonous* [see Deadly nightshade])
Joe-pye weed	Eupatorium purpureum	Dried root said to prevent formation of gallstones
Joshua tree	Yucca brevifolia	Cortisone and estrogenic hormones made from sapogenins produced in the roots
Jujube	Ziziphus jujuba	Fruit extracts shown to promote restful sleep and to aid in balancing irregular heartbeat
Juniper	Juniperus spp.	Tea of "berries" drunk by Zuni Indian women to relax muscles following childbirth (*Caution: Internal consumption can interfere with absorption of iron and other minerals*)
Kansas snakeroot	Echinacea angustifolia	Dried roots used as antiseptic in treatment of sores and boils, periodontal disease, and sinus drainage problems
Kava kava	Piper methysticum	Leaf tea used as a sedative, as a muscle relaxant, and as a pain reliever
Kirilow's cucumber	Trichosanthes kirilowii	Used to inhibit mucous production in the lungs
Lactobacillus	Lactobacillus acidophilus, L. rhamnosus, L. salivarius, and others	Lactobacilli normally populate the gastrointestinal tract, where they function in various ways to boost the immune system and destroy pathogenic bacteria; antibiotics destroy these useful bacteria, and repopulation is facilitated by ingestion of lactobacilli capsules or products such as yogurt, which contain the useful bacteria

Table A3.4 — Plants, Algae, Fungi, and Bacteria Associated with Medicinal Uses

Organism	Scientific Name	Uses
Lemon balm	*Melissa officinalis*	Leaf extracts and oils used for colds
Licorice	*Glycyrrhiza glabra*	Rhizomes source of licorice used in cough drops and for the soothing of inflamed mucous membranes; stimulates interferon production; relieves allergic symptoms; boosts immune system; deglycyrrhizinated licorice increases protection of upper digestive tracts by augmenting the mucous coating (**Caution: Undeglycyrrhizinated licorice can elevate blood pressure**)
Ligusticum	*Ligusticum wallichii*	Extract has been demonstrated to relax blood vessels
Lily of the valley	*Convallaria majalis*	All parts of plant contain a heart stimulant similar to digitalis; used to control irregular heartbeat (**Caution: Plants are poisonous**)
Lilyturf plant	*Ophiopogon japonicus*	Root extracts said to aid in diluting thick mucous secretions in the lungs
Lobelia	*Lobelia inflata*	Drug lobeline sulfate obtained from dried leaves; drug used in preparations to aid in cessation of smoking and in treatment of respiratory disorders (**Caution: Has effects similar to those of nicotine; more than 10 grams [one-third ounce] of dried plant can produce a coma**)
Lycopodium serratum (see Chinese club moss)		
Maca	*Lepidium meyenii*	Root extracts said to elevate testosterone levels and improve sexual performance in men
Madagascar periwinkle	*Catharanthus roseus*	A semisynthetic extract (vinpocetine) derived from vincamine produced by this plant said to be a significant memory enhancer
Magnolia vine	*Schisandra chinensis*	Plant extracts contain a powerful antioxidant that appears to protect healthy tissues (liver in particular) from damage caused by higher-than-normal blood sugar levels. Synergistic effect when combined with ginseng
Ma huang (see Ephedra)		
Maitake mushrooms	*Grifola frondosa*	A substance (beta-glucan) produced by these mushrooms evidently stimulates the production of cells that aid in the inhibition of cancer cells
Malabar kino	*Pterocarpus marsupium*	Leaf extracts contain epicatechin, which promotes oxygen uptake and better processing of sugar by body tissues
Mandrake	*Mandragora officinarum*	Extracts of plant used in folk medicine as a painkiller (drugs hyoscyamine, podophyllin, and mandragorin have been isolated; podophyllin used experimentally in treatment of paralysis)
Mangosteen	*Garcinia mangostana*	Fruit acid is believed to aid in weight reduction
Manroot	*Marah* spp.	Native Americans used oil from seeds to treat scalp problems and the crushed roots for relief from saddle sores
Marginal fern	*Dryopteris marginalis*	Rhizomes contain oleoresin used in expulsion of tapeworms from the intestinal tract
Marijuana	*Cannabis sativa*	Tetrahydrocannabinol obtained from resinous hairs in inflorescences; ancient medicinal drug of China
Mayapple	*Podophyllum peltatum*	Podophyllin obtained from rhizomes used experimentally in treatment of paralysis; dried rhizome powder used on warts (**Caution: Plant is poisonous, and extracts are very irritating to the skin**)
Maypop	*Passiflora incarnata*	Dried leaves used as sedative; Native Americans used juice as treatment for sore eyes
Melia	*Melia toosendan*	Used in traditional Chinese medicine to relieve joint pain
Membranous milk vetch	*Astragalus membranaceus*	Extracts strengthen the immune system, especially that of the upper respiratory tract; promote interferon production and repair of damaged bronchial tubes; there is evidence it can counter bone loss (osteoporosis) resulting from extended use of corticosteroids (**Caution: Some other *Astragalus* spp., also known as milk vetch, are toxic**)
Mesquite	*Prosopis glandulosa*	Native Americans mixed dried leaf powder with water and used liquid to treat sore eyes
Mexican yam	*Dioscorea floribunda*	Tuberous roots produce up to 10% diosgenin, a precursor of progesterone and cortisone, and are a source of DHEA (dihydroepiandrosterone), a complex hormone naturally produced by humans; DHEA levels decline with aging; there is some evidence that controlled DHEA supplementation in older persons retards some aspects of aging
Milk thistle	*Silybum marianum*	Silymarin extracted from plants has antioxidant properties that appear to be especially beneficial to the liver
Milk vetch (see Astragalus)		

Table A3.4 Plants, Algae, Fungi, and Bacteria Associated with Medicinal Uses

Organism	Scientific Name	Uses
Milkweed	*Asclepias syriaca*	Quebec Indians promoted temporary sterility by drinking infusion of pounded roots
Mistletoe	*Phoradenron flavescens*	Native Americans reported to use small amounts as a contraceptive and sedative (**Caution: Plants are toxic and should not be taken internally**)
Monkshood	*Aconitum napellus*	Source of aconite once used in treatment of rheumatism and neuralgia (**Caution: Plant is highly toxic**)
Mormon tea (see Ephedra)		
Muira puama	*Ptychopetalum olacoides*	Leaf extracts said to stimulate hormone production and improve circulation; some consider it to be an aphrodisiac
Mukul myrrh	*Commiphora mukul*	Resin from tree stabilizes cholesterol levels and reduces osteoarthritis pain
Mulberry (red)	*Morus rubra*	Rappahannock Indians applied milky latex of leaf petioles to scalp to control ringworm
Mulberry (white)	*Morus alba*	Bark extract said to function as a bronchodilator
Mullein	*Verbascum thapsus*	Native Americans smoked leaves for respiratory ailments and asthma; flowers once widely used in cough medicines
Nettle (see Stinging nettle)		
Noni	*Morinda citrifolia*	Used in treatment of diabetes, high blood pressure, kidney disorders, and other ailments
Oats	*Avena sativa*	Extract from green oat seeds said to enhance both physical and sexual health
Olive	*Olea europaea*	Leaf extract contains oleuropein (calcium elenolate), which is a wide-spectrum bactericide and a virucide; it evidently enhances the production of phagocytes, thereby strengthening the immune system
Onion	*Allium* spp.	Cheyenne Indians applied bulbs in poultice to boils; juice and olive oil used to cure earaches; can lower blood pressure and help dissolve blood clots; the active principle, allicin, is also produced by garlic
Ophiopogon japonicus (see Lilyturf plant)		
Opium poppy	*Papaver somniferum*	Morphine and codeine obtained from latex of immature fruits
Oregon grape	*Berberis aquifolium*	Bark tea drunk by Native Americans to settle upset stomach; used in strong doses for treatment of venereal diseases
Pacific yew	*Taxus brevifolia*	Taxol, a promising anticancer agent, is extracted from bark
Panax ginseng	*Panax pseudoginseng, Panax ginseng*	Root extract strengthens respiratory immune system, resulting in reduction of respiratory infections; strong antioxidant
Pansy (wild)	*Viola* spp.	Plants ground up and applied to skin sores or inflammations
Papaya	*Carica papaya*	Exudate of scarified unripe fruit is source of *papain* (a protein that is used to digest ruptured back discs and to facilitate digestion of food, as a meat tenderizer, for termite control, and for reduction of cloudiness in beer); *papain*, which is also believed to have antibiotic properties, may be used with *bromelain* from pineapples and trypsin to facilitate breakdown of cardiovascular plaque
Parsley	*Petroselinum crispum*	Richer in vitamin C than citrus fruits; inhibits proliferation of tumor cells; suppresses halitosis; general organ tonic
Pau d'arco	*Tabebuia heptaphylla*	General immune system booster
Peanut	*Arachis hypogoea*	Reservatrol extracted from peanut and mulberry plants said to be effective in inhibiting several types of cancers
Pennyroyal	*Mentha pulegium*	Native Americans used leaf tea in small amounts for relief of headaches and flatulence and to repel chiggers (**Caution: Pennyroyal is toxic in larger amounts**)
Peppermint	*Mentha piperita*	Peppermint oil is used to alleviate symptoms of respiratory infections and inflammation
Perilla	*Perilla frutescens*	Seeds are the source of perilla oil, which is exceptionally rich in omega-3 fatty acids essential to cardiac health
Persimmon	*Diospyros virginiana*	Liquid from boiled fruit used as an astringent; fruits with high beta-carotene content; leaves have high vitamin C content
Peruvian balsam	*Myroxylon balsamum*	Resin obtained from scorched or incised tree trunks is used as an antiseptic on burns, wounds, and hemorrhoids
Peyote	*Lophophora williamsii*	Alcoholic extract of plant used as an antibiotic
Pine	*Pinus* spp.	Pycnogenols extracted from bark have powerful antioxidant properties

Table A3.4 — Plants, Algae, Fungi, and Bacteria Associated with Medicinal Uses

Organism	Scientific Name	Uses
Pineapple	*Ananas comosus*	Bromelain extracted from pineapple decreases clumping of blood platelets and fibrin, thereby improving circulation; bromelain also accelerates healing and can relieve pain, all without the side effects of aspirin, which is widely used for the same purposes; repeatedly chewing or holding fresh pineapple in the mouth may cure mouth ulcers
Pinkroot	*Spigelia marilandica*	Powdered root very effective in expulsion of roundworms from intestinal tract
Pipssisewa	*Chimaphila umbellata*	Native Americans steeped plant in water and used liquid to draw out blisters
Pitcher plant	*Sarracenia purpurea*	Native Americans used root widely as smallpox cure (records indicate it was effective)
Plantain	*Plantago ovata* and other *Plantago* spp.	(Not related to banana-like plantains.) Seed husks (known as psyllium) absorb water and are widely used in bulking laxatives; said to lower LDL cholesterol levels
Pleurisy root	*Asclepias tuberosa*	Liquid from roots boiled in water used in treatment of respiratory problems
Polypore fungus	*Grifola umbellata*	All parts enhance kidney and bladder function; also believed to have anticancer immune system–boosting properties
Prickly ash	*Zanthoxylum americanum*	Bark and berries widely used by Native Americans for toothache (pieces inserted in cavities); liquid infusion drunk for venereal diseases
Psoralea	*Psoralea corylifolia*	Flavonoids used in Chinese medicine to facilitate relief of urinary tract problems
Pumpkin	*Cucurbita pepo*	Seed oil used to promote prostate health
Puncture vine	*Tribulus terrestris*	Plant extracts believed to elevate testosterone levels and promote muscle gain in men and to elevate estrogen levels in women
Purple coneflower	*Echinacea purpurea*	Plant extracts used to boost the immune system
Pygeum	*Pygeum africanum*	Bark extracts used to promote shrinkage of benign swelling of the prostate gland in men; there is evidence that a combination of pygeum and stinging nettle (*Urtica dioica*) can significantly reduce urgency for night urination
Quassia	*Picraea excelsa*, *Quassia amara*	Wood extracts used as pinworm remedy and as insecticides
Rauvolfia[1]	*Rauvolfia serpentina*	Reserpine obtained from roots; drug used in treatment of mental illness and in counteracting effects of LSD
Rauvolfia[1] *yunnanensis* (see Yunnan rauvolfia)		
Red-rooted sage	*Salvia miltiorhiza*	Plant extracts elevate blood oxygen content and are used to enhance blood circulation, particularly in the lungs; inhibits blood platelet clumping
Red yeast	*Monascus purpureus*	This yeast is cultured on rice; the combination improves circulation and balances cholesterol levels
Rehmannia	*Rehmannia glutinosa*	Experiments with animals indicates efficacy in strengthening kidney function and in lowering blood pressure
Rice (brown)	*Oryza sativa*	Inositol hexaphosphate (IP6), a B vitamin that is produced by rice, has been shown to control growth of cancer cells
Rye	*Secale cereale*	An Australian patented ryegrass extract known as Oralmat is proving to be effective in treating asthma, allergies, and other disorders, without the side effects of steroids
Saffron (meadow)	*Colchicum autumnale*	Drug colchicine from corms used in past for treatment of gout and back disc problems, but now mostly used experimentally to induce doubling of chromosome numbers in plants
St. John's wort	*Hypericum perforatum*	Extracts used in the treatment of depression; boosts serotonin production in the brain; the serotonin suppresses cravings for carbohydrates and tends to promote normal sleep patterns (Note: St. John's wort can interfere with the normal metabolic activities of some prescription medications, and its use should be supervised by informed personnel)
Salvia miltiorhiza (see Red-rooted sage)		
Sarsaparilla	*Aralia nudicaulis*	Cough medicines made from roots
Sassafras	*Sassafras albidum*	Tea of root bark used to induce sweating; used externally as a liniment

[1]Frequently misspelled *Rauwolfia*.

Table A3.4 Plants, Algae, Fungi, and Bacteria Associated with Medicinal Uses

Organism	Scientific Name	Uses
Saw palmetto	*Serenoa repens*	Berry extracts clinically demonstrated to aid in shrinkage of benign swelling of prostate gland, increase urine flow, and normalize frequency of urination in men
Schisandra chinensis (see Magnolia vine)		
Scutellaria baicalensis (see Asian skullcap)		
Self-heal	*Prunella vulgaris*	Native Americans applied plants in poultices to boils; plant glucosides said to tone blood vessels
Seneca snakeroot	*Polygala senega*	Liquid from bark boiled in water applied to snakebites; taken internally as an abortifacient; used in a cough remedy
Senna	*Cassia senna* and other spp.	Leaf extract used as a laxative or purgative
Siberian ginseng	*Eleutherococcus senticosus*	Liquid extract of rhizome and roots used as an immune system and stamina booster; in some individuals, it appears to counteract chronic fatigue syndrome
Sicklepod	*Cassia obtusifolia*	Plant extracts said to lower both blood pressure and LDL cholesterol levels
Skeleton weed	*Lygodesmia juncea*	Widely used by Native American women to increase milk flow
Skullcap	*Scutellaria laterifolia*	Dried plant used as an anticonvulsive in treatment of epilepsy and as a sedative (*see also* Asian skullcap)
Slippery elm	*Ulmus fulva*	Dried inner bark, which contains an aspirin-like substance, used to soothe inflamed membranes
Southern tsangshu	*Atractylodes lancea, Atractylodes macrocephala*	Plant extracts used as a diuretic; also used to balance blood sugar levels and promote spleen health
Soy	*Glycine max*	Isoflavones from plants (especially fruits) have medicinal value; ipriflavone appears to inhibit development of osteoporosis and increase bone density; genistein apparently diminishes production of cellular stress protein and inhibits the growth of human prostate cancer cells; reduces symptoms of menopause
Spicebush	*Lindera benzoin*	Berries, buds, and bark brewed for tea used to reduce fevers
Spruce	*Picea* spp.	Cree Indians ate small, immature female cones for treatment of sore throat; spruce leaf oil and spruce shoots used in Europe to alleviate cold, bronchitis, and fever symptoms
Squills	*Urginea maritima*	Bulbs of red variety are source of a heart stimulant; bulbs of white variety are widely used as a rodent killer
Stevia	*Stevia rebaudiana*	Stevia extracts are 30–100 times sweeter than sugar; used as a sweetener by diabetics and others needing to reduce their sugar intake
Stinging nettle	*Urtica dioica*	Used in treatment of allergic disorders and inflammatory conditions of the lungs; the roots are rich in vitamin C, which apparently inhibits breakdown of testosterone, consequently increasing testosterone levels in the body (*see also* Pygeum)
Stoneseed	*Lithospermum ruderale*	Shoshoni women reported to have drunk cold-water infusion of roots daily for 6 months to ensure permanent sterility (experiments with mice suggest substance to the reports)
Strophanthus	*Strophanthus* spp.	Seeds are major source of cortisone and also source of a heart stimulant
Strychnine plant	*Strychnos nox-vomica*	Strychnine extracted from seeds widely used as an insect and animal poison and as the principal ingredient in blowgun darts used by South American aborigines; minute amounts stimulate the central nervous system and relieve paralysis
Sumac	*Rhus* spp. (especially *R. glabra*)	Native Americans applied leaf decoction as a remedy for frostbite; fruits and liquid made from leaves applied to poison ivy rash and gonorrhea sores; root chewed for treament of mouth ulcers
Sweet flag	*Acorus calamus*	Boiled root applied to burns; root chewed for relief of colds and toothache
Sweet gum	*Liquidambar styraciflua*	Bud balsam used to treat chigger bites; balsam also used in insect fumigating powders
Sword fern	*Polystichum munitum*	Boiled rhizome used by Native Americans to treat dandruff; sporangia and spores applied to burns
Tamarind	*Tamarindus indica*	Fruit pulp used as laxative
Tinospora	*Tinospora cordifolia*	Reported to boost the immune system
Tomato	*Solanum esculentum*	Rich source of lycopene, which is involved with health of the prostate gland and eyes

Table A3.4 — Plants, Algae, Fungi, and Bacteria Associated with Medicinal Uses

Organism	Scientific Name	Uses
Tree peony	*Paeonia suffruticosa*	Bark extract said to reduce production of blood platelets
Trichosanthes kirilowii (see Kirilow's cucumber)		
Turmeric	*Curcuma longa*	Rhizome extracts appear to lower LDL cholesterol levels, prevent blood clots, suppress cancer proliferation, and reduce joint pain
Uncaria	*Uncaria rhynchophilla*	Glucoside extracts used in China to lower blood pressure
Uzara	*Xysmalobium undulatum*	Root extracts used to control diarrhea (***Caution:*** *Uzara glycosides may react with other glycosides; drug should be taken only at recommended doses and not with other medications*)
Valerian	*Valeriana septentrionalis*	Pulverized plant applied to wounds; extracts taken internally have sedative effect and are used to treat insomnia
Velvet bean	*Mucuna* spp.	Seeds contain L-dopa used in treatment of Parkinson's disease
Virginia snakeroot	*Aristolochia serpentaria*	Native Americans used tea of plant for reducing high fevers
Wahoo	*Euonymus atropurpurea*	Bark steeped in water; liquid has digitalis-like effect on heart
Walnut (English)	*Juglans regia*	Extracts have been demonstrated to kill or inhibit a wide range of pathogenic microorganisms and to reduce LDL cholesterol levels
Watercress	*Rorippa nasturtium-aquaticum*	Some evidence that daily consumption retards development of lung cancer in smokers
Water plantain	*Alisma plantago-aquatica*	Rhizome extracts have diuretic properties; believed to improve bladder and kidney function
Western wallflower	*Erysimum capitatum*	Zuni Indians ground plant with water and applied it to skin to prevent sunburn
White mulberry	*Morus alba*	Fruit believed to improve kidney and liver function; also alleviates respiratory problems including asthma and mucous production
Willow	*Salix* spp.	Chickasaw Indians snuffed infusion of roots as a remedy for nosebleed; Pomo Indians boiled bark in water and applied liquid for relief of skin itches; fresh inner bark contains salicin, an aspirin-like compound used to reduce fevers
Wintergreen	*Gaultheria procumbens*	Oil from leaves used as a folk remedy for body aches and pains
Witch hazel	*Hamamelis virginiana*	Oil distilled from twigs and leaves used primarily as an external astringent that staunches bleeding
Wormseed	*Chenopodium ambrosioides*	Oil from seeds used to expel intestinal worms
Wormwood	*Artemisia* spp.	Yokia Indians made tea from leaves to treat bronchitis; other Native Americans used tea as a cold remedy
Yarrow	*Achillea millefolium*	Native Americans used plant infusion for treating wounds, earaches, and burns; infusion drunk to relieve upper respiratory tightness; fresh leaf inserted in nostril to staunch nosebleed
Yellow lady's slipper	*Cypripedium calceolus*	Dried root used for relief of insomnia and as sedative
Yellow nut grass	*Cyperus esculentus*	Paiute Indians pounded tubers with tobacco leaves and applied mass in wet dressing for treatment of athlete's foot
Yerba santa	*Eriodictyon californicum*	Native Americans smoked leaves or drank leaf tea for treatment of colds or asthma
Yucca	*Yucca* spp.	Plants produce saponins used in birth control preparations and to treat inflammation and other conditions that might otherwise be treated with steroids such as cortisone
Yunnan rauvolfia	*Rauvolfia yunnanensis*	Plant extracts said to lower blood pressure
Zhi-mu (see Chinese lily)		

Memory Problems or Enhancement

Memory is associated with an adequate supply of oxygen to the brain and the proper functioning of chemicals involved in messages transmitted via nerves. The following plants or plant substances may improve blood flow to the brain (thereby increasing the oxygen supply) and either protect or augment *acetylcholine,* which is the chemical involved in transmissions between nerve cells.

1. *Ginkgo biloba* (**Maidenhair Tree**)

   The leaves of this tree contain chemicals (e.g., specific terpenes, lactones, flavonoids) unknown in other plants. Numerous studies, particularly in Europe, indicate that *Ginkgo* extracts relax brain capillaries and thus improve blood flow. Clinical trials have shed light on the ability of *Ginkgo biloba* to enhance memory in healthy adults. The effects on memory enhancements, though, have not been conclusively demonstrated. *Ginkgo* extracts normally should not be taken concurrently with blood thinners such as coumadin.

2. *Huperzia serrata* (*= Lycopodium serratum*)

   This Chinese club moss produces an alkaloid known as huperzine A. Huperzine A inhibits an enzyme that destroys acetylcholine, which is essential to nerve impulse transmission.

3. **Lecithin**

   This well-known food product, which is derived commercially mostly from soybeans, is broken down and converted to acetylcholine in the brain. It is most effective in granular form.

4. *Centella asiatica* (**Gotu kola**)

   Gotu kola is said to improve brain function, partly as a result of strengthening veins.

5. **Phosphatidylserine**

   This fatty acid, whose absorption by the brain requires vitamin B_{12} and bioflavonoids, tends to block stress hormones that impair memory. It is believed to have potential to reverse cognitive impairment in the elderly.

6. **Dimethylaminoethanol (DMAE)**

   DMAE plays a role in the production of acetylcholine and helps the brain function more efficiently.

7. **Vinpocetine**

   Vinpocetine is derived from an extract of periwinkle plants native to Madagascar. Experimental studies have shown vinpocetine significantly improves short-term memory, evidently by enhancing blood flow to areas of the brain needing it most, and by increasing the capacity of red blood cells to carry oxygen. *Caution:* Long-term studies are needed to determine if repeated use could have any negative side effects.

8. **Other Factors or Nutrients Involved in Memory**

   Higher-than-normal amounts of cholesterol and triglycerides in the blood can interfere with an adequate supply of nutrients to the brain (See "Cholesterol-Lowering Plants or Plant Derivatives" in this chapter). Exposure to free radicals, low blood sugar, and poor diet and/or exercise can also contribute to memory problems. It is important to have an adequate intake of manganese (5 mg daily), zinc (50 mg daily), vitamins A, B, C, and E; other antioxidants such as COQ10, pycnogenol, SOD (superoxide dismutase), grape seed extract; and the amino acids L-glutamine, L-pheynlalanine, and L-tyrosine. Other plants or plant derivatives believed to aid memory include blue cohosh, *Gynostemma,* anise, piracetam, and rosemary.

References

Allain, H., et al. 1993. Effect of two doses of *Ginkgo biloba* extract on the dual-decoding test in elderly subjects. *Clinical Therapeutics* 15(3): 549–58.

Joseph, J., et al. 1998. Long-term dietary strawberry, spinach or Vitamin E supplementation retards the onset of age-related neuronal signal transduction and cognitive behavioral defects. *Journal of Neuroscience* 18(19): 8047–55.

Mrak, R., et al. 1997. Aging-associated changes in the human brain. *Journal of Neuropathology and Experimental Neurology* 56(12): 1269–75.

Skolnick, A. 1997. Old Chinese herbal medicine used for fever yields possible new Alzheimer's disease therapy. *Journal of the American Medical Association* 227(10): 776.

Wang, Z., et al. 1994. A double-blind control study of huperzine A and piracetam in patients with age-associated memory impairment and Alzheimer's disease. *Neuropsychopharmacology* 10(35): 7635.

Cholesterol-Lowering Plants or Plant Derivatives

1. *Allium sativum* (**Garlic**)

   Allicin, which is produced by garlic and onions, has been demonstrated to have several beneficial effects on human health, including the capacity to significantly lower blood cholesterol levels. Deodorized garlic powder is available in capsule form.

2. *Crataegus oxycantha* (**Hawthorn**)

   Hawthorn plant extracts have been shown to regulate blood levels of cholesterol and reduce blood pressure.

3. *Docosahexaenoic Acid* (**DHA**)

   There is evidence that DHA can play a role in balancing cholesterol and triglyceride levels in the blood.

4. *Commiphora mukul* (**Mukul myrrh**)

   This small tree, native to Arabia and India, produces a resin known as *guggul* that lowers LDL cholesterol and other fat levels in the blood by stimulating the thyroid gland while at the same time raising HDL levels.

5. *Cassia obtusifolia* (**Sicklepod**)

   Extracts of this plant have been shown to lower cholesterol levels.

6. *Linum* spp. (**Flax**)

   Including flaxseed in one's daily diet has been shown to balance triglyceride and cholesterol levels in the blood.

7. *Monascus purpureus* (**Red yeast rice**)

   Clinical trials involving the consumption of red yeast rice have demonstrated the rice's ability to significantly lower total serum cholesterol levels in humans.

References

Addler, A. J., et al. 1997. Effect of garlic and fish oil supplementation on serum lipid and lipoprotein concentrations in hypercholesterolemic men. *Nutrition Research* 65(2): 445–50.

Heber, D., et al. 1999. Cholesterol-lowering effects of proprietary Chinese red yeast rice supplement. *American Journal of Clinical Nutrition* 69(2): 231–36.

Pearson, T., et al. 2000. The lipid treatment assessment project (L-TAP): A multicenter survey to evaluate the percentages of dislipodemic patients receiving lipid-lowering therapy and achieving low-density lipoprotein cholesterol goals. *Archives of Internal Medicine* 160(4): 459–67.

Wang, J., et al. 1997. Multicenter clinical trial of serum lipid-lowering effects of *Monascus purpureus* rice preparation from traditional Chinese medicine. *Current Therapeutic Research* 58(12): 964–66.

Hallucinogenic Plants

Although a few hallucinogenic substances produced by animals have been isolated and some have been synthesized, the majority of known hallucinogens are produced by plants. Table A3.5 is not a complete list, but it includes the better-known sources. The reader is referred to the "Additional Reading" section for further information.

Spice Plants

The word *spice* describes any aromatic plant or part of a plant used to flavor or season food; spices are also used to add scent or flavor to manufactured products (Table A3.6). Although spices have no nutritional value, they add a pleasurable zest to meals, and before food preservation was possible, they helped make palatable food that was still edible but unappealing.

The value placed on spice plants was responsible for changing the course of Western civilization as a principal motive behind the voyages of discovery.

Dye Plants

In the recent and the ancient past, dyes from many different plants were used to color cotton, linen, and other fabrics. Since the middle of the 19th century, however, natural dyes have been almost completely replaced in industry by synthetic dyes, and today, the use of natural dyes is largely confined to individual hobbyists.

Any reader interested in experimenting with natural dyes is encouraged not only to choose those plant materials included in Table A3.7, but to try any local plants available. The experimenter will soon find that quite unexpected colors may be derived from plants, as the colors of fresh flowers, bark, or leaves often bear little relationship to the colors of the dyes. For methods of dyeing, see references in the "Additional Reading" section.

Table A3.5 Hallucinogenic Substances Produced by Plants and Fungi

Organism	Scientific Name	Part Used	Principal Active Substance
Ajuca	*Mimosa hostilis*	Roots	Nigerine
Belladonna	*Atropa belladonna*	Leaves	Hyoscyamine, scopolamine
Caapi	*Banisteriopsis caapi*	Wood	Harmine
Canary broom (Genista of florists, but true Genista is a broom genus that differs from Cytisus in that Cytisus spp. have a seed appendage whereas those of Genista do not)	*Cytisus canariensis*	Seeds	Cytisine
Catnip	*Nepeta cataria*	Leaves	Nepetalactone
Cohoba	*Piptadenia peregrina, P. macrocarpa*	Seeds, pods (snuff)	Tryptamines
Coral bean	*Erythrina* spp.	Seeds	Unknown
Cubbra borrachera	*Brugmansia* spp.	Leaves	Scopolamine
Ergot fungus	*Claviceps purpurea*	Rhizomorph	Ergine (LSD)
Fly agaric	*Amanita muscaria*	Mushroom cap	Ibotenic acid, muscimol
Henbane	*Hyoscyamus* spp.	Leaves	Hyoscyamine, scopolamine
Iboga	*Tabernanthe iboga*	Root bark	Ibogaine
Jimson weed	*Datura* spp.	All parts	Scopolamine
Kava kava	*Piper methysticum*	Root (large amounts of beverage produce hallucinations)	Myristicin-like compound
Mace	*Myristica fragrans*	Aril of seed	Myristicin
Mescal bean	*Sophora secundiflora*	Seeds	Cytisine
Morning glory	*Ipomoea violacea*	Seeds	Ergine
Nutmeg	*Myristica fragrans*	Seeds	Myristicin
Ololiuqui	*Rivea corymbosa*	Seeds	Turbicoryn
Peyote	*Lophophora williamsii*	Stems	Mescaline
Psilocybe and related mushrooms	*Psilocybe* spp., *Conocybe* spp., *Panaeolus* spp., and others	All parts	Psilocybin, psilocin
Rape dos Indios	*Maquira sclerophylla*	Dried plant (snuff)	Unknown
San Pedro	*Trichocereus pachanoi*	Stems	Mescaline
Sassafras	*Sassafras albidum*	Root bark (large amounts of tea)	Safrole
Sweet flag	*Acorus calamus*	Dried root	Asarone, β-asarone
Syrian rue	*Peganum harmala*	Seeds	Harmine
Vygie	*Mesembryanthemum expansum*	All parts	Mesebrine
Wood rose	*Argyreia nervosa*	Seeds	Ergoline alkaloids
Yakee (Parica)	*Virola* spp.	Resin from inner surface of freshly removed bark (snuff)	Tryptamine
Yohimbe	*Corynanthe* spp., *Pausinystalia johimbe*	Bark	Yohimbine

Table A3.6 Plants Used to Season or Flavor

Spice	Scientific Name of Plants	Part Used; Remarks	Principal Source
Allspice	Pimento officinalis	Powdered dried fruit	Jamaica
Almond	Prunus amygdalus	Oil from seed used for flavoring baked goods	Mediterranean; U.S.
Angelica	Angelica archangelica	Stems candied; oil from seeds and roots used in liqueurs	Europe; Asia
Anise	Pimpinella anisum	Oil distilled from fruits used for flavoring	Widely cultivated
Arrowroot	Maranta arundinacea	Powdered root used in milk puddings, baked goods	South America
Asafoetida	Ferula asafoetida	Powdered gum from stems and roots used in minute quantities with fish	Middle East
Balm (Melissa)	Melissa officinalis	Oil from leaves used in beverages; leaves used as food flavoring	U.S.; Mediterranean
Basil	Ocimum basilicum	Leaves used in meat dishes, soups, sauces	Mediterranean
Bay	Laurus nobilis	Leaves used in soups, sauces	Europe
Bell pepper	Capsicum frutescens	Dried, diced fruit used in chip dips, salad dressings	Widely cultivated
Bergamot	Monarda didyma	Leaves used with pork (*Note:* A perfume oil obtained from a variety of orange—*Citrus aurantium* var. *bergamia*—is also called bergamot)	North America (*Monarda*); Italy (*Citrus*)
Black pepper	Piper nigrum	Dried fruits used as a condiment	India; Indonesia
Borage	Borago officinalis	Leaves used as a beverage flavoring	England
Burnet	Sanguisorba minor	Used in soups and casseroles	Eurasia
Calamus	Acorus calamus	Powdered rhizome used for flavoring	Europe; Asia; North America
Capers	Capparis spinosa	Flower buds used for flavoring relishes, pickles, sauces	Mediterranean
Caraway	Carum carvi	Seeds used in breads, cheeses; seed oil used in the liqueur kümmel	North America; Europe
Cardamon	Elletaria cardamomum	Dried fruit and seeds used for flavoring baked goods (*Note:* Several false cardamons—*Amomum* spp.—are sold commercially)	India; Sri Lanka; Central America
Cassia	Cinnamomum cassia	Powdered bark used as cinnamon substitute	Southeast Asia
Cayenne pepper	Capsicum spp.	Powdered dried fruits used in chili powder, Tabasco sauce	American tropics
Celery	Apium graveolens	Seeds used in celery salt, soups	Europe; U.S.
Chervil	Anthriscus cerefolium	Used as a parsley substitute	Europe; Near East
Chicory	Cichorium intybus	Ground, dried root added to coffee	Mediterranean
Chives	Allium schoenoprasum	Leaves, bulbs used with sour cream, butter	Widely cultivated
Chocolate	Theobroma cacao	Ground seeds used for flavoring	Africa; South America
Cilantro	Coriandrum sativum	Leaves used in avocado dip and with poultry	Europe
Cinnamon	Cinnamomum zeylanicum	Ground bark used for flavoring baked goods; oil from leaves used as flavoring, clearing agent	Seychelles; Sri Lanka
Citrus	Citrus spp.	Fruits, especially rinds, source of flavoring oil	Mediterranean; South Africa; U.S.
Cloves	Syzygium aromaticum	Dried flower buds used to flavor cooked fruits, toothpaste, candy	Moluccas (Spice Islands)
Coffee	Coffea arabica	Roasted seeds source of mocha-coffee flavoring	Tropics
Coriander	Coriandrum sativum	Ground seed used in German frankfurters, curry powders	Mediterranean
Costmary	Chrysanthemum balsamita	The leaves used sparingly in salads add a mint flavor	Europe, Asia
Cubebs	Piper cubeba	Dried fruits used as seasoning	East Indies
Cumin	Cuminum cyminum	Ground seed used with meats, pickles, cheeses, curry	Mediterranean
Curry	A blend of parts of plants of several different spp.	A spicy condiment containing several ingredients, such as turmeric, cumin, fenugreek, and zedoary	India
Dill	Anethum graveolens	Seeds used in pickling brines; leaves used for seasoning meat loaves, sauces	Europe; Asia
Dittany	Origanum dictamnus	Leaves used as seasoning for poultry, meats	Crete
Eucalyptus	Eucalyptus spp.	Oil from leaves used in toothpastes, flavoring agents	Australia
Fennel	Foeniculum vulgare	Seeds used in baked goods	Europe
Fenugreek	Trigonella foenum-graecum	Oil distilled from seeds used in pickle, chutney, curry powders, imitation maple flavoring	Widely cultivated
Filé (see Sassafras)			

Table A3.6 — Plants Used to Season or Flavor

Spice	Scientific Name of Plants	Part Used; Remarks	Principal Source
Fruit-scented sage	*Salvia dorisiana*	Plant used with beef and fish; it adds a grapefruit-pineapple flavor to the meat	Honduras
Garlic	*Allium sativum*	Fresh or dry bulbs used for meat seasonings	Widely cultivated
Ginger	*Zingiber officinale* and others	Dried rhizomes used for flavoring many foods and drinks	India; Taiwan
Grains of paradise	*Afromomum melegueta*	Seeds used to flavor beverages and medicines	West Africa
Hops	*Humulus lupulus*	Dried inflorescences of female plants used in brewing beer	Europe; North America
Horseradish	*Rorippa armoracia*	Grated fresh root used as a condiment	Europe; North America
Juniper	*Juniperus* spp.	"Berries" used to season beef roasts, poultry, sauces	North America
Lemon balm	*Melissa officinalis*	Leaves give a lemon-mint flavor to stews and desserts	Southern Europe
Licorice	*Glycyrrhiza glabra*	Dried rhizome and root used to flavor pontefract cakes, candies	Middle East
Lovage	*Ligusticum scoticum*	Stems candied; seeds used in pickling sauces; celery substitute	Europe
Mace	*Myristica fragrans*	Aril of seed used for flavoring beverages, foods	Grenada; Indonesia; Sri Lanka
Marigold	*Tagetes* spp.	Petals substituted for saffron in rice dishes, stews	Widely cultivated
Marjoram	*Origanum hortensis*	Leaves used in stews, dressings, sauces	Mediterranean
Mugwort	*Artemisia douglasiana*	Fatty meat flavored with leaves	West Coast of North America
Mustard	*Brassica* spp.	Ground seeds used in meat condiment	Europe; China
Nasturtium	*Tropaeolum majus*	Flowers, seeds, leaves used in salads	Widely cultivated
Nutmeg	*Myristica fragrans*	Seeds used for flavoring foods, beverages	Grenada; Indonesia; Sri Lanka
Oregano	*Origanum vulgare* and others	Leaves used as seasoning with poultry, meats	Europe
Paprika (see *Cayenne pepper*)			
Parsley	*Petroselinum crispum*	Leaves used as meat garnish and flavoring in sauces	Widely cultivated
Peppermint	*Mentha piperita*	Oil from leaves used for food, drink, toothpaste flavoring (much commercial menthol is derived from *Mentha arvensis* grown in Japan)	U.S.; Russia
Pimiento	*Capsicum* spp.	Bright red fruits of a cultivated variety of pepper used in stuffing olives and in cold meats, cheeses	Central and South America
Poppy	*Papaver somniferum*	Seeds used in baking	Widely cultivated
Rosemary	*Rosmarinus officinalis*	Oil from leaves used in perfumes, soaps	Mediterranean
Rue	*Ruta graveolens*	Flavoring for fruit cups, salads	Europe
Saffron	*Crocus sativus*	Dried stigmas used to flavor oriental-style dishes	Spain; India
Sage	*Salvia officinalis*	Leaves used in poultry and meat dressings	Yugoslavia
Salad burnet	*Poterium sanguisorba*	Leaves impart a cucumber-like flavor to salads	Europe; W. Asia
Sarsaparilla	*Smilax* spp.	Roots are source of flavoring for beverages, medicines	American tropics
Sassafras	*Sassafras albidum*	Bark and wood yield flavoring for beverages, toothpaste, gumbo	U.S.
Savory (summer)	*Satureia hortensis*	Leaves used in green bean and bean salads, lentil soup, with fish	Mediterranean
Savory (winter)	*Satureia montana*	Leaves used as seasoning in stuffings, meat loaf, stews	Europe
Scallion	*Allium fistulosum*	Leaves used in wine cookery, soups	Widely cultivated
Sesame	*Sesamum indicum*	Seeds used in baking	Asia
Shallot	*Allium ascalonicum*	Bulbs, leaves used in Colbert butter, wine cookery	Widely cultivated
Southernwood	*Artemisia abrotanum*	Leaves used to flavor cakes	Europe
Star anise	*Illicium verum*	Fruits used in candy and cough drops	China
Stonecrop	*Sedum acre*	Dried leaves (ground) used as pepper substitute	Europe
Sweet cicely	*Osmorhiza* spp.	Leaves have sweet, slight anise flavor; used to flavor dishes and baked goods that incorporate cooked fruits	North America; E. Asia
Sweet woodruff	*Galium odoratum*	Plants used to flavor fruit punches and strawberries	Europe; N. Africa; Asia

Table A3.6 Plants Used to Season or Flavor

Spi	Scientific Name of Plants	Part Used; Remarks	Principal Source
Tansy	*Tanacetum vulgare*	Leaves used in small amounts to flavor baked goods, pancakes, and puddings	Europe; Asia
Tarragon	*Artemisia dracunculus*	Leaves and flowering tops used in pickling sauces	Europe
Thyme	*Thymus vulgaris*	Leaves used in meat and poultry dishes, soups, sauces	Widely cultivated
Tonka bean	*Dipteryx* spp.	Seeds source of flavoring for tobacco; vanilla substitute (now largely synthesized)	American tropics
Turmeric	*Curcuma longa*	Rhizomes powdered and used in curry powders, meat flavoring	India; China
Vanilla	*Vanilla planifolia*	Flavoring extracted from fruits; used in foods, drinks	Malagasay Republic
Wintergreen	*Gaultheria procumbens*	Oil from leaves, bark used as flavoring for confections, toothpaste	U.S.
Zedoary	*Curcuma zedoaria*	Dried rhizome used in liqueurs, curry powders	India

Table A3.7 Plant and Fungal Sources of Natural Dyes

Organism or Dye	Scientific Name of Source	Remarks
Acacia	*Acacia* spp.	Brown dyes from bark and fruits
Alder	*Alnus* spp.	Brownish dyes from bark
Alkanet	*Alkanna tinctoria*	Red dye from roots
Annatto	*Bixa orellana*	Yellow or red dye from pulp surrounding seeds
Bamboo	*Bambusa* spp.	Light green dye from leaves
Barberry	*Berberis vulgaris*	Grayish dye from leaves
Barwood	*Baphia nitida*	Purplish dyes from wood
Bearberry	*Arctostaphylos uva-ursi*	Yellowish dye from leaves
Bedstraw	*Galium* spp.	Light reddish brown dyes from roots
Birch	*Betula* spp.	Light brown to black dyes from bark
Black cherry	*Prunus serotina*	Red dye from bark; gray to green dyes from leaves
Black walnut	*Juglans nigra*	Rich brown dye from bark; brown dye from walnut hulls
Bloodroot	*Sanguinaria canadensis*	Red dye from rhizomes
Blueberry	*Vaccinium* spp.	Blue to gray dye from mature fruits (tends to fade)
Bougainvillea	*Bougainvillea* spp.	Light brownish dyes from floral bracts
Brazilwood	*Caesalpinia* spp.	Reddish dyes from wood
Buckthorn	*Rhamnus* spp.	Green dyes from fruits
Buckwheat	*Fagopyrum esculentum*	Blue dye from stems
Buckwheat (wild)	*Eriogonum* spp.	Dark gold, pale yellow, and beige dyes from stems and flowers
Buffaloberry	*Shepherdia argentea*	Red dye from fruit
Butternut	*Juglans cinerea*	Yellow to grayish brown dyes from fruit hulls
Cocklebur	*Xanthium strumarium*	Dark green dye from stems and leaves
Coffee	*Coffea arabica*	Light brown dye from ground, roasted seeds
Cudbear (Archil)	*Rocella* spp. (lichen)	Red dye obtained by fermentation of thallus
Cutch	*Acacia* spp.; *Uncaria gambir*	Brown to drab green dyes from stem gums
Dock	*Rumex* spp.	Light brown dyes from stems and leaves
Dogwood	*Cornus florida*	Red dye from bark; purplish dye from root
Doveweed	*Eremocarpus setigerus*	Light to olive green dye from entire plant
Dyer's rocket	*Reseda luteola*	Orangish dye from all parts
Elderberry	*Sambucus* spp.	Blackish dye from bark; purple, blue, or dark brown dyes from fruits

Table A3.7 Plant and Fungal Sources of Natural Dyes

Organism or Dye	Scientific Name of Source	Remarks
Eucalyptus	*Eucalyptus* spp.	Beige dyes from bark
Fennel	*Foeniculum vulgare*	Yellow dyes from shoots
Fig	*Ficus carica*	Green dyes from leaves and fruits
Fustic	*Chlorophora tinctoria*	Yellow, bright orange, and greenish dyes from heartwood
Gamboge	*Garcinia* spp.	Yellow dye from resins that ooze from cuts made on stems
Giant reed	*Arundo donax*	Pale yellow dye from leaves
Grape	*Vitis* spp.	Bright yellow to olive green dyes from leaves
Hawthorn	*Crataegus* spp.	Pink dye from ripe fruits
Hemlock	*Tsuga* spp.	Reddish brown dye from bark
Henna	*Lawsonia inermis*	Orange dye from shoots and leaves
Hickory	*Carya tomentosa*	Yellow dye from bark
Hollyhock	*Althaea rosea*	Purplish black dye from flower petals
Horsetail	*Equisetum* spp.	Tan dyes from all green parts
Indigo	*Indigofera tinctoria*	Bright blue dyes from leaves
Kendall green (see *Woadwaxen*)		
Larkspur	*Delphinium* spp.	Blue dyes from petals
Lichens	Many genera and species	Many lichens yield (with various mordants) brilliant shades of yellows, golds, and browns
Litmus	*Rocella tinctoria*	Widely used pink-to-blue pH indicator dye from thallus
Logwood	*Haematoxylon campechianum*	Dark blue-purple dye from heartwood (widely used for staining tissues in microscope slides)
Lokao	*Rhamnus* spp.	Green dye from wood
Lupine	*Lupinus* spp.	Greenish dyes from flowers
Madder	*Rubia tinctorium*	Bright red dye from roots
Madrone	*Arbutus menziesii*	Brown dye from bark
Manzanita	*Arctostaphylos* spp.	Beige to dull yellow dyes from dried fruits
Maple	*Acer* spp.	Pink dye from bark
Marsh marigold	*Caltha palustris*	Yellow dye from petals
Milkweed	*Asclepias speciosa*	Pale yellow dyes from leaves
Morning glory	*Ipomoea violacea*	Gray-green dye from blue flowers
Mullein	*Verbascum thapsus*	Gold dyes from leaves
Oak	*Quercus* spp.	Yellow dye from bark
Onion	*Allium cepa*	Reddish brown dyes from dry outer bulb scales of red onions; yellow dyes from similar parts of yellow onions
Oregon grape	*Berberis aquifolium*	Yellow dyes from roots
Osage orange	*Maclura pomifera*	Yellow, gray, and green dyes from fruits; yellow-orange dye from wood
Peach	*Prunus persica*	Green dyes from leaves
Poke	*Phytolacca americana*	Red dyes from mature fruits
Pomegranate	*Punica granatum*	Dark gold dye from fruit rinds
Prickly lettuce	*Lactuca serriola*	Green dye from leaves
Privet	*Ligustrum vulgare*	Yellow-green dye from leaves; deep gray dye from berries
Quercitron	*Quercus velutina*	Bright yellow dye from bark
Rhododendron	*Rhododendron* spp.	Tan dyes from leaves
Safflower	*Carthamnus tinctorius*	Reddish dye from flower heads
Saffron	*Crocus sativus*	Powerful yellow dye from stigmas
Sage	*Salvia officinalis*	Yellow dye from shoots
St. John's wort	*Hypericum* spp.	Light brownish dyes from leaves
Sandalwood	*Pterocarpus santalinus*	Red dye from wood
Sappanwood	*Caesalpinia sappan*	Red dye from heartwood
Sassafras	*Sassafras albidum*	Orange brown dye from bark
Scotch broom	*Cytisus scoparius*	Yellow dye from all parts of plant
Smoke tree	*Cotinus coggyria*	Orange-yellow dye from wood (dye sometimes called "young fustic")
Smooth sumac	*Rhus glabra*	Grayish brown dye from bark
Tansy	*Tanacetum* spp.	Yellow and green dyes from leaves

Table A3.7 Plant and Fungal Sources of Natural Dyes

Organism or Dye	Scientific Name of Source	Remarks
Toyon	*Heteromeles arbutifolia*	Reddish brown dyes from leaves
Turmeric	*Curcuma longa*	Orangish dye from rhizome
Woad	*Isatis tinctoria*	Blue dye from leaves
Woadwaxen	*Genista tinctoria*	Yellow dye from all parts
Yerba santa	*Eriodictyon californicum*	Rich dark brown dyes from leaves

Additional Reading

Berry, A. 1995. *Know your spices: An alphabetical guide to your spice rack.* Franklin, TN: Runaway Press.

Bliss, A. 1994. *North American dye plants*, rev. ed. New York: Loveland, CO: Interweave Press.

Buchanan, R. 1995. *A dyer's garden: From plant to pot. Growing dyes for natural fibers.* Loveland, CO: Interweave Press.

Elias, J. 1999. *The A to Z guide to healing herbal remedies.* New York: Random House Value Publishing.

Foster, S. and J. A. Duke. 2014. *Peterson Field Guide to Medicinal Plants and Herbs of Central North America,* 3d ed. Boston: Houghton Mifflin Harcourt.

Graedon, J., and T. Graedon. 2001. *The people's pharmacy guide to home and herbal remedies.* New York: St. Martin's Press.

Heatherly, A. N. 1998. *Healing plants: A medicinal guide to native North American plants and herbs.* New York: Lyons Press.

Kallas, J. 2010. *Edible wild plants: Wild foods from dirt to plate.* Layton, UT: Smith, Gibbs.

Null, G. 1998. *The complete encyclopedia of natural healing.* New York: Kensington.

Sanecki, K. N. 1998. *Growing and using herbs.* New York: Sterling.

Schneider, E. 1998. *Uncommon fruits and vegetables.* New York: William Morrow & Co, Inc.

Schultes, R. E., and A. Hoffman. 1980. *The botany and chemistry of hallucinogens,* 2d ed. Springfield, IL: Charles C. Thomas.

Vogel, V. J. 1990. *American Indian medicine.* Norman, OK: University of Oklahoma Press.

For a list of useful plants, edible tropical and uncommon fruits, along with pertinent comments about them, check the following website: http://www.mhhe.com/stern15e, by clicking on Student Edition, then "Useful Plants."

APPENDIX 4
Houseplants and Home Gardening

Growing Houseplants
    Water
    Containers
    Temperatures
    Light
    Humidity
Common Houseplants
    Water
    Temperature
    Light
    Humidity
    Potting Medium
Growing Vegetables
    General Tips on Vegetable Growing
        Seed Germination
        Transplanting
        Cutworms
        Protection against Cold
        Watering
        Fertilizers
        Pests and Diseases

Common Vegetables and Their Nutritional
    Values
    Asparagus
    Beans
    Beets
    Broccoli
    Cabbage
    Carrots
    Cauliflower
    Corn
    Cucumber
    Eggplant
    Lettuce
    Onion
    Peas
    Peppers
    Potatoes
    Spinach
    Squash
    Tomatoes

Pruning
    Fruit Trees
    Grapevines
    Roses
    Raspberries, Blackberries,
        and Their Relatives
    Bonsai
Major Types of Grafting
    Whip, or Tongue, Grafting
    Splice Grafting
    Cleft Grafting
    Side Grafting
    Approach Grafting
    Bridge Grafting
    Bud Grafting (Budding)
    Root Grafting
    Double-Working
Additional Reading

Growing Houseplants

If sales volume is an indication, houseplants have never been more popular in the United States than they are now. Many are easy to grow and will brighten windowsills, planters, and other indoor spots for years if a few simple steps are followed to ensure their health and vigor.

Water

Houseplants are commonly overwatered, resulting in poor soil aeration and root rot (see Table A4.2). As a rule, the surface of the potting medium should be dry to the touch before watering, but the medium should not be allowed to dry out completely unless the plant is dormant. Care should be taken, particularly during the winter, that the water is at room temperature. Rain water, if available, is preferred over tap water, particularly if the water has a high mineral content or is chlorinated. Broad-leaved plants should periodically have house dust removed with a damp sponge (never use detergents to clean surfaces—they remove protective waxes). Many plants benefit from a daily misting with water, particularly in heated rooms.

Containers

Over time, plants may develop too extensive a root system for the pots in which they are growing (commonly called *becoming root bound*). Nutrients in the potting medium may become exhausted, salts and other residues from fertilizers and water may build up to the point of inhibiting growth, or the plants themselves may produce substances that accumulate until they interfere with the plant's growth. To resolve these problems, the plants should be periodically repotted and, if necessary, divided at the time of repotting.

Temperatures

Most houseplants don't thrive where the temperatures are either too high or too low (see Table A4.1). In general, they tend to do best with minimum temperatures of about 13°C (55°F) and maximum temperatures of about 29°C (84°F). Many houseplants that prefer warmer temperatures while actively growing also benefit from a rest period at lower temperatures after flowering.

Light

Next to overwatering, insufficient light most commonly contributes to the decline or death of houseplants (see Table A4.2). This does not mean that houseplants do better in direct sunlight—such light frequently damages them; filtered sunlight is usually better for the plant than the light available in the middle of the room. Some plants can also thrive in artificial light of appropriate quality. Ordinary incandescent bulbs have too little light of blue wavelengths, and ordinary fluorescent tubes emit too little red light. A combination of the two, however, works very well. Generally, the wattage of the incandescent bulbs should be only one-fourth that of the fluorescent tubes in such a combination. Several types of fluorescent tubes specially balanced to imitate sunlight are also available. LED plant grow lights offer energy efficiency and low heat output.

Humidity

Dry air is hard on most houseplants. The level of humidity around the plants can be raised by standing the pots in dishes containing gravel or crushed rock to which water has been added. The humidity level can also be raised through the use of humidifiers, which come in a variety of sizes and capacities. Daily misting, as mentioned previously, can also help.

Common Houseplants

Explanation of symbols given with each plant:

Water

 = needs little water (applies primarily to cacti and succulents; these plants store water in such a way that the soil can be completely dry for a week or two without them being adversely affected)

 = water regularly but not excessively; wait until the potting medium surface is dry to the touch before watering

 = immerse pot in water for a few minutes each week and water frequently, never allowing the potting medium to become dry; do not, however, leave the base of the pot standing in water

 = little to regular

 = regular to frequent

Minerals in hard water are taxing on houseplants and commercial water softeners do not improve water for the plants. Use rain water or filtered water if at all possible; otherwise repot more often.

Temperature

 = cool. Maximum 13°C–16°C (55°F–61°F); minimum 5°C–7°C (41°F–45°F)

 = cool to medium. Maximum 18°C–21°C (65°F–70°F); minimum 10°C–13°C (50°F–55°F)

 = medium. Maximum 22°C–29°C (72°F–84°F); minimum 14°C–17°C (57°F–63°F)

 = medium to warm. Maximum 30°C–32°C (86°F–89°F); minimum 18°C–20°C (64°F–68°F)

 = warm. Maximum 33°C–37°C (91°F–99°F); minimum 21°C–24°C (69°F–75°F)

Many houseplants are native to the tropics, where they thrive under year-round warm temperatures, while cacti and succulents thrive in warm summers and cool winters, with wide daily temperature fluctuations. The closer one is able to imitate a plant's natural environment, the better the plant will grow (see Table A4.1).

Light

 = needs shading or indirect daylight

 = prefers bright light but needs to be screened from direct sunlight

 = prefers direct sunlight

 = prefers shading to bright

 = prefers bright to direct light

As mentioned previously, improper lighting is second only to overwatering as a cause of problems for houseplants. They are frequently given too little light or occasionally too much. A south-facing windowsill may be ideal for certain plants in midwinter but excessively bright in midsummer; conversely, a north-facing windowsill may have enough light for certain plants in midsummer but not in midwinter. Adjustable screens permit manipulation of daylight to suit the plants involved.

Humidity

 = tolerates dry air

 = tolerates dry to regular

 = will tolerate the air in most houses provided it is mist-sprayed occasionally

 = needs regular to humid

 = needs high humidity; use a humidifier if possible

Virtually all plants with or symbols benefit from having a pan of gravel with water beneath the pot.

Potting Medium

 = requires a porous, slightly acidic medium that drains immediately

 = requires a loam that is slightly alkaline (e.g., a mixture of sand and standard commercial potting medium)

 = requires a peaty potting mixture and acidic fertilizer

Appendix 4

Table A4.1 — Environments Suitable for Common Houseplants

Plant	Scientific Name	Environmental Requirements	Remarks
Abutilon	*Abutilon* spp. and hybrids		Avoid rangy growth by pinching out branch tip buds
Aechmea	*Aechmea fasciata*		See **Bromeliad**; produces side shoots that should be propagated as main plant dies after flowering
African lily	*Agapanthus* spp.		Do not repot until pot is full; keep cool in winter
African violet	*Saintpaulia* spp.		Let rest under cooler conditions after flowering; dislikes cold water
Agave	*Agave* spp.		Keep cool and dry in winter
Algerian ivy	*Hedera canariensis*		Resembles a larger-leaved English ivy; leaves often variegated
Aloe	*Aloe* spp.		Keep cool and dry in winter
Aluminum plant	*Pilea cadierei*		Plants do not usually survive long in houses
Amaryllis	*Amaryllis* spp.		Let leaves die back in fall; put bulb in cool, dark place until early spring; then repot, water, and fertilize weekly
Amazon lily	*Eucharis grandiflora*		Needs warm greenhouse; flowers strong-scented
Anthurium	*Anthurium* spp.		If it grows but doesn't flower, try putting it in a cooler location for a few weeks
Aphelandra	*Aphelandra squarrosa*		Mist-spray frequently; fertilize regularly
Arabian violet	*Exacum affine*		Needs well-fertilized soil with humus content
Aralia (see Fatsia)			
Asparagus fern	*Asparagus plumosus*		Not a true fern; repot annually; fertilize weekly
Aspidistra	*Aspidistra* spp.		Sometimes called "cast iron plant" because it can stand neglect
Aucuba	*Aucuba japonica*		Must be kept cool in winter
Avocado	*Persea* spp.		Easily propagated from seed; provides good greenery but will not produce fruit indoors
Azalea	*Rhododendron* spp.		Needs acid fertilizer; avoid warm locations
Bamboo (dwarf)	*Bambusa angulata*		Needs good air circulation, bright light
Basket grass	*Oplismenus hirtellus*		Hanging-basket cultivated variety sometimes sold as *Panicum variegatum*

Table A4.1 — Environments Suitable for Common Houseplants

Plant	Scientific Name	Remarks
Begonia	*Begonia* spp.	Easily propagated from leaves; repot regularly
Bilbergia	*Bilbergia* spp.	See **Bromeliad**; tough plant that can tolerate some neglect
Birdcatcher plant	*Pisonia umbellifera*	Sticky exudate on fruits attracts birds in the plant's native habitat of New Zealand; strictly a foliage plant in houses
Bird of paradise plant	*Strelitzia reginae*	Can be grown outdoors in milder climates
Bird's nest fern	*Asplenium nidus*	Produces a spongelike mass of roots at base; requires much water and regular fertilizing
Black-eyed Susan	*Thunbergia alata*	Annual climbing vine; grow from seed
Bloodleaf	*Iresine herbstii*	Easily propagated
Boston fern	*Nephrolepis exaltata*	Needs regular watering and fertilizing
Brake	*Pteris cretica*	See **Ferns** for propagation notes
Bromeliads	Many species	These plants absorb virtually all their water and nutrients through their leaves; they should not be placed in regular potting soil (regular potting soil usually kills them) nor should they be watered with high-calcium-content water; they produce offshoots that should be propagated as main plant dies after flowering
Cacti	Many species	Contrary to popular belief, these slow-growing plants should not be grown in pure sand; add some humus to potting mixture; withhold water and keep cool in winter
Caladium	*Caladium* spp.	Must have high humidity; keep root at no less than 18°C (65°F) in pot during winter
Calceolaria	*Calceolaria crenatiflora*	Discard after flowering
Calla lily	*Zantedeschia aethiopica*	After flowering, allow plant to dry up; repot in fall and start over
Cape jasmine	*Gardenia jasminoides*	Night temperatures below 22°C (72°F) needed to initiate flowering; needs cool temperatures in winter
Carpet plant	*Episcia lilacina*	Needs minimum of 70% humidity
Carrion flower	*Stapelia* spp.	Cactuslike plants with foul-smelling flowers

Table A4.1 — Environments Suitable for Common Houseplants

Plant	Scientific Name	Remarks
Century plant (see Agave)		
Chinese evergreen	Aglaonema costatum	Needs warm temperatures and much water all year
Chrysanthemum	Chrysanthemum spp.	Plants may be artificially dwarfed through use of chemicals; flowering initiated by short days
Cineraria	Senecio x hybridus (other plants called cinerarias include Chrysanthemum ptarmiciflorum, Senecio vira-vira, and Senecio cineraria)	Needs cool temperatures; discard after flowering
Cliff brake	Pellaea rotundifolia	Hanging-basket fern; needs minimum temperature above 10°C (50°F)
Coffee	Coffea arabica	Handsome foliage plant that will produce fruit if self-pollinating variety is obtained
Coleus	Coleus blumei, C. hybridus, and others	To control size, restart plants from cuttings annually
Columnea	Columnea spp.	See **African violet** for culture conditions
Copperleaf	Acalypha wilkesiana	Seldom survives average house environment for long
Corn plant	Dracaena massangeana	Uses much water when large; easy to grow
Croton	Codiaeum spp.	Needs constant high humidity and bright light
Crown of thorns	Euphorbia milii and E. splendens	Deviation from watering routine may result in loss of leaves, but plant generally recovers
Cyclamen	Cyclamen spp.	Fertilize weekly; keep cool; withhold water after flowering for few weeks, then start over
Donkey tail	Sedum morganianum	Keep cool and dry in winter
Dracaena	Dracaena spp.	Many kinds—all easy to grow and tolerant of some neglect
Dumbcane	Dieffenbachia spp.	Needs regular fertilizing; keep away from small children (poisonous)
Dwarf banana	Musa cavendishii	Will produce small edible bananas if given enough light, water, and humidity
Dwarf cocos palm	Microcoleum weddelianum	Keep temperature above 18°C (65°F) at all times
Echeveria	Echeveria spp.	Keep cool in winter
English ivy	Hedera helix	Needs cool temperatures to grow at its best but will usually survive higher temperatures

Table A4.1 — Environments Suitable for Common Houseplants

Plant	Scientific Name	Environmental Requirements	Remarks
False aralia	*Dizygotheca elegantissima*		Benefits from frequent mist-spraying
Fatshedera (= hybrid between *Fatsia* and *Hedera*)	*Fatshedera lizei*		Climbing shrub
Fatsia	*Fatsia japonica*		Also called *Aralia*
Ferns	Many species		Water regularly; propagate from spores or rhizomes
Figs:			
Climbing fig	*Ficus pumila*		Damp-sponge leaves regularly
Fiddleleaf fig	*Ficus lyrata*		Damp-sponge leaves regularly
Weeping fig	*Ficus benjamina*		Damp-sponge leaves regularly
Fingernail plant	*Neoregelia* spp.		See **Bromeliads**; name from red tips of leaves
Fittonia	*Fittonia* spp.		Strictly terrarium plants—humidity too low elsewhere
Flame violet	*Episcia cupreata*		Add charcoal and peat to potting medium
Flowering maple	*Abutilon pictum*		Needs bright light to flower
Fuchsia	*Fuchsia* spp.		Soil must be alkaline
Gardenia (*see* Cape jasmine)			
Geranium	*Pelargonium* spp.		Make cuttings annually and discard parent plants each fall; keep cool through winter; available with various scents, including orange, rose, or coconut
Gloxinia	*Sinningia speciosa*		Fertilize heavily and water frequently; after flowering, withhold water and keep bulb cold for few weeks
Goldfish plant	*Columnea* spp.		Pot in mixture of leaf mold, fern bark, peat moss, and charcoal; use only rain water or filtered water
Goosefoot plant (not related to true goosefoots in the Goosefoot family—Chenopodiaceae)	*Nephthytis* spp.		Soil should have high organic content and be kept evenly moist; humidity must be maintained
Grape ivy	*Rhoicissus rhomboidea*		Tolerates low light better than most plants
Gynura (*see* Velvetleaf)			
Hart's tongue fern	*Phyllitis scolopendrium*		Doesn't thrive if water is of poor quality
Haworthia	*Haworthia* spp.		Aloe-like plants that need minimum temperatures above 10°C (50°F) in winter
Hen and chickens	*Sempervivum tectorum* and hybrids		Keep cool in winter

Appendix 4

Table A4.1 — Environments Suitable for Common Houseplants

Plant	Scientific Name	Environmental Requirements	Remarks
Hibiscus	*Hibiscus rosa-sinensis*		Fertilize weekly; does better outdoors in mild climates
Hippeastrum (see Amaryllis)			
Holly fern	*Cyrtomium falcatum*		Relatively tough fern; keep cool in winter
Houseleek	*Sempervivum* spp.		Keep cool in winter
Hydrangea	*Hydrangea* spp.		Prune after flowering; keep cool in winter; blue-flowering plant can be converted to pink by changing the soil to acid, and vice versa (**Caution: Leaves and buds are poisonous**)
Hypocyrta	*Nematanthus* spp.		Plant several in a hanging basket
Hypoestes (see Pink polka dot plant)			
Impatiens	*Impatiens* spp.		Exceptionally easy to propagate from cuttings
Ivy-arum	*Epipremnum aureum*		Can tolerate some neglect
Jade plant	*Crassula argenta* and others		Keep cool in winter
Jamaican pansy	*Achimenes* spp.		Requires treatment similar to that for **Gloxinia**
Kaffir lily	*Clivia miniata*		Save the plant's energy by removing flowers as they wither
Kalanchoë	*Kalanchoë* spp.		Withhold water and fertilizer for a few weeks after flowering, then repot and start over
Lantana	*Lantana camara*		Fertilize twice a month; can be espaliered
Madagascar jasmine (see Stephanotis)			
Maidenhair fern	*Adiantum* spp.		Mist-spray regularly
Meyer fern (not a true fern)	*Asparagus densiflora* var. *meyeri*		Fertilize regularly; repot annually
Moneywort	*Lysimachia nummularia*		Hanging-pot plant; needs bright light
Moonstones	*Pachyphytum* spp.		Keep cool in winter
Moses in the cradle	*Rhoeo* spp.		Can tolerate some neglect
Mother-in-law's tongue (see Sansevieria)			
Mother of thousands	*Saxifraga* spp.		Also called *Saxifrage;* hanging plant; plantlets formed on runners can be removed and grown separately
Neanthe palm	*Chamaedorea elegans*		Sometimes also called *Parlor palm;* stays less than 1 meter tall

Table A4.1 Environments Suitable for Common Houseplants

Plant	Scientific Name	Environmental Requirements	Remarks
Nerine	*Nerine* spp.		Water freely until flowering completed, then rest bulbs for a few months
Norfolk Island pine	*Araucaria* spp.		Needs cold temperatures 2°C–3°C (36°F–38°F) in winter
Octopus tree	*Schefflera octophylla*		Does best under cool conditions
Oleander	*Nerium oleander*		Keep pot cool in winter for better flowering; keep away from small children (very poisonous)
Orchid	Thousands of species	No single set of environmental conditions applies	Contrary to popular belief, the common *Cattleya* and related orchids do not need high temperatures and humidity; most can get along with a minimum temperature of 13°C (56°F) at night and a minimum humidity of 40%; most need bright light; they should never be placed in soil; pot them in sterilized pots with chips of fir bark or shreds of tree fern bark (*see* "Additional Reading" at the end of this appendix for culture references)
Oxtongue	*Gasteria* spp.		Can tolerate some neglect; needs a cool and relatively dry winter to flower
Palms	Many species		Use deep pots; fertilize regularly
Parlor palm	*Chamaedorea elegans* (*see also* **Neanthe palm**)		Palm is relatively easy to grow
Peperomia	*Peperomia* spp.		Many varieties; keep warm, humid; fertilize regularly
Persian violet	*Exacum affine*		Needs good air circulation
Philodendron	*Philodendron* spp.		Relatively tough plants; repot each spring
Piggyback plant	*Tolmiea menziesii*		Plantlets formed on leaves can be separated and propagated
Pilea (*see* Aluminum plant)			
Pineapple	*Ananas comosus*		*See* **Bromeliads**; easily grown from the top of a pineapple; if plant has not flowered after one year, enclose it in a plastic bag with a ripe apple for a few days (ethylene from apple should initiate flowering); no temperature below 15°C (59°F)
Pink polka dot plant	*Hypoestes sanguinolenta*		Susceptible to diseases and pests; usually does not last long as a houseplant
Pittosporum	*Pittosporum* spp.		Put several cuttings in one pot for bushy appearance

Table A4.1 — Environments Suitable for Common Houseplants

Plant	Scientific Name	Environmental Requirements	Remarks
Pleomele (see Dracaena)			
Poinsettia	Euphorbia pulcherrima		After flowering, let plant dry under cool conditions until it loses its leaves, place in total darkness for a month, then restart
Prayer plant	Maranta leuconeura		Name derived from fact that leaves fold together in evening
Primrose	Primula spp.		Needs much water; does well outside in cool weather
Pteris (see Brake)			
Purple tiger	Calathea amabilis		Use pots that are broader than deep
Rosary plant	Ceropegia woodii		Hanging-pot plant whose potting medium must drain well or plant will not survive
Rubber plant	Ficus elastica		Do not overwater!
Sago palm	Cycas revoluta		Very slow growing (not a palm but a gymnosperm); never allow to dry out, but be certain water drains
Sansevieria	Sansevieria spp.		Perhaps the toughest of all houseplants—nearly indestructible
Satin pothos	Scindapsus pictus		Basket or pot plant
Schefflera	Schefflera spp.		Plants can get quite large if growth is not controlled
Screw pine	Pandanus spp.		If given space, can become large; mist-spray often
Selaginella (see Spike moss)			
Sensitive plant	Mimosa pudica		Leaves fold when touched; does not usually last more than a few months in most houses
Sentry palm	Howea fosterana		Palm that is easy to grow
Setcreasea	Setcreasea pallida		Do not overwater
Shrimp plant	Beloperone guttata		Winter temperatures should be above 15°C (59°F)
Spathe flower	Spathiphyllum spp.		Prefers warm winters and even warmer summers; potting soil should have high organic content
Spider plant	Chlorophytum comosum		Plantlets formed at tips of stems can be propagated separately
Spiderwort	Tradescantia spp.		Easy to grow; do not overwater
Spike moss	Selaginella spp.		Can become a weed in greenhouses

Table A4.1 Environments Suitable for Common Houseplants

Plant	Scientific Name	Environmental Requirements	Remarks
Splitleaf philodendron	*Monstera deliciosa*		Plant adapts to various indoor locations quite well
Sprenger fern (not a true fern)	*Asparagus densiflora* var. *sprengeri*		Fertilize weekly; repot annually
Staghorn fern	*Platycerium* spp.		Tough plant; immerse in water weekly
Stephanotis	*Stephanotis floribunda*		Use very little fertilizer; keep cool in winter but water sparingly
Stonecrop	*Sedum* spp., *Crassula* spp.		Keep cool in winter
Stove fern	*Pteris cretica*		Water and fertilize regularly
String-of-pearls	*Senecio rowleyanus*		Keep cool in winter
Stromanthe	*Stromanthe* spp.		Needs warm, shaded environment
Sundew	*Drosera* spp.		Sterilize pots; grow only on sphagnum moss (plants die soon in soil)
Syngonium	*Syngonium podophyllum*		Repot annually in spring
Tillandsia	*Tillandsia* spp.		See **Bromeliads**; best-known species is called *Spanish moss*, which does best suspended from wires or other plants
Ti plant	*Cordyline terminalis*		Needs high humidity; seems to do better with other plants in pot
Treebine	*Cissus antarctica*		Plant does not do well in acidic potting medium
Umbrella plant	*Cyperus* spp.		One of very few plants that need to stand in water
Velvetleaf	*Gynura sarmentosa*		Gets "stringy" but is easily restarted from cuttings; flowers have unpleasant odor
Venus's flytrap	*Dionaea muscipula*		Sterilize pots; grow only in sphagnum moss; repot annually
Vriesia	*Vriesia* spp.		See **Bromeliads**
Wandering Jew	*Zebrina pendula*		Easy to grow; do not overwater
Wax plant	*Hoya* spp.		Climber; leave pot in one place—does not like to be moved

Zebrina (*see* Wandering Jew)

Table A4.2 — Common Ailments of Houseplants

Problem Symptoms	Possible Causes
Wilting or collapse of whole plant	Lack of water; too much heat; too much water or poor drainage resulting in root rot
Yellowish or pale leaves	Insufficient light; too much light; microscopic pests (especially spider mites); too much or too little fertilizer
Brown, dry leaves	Humidity too low; too much heat; poor air circulation; lack of water
Tips and margins of leaves brown	Mineral content of water; drafts; too much sun or heat; too much or too little water
Ringed spots on leaves	Water too cold
Leaves falling off	Improper watering or water too cold; excessive use of fertilizer or wrong fertilizer; too much sun or, if lower leaf drop only, too little light
Stringy growth	Needs more light; too much fertilizer
Base of plant soft or rotting	Overwatering
No flowers or flower buds drop	Too much or too little light; night temperatures too high
Water does not drain	Drain hole plugged; potting mixture has too much clay
Mildew present	Fungi present—arrest with sulfur dust

Common Pests	Controls
Aphids	Wash off under faucet or spray with soapy (not detergent) water; pyrethrum or rotenone sprays also effective
Mealybugs	Remove with cotton swabs dipped in alcohol; spray with Volck oil
Scale insects	Remove by hand; spray with Volck oil
Spider mites	Use sprays containing small amounts of xylene (act as soon as possible—spider mites multiply very rapidly)
Thrips	Spray with pyrethrum/rotenone or Volck oil sprays
White flies	Spray with soapy water or pyrethrum/rotenone sprays every 4 days for 2 weeks (only the adults are susceptible to the sprays)

For additional controls, see Appendix 2, "Biological Controls."

Growing Vegetables
General Tips on Vegetable Growing

Seed Germination

Many vegetable plants can be direct seeded into the garden. Carefully follow the directions on the seed packet for depth and spacing requirements. For some vegetables, especially in northern states, the growing season is not long enough for direct seeding. In that case, seeds can be started indoors. It is best to sow the seeds in commercial potting mix by sprinkling them on the surface and covering them lightly with potting mix or vermiculite. When seedlings have reached the first true leaf stage, thin out the weakest plants. Make sure the container has drainage holes. Supplementary light is usually needed for adequate seedling growth in the spring.

Transplanting

Seedlings can be transplanted into the garden when they have established root systems, the soil is warm enough, and the danger of frost is past. Before seedlings are put into a garden, they should be hardened off. For a few hours a day, place them outside; bring them in at night. Increase to several hours over the course of a week. Then, the seedlings are ready for transplanting. Make a hole in the ground, place the plant in at the same level as it grew in the pot, add soil around the plant, and press down on the soil. It is important for the roots to make contact with the soil. Make sure to water well, especially while the transplant is becoming established. A layer of straw or other mulch will help to retain moisture and prevent weed seeds from germinating.

Cutworms

Cutworms forage at or just beneath the surface of the soil. Their damage to young seedlings can be minimized by a collar placed around each plant. Tuna cans with both ends removed make effective collars when pressed into the ground to a depth of about 1 to 2 centimeters (0.4 to 0.8 inch).

Protection against Cold

Some vegetables, such as peas and spinach, grow best under cool spring temperatures and can tolerate light frosts. Others, such as tomatoes and corn, require warm soil and are easily damaged by frost. It is important to read package labels to determine optimum temperatures for growth.

Watering

Proper watering promotes healthy growth. It is much better to water an area deeply every few days than to wet the surface daily. Shallow daily watering promotes root development near the surface, where midsummer heat can damage the root system. Conversely, too much watering can leach minerals out of the topsoil and can lead to poor root growth if the soil is not allowed to drain between waterings.

Fertilizers

Composted manures, bone and blood meals, and other organic fertilizers are preferred by many gardeners, as they release nutrients more slowly than commercial fertilizers. Regular application of a dilute general plant fertilizer is also typically sufficient for most vegetables. Sometimes specialized fertilizers that enhance flower and fruit formation can be useful. Calcium is especially important in preventing blossom end rot, a common problem with tomatoes. Slow release fertilizer pellets can provide a steady supply of nutrients across the growing season.

Pests and Diseases

The best way to control pests and diseases is to keep plants healthy and prevent the accumulation of inoculum. Remove dead leaves, thin properly to allow air movement between plants, and dispose of diseased plant parts. Synthetic pesticides can be very effective when used properly. Biological pest controls are discussed in Appendix 2.

Common Vegetables and Their Nutritional Values

Note: The nutritional values (NV) given are per 100 grams (3.5 ounces), edible portion, as determined by the United States Department of Agriculture.

Asparagus

NV (spears cooked in water): 20 calories; protein, 2.2 gm; fats, 0.2 gm; vit. A, 900 I.U.; vit. B_1, 0.16 mg; vit. B_2, 0.18 mg; niacin, 1.5 mg; vit. C, 26 mg; fiber, 0.7 gm; calcium, 21 mg; phosphorus, 50 mg; iron, 0.6 mg; sodium, 1.0 mg; potassium, 208 mg

Asparagus can be started from seed, but time until the first harvest can be reduced by a year or two if planting begins with 1-year-old root clusters of healthy, disease-resistant varieties. Asparagus requires little care if appropriate preparations are made before planting in the permanent location. Seeds should be sown sparsely, and the seedlings thinned to about 7.5 centimeters (3 inches) apart. Before transplantation the following spring, dig a trench 30 to 60 centimeters (12 to 24 inches) deep and about 50 centimeters (20 inches) wide in an area that receives full sun, usually along one edge of the garden. If the soil is heavy, place crushed rock or gravel on the bottom of the trench to provide good drainage. Add a layer of steer manure about 10 centimeters (4 inches) thick, followed by about 7.5 centimeters (3 inches) of rich soil that has been prepared by thorough mixing with generous quantities of steer manure and bone meal. Place root clusters about 45 centimeters (18 inches) apart in the trench and cover with about 15 centimeters (6 inches) of prepared soil (be sure not to allow root clusters to dry out). As the plants grow, gradually fill in the trench. If 1-year-old roots are planted, wait for 2 years before harvesting tips; if 2-year-old roots are planted, some asparagus may be harvested the following year. In all cases, no harvesting should be done after June so that the plants may build up reserves for the following season. Cut shoots below the surface but well above the crown before the buds begin to expand. Cut all stems to the ground after they have turned yellow later in the season.

Beans

String, or Snap, Beans

NV (young pods cooked in water): 25 calories; protein, 1.6 gm; fats, 0.2 gm; vit. A, 540 I.U.; vit. B_1, 0.07 mg; vit. B_2, 0.09 mg; niacin, 0.5 mg; vit. C, 12 mg; fiber, 1.0 gm; calcium, 50 mg; phosphorus, 37 mg; iron, 0.6 mg; sodium, 4 mg; potassium, 151 mg

String, or snap, beans are warm-weather plants, although they can be grown almost anywhere in the United States. Wait until all danger of frost has passed and the soil is warm. Prepare the soil, preferably the previous winter, by digging to a depth of 30 centimeters (12 inches) and mixing in aged manure and bone meal. Pulverize the soil just before sowing; if soil has a low pH, add lime. Plant seeds thinly in rows about 40 to 50 centimeters (16 to 20 inches) apart; thin plants to 10 centimeters (4 inches) apart when the first true leaves have developed. Beans respond unfavorably to a very wet soil—do not overwater! In areas with hot summers, beans also prefer some light shade, particularly in mid-afternoon. In most soils, as the bean plants grow, nitrogen-fixing bacteria invade the roots and supplement the nitrogen supply. If a legume has not been planted in the garden before, it may be helpful to add Rhizobia to the soil. Early vigorous growth can be enhanced by inoculating the seeds with such bacteria, which are available commercially in a powdered form. To maintain a continuous supply of green beans, plant a new row every 2 to 3 weeks during the growing season until 2 months before the first predicted frost. Cultivate regularly to control weeds, taking care not to damage root systems. Do not harvest or work with beans while the plants are wet, as this may invite disease problems.

Pole Beans

NV similar to those of string beans

Soil preparation and cultivation are the same as for string beans. Plant beans in hills around poles that are not less than 5 centimeters (2 inches) in diameter and at least 2 meters (6.5 feet) tall. As beans twine around their supports, it helps to tie them to the support with plastic tape as they grow. If harvested before the pods are mature, pole beans will produce over a longer period of time than bush varieties.

Lima Beans

NV (immature seeds cooked in water): 111 calories; protein, 7.6 gm; fats, 0.5 gm; vit. A, 280 I.U.; vit. B_1, 0.18 mg; vit. B_2, 0.10 mg; niacin, 1.3 mg; vit. C, 17 mg; fiber, 1.8 gm; calcium, 47 mg; phosphorus, 121 mg; iron, 2.5 mg; sodium, 1.0 mg; potassium, 422 mg

Lima beans take longer to mature than other beans and are more sensitive to wet or cool weather. They definitely need warm weather to do well. Prepare soil and cultivate as for string beans.

Soybeans

NV (dry, mature seeds): 403 calories; protein, 34.1 gm; fats, 17.7 gm; vit. A, 80 I.U.; vit. B_1, 1.10 mg; vit. B_2, 0.31 mg; niacin, 2.2 mg; vit. C;[1] fiber, 4.9 gm; calcium, 226 mg; phosphorus, 554 mg; iron, 8.4 mg; sodium, 5.0 mg; potassium, 1,677 mg

Prepare soil and cultivate as for string beans.

Broad (Fava) Beans

NV (dry, mature seeds): 338 calories; protein, 25.1 gm; fats, 1.7 gm; vit. A, 70 I.U.; vit. B_1, 0.5 mg; vit. B_2, 0.3 mg; niacin, 2.5 mg; vit. C;[1] fiber, 6.7 gm; calcium, 47 mg; phosphorus, 121 mg; iron, 2.5 mg; sodium, 1.0 mg; potassium, 422 mg

Unlike other beans, broad beans need cool weather for their development. Sow as early as possible (in mild climates, they may be sown in the fall, as they can withstand light frosts). Since the plants occupy a little more space than bush beans, plant in rows about 0.9 to 1 meter (3 feet or more) apart and thin to about 20 centimeters (8 inches) apart in the rows. After the first pods mature, pinch out the tips to promote bushier development. To most palates, broad beans do not taste as good as other types of beans.

Beets

NV (cooked in water): 32 calories; protein, 1.1 gm; fats, 0.1 gm; vit. A, 20 I.U.; vit. B_1, 0.03 mg; vit. B_2, 0.04 mg; vit. C, 6 mg; fiber, 0.8 gm; calcium, 14 mg; phosphorus, 23 mg; iron, 0.5 mg; sodium, 43 mg; potassium, 208 mg

Beets will grow in a variety of climates but do best in cooler weather. They can tolerate light frosts and can be grown on a variety of soil types, although they prefer a sandy loam supplemented with well-aged organic matter. As with any bulb or root crop, they develop best in soil that is free of rocks and lumps.

Beet "seeds" are really fruits containing several tiny seeds. Plant them in rows 40 to 60 centimeters (16 to 24 inches) or more apart and thin to about 10 centimeters (4 inches) apart in the rows after germination. After harvesting, the beets will keep in cold storage for up to several months. The leaves, if used when first picked, make a good substitute for spinach.

Broccoli

NV (spears cooked in water): 26 calories; protein, 3.1 gm; fats, 0.3 gm; vit. A, 2,500 I.U.; vit. B_1, 0.09 mg; vit. B_2, 0.2 mg; niacin, 0.8 mg; fiber, 1.5 gm; calcium, 88 mg; phosphorus, 62 mg; iron, 0.8 mg; sodium, 10 mg; potassium, 267 mg

Broccoli is a cool-weather plant that will thrive in any well-prepared soil, providing that it has not been heavily fertilized just prior to planting (fresh fertilizer promotes rank growth). The plants can stand light frosts and are planted in both the spring and fall in areas with mild climate. Although broccoli may continue to produce during the summer, most growers prefer not to keep the plants going during warm seasons because of the large numbers of pest insects they may attract.

Sow seeds indoors and transplant outdoors after danger of killing frosts has passed. Place plants about 0.9 to 1 meter (3 feet or more) apart and keep well watered. Keep area weeded and pests under control. Harvest heads (bundles of spears) while they are still compact. Smaller heads will develop very shortly after the first harvest; if these are removed regularly, the plants will continue to produce for some time, although the heads become smaller as the plants age.

Cabbage

NV (raw): 24 calories; protein, 1.3 gm; fats, 0.2 gm; vit. A, 130 I.U.; vit. B_1, 0.05 mg; vit. B_2, 0.05 mg; niacin, 0.3 mg; vit. C, 47 mg; fiber, 1.0 gm; calcium, 49 mg; phosphorus, 29 mg; iron, 0.4 mg; sodium, 20 mg; potassium, 233 mg

Growth requirements of cabbage are similar to those of broccoli.

Carrots

NV (raw): 42 calories; protein, 1.1 gm; fats, 0.2 gm; vit. A, 11,000 I.U.; vit. B_1, 0.6 mg; vit. B_2, 0.5 mg; niacin, 0.6 mg; vit. C, 8 mg; fiber, 1.0 gm; calcium, 37 mg; phosphorus, 36 mg; iron, 0.7 mg; sodium, 47 mg; potassium, 341 mg

Carrots are hardy plants that can tolerate a wide range of climate and soils, but the soil must be well prepared, free of rocks and lumps, and preferably not too acidic. The seeds are slow to germinate. Plant in rows a little more than

---

[1] Values not available.

30 centimeters (12 inches) apart and thin seedlings to about 5 centimeters (2 inches) apart in the rows. Weed the rows regularly until harvest. Carrots keep well in below-ground storage containers when freezing weather arrives.

Cauliflower

NV (cooked in water): 22 calories; protein, 2.3 gm; fats, 0.2 gm; vit. A, 60 I.U.; vit. B_1, 0.09 mg; vit. B_2, 0.08 mg; niacin, 0.6 mg; vit. C, 55 mg; fiber, 1.0 gm; calcium, 21 mg; phosphorus, 42 mg; iron, 0.7 mg; sodium, 9 mg; potassium, 206 mg

Growth requirements of cauliflower are similar to those of broccoli except that heavier fertilizing is required. As cauliflower heads develop, protect them from the sun by tying the larger leaves over the tender heads. Harvest while the heads are still solid.

Corn

NV (cooked sweet corn kernels): 83 calories; protein, 3.2 gm; fats, 1.0 gm; vit. A, 400 I.U. (yellow varieties; white varieties have negligible vit. A content); vit. B_1, 0.11 mg; vit. B_2, 0.10 mg; niacin, 1.3 mg; vit. C, 7 mg; fiber, 0.7 gm; calcium, 3 mg; phosphorus, 89 mg; iron, 0.6 mg; sodium, trace; potassium, 165 mg

There are several types of corn (e.g., popcorn, flint corn, dent corn), but sweet corn is the only type grown to any extent by home gardeners. It can be grown in any location where there is at least a 10-week growing season and warm summer weather.

Corn does best in a fertile soil that should be prepared by mixing with compost and liberal amounts of chicken manure or fish meal. Since corn is wind-pollinated, it can be helpful to grow the plants in several short rows at right angles to the prevailing winds rather than in a single long row. For best results, use only fresh seeds and plant in rows 60 centimeters (24 inches) apart for dwarf varieties and 90 centimeters (36 inches) apart for standard varieties. Thin to 20 to 30 centimeters (8 to 12 inches) apart in the rows after the plants have produced three to four leaves. Cultivate frequently to control weeds. The corn is ready to harvest when the silks begin to wither.

Cucumber

NV (raw, with skin): 15 calories; protein, 0.9 gm; fats, 0.1 gm; vit. A, 250 I.U.; vit. B_1, 0.03 mg; vit. B_2, 0.04 mg; niacin, 0.2 mg; vit. C, 11 mg; fiber, 0.6 mg; calcium, 25 mg; phosphorus, 27 mg; iron, 1.1 mg; sodium, 6 mg; potassium, 160 mg

Until drought- and disease-resistant varieties were developed in recent years, cucumbers were considered rather temperamental plants to grow. The newer varieties are no more difficult to raise than those of most other common vegetables.

The soil should be a light loam—neither too heavy nor too sandy. It should be mixed with well-aged manure and compost and heaped into small mounds about 2 meters (6.5 feet) apart. Five to six seeds should be planted in each mound about 2.5 centimeters (1 inch) below the surface in the middle of the spring. When the plants are about 1 decimeter (4 inches) tall, thin to three plants per mound. Cultivate regularly, and to promote continued production, pick all cucumbers as soon as they attain eating size.

Eggplant

NV (cooked in water): 19 calories; protein, 1.0 gm; fats, 0.2 gm; vit. A, 10 I.U.; vit. B_1, 0.05 mg; vit. B_2, 0.04 mg; niacin, 0.5 mg; vit. C, 3 mg; fiber, 0.9 gm; calcium, 11 mg; phosphorus, 21 mg; iron, 0.6 mg; sodium, 1 mg; potassium, 150 mg

Eggplant is strictly a hot-weather plant that is sensitive to cold weather or dry periods and needs heavy fertilizing. Since seedling development is initially slow, plant the seeds indoors about 2 months before the plants will be set out, which should be about 5 to 6 weeks after the last average date of frost.

Eggplants do best in enriched sandy soils that are supplemented with additional fertilizer once a month. Never permit them to dry out. Place the seedlings about 70 to 80 centimeters (28 to 32 inches) apart in rows 0.9 to 1 meter (3 feet or more) apart. Some staking of the plants may be desirable. The fruits are ready to harvest when they have a high gloss. They are still edible after greenish streaks appear and the gloss diminishes, but they are not as tender at this stage.

Lettuce

NV (crisp, cabbage-head varieties): 13 calories; protein, 0.9 gm; fats, 0.1 gm; vit. A, 330 I.U.; vit. B_1, 0.06 mg; vit. B_2, 0.06 mg; niacin, 0.3 mg; vit. C, 6 mg; fiber, 0.5 gm; calcium, 20 mg; phosphorus, 22 mg; iron, 0.5 mg; sodium, 9 mg; potassium, 175 mg

NV (leaf varieties): 18 calories; protein, 1.3 gm; fats, 0.3 gm; vit. A, 1,900 I.U., vit. B_1, 0.05 mg; vit. B_2, 0.08 mg; niacin, 0.4 mg; vit. C, 18 mg; calcium, 68 mg; phosphorus, 25 mg; iron, 1.4 mg; sodium, 9 mg; potassium, 264 mg

This favorite salad plant comes in a wide variety of types and forms, all of which do better in cooler weather, although a few of the leaf types (e.g., oak leaf) can tolerate some hot periods. As long as the individual plants are given room to develop and the soil is not too acidic, most varieties can be grown on a wide range of soil types.

Since lettuce can stand some frost, sow the seeds outdoors as early in the spring as the ground can be cultivated. Do not cover the seeds with more than a millimeter or two of soil—they need light to germinate. Mix the soil with a well-aged manure and a general-purpose fertilizer a week or two before sowing. Plant seedlings about 30 centimeters (12 inches) apart in rows 30 to 40 centimeters (12 to 16 inches) apart. For best results, do not allow the soil to dry out, and plant only varieties suited to local conditions. The most common crisp, cabbage-head varieties found in produce markets will not form heads in hot weather, and many others will bolt (begin to flower) during hot weather and longer days. Cultivate weekly between rows to promote rapid growth and to control weeds.

Onion

NV (raw): 38 calories; protein, 1.5 gm; fats, 0.1 gm; vit. A, 40 I.U. (yellow varieties only); vit. B_1, 0.03 mg; vit. B_2, 0.04 mg; niacin, 0.2 mg; vit. C, 10 mg; fiber, 0.6 gm; calcium, 27 mg; phosphorus, 36 mg; iron, 0.5 mg; sodium, 10 mg; potassium, 157 mg

These easy-to-grow vegetables do best in fertile soils that are free of rocks and lumps, are well drained, and are not too acidic or sandy.

Onions may take several months to mature from seed. The viability of the seed decreases rapidly after the first year. Bulb formation is determined by day length rather than by the total number of hours in the ground. Because of these characteristics of onions, most gardeners prefer to purchase *sets* (young plants that already have a small bulb) from commercial growers, although green or bunching onions are still easily grown from seed.

Plant the sets upright 6 to 7.5 centimeters (2.5 to 3 inches) apart in rows and firm in place. Except for weeding, watering, and occasional shallow cultivation, they will need little care until harvest about 14 weeks later. The onions are mature when the tops fall over. After they are pulled from the ground, allow them to dry in the shade for 2 days. Then remove the tops 2 to 3 centimeters (about 1 inch) above the bulbs, and spread out the bulbs to continue curing for 2 to 3 more weeks. After this, store them in sacks or other containers that permit air circulation until needed.

Peas

NV (green, cooked in water): 71 calories; protein, 5.4 gm; fats, 0.4 gm; vit. A, 540 I.U.; vit. B_1, 0.28 mg; vit. B_2, 0.11 mg; niacin, 2.3 mg; vit. C, 20 mg; fiber, 2.0 gm; calcium, 23 mg; phosphorus, 99 mg; iron, 1.8 mg; sodium, 1 mg; potassium 196 mg

Peas are strictly cool-weather plants that generally produce poorly when the soil becomes too warm. Seeds should be planted in the fall or very early spring. As is the case with beans, peas receive a better start if the seeds are inoculated with nitrogen-fixing bacteria (see discussion of beans) at planting time. Prepare the soil by mixing thoroughly with liberal amounts of aged manure and bone meal. Plant the seeds about 2.5 centimeters (1 inch) deep in heavy soil or 5 centimeters (2 inches) deep in light, sandy soil, about 2.5 centimeters (1 inch) apart in single rows for dwarf bush varieties or in double files 15 centimeters (6 inches) apart for standard varieties, with intervals of 80 to 90 centimeters (32 to 36 inches) between the rows. After germination, thin the plants to 10 centimeters apart. Place support wires, strings, or chicken wire between the rows at the time of planting; peas will not do well without such support.

Green peas should be picked while still young and then cooked or frozen immediately, as the sugars that make them sweet are converted to starch within 2 to 3 hours after harvest.

Peppers

NV (raw sweet or bell peppers): 22 calories; protein, 1.2 gm; fats, 0.2 gm; vit. A, 420 I.U.; vit. B_1, 0.08 mg; vit. B_2, 0.08 mg; niacin, 0.5 mg; vit. C, 128 mg; fiber, 1.4 gm; calcium, 9 mg; phosphorus, 22 mg; iron, 0.7 mg; sodium, 13 mg; potassium, 213 mg

Peppers, like eggplants, are strictly hot-weather plants for most of their growing season, but unlike eggplants, they actually do better toward the end of their season if temperatures have moderated somewhat. Sweet (also called bell) peppers are closely related and have similar cultural requirements.

Plant seeds indoors 8 to 10 weeks before the outdoor planting date, which is generally after the soil has become thoroughly warm. They will grow in almost any sunny location in a wide variety of soils, but to obtain the large fruits seen in produce markets, the plants need to be fertilized heavily and watered regularly. Plant seedlings 50 to 60 centimeters (20 to 24 inches) apart in rows that are 60 to 90 centimeters (24 to 36 inches) apart. Sweet peppers can be harvested at almost any stage and are still perfectly edible after they have turned red.

Potatoes

NV (baked in skin): 93 calories; protein, 2.6 gm; fats, 0.1 gm; vit. A, trace; vit. B_1, 0.10 mg; vit. B_2, 0.04 mg; niacin, 1.7 mg; vit. C, 20 mg; fiber, 0.6 gm; calcium, 9 mg; phosphorus, 65 mg; iron, 0.7 mg; sodium, 4 mg; potassium, 503 mg

Potatoes grow best in a well-drained soil that has had compost or well-aged manure added to it. They are subject to several diseases, and it is advisable to use disease-free seed potatoes purchased from a reputable dealer. Two to 3 weeks before the average date of the last spring frost, plant the seed potatoes whole or cut into several pieces, making sure that each piece has at least one eye and is large enough to supply sufficient nutrients. Place the potato pieces about 30 centimeters (12 inches) or more apart at a depth of about 12 to 15 centimeters (5 to 6 inches) in rows 0.9 to 1 meter (about 3 feet) apart. Later plantings are feasible. Spread a thick mulch (e.g., straw) over the area after planting to keep soil temperatures down and to retain soil moisture.

Potatoes are ready for harvest when the tops start turning yellow, but they may be left in the ground for several weeks after that if the soil is not too wet. After harvest, wash the potatoes immediately and place them in a dry, cool, dark place until needed. If left exposed to light, the outer parts of the potato turn green; poisonous substances are produced in these tissues, and such potatoes should be discarded.

Spinach

NV (raw): 26 calories; protein, 3.2 gm; fats, 0.3 gm; vit. A, 8,100 I.U.; vit. B_1, 0.10 mg; vit. B_2, 0.20 mg; niacin, 0.6 mg; vit. C, 51 mg; fiber, 0.6 gm; calcium, 93 mg; phosphorus, 51 mg; iron, 3.1 mg; sodium, 71 mg; potassium, 470 mg

Spinach is a cool-season crop that goes to seed as soon as the days become long and warm. It should be planted in the fall or early spring. If protected by straw or other mulches, it will overwinter in the ground in most areas and be ready for use early in the spring. Spinach has a high nitrogen requirement and reacts negatively to acidic soils. It is otherwise easy to grow. Mix the soil thoroughly with aged manure and bone meal before planting. Plant seedlings 3 to 5 centimeters (1 to 2 inches) apart in rows 40 to 50 centimeters (16 to 20 inches) apart. Keep the plants supplied with adequate moisture and their growing area free of weeds. Harvest the whole plant when a healthy crown of leaves develops.

Squash

NV (cooked zucchini): 12 calories; protein, 1.0 gm; fats, 0.1 gm; vit. A, 300 I.U.; vit. B_1, 0.05 mg; vit. B_2, 0.08 mg; niacin, 0.8 mg; vit. C, 9 mg; fiber, 0.6 gm; calcium, 25 mg; phosphorus, 25 mg; iron, 0.4 mg; sodium, 1 mg; potassium, 141 mg

All varieties of squash are warm-weather plants, and all are targets of a variety of pests. Thorough preparation of the soil before planting pays dividends in production and in the health of the plants. Mix compost and aged manure with the soil and heap the soil in small hills about 1.2 meters (4 feet) apart from one another. Plant four to five seeds in each hill and thin the seedlings to three after they are about 10 centimeters (4 inches) tall. Summer squashes (e.g., zucchini) mature in about 2 months, while winter squashes (e.g., acorn) can take twice as long to mature. Summer squashes should be harvested while very young—they can balloon, seemingly overnight, into huge fruits. Winter squashes should be harvested before the first frost; only clean, undamaged fruits will store well. Keep such squashes laid out in a cool, dry place until use and not piled on top of one another. Check them occasionally for the development of surface fungi.

Tomatoes

NV (raw, ripe): 22 calories; protein, 1.1 gm; fats, 0.2 gm; vit. A, 900 I.U.; vit. B_1, 0.06 mg; vit. B_2, 0.04 mg; niacin, 0.7 mg; vit. C, 23 mg; fiber, 0.5 gm; calcium, 13 mg; phosphorus, 27 mg; iron, 0.5 mg; sodium, 3 mg; potassium, 244 mg

These almost universally used fruits are easy to grow, providing one understands a few basic aspects of their cultural requirements:

(1) Tomato fruit set is inhibited by high or low temperatures when the plants are flowering. For the earliest yields, seeds may be germinated indoors several weeks before the plants are to be placed outside, but little is accomplished by transplanting before the night temperatures begin to remain above 14°C (57°F).

(2) Tomatoes require considerably more phosphorus than nitrogen from any fertilizers added to the soil where they are to be grown. Many inexperienced gardeners make the mistake of giving the plants lawn or general-purpose fertilizers that are proportionately high in nitrogen. As a result, the plants may grow vigorously but produce very few tomatoes. Give tomatoes bone meal, tomato food (Magamp is an excellent commercial slow-release preparation), or steer manure mixed with superphosphate.

(3) Tomato plants seem more susceptible than most to soil fungi and to root-knot nematodes. The damage caused by these organisms may not become evident until the plants begin to bear. Then the lower leaves begin to wither, and yellowing progresses up the plant or there seems to be a general loss of vigor and productivity. Using disease- and nematode-resistant varieties (usually indicated by the letters V, F, and N on seed packets) is by far the simplest method of controlling these problems.

(4) Many garden varieties of tomatoes need to be staked to keep fruits off the ground where snails and other organisms can gain easy access to them. When using wooden stakes, be sure they have a diameter of 5 centimeters (2 inches) or more, and tie the plants securely to the stakes with plastic tape. Thinner stakes are likely to break or collapse when the plants grow to a height of 2 meters (6.5 feet) or more. Some growers prefer to use heavy wire tomato towers instead of stakes.

(5) Hornworms and tomato worms almost invariably appear on tomato plants during the growing season. They can virtually strip a plant and ruin the fruits if not controlled. Fortunately, control is simple and highly effective with the use of *Bacillus thuringiensis*, which is discussed in Appendix 2.

(6) The eating season for garden tomatoes can be extended for about 2 months past the first frost if all the green tomatoes are picked before frost occurs. Place the tomatoes on sheets of newspaper on a flat surface in a cool, dry place, where they will ripen slowly a few at a time. Generally, the taste of tomatoes ripened this way is superior to that of hothouse tomatoes sold in produce markets. Be sure when picking the green tomatoes to handle them very gently, as they bruise very easily and molds quickly develop in the bruised areas.

Pruning

A good gardener or orchardist makes a habit of pruning trees, shrubs, and other plants regularly for a variety of reasons. He or she may wish to improve the quality and size of the fruits and flowers, restrict the size of the plants, keep the plants healthy, shape the shrubbery, or generally get more from the plants.

Except for spring-flowering ornamental shrubs, which should be pruned right after flowering, most maintenance pruning is done in the winter when active growth is not taking place. It usually involves removal of portions of stems, but it can also involve roots. When a terminal bud is removed, the axillary or lateral bud just below the cut will usually develop into a branch, and a bushier growth will result. Some gardeners pinch off terminal buds routinely to encourage such growth. The following sections provide a few generalities and specifics pertaining to several types of plants.

Fruit Trees

When young fruit trees are first planted, any damaged roots and all except four or five stems should be pruned off. The remaining stems should be cut back so that there is one central leader about 1 meter (3 feet) tall, with shorter side branches facing in different directions (Fig. A4.1). When cutting the stems be careful to cut in such a way that the axillary or lateral bud just below the cut is facing outward. Each succeeding year, prune back new growth to within a few centimeters of the previous year's cut. Any dead or diseased branches should be removed, and stems that have grown so that they are rubbing against each other should be pruned. The central leader should be cut out the second year in peach trees so that the interior of the tree is left relatively open. Any sucker shoots that develop from the base or along the trunk of the tree should be regularly removed.

Grapevines

There are several methods of pruning grapevines, depending on the type of vine and the circumstances under which the vines are being grown. In general, grapevines should be pruned heavily in late winter for best fruit production. After a central trunk has been allowed to develop, each shoot, regardless of its length, should be cut back so that no more than three axillary buds remain. Exceptions to this rule involve situations in which the vines are trained on arbors or wires, when the shoots initially may be allowed to grow longer. Even then, however, after the desired training has taken place (Fig. A4.2), pruning should be heavy for best results.

Figure A4.1 A young peach tree. (a) Before pruning. (b) After pruning.

Figure A4.2 A grapevine. (a) Before pruning. (b) After pruning.

Roses

Rose bushes should be pruned heavily—they will recover! In general, new stems should be cut back to within 10 to 20 centimeters (4 to 8 inches) of their point of origin, with care taken that the top remaining axillary bud of each stem is pointing outward. This promotes growth that leaves the center open for better air circulation. Any dead or diseased canes should be removed and the number of remaining canes limited to three or four per plant.

Raspberries, Blackberries, and Their Relatives

Berry canes are biennial. They are produced from the base the first year, branch during the summer, and usually produce fruit on the branches the second year, although in milder climates they may also produce fruit on the first year's growth. The canes die after the second year.

Old, dead canes should be removed, and all but three or four canes developing from each crown should be pulled out when the ground is soft. New canes should be cut back to lengths of 1 meter (3 feet) or less in the spring. Branches of one-year-old canes should be cut back in early spring to lengths of about 30 centimeters (12 inches) for larger fruits.

Bonsai

Container-grown trees that are dwarfed through the constant careful pruning of both roots and stems, the manipulation of soil mixtures, and the weighting of branches are called **bonsai** (Fig. A4.3). Some of these dwarfed trees attain ages of 50 to 75 years or more and may be less than 1 meter (3 feet) tall. Bonsai is an art that requires knowledge of the environmental requirements and tolerances of individual species.

In general, bonsai growers pinch out new growth above a bud every few days during a growing season but never prune when a plant is dormant. Refer to the Additional Reading section at the end of Appendix 4 for more information on the subject.

Major Types of Grafting

Whip, or Tongue, Grafting

Whip, or *tongue, grafting* is widely used for relatively small material—that is, wood between 0.7 and 1.25 centimeters (0.25 and 0.5 inch) in diameter. The stock and scion (rooted portion and aerial portion, respectively) should be nearly identical in diameter to bring about maximum contact between the cambia. The scion should contain two or three buds, and the cuts on both the stock and scion should be made in an internode. As shown in Figure A4.4, a smooth tangential cut about 5 centimeters (2 inches) long is made with a sharp sterilized knife at the bottom of the scion and at the top of the stock. The angles of both cuts should also be as nearly identical as possible, and there should be no irregularities or undulations in the surfaces (such as those caused by a dull cutting instrument). A second cut is then made in both the stock and scion about one-third of the distance from the tip of the cut surfaces. This cut is made back into the wood, nearly parallel to the first cut so that it forms a little tongue. The scion is then inserted into the stock as tightly as possible, taking care not to force a split. In addition, the bottom edge of the scion should not protrude past the bottom of the cut of the stock. The process is completed by binding the materials and adding grafting wax.

If the stock and scion are not identical in diameter, it is still possible to obtain a graft if care is taken to bring the cambia in close contact along one edge of the cuts (Fig. A4.5).

Splice Grafting

Splice grafting is sometimes used with plants where the pith is extensive. It is essentially the same as whip grafting, except that no second cut is made to form a tongue.

Cleft Grafting

Cleft grafting is a widespread method of grafting used routinely when the stock has a considerably greater diameter than that of the scion or scions. The stock branch or trunk

Figure A4.6 Stages in cleft grafting. (a) The stock is cut transversely and a meat cleaver or heavy knife is driven into the wood to make a vertical cut or split. (b) A wedge is temporarily inserted into the vertical cut to keep it open. (c) Scions are inserted into the cut in the vicinity of the cambium, and the wedge is removed. (d) The exposed surfaces are sealed with grafting wax.

may turn to the bark of trees and gnaw off a band of tissue, sometimes a decimeter or two wide. The damage usually extends through the phloem and cambium, as these tissues are the most palatable to the animals. This stripping of tissues is frequently referred to as **girdling** a tree and, if left untreated, will result in its death through starvation of the roots, since the phloem cannot conduct food past the damaged area.

The tree can often be saved, however, through *bridge grafting,* particularly if it is done in the early spring just as new growth is beginning. The scions should be cut from dormant twigs of the same tree and kept in a refrigerator until needed. The damaged area should be cleaned and prepared by cutting out any remaining dead or tattered tissues. The scions are then inserted above and below the girdle, about 5.0 to 7.5 centimeters (2 to 3 inches) apart, with the natural bottom ends facing down and the tip ends up (Fig. A4.10). The graft will not succeed if a scion is put in upside down.

Bud Grafting (Budding)

Bud grafting, or **budding,** is a form of grafting that utilizes a single bud. The method is widely used in commercial nurseries, partly because a single team of two to three workers can produce over a thousand such grafts a day and also because frequently better than 95% of the grafts are successful.

Budding is usually done in the summer when the season's axillary buds are mature and while the sap of the stocks is flowing freely. Budding is generally most successful when plump leaf buds (not flower or mixed buds) of the current season's growth are grafted to healthy stocks that are 2 to 3 years old.

Figure A4.7 How a side graft is made. A tangential cut is made on the side of the stock, and a prepared scion is inserted.

Figure A4.8 An approach graft. Two independently growing plants are prepared and grafted together as shown. (a) Smooth slanting vertical cuts identical in length and depth and at the same height are made on both stems; the cuts, however, go in opposite directions. (b) The two parts are wedged together. (c) The united area is tied and sealed.

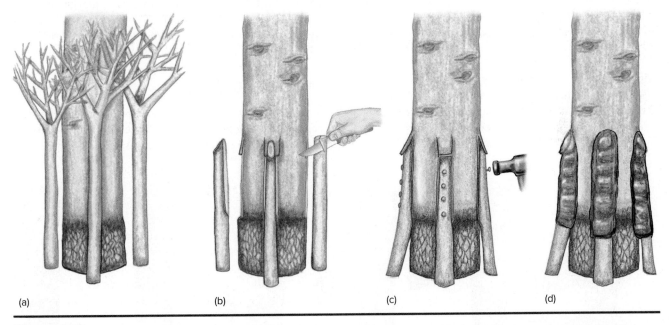

Figure A4.9 Inarching. (a) Established seedlings, which had previously been planted around the base of the tree, are cut vertically at their tips. (b) Vertical slots of similar length are made in the tree adjacent to the seedling tips. (c) The seedling tips are nailed into the slots. Growth of the seedling bases may be very rapid if the inarching is successful. (d) Sealing the grafts lessens chance of disease.

The stock is prepared by removing all the leaves and side branches below and in the vicinity of the point at which the graft is to be made. Then a T-shaped incision is made through the bark with a sterilized, razor-sharp knife. The transverse cut should be roughly 1.25 centimeters (0.5 inch) wide and the vertical cut about 2.5 to 3.0 centimeters (1 to 1.2 inches) long. Both cuts should be no deeper than the cambium, and the bark should peel back easily at the junction of the two cuts.

A *bud stick* is prepared by removing all leaf material except for 1.25 centimeters (0.5 inch) of the petioles, which are left to serve as handles. Then a bud is carefully cut from the stick so that an oval piece of tissue about 1.75 centimeters (0.75 inch) in diameter surrounds it. If the bud separates from the tissue, it should be discarded and another one cut. The bud and its oval shield are next inserted into the T-shaped cut of the stock and the flaps of bark are folded over the shield, leaving the bud exposed. Use flat rubber

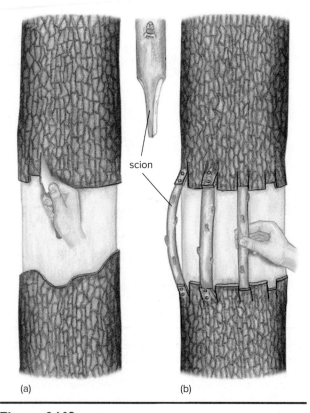

Figure A4.10 Bridge grafting. Scions of small diameter are cut at an angle at each end and inserted into prepared slots cut above and below the girdled area.

strips to tie the T shut so that only the bud remains visible. After growth begins, the stock is cut off just above the bud. The stock should not be cut off any earlier because the bud derives benefit from the transport of substances up and down the stock (Fig. A4.11).

Bud grafting can also be done in the spring, as soon as possible after growth of the stock begins but before the bud sticks become active. The bud material is frequently cut and stored in a refrigerator before growth begins. In areas with long growing seasons, bud grafting may also be done in early summer if the bark still peels back easily, but it should not be done any later because a young tree needs to produce sufficient growth before fall to be healthy and vigorous the following season.

In thick-barked trees, such as the Pará rubber tree and some of the nut trees, a rectangular patch of bark is cut out of the stock and replaced with a similar patch containing a bud. Normally, the patch is not more than 5 centimeters (2 inches) in diameter. This method is slower than the other budding methods described but generally gives much better results in species with thick bark.

Root Grafting

Whip, or tongue, grafts with roots or pieces of roots used for stocks are sometimes used in the propagation of apple, pear, and other fruit trees. After the grafts have been made, they are usually stored in a cool, damp place for about 2 months and then refrigerated until early spring, when they are planted before growth starts. After growth begins, they are checked to make sure the scion is not producing its own roots, or the advantages of the original rootstock will be lost. These advantages may include disease resistance or dwarfing.

Double-Working

In *double-working* grafts, an *interstock* consisting of a stem segment varying in length from 2.5 centimeters to 30.0 centimeters (1 to 12 inches) or more is grafted between the

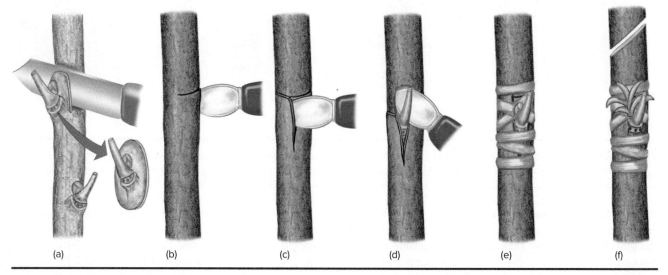

Figure A4.11 Budding (bud grafting). Leaves and side branches are removed from the stock below the point at which the graft is to be made. A bud stick is prepared by removing all leaf material except a short portion of each petiole. (a) A bud, with an oval patch of tissue surrounding it, is cut from the bud stick. (b) A T-shaped incision is made in the bark of the stock. (c) The bark is pried up slightly at the corners. (d) The bud and its surrounding tissue are inserted into the T-shaped incision on the stock. (e) The flaps of bark are folded over the tissue, leaving the bud exposed, and the T is tied shut with flat rubber strips. (f) After growth begins, the stock is cut off just above the bud.

stock and scion (i.e., three sections of stem are used for two grafts). Double-working grafts are used for special purposes. One such purpose is dwarfing, usually achieved by using special combinations of materials but sometimes by this method of grafting. A young tree is cut off above the ground, an additional segment is cut to serve as the interstock, and the stock, interstock, and top (scion) are grafted together with the interstock inverted. This method will work only with certain varieties, as inversion normally effectively blocks the flow of materials up and down the stem. Other purposes of double-working grafts—in which the interstock is not inverted—include the influencing of growth so as to promote greater flower production than would otherwise occur, providing a disease- or cold-resistant trunk, and circumventing graft incompatibility (the failure of grafts to form permanent unions). With regard to this last purpose, if scions of certain varieties will not produce permanent unions when grafted to stocks of another variety but will form good grafts with a third variety, it may be possible to graft the third variety to the stock so that it can function as an interstock, thereby circumventing the problem.

Additional Reading

Adams, C. F. 1981. *Nutritional value of American foods in common units.* Washington, DC: Government Printing Office.

Alexander, D. E., and W. J. Lewis. 2009. *Grafting and budding: A practical guide for fruit and nut plants.* Collingwood, Victoria, Australia: CSIRO Publishing.

Bradley, F. M., E. Phillips, and B. Ellis (Eds.). 2009. *Rodale's ultimate encyclopedia of organic gardening: The indispensable green resource for every gardener.* Emmaus, PA: Rodale Press, Inc.

Brickell, C. 1999. *American Horticultural Society: Pruning and training.* New York: DK Publishing.

Brickell, C., and J. D. Zuk (Eds.). 1996. *American Horticultural Society A to Z encyclopedia of garden plants.* New York: DK Publishing.

Bubel, N. 1998. *The new seed-starters handbook.* Emmaus, PA: Rodale Press.

Garner, R. J. 2003. *The grafter's handbook.* London: Orion Publishing, Limited.

Hartmann, H. T., D. E. Kester, and F. T. Davies, Jr. 1997. *Plant propagation: Principles and practices,* 6th ed. Upper Saddle River, NJ: Prentice-Hall.

Hessayon, D. G. 1998. *The house plant expert.* New York: Sterling.

Hill, L. 1992. *Fruits and berries for the home garden,* rev. ed. Pownal, VT: Garden Way Publishing.

Jeavons, J. 1995. *How to grow more vegetables,* 5th ed. Berkeley, CA: Ten Speed Press.

Kramer, J. 1992. *Know your houseplants.* New York: Lyons Press.

Little, B. 2006. *The dictionary of vegetable gardening.* Chicago, IL: Silverleaf Press.

Norman, K. 1998. *Essential bonsai.* New York: Anness Publishing.

Sunset Editors. 1999. *Basic gardening illustrated.* Menlo Park, CA: Lane Publishing.

Sunset Editors. 2001. *Northeast garden book.* Menlo Park, CA: Lane Publishing.

Sunset Editors. 2001. *Western garden book.* Menlo Park, CA: Lane Publishing.

APPENDIX 5
Metric Equivalents and Conversion Tables

Metric System of Measurement

Application	International System of Unitse	English System Equivalents
Length	Kilometer	0.62137 mile
	Meter	39.37 inches
	Centimeter	0.3937 inch
	Millimeter	0.03937 inch
	Micrometer	0.00003937 inch
	Nanometer	0.00000003937 inch
	Angstrom	0.0000000003937 inch
Mass (Weight)	Metric ton	2,200 pounds
	Kilogram	2.2 pounds
	Gram	0.03527 ounce
	Milligram	0.00003527 ounce
Volume	Liter	1.06 quart
	Milliliter	0.00106 quart
	Cubic meter	35.314 cubic feet
	Cubic centimeter	0.061 cubic inch
Area	Hectare	2.471 acres
Temperature	To convert Celsius to Fahrenheit, multiply the Celsius figure by 9, divide the total by 5, and add 32.	

Conversion Table

To Convert	To	Multiply by
Millimeters	Inches	0.039
Centimeters	Inches	0.39
Inches	Centimeters	2.54
Feet	Centimeters	30.48
Ounces	Grams	28.35
Pounds	Grams	453.6
Pounds	Kilograms	0.4536
Grams	Ounces	0.035
Kilograms	Pounds	2.205
Fluid ounces	Milliliters	29.57
Quarts	Liters	0.9463
Milliliters	Fluid ounces	0.03
Liters	Quarts	1.057

Temperature Conversion Scale

To convert Fahrenheit to Celsius use the following formula:
$°C = 5/9(°F - 32)$

To convert Celsius to Fahrenheit use the following formula:
$°F = (9/5 °C) + 32$

APPENDIX 6

Periodic Table of the Elements

REPRESENTATIVE ELEMENTS

Legend:
- Metals (main group)
- Metals (transition)
- Metals (inner transition)
- Metalloids
- Nonmetals

Key:
1 — Atomic number
H — Symbol
Hydrogen
1.008 — Atomic mass

Period	IA (1)	IIA (2)		IIIB (3)	IVB (4)	VB (5)	VIB (6)	VIIB (7)	—VIIIB— (8)	(9)	(10)	IB (11)	IIB (12)	IIIA (13)	IVA (14)	VA (15)	VIA (16)	VIIA (17)	VIIIA (18)
1	1 H Hydrogen 1.008																		2 He Helium 4.003
2	3 Li Lithium 6.941	4 Be Beryllium 9.012												5 B Boron 10.81	6 C Carbon 12.01	7 N Nitrogen 14.01	8 O Oxygen 16.00	9 F Fluorine 19.00	10 Ne Neon 20.18
3	11 Na Sodium 22.99	12 Mg Magnesium 24.31							TRANSITION ELEMENTS					13 Al Aluminum 26.98	14 Si Silicon 28.09	15 P Phosphorus 30.97	16 S Sulfur 32.07	17 Cl Chlorine 35.45	18 Ar Argon 39.95
4	19 K Potassium 39.10	20 Ca Calcium 40.08		21 Sc Scandium 44.96	22 Ti Titanium 47.88	23 V Vanadium 50.94	24 Cr Chromium 52.00	25 Mn Manganese 54.94	26 Fe Iron 55.85	27 Co Cobalt 58.93	28 Ni Nickel 58.69	29 Cu Copper 63.55	30 Zn Zinc 65.39	31 Ga Gallium 69.72	32 Ge Germanium 72.61	33 As Arsenic 74.92	34 Se Selenium 78.96	35 Br Bromine 79.90	36 Kr Krypton 83.80
5	37 Rb Rubidium 85.47	38 Sr Strontium 87.62		39 Y Yttrium 88.91	40 Zr Zirconium 91.22	41 Nb Niobium 92.91	42 Mo Molybdenum 95.94	43 Tc Technetium (98)	44 Ru Ruthenium 101.1	45 Rh Rhodium 102.9	46 Pd Palladium 106.4	47 Ag Silver 107.9	48 Cd Cadmium 112.4	49 In Indium 114.8	50 Sn Tin 118.7	51 Sb Antimony 121.8	52 Te Tellurium 127.6	53 I Iodine 126.9	54 Xe Xenon 131.3
6	55 Cs Cesium 132.9	56 Ba Barium 137.3		57 La Lanthanum 138.9	72 Hf Hafnium 178.5	73 Ta Tantalum 180.9	74 W Tungsten 183.9	75 Re Rhenium 186.2	76 Os Osmium 190.2	77 Ir Iridium 192.2	78 Pt Platinum 195.1	79 Au Gold 197.0	80 Hg Mercury 200.6	81 Tl Thallium 204.4	82 Pb Lead 207.2	83 Bi Bismuth 209.0	84 Po Polonium (209)	85 At Astatine (210)	86 Rn Radon (222)
7	87 Fr Francium (223)	88 Ra Radium (226)		89 Ac Actinium (227)	104 Rf Rutherfordium (261)	105 Db Dubnium (262)	106 Sg Seaborgium (266)	107 Bh Bohrium (262)	108 Hs Hassium (265)	109 Mt Meitnerium (266)	110 Ds Darmstadtium (281)	111 Rg Roentgenium (272)	112 Cn Copernicium (285)	113 Uut Ununtrium (284)	114 Fl Flerovium (289)	115 Uup Ununpentium (288)	116 Lv Livermorium (293)	117 Uus Ununseptium (293)	118 Uuo Ununoctium (294)

INNER TRANSITION ELEMENTS

6 Lanthanides	58 Ce Cerium 140.1	59 Pr Praseodymium 140.9	60 Nd Neodymium 144.2	61 Pm Promethium (147)	62 Sm Samarium 150.4	63 Eu Europium 152.0	64 Gd Gadolinium 157.3	65 Tb Terbium 158.9	66 Dy Dysprosium 162.5	67 Ho Holmium 164.9	68 Er Erbium 167.3	69 Tm Thulium 168.9	70 Yb Ytterbium 173.0	71 Lu Lutetium 175.0
7 Actinides	90 Th Thorium 232.0	91 Pa Protactinium (231)	92 U Uranium 238.0	93 Np Neptunium (237)	94 Pu Plutonium (242)	95 Am Americium (243)	96 Cm Curium (247)	97 Bk Berkelium (247)	98 Cf Californium (251)	99 Es Einsteinium (252)	100 Fm Fermium (257)	101 Md Mendelevium (258)	102 No Nobelium (259)	103 Lr Lawrencium (260)

The 1-18 group designation is recommended (International Union of Pure and Applied Chemistry) but not widely used. This text uses standard U.S. notation for groups (IA-VIIA and IB-VIIIB.)

Glossary

a

abscisic acid (ab-siz′ik as′id) (ABA) a growth-inhibiting hormone of plants; it is involved with other hormones in dormancy

abscission (ab-sizh′un) the separation of leaves, flowers, and fruits from plants after the formation of an abscission zone at the base of their petioles, peduncles, and pedicels

achene (uh-keen′) a single-seeded fruit in which the seed is attached to the pericarp only at its base

acid (as′id) a substance that dissociates in water, releasing hydrogen ions

active transport (ak′tiv trans′port) the expenditure of energy by a cell in moving a substance across a plasma membrane against a diffusion gradient

adenine (ad′n-een) a nucleic acid component that base-pairs with thymine in DNA

adventitious (ad-ven-tish′uss) said of buds developing in internodes or on roots, or of roots developing along stems or on leaves

aerial root (air′ee-uhl root) root formed above the ground

aerobic respiration (air-oh′bik res-puh-ray′shun) respiration that requires free oxygen

agar (ah′gur) a gelatinous substance produced by certain red algae and also a few brown algae; it is often used as a culture medium, particularly for bacteria

aggregate fruit (ag′gruh-git froot) a fruit derived from a single flower having several to many pistils

air layering (air lay′urr-ing) an asexual plant propagation technique whereby aerial stems are induced to form roots. The rooted portion of the stem is then cut and planted

algin (al′jin) a gelatinous substance produced by certain brown algae; it is used in a wide variety of food substances and in pharmaceutical, industrial, and household products

allele (uh-leel′) one of at least two alternative forms of a gene

alternate (awl′ter-neyt) placed at different heights or distance from one another, as leaves can be arranged on a stem

alternation of generations (ol-tur-nay′shun uv jen-ur-ay′shunz) alternation between a haploid gametophyte phase and a diploid sporophyte phase in the life cycle of sexually reproducing organisms

amino acid (ah-meé noh as′id) one of the organic nitrogen-containing units from which proteins are synthesized; there are about 20 in all proteins

amino group (ah-mee′noh groop) the –NH$_2$ portion of an amino acid

amyloplast (am′uh-loh-plast) a colorless, starch-forming plastid found in roots and involved in gravity perception

anabolism (uh-nab′uh-liz-uhm) metabolic reactions in which chemical bonds are formed

anaerobic respiration (an-air-oh′bik res-puh-ray′shun) respiration in which the hydrogen removed from the glucose during glycolysis is typically combined with an inorganic ion (in the absence of oxygen)

anaphase (an′uh-feyz) stage of mitosis or meiosis in which sister chromatids (or homologous chromosomes in meiosis I) separate

aneuploid (an′you-ploid) an aberration in normal chromosome number in which one or more extra chromosomes are present or one or more chromosomes are missing

angiosperm (an′jee-oh-spurm) a plant whose seeds develop within ovaries that mature into fruits

annual (an′you-ul) a plant that completes its entire life cycle in a single growing season

annual ring (an′you-ul ring) a single season's production of xylem (wood) by the vascular cambium

annulus (an′yu-luss) a specialized layer of cells around a fern sporangium; it aids in spore dispersal through a springlike action; also a membranous ring around the stipe of a mushroom

anther (an′thur) the pollen-bearing part of a stamen

antheridiophore (an-thur-id′ee-oh-for) a stalk that bears an antheridium

antheridium (pl. **antheridia**) (an-thur-id′ee-um; pl. an-thur-id′ee-ah) the male gametangium of certain algae, fungi, bryophytes, and vascular plants other than gymnosperms and angiosperms

anthocyanin (an-thoh-sy′ah-nin) a water-soluble pigment found in cell sap; anthocyanins vary in color from red to blue

antibiotic (an-tee-by-ot′ik) a substance produced by a living organism that interferes with the normal metabolism of another living organism

anticodon (an-tee-koh′don) the three-nucleotide sequence in a tRNA molecule that base-pairs with the complementary mRNA codon for the amino acid carried by that specific tRNA

apical dominance (ay′pi-kul dom′in-uns) suppression of growth of lateral buds by hormones

apical meristem (ay′pi-kul mair′i-stem) a meristem at the tip of a shoot or root

apomixis (ap-uh-mik′sis) reproduction without fusion of gametes or meiosis in otherwise normal sexual structures

aquifer (ak′wuh-fer) a layer of porous rock or sand found underground and containing water that can be brought to the surface via wells

archegoniophore (ahr-kuh-goh′nee-oh-for) a stalk bearing an archegonium

archegonium (pl. **archegonia**) (ahr-kuh-goh′nee-um; pl. ahr-kuh-goh′nee-ah) the multicellular female gametangium of bryophytes and most vascular plants other than angiosperms

aril (air′il) an often brightly colored appendage surrounding the seed of certain plants (e.g., yew)

artificial selection (ahr-tuh-fish′uhl si-lek′shuhn) intentional breeding for desirable traits

ascomycetes (as-kuh-mayh′seets) sac fungi that are members of the Phylum Ascomycota

ascus (pl. **asci**) (as′kus; pl. as′eye) one of often numerous, frequently fingerlike hollow structures in which the fusion of two haploid nuclei is followed by meiosis; a row of ascospores (usually eight) is ultimately produced in each ascus on or within the sexually initiated reproductive bodies of cup (sac) fungi

asexual reproduction (ay-seksh′yule ree-proh-duk′shun) any form of reproduction not involving the union of gametes

assimilation (uh-sim-i-lay′shun) cellular conversion of raw materials into protoplasm and cell walls

atom (at′um) the smallest individual unit of an element that retains the properties of the element

ATP (ay-tee-pee) adenosine triphosphate, a molecule with three phosphate groups found in all living cells; it is the principal vehicle for energy storage and exchange in cell metabolism

autoclave (aw′tuh-kleyv) a pressure cooker used to sterilize materials

autotrophic (aw-toh-troh′fik) descriptive of an organism capable of sustaining itself through conversion of inorganic substances to organic material

auxin (awk′sin) a growth-regulating substance produced either naturally by plants or synthetically

axil (ak′sil) the angle formed between a twig and the petiole of a leaf; normally the site of an *axillary bud* (also called *lateral bud*)

b

backcross (bak′kross) a cross involving a hybrid and one of its parents

bacteriophage (bak-teer′ee-oh-fayj) a virus whose host is a bacterium

bark (bahrk) tissues of a woody stem between the vascular cambium and the exterior

base (bayss) a substance that dissociates in water, releasing hydroxyl (OH$^-$) ions

basidiomycetes (buh-sid-ee-oh-mayh′seets) club fungi that are members of the Phylum Basidiomycota

basidiospore (buh-sidd′ee-oh-spor) a spore produced on a basidium

basidium (pl. **basidia**) (buh-sid′ee-um; pl. buh-sid′ee-ah) one of usually numerous, frequently club-shaped hollow structures in which the fusion of two haploid nuclei is followed by meiosis, the four resulting nuclei becoming externally borne basidiospores; basidia are produced on or within sexually initiated reproductive bodies of the club fungi (e.g., mushrooms, puffballs)

behavioral evolution (bih-heyv′yer-al ev-oh-loo′shun) study of the effect of genetic variation on behavioral traits

berry (bair′ee) a thin-skinned fruit that usually develops from a compound ovary and commonly contains more than one seed

biennial (by-en′ee-ul) a plant that normally requires two seasons to complete its life cycle, the first season's growth being strictly vegetative

bioinformatics (by-oh-in-for-mat′-ics) the combination of biology, statistics, and computer science to analyze the huge data sets generated by DNA and RNA sequencing efforts

biological controls (by-oh-loj′i-kull kun-trohlz′) the use of natural enemies and inhibitors in combating insect pests and other destructive organisms

biomass (by-oh-mass) the total mass of living organisms present

biome (by′ohm) similar biotic communities considered on a worldwide scale (e.g., desert biome, grassland biome)

biotechnology (by-oh-tek-nol′oh-jee) the manipulation of organisms, tissues, cells, or molecules for specific applications primarily intended for human benefit

biotic community (by-ot′ik kuh-myu′nit-ee) an association of plants, animals, and other organisms (e.g., woodland)

blade (blayd) the conspicuous flattened part of a leaf (also called *lamina*) or seaweed

bond (bond) a force that holds atoms together

bonsai (bon-sy′) container-grown plants (usually trees) that have been dwarfed artificially through skillful pruning and manipulation of the growing medium

botanist (bot′an-ist) someone who studies or specializes in botany

botany (bot′an-ee) science involving the study of plants

botulism (bot′yu-lizm) poisoning from consumption of food infected by botulism bacteria

bract (brakt) a structure that is usually leaflike and modified in size, shape, or color

bristlecone pine (bris′uhl-kohn pahyn) a tree species—known to be the oldest living organism on earth

bryophyte (bry′oh-fyt) a photosynthetic, terrestrial, aquatic, or epiphytic, embryo-producing plant without xylem and phloem (e.g., mosses, liverworts, hornworts)

budding (budd′ing) a form of asexual reproduction in which a new cell develops to full size from a protuberance arising from a mature cell, as in yeasts

bud gap (buhd gap) thumbnail-shaped opening in the cylinder of vascular tissue

bulb (buhlb) an underground food-storage organ that is essentially a modified bud consisting of fleshy leaves that surround and are attached to a small stem

bundle scar (bun′dul skahr) a small scar left by a vascular bundle within a leaf scar when the leaf separates from its stem through abscission

bundle sheath (bun′dul sheeth) the parenchyma and/or sclerenchyma cells surrounding a vascular bundle

C

C_3 plant (see three plant) plant in which the first isolated product of the light-independent reactions is 3-phosphoglyceric acid

C_4 plant (see fawr plant) plant in which the first isolated product of the light-independent reactions is oxaloacetic acid

callose (kal′ohs) a complex carbohydrate that develops in sieve tubes following an injury; it is commonly associated with the sieve areas of sieve tube members

callus (kal′uss) undifferentiated tissue that develops around injured areas of stems and roots; also the undifferentiated tissue that develops during tissue culture

Calvin cycle (kal′vin sy′kuhl) see *light-independent reactions*

calyptra (kuh-lip′truh) tissue from the enlarged archegonial wall of many mosses that forms a partial or complete cap over the capsule

calyx (kay′liks) collective term for the sepals of a flower

cambium (kam′bee-um) a meristem producing secondary tissues; see *vascular cambium, cork cambium*

capillary water (kap′i-lair-ee waw′tuhr) water held in the soil against the force of gravity; capillary water is available to plants

capsule (kapp′sool) a dry fruit that splits in various ways at maturity, often along or between carpel margins; also the main part of a sporophyte in which different types of tissues develop

carbohydrate (kahr-boh-hy′drayt) an organic compound containing carbon, hydrogen, and oxygen, with twice as many hydrogen as oxygen atoms per molecule

carboxyl group (khar-bok′sil groop) the — COOH portion of an amino acid

carpel (kahr′pul) an ovule-bearing unit that is a part of a pistil

caryopsis (kare-ee-op′siss) a dry fruit in which the pericarp is tightly fused to the seed; it does not split at maturity

Casparian strip (kass-pair′ee-un strip) a band of suberin around the radial and transverse walls of an endodermal cell

catabolism (kuh-tab′uh-liz-uhm) metabolic reactions which require chemical bonds to be broken

cell (sel) the basic structural and functional unit of living organisms; in plants, it consists of protoplasm surrounded by a cell wall

cell biology (sel by-ol′uh-jee) the biological discipline involving the study of cells and their functions

cell cycle (sel sy′kul) a sequence of events involved in the division of a cell

cell division see *cytokinesis*

cell membrane (sel mem′brayn) see *plasma membrane*

cell plate (sel playt) the precursor of the middle lamella; it forms at the equator during telophase

cell sap (sel sap) the liquid contents of a vacuole

cell theory (sel thee-u-ree) a theory stating that all living organisms are composed of cells

cellular respiration (sel′yuh-ler res-puh-rey′shuhn) metabolic process that involves the breakdown of carbohydrates to release energy

cell wall (sel wawl) the relatively rigid boundary of cells of plants and certain other organisms

central cell nuclei (sen-truhl sell new′klee-eye) nuclei, frequently two in number, that unite with a sperm in an embryo sac, forming a primary endosperm nucleus

centromere (sen′truh-meer) the dense, constricted portion of a chromosome to which a spindle fiber is attached

character state (kar′ik-ter steyt) value or form of a feature used to classify an organism

chemiosmosis (kem-ee-oz-moh′siss) a theory that energy is provided for phosphorylation by protons being "pumped" across inner mitochondrial and thylakoid membranes

chiasma (pl. **chiasmata**) (kyaz′mah; pl. ky-az′mah-tah) the X-shaped configuration formed by two chromatids of homologous chromosomes as they remain attached to each other during prophase I of meiosis

chlorenchyma (klor-en′kuh-mah) tissue composed of parenchyma cells that contain chloroplasts

chlorophyll (klor′uh-fil) green pigments essential to photosynthesis

chloroplast (klor′uh-plast) an organelle containing chlorophyll, found in cells of most photosynthetic organisms

chromatid (kroh′muh-tid) one of the two strands of a chromosome; they are united by a centromere

chromatin (kroh′muh-tin) a readily staining complex of DNA and proteins found in chromosomes

chromoplast (kroh′muh-plast) a plastid containing pigments other than chlorophyll; the pigments are usually yellow to orange

chromosome (kroh′muh-sohm) a body consisting of a linear sequence of genes and composed of DNA and proteins; chromosomes are found in cell nuclei and appear in contracted form during mitosis and meiosis

chytrid (kahy′trid) simple, alga-like organism that is a member of the Phylum Chytridiomycota

cilium (pl. **cilia**) (sil′ee-um; pl. sil′ee-uh) a short hairlike structure usually found on the cells of unicellular aquatic organisms, normally in large numbers and arranged in rows; the most common function of cilia is propulsion of the cell

circadian rhythm (sur-kay′dee-an rith′um) a mostly daily rhythm of growth and activity found in living organisms

citric acid cycle (sit′rik as-id sy′kul) a complex series of reactions following glycolysis in aerobic respiration that involves ATP, mitochondria, and enzymes and that results in the combining of free oxygen with protons and electrons from pyruvic acid to make water

cladistics (kluh-dis′tiks) analysis of shared features
cladogram (klad′uh-gram) straight-line diagram used to portray evolutionary relationships
cladophyll (klad′uh-fil) a flattened stem that resembles a leaf; also called *phylloclade*
class (klas) a category of classification between a phylum and an order
climax community (kly′maks kuh-myoo′ ni-tee) group of organisms living in an ecosystem at its peak equilibrium
climax vegetation (kly′maks vej-uh-tay′shun) vegetational association that perpetuates itself indefinitely at the culmination of ecological succession
cloning vector (kloh′ning vek′torr) a DNA molecule that can replicate and transfer DNA between cells
club fungi (kluhb fuhn′-jahy) member of class basidiomycetes with a simple or branched club-shaped sporophore
club moss (kluhb maws) member of the genus *Lycopodium* with structure resembling mosses and reproducing by spores
codon (koh′donn) the sequence of three nucleotides in an mRNA molecule that constitutes the code for a specific amino acid or a stop signal in protein synthesis; it is complementary to an anticodon
coenocytic (see′no-sitt-ik) multinucleate, the nuclei not individually separated from one another by crosswalls, as in the hyphae of water molds
cohesion-tension theory (koh-hee′zhun ten′shun thee′uh-ree) theory that explains the rise of water in plants through a combination of cohesion of water molecules in vessels and tracheids and tension on the water columns brought about by transpiration
coleoptile (koh-lee-op′tul) a protective sheath surrounding the emerging shoot of seedlings of the Grass Family (Poaceae) (e.g., corn, wheat)
coleorhiza (koh-lee-uh-ry′zuh) a protective sheath surrounding the emerging radicle (immature root) of members of the Grass Family (Poaceae) (e.g., corn, wheat)
collenchyma (kuh-len′kuh-muh) tissue composed of cells with unevenly thickened walls
colloid (kol′oyd) a substance consisting of a medium in which fine particles are permanently dispersed
community (kuh-myu′nit-ee) a collective term for all the living organisms sharing a common environment and interacting with one another
companion cell (kum-pan′yun sel) a specialized cell derived from the same parent cell as the closely associated sieve tube member immediately adjacent to it (in angiosperm phloem)
compost (kom′post) a mixture of decomposed organic matter, particularly decomposed plant materials
compound (kom′pownd) a substance whose molecules are composed of two or more elements
compound leaf (kom′pownd leef) a leaf whose blade is divided into distinct leaflets
conidium (pl. **conidia**) (kuh-nid′ee-um; pl. kuh-nid′ee-uh) an asexually produced fungal spore formed outside of a sporangium
conifer (kon′i-fur) a cone-bearing tree or shrub

conjugation (kon-juh-gay′shun) a process leading to the fusion of isogametes in algae, fungi, and protozoa; also the means by which certain bacteria exchange DNA
conjugation tube (kon-juh-gay′shun t(y)oob) a tube permitting transfer of a gamete or gametes between adjacent cells, as in *Spirogyra* or desmids
consumers (kon-soo′muhrz) organisms that feed on producers
cork (kork) tissue composed of cells whose walls are impregnated with suberin at maturity; the outer layer of tissue of an older woody stem; produced by the cork cambium
cork cambium (kork kam′bee-um) a narrow cylindrical sheath of cells between the exterior of a woody root or stem and the central vascular tissue; it produces *cork* to its exterior and *phelloderm* to its interior; it is also called *phellogen*
cork cell (kork sel) cell produced to the outside of the cork cambium
corm (korm) a vertically oriented, thickened food-storage stem that is usually enveloped by a few papery, nonfunctional leaves
corolla (kuh-rahl′uh) collective term for the petals of a flower
cortex (kor′teks) a primary tissue composed mainly of parenchyma; the tissue usually extends between the epidermis and the vascular tissue
cotyledon (kot-uh-lee′dun) an embryo leaf ("seed leaf") that usually either stores or absorbs food
covalent bond (koh-vay′luhnt bond) a force provided by pairs of electrons that travel between two or more atomic nuclei, holding atoms together and keeping them at a stable distance from each other
crossing-over (kross′ing oh′vur) the exchange of corresponding segments of chromatids between homologous chromosomes during prophase I of meiosis
cross-pollinating (kross pol′uh-neyt-ing) a plant species that reproduces mainly by outcrossing
crown division (krown duh-vizh′unn) the asexual production of multiple plants by division of the base of a stem (crown)
cryptochrome (krip′toh-krome) light-sensitive pigment associated with circadian rhythms
cryptomonads (krip-tom′un-nad) group of asymmetrical, somewhat flattened, unicellular, marine and freshwater algae with two flagella
cuticle (kyut′i-kul) a waxy or fatty layer of varying thickness on the outer walls of epidermal cells
cutin (kyu′tin) the waxy or fatty substance of which a cuticle is composed
cutting (kutt′ing) any vegetative plant part used for asexual propagation
cycad (sy′kad) member of the Phylum Cycadophyta that has the appearance of a hybrid between a tree fern and a palm
cyclosis (sy-kloh′sis) the flowing or streaming of cytoplasm within a cell
cytochrome (sy′toh-krohm) iron-containing protein involved in molecule transfer in an electron transport system
cytogenetics (sy′toh-juh-net-iks) the study of the genetic effects of chromosome structure and behavior

cytokinesis (sy-toh-kuh-nee′sis) division of a cell, usually following mitosis
cytokinin (syt-uh-ky′nin) a growth hormone involved in cell division and several other metabolic activities of cells
cytology (sy-tol′uh-jee) see *cell biology*
cytoplasm (sy′tuh-plazm) the protoplasm of a cell exclusive of the nucleus
cytoplasmic streaming (sy-tuh-plaz′mik streem′ing) see *cyclosis*
cytosine (sahy′tuh-seen) a component of nucleic acids that base-pairs with guanine in DNA
cytoskeleton (sy-toh-skel′uh-ton) a network of microtubules and microfilaments involved in movement within a cell
cytosol (sy′toh-sol) fluid, living part of a cell; organelles are distributed within it

d

dark reactions (dahrk ree-ak′shunz) see *light-independent reactions*
daughter cell (daw′ter sel) independent cell after completion of mitosis or meiosis
daughter chromosome (daw′ter kroh′muh-sohm) chromatid that becomes a chromosome after separation from a duplicated DNA molecule
day-neutral plant (day new′trul plant) a plant that is not dependent on specific day lengths for the initiation of flowering
deciduous (duh-sij′yu-wuss) shedding leaves annually
decomposer (dee-kuhm-poh′zur) organism (e.g., bacterium, fungus) that breaks down organic material to forms capable of being recycled
dedifferentiate (dee-diff-urr-en′shee-ayt) to become less specialized (usually pertains to cells)
deletion (dih-lee′shuhn) in genetics, the loss of one or more base pairs of a DNA molecule
deoxyribose (dee-ok-si-rahy′bohs) deoxygenated form of ribose that is found in DNA
dessication (des-i-keyt′shun) process of removing moisture
determinant growth (dee-tur′men-ant grohth) growth that stops after an organ such as a flower or leaf stops after a plant has reached a certain size.
development (duh-vel′up-ment) changes in the form of a plant resulting from growth and differentiation of its cells into tissues and organs
diatom (dahy′uh-tom) a microscopic, unicellular alga of the phylum Chrysophyta
dicotyledon (dy-kot-uh-lee′dun) a class of angiosperms whose seeds commonly have two cotyledons; frequently abbreviated to *dicot*
dictyosome (dik′tee-oh-sohm) an organelle consisting of disc-shaped, often branching hollow tubules that function in accumulating and packaging substances used in the synthesis of materials by the cell
differentially permeable membrane (dif-uh-rensh′uh-lee pur′mee-uh-bul mem′brayn) a membrane through which different substances diffuse at different rates; see *semipermeable membrane*

differentiation (dif-uh-ren-shee-ay′shun) the change of a relatively unspecialized cell to a more specialized one (e.g., the change of a cell just produced by a meristem to a vessel member or fiber)

diffusion (dif-fyu′zhin) the random movement of molecules or particles from a region of higher concentration to a region of lower concentration, ultimately resulting in uniform distribution

digestion (duh-jes′jin) an enzyme-controlled conversion of complex, usually insoluble substances to simpler, usually soluble substances

dihybrid cross (dy-hy′brid kross) a cross involving two different pairs of genes and heterozygous parents

dikaryotic (dy-kair-ee-ot′ik) having a pair of nuclei in each cell or a type of the mycelium in club fungi

dioecious (dy-ee′shuss) having unisexual flowers or cones, with the male flowers or cones confined to certain plants and the female flowers or cones of the same species confined to different plants

diploid (dip′loyd) having two sets of chromosomes in each cell; the $2x$ chromosome number characteristic of the sporophyte generation

disinfest (diss′in-fest) the removing of surface contaminants from a plant surface

diuretic (dy-yu-ret′ik) a substance tending to increase the flow of urine

division (duh-vizh′un) the largest undivided category of classification of organisms within a kingdom; considered synonymous with *phylum*

DNA (dee′en-ay) standard abbreviation of deoxyribonucleic acid, the carrier of genetic information in cells and viruses

DNA polymerase (dee′en-ay pol′uh-muh-reys) enzyme responsible for adding nucleotides during DNA synthesis

DNA synthesizer (dee-en-ay sin′thuh-size-urr) a machine that creates specific DNA sequences

domain (doh-meyn′) one of three super-kingdoms in the classification of life

domesticate (duh-mes′ti-keyt) process by which plants have been genetically altered to meet specific needs of humans

dominance (dom′in-uns) a condition in which one allele of a gene (dominant allele) masks the phenotypic expression of another allele (recessive allele)

dominant (dom′uh-nuhnt) one of a pair of alleles that masks expression of the other

dormancy (dor′man-see) a period of growth inactivity in seeds, buds, bulbs, and other plant organs even when environmental conditions normally required for growth are met

double fertilization (dub′ul fur-tl-uh-zey′shuhn) event in angiosperms in which one sperm fertilizes the egg to produce a zygote and another sperm fertilizes the central cell to produce endosperm tissue

double fusion (dub′ul fu′shun) the more or less simultaneous union of one sperm and egg (forming a zygote) and union of another sperm and central cell nuclei (forming a primary endosperm nucleus) that occur in the megagametophyte of flowering plants

drupe (droop) a simple fleshy fruit whose single seed is enclosed within a hard endocarp

e

ecology (ee-kol′uh-jee) the biological discipline involving the study of the relationships of organisms to each other and to their environment

economic botany (ek-uh-nom′ik bot′an-ee) study of commercial uses of plants

ecosystem (ee′koh-sis-tim) a system involving interactions of living organisms with one another and with their nonliving environment

egg (eg) a nonmotile female gamete

elater (el′uh-tur) a straplike appendage (usually occurring in pairs) attached to a horsetail (*Equisetum*) spore ; also, a somewhat spindle-shaped sterile cell occurring in large numbers in liverwort sporangia; both types of elaters facilitate spore dispersal

electron (ee-lek′tron) a negatively charged particle of an atom

electron microscope (ee-lek′tron mahy′kruh-skohp) instrument that uses a beam of electrons to obtain images of a specimen

element (el′uh-mint) one of more than 100 types of matter, most existing naturally but some human-made, each of which is composed of one kind of atom

embryo (em′bree-oh) an immature sporophyte that develops from a zygote within an ovule or archegonium after fertilization

enation (ee-nay′shun) one of the tiny, green leaflike outgrowths on the stems of whisk ferns (*Psilotum*)

endocarp (en′doh-kahrp) the innermost layer of a fruit wall

endodermis (en-doh-dur′mis) a single layer of cells surrounding the vascular tissue (stele) in roots and some stems; the cells have Casparian strips

endoplasmic reticulum (en-doh-plaz′mik ruh-tik′yu-lum) also known as ER. A complex system of interlinked, double-membrane channels subdividing the cytoplasm of a cell into compartments; parts of it are lined with ribosomes

endosperm (en′doh-spurm) a food-storage tissue that develops through divisions of the primary endosperm nucleus; it is digested by the sporophyte after germination in some species (e.g., corn) or before maturation of the seed in other species (e.g., beans)

endosymbiotic theory (endoh-sim′by-ot′ik thee-u-ree) the theory that mitochondria and chloroplasts were free-living bacteria that became incorporated in cells

energy (en′ur-jee) the capacity to do work; some forms of energy are heat, light, and kinetic

enzyme (en′zym) one of numerous complex proteins that speeds up a chemical reaction in living cells without being used up in the reaction (i.e., it catalyzes the reaction)

epicotyl (ep′uh-kaht-ul) the part of an embryo or seedling above the attachment point of the cotyledon(s)

epidermis (ep-uh-dur′mis) the exterior tissue, usually one cell thick, of leaves, young stems and roots, and other parts of plants

epigynous (ee-pidj′uh-nuss) having flower parts attached above the ovary

epiparasite (ep′uh-par-uh-sahyt) a parasite that feeds upon another parasite

epiphyte (ep′uh-fyt) an organism that is attached to and grows on another organism without parasitizing it

ergotism (ur′got-izm) a disease resulting from consumption of goods made with flour containing ergot fungus

essential element (eh-sen′shul el′uh-mint) one of 18 elements generally considered essential to the normal growth, development, and reproduction of most plants

ethnobotany (eth-noh-bot′an-ee) study of relationships between plants and people

ethylene (eth′uh-leen) a simple, naturally produced, gaseous hormone that inhibits plant growth and promotes the ripening of fruit

etiolation (ee-tee-oh-lay′shun) a condition characterized by long internodes, poor leaf development, and pale, weak appearance due to a plant's having been deprived of light

eukaryotic (yu-kair-ee-ot′ik) pertaining to cells having distinct membrane-bound organelles, including a nucleus with chromosomes

eutrophication (yu-troh-fuh-kay′shun) the gradual enrichment of a body of water through the accumulation of nutrients, resulting in a corresponding increase in algae and other organisms

evolution (ev-oh-loo′shun) the accumulation of genetic changes in populations of living organisms through many generations

evolutionary developmental biology (ev-oh-loo′shun-ner-ee dih-vel′uhp-muhnt-al bahy-ol′uh-jee) study of genetic variation on traits that affect survival and reproduction

evolutionary ecology (ev-uh-loo′shun-ner-ee ee-kol′uh-jee) study of ecological features that affect evolution of species and ecosystems

evolutionary genetics (ev-uh-loo′shun-ner-ee juh-net′iks) use of molecular and transmission genetics to understand origins of genetic variability

evolutionary paleontology (ev-uh-loo′shun-ner-ee pey-lee-uhn-tol′uh-jee) use of fossil records to examine large-scale evolutionary change

evolutionary physiology and morphology (ev-oh-loo′shun-ner-ee fiz-ee-ol′uh-jee and mor-fol′uh-jee) study of adaptation through biochemical, physiological, and anatomical changes

exine (ek′syne) the outer layer of the wall of a pollen grain or spore

exocarp (ek′soh-kahrp) the outermost layer of a fruit wall

explant (eks′plant) an excised portion of leaf or stem tissue used for tissue culture

extranuclear DNA (ex-truh-nyu′klee-ahr dee′en-ay) DNA found outside the nucleus, typically in plastids and mitochondria

ex vitro (eks vee′tro) tissue culture technique in which microshoots are rooted in soil and treated as cuttings

eyespot (eye′spot) a small, often reddish structure within a motile unicellular organism; it appears to be sensitive to light (also called *stigma*)

f

F_1 (first filial generation) (eff wun) the offspring of a cross between two parent plants

F_2 (second filial generation) (eff too) the offspring of the F_1 generation

FAD (eff-ay-dee) flavin adenine dinucleotide, a hydrogen acceptor molecule involved in the Krebs cycle of respiration and in photosynthesis

$FADH_2$ (ef′ey-dee-eych too) acronym for a high-energy compound called reduced flavin adenine dinucleotide

family (famm′uh-lee) a classification category between genus and order

fat (fat) an organic compound containing carbon, hydrogen, and oxygen but with proportionately much less oxygen than is present in a carbohydrate molecule

Fd (ef dee) abbreviation for an iron-sulfur acceptor molecule called ferredoxin

fermentation (fur-men-tay′shun) respiration in which the hydrogen removed from the glucose during glycolysis is transferred back to pyruvic acid, creating substances such as ethyl alcohol or lactic acid

fertilization (fur-til-i-zay′shun) formation of a zygote through the fusion of two gametes

fiber (fy′bur) a long, thick-walled cell whose protoplasm often is dead at maturity

filament (fil′uh-mint) threadlike body of certain bacteria, algae, and fungi; also the stalk portion of a stamen

first filial (F_1) (furst fil′ee-uhl (eff wun)) first generation after the parental generation

fission (fish′un) the division of cells of bacteria and related organisms into two new cells

flagellum (pl. **flagella**) (fluh-jel′um; pl. fluh-jel′uh) a fine, threadlike structure protruding from a motile unicellular organism or the motile cells produced by multicellular organisms; functions primarily in locomotion

floret (flor′et) a small flower that is a part of the inflorescence of members of the Sunflower Family (Asteraceae) and the Grass Family (Poaceae)

florigen (flor′uh-jen) one or more hormones once thought from circumstantial evidence to initiate flowering but which have never been isolated or proved to exist

flower (flou′er) plant structure that contains reproductive organs and associated tissues

follicle (foll′uh-kuhl) a dry fruit that splits along one side only

food chain (food chayn) a natural chain of organisms of a community wherein each member of the chain feeds on members below it and is consumed by members above it, with autotrophic organisms (*producers*) being at the bottom; interconnected food chains are referred to as *food webs*

foot (foot) the basal part of the embryo of bryophytes and other plants; it is attached to and absorbs food from the gametophyte

fossil (fos′ul) the remains or impressions of any natural object that has been preserved in the earth's crust

frond (frond) a fern leaf; term occasionally also applied to palm leaves

fruit (froot) a mature ovary usually containing seeds; term also somewhat loosely applied to the reproductive structures of groups of plants other than angiosperms

fucoxanthin (fyu-koh-zan′thin) a brownish pigment occurring in brown and other algae

g

gametangium (pl. **gametangia**) (gam-uh-tan′jee-um; pl. gam-uh-tan′jee-ah) any cell or structure in which gametes are produced

gamete (gam′eet) a sex cell; one of two cells that unite, forming a *zygote*

gametophore (guh-me′toh-for) a stalk on which a gametangium is borne

gametophyte (guh-me′toh-fyte) the haploid (*n*) gamete-producing phase of the life cycle of an organism that exhibits alternation of generations

gemma (pl. **gemmae**) (jem′uh; pl. jem′ee) a small outgrowth of tissue that becomes detached from the parent body and is capable of developing into a complete new plant or other organism; gemmae are produced in cuplike structures on liverwort thalli and are also produced by certain fungi

gene (jeen) a unit of heredity; part of a linear sequence of such units occurring in the DNA of chromosomes

gene bank (jeen bank) a collection of plants or seeds maintained for their germ plasm

generative cell (jen′uh-ray-tiv sel) the cell of the male gametophyte of angiosperms that divides, producing two sperms; also, the cell of the male gametophyte of gymnosperms that divides, producing a *sterile cell* and a *spermatogenous cell*

generative nucleus (jen′er-uh-tiv new′klee-uss) a pollen grain nucleus that divides by mitosis to produce sperm cells

genetic drift (juh-net′ik drift) a change in the genetic makeup of a population that may take place by chance alone

genetic engineering (juh′net′ik en-juh-neer′ing) the introduction, by artificial means, of genes from one form of DNA into another form of DNA

genetics (juh-net′iks) the biological discipline involving the study of heredity

genome (jee′nohm) the sum total of DNA in an organism's chromosomes

genomics (jee′nohm-iks) the study of genomes and their functions

genotype (jeen′oh-typ) the genetic constitution of an organism; may or may not be visibly expressed, as contrasted with phenotype

genus (pl. **genera**) (jee′nus; pl. jen′er-ah) a category of classification between a family and a species

germination (jur′min-ay-shun) the beginning or resumption of growth of a seed or spore

germ-line mutation (jurm′lyn myu-tay′shun) a mutation in a cell from which gametes are derived; the mutation can be passed on to offspring

germplasm (jurm plaz′im) the sum total of all the genes of a species or group of organisms

gibberellin (GA) (jib-uh-rel′in) one of a group of plant hormones that have a variety of effects on growth; they are particularly known for promoting elongation of stems

gill (gil) one of the flattened plates of compact mycelium that radiate out from the stalk on the underside of the caps of most mushrooms

ginkgo (ging′koh) member of the Phylum Ginkgophyta that has fan-shaped leaves and fleshy seeds

girdling (gurd′ling) the removal of a band of tissues extending inward to the vascular cambium on the stem of a woody plant

gland (gland) a small body of variable shape and size that may secrete certain substances but that also may be functionless

glycine (gly′seen) the most simple amino acid, with H as the R group

glycolysis (gly-kol′uh-sis) the initial phase of all types of respiration in which glucose is converted to pyruvic acid without involving free oxygen

glyoxisome (gly-ox′i-zome) organelle that contains enzymes for the conversion of fats into sugars

gnetophyte (neet′oh-fite) member of the Phylum Gnetophyta that is classified with gymnosperms but has angiosperm-like characteristics, such as complex vascular tissue

graft (graft) the union of a segment of a plant, the *scion*, with a rooted portion, the *stock*

grain (grayn) see *caryopsis*

granum (pl. **grana**) (gra′num; pl. gra′nuh) a series of stacked thylakoids within a chloroplast

gravitational water (grav-uh-tay′shun-ul waw′tuhr) water that drains out of the pore spaces of a soil after a rain

gravitropism (grav-uh-troh′pism) growth response to gravity

green algae (green al′jee) members of the Phylum Chlorophyta characterized, among other things, as having both chlorophylls *a* and *b*

ground meristem (grownd mair′i-stem) meristem that produces all the primary tissues other than the epidermis and stele (e.g., cortex, pith)

growing degree days (groh′-ing de-gree days) (GDD) a measure of heat accumulation used to predict plant development

growth (grohth) progressive increase in size and volume through natural development

guanine (gwah-neen) nucleic acid component that base-pairs with cytosine in DNA

guard cell (gahrd sel) one of a pair of specialized cells surrounding a stoma

guttation (guh-tay′shun) the exudation from leaves of water in liquid form due to root pressure

gymnosperm (jim′noh-spurm) a plant whose seeds are not enclosed within an ovary during their development (e.g., pine tree)

h

half-life (haf lyf) the amount of time it takes for a radioactive element to lose half of its radioactivity

haploid (hap′loyd) having one set of chromosomes per cell, as in gametophytes; also referred to as having x chromosomes (as contrasted with $2x$ chromosomes in the diploid cells of sporophytes)

Hardy-Weinberg law (hahr′dee wine′burgh law) specifies the criteria for genetic equilibrium in large populations

haustorium (pl. **haustoria**) (haw-stor′ee-um; pl. haw-stor′ee-uh) a protuberance of a fungal hypha or plant organ such as a root that functions as a penetrating and absorbing structure

heartwood (hahrt′wood) nonliving, usually darker-colored wood whose cells have ceased to function in water conduction

heirloom variety (air′loom vuh-rye′it-ee) a previously popular plant variety that is currently being maintained because of certain desirable qualities

herbaceous (hur-bay′shuss or ur-bay′shuss) referring to nonwoody plants

herbarium (pl. **herbaria**) (hur-bair′ee-um or ur-bair′ee-um; pl. hur-bair′ee-uh) a collection of dried, pressed plant specimens, usually mounted on paper and provided with a label that gives collection information and an identification

heterocyst (het′uh-roh-sist) a transparent, thick-walled, slightly enlarged cell occurring in the filaments of certain cyanobacteria

heterosis (hett-urr-oh′sis) hybrid vigor; superior qualities of heterozygous offspring as compared with those of their homozygous parents

heterospory (het-uh-ross′por-ee) the production of both microspores and megaspores

heterotrophic (het-ur-oh-troh′fick) incapable of synthesizing food and therefore dependent on other organisms for it

heterozygous (het-uh-roh-zy′guss) having two different alleles at the same locus on homologous chromosomes

holdfast (hold′fast) attachment organ or cell at the base of the thallus or filament of certain algae

homologous chromosomes (hoh-mol′uh-guss kroh′muh-sohmz) pairs of chromosomes that associate together in prophase I of meiosis; each member of a pair is derived from a different parent

homozygous (hoh-moh-zy′guss) having two identical alleles at the same locus on a pair of homologous chromosomes

hormone (hor′mohn) an organic substance generally produced in minute amounts in one part of an organism and transported to another part of the organism where it controls or affects growth and development

human evolution (hyoo′muhn ev-oh-loo′shun) use of all evolutionary disciplines to study genetic variation in both modern and historical human populations

hybrid (hy′brid) heterozygous offspring of two parents that differ in one or more inheritable characteristics

hybrid seed (hy′brid seed) mature ovule from genetically different parents

hydathode (hy′duh-thohde) structure at the tip of a leaf vein through which water is forced by root pressures

hydrogen bond (hy′druh-jin bond) attraction between a positively charged hydrogen atom and a negatively charged atom in polar molecules

hydrolysis (hy-drol′uh-sis) the breakdown of complex molecules to simpler ones as a result of the union of water with the compound; the process is usually controlled by enzymes

hydroponics (hahy-druh-pon′iks) technique used to grow plants using mineral nutrient solutions without soil

hygroscopic water (hy-gruh-skop′ik waw′tuhr) water that is chemically bound to soil particles and therefore unavailable to plants

hypha (pl. **hyphae**) (hy′fuh; pl. hy′fee) a single, usually tubular, threadlike filament of a fungus; *mycelium* is a collective term for hyphae

hypocotyl (hy-poh-kot′ul) the portion of an embryo or seedling between the radicle and the cotyledon(s)

hypodermis (hy-poh-dur′mis) a layer of cells immediately beneath the epidermis and distinct from the parenchyma cells of the cortex in certain plants

hypogynous (hi-podj′un-nuss) having flower parts attached below the ovary

hypothesis (hy-poth′uh-sis) a postulated explanation for some observed facts that must be tested experimentally before it can be accepted as valid or discarded if it proves to be incorrect

i

imbibition (im-buh-bish′un) adsorption of water and subsequent swelling of organic materials because of the adhesion of the water molecules to the internal surfaces

imperfect fungus (im-pur′fikt fuhng′guhs) any fungus for which a sexual stage has not been observed and that is a member of the Phylum Deuteromycota

inbreeding (in′breed-ing) mating between individuals with a common ancestry

inbreeding depression (in′breed-ing dee-presh′unn) poor performance and low fertility of inbred individuals

incomplete dominance (in′kom-pleet dom′in-uns) a condition in which the heterozygous phenotype is intermediate to the two homozygous phenotypes as a result of one allele only partly masking another allele

increment borer (in′kruh-muhnt bawr′er) drill that is inserted into a tree to determine its widest girth

indehiscent fruit (in′dee-his-ent froot) a plant ovary that retains its seeds and does not open after ripening

indeterminant growth (in-dee-tur′men-ant grohth) growth in which new tissues are added indefinitely

indusium (pl. **indusia**) (in-dew′zee-um; pl. in-dew′zee-uh) the small, membranous, sometimes umbrella-like covering of a developing fern sorus

inferior ovary (in-feer′ee-or oh′vuh-ree) an ovary to which parts of the calyx, corolla, and stamens have become more or less united so they appear to be attached at the top of it

inflorescence (in-fluh-res′ints) a collective term for a group of flowers attached to a common axis in a specific arrangement

inorganic (in-or-gan′ik) descriptive of compounds having no carbon atoms

integument (in-teg′yu-mint) the outermost layer of an ovule; usually develops into a seed coat; a gymnosperm ovule usually has a single integument, and an angiosperm ovule usually has two integuments

intermediate-day plant (in-tur-me′dee-ut day plant) a plant that has two critical photo periods; it will not flower if the days are either too short or too long

Internet (in′ter-net) a network of computers that enables users to search for information throughout the World Wide Web

internode (in′tur-nohd) a stem region between nodes

interphase (in′ter-feyz) period in which the cell is performing its intended function and not dividing

inversion (chromosomal) (in-verzh′un) a chromosome rearrangement as a result of a segment having been removed, rotated 180°, and then reinserted

in vitro (in vee′troh) "in glass;" growing or being maintained on artificial media, usually in glass test tubes or flasks

ion (eye′on) a molecule or atom that has become electrically charged through the loss or gain of one or more electrons

ionic bond (eye-on′ic bond) attraction between oppositely charged ions

isogamy (eye-sog′uh-me) sexual reproduction in certain algae and fungi having gametes that are alike in size

isotope (eye′suh-tohp) one of two or more forms of an element that have the same chemical properties but differ in the number of neutrons in the nuclei of their atoms

k

kinetochore (kuh-net′uh-kor) a specialized protein complex that develops on the vertical face of a centromere during late prophase; spindle fibers are attached to it

kingdom (king′dum) the highest category of classification (e.g., Plant Kingdom, Animal Kingdom)

knot (not) a portion of the base of a branch enclosed within wood

l

lamina (lam′uh-nuh) see *blade*
laminar flow hood (lam′uh-ner flow hood) device used to filter air to prevent introduction of contaminants during tissue culture
lateral bud (lat′uh-rul bud) see *axil*
lateral meristem (lat′uh-rul mair′uh-stem) region in a plant that produces tissue that increases the girth of roots and stems
laticifer (luh-tis′uh-fur) specialized cells or ducts resembling vessels; they form branched networks of latex-secreting cells in the phloem and other parts of plants
law of dominance (law uhv dom′in-uns) principle stating that for any given pair of alleles, one may mask expression of the other
law of independent assortment (law uhv in-di-pen′duhnt uh-sawrt′muhnt) principle stating that individual hereditary factors assort independently during gamete production
law of unit characters (law uhv yoo′nit kar′ ikters) principle stating that factors, always occurring in pairs, control the inheritance of traits
leaf (leef) a flattened, usually photosynthetic structure arranged in various ways on a stem
leaf gap (leef gap) a parenchyma-filled interruption in a stem's cylinder of vascular tissue immediately above the point at which a branch of vascular tissue (*leaf trace*) leading to a leaf occurs
leaflet (leef′lit) one of the subdivisions of a compound leaf
leaf scar (leef skahr) the suberin-covered scar left on a twig when a leaf separates from it through abscission
leaf trace (leef trays) see *leaf gap*
legume (leg′yoom) a dry fruit that splits along two "seams," the seeds being attached along the edges
lenticel (lent′uh-sel) one of usually numerous, slightly raised, somewhat spongy groups of cells in the bark of woody plants; lenticels permit gas exchange between the interior of a plant and the external atmosphere
leucoplast (loo′kuh-plast) a colorless plastid commonly associated with starch accumulation
light-dependent reactions (lyt-dee-pen-dent ree-ak′shunz) a series of chemical and physical reactions through which light energy is converted to chemical energy with the aid of chlorophyll molecules; in the process, water molecules are split, with hydrogen ions and electrons being produced and oxygen gas being released; ATP and NADPH also are created
light-independent reactions (lyt in-dee-pendent ree-ak′shunz) a cyclical series of chemical reactions that utilizes carbon dioxide and energy generated during the light-dependent reactions of photosynthesis, producing sugars, some of which are stored as insoluble carbohydrates, while others are recycled; the reactions are independent of light and occur in the stroma of chloroplasts
light microscope (lyt mahy′kruh-skohp) device that uses light and a series of lenses to obtain images of a specimen
lignin (lig′nin) a polymer with which certain cell walls (e.g., those of wood) become impregnated
ligule (lig′yool) the tiny, tongue-like appendage at the base of a spike moss (*Selaginella*) or quillwort (*Isoetes*) leaf
linked genes (linked jeens) genes located on the same chromosome
lipid (lip′id) a general term for fats, fatty substances, and oils
locule (lok′yool) a cavity within an ovary or a sporangium
locus (loh′kuss) the position of a gene on a chromosome
long-day plant (long-day plant) a plant in which flowering is not initiated unless exposure to more than a critical day length occurs

m

map unit (map you′nit) a unit of measure equivalent to 1% recombination
mass-flow hypothesis (mass flo hy-poth′uh-sis) see *pressure-flow hypothesis*
mass selection (mass suh-lek′shun) a plant breeding technique in which seeds of plants in a population are used to create each generation
maternal inheritance (muh-terr′nal in-hair′itans) inheritance in which the female gamete contributes extranuclear genes to the offspring
megagametophyte (meg-uh-ga-mee′toh-fyt) the female gametophyte of angiosperms, which, in approximately 70% of the species investigated, contains eight nuclei
megaphyll (meg′uh-fill) a leaf having branching veins; it is associated with a *leaf gap*
megasporangium (meg-uh-spor′an-jee-um) a sporangium in which megaspores are formed
megaspore (meg′uh-spor) a spore that develops into a female gametophyte (megagametophyte)
megasporocyte (meg uh-spor′oh-syt) a diploid cell that produces megaspores upon undergoing meiosis
megasporophyll (meg-uh-spor′uh-fil) a leaf, usually reduced in size, that produces megaspores following meiosis
meiocyte (my′oh-syt) see *sporocyte*
meiosis (my-oh′sis) the process of two successive nuclear divisions through which segregation of genes occurs and a single diploid (2*x*) cell becomes four haploid (*x*) cells
mericloning (mair′i-kloh-ning) see *micropropagation*
meristem (mair′uh-stem) a region of undifferentiated cells in which new cells arise
meristematic (mair′uh-stem-at′ik) consisting of or having a region of undifferentiated cells
mesocarp (mez′uh-karp) the middle region of the fruit wall that lies between the exocarp and the endocarp
mesophyll (mez′uh-fil) parenchyma (chlorenchyma) tissue between the upper and lower epidermis of a leaf
metabolism (muh-tab′uh-lizm) the sum of all the interrelated chemical processes occurring in a living organism
metaphase (met′uh-feyz) stage of mitosis or meiosis in which sister chromatids (or homologous chromosomes in metaphase I) line up along the equatorial plane
microbody (my′kro-bod-ee) a small, spherical organelle which contains special enzymes and is bounded by a single membrane
microfilament (my′kroh-fil′uh-mint) a protein filament involved with cytoplasmic streaming and with contraction and movement in eukaryotic cells
microphyll (my′kroh-fil) a leaf having a single unbranched vein not associated with a *leaf gap*
micropropagation (my-kroh-prop-uh-gay′shun) propagation of plants *in vitro*
micropyle (my′kroh-pyl) a pore or opening in the integuments of an ovule through which a pollen tube gains access to an embryo sac or archegonium of a seed plant
microshoot (my′kroh-shoot) one of several to many shoots produced by a plant growing *in vitro*
microsporangium (my-kro-spor-anj′ee-um) a sporangium in which microspores are formed
microspore (my′kroh-spor) a spore that develops into a male gametophyte (microsporocyte)
microsporocyte (my′kroh-spor′oh-syt) a diploid cell that produces microspores upon undergoing meiosis
microsporophyll (my-kroh-spor′uh-fil) a leaf, usually reduced in size, on or within which microspores are produced
microtubule (my′kroh-t(y)oo-byul) an unbranched, tubelike, proteinaceous structure commonly found inside the plasma membrane where it apparently regulates the addition of cellulose to the cell wall
middle lamella (mid′ul luh-mel′uh) a layer of material, rich in pectin, that cements two adjacent cell walls together
midrib (mid′rib) the central (main) vein of a pinnately veined leaf or leaflet
midvein (mid′vayn) primary region of vascular tissue, usually spanning the center of a leaf
mitochondrion (pl. **mitochondria**) (my-toh-kon′dree-un; pl. my-toh-kon′dree-uh) an organelle containing enzymes that function in the citric acid cycle and the electron transport chain of aerobic respiration
mitosis (my-toh′sis) nuclear division, usually accompanied by cytokinesis, during which the chromatids of the chromosomes separate and two genetically identical daughter nuclei are produced
molecular evolution (muh-lek′yuh-ler ev-oh-loo′shun) study of evolutionary changes in DNA in relation to gene structure, organization, and control of expression
molecular genetics (muh-lek′yuh-ler juh-net′iks) study of gene expression and transmission through investigations of nucleic acid sequences
molecule (mol′uh-kyul) the smallest unit of an element or compound retaining its own identity; consists of two or more atoms

monocotyledon (mon-oh-kot-uh-lee′dun) a class of angiosperms whose seeds have a single cotyledon; commonly abbreviated to *monocot*

monoecious (moh-nee′shuss) having unisexual male flowers or cones and unisexual female flowers or cones both on the same plant

monohybrid cross (mon-oh-hy′brid kross) a cross involving a single pair of genes and heterozygous parents

monokaryotic (mon-oh-kair-ee-ot′ik) having a single nucleus in each cell or unit of the mycelium in club fungi

monomer (mon′oh-mur) a simple individual molecular unit of a polymer

motile (moh′tul) capable of independent movement

multiple fruit (mul′tuh-pul froot) a fruit derived from several to many individual flowers in a single inflorescence

mushroom (mush′room) a sexually initiated phase in the life cycle of a club fungus, usually consisting of an expanded *cap* and a *stalk (stipe)*

mutagen (mew′tuh-jenn) an agent that causes a mutation to occur

mutate (myu-teyt) to cause a heritable change in a DNA sequence

mutation (myu-tay′shun) a heritable change in a gene or chromosome

mycelium (my-see′lee-um) a mass of fungal hyphae

myco-heterotroph (my-co-het′-ur-oh-trohf) a plant that obtains its carbon from a fungus rather than from photosynthesis

mycorrhiza (pl. **mycorrhizae**) (my-kuh-ry′zuh; pl. my-kuh-ry′zee) a symbiotic association between fungal hyphae and a plant root

n

n (en) a designation for the gametophyte generation

2n (too-en) a designation for the sporophyte generation

NAD (en-ay-dee) nicotinamide adenine dinucleotide, a molecule that during respiration temporarily accepts electrons whose negative charges are balanced by also accepting protons and thereby hydrogen atoms

NADH (en′ay-dee-eych) acronym for a high-energy compound called reduced nicotinamide adenine dinucleotide

NADP (en-ay-dee-pee) nicotinamide adenine dinucleotide phosphate, a high-energy storage molecule that temporarily accepts electrons from photosystem I in the light reactions of photosynthesis

NADPH (en′ay-dee-pee-eych) acronym for a high-energy compound called reduced nicotinamide adenine dinucleotide phosphate

nastic movement (nass′tik moov′mint) a nondirected movement of a flat organ (e.g., petal, leaf) in which the organ bends alternately up and down

natural selection (nach′er-uhl si-lek′shuhn) tendency of populations to adapt to the environment to enhance survival

neutron (new′tron) an uncharged particle in the nucleus of an atom

node (nohd) region of a stem where one or more leaves are attached

noncoding DNA (non-koh′ding dee-en-ay) DNA that does not code for a gene

nucellus (new-sel′us) ovule tissue within which an embryo sac develops

nuclear envelope (new′klee-ur en′vuh-lohp) a porous double membrane enclosing a nucleus

nucleic acid (new-klay′ik as′id) see *DNA, RNA*

nucleolus (pl. **nucleoli**) (new-klee′oh-luss; pl. new-klee′oh-ly) a somewhat spherical body within a nucleus; contains primarily RNA and protein; there may be more than one nucleolus per nucleus

nucleotide (new′klee-oh-tyd) the structural unit of DNA and RNA

nucleus (new′klee-uss) the organelle of a living cell that contains chromosomes and is essential to the regulation and control of all the cell's functions; also, the core of an atom

nut (nutt) a one-seeded, dry fruit with a hard, thick pericarp; a nut develops with a cup or cluster of bracts at the base

nutrient (noo′tree-uhnt) a substance that furnishes the elements and energy for the organic molecules that are the building blocks from which an organism develops

o

oil (oyl) a fat in a liquid state

oogamy (oh-og′uh-mee) sexual reproduction in which the female gamete, or egg, is nonmotile and larger than the male gamete, or sperm, which is motile

oogonium (pl. **oogonia**) (oh-oh-goh′nee-um; pl. oh-oh-goh′nee-ah) a female sex organ of certain algae and fungi; it consists of a single cell that contains one to several eggs

operculum (oh-per′kyu-lum) the lid or cap that protects the peristome of a moss sporangium

opposite (op′uh-zit) placed at same heights or distance from one another, as leaves can be arranged on a stem

orbital (or′buh-till) a volume of space in which a given electron occurs 90% of the time

order (or′dur) a category of classification between a class and a family

organelle (or-guh-nel′) a membrane-bound body in the cytoplasm of a cell; there are several kinds, each with a specific function (e.g., mitochondrion, chloroplast)[1]

organic (or-gan′ik) pertaining to or derived from living organisms and pertaining to the chemistry of carbon-containing compounds

osmosis (oz-moh′sis) the diffusion of water or other solvents through a differentially permeable membrane from a region of higher concentration to a region of lower concentration

osmotic potential (oz-mot′ik puh-ten′shil) a measure of the potential of water to move from one cell to another as influenced by solute concentration

outcrossing (out′kross-ing) cross-pollination between individuals of the same species

ovary (oh′vuh-ree) the enlarged basal portion of a pistil that contains an ovule or ovules and usually develops into a fruit

ovule (oh′vyool) a structure of seed plants that contains a female gametophyte and has the potential to develop into a seed

oxidation (ok-si-dey′shuhn) process involving the loss of one or more electrons

oxidation-reduction reactions (ok-suh-day′shun ree-duk′shun ree-ak′shunz) chemical reactions involving gain or loss of electrons to or from a compound

p

P$_{680}$ (pee siks-ey′tee) special reaction center molecule of photosystem II

P$_{700}$ (pee sev′uhn huhn′drid) special reaction center molecule of photosystem I

palisade mesophyll (pal-uh-sayd′ mez′uh-fil) mesophyll having one or more relatively uniform rows of tightly packed, elongate, columnar parenchyma (chlorenchyma) cells beneath the upper epidermis of a leaf

palmately compound; palmately veined (pahl′mayt-lee kom′pownd; pahl′-mayt-lee vaynd) having leaflets or principal veins radiating out from a common point

papilla (pl. **papillae**) (puh-pil′uh; pl. puh-pill′ay) a small, usually rounded or conical protuberance

parenchyma (puh-ren′kuh-muh) thin-walled cells varying in size, shape, and function; the most common type of plant cell

parental generation (P) (puh-ren′tl jen-uh-rey′shuhn) (pee) original organisms involved in making crosses

parental type (pah-renn′tuhl typ) an offspring with the same combination of alleles as one of its parents

parthenocarpic (par-thuh-noh-kar′pik) descriptive of fruits that develop from unfertilized ovaries; the resulting fruit is, therefore, usually seedless

particle gun (pahrt′ik-kuhl gunn) a machine capable of changing the genetic makeup of plant tissue by shooting DNA-coated particles into it

passage cell (pas′ij sel) a thin-walled cell of an endodermis

pectin (pek′tin) a water-soluble organic compound occurring primarily in the middle lamella; when combined with organic acids and sugar, it becomes a jelly

pedicel (ped′i-sel) the stalk of an individual flower or the branchlets of smaller stalks of a flower

peduncle (pee′dun-kul) the main stalk supporting an inflorescence

PEP carboxylase (pee′ee-pee kahr-bok′ suh-leys) an enzyme involved in the C$_4$ photosynthetic pathway that combines carbon dioxide with phosphoenolpyruvate

---

[1]Ribosomes, which are considered organelles, are an exception in that they are not bounded by a membrane.

peptide bond (pep′tyd bond) the type of chemical bond formed when two amino acids link together in the synthesis of proteins

perennial (puh-ren′ee-ul) a plant that continues to live indefinitely after flowering

perforation plate (per′for-ey-shun playt) the end walls of xylem vessel elements with openings through which water flows

perianth (pari′ee-anth) the calyx and corolla of a flower

pericarp (per′uh-karp) collective term for all the layers of a fruit wall

pericycle (per′uh-sy-kul) tissue sandwiched between the endodermis and phloem of a root; often only one or two cells wide in transverse section; the site of origin of lateral roots

periderm (pair′uh-durm) outer bark; composed primarily of cork cells

perigynous (purr-idj′uh-nuss) having flower parts attached around the ovary; the flower parts are usually attached to a cup

peristome (per′uh-stohm) one or two series of flattened, often ornamented structures (*teeth*) arranged around the margin of the open end of a moss sporangium; the teeth are sensitive to changes in humidity and facilitate the release of spores

peroxisome (per-ox′i-zome) organelle that contains enzymes for breakdown of toxic substances and, in plants, participates in photorespiration

petal (pet′ul) a unit of a corolla; it is usually both flattened and colored

petiole (pet′ee-ohl) the stalk of a leaf

$P_{far-red}$, or P_{fr} (pee-far-red or pee-ef-ahr) a form of phytochrome; see *phytochrome*)

pH (pee-aitch) a symbol of hydrogen ion concentration indicating the degree of acidity or alkalinity

pH scale (pee-aitch′hrskeyl) degrees of acidity or alkalinity on specific scale

phage (fayj) see *bacteriophage*

phelloderm (fel′uh-durm) tissue produced to the inside of the cork cambium

phellogen (fel′uh-jun) see *cork cambium*

phenotype (fee′noh-typ) the physical appearance of an organism

pheromone (fer′uh-mohn) something produced by an organism that facilitates chemical communication with another organism

phloem (flohm) the food-conducting tissue of a vascular plant

photon (foh′ton) a unit of light energy

photoperiodism (foh-toh-pir′ee-ud-izm) the initiation of flowering and certain vegetative activities of plants in response to relative lengths of day and night

photorespiration (foh-toh-res-puh-rey′shuhn) alternate photosynthetic process that prevents photoxidative damage in C_3 plants

photosynthesis (foh-toh-sin′thuh-sis) the conversion of light energy to chemical energy; water, carbon dioxide, and chlorophyll are all essential to the process, which ultimately produces carbohydrate, with oxygen being released as a by-product

photosynthetic unit (foh-toh-sin-thet′ik yew′nit) one of two groups of about 250 to 400 pigment molecules each that function together in chloroplasts in the light reactions of photosynthesis; the units are exceedingly numerous in each chloroplast

photosystem (foh′toh-sis-tum) collective term for a specific functional aggregation of photosynthetic units

phragmoplast (frag′mo-plast) a complex of microtubules and endoplasmic reticulum that develops during telophase of mitosis

phytochelatin (fahy-toh-kee′la-tin) polypeptide that sequesters and detoxifies heavy metals

phytochrome (fahy′tuh-krohm) protein pigment associated with the absorption of light; it is found in the cytoplasm of cells of green plants and occurs in interconvertible active and inactive forms ($P_{far\ red}$ and P_{red}); it facilitates a plant's capacity to detect the presence (or absence) and duration of light

phytoremediation (fahy-toh-ri-mee-dee-ey′shuhn) use of plants to facilitate removal of toxins from the soil

pilus (pl. **pili**) (py′lis; pl. py′lee) the equivalent of a conjugation tube in bacteria

pinna (pl. **pinnae**) (pin′uh; pl. pin′ee) a primary subdivision of a fern frond; the term is also applied to a leaflet of a compound leaf

pinnately compound; pinnately veined (pin′ayt-lee kom′pownd; pin′ayt-lee vaynd) having leaflets or veins on both sides of a common axis (e.g., rachis, midrib) to which they are attached

pistil (pis′tul) a female reproductive structure of a flower, composed of one or more carpels and consisting of an ovary, style, and stigma

pit (pit) a more or less round or elliptical thin area in a cell wall; pits occur in pairs opposite each other, with or without shallow, domelike borders

pith (pith) central tissue of a dicot stem and certain roots; it usually consists of parenchyma cells that become proportionately less of the volume of woody plants as cambial activity increases the organ's girth

plankton (plank′ton) free-floating aquatic organisms that are mostly microscopic

plant anatomy (plant uh-nat′uh-mee) the botanical discipline that pertains to the internal structure of plants

plant breeding (plant bree′ding) accelerated plant evolution guided by humans rather than nature

plant community (plant kuh-myu′nuh-tee) an association of plants inhabiting a common environment and interacting with one another

plant ecology (plant ee-kol′uh-jee) the science that deals with the relationships and interactions between plants and their environment

plant geography (plant jee-og′ruh-fee) the botanical discipline that pertains to the broader aspects of the space relations of plants and their distribution over the surface of the earth

plant morphology (plant mor-fol′uh-jee) the botanical discipline that pertains to plant form and development

plant physiology (plant fiz-ee-ol′uh-jee) the botanical discipline that pertains to the metabolic activities and processes of plants

plant systematics (plant sis-tuh-mat′iks) the study of the diversity of organisms and the relationships among them

plant taxonomy (plant tak-son′uh-mee) the botanical discipline that pertains to the classification, naming, and identification of plants

plasma membrane (plaz′muh mem′brayn) the outer boundary of the protoplasm of a cell; also called *cell membrane*, particularly in animal cells

plasmid (plaz′mid) one of up to 30 or 40 small, circular DNA molecules usually present in a bacterial cell

plasmodesma (pl. **plasmodesmata**) (plaz-muh-dez′muh; pl. plaz-muh-dez′muh-tah) minute strands of cytoplasm that extend between adjacent cells through pores in the walls

plasmodium (pl. **plasmodia**) (plaz-moh′ dee-um; pl. plaz-moh′dee-ah) the multinucleate, semiviscous liquid, active form of slime mold; it moves in a "crawling-flowing" motion

plasmolysis (plaz-mol′uh-sis) the shrinking in volume of the protoplasm of a cell and the separation of the protoplasm from the cell wall due to loss of water via osmosis

plastid (plas′tid) an organelle associated primarily with the storage or manufacture of carbohydrates (e.g., leucoplast, chloroplast)

plumule (ploo′myool) the terminal bud of the embryo of a seed plant

pneumatophore (noo-mat′oh-for) spongy root extending above the surface of the water, produced by a plant growing in water; pneumatophores facilitate oxygen absorption

pole (pohl) an invisible focal point toward each end of a cell from which spindle fibers extend in arcs during mitosis or meiosis

pollen cone (pahl′un kohn) male strobilus that produces microsporangia

pollen grain (pahl′un grayn) a structure derived from the microspore of seed plants that develops into a male gametophyte

pollen tube (pahl′un t(y)oob) a tube that develops from a pollen grain and conveys the sperms to the female gametophyte

pollination (pahl-uh-nay′shun) the transfer of pollen from an anther to a stigma

pollinium (pl. **pollinia**) (pah-lin′ee-um; pl. pah-lin′ee-ah) a cohesive mass of pollen grains commonly found in members of the Orchid Family (Orchidaceae) and the Milkweed Family (Asclepiadaceae)

polymer (pahl′i-mur) a large molecule composed of many monomers

polymerase (poh-limm′err-ace) an enzyme that creates a polymer (e.g., DNA polymerase synthesizes DNA)

polypeptide (pahl-ee-pep′tide) a chain of amino acids

polyploid (pahl′i-ployd) having more than two complete sets of chromosomes per cell

pome (pohm) a simple fleshy fruit whose flesh is derived primarily from the receptacle

population (pop-yew-lay′shun) a group of organisms, usually of the same species, occupying a given area at the same time

PQ (pee′kyoo) acronym for an electron acceptor in the light-dependent reactions called plastoquinone

P$_{red}$ or **P$_r$** (pee-red or pee-ahr) a form of phytochrome; see *phytochrome*

pressure-flow hypothesis (presh′ur floh hy-poth′uh-sis) the theory that food substances in solution in plants flow along concentration gradients between the sources of the food and *sinks* (places where the food is utilized)

pressure potential (presh′er puh-ten′shuhl) turgor that builds up against the cell wall as a result of water entering the vacuole

prickle (prik′uhl) a pointed outgrowth from an epidermis or cortex beneath the epidermis

primary consumer (pry′mer-ree kon-soo′mur) organism that feeds directly on producers

primary tissue (pry′mer-ee tish′yu) a tissue produced by an apical meristem (e.g., epidermis, cortex, primary xylem and phloem, pith)

primordium (pry-mord′ee-um) an organ or structure (e.g., leaf, bud) at its earliest stage of development

procambium (proh-kam′bee-um) a primary meristematic tissue that differentiates into primary xylem and phloem

producer (pruh-dew′sur) an organism that manufactures food through the process of photosynthesis

prokaryotic (proh-kair-ee-ot′ik) having a cell or cells that lack a distinct nucleus and other membrane-bound organelles (e.g., bacteria)

promoter region (proh-moh′turr ree′jin) the DNA sequence to which RNA polymerase binds to initiate transcription

propagate (prop′uh-geyt) to multiply by any process of reproduction

propagule (prop′uh-gyool) structure acting as an agent of reproduction

prophase (pro′feyz) stage of mitosis or meiosis in which chromosomes become thick and visible

proplastid (proh-plas′tid) a tiny, undifferentiated organelle that can duplicate itself and that may develop into a chloroplast, leucoplast, or other type of plastid

protein (proh′tee-in or proh′teen) a polymer composed of many amino acids linked together by peptide bonds

protein sequencer (proh′tee-in or proh′teen see′kwens-urr) a machine that reveals the sequence of amino acids in a protein

prothallus (pl. **prothalli**) (proh-thal′us; pl. proh-thal′eye) the gametophyte of ferns and their relatives; also called *prothallium*

protoderm (proh′tuh-durm) the primary meristem that gives rise to the epidermis

proton (proh′ton) a positively charged particle in the nucleus of an atom

protonema (proh-tuh-nee′muh) a green, usually branched, threadlike or sometimes platelike growth from a bryophyte spore; it gives rise to "leafy" gametophytes

protoplasm (proh′tuh-plaz-uhm) living part of a cell

protoplast (proh′toh-plast) the unit of protoplasm within a plant cell wall

protoplast fusion (proh′toh-plast few′zhuhn) the process of combining *in vitro* two protoplasts in one cell

pruning (proon′ing) removal of portions of plants for aesthetic purposes, for improving quality and size of fruits or flowers, or for elimination of diseased tissues

Punnett square (puhn′it skwair) diagram used to determine genotypes produced by a cross

pure-line selection (pyoor lahyn si-lek′shuhn) choosing the best and most uniform of organisms for subsequent generations of a self-pollinated crop

purine (pyoor′een) constituent of DNA that may be either adenine or guanine

pyrenoid (py′ruh-noyd) a small body found on the chloroplasts of certain green algae and hornworts; pyrenoids are associated with starch accumulation; they may occur singly on a chloroplast, or they may be numerous

pyrimidine (pahy-rim′i-deen) constituent of DNA that may be either cytosine or thymine

pyruvic acid (py-roo′vik as′id) the organic compound that is the end product of the glycolysis phase of respiration

q

quantitative trait (kwan′tuh-tay-tiv trait) a trait controlled by several genes and influenced by the environment; it is usually measured on a continuous scale

quantitative trait loci (kwon′ti-tey-tiv treyt loh′sahy) genes that control the quantity of a trait that is not completely determined by any one gene acting alone

quiescence (kwy′ess-ens) a state in which a seed or other plant part will not germinate or grow unless environmental conditions normally required for growth are present

r

R group (ahr groop) an atom or group of atoms, bonded to the central carbon, that determines the identity of an amino acid

rachis (ray′kiss) the axis of a pinnately compound leaf or frond extending between the lowermost leaflets or pinnae and the terminal leaflet or pinna (corresponds with the midrib of a simple leaf)

radicle (rad′i-kuhl) the part of an embryo in a seed that develops into a root

ray (ray) radially oriented tiers of parenchyma cells that conduct food, water, and other materials laterally in the stems and roots of woody plants; they are generally continuous across the vascular cambium between the xylem and the phloem; the portion within the wood is called a *xylem ray,* while the extension of the same ray in the phloem is called a *phloem ray*

receptacle (ree-sep′tuh-kuhl) the commonly expanded tip of a peduncle or pedicel to which the various parts of a flower (e.g., calyx, corolla) are attached

recessive (ree-ses′iv) a condition in which the phenotypic expression of one allele of a gene is masked by the phenotypic expression of another (dominant) allele

recombinant DNA (ree-komm′bin-int dee-en-ay) a molecule created *in vitro* containing DNA from at least two organisms

recombinant type (ree-komm′bin-int typ) an individual offspring that due to recombination has a combination of alleles different from either of its parents

reduction (ri-duhk′shuhn) process involving the gain of one or more electrons

red tide (red tyd) the marine phenomenon that results in the water becoming temporarily tinged with red due to the sudden proliferation of certain dinoflagellates that produce substances poisonous to animal life and humans

region of cell division (ree′juhn uhv sel′ dih-vizh-uhn) area of actively dividing cells at an apical meristem

region of differentiation (ree′juhn uhv diff-uhr-en-tee-a′shun) area of maturing cells near an apical meristem

region of elongation (ree′juhn uhv ih-lawng-gey′shuhn) area near an apical meristem in which cells grow in length and width

reproduction (ree-proh-duk′shun) the development of new individual organisms through either sexual or asexual means

resin canal (rez′in kuh-nal′) a tubular duct of many conifers and some angiosperms that is lined with resin-secreting cells

respiration (res-puh-ray′shun) the cellular breakdown of sugar and other foods, accompanied by release of energy; in aerobic respiration, oxygen is utilized

restriction enzyme (ruh-strikt′shunn en′zym) an enzyme capable of severing a DNA molecule at a specific site

rhizoid (ry′zoyd) a delicate root- or root-hair-like structure of algae, fungi, the gametophytes of bryophytes, and certain structures of a few vascular plants; functions in anchorage and absorption but has no xylem or phloem

rhizome (ry′zohm) an underground stem, usually horizontally oriented, that may be superficially rootlike in appearance but that has definite nodes and internodes

ribose (rahy′bohs) sugar that is found in RNA

ribosome (ry′boh-sohm) a granular particle composed of two subunits consisting of RNA and proteins; ribosomes lack membranes, are the sites of protein synthesis, and are very numerous in living cells

RNA (ar-en-ay) the standard abbreviation for *ribonucleic acid,* an important cellular molecule that occurs in three forms, all involved in communication between the nucleus and the cytoplasm and in the synthesis of proteins

RNA polymerase (ahr-en-ey pol′uh-muh-reys) enzyme responsible for adding nucleotides to RNA

herbivores, 477
seed dispersal by, 135–36
of the taiga, 500–501
vs. plants, 48, 279–80
Anise, 132, 463
Annuals, 87, 123, 479, 506–7
Annual tree rings, 7, 82, 87, 89, 410
Annulus, 357, 400, 401
Ant colonies in plants, 111
Antenna pigments, 169
Antheridia
of ascomycetes, 349–51, 354, 356
of brown algae, 329
of ferns, 400, 401
of ground pines, 390, 391
of horsetails, 397
of liverworts, 376
of mosses, 378
of quillworts, 392
of spike mosses, 391
of whisk ferns, 386, 388
Antheridiophores, 373
Anthers, 124, 429, 466, 469
Anthocerophyta, 371, 377, 383
Anthocyanine pigments, 40, 41, 115, 176, 206
Anthophyta, 428
Anthrax, 298, 300, 301
Anthuriums, 402
Antibiotic production, natural, 306
Antibiotics, 299, 345, 364, 367
development of resistance to, 263
Anticodon, 228
Antipodals, 429, 432–34
Antiseptic mosses, 371
Ants, 76, 77, 111, 135, 363, 478, 507
Aphids
feeding, 154
stylets, 154
as virus vectors, 311
Apiaceae, 132, 463
Apical dominance, 196, 197
Apical meristems
mutation in, 52–53
primary growth and, 53
in roots, 66
in stems, 84–87
Apomixis, 272, 434
Apothecium, 351
Apples, 195, 229, 257, 448
aerial stems (suckers), 70
pomes, 128
seeds of, 137–41, 409
Aquaculture, 337, 490
Aquatic plants
adaptation to environment, 475
algae associated with, 322
insect-trapping, 114–15

intercellular spaces in, 54
leaf modification in, 153
leaves of, 109
roots of, 64
xylem in, 57
Aquatic system, 477, 486
Aquifers, 480, 491
Archaea (archaebacteria), 280, 282, 292
Archaebacteria
characteristics of, 307–9
Archegonia
of conifers, 415–16
of ferns, 400–2
of ground pines, 390, 391
of liverworts, 376
of mosses, 377
of quillworts, 393
of spike mosses, 391, 392
of whisk ferns, 386–88
Archegoniophores, 374, 376
Arctic tundra, 139, 140, 499, 500
Argon, 163
Arid regions, 69, 109, 166, 174, 254
mosses in, 381
photorespiration in, 174
photosynthesis in, 175–78, 506
soils development, 77
Aril, 415, 424
Aristotle, 264, 276
Arms, of root xylem, 68
Artificial selection, 262. *See also* Breeding
Asci, 350–52
Ascocarp, 351
Ascogonia, 350–52
Ascoma, 351–53
Ascomycetes, 349, 350, 354, 356
Ascorbic acid, 303
Ascospores, 351–52, 366
Asexual plant propagation, 255–60
Asexual reproduction, 213
of ascomycetes, 350
of bacteria, 293
of brown algae, 327
of chytrids, 347–48
of club fungi, 355–60
of coenocytic fungi, 348–49
of green algae, 318, 321, 323
of ground pines, 390
of hornworts, 377
of horsetails, 395
of lichens, 365–67
of liverworts, 374
of *Oedogonium,* 322
of *Ulothrix,* 320
of water molds, 342
of yeasts, 351
of yellow-green algae, 325

Ash trees, 91, 502
Asparagus, 95, 448, 467
Aspartic acid, 176
Aspen trees, 115, 487, 500, 502
Aspergillus, 364–65
Assimilation, 13, 182
Asteraceae, 465–67
Asters, 44, 205, 502
Atmosphere, 162, 163. *See also* Air pollution
carbon dioxide in, 163–64, 184, 480–81
nitrogen in, 163, 481–83, 485
water in, 480
Atole, 459
Atomic mass, 14
Atomic number, 14, 15
Atoms
ATP, 148, 162–63, 176
bonds between, 17–18
characteristics of, 13–15
combinations of, 15–16
electrical balance, 14
energy levels of, 14
isotopes, 14–15
movement of hormones and, 189
in photosynthesis, 165–67
production of, 171–72
in respiration, 177
valence of, 16
ATP, 148, 162–63, 176
ATPase, 171, 181
Atropine, 116, 462
Autoclave, 258, 259
Autofusion, 249
Autograph tree, 131
Autopolyploid, 272
Autotrophic bacteria, 296–97
Autumnal change in leaves, 115
Autumn leaves, 102, 115
Auxins, 189–91, 196, 209
cytokinins and, 193
dormancy control and, 208–9
movement in plants and, 198, 199
rooting stimulation and, 256
senescence delay and, 197
Auxospores, 327
Available water, 79, 80
Avian flu, 312
Avocados, 127, 453
Axil
in tiger lilies, 96
in woody twig, 83
Axillary buds
hormonal suppression of, 196
in potatoes, 95
in woody twig, 83

b

Bacillariophyceae, 326–27
Bacilli, 295, 296
Bacillus popilliae, 302
Bacillus thuringiensis (Bt), 252, 302, A21
Bacillus thuringiensis var. *israelensis* (BtI), 302
Backcross, 237
Bacteria
anaerobic respiration in, 177, 179
antibiotic-resistant, 300
archaea and viruses, 291–314
autotrophic, 296–97
beneficial, 292–93, 303
carbon cycle, role of, 480–81
cells of, 31, 292, 309–15
classification of, 280–85, 293–95
communication between, 297
complexity of organization, 12
in composting, 297–98
diseases caused by, 298–99
distribution of, 292
DNA in, 12
enzymes produced by, 24
forms of, 295, 303
galls caused by, 193
genome of, 224
in geologic time scale, 389
gram-negative and gram-positive, 295
in hardpan soil, 483
heterotrophic, 297
inhibitory chemicals in, 478
interferon synthesis and, 311
kingdoms of, 280
luminescent, 292
methane produced by, 308, 489–90
mitochondria origins and, 239
movement of, 295, 298
mutualism with plants, 254
nitrogen cycle role of, 481–83
photosynthetic, 168, 171, 293, 296, 304, 308
photosynthetic pigments in, 166
reproduction in, 293
role in ecosystem, 475–77
size of, 295
slime molds feeding on, 339
thermophilic, 309
true, 295–97
use in agriculture, 254, 302–3, A20–A21
use in bioremediation, 302
use in recombinant technology, 250
viruses attacking, 310, 313

Index

Bacteriochlorophyll pigments, 296
Bacteriophages, 296, 310
Balance of Nature, 477
Bald cypress, 70, 97, 365, 420, 422, 502
Balsam fir, 422
Balsamroot, 465
Balsa wood, density, 96
Bamboo orchids, 469
Bamboos, 203, 448
Bananas, 128, 133, 195, 232, 434
Banner petal, 456
Banyan trees, 70, 71
Baobab, 449
Barberry plants, 202, 360
Barberry spines, 111
Bark, 91
  of conifers, 408–9, 410–16
  of ferns, 403
  outer, 55
Barley, 93, 153, 223, 448, 466
  ergotized, 352
Barrel cactus, 267, 459, 460
Base, of nucleotide, 222, 223
Bases (alkaline compounds), 18
Basidia, 355, 357
Basidioma, 357–60
Basidiomycetes, 355–60
Basidiospores, 357
Basswood, 502
  durability, 96
  rays, 90
Bateson, W., 239, 241
Bats, pollination by, 439
Bay trees, 453
Beadle, George, 231
Beano, 24
Beans, 205, 244, 449
  endosperm in, 434
  gibberellins applied to, 192
  movements in, 198, 202
  mutualism with bacteria, 254
  plants, deficiency symptoms, 157
  seeds of, 254
Beech fern, 401
Beech tree, 91
Beefsteak morel, 352
Bees, pollination by, 270
Beetles, pollination by, 438
Beets, 69, 205, 448
  storage tissues in, 176
Begonia, 255
Beijerinck, Martinus, 310
Belladonna, 116, 448, 462
Belt, Thomas, 478
Bermuda grass, 199
Berries, 127
Beta carotene, 169
Betacyanin, 115
Beta sheet polypeptide, 23
Betel nut, 449

Biennial flowers, 123
Biennials, 479
Big bang reproduction strategy, 479
Binary fission, in bacteria, 294, 295
Binnig, Gerd, 30
Binomial nomenclature, 278, 285
Bioassays, 167, 189, 190
Biodiversity, loss of, 491–95
Bioinformatics, 8, 224
Biological controls, A20–A25
  fungi as, 364–65
  insects as, A20–A22, A24
  viruses as, 312–13, A21
Bioluminescence, 203, 317, 334
Biomass, 475
  plants as percentage of, 3
Biomes, 8, 487, 488, 498–508
Bioremediation, 302
Biotechnology, 264
Bioterrorism, 301, 309
  viral, 309
Biotic community, 474, 475
Biotic factors, 474, 475
Bipinnately compound leaves, 102
Birch trees, 499, 500, 502
  lenticels, 86
  transpiration rate, 149
Bird flu virus, 312
Birds
  nests of, plants used in, 372, 403
  pollination by, 437–42
  of the taiga, 501
Bird's foot fern, 404
Bird's nest fern, 398
Bird's nest fungi, 345, 355, 359–60
Bivalent, 215
Blackberries, 127, 132, 133, 256, 455, 456
Black bread molds, 348–49
Black chokecherry, 456
Black-eyed pea seeds
  effect of water imbibition, 147
Black locust tree, 58, 96, 457
Black locust tree phloem, 58
Black mangrove, 70
Black raspberry cane, 106
Black stem rust of wheat, 360, 361
Black walnut tree, 98, 104, 477
Bladders, of brown algae, 327–29
Bladderworts, 114–15, 201
Blades
  of brown algae, 327
  leaf, 83
Bleeding hearts, 428
Bloodroot, 453, 502

Blooms, algal-bacterial, 306, 331
Blowouts, in tundra, 499
Blueberries, 449
Blue-eyed Mary, 237
Blue-green bacteria. *See* Cyanobacteria
Boletes, 358, 359
Bonds, chemical, 17–18, 22
Bongard, August, 285
Bordeaux mixture, 343
Bordered pits
  torus, 56
  tracheids, 56
Boreal forests, 492, 500
Borlaug, Norman, 246, 247
Boron
  deficiency symptoms, 157
  function in plants, 156
Boston fern, 260, 402
Boston ivy, 71, 96
Botany
  defined, 6
  economic, 8–10
Botulism, 226, 300, 311
Bowstring hemps, 467
Bowstrings
  use of leaves for, 116
Box Family (Buxaceae), 458
Boysenberries, 256
Bracken fern, 403
Bracket fungi, 340, 345, 355, 358
Bracts, 112, 124, 466
Branch roots, 64
Brassicaceae, 130, 267, 454–55
Brassinosteroids, 188, 195, 196
Bread molds, 348–49
Bread wheat, 216, 273
Breeding, plant, 246–54
  of resistant varieties, 475, 477, A22
  with sexually compatible germplasm, 246–48
  with sexually incompatible germplasm, 248–54
Breil, David, 372
Bridal wreath flowers, 125
Bristlecone pines, 86, 197, 410
British soldier lichens, 345
Broccoli, 448, 454
Bromeliads, 176, 403, 507
Broomrapes, 72, 429
Brown algae, 325, 327–29, 338, 343
Brownian movement, 144
Brown, Robert, 28
Brucellosis, 300
Bryophyta. *See* Mosses
Bryophytes, 371–73
  ancestors of, 386
  in geologic time scale, 388
  preserving, 442
Bt toxin, 252, 302, A21

Bubonic plague, 301
Buchner, Eduard, 29
Buckeye, 502
Buckwheat, 448
Budding, 350, 351
Buds
  axillary, 196, 197
  bud scale scars in, 84
  dormancy of, 209
  flowers, 124
  gaps, 86
  hormones in, 189, 194
  of mosses, 379, 380
  primordia, 102
  scale scars, 84
  sensitivity to day length, 208
  terminal, 196–97
Bulbils, 390
Bulblets, 96
Bulbs, 95
  contractile roots in, 198
  dormancy of, 206
  propagation of, 256
  storage, 110
Bulliform cells, 107, 203
Bundle scars, 84, 116
Bundle sheath, 107, 175, 176
Burnam, Charles, 231
Bush monkey flower, 202
Butcher's bloom, 95
Buttercup Family, 276, 450–52
Buttercup root, 151
Buttercups, 451
Butterflies, pollination by, 438–41
Button, of club fungi, 357
Butyl alcohol, 303
Buxaceae, 458

C

Cabbage, 192, 267, 302, 341, 448, 454
Cacao tree, 447
Cacti, 54, 64, 95, 267, 457–60, 474, 506
  CAM photosynthesis in, 176
  as houseplants, 459
Caffeine, 470–71
Calabazilla plant, 70
Calcium, 16
  deficiency symptoms, 157
  function in plants, 156
  in thigmomorphogenesis, 194
Calcium chloride, 16
Calcium ions, in roots, 66
Calendulas, 205
California bay tree, 453
California nutmeg, 415
California poppy root, 64
California soaproot, 468
Callose, 13, 57, 58
Callus, 13

Calmodulin, 194
Calvin cycle, 166, 170, 172–74
Calvin, Melvin, 166, 172, 458
Calyptra, 376, 378, 380
Calyx, 466
   flower, 124
Cambial cylinder, 53
Cambium, 43. See also Vascular cambium
   in dicots, 87–88
   phloem cells, 56–58
Camelina oil, 455
Camphor tree, 451
CAM photosynthesis, 110, 153, 176, 475, 506
Canada balsam, 422
Canada thistle, 70
Canal, of moss archegonia, 378, 379
Cancers
   aflatoxin causing, 365
   medicines derived from plants, 263
   plant hormones and, 195
   tobacco use causing, 462
   viruses causing, 311–12
Candolle, Alphonse Pyramus de, 448
Cannel coal, 397, 407
Canola
   oil, 455
   transgenic, 252
Cantaloupes, 464
Capillary action, 16, 80, 150
Capillary water, 78, 79
Capillitium, 340
Capsicum, 462
Capsules
   dry fruits, 129–32
   of liverworts, 376
   of mosses, 377–82
Caraway, 463
Carbohydrates, 19–21, 35, 38
   conversion to carbon dioxide, 480–81
   disaccharides, 20
   growth and, 172
   monosaccharides, 20
   photosynthesis and, 161, 162, 172–77, 182–83
Carbon, 15
   in carbohydrate molecules, 19
   in compost pile, 297
   in natural cycles, 479–81
   orbitals of, 17
   radioactive, 266
Carbon cycle, 480–81
Carbon dioxide (CO_2)
   in atmosphere, 163–64, 184, 481
   in coal, 405
   in fermentation, 354
   global warming and, 488
   ice ages, cycles of, 488
   in photosynthesis, 102–3, 162–63, 166–67, 172–76, 182
   plants' role in removal, 2
   in respiration, 176, 177, 184
   storage in desert plants, 153
   transpiration rates and, 153
Carbon-fixing and reducing reactions, 166
Carboniferous period, 392, 396, 402
Carbon monoxide, from algae, 323
Carbon skeletons, 40
Carboxyl, 21
Carboxylase, 174
Carboxyl group, 23
Cardamom, 449
Carinal canals, 395–96
Carnations, 205
Carnivorous plants, 2. See also Insect-trapping plants
Carob, 456
Carotenes, 41
Carotenoid pigments, 40, 41, 115, 166
   absorption spectrum of, 167
   hormone synthesized from, 193
   as vitamin A precursors, 188
   yellow, 115
Carpels, 428, 435
   flower, 125
Carpetweed Family, 111
Carpogonia, 330
Carposporangia, 331
Carpospores, 331
Carrageenan, 338
Carrière, Élie, 285
Carrion flowers, 438
Carrot, 69, 448, 479
   chromosomes of, 231–32
   Family, 450, 463
   flowers, 125
   storage tissues in, 176
Cashew nuts, 449
Casparian strips, 67–68, 151
Cassava, 449, 457
Cassia, 451
Castor bean, 449, 458
Casts, fossil, 405
Catabolism, 161
Catalpa, durability, 96
Catalysts, 16, 29
Catastrophism, 264
Caterpillar nematode, A21
Cattails, 79, 428, 486
Cedar-apple rust, 360
Cedar trees, 96, 360, 424
Celery, 192, 254, 463, 479
Cell biology, 8
Cell biology (cytology), 8
Cell model, 31–32
Cells, 12, 27–48
   activities of, and role of water in, 150
   amyloplasts, 40
   of bacteria, 31, 292, 308–9
   cell plate, telophase, 47–48
   cell sap, 41
   cell signaling, 189
   cell theory, 29
   chemical components, 19
   chemical reactions in, 24
   chlorophyll, 40
   chloroplasts, 31, 36, 38
   chromoplasts, 40
   cilia, 42
   communication between, 34
   components, 34
   constraint on volume, 31–32
   of cyanobacteria, 303
   cycle, 42
   cytosol, 31
   dedifferentiation of, 255
   differentiation of, 27, 188, 191, 225–26
   digestion within, 182
   division, 193, 213
   elaioplasts, 40
   endoplasmic reticulum, 36–38
   energy storage, 21
   enzymes, 16
   epidermal, 33
   eukaryotic vs. prokaryotic, 31, 38, 293
   flagella, 42
   flow of information in, 229, 231
   gene function, 24
   glycoproteins, 33
   Golgi apparatus, 38
   growth and, 188
   membranes, 12, 145
   microbodies, 40–41
   middle lamella, 33
   movement within, 42
   organelles, 31
   origins of, 35
   osmotic pressure, 35
   phragmoplasts, 42
   pH regulation, 41
   plasma membrane, 31
   plates, 371
   protein content, 22
   protein production, 39
   protoplasm, 31, 34
   reproduction, 42–48
   sap, composition of, 41
   shapes, 33
   size, 31
   structure of, 31–32
   study of, 28–29
   totipotency of, 258
   turgor, 146
   vacuolar membranes (tonoplasts), 41–42
   vacuoles, 41–42
   water content of, 149, 179–80, 480
Cellular respiration, 40, 161
Cellular slime molds, 341
Cellulose, 20, 96
   in cell walls, 33
   in chytrids, 347
   in dinoflagellates, 332
   microfibrils, 33
   production of, 34
   in roots, 68
   water imbibition and, 147, 151
   in water molds, 343
Cellulosic ethanol, 5
Cell wall, 12, 31, 33
   absence in plasmodia, 339
   in cell model, 32
   cellulose, 20
   new, in meiosis, 216
   permeability, 34
   strength, 33
Cenozoic era, 435
Centers of Origin (Vavilov), 245
Central America, plant origins in, 449
Central cell nuclei, in flowering plants, 429, 433, 434
Central Dogma of Molecular Genetics, 229, 231
Centrioles
   cellular reproduction, 44
   plants vs. animals, 48
Centromeres, 43–45, 214, 216
   anaphase, 46
   metaphase, 46
Cesalpino, Andrea, 362
Chain fern, 404
Chamberland, Charles, 309
Chamise, 487
Chaparral, 487, 488
Character state, 288
Chargaff, Erwin, 223
Charophyta (stoneworts), 334
Chayote, 449
Chemical bonds, 17–18
Chemical formulas, 16
Chemical reactions, 19
   in cells, 24
   role of enzymes in, 24, 29
Chemiosmosis, 171–72, 179, 181
Chemoautotrophic bacteria, 297
Chemotaxic response, 204
Chemotropism, 201
Chernobyl disaster, 367, 493
Cherries, 70
Cherry trees, 449
   lenticels, 60, 86
Chestnut blight, 355, 502
Chia, 461–62
Chiasma (chiasmata), 214
Chicle tree, 92

Chicory, 465, 466
Chilghoza pine, 420
Chinese region of origin, 448
Chisholm, Sally, 307
Chitin, 346
Chlamydia, 300
Chlamydomonas, 319–20
Chlorella, 337
    green algae, 323
    oxygen generation by, 323
Chlorenchyma, 107
    cells, 374
Chlorine, 19, 157
Chlorobium chlorophyll, 297
Chlorofluorocarbons (CFCs), 488, 490
4-Chloroindoleacetic acid, 191
Chlorophyll, 39–40, 163
    in algae, 319, 326, 331, 333, 335
    in bacteria, 296, 304
    destruction of, 165
    green, 115
    light absorption by, 167
    types of, 165–66
Chlorophyll a, 165–66
    absorption spectrum of, 167
    in algae, 331
    bacteriochlorophyll compared to, 296
    in cyanobacteria, 304
    in diatoms, 325
    in dinoflagellates, 332
    in photosynthetic unit, 168
    in prochlorobacteria, 307
Chlorophyll b, 165–66
    absorption spectrum of, 167
    in algae, 319
    in photosynthetic unit, 168
    in prochlorobacteria, 307
Chlorophyll c, 166, 333
Chlorophyll c_1 and c_2, in diatoms, 326
Chlorophyll d, 166, 331
Chlorophyll e, 166
Chlorophyta. *See* Green algae
Chloroplasts, 36, 39, 107
    biochemical reactions in, 176
    CAM photosynthesis, 176
    carotene pigments in, 189
    in cell model, 32
    in cryptomonads, 334
    in cyanobacteria, 304
    in diatoms, 326
    in dinoflagellates, 332
    DNA in, 238
    in euglenoids, 279, 331
    genomes of, 239
    in green algae, 318–23
    in haptophytes, 334
    hormones and development of, 193–94
    in hornworts, 377
    inner membrane, 40
    in Kranz anatomy leaves, 174
    in liverworts, 373
    in mosses, 379
    number in cells, 38
    in *Oedogonium*, 322
    origins of, 31, 305, 307
    in photorespiration, 174
    in photosynthesis, 163, 166, 183
    photosynthetic units in, 168
    in red algae, 331
    size and shape, 38
    in *Spirogyra*, 321
    structure, 39
Chloroxybacteria, 293, 296
Cholesterol, 21, 303, 362
Cholesterol-lowering plants or derivatives, A43–A44
Cholla cactus, 459
Chromatids, 43, 215, 216
Chromatin, 36, 222
Chromium, in soil, 475
Chromophyta, 325–29, 334
Chromoplasts
    color, 39
    DNA in, 39
Chromosomes, 36
    in bacterial cells, 293
    changes in structure and number, 231–32
    crossover of, 214
    deletion in, 268
    DNA molecules in, 222–23
    in gametophyte generation, 217
    genetic information in, 222
    genetic mapping of, 235, 239
    homologous, 213–16
    inversions of, 231, 268
    locus of genes on, 234, 241
    during meiosis, 212, 213, 216
    study of, 231
    translocations of, 231, 268
    transposable genetic elements in, 221, 238
Chronic wasting disease, 314
Chrysanthemums, 192, 205, 465
Chrysophyceae (golden-brown algae), 325
Chytridiomycota, 347–48
Chytrids, 347–48
Cilia, 42, 204
Cinnamon, 451
Cinnamon fern, 398
Circadian rhythms, 202–3, 206
Citric acid, 303, 364
Citric acid cycle, 176–78, 180, 181
Citronella oil, 467
Citrus fruits, 137, 195, 449
Clades, 279, 284
Cladistics, 7, 288
Cladograms, 288, 289, 318
Cladophora, green algae, 325
Cladophyll, 82, 95
Clamp connections, 356
Clary's annual sage
    bracts, 112
    flowers, 112
Class Bacteriae, 296
Class Cyanobacteriae, 303
Classes, 280
Classification systems
    for bacteria, 280–82, 285, 295
    binomial, 277
    categories in, 280–85
    DNA sequence data and, 289, 347
    early, 276–79
    for flowering plants, 289, 428, 435–37, 449–50
    for fungi, 346
    International Code, 278–79
    keys in, 285
    kingdoms in, 279–80
    phylogenetic nomenclature, 279
    species in, 278, 285–89
Class Prochlorobacteria, 307
Clay, 78
Cleistothecium, 351
Clematis, 451
Clematis rachises, tendrils, 109–10
Climacteric fruits, 195
Climate. *See also* Global warming
    adaptation, leaves, 108–12
    biomes and, 499, 505
    conditions, annual rings, 89
    data from annual tree rings, 6, 84
Climax communities, 485–86
Clinostat, 200
Cloned genes, 7
Cloning vector, 249
Closed system, 7, 323
Clostridium botulinum, 300
Clostridium difficile, 300
Clover, 448, 457
Clover fern, 401
Club fungi, 351, 355–60
Club mosses, 87, 378, 386, 388–90
    ancient relatives of, 392
    human relevance of, 392–94
Coal
    energy stored in, 177
    formation, 99
    origins of, 4, 397, 405, 407, 479
Coastal forests, 502–3
Coastal redwood, 92, 422
Cobalt, 19, 156
Cocaine, 116, 449
Cocci, 295, 296
Coccidiomycosis, 365
Coccolithophores, 334
Cocklebur, 147, 207, 209
Coconuts, 449
    fruit, 127
    milk, 193
    palms, 409
Codeine, 453, 454
Codons, 228–30
Coenocytic hyphae, 342
Coenocytic mycelia, 348
Coenocytic true fungi, 348–49
Coenozygote, 349
Coenzyme A (CoA), 177, 180, 192
Coenzymes, vitamins as, 188
Coffee, 5, 209, 449, 470–71
Coffee berries, 5
Cohesion-tension theory, 150–51
Colchicine, 467
Coleochaete, 317
Coleoptiles, 137, 189
    gravitropism and, 199
    phototropism and, 198–99
Coleorhiza, 137
Coleus, 199, 255, 461
    stem, section, 85
Collenchyma tissue, 54
Colloid, 483
Colloidal materials, 147
Colloidal particles, 78
Color change in leaves, 115, 164–66
Colors, of flowers
    genetic determination of, 234, 237–38
    role in pollination, 438–39
Columbine, 220, 451, 452
Column, of orchids, 469
Common and scientific names, A1–A19
Common names, of plants, 276
Communities, plant, 475, 483
    ecological balance in, 475–77, A20
    human impact on, 483, 488–91
Comose seed, 134
Companion cells, 85
    phloem, 56–58
Companion planting, A22, A24–A25
Compass plant, 109, 110
Competitive species, 479
Complete flowers, 436
Complex tissues, 55–61
Composting, 297–98, 363

Compound
  compound ovaries, flower, 125
  defined, 15
  flower, 126
  leaf structure, 102
  light microscopes, 29, 30
  ovary, 436–37
  pistils, 436–37
Compression fossils, 405
Conceptacles, of brown algae, 329
Condensation reaction, 20
Conducting sclerenchyma (xylem), 54–55
Cone-bearing trees, 87–88, 91
Cone, female cycad, 408
Coneflower, 125, 438, 479
Confocal scanning microscopy, 29
Conidia, 350, 351, 356, 363–65
Conidiophores, 350, 365
Conifer roots, 68
Conifers, 386, 408–16
  in boreal forests, 492, 500
  climate change and, 502
  and flowering plants, evolution of, 428–32
  human relevance of, 419–24, 425
  in primary succession, 486
  resin canals in, 410, 411, 420
  resin secretion in, 410
  wood of, 415
Conjugation, 293, 295, 321, 349
Conjugation tubes, 321, 322
Conservation
  biology, 263
  efforts, 3, 227
Consumers, in ecosystem, 476, 478
Contractile roots, 71–72, 198
Contraction movements, 198
Convergent evolution, 267
Cooling, water needed for, 150
Copals, 456
Copeland, Herbert F., 279, 280
Copper, 13, 19, 171
Coprolites, 406
Coralline algae, 330
Coral tree, 456
Cork cambium, 43, 53, 59, 60, 69, 85, 92, 193, 403
  in roots, 69
Cork cells, 28, 59, 86
Corms, 95
  propagation of, 257
  of quillworts, 392, 393
Corn, 70, 93, 138, 449, 466
  ancestor of, 265
  carbon absorption by, 163
  corn rust, 360
  C_4 photosynthesis in, 175

diploid, 272
diseases of, 360
endosperm in, 433
epigenetic alterations in, 265
ethanol production from, 5
hormones produced by, 191
hybrid, 248, 254
leaf cells, 38
leaf section, 174
leaves of, 191
plants, transpiration rate, 149
root growth, 150
seed germination in, 254
transgenic, 252–54
transposable genetic elements in, 221–22
vascular bundle, 93
Corolla, 466
  flower, 124–26
Corpse flower, 121–23
Correns, Carl, 232
Cortex, 85, 88
  in roots, 66, 67
Cotton
  plants, 5, 205, 232, 449
  plants, transgenic, 251
  transgenic, 252
Cottonwood, durability, 96
Cotyledons, 87, 137, 193
Coulter pines, 412
Covalent bonds, 35
  amino acids, 22
  nonpolar, 17
  polar, 17
Cowslips, 451
Cox, Jeff, A21
C_3 plants, 174–76, 481
C_4 plants, 174–76, 481
Cranberries, 80
Crassulacean acid metabolism (CAM)
  photosynthesis, 176
Cretaceous period, 392, 423, 435
Crick, Francis, 221, 223
CRISPR/Cas, SSN, 253
Cristae, 40
Crista, in mitochondrion model, 40
Crocus, 95
Crop plants
  breeding of, 246–54
  cross-pollinating, 247
  domestication of, 244–45
  endosperm in, 433–34
  germplasm of, 248
  global warming and, 164
  in Grass Family, 466
  in Legume Family, 456
  loss of biodiversity and, 491
  major, 244
  nitrogen loss from harvesting of, 482–83

origins of, 448–49, 491
polyploidy in, 271–72
propagation of, 254–60
rotation of, 491
self-pollinating, 246
temperature and growth of, 207–8
transgenic, 249–52, 264, 302
viruses and loss of, 310
wild relatives of, 245
Cross, Brian, 192
Crossing-over, 214–15, 231, 239, 241
Cross-pollinating plants, 246
Crown division, 255
Crown gall, 250
Crown-of-thorns, 458, 471
Croziers, 351, 399, 400
Cruciferae, 454
Crustose lichens, 366, 367
Cryptochromes, 206, 210
  lichens and, 304
Cryptomonads (Cryptophyta), 334
Crystal, in model cell, 32
Cucumbers
  carpels of, 435
  seeds of, 254
Cucurbitaceae. *See* Pumpkin Family
Cultivar, 279
Cultivated plants
  cultigen, 278
  diversity of, 449
Cuticles, 58, 67, 84
  of pines, 409
  plant evolution and, 371, 386
  of in vitro grown plants, 259
Cutin, 22, 58
  leaves, 105
Cuttings, 255
Cuvier, Georges, 264
Cyanobacteria, 168, 171, 179, 204, 293, 304–7
  after fire, 488
  blue-green bacteria, 297
  evolution of, 168, 171
  in ferns, 404
  in hornworts, 377
  in lichens, 304, 365
  representative, 305
Cyanobacterins, 306
Cyanophages, 306
Cyanophycin, 304
Cyanosis, 492
Cyathium, 457, 458
Cycadeoids, 417
Cycad, female cone of, 408
Cycadophyta, 49, 416–17
Cycads, 409, 416–17, 424
Cyclamens, 205
Cyclic electron flow, 171

Cyclic photophosphorylation, 171
Cyclosis, 42, 154
Cyclosporine, 364
Cypress trees, 365, 420, 422. *See also* Bald cypress
Cysteine, 176
Cystocarps, 331
Cytochrome complex, 171, 179
Cytochromes, 171, 179, 181
Cytogenetics, 231–32
Cytokinesis, 43
Cytokinins, 193, 195, 197, 209
Cytoplasm, 12, 19, 31, 145
  in cell model, 32
  cytoplasmic streaming, 42
  effect of plasmolysis on, 146
  movement within, 42
  of pollen grain, 432–33
  protein synthesis in, 34
  respiration in, 176–78
Cytosine, 222, 223, 225
Cytoskeleton, structure, 42
Cytosol, 31

d

2, 4-D, 191
Daffodils, 467
Dahlias, 206, 465
Damask roses, 455
Dammar, 420
Dandelions, 69, 92, 205, 272
  adaptability, 12
  flowers of, 465
  reproduction of, 272
  susceptibility to hormones, 191
  uses of, 466
Darwin, Charles, 189, 197, 262, 264–66, 268–69, 273, 292, 428
  On the Origin of Species, 262, 265, 266
  plant movement studies of, 197
Darwin, Francis, 189
Data, 6
Daughter cells (mitosis), 43
Daughter chromosomes
  anaphase, 46
  mitosis, 43
  telophase, 46–48
Da Vinci, Leonardo, 264
Dawn redwood, 421
Day length, sensitivity to, 205
Daylily, 254
Day-neutral plants, 205, 462
DDT, 338, 355
Death angel mushrooms, 357, 360
De Candolle, Alphonse, 448
Decay organisms, 96

Index

Fats—Cont.
  polyunsaturated, 22
  saturated, 21
  unsaturated, 22
Fatty acids, 21
  chains, 40
  source, 21
Feeder roots, 64
Female cycad cone, 408
Fermentation, 177–79, 354
Ferns, 83, 386, 398–407
  fossil fuels and, 406
  fossil relatives of, 402
  in geologic time scale, 388
  as houseplants, 398, 402
  human relevance of, 402–4
  mineral requirements, 156
  reproduction of, 399–402
  rhizomes in, 403, 406
  structure and form of, 398
  taxic movement in, 204
  vs. flowering plants, 409
Ferredoxin (Fd), 171
Fertilization, 125, 218, 432–34
Fertilizers
  algae used as, 338
  compost used as, 297–98
  cyanobacteria used as, 305
  ferns used as, 404
  Green Revolution and, 247–47
  inorganic, 483
  from methane digester, 309
  use in plant propagation, 255
  water contamination by, 492
Fetal testing, 227
Fibers, 87, 162
Fibrous roots, 64, 65
Fiddleheads, 399, 400, 403
Field capacity, 80
Fig, 134
Figwort Family, 478
Filaments
  of bacteria, 292, 304
  flower, 125
  of fungi, 347
  of grasses, 466
  of mosses, 378
  of *Oedogonium,* 322
  of red algae, 330
  of *Spirogyra* (watersilk), 322
  of *Ulothrix,* 320, 321
Filaree fruitlet, 137
Fir, durability, 96
Fires
  annual rings, 89
  carbon dioxide released by, 481
  diversity of species and, 483, 484
  nitrogen loss from, 482
  role in chaparral biome, 508

  role in ecosystem, 484, 487–88
  seed germination and, 414, 487, 503, 508
  trees resistant to, 487, 503
Fireweed, 272
First filial (F_1) generation, 233
First law of thermodynamics, 18
Fir trees
  apical dominance in, 197
  biomes of, 500, 503
  Douglas, 422, 502, 503
  use by humans, 422
Fish
  algal-bacterial blooms and, 306
  farming, 336
  flashlight, 292
  mosquito control using, A21
  water molds on, 342
Fission, in bacteria, 293
Flaccid cell, 146
Flagella, 42, 204
  of bacteria, 295
  of brown algae, 326, 335
  of *Chlamydomonas,* 320
  of chytrids, 347
  of cryptomonads, 334
  of dinoflagellates, 333
  of euglenoids, 279, 332, 334
  of green algae, 319–20, 324, 336
  of haptophytes, 334
  of *Oedogonium,* 322
  of slime molds, 340
  of water molds, 342
Flavin adenine dinucleotide (FAD), 171, 179, 180
Flax, 448, 467
Fleming, Sir Alexander, 364
Flies, A21
  pollination by, 438
Floating liverworts, 378
Flora, 277
Flora of Holstein (Wiggers), 277
Florets, 465
Floridean starch, 331
Flowering plants, 428–37
  classification of, 289, 427, 435, 449–50
  evolution of, 433, 435, 437–38, 443, 449
  fertilization in, 432–34
  in geologic time scale, 388
  humans relevance of, 450–70
  life cycle of, 430
  photoperiodism in, 205–7
  phototropism in, 198–99
  pollination of, 432, 437–42
  polyploidy in, 271
  preservation methods, 442–45
  seeds of, 409

  at vernal pools, 503, 504
  vs. ferns, 409
Flower pot leaves, 111
Flowers
  abnormalities in, viruses and, 311
  advanced, 436
  annuals, 123
  biennials, 123
  bilateral symmetry vs. irregularity in, 438
  complete vs. incomplete, 436, 437
  corpse flower, 121–22
  diversity of, 120–23
  epiphytes, 122
  ethylene produced by, 193
  evolution of, 433, 435, 437, 438
  form and structure, 124–26
  fruit development from, 429
  fruits and seeds, 120–41
  gibberellins and growth in, 191–93
  habitat variations, 122–23
  maturation process, 123
  perennials, 123
  perfect vs. imperfect, 437
  preservation methods, 442–44
  primitive, 435–36
  structure of, 435–37
  transgenic, 252–54
  turgor contact movements in, 202, 203
Flower structure, 124–34
Fluorescence, 168. See also Bioluminescence
Fluorescence microscopy, 29
Fly agaric mushroom, 361
Foliose lichens, 366
Follicles, 129
Food (human)
  algae products, 337, 338
  bacterial contamination of, 299–300
  balancing amino acids, 22–23
  beneficial bacteria in, 300, 303
  from Buttercup Family, 450–51
  cacti used for, 458–59
  from Carrot Family, 463
  coffee, 5, 470
  coffee beans, 5
  conifers and, 419–24
  cyanobacteria as, 306
  fats in, 21
  flowering plants as, 433–34
  fungi used for, 350, 351, 364
  genetically modified products, 252
  from grasses, 123
  from Grass Family, 466–67

  horsetails used for, 396
  from Laurel Family, 451–53
  leaves, use of, 116–18
  from Legume Family, 456–57
  legume seed deficiency, 23
  lichens as, 365
  from Lily Family, 467–68
  mushrooms as, 361–62
  from Mustard Family, 454–55
  Penicillium molds and, 364
  photosynthesis and, 171
  pines and, 419–20
  population growth and, 491
  protein sources in, 23
  from Pumpkin Family, 463–64
  quillwort corms as, 392–93
  refrigeration of, 179
  from Rose Family, 456
  safety of, PCR screening techniques, 226
  shelf life of, hormones extending, 196
  from Spurge Family, 457
  starches, 20
  storage in low oxygen environment, 180
  from Sunflower Family, 465–66
  vegetarian diets, 477, 491
  viral contamination of, 311
  wild plants and fungi, A26
Food (plant)
  storage of, 21
  transport of, 152, 154–56
Food chains, 475
  algae in, 319, 335
  cyanobacteria in, 305, 335
  dinoflagellates in, 333
Food webs, 476, 495
Foot
  of hornworts, 377
  of liverworts, 376
  of whisk ferns, 386–87
Ford, Clellan, 448
Foreign species, invasion of, 493
Forensics, 226
Forensic science, 7
Forests, 474
  boreal, 492, 500
  coastal, 501–2
  destruction of, 494, 508
  destruction of tropical, 2
  fires in, 483, 484, 487–88, 502
  mountain, 502, 503
  temperate deciduous, 499
  tropical rain, 2, 184, 507–8
Forsythia, 255
Fossil fuels
  acid depositions and, 492

alternatives to, 4, 488
carbon dioxide released by, 163, 184, 480–81
energy stored in, 161, 177, 179
fertilizers produced from, 248
as nonrenewable resource, 404
origins of, 397, 404, 405
Fossils, 405–6
of algae, 317, 319
amber, 420
of bacteria, 292
early plant features in, 370
as evidence for evolution, 264–65
of ferns and fern ancestors, 386, 402
of flowering plants, 433–34
living, 423
study of, 264
Foxgloves, 88, 479
Fragmentation, reproduction by, 321–22, 377, 382
Frankincense, 91
Franklin, Rosalind, 223
Fronds, of ferns, 398–400, 402–4
Fructokinase, 180
Fructose, 20, 465
in fermentation, 177
in respiration, 177
Fruit, 125–37
ABA in, 193
animal dispersal, 135–36
botanical, 126
controlled ripening of, 194–96
dehiscent, 129
development from flowers, 430
dispersal, 133–37
harvested, respiration in, 179
indehiscent, 131
irregularities in, viruses and, 311
parthenocarpic, 126, 128
seed dispersal and, 133–37
seedless, 191, 192, 232, 434
wind dispersed, 134–35
Fruit regions, 126
Fruticose lichens, 366, 367
FT (gene), 205
Fuchsias, 211, 439
Fucoxanthin, 325–27
Fuel
wood use, 96, 99
from yareta plants, 116
Fumitory Family (Fumariaceae), 276
Functions of secondary tissues, 85

Fungal hyphae, 73
Fungi
bryophytes associated with, 371–73
chytrids, 347–48
classification of, 283
club (basidiomycetes), 355–60
coenocytic, 348–49
diseases caused by, 348, 352, 354–55, 363
enzymes derived from, 24
fire and decrease in, 487–88
forms of, 346
human and ecological relevance of, 346
imperfect (deuteromycetes), 363–64
inhibitory chemicals in, 478
in lichens, 365
mycorrhizal, 347, 364, 472, 487
in nitrogen cycle, 481–83
phototropism in, 198
plant diseases caused by, 298
preservation techniques, 442–45
role in ecosystem, 475–77
and roots, 73
sac (ascomycetes), 350
vs. Protista, 345–47
Fungicides, 254, 255, 343
Funori, 338

g

GA. See Gibberellins (GA)
Galls, bacteria and, 193, 250
Gametangia, 329, 342, 349, 371, 378, 386
Gametes, 213, 217
chromosomes in, 213
in green algae, 320–22
haploid, 216
in Mendelian genetics, 234
Gametophores, 373, 374
Gametophyte generation (n), 217, 218, 435
Gametophytes, 217
of bryophytes, 370–73
evolution of, 387
of ferns, 399, 401, 402
of flowering plants, 429, 433, 434
of ground pines, 389–90, 391
of gymnosperms, 408–11, 415
of hornworts, 377
of horsetails, 397
of liverworts, 373–74, 376
of mosses, 377–78, 379
of quillworts, 392
of spike mosses, 390, 392
of whisk ferns, 387, 388

Gane, R., 194
GA3P, 173–75, 177, 178, 180
Garbanzo beans, 204. See Chickpeas
Garden bean, 138
Garden camellia, leaves, 118
Gardenia, 191
Gardening
biological controls in, A20
companion planting in, A22
mulch use in, 483
slug control in, 336
Garden pea, tendrils, 110
Garlic, 448, A22
Garner, Wightman W., 205, 206
Gas chromatography, 189
Gas gangrene, 301
Gauthier, Jacques, 279
Gelatinous sheaths, of cyanobacteria, 304
Gemma (gemmae), 374
Gemmae cups, 374
Genera, 277, 280
Generative cell, 413
Generative nucleus, 432
Gene(s), 25
banks, 248
in chloroplast and mitochondrial genomes, 239
cloning, by bacteria, 249–50
defined, 7, 25, 188, 223
developmental control, 265
in DNA molecules, 24
edited plants, 253
environment and expression of, 188, 225, 228
enzymes and, 188
epigenetic control of, 264–65
flow, 268
homeobox, 265
hormones and, 188
incomplete dominance of, 237
interactions among, 237–38
jumping, 220–21
law of independent assortment of, 235
linked, 239–41
locus on chromosome, 234, 237
mutations within, 221
promoter region in, 228
regulatory, 265
sexual reproduction and recombination of, 214, 218
splicing, 248–52
transfer, 8, 193
unlinked, 235, 237
Genetic code, 228–29
Genetic drift, 269

Genetic engineering, 8, 249–50, 252, 253, 262
biological pest control and, 302
increased crop production, 491
Genetic equilibrium, 241
Genetic mapping, 235, 239
Genetic recombination, in bacteria, 293, 305
Genetics, 8
cytogenetics, 231
Mendelian, 232–38
molecular, 220–41
population, 264
Genetic variability, and plant breeding, 247
Genome, 224, 228
Genome editing, 253
Genomics, 8, 214
Genotype, 234
Genus, in binomial system, 277, 285
Geographic isolation, and evolution, 269–70
Geography, plant, 8, 448
Geologic time scale, 388
Geomagnetotropism, 201
Geranium Family, 204, 255
Germination, 137–39
conditions necessary for, 208
enzymes and, 209
fire and, 415, 488, 508
force exerted by, 147, 484
hormones and, 192, 194, 197
phytochrome pigments and, 206
role of water imbibition, 147
temperature and, 208
water and, 254
Germ-line modifications, 253
Germ-line mutations, 229–30
Germplasm, 248
Giant sequoia trunks, 82
Giant water lily leaves, 103
Gibberellins (GA), 191–93, 197
dormancy control and, 208, 209
limitations on, 192
seed germination, 197, 209
senescence delay, 197
Gibbons, Euell, 419
Gills, mushroom, 357, 358
Ginger, 448
Ginkgoes, 408, 409
human relevance of, 424
leaves of, 409
venation, 103
Ginkgophyta, 409, 416
Ginseng, 448
Girdling, 256, 478
Gladiolus, 95
Glands, 59
leaf, 105

Index

Glass, in grasses, 117
Gliding movements, 204
Global warming, 2, 163–64, 184, 484
   alpine flora as indication of, 505
   human contributions to, 487
Gloeocapsa, 305
Glucose, 20, 24
   in fermentation, 177
   in photosynthesis, 161, 163, 183
   in respiration, 176–81
Glume, 466
Glutamine, 176
Glyceraldehyde 3-phosphate (GA3P), 174, 177, 180
Glycerol, 21
Glycine, 23
Glycolysis, 176–81
Glycoproteins, 33
Glyoxisomes, 41
Gnetophyta, 409, 416
Gnetophytes, 409, 416
Gnetum, 418, 424
Golden-brown algae, 325
Golden chain tree, 91
Golden rice, 252
Goldenrods, 205, 502
Goldenseal, 451
Goldenweed, 36
Golgi apparatus, 38
Golgi body, 38
Golgi, Camillo, 38
Gonorrhea, 300
Goober peas, 127
Gooseberries, 128, 360, 487
Gore, Al, 489
Gourds, 448, 464
G_1 period, 43
G_2 period, 43
   cellular reproduction, 43
   in growth, 193
Grafting, 83, 256–57
   approach, A70, A72
   bridge, A70–A71, A73
   bud, A71–A73
   cleft, A68, A71
   double-working, A73–A74
   inarching, A70, A72
   root, A73
   side, A70
   splice, A68
   whip of tongue, A68
Grains
   domestication of, 244–45
   ergot fungus on, 352
   fruit, 132
   natural stratification, 208
Gram, Christian, 295
Gram-negative bacteria, 295
Gram-positive bacteria, 295
Grand Rapids lettuce, 206
Granum (grana), 39

Grapes, 96
   downy mildew of, 343
   flowers, 125
   GA use in, 192
   as nonclimacteric fruits, 195
   seedless, 192, 434
Grasses, 428, 450, 466–67, 468
   adaptation to fire, 487
   bending toward light source, 189
   biomes of, 499
   carbon-fixing in, 166
   endosperm in, 433
   environment of, 474
   flowers of, 428, 466
   in geologic time scale, 388
   glass teeth, 117
   as herbaceous perennials, 479
   internodes, 93
   leaves of, 466
   photoperiodism in, 205
   roots, 64
   water conservation movements in, 203, 204
Grassland, 503–6
Grass leaf cross section, 108
Gravitropism, 65, 196, 199–201
Green algae, 38, 318–25
   lichens and, 365
   in plant evolution, 371
   use for food, 337
   vs. plants, 371
Greenbriers, 95
Greenhouse effect, 163, 184, 488
Greenhouse stores, 382
Green manure, 491
Green plant phylogeny, 318
Green Revolution, 246–47
Green sulfur bacteria, 296
Grew, Nehemiah, 7, 28, 150
Grooves, in dinoflagellates, 332
Ground meristem, 53, 66, 84
Ground pines, 388–94
Growth
   accelerators, 193–94
   carbohydrates and, 172
   definition of, 188
   determinate, 69
   determinate *vs.* indeterminate, 188
   environmental influences on, 478
   essential elements required, 156
   factors required for, 156–57
   hormones and, 188–96
   indeterminate, 69
   inhibitor, 137–38, 193
   movements, 197
   nutrients and, 188

photoperiodism and, 205–7
retardants, 192
signal transduction pathway and, 188
temperature and, 207–8
of tissues, 52–61
variation in form, 12
Guanine, 222, 223, 225
Guard cells, 59, 151
   functions, 106
   leaves, 103, 104, 106
Guava, 449
Guillain-Barré syndrome, 312
Gullet, 280, 332, 334
Guttation, 103, 153, 154
Gymnosperms, 408
   conifers, 410–16
   cycads, 409, 416, 417
   ginkgoes, 409, 416–17
   gnetophytes, 409, 418–19
   human relevance of, 419–24
   phyla of, 416
   seeds of, 408
   vs. angiosperms, 427, 428
Gynophore, 457
Gypsy moth caterpillars, 313

h

Haberlandt, Gottlieb, 193
Habitats of flowers, 122–23
Haeckel, Ernst, 279, 474
Hairs, epidermal in stomatal crypts, 153
Hairy cap mosses, 378, 381
Hales, Stephen, 150
Half-life, 266
Hallucinogenic plants, A44–A45
Halons, 490
Haploid cells, 216, 217
Haptonema, 334
Haptophytes, 334
Hard maple, uses, 98
Hardpan soil, 483
Hardwood, 91, 397
   tree, water use, 149
Hardy, G. H., 241
Hardy-Weinberg law, 241, 246, 269
Harlan, Jack R., 448
Haustoria, 72, 363
Hawaiian tree fern, 388
Heart disease, 21
Heartwood, 90
Heath Family, 125
Hedgehog cactus, 459
Heirloom varieties, 247
Heliotropism, 203
Helium, 17, 163
Hemberg, Torsten, 193
Heme of hemoglobin, 165
Hemicellulose, 33
   chains, in cell model, 33

Hemlock, 463, 502
   use by humans, 424
   uses, 97
   water, 463, 464
Hemp, 93, 449
Hepatica, 452, 502
Hepaticophyta, 371, 373–76
Herbaceous dicots, 52, 87
Herbaceous dicotyledons, 87–88
Herbaceous plants
   cell turgor and water in, 149
Herbaria, 442–45
Herbicides
   GA counteracting, 192
   hormones used as, 192
   plants resistant to, 251
Herbivores, 477, 486
Hereditary variations, 268
Hesperidium, fruit, 128
Heterocyst, 304
Heterosis, 247
Heterospory, 391, 429
Heterotrophic bacteria, 292, 296
Heterozygous plants, 238, 246
Hibiscus, 120
Hickory, 502
   uses, 98
Hill reaction, 166
Hill, Robin, 166
H^+ ions, 18
Hips, rose, 455
HIV, 313
Hobbs, R. J., 495
Hogg, John, 279
Holdfast
   of algae, 320, 325, 327
   of ulothrix, 320
Holly, 194
Homeobox genes, 265
Homologies, 267
Homologous chromosomes, 213–15
Homozygous plants, 234, 246
Honey, botulism spores in, 300
Honeydew (aphid), 154
Honey locust, 95
Hooker Oak, age, 90
Hooke, Robert, 28–29, 33
Hooker, Sir Joseph, 90
Horehound, 461
Hormones, plant, 188
   chemical structures of, 189
   dormancy and, 208
   interactions among, 196–97
   measurement of, 189
   movement in plants and, 197–99
   movement of, 188–96
   seedless fruits and, 434
   types of, 188–96
   vs. enzymes, 189
Hormones, synthetic, 191
Hornworts, 370, 371, 373, 377

Index

Horseradish, 70, 448, 454
Horsetails, 385, 386, 394
   ancient relatives of, 392, 396
Hot springs, bacteria in, 292, 301, 304, 308
Houseplants
   cacti, 459
   ferns, 398, 403
   gas heating and problems with, 195
   growing
      common ailments of, A61
      environments suitable for, A53–A60
      humidity, A52
      light, A51–A52
      temperatures, A51
      water, A51
   orchids, 469
   propagation by cuttings, 255
   rubber, 457
Huber, Robert, 171
Human food. *See* Food (human)
Human Genome Project, 224
Human immunodeficiency virus (HIV), 313
Humans
   algae relevance to, 335
   archaebacteria relevance to, 308–9
   bacterial diseases affecting, 299
   bacteria useful to, 301–3
   bryophytes relevance to, 382
   chromosomes of, 36
   club mosses used by, 396
   conifers' relevance to, 418–24
   cryptochromes in, 206
   cyanobacteria relevance to, 305–7
   dependence on plants, 4–6, 160, 168
   diet, sources of nutrients, 4
   Douglas fir trees (food), 92
   extinction, prevention of, 4, 6
   ferns, relevance to, 402–4
   flowering plants' relevance to, 448–50, 455
   food from plants (*See* Food (human))
   fungi relevance to, 349, 351–55, 360–67
   gymnosperm relevance of, 419–24
   horsetails and scouring rushes used by, 394–97
   impact on environment, 2, 474, 483, 488–91
   impact on plant communities, 483, 488–91
   leaves, use for landscaping, 116–18
   lichens' relevance to, 365–67
   population growth, 4, 489
   population size, constraints on, 162
   quillworts used by, 392
   as secondary consumers, 476
   slime molds' relevance to, 340–41
   use of wood, 82, 96
   viruses' relevance to, 311–13
   water use *vs.* plants, 149–51
Hummingbirds, 403, 438
Humus, 297
   formation, 382
Hunter-gatherer societies, 244, 262
Hutton, John, 266
Hyacinth bean, solar orientation, 102
Hyacinths, 232
Hybrid fusion, 249
Hybrids, 247–49, 254
Hybrid seed, 247
Hydrodictyon (water nets), 323
Hydrogen, 13–14
   in atmosphere, 163
   bacteria, 296–97
   bonds, 17–18
   in carbohydrate molecules, 19–21
   in cells, 19
   in DNA molecule, 221, 223
   importance in biology, 17
   ions, 18
   in lipid molecules, 21
   molecular structure, 15–16
   in oxidation-reduction reactions, 161–62
   in polymer formation, 19–21
   in protein molecules, 22–23
   strength of, 17
Hydroids, 372, 378
Hydrolysis, 19, 22, 24
Hydrophytes, 475
Hydroponics, 157
Hydrotropisms, 200
Hydroxyl ions, 18, 21
Hygroscopic water, 78
Hymenium, 351, 352
Hyoscyamine, 462
Hyperaccumulating plants, 79
Hyphae, 342, 343, 346–48
   of ascomycetes, 350
   of club fungi, 355, 356
   of lichens, 366
   of water molds, 341–43
Hypocotyl, 137
Hypodermis, 109, 410, 411
Hypogeous germination, 137
Hypogynous flower parts, 436, 437
Hypothesis, scientific, 6

i

IAA (indoleacetic acid), 191
IBA (indolebutyric acid), 191
ICBN *vs.* ICNCP, 278
Ice ages, cycles of, 488
Ice plants, 205
Ichneumon wasps, A21
Ignis fatuus, 308
Imbibition, 144, 147
   pressure potential of, 146
Immunization, 301, 312
Imperfect flowers, 437
Imperfect fungi, 363–64
Imprint fossils, 405
Inbreeding, 247
   depression, 247
Inca agriculture, 79
Incense cedars, 424
Incomplete dominance, 237, 238
Incomplete flowers, 436, 437
Incremental borer, 87
Indehiscent dry fruits, 131
Indeterminate growth, 69, 83, 188
Indian mustard, 79
Indian pipe, 73
Indian warrior, 72
Indicator species, 479
Indusia, of ferns, 400
Inferior ovaries, 436
   flower, 127
Inflorescences, 132, 133, 439, 456
   cyathium, 457
   flower, 125
   types, 125
Influenza viruses, 311
Ingen-Housz, Jan, 167
Inheritance. *See also* Gene(s)
   Mendel's model of, 233–34
Initials (cambium cells), 53
Injury
   plant response, 13
   reaction of leaves, 104
   response to, 59
Inner membrane (nuclear)
   in drawing of nucleus, 37
Inorganic molecules, 19
Inositol, in growth medium, 257
Insect control, viruses and, 313
Insecticides
   from chrysanthemums, 465
   nicotine used in, 462
   from plants, 118
   plants producing, A21
   resistance to, 262, A20
   toxins from, 302
Insect-resistant plants, 249, 250
Insects
   beneficial, A20
   biological control of, A20–A21, A24
   invasive species control and, 493
   mutualism with plants, 478
   plants mimicking, 261
   pollination by, 437–42
   prehistoric, 421
   in tropical rain forests, 507
   as vectors for viruses, 311
Insect-trapping plants, 113–15
   Bladderworts, 114
   digestion in, 183
   leaves, 112–14
   pitcher plants, 113–15
   sundew plants, 113–14
   Venus's flytraps, 114
Institute, 248
Integument
   of angiosperms, 428, 433
   of gymnosperms, 408–9, 412, 415
Interbreeding, 269, 288
Intercalary meristems, 53, 93
Intercellular space, in cell model, 32
Interferon, 311
Intermediate-day plants, 205
International Botanical Congress, 278, 280
International Code of Botanical Nomenclature (ICBN), 278
International Code of Nomenclature for Cultivated Plants (ICNCP), 278–79
Internet resources
   plant biology, 9
Internode
   grasses, 93
   stem, 83
Interphase, 42–43
Invasive species, 263–64, 493
Inversions, of chromosomes, 231, 268
Invertase, 354
In vitro plant propagation, 257–60
Iodine, 157, 338
Ionic bonds, 17–18
Ions, 17, 149
   chlorine, 17
   magnesium, 17
   sodium, 17

Index

Irises, 1, 95
Irish moss, 338
Irish potato famine, 338, 343
Irish potato plants, 95
Irish valentines, 401
Iron, 19, 157
   availability, 80
   deficiency symptoms, 157
   in plants, 171
   in soil, 475
Iron bacteria, 297
Iron-sulfur acceptor molecule (Fe-S), 171
Ironwood, density, 96
Isidia, 366
Isoetes, 385, 392
Isogamy, 321
Isolation barriers, and evolution, 269
Isomer, 20
Isotope, 14
   radioactive, 15

J

Jackfruit, 449
Jack pines, 415
Janzen, Daniel, 478
Jasmonates, 195
Jeffrey pine tree, 196
Jelly fungi, 355, 362
Jenner, Edward, 309
Jequirity beans, 24, 457
Jerusalem artichoke, 206, 465
Jesty, Benjamin, 309
Jimson weed, 462
Joint firs, 418, 424
Jojoba, 458
Judd, Walter, 289
Jumping gene, 220–21
Junipers, 255, 416, 422, 424
Jurassic period, 435

K

Kaffir lily leaf, 59
Kalanchoë, 256
Kauri pines, 420
Keel (flower), 456
Kelp, 316, 318, 325, 327, 328, 338
Keys, to classification, 285
Kinetic energy, 19
Kinetin, 193
Kinetochore, 43
Kingdom Fungi, 346–67
Kingdom Protista, 304, 316–43, 347
Kingdoms, of organisms, 279, 281, 282
Knapweed, 477
Knobcone pines, 415
Knots, 96–98
Koch, Robert, 301
Koch's Postulates, 301
Krakatau, 485
Kranz anatomy of leaves, 174
Krebs cycle, 177, 180–81
Krebs, Hans, 177
Kudzu, 263, 493
Kurosawa, Eiichi, 191

L

Lacewings, A20
Lactic acid, 178, 179, 303
Lactobacillus acidophilus, 303
Ladybugs, A20
Lady fern, 404
Lady's slipper, 427
Lake, James, 280, 308
Lakes
   algae in, 306–8
   eutrophication of, 486, 492
   methane bacteria in, 308
   succession in, 485
Laki volcano, 90
Lamarck, Jean Baptiste de, 28, 264
Lamb's ear mint, 461
Lamiaceae. *See* Mint Family
Lamina (blade), 102
Laminar flow hood, 258, 259
Laminarin, 326, 328
Land plants, origins of, 317, 371, 428
Land restoration, 495
Landscaping
   use of leaves in, 116–18
Lanjouw, Joseph, 278
Lantana, 439
Larch trees, 91, 360, 421
Larkspur, 206, 451
Lateral adventitious roots, 65
Lateral buds, 196–97
Lateral meristems, 53, 84
   cork cambium, 53
   vascular cambium, 53
Lateral (branch) roots, 68
Latex, 2, 92, 457
Laticifers, 92
Laurel Family (Lauraceae), 450–53
Lavender, 60, 116, 419, A22
Law of dominance, 234
Law of independent assortment, 235
Law of segregation, 235
Law of unit characters, 234
Layering, propagation by, 256
Leaching by rain, 77
Lead, 266
Leaflets, 102
Leaf structure. *See also* Leaves
   aquatic adaptation, 109
   arid regions, 109
   axil, 84, 96
   axillary bud, 84, 102
   bipinnately compound, 102
   blade, 83
   bracts (floral leaves), 112
   bud primordia, 84
   bulliform cells, 107
   bundle sheath, 107
   compound, 102
   cuticle, 84
   dicotyledons, 87
   epidermis, 84, 102, 104
   flower-pot leaves, 111
   form, 101
   gaps, 85, 102
   glands, 105
   Kranz anatomy of, 174
   lamina (blade), 102
   lateral meristem, 84
   leaf gaps, 84, 102
   leaflets, 102
   leaf scars, 84
   lichens on, 366
   midvein, 103
   monocotyledons, 87
   movements of, 198, 201, 202, 457
   netted venation, 103
   nodes, 53, 84, 103
   opposite, 103, 104
   palmately compound, 102
   palmately veined, 103
   petiole, 102
   phyllotaxy, 103
   of pines, 153, 410
   pinnately compound, 102
   pinnately veined, 103
   prickles, 110
   primordia, 84, 102
   procambium, 84
   propagation using, 255
   protective layer, 116
   of quillworts, 392
   rachis, 102
   reaction to injury, 104
   reaction to pollution, 105
   reproductive leaves, 112
   respiration, 103
   role in transpiration, 153
   scars, 84
   sensitivity to day length, 206
   separation layer, 116
   sessile, 102
   shade leaves, 108
   simple, 102
   size and form, 103–7
   as solar panels, 102
   solar tracking by, 203
   of spike mosses, 393
   spines, 110
   stipules, 83–84, 102
   stipule scars, 84
   stomata, 103–6
   tendrils, 109–10
   terminal bud, 83
   thorns, 110
   trace, 84
   transpiration, 103
   types, 103–5
   variegated, 53, 239
   veins, 103
   venation, 103
   vessel elements, 85
   water conservation movements in, 203
   whorled, 103
   window leaves, 111, 112
Leafy liverworts, 376
Leaves, 101–18. *See also* Leaf structure
   abnormalities in, viruses and, 311
   abscission of, 103, 191, 193, 194, 196
   adaptation, 108–10, 153, 474
   autumnal color changes, 165
   color change, 115
   of desert plants, 459, 506
   ethylene produced by, 193
   evolution, 108
   of ferns, 398–99, 401
   functions, 103
   of ginkgoes, 409, 416–17
   of grasses, 466
   of ground pines, 388–89
   hormones and aging in, 192–93
   hormones in, 191, 193
   of horsetails and scouring rushes, 394
   human use, 116–18
   insect-trapping leaves, 113–15
   lifespan, 102
   waste disposal, 105
Leeuwenhoek, Anton van, 28
Legionnaire disease, 300
Legume Family, 23, 450, 456–57
   bacteria and, 254
   domestication of, 245
   endosperm in, 433, 434
   flowers in, 456
   fruits in, 456
   movements in, 197, 202
   seeds in, 198, 209
Lemma, 466
Lemon grass oil, 467
Lenticels, 61, 86
Lentils, 448
Lentinacin, 362
Leptoids, 372
Lettuce, 205, 206, 254, 448, 465
Leucoplasts, 40
Lewin, Ralph A., 307
Lichens, 345, 365–67, 499

and primary succession, 484
symbiotic relationship
in, 305
Licorice fern, 404
Life. *See also* Living organisms
dependence on chloroplasts,
38
difficulty defining, 12
elements of, 13–16
histories, 479
photosynthesis-respiration
balance and, 160, 161
water and, 480
Life cycles, classification
schemes, 218
Light
bending toward, 189, 198
characteristics of, 167
mushroom growth and, 362
photoperiodism, 205–6
photosynthesis and,
160–67
role in ecosystem, 475, 477
transpiration rates and, 152
turgor movements in
response to, 201
Light-dependent reactions, in
photosynthesis, 166–72
Light-independent reactions, in
photosynthesis, 166–67
Light microscopes, 29–31
Lignin, 33, 182
in algae, 317
in cell model, 35
content, wood, 96
human use of, 182
in roots, 67–68
Lignum vitae, density, 96
Ligules, 391–93
Lilac flowers, 125
Liliopsida, 123, 428
Lily
bulbs, 71–72
leaves, 110
Lily Family, 450, 467–68
megagametophytes of, 432,
433
meiosis in, 213
polyploidy in, 232
Lima beans, 449
Liming, 80
Limonoid sprays, A21
Linden tree
cross section, 88
fibers, 56
Lindstrom, D. L., 488
Linked genes, 239, 241
Linnaeus, Carolus *(Species
Plantarum),* 7, 277–78
Lipids, 21–22, 35
fats, 21
oils, 21
waxes, 22
Litchi, 448

Live oak
after acorn imbibed
water, 148
Liverworts, 370, 371, 373–86
Living organisms
adaptation, 13
attributes, 12
cells of, 20, 28, 29
chemical reactions in, 19
complexity, 12
elements of life, 13–25
energy in cells of, 19
energy requirements of, 161
environmental adaptation, 13
nitrogen in, 482
number of cells, 31
water in, 480
Living rock cacti, 459
Lobeline sulfate, from leaves,
117
Locoweed, 456
mineral requirements, 156
Locules, 130
Locus, on chromosome,
234, 237
Lodgepole pines, 415, 420, 484,
487
Long-day plants, 205
Lower cortex, of lichens, 366
LSD, 353
Luffa, 464
Lumen, thylakoid, 170–71
Lunularic acid, 374
Lupines, 277, 448, 457,
485, 508
seeds of, 209
Lycophyta, 386, 388–93
Lyell, Charles, 266
Lyme disease, 301
Lyngbya, 306
Lysosomes, digestive
activities, 41

m

MacAyeal, D. R., 488
Macroevolution, 269–71
Macromolecules, 19
Macronutrients, 156
Magnesium, 15, 17, 157
in cells, 20
deficiency symptoms, 157
in plants, 165, 474
in soil, 475
Magnolia, 130, 428
Magnoliophyta, 428, 429
Magnoliopsida, 123, 428
Maidenhair ferns, 417
Malaysian tree frog, pitcher
plant, 113
Male fern, 404
Malpighi, Marcello, 7,
28, 150
Maltase, 24

Malthus, Thomas, 266
Maltose, 24
Manganese, 19
availability, 80
deficiency symptoms, 157
Mangoes, 195, 449
Mangroves, 149
salt glands in, 149
Mannitol, 149
Manroot, 69, 198, 471
Maple leaf cross section, 108
Maples, 91, 115, 502
Map unit, 239
Marchantia, 374
Marchant, N., 374
Marigolds, 232, 465, A22
Marijuana, 60, 117
Mass, 12
Mass-flow hypothesis, 154
Mass selection, 247
Maté, 118
Maternal inheritance, 239
Matrix, in mitochondrion
model, 40
Matter, 13
characteristics of, 13–15
states of, 13
units of, 13
Maxwell, James, 167
McClintock, Barbara,
221, 231
MCPA, 191
Meadowfoams, 458
Meadow saffron, 467
Mechanical isolation, 270
Medicinal plants, A26,
A32–A42
Medicine
from algae, 337
from Buttercup Family, 450
from conifers, 423
evolutionary, 263
from fungi, 354, 364
horsetails used for, 396
from Lily Family, 467
from Mint Family, 459
from Nightshade Family,
462, 463
plant origins, 3–4, 8, 448
from Poppy Family, 453
secondary compounds of
plants and, 182
from stems, 82
Medulla, of lichens, 365
Megagametophytes, 429,
432–34
Megaphylls, 386, 398
Megasporangia, 391, 412
Megaspores
of conifers, 410, 412, 415
of ferns, 398
of flowering plants, 428, 431
of quillworts, 392
of spike mosses, 390

Megasporocytes
of conifers, 415
of flowering plants, 435, 437
of quillworts, 392
of spike mosses, 390
Megasporophylls, 391, 412
Meiosis, 212, 213. *See also*
Sexual reproduction
alternation of generations
and, 216
crossing-over during,
215, 216
phases of, 213
vs. mitosis, 213, 216, 217
Melonette, 465
Melons, 195, 216, 435
Membranes
cristae, 40
differentially permeable, 146
in mitochondrion model, 40
synthesis, 36
Memory-enhancing substances,
A43
Mendel, Gregor, 8, 232, 238,
241, 265
Mendelian genetics, 232
Menthol, 420, 461
Mericloning, 469
Meristematic regions
creating new, 255
Meristematic tissues, 52–54
apical meristems, 52
ground meristem, 52
procambium, 53
protoderm, 53
Meristems, 43
auxins in, 189
collenchyma tissue,
54–55
complex tissue, 55
conducting sclerenchyma
(xylem), 55–56
epidermis, 58–59
parenchyma tissue, 54
periderm, 59
phloem tissue, 56–58
phytochromes in, 206
plantlets developed from,
259
sclerenchyma tissue, 54
tissue production, 53–61
xylem (conducting
sclerenchyma),
55–56
Merkus pine, 410
Mermaid's wineglass
Acetabularia, 324
cell size, 31
Mescal beans, 457
Mescaline, 459
Mesocarp (flesh), fruit, 126
Mesophyll, 106, 107
of gymnosperms, 409, 411
leaf, 103, 104

Mesophyll—Cont.
    leaves, 108, 109
    of pines, 410
    during transpiration, 151
Mesophyll cell
    chloroplasts, 38, 39
    role of water in, 149, 151
Mesozoic era, 417
Mesquite roots, 64
Messenger RNA (mRNA), 228–30
Metabolic reactions, 34
Metabolism, 19, 34, 181,
    secondary. *See also*
        Photosynthesis;
        Respiration
    assimilation, 13
    defined, 13, 14, 161
    digestion, 13
    enzymes in, 161, 162
    photosynthesis, 13
    plant, 160–86
    respiration, 13
    secondary, 181–84
Metabolites, 34
Metal absorption, 79
Metal-Munching plants, 79
Metals
    photoelectric effect in, 167
    plant growth and, 474–81
Metaphase (mitosis), 45–46, 213, 215
    centromeres, 43, 45–46
    equator, 45, 46
    features of, 45, 46
    sister chromatids, 43, 46
    spindle, 45, 46
Metaphase I (meiosis), 213, 215–17
Metaphase II (meiosis), 213–16
Methane, 17, 489–90
    bacteria, 308, 309
    digester, 309, 310
    global warming and, 488
    in natural gas, 308, 309
    production of, as true anaerobic respiration, 179
Methionine, 24, 176, 194, 228–30, 253, 254
Methuselah, bristlecone pine, 86, 87
Methylation, 264, 265
2-Methyl-4-chlorophenoxyacetic acid (MCPA), 191
Methylene chloride, 471
Metric equivalents and conversions, A75–A76
Mevalonic acid pathway, 182
Mexican jumping bean, 458
Mexican poppies, 204
Michel, Hartmut, 171

Microbodies, cellular, 40–41
Microcysti, 305
Microcystis, 305
Microevolution, 267–71, 267–73
Microfilaments, 42, 43
    size, 42, 43
    structure, 42
Microgametophyte, 433
Micronutrients, 156
Microorganisms, 22, 24, 29
    importance of, 4
Microphylls, 386, 389, 392–94
Micropropagation, 257–60, 338
Micropyle, 137
    of conifers, 412, 413, 415, 416
    of flowering plants, 428, 429, 432, 437
    of gnetophytes, 418
Microscopes
    compound, 29
    confocal scanning microscopy, 29
    dissecting, 29
    electron, 29
    fluorescence microscopy, 29
    light, 29
    phase-contrast, 29
    resolution of, 29
    scanning electron, 29–30, 46–47
    scanning tunneling, 30
    secondary electrons, 46
    transmission electron, 29, 46
Microshoots, 258, 259
    tissue production, 53–61
    xylem (conducting sclerenchyma), 55–57
Mineral components, classification, 78
Mint Family
    biological pest control, A22
Mites, A21
Monocotyledonous stems, 92–93
    composition and structure, 93
    sclerenchyma cells, 93
Monocotyledons, 88
Monoecious flowering plants, 437
Monohybrid cross, 234–35
Monokaryotic mycelia, 356
Monomers, 19
Monosaccharides, 20
Monosodium glutamate, 154
Monterey pines, 420
Morels, 351
Morgan, T. H., 239
Mormon tea, 354
Morning glory, 198
Morphine, 453

Morphology, plant, 8, 288
Mosquito control
    bacteria and, 302
    citronella oil and, 467
    mustard plants and, 454
Mosquitoes, pollination by, 441
Mosquito fern, 404
Mosquito fish, A21
Mosses, 371, 377–82
    antiseptic properties of, 382
    daylight required by, 163
    habitats of, 372
    liverworts compared to, 374
    primary succession and, 484
Moss rose, 202
Moths, pollination by, 438, 440, 441
Mountain forests, 502, 503
Movement
    of algae, 318, 326
    of auxins, 189–93
    of bacteria, 296
    within cell, 12
    of cyanobacteria, 305
    of dinoflagellates, 333
    of euglenoids, 332
    of leaves, 198, 457
    of living things, 13
    molecular, 15
    plant, 197–204
    of slime molds, 339–41
Moyle, Susan, 373
Mucilage, in hornwarts, 377
Mucilaginous lubricant, 65
Mucilaginous sheaths, of cyanobacteria, 306
Muir, John, 494
Mulch, 483
Mule ears, 465
Mullis, Kary, 226
Multiple fruits, 132, 133
Multiplication medium, 258
Mung beans, 449, 457
Muramic acid, 286
Mushrooms, 346, 351, 355
    fairy rings of, 358
    life cycle of, 357
    phototropism in, 199
    poisoning by, 360–63
    structure of, 357
    use for food, 361–62
Mustard Family, 130, 267, 450, 454
Mustard weed, 253
Mutagenesis, 253
Mutagens, 229
Mutations, 25, 221, 225, 229–31
    in apical meristem, 52–53
    DNA replication and, 224–29
    evolution and, 214, 231, 268
    germ-line, 229–30

pesticide use and, A20
point, 268
silent, 229, 268
somatic, 229
transposable elements and, 221
in viruses, 312, 313
Mutualism, 478. *See also* Symbiosis
Mutualistic association, 73
Mycelium
    of ascomycetes, 350
    of club fungi, 355
    of coenocytic fungi, 347
    dikaryotic, 356
    fairy rings of, 358
    monokaryotic, 356
    of water molds, 343
Mycobacterium tuberculosis, 298
Myco-heterotroph, 73
Mycologists, 346
Mycophagists, 346
Mycoplasmas, 301
Mycorrhizae, 73
Mycorrhizal fungi, 347, 364, 387, 410, 478, 485, 492
    acidity of soil and, 492
    bryophytes and, 372
Myrrh tree, 91
Myxamoebae, 340

n

NAA, 191
NAD, 177, 179–81, 185, 186
NADH, 177–81, 185, 186
NADP, 166, 171, 172, 185
NADPH, 162, 166–69, 171–73, 185
Narcotics. *See* Drugs
Nastic movements, 198, 202
Nasturtiums, 75, 205
National parks, 494, 503, 504
National Plant Germplasm System, 248
Native American plant use
    Buttercup Family, 450–51
    Cacti Family, 458–59
    Carrot Family, 463
    club mosses, 392
    conifers, 419–22, 425, 428, 502, 503
    ferns, 398, 402–4, 405
    horsetails and scouring rushes, 386, 394, 396–97
    Laurel Family, 451–53
    Lily Family, 467–68
    Mint Family, 459–62
    Mustard Family, 454–55
    Nightshade Family, 462–63
    Poppy Family, 453–54

Pumpkin Family, 463–65
Rose Family, 455–56
stoneseed, 448
Sunflower Family, 465–66
wild rice, 469
Natural cycles, 479
Natural gas, methane in, 309
Natural grafting, in roots, 69
Natural selection, 13, 262
Darwin on, 265–68
vs. human selection, 246
Navel oranges, 192
Near-eastern region of origin, 448
Neck, of moss archegonia, 376, 378
Nectar, 438
Neljubow, Dimitry, 194
Nematodes, 363, 465, A21
Nematode-trapping fungi, 363
Neon, 163
Netted venation, 103
Nettles, cell size, 31
Neutrons, 14
Newton, Sir Isaac, 167
Niacin, 166
Nicotine, 75, 118, 462
Nicotinic acid, in growth medium, 257
Nightshade Family, 462–63
Nitrates, 176, 492
Nitric acid, 492
Nitric oxide, 195
Nitrifying bacteria, 297
Nitrogen, 13–14, 156
in atmosphere, 163, 482, 483
in cells, 19
in compost pile, 297
deficiency symptoms, 158
in plants, 176, 195
in protein molecules, 22, 482
radioactive, 266
in soil, 482
Nitrogen cycle, 481–83
Nitrogen-fixing bacteria, 297, 482, 483
cyanobacteria compared to, 303, 307
mutualistic associations with, 254
viruses attacking, 310–11
Nitrogen gas, in food storage, 180
Nitrogenous bases
in DNA, 222, 223
in RNA, 227–28
Nitrosamines, in tobacco, 117
Noble, David, 423
Nodding movements of plants, 198
Nodes, 53, 84

Nominalistic species concept, 289
Noncoding DNA,
Noncyclic photophosphorylation, 169, 171
Nonpolar covalent bonds, 17
Nori, 338
North American region of origin, 449
Noscapine, 454
Nostoc, 304, 306
Nucellus, of gymnosperms, 409, 412, 413
Nuclear envelope, 36
cellular reproduction, 42–45
telophase, 46–48
Nuclear pores
in drawing of nucleus, 37
Nuclear radiation, 367
Nucleic acids, 24–25
Nucleoid, in bacterial cells, 293
Nucleoli, telophase, 46–48
Nucleolus, 29, 36
cellular reproduction, 44
Nucleomorph, of cryptomonads, 334
Nucleoplasm, 36
Nucleoside triphosphates, 226
Nucleotides
DNA, 24, 222, 223, 225
RNA, 227–28
Nucleus (atomic), 14
Nucleus (cell), 12, 31
in algae, 319
chromatin, 36
in coenocytic fungi, 350
of cryptomonads, 334
in dinoflagellates, 333
discovery of, 29
DNA in, 223
in drawing and micrograph, 37
functions, 35–37
in green algae, 318–19, 322–25
influence on form of cell, 324
size, 36
in slime molds, 341
Nurse plants, 477
Nutations, 197
Nutmeg, 415, 449
Nutrients, 188
availability, 80
plant needs, 4
Nuts, fruit, 132
Nut trees, propagation of, 256

O

Oak trees, 501, 502
Oak wilt, 355
Oats, 93, 232, 466
coleoptiles of, 189–90
etiolated seedlings of, 206
Occam's razor, 288
Oceans
acidity of, 481
algae in, 307, 319
bacteria in, 307, 308
carbon balance and, 163, 184
carbon in, 480–81
cyanobacteria in, 304
photosynthesis in, 162
pollution of, 163
red tide in, 332–34
Oedogonium, life cycle of, 322
OH⁻ ions, 18
Oil (petroleum)
energy stored in, 162, 176
origins of, 4, 336
Oil palm, 449
Oils, 21, 459–62, 474
from plants, 116
in wood, 96
Oil synthesis, in elaioplasts, 40
Okra, 449
Oldest living species, 87
Olibanum tree, 91
Oligosaccharins, 196
Olives, 128, 362
Onion, 95, 285, 449
bulbs, 23
cellular reproduction, 43, 44
leaves, 110
poisonous substances in, 468, A26
seeds of, 254
storage organs of, 23
Oogamy, 322
Oogonia
of brown algae, 329
of *Oedogonium*, 322, 323
of water molds, 342
Oomycota, 341–43
Operculum, of mosses, 379
Opioid epidemic, in the United States, 2
Opium poppies, 2, 92, 448, 453, 454
Opposite leaf structure, 104
Oppositely arranged, 84
Oranges, 192, 216, 229, 249, 434
Orbitals, of electrons, 14, 15, 17
Orchid Family, 70, 95, 456, 468–70
CAM photosynthesis in, 176
endangered members of, 427
flowers of, 436, 468, 469

growing on tree ferns, 402, 403
as houseplants, 77, 116
micropropagation of, 258
mycorrhizal fungi associated with, 478
pollination mechanisms in, 270, 437–42
pollinia in, 270
roots, 70–71
seeds of, 469
in tropical rain forests, 507
turgor contact movements in, 202
underground flowering, 121
Ordering complexity, 218
Orders, 280
Organelles, 29, 31, 34, 40–42, 48
Organic evolution, 262
Organic molecules, 19
Organic solutes, 149
Organisms
common and scientific names, A1–A19
Organ-pipe cactus, 460
Organs, plant, 52
Origin
of coal, 4–5
of oil, 4–5
Origin of the Species, The (Darwin), 262, 265, 267
Ornamental mint, 59
Osage orange, durability, 96
Oscillatoria, 304
Osmometer, 145
Osmosis, 35, 145–46
in cells, 41
water transport in plants, 150, 151
Osmotic potential, 145
Osmotic pressure, 35, 145
Ostrich fern, 398
Outcrossing, 247
Outer membrane (nuclear)
in drawing of nucleus, 37
Ovary
compound, 436
development into fruit, 434
flower, 125
of flowering plants, 428, 429, 434–35, 437
inferior, 436, 437
simple, 436
superior, 436, 437
Overproduction, 268
Ovule
development into seed, 433
of flowering plants, 125, 428, 429
of gymnosperms, 409, 412, 413, 415
Oxaloacetic acid, 175, 178, 180

Oxidation, 162
Oxidation-reduction reactions, 162, 171
Oxidative phosphorylation, 172, 181
Oxygen
   in aerobic respiration, 177–79, 181
   algae and production of, 6, 337
   atomic structure, 14–15
   in carbohydrate molecules, 20
   in cells, 19
   Chlorella and production of, 323–25
   cyanobacteria and production of, 304–6
   isotopes of, 13–15
   life and, 162
   in lipid molecules, 21
   molecular structure, 15–16
   in oxidation-reduction reactions, 162
   in photosynthesis, 102–3
   production by plants, 4
   in protein molecules, 22
   release in photosynthesis, 162–66, 171, 181
   in water, 70–71
Oxygenase, 174
Oxygen-evolving complex (OEC), 171
Oyster mushrooms, 346
Ozone, 305, 490

P

P_{680}, 169–71
P_{700}, 169–71
PAA (phenylacetic acid), 191
Pacific bleeding heart, 136
Pacific Islands, plant origins in, 449
Pacific yew tree, 424
Palea, 466
Paleobotany, 7
Paleozoic era, 107
Paleozoic evolution, 107
Palisade layer, leaves, 109
Palisade mesophyll, 107, 176
Palmately compound leaves, 102
Palmately veined leaves, 103, 104
Palm trees, 93
Panama hats
   use of leaves for, 116
Pandemic, 312
Pantothenic acid, 180
Papaveraceae. *See* Poppy Family
Papaverine, 454
Papillae, 322
Papovaviruses, 310
Papovaviruses, in human wart, 311

Parallel veins (monocot), 105, 107
Paramylon, 332
Paraphyses, 378, 382
Pará rubber trees, 92, 457, 458
Parasites, 478
   bacteria, 297
   flowering plants, 429
   fungi, 359, 362–63, 372
   lichens, 365
   viruses, 310–11
Parenchyma cells, 65, 67, 92
   hormone transport and, 193, 201
   of liverworts, 374
Parenchyma tissue, 54, 84
Parental generation (P), 233
Parental type offspring, 241
Parsimony, 288
Parsley, 448, 463, 471
Parsnips, 463
Parthenocarpic development, 126, 128, 133, 434
Partially parasitic plants, 72
Particle gun, 250, 251
Passage cells, in roots, 68
Pasteur, Louis, 29, 309
Pauling, Linus, 223
PCBs, 162
PCR (polymerase chain reaction), 226–27
PCR (photosynthetic carbon reduction) pathway, 172–76
Peach flowers, 125
Peach leaf curl, 355
Peach trees, 355
Peanuts, 127, 129, 232, 276, 365, 449
Pea plants, 448
   axillary buds in, 197
   crosses between, 234–37
   domestication of, 245
   endosperm in, 433
   flowers, 125
   genetic map of, 240
   Mendel's studies of, 232–34, 237
   mutualism with bacteria, 255
   pod, 428
   sleep movements in, 203
Pear trees, 31, 449
Peat mosses, 372, 378, 382, 486
Pectin, 33
Peduncle, flower, 124–26
Pellicles, 332
Penicillium, 5, 362–68
PEP (phosphoenolpyruvate), 175
PEP carboxylase, 175, 176
Pepos, fruit, 128
Pepper Family, 58, 449
   thermoperiodism in, 207
Peppermint, 277, 459, 461

Peptide bonds, 23
Perception phase (tropism), 198
Percolation, 480
Perennial flowers, 121
Perennial grasses, 83
Perennials, 479, 499
   of red algae, 330–31
Pericycle, in roots, 68
Periderm, 55, 59, 92
Perigynous flower parts, 436
Peristomes, 379, 383
Perithecium, 351
Periwinkles, 263
Permafrost, 499–500
Permanent wilting point, 80
Peroxisomes, 40, 41, 174
Persimmons, 448
Pesticides, 492, 496, A20. *See also* Biological controls; Insecticides
Petals
   flower, 124
   movements of, 198, 202
Petioles, 102
   heliotropic movements of, 203
   stem, 84
Petrifactions, 405–6, 407
Petunias, 462
Peyote, 449, 459, 460
$P_{far-red}$ (P_{fr}), 206
3PGA, 166, 173
PGR (plant growth regulator), 195
pH. *See also* Acidity
   mycorrhizae, 80
Phaeophyceae, 327
Phages, 310–13
Phase-contrast microscopy, 29
Phenolics, 182
Phenology, 479
Phenotype, 234
   environmental influence on, 238, 265–68
   intermediate, 237, 238
Phenylacetic acid (PAA), 191
Pheophytin (Pheo), 171
Pheromones, 350, A21
Phloem, 88, 93
   absence in bryophytes, 372
   absence in mosses, 378
   in conifers, 410
   evolution of, 386
   food transport and, 153–54
   hormone transport and, 192, 193
   of horsetails and scouring rushes, 394
   in leaf, 107, 109
   phloem-loading, 155
   rays, 90
   sieve tube members, 57
   sugar content in, 419
   synopsis, 156

   tissue, 56–58
   in whisk ferns, 387
Phosphate group, 22
   of nucleotide, 222, 223
Phosphates, 157
Phosphoenolpyruvate. *See* PEP
3-Phosphoglyceric acid (3PGA), 166, 173, 175
Phosphoglycolic acid, 174
Phospholipids, 22, 35
   in plasma membrane model, 36
Phosphorescence, 168
Phosphorus, 19, 479
   in cells, 19
   in plants, 478
Phosphorylation, 180, 181
Phospodiester bonds
   in DNA, 222
Photoelectric effect, 167
Photolysis, 171
Photons, 166, 167
Photooxidation, 165
Photoperiodism, 205–7
Photophosphorylation, 171
Photoreceptors, in algae, 326
Photorespiration, 41, 165, 174, 175
Photosynthesis, 13, 162–76
   bacteria and, 166, 168, 171, 296–97
   CAM, 153, 176, 475, 506
   C_3 form of, 165
   C_4 form of, 165
   chlorella and, 323–25
   in chloroxybacteria, 293
   conversion process of, 41
   C_3 pathway of, 174
   C_4 pathway of, 174–76
   in cyanobacteria, 293, 296, 303–7
   in desert plants, 153
   dinoflagellates and, 333
   effects of temperature on, 165
   energy-storing process of, 163
   energy transfer in, 161–62, 166
   in euglenoids, 279
   first products of, 162
   in flowering plants, 428–29
   global warming and, 184
   glucose production in, 20
   in gnetophytes, 418–19
   in horsetails and scouring rushes, 394
   in lichens, 365
   light-dependent reactions in, 166–67, 169, 172
   light-independent reactions in, 166, 167, 172–76
   in mosses, 372, 378
   process, 102–6

Index

process location, 107
respiration and, 161, 182
role of water in, 149–51
in salt bacteria, 308
stipules, 110
study of, history of, 167
vs. respiration, 182
water in, 150
in whisk ferns, 387
Photosynthetic carbon reduction (PCR) pathway, 172–78
Photosynthetic roots, 70
Photosynthetic units, 166, 168, 171
Photosystem I, 168–71
Photosystem II, 168–71
Phototaxic response, 204
Phototropism, in plants, 198–99
Phragmoplast, 42, 371
telophase, 47–48
pH scale, H^+ measurement, 18
Phycobilins, 166, 304, 331
Phycocyanin, 304
Phycoerythrin, 304
Phycologists, 326
Phyla, 280
Phyllotaxy, 103
PhyloCode, 279
Phylogenetic nomenclature, 279
Phylogenetics, 288
Phylum
 Anthocerophyta, 371, 377
 Ascomycota, 349–51
 Bacteriophyta, 296–97
 Basidiomycota, 355–60
 Bryophyta, 371, 377–82
 Charophyta, 334–35
 Chlorophyta, 319–25
 Chromophyta, 325–29
 Chytridiomycota, 347–65
 Cryptophyta, 334
 Cycadophyta, 409, 416, 417, 425
 Deuteromycetes, 363–64
 Dictyosteliomycota, 341
 Dinophyta, 332–33
 Equisetophyta (horsetails and scouring rushes), 386, 394–97
 Euglenophyta, 331–32
 Ginkgophyta, 409, 416
 Gnetophyta, 409, 416, 418–19
 Hepaticophyta, 371, 373–77
 Lycophyta, 388–93
 Lycophyta (club mosses and quillworts), 383–93
 Magnoliophyta, 428–36
 Myxomycota, 339–40
 Oomycota, 341–43
 Pinophyta, 410–15

Polypodiophyta, 398–404
Polypodiophyta (ferns), 386, 398–404
Prymnesiophyta, 334
Psilotophyta (whisk ferns), 386–88
Rhodophyta, 330
Zygomycota, 348–49
Physiology, plant, 7
Phytoalexins, 477
Phytochelatins, 79
Phytochromes, 206
Phytoremediation, 79
Pigments
 absorption spectra of, 167–68
 in bacteria, 297, 305, 307
 in brown algae, 327
 in cell sap, 42
 in cyanobacteria, 304
 in diatoms, 326
 in dinoflagellates, 333
 in ferns, 403
 in fungi, 349
 in green algae, 319
 in haptophytes, 334
 photoperiodism and, 206
 photosynthetic, 165–66
 in prochlorobacteria, 307
 in red algae, 331
 in yellow-green algae, 325
Pignolias, 420
Pileus, of mushroom, 357
Pili, 293
Pilobolus, dung fungus, 349
Pilus (pili), 293, 295
Pineapples, 176, 196, 208, 449
Pine cones, 409, 416
Pinedrops, 72
Pine leaves, leaf modification in, 153
Pine needle cross section, 110
Pines, 410–15
 apical dominance in, 197
 biomes of, 500, 502
 bristlecone, 410
 genome of, 224
 human relevance of, 420–24
 leaf modifications in, 410
 life cycle of, 414
 reproduction of, 410–15
 resin canals in, 92, 410, 411, 420
 resin secretion in, 410, 420–22
 rusts and, 359
 seeds of, 409, 420
 wood of, 411, 420
Pine trees, 87, 109, 153
Pinnae, of ferns, 399
Pinnately compound leaves, 102

Pinnately veined leaves, 103
Pinophyta, 409–16. *See also* Conifers
Pinyons, 420
Piperaceae. *See* Pepper Family
Pistachio, 448
Pistils, 125, 428
 compound, 435, 437
 of orchids, 469
 in primitive flowers, 428, 436
Pitcher plant leaves, 113
Pitcher plants, 113
Pitch pine, 420
Pith, 84–87
 hormonal effects on, 197
 of horsetails and scouring rushes, 394, 395
 of pines, 411
 in roots, 66
Pits, tracheid, 55–56
Plain-sawed wood, 96
Plankton, 304, 318, 326, 334, 337
Plantains, 191, 276
Plant biology, Internet resources, 10
Plant Kingdom, 279–80, 371
Plant life cycles, 435
Plant press, 442
Plants. *See also* Plant structures
 cells of *vs.* animal cells, 48
 distinguished from animals, 279–80
 ecology, 8, 72
 forensic science, 7
 human uses, 4–5, 9
 identification, 83–84
 medicines from, 4, 8
 names, 276–89
 organs, 52
 pests, biological control, 302–3
 transition from water to land, 317
 use of term, 371
Plant structures. *See also* Plants
 apical meristem, 84
 dicotyledons, 87
 environmental adaptation, 60
 monocotyledons, 87
 shoot system, 83
Plant systematics, 7
Plant taxonomy, 7
Plant viruses, 310
Plant viruses, naming, 310
Plaque, dental, 298
Plasma membranes, 31
 cellulose production, 35
 in euglenoids, 332
 function of, 34–35
 model of, 36
 in plant cell, 32
 structure of, 35
 thickness of, 34

Plasmids, 249–50, 292
Plasmodesma, 311
Plasmodesmata, 34
 in roots, 68
 telophase, 47–48
Plasmodia, 339–41
Plasmodial slime mold, life cycle, 341
Plasmolysis, 146
 mangroves, 149
Plastids, 38–39, 193
 plants *vs.* animals, 48
Plastocyanin, 171
Plastoquinone (PQ), 171
Plato, 276
Pleated sheet polypeptide, 23
Pliny, 454
Plumbing and storage, plant, 55–56
Plumule, 137
Pneumatophores, 70
Pneumonia, bacterial, 301
Poaceae. *See* Grasses
Podocarps, 415, 424, 426
Poinsettia, 205, 458
 bracts, 112
 flower, 112
Point mutation, 268
Poison ivy/oak, A26
Poisonous organisms, A31
Poisonous plants and fungi, A26
Poisons, plant. *See also* Toxins
 in Buttercup Family, 450–51
 in Cacti Family, 459
 in Carrot Family, 463
 in club fungi (mushrooms), 360–63
 in conifers, 424
 in ferns, 404
 in horsetails, 396
 jequirity beans, 24
 in Legume Family, 24, 457
 in Lily Family, 467
 in Nightshade Family, 462, 463
 in Spurge Family, 458
Polar covalent bonds, 17
Polarity, imbibition and, 144, 147
Polarity of molecules, 16
Polar molecules, 147
Polar movement, 191
Poles, cellular reproduction, 42–47
Poles, of cells, 215, 216
Pollen
 cones, 409, 412
 drops, 413
 hormones, 125–26
 tube, 413, 425, 429, 433

Pollen grains
  apertures, 431–32
  flower, 125
  of flowering plants, 429, 431–33
  of gymnosperms, 409, 411–13
  sperm cells, 432
Pollination, 432, 437–42
  by carrion beetles, 121
  by tiny flies, 121
  by the wind, 122
Pollinia, 270, 439–41, 469
Pollution. *See also* Air pollution
  bioremediation of, 302
  carbon dioxide accumulation, 163
  effect on leaves, 105
  impact on oceans, 163
  lichens' sensitivity to, 367
  public awareness of, 492
  of water, 491, 492
Polyamines, 195
Polychlorinated biphenyls (PCBs), 162
Polymerase chain reaction (PCR), 226, 227
Polymers, 19
  amino acids, 22–24
  carbohydrates, 19–21
  lipids, 21–22
  polypeptides, 22–23
  proteins, 22–23
Polypeptide, 22–23
  beta sheet, 23
  pleated sheet, 23
  quaternary structure, 23
  tertiary structure, 23
Polypeptide chains, 22
Polyploidy, 232, 271–73
Polypodiophyta, 386, 398–404
Polysaccharides, 20–21, 38, 162
Polyunsaturated fat, 22
Pomes, 128–30, 455
Pomes (apples), 129
Ponderosa pine, 420, 487, 502
Poor man's pepper, 208
Poplar leaf spot, 360
Poplars, 277, 360, 362
Poppy Family, 448, 453–54. *See also* Opium poppies
  pollen grain of, 453
Population ecology, 483
Population genetics, 265
Population growth, human, 4, 273, 491
Populations (plant), 475, 495
Pores
  of hornworts, 377
  of leaves, 152–54
  of liverworts, 374
  in nuclear envelope, 36, 37
Portabella mushrooms, 362

Postzygotic barriers, 269
Potassium, 19, 156
  in cells, 19
  deficiency symptoms, 158
  ions, role in guard cells, 151–54
Potato plants, 244, 246, 248, 449, 457, 462, 463
  ethylene application and, 196
  eyes, 95
  gene-edited variety, 253
  late blight of, 343
  leaves of, 232
  as long-day plants, 205
  nitrogen storage in, 176
  poisons produced in, 463
  polyploidy in, 232
  powdery scab of, 341
  propagation of, 258
  storage organs of, 23–24
  tubers, 23, 95, 246
  vine, tendrils, 110
Potential energy, 19
Powderpuff flower, 456
PQ (plastoquinone), 171
Prairies, 503
Prayer plants, 202, 203
Praying mantis, A20
Precipitation, 483. *See also* Acid rain
P_{red} (P_r), 206
Preprophase band, 43, 45
Pressure-flow hypothesis, 154–56
Pressure potential of imbibition, 148
Prezygotic barriers, 269
Prickles, leaf, 110
Prickly pear cactus, 95, 449, 459, 460
Priestley, Joseph, 167
Primary cell wall, flexible, 33–34
Primary consumers, in ecosystem, 476, 478
Primary growth, 53
Primary meristems, 53
Primary phloem
  cells, 84
  in roots, 66, 67, 69
  in stems, 84
Primary roots, 65
Primary structure of protein, 23
Primary succession, 484–86
Primary thickening in monocots, 124
Primary tissues, 53, 84
Primary veins
  dicots, 103
  leaves, 103
  monocots, 103
Primary xylem
  cells, 84

  in roots, 66, 67, 69
  in stems, 84
Primers, 226
Primordia, 102
Primordium, flowers, 124
Primrose flowers, 125
Principle, scientific, 6
Prions, 313, 314
Procambium, 53, 84
  in roots, 66
Prochlorobacteria, 304, 307, 314
Prochloron, 307
Prochlorophyta, 307
Prochlorothrix, 307
Producers, in ecosystem, 475, 477
Progametangia, 349
Prokaryotes, classification of, 280
Prokaryotic cells, 31, 293
Prokaryotic reticulum, 37, 38
Proline, 149
Promoter region, 228
Propagation, 254–64
  by cuttings, 255
  with seeds, 254–55
  using adventitious roots, 256
Propagules, 248, 249
Prophase (mitosis), 43–45, 213
Prophase I (meiosis), 213–19
Prophase II (meiosis), 214, 216, 217
Proplastids, 176
Prop roots, 70
Prostele, 87
Protectant, seed, 254, 255
Protective layer, leaf, 116, 117
Proteins, 22–24, in plasma membrane
  algae as source of, 337
  in chromosomes, 222
  coats, of viruses, 310, 311
  digestion of, 182, 183
  enzymes, 23, 24
  in grass mowings, 467
  model, 36
  molecular structure, 22
  nitrogen in, 22, 482
  from plant leaves, 118, 119
  primary structure of, 23
  production in chloroplasts, 40
  secondary structure of, 23
  sequence, 249, 250
  in shiitake mushrooms, 362
  storage in plants, 23
  synthesis, 39, 40
  synthesis in cells, 34, 35, 37–38, 223, 224, 228–33, 237, 239
  transport in plants, 148, 149
  yeasts as source of, 354

Prothalli, of ferns, 399–402
Protista, 281–83, 317–18
  vs. fungi, 346–53
Protoctista, 279, 280
Protoderm, 53, 84
  in roots, 66
Protonema, 373, 376, 378, 379, 382
  liverwort, 374
Proton pump, 148
Protons, 14
Protoplasm, 31, 34
  in slime molds, 339
Protoplastids (organelles), 40
Protoplasts
  fusion, 249–54
  in green algae, 322
  in *Spirogyra* (watersilk), 322
Protozoans, 317, 334, 339, 363
Pruning
  bonsai, A68
  fruit trees, A67
  grapevines, A67
  raspberries and blackberries, A68
  roses, A68
Prymnesiophyta, 334
Pseudoplasmodium, 341
Psilophytes, 387
Psilotophyta, 386–88
Psyllium, 448
Pteridosperms, 409, 428
Puccoons, 448
Puffballs, 355, 358
Pulp, 98
Pulque, from *Agave*, 118
Pulvini, 201–3
Pumpkin Family, 69, 128, 136, 139, 449, 463–65
  flowers of, 196, 437, 464
  tendrils, 109
Punctuated equilibrium model, 269, 276
Puncture vine fruit, 135, 139
Punnett, R. C., 239, 241
Punnett square, 236, 237
Pure-line selection, 246, 247
Purines, 222, 223
Purple nonsulfur bacteria, 296, 297
Purple sulfur bacteria, 296, 297
Pygmy mosses, 372
Pyrenoids, 320–22, 377
Pyrethrum, 465
Pyrimidines, 222, 223
Pyrodictium, 308, 309
Pyruvic acid, 177–81
  formation, 180
  of red algae, 330
  repeated, 479
  of slime molds, 339, 340
  of spike mosses, 390–92
  of *Ulothrix,* 320

of viruses, 311–13
vs. survival, 479
of water molds, 341–43
of yeasts, 351
of yellow-green algae, 325
Pyruvic acid
 formation, 180

q

Quantitative trait loci (QTLs), 239
Quantitative traits, 238–39
Quantum leap theory, 14
Quarks, 14
Quaternary period, 388
Quaternary structure
 polypeptide, 23
Quartersawed wood, 96, 98
Queiroz, Kevin de, 279
Quiescence, 208–9
Quillworts, 385, 386, 388–94, 410
 life cycle of, 393
Quinoa, 449

r

Rabies, 309
Rachis, 399
Radially cut wood, 96
Radicle, 64, 137, 209
Radioactive elements, 266, 367
Radioactive isotope, 15
Radioactive wastes, 493
Radishes, 36, 67, 130, 454
 root hairs, 67
Rain forests, 2
Rafflesia flower, 121, 122
Ragweeds, 205
Rainfall, annual rings, 89
Ramblers, 96
Ranunculaceae, 134, 276, 450–51
Raphe, of diatoms, 326
Raspberries, 96, 110, 255, 455
Ray cells, 55–57, 90
Ray initials, 56
Reaction-center molecules, 169, 171
Receptacles
 of brown algae, 329
 flower, 124
Receptors, hormones and, 188
Recessive allele, 234, 237, 238, 241, 247
Recombinant DNA technology, 249
Recombinant type offspring, 241
Recombination hotspots, 214
Recycling, 5, 42, 296, 346, 475, 492

Red algae, 166, 257, 330–31, 338, 378
Redbuds, 270
Red cedar, 424, 502
Red mulberry, durability, 96
Red pepper, 40, 449
Red spruce, uses, 97
Red tide, 332–33
Reduction division, in meiosis, 214, 215–16
Redwood sorrel, 202
Redwood trees, 2, 31, 33, 416, 421, 422, 503
 fire-resistance of, 487
 size, 2, 3, 31, 416
Regulatory genes, 265
 mutations in, 268
Reindeer moss lichens, 367
Reproduction
 of algae, 318, 319–20
 of ascomycetes, 349–51
 of bacteria, 292–95
 big bang strategy for, 479
 of brown algae, 327–29
 of bryophytes, 371
 of cells, 213
 characteristics of, 12
 of *Chlamydomonas,* 320
 of chytrids, 347
 of club fungi, 356
 of conifers, 410–15
 of cryptomonads, 334
 of diatoms, 327
 of dinoflagellates, 334
 essential nature of, 12–13
 of euglenoids, 332
 of ferns, 399–402
 of flowering plants, 430
 of fungi, 346
 of green algae, 319–22
 of ground pines, 389–92
 of hornworts, 377
 of horsetails, 395–96
 of imperfect fungi, 363–65
 of lichens, 365–66
 of liverworts, 374–76
 of mosses, 378–80
 of quillworts, 392
 of red algae, 330
 repeated, 479
 of slime molds, 341
 of spike mosses, 390–92
 vs. survival, 479
 of *Ulothrix,* 320
 of viruses, 311
 of water molds, 341–43
 of yeasts, 350
 of yellow-green algae, 325
Reproductive isolation, 270
Reproductive leaves, 101, 112
Resin, 410, 411, 421
 injury response, 420
 use by humans, 422

Resin canals, 91, 92, 410, 411, 420
Resistant varieties, 311, 360, 477, A22
Respiration, 13, 92, 103, 176–81
 in carbon cycle, 481
 defined, 176
 energy transfer in, 162
 forms of, 176–81
 global warming and, 164
 photosynthesis and, 161, 182
 temperature and, 179, 184
Restoration ecology, 495
Restriction enzymes, 249
Resurrection plant, 392
Reticulate venation, 103
Retroviruses, 313
Retting, 93
R groups, 23
Rhizines, of lichens, 366
Rhizoids, 348
 of liverworts, 374, 376
 of mosses, 377, 380
 of whisk ferns, 387, 388
Rhizomes, 94
 of ferns, 404
 of ground pines, 389
 of horsetails, 395
 propagation of, 257
 of whisk ferns, 385, 386
Rhizopus, 348–49, 368
Rhoades, Marcus, 231
Rhodophyta, 330–31
Rhodopsin, 303, 308
Rhythmic cell division, 66
Ribonucleic acid. See RNA
Ribose, 226
Ribosomal RNA (rRNA), 228
Ribosomes
 assembly, 37–38
 in cell model, 32
 composition of, 37–38
 in eukaryotic cells, 38
 in prokaryotic cells, 37, 293
 protein content, 37
 shape, 37
 during transcription, 228
 during translation, 228–30
Rice, 93, 448, 449, 466
 endosperm in, 433, 434
 golden, 252
 vs. wild rice, 469
Rice cutgrass, margin, 117
Rice-paper plants, 70
Rickettsias, 301
Ringworm, 365
RNA, 24, 36
 cytokinins in, 193
 DNA compared to, 227
 molecular structure, 24
 polymerase, 228
 in ribosomes, 37, 38

transcription and, 228, 231
translation and, 228–29, 230, 231
in viroids, 313–14
in viruses, 311–13
Rock cress plants, 360
Rock mosses, 484
Rocks
 primary succession, 484
 splitting by seeds, 147–48, 484
Rock tripe, 367
Rockweed, 329
Rohrer, Heinrich, 30
Roosevelt, Theodore, 494
Root cap, 65–66, 199
Root hairs, 65
Root hair zone, 66
Rooting powder, 255–56
Rootlets, 70
Root pressure, 103, 150
Roots. *See also* Adventitious roots
 acidity of soil and, 492
 adventitious roots, 64–65
 aerial, 58
 branch roots, 64
 of cacti, 506
 of club mosses, 388
 contractile, 71, 198
 cytokinins synthesized in, 193, 197
 development, 64–65
 dicotyledonous plants, 65
 embryo, 64
 ethylene produced by, 194
 ex vitro propagation by, 259
 feeder roots, 64
 fibrous roots, 64
 function, 64
 fungi, 73
 geomagnetotropism in, 201
 gravitropism in, 199, 200
 hormones and, 191
 human food use, 75–76
 monocotyledonous plants, 65
 natural grafting in, 69
 nitrogen storage in, 176
 phototropism in, 198–99
 plant, 52
 primary meristem, 65
 primary roots, 65
 propagation by, 254–60
 radicle, 64
 root hairs, 65
 secondary roots, 65
 size and number, 64
 soils, 64–80
 structure, 64
 surface area, 65–67
 taproot, 64
 transpiration and growth of, 151
 velamen, 58

Index

Root section, 65
Rootstock, 256
Root tip, section, 65
Ropes, use of leaves for, 116
Rose Family (Rosaceae), 455
  as day-neutral plants, 205
  edible parts of, 457
  propagation of, 255
Rose prickles, 110–11
Rosewood, 456
Rosin, 420
Rosy periwinkle, 263
Rough endoplasmic reticulum, 36–37
Roundup, 252
Rubber, 92
Rubber trees, 92, 457, 458
Rubisco, 173–175
RuBP, 166
Ruellias, leaf modification in, 153
Runners and stolons, 95
Russula mushroom., 356
Rust, life cycle, 360
Rusts, 355, 360
Rutabagas, 75
Rye, 93, 231, 352, 466
Ryegrass roots, 67

S

Sac fungi, 349–51, 365
Safflower, 449
Saffron, 467
Sagebrush, 465, 477
Saguaro, 459
Salicylic acid, 195, 197
Salmonella, 226, 299
Salsify, 69, 465
Salt bacteria, 308
Salt formation (chemical), 18
Salvias, 205, 461
Samaras, fruit, 132
Sand, use in flower drying, 444
Sandy soils, 79–80
Sansevieria, 255, 467, 468
Saprobes, 296, 347
Saprophytes, 73, 429
Sarsaparilla, 467
Savanna, 508
Saxifrages, 95
Scales, 334, 374
Scanning electron microscope, 33, 46, 60
Scarification, 137
Schizocarp, fruit, 132–33
Schleiden, Matthias, 29
Schwann, Theodor, 29
Science
  defined, 6
  vs. beliefs, 273
Scientific method, 6
Scientific names, 276, 285

Scientific theory, 6
Scion, 257
Sclereids, 54
Sclerenchyma
  cells, monocot, 93
  tissue, 54–55
Scopolamine, 116, 462
Scouring rushes, 386, 394, 396
Screw pines, 71
Scrophulariaceae, 478
Sea lettuce, 325, 337
Sea level rise, 489, 490
Seasoning wood, 96
Seasons, plant responses to, 205–6
Sea squirts, 307
Seaweeds, 318, 327, 330
  human relevance of, 341
Secondary cells, 182
Secondary cell wall, 32, 33
Secondary compounds, 182–83
Secondary consumers, 476, 477
Secondary electrons (microscopy), 46
Secondary growth in roots, 69
Secondary meristem, monocots, 93
Secondary metabolism, 181, 182
Secondary phloem, 85, 86, 89
Secondary roots, 65
Secondary structure of protein, 23
Secondary succession, 484, 487–88
Secondary tissues, 53, 65–66, 88
Secondary xylem, 85, 86, 88
Second filial (F_2) generation, 233
Second law of thermodynamics, 19
Sedges, 486, 499
  adaptation, 136
Seed coat
  of angiosperms, 427, 428, 432
  of gymnosperms, 409, 411, 412, 415
Seed cones, 412, 413, 415
Seedless fruits, 191, 232, 435
Seedless vascular plants
  ancient relatives of, 386, 387
  club mosses, 378, 386, 389–90
  ferns, 385, 398–404
  ground pines, 388–94
  horsetails and scouring rushes, 386, 394–97
  quillworts, 386, 392
  spike mosses, 390–91
  vs. seed plants, 409
  whisk ferns, 386–87
Seeds, 137–41. See also Germination

  absence in primitive vascular plants, 386
  animal dispersal, 135–36
  composition, 23
  cytokinins synthesized in, 193
  dispersal, 135–36, 245
  dormancy of, 208–9
  effect of water imbibition, 147
  ethylene produced by, 194
  evolution of, 411
  explosive release of, 204
  flower, 124–26
  of flowering plants, 409
  force exerted by, 484
  force generated by germination, 147
  of ginkgoes, 416, 417
  of gymnosperms, 408, 415
  hormones produced by, 193
  hybrid, 247
  leaves, 137
  oldest known, 409
  of orchids, 468–69
  of pines, 409, 419
  production without fertilization, 272
  propagation by, 254–55
  proteins in, 23
  quiescence of, 208
  respiration in, 179
  sexual reproduction and, 211
  survival value of, 409, 411
  viability of, 254
  wet, 180, 366
  wind dispersed, 134
Seed Savers Exchange, 248
Selaginella (spike mosses), 388–91
Selenium, 157
Self-fertile plants, 232
Self-pollinating plants, 246
Semiconservative replication, 225
Semipermeable membranes, 145
Senebier, Jean, 167
Senescence, 196
Sensitive plants, 201, 202, 457
Sepals, flower, 124
Separation layer, leaf, 116
Sequence-specific nucleases (SSNs), 253
Sequoia trees, 3
Serpentine soils, 475
Sesame, 449
Sessile leaves, 102
Seta, 376, 379, 380
Sexual hybrids, 249
Sexually transmitted diseases, 300

Sexual reproduction, 211. See also Meiosis
  of ascomycetes, 350
  of brown algae, 327–29
  of bryophytes, 370–73
  of chytrids, 347–48
  of club fungi, 355–60
  of coenocytic fungi, 349
  of conifers, 410–16
  of ferns, 399–402
  of flowering plants, 427–32
  of fungi, 346
  of green algae, 322
  of ground pines, 388–94
  of hornworts, 377
  of horsetails, 395–96
  of lichens, 365–67
  of liverworts, 374–76
  of mosses, 378–79
  new gene combinations, 218
  of quillworts, 392
  of red algae, 330–31
  seed propagation, 254–55
  of spike mosses, 390–92
  of stoneworts, 334–35
  of water molds, 341–43
  of whisk ferns, 386, 387
  of yeasts, 351
  of yellow-green algae, 325
Seychelles Island palm leaves, 103
Shade leaves, structure, 108
Shelf fungi, 355, 356, 358
Shepherd's purse, 454
Shiitake mushrooms, 362
Shikimic acid pathway, 182
Shoot system, 83
Short-day plants, 205
Shull, G. H., 248
Siberian white pine, 420
Sierra Club, 494
Sieve cells, 58
Sieve tube members, 56, 57, 85
Sieve tube network, 155
Signal transduction pathway, 188
Silent mutations, 229, 268
Silibinin, 360
Silica, 326, 395, 397, 405, 443–44
Silicles, 130–31, 454
Silicon, 78, 157, 327
Siliques, 130–31, 454
Silurian period, 388
Silurian period fossils, 388
Silymarin, 360, 361
Simple fleshy fruits, 126–28
Simple flower, 127
Simple leaf structure, 102
Simple ovary, 436
Simple tissues, 54–55
Sink

for food flow in plants, 154–57
Sipe, M., A21
Siphonosteles, 87
Sisal, 93
Sister chromatids, metaphase, 46
Sitka spruces, 421
Skototropism, 201
Slash pine, 420
Sleep movements, 198, 202–3
Slime molds, 204, 280, 317, 339–40
   cellular, 341
   human relevance of, 341
   plasmodial, 339–41
Slugs, 336, 341
Smallpox, 309
Small stem cross section, 90
Smooth endoplasmic reticulum, 37
Smuts, 355, 359
Snake plant, 255
Snapdragons, 205, 232, 237, 238, 478
Sneeze, stop-action photo, 299
Snowplant, 429
Socrates, 463
Sodium, 156
   in cells, 19
Sodium chloride, 15, 17, 18, 300
Sodium hydroxide, 18
Softwood, 91, 97
Soils
   acidified, 492
   after fire, 487
   alkalinity, counteracting, 80
   available water, 80
   bacteria in, 292, 300, 301
   capillary water, 78
   clay, 78–79
   effect of climate, 77
   effect of topography, 77
   erosion of, 483, 490–91
   excess air, 80
   field capacity, 79–80
   formation of, 484
   gravitational water, 78
   hardpan, 483
   hygroscopic water, 78
   impact on plant species, 474
   Inca treatments, 79
   mosses and, 381
   nitrogen in, 481–83
   nutrient availability, 80
   organisms in, 77
   permafrost, 499–500
   permanent wilting point, 80
   pH (acidity or alkalinity), 80
   productive agricultural, 78
   sandy, 80
   structure, 77, 78
   texture and composition, 77–78

Solanaceae, 450, 462–63
Solanine, 462, 463
Solar energy, 18, 163, 176
Solar panels, leaves as, 101–2
Solar tracking, 203
Solvents, 145, 303
Somatic hybrids, 249
Somatic mutations, 229
Soredia, 366
Sorghum, 174, 205, 449, 467
Sori, of ferns, 385, 400
Sorrel, 202, 360
South African orchid, 261
South America, plants from, 449
South Asia, plants from, 449
Soybeans, 205, 448
   transgenic, 251
Spanish moss, 123
Specialized stems, 94
Speciation, 231
Species, 267. *See also* Endangered species
   classification systems for, 276, 285–88
   competitive, 479
   evolution of, 267, 269–71
   extinction of, 263, 264, 491
   indicator, 479
   invasive, 263, 493–95
   life histories of, 479
   scientific names for, 276
   stress-tolerant, 479
   sympatric, 270, 271
   total number, 8
Species Plantarum (Linnaeus), 7, 277, 278, 285, 289
Specific epithet, 285
Sperm, 213
   of moss, 377
Spermatangia, 330, 331
Spermatia, 330–31
Spermatogenous cell, 413
Sperm cells
   in flowering plants, 430, 433
Sphagnum moss, 257, 371, 372
S phase, 224
Spice plants, A44, A46–A48
Spices, 4, 451
   from stems, 99
   use of leaves for, 116–18
Spike mosses, 388–94
Spinach, 205
Spindle fibers, 215, 216
   cellular reproduction, 42–48
   telophase, 46–48
Spindles, 42–46
Spines, 96, 110
Spiraling movements of plants, 197
Spirilla, 295
Spirogyra, algae, 321
Spirulina, 305, 306

Sponges, 339
Spongy layer, leaves, 107, 109
Spongy mesophyll, 106, 107
Spontaneous generation theory, 29
Sporangia
   of ferns, 400, 401
   of fungi, 349, 350
   of ground pines, 389, 391
   of horsetails, 394–95
   plant evolution and, 371
   of slime molds, 340–42
   of spike mosses, 390–92
   of whisk ferns, 386, 387
Sporangiophores, 348–50, 395
Spore print, 358
Spores, 217
   of club fungi, 355–56
   of coenocytic fungi, 348
   of ferns, 385, 399, 402
   of ground pines, 389, 391
   of horsetails, 395–97
   of liverworts, 373, 376
   of mosses, 377
   of quillworts, 392
   of spike mosses, 390, 392
   vs. seeds, 409
   of whisk ferns, 386
Sporocytes, 217, 218
   of ferns, 399
   of ground pines, 389
   of horsetails, 395, 397
   of liverworts, 376
   of mosses, 379
Sporophylls, 389, 390
Sporophyte generation (2n), 217, 218, 435
Sporophytes, 217, 218, 386
   of bryophytes, 371, 373
   of club mosses, 388
   of ferns, 399, 402
   of flowering plants, 429
   of ground pines, 390, 391
   of gymnosperms, 409
   of hornworts, 377
   of liverworts, 376
   of mosses, 377–80
   of quillworts, 392
   of spike mosses, 390
   of whisk ferns, 386–87
Sport (mutation), 229
Spotted fever, 301
Spotted spurge, 457
Sprays, liquified pests, A21–A22
Spring wood, 87
Spruces, 196, 421–22, 500
Spurge Family, 92, 457–58
Squash, 188, 437, 449
Squills, 467
Squirrel corn, 270, 271, 276
Squirting cucumber, 204
Stamens

flower, 124
   in legume flowers, 456
   of orchids, 469
   in primitive flowers, 435
   turgor contact movements in, 202
Stanley, Wendel, 309, 310
St. Anthony's fire, 353
Staphylococcus, 299
Starches, 19–21, 162, 183
   in dinoflagellates, 333
   energy storage in, 163
   in plants and green algae, 371
   synthesis, 40
Statospores, 325–26
Stele, 87
Stems, 82–99
   alternately arranged, 83
   of cacti, 458–59
   of climbing plants, 200
   of club mosses, 388
   of desert plants, 506
   grafting, 83
   gravitropism in, 199
   of ground pines, 388
   herbaceous *vs.* woody, 88
   hormones and growth in, 193
   of horsetails and scouring rushes, 394
   internode, 83
   layering of, 256
   node, 83
   oppositely arranged, 84
   phytochrome pigments and, 206
   plant, 52
   shoot system, 83
   twining of, 198
   of whisk ferns, 386
   whorled, 84
Stereomicroscopes, 29, 30
Stereoscopic view, leaf, 105
Sterigmata, 357
Stern, Kingsley, ix
Steward, F. C., 188
St. Helens, Mount, 485
Stigma, 124, 125, 465, 466, 469
Stimuli, plant response, 11–13
Stinkhorns, 355
Stipe, 327, 328, 357
Stipules, 83–84, 96, 102
   photosynthesis, 109
   scars, 84
Stodola, Frank, 192
Stomata, 58, 86, 103
   of CAM photosynthesis plants, 176, 506
   evolution of, 386
   global warming and, 164
   of hornworts, 377

Index

Stomata—Cont.
  of horsetails and scouring rushes, 394
  leaves, 104–6
  light intensity and, 165
  of mosses, 378–79
  in photorespiration, 174
  of pines, 408, 410
  recessed, 153
  role in transpiration, 146
  transpiration and, 149–51
  of *in vitro* grown plants, 259
Stomatal apparatus, 103, 151
Stomatal crypts, 153
Stonecrops, 176, 204
Stone fruits, 127, 448, 455
Stone pine, 420
Stoneseed, 448
Stoneworts (Charophyta), 334
Storage proteins, 23
Stratification, 208
Strawberries, 95, 127, 132, 205, 212, 216, 217, 232
Streptococcus mutans, 298
Stress-tolerant species, 479, 492
Strobili, 409
  of conifers, 409, 411
  of cycads, 409, 416, 417
  flowers compared to, 428
  of gnetophytes, 416, 418
  of ground pines, 389, 391
  of horsetails, 395, 397
  of spike mosses, 392
Stroma, 39, 166, 172
Structural gene, mutations in, 268
Struggle for existence, 268
Style, 125, 466
Subatomic particles, 14
Subculturing, 260
Suberin, 22, 59, 60, 85
  leaf, 116
  in roots, 67–68
Subsidiary cells, 152
Substrate, 24
Succession, 382, 483–88, 495
Succulent leaves, 110, 112
Suckers, 70
Sucrose, 20, 155, 162, 182
Sugar beets, 20
Sugarcane, 20, 174, 192, 232, 244, 449, 466, 467
Sugar cleavage, 180
Sugar maple, fall colors, 115
Sugar pines, 412, 502
Sugars
  active transport in plants, 155
  basic forms of, 20
  in DNA molecule, 221, 222
  energy storage in, 162
  osmotic water movement, 156
  in phloem, 419
  in photosynthesis, 162, 182
  in respiration, 177, 179
  in RNA molecule, 228
Sulfates, 176
Sulfide, 176
Sulfolobus bacteria, 309
Sulfur, 15, 156
  in cells, 19
  deficiency symptoms, 157
Sulfur bacteria, 296
Sulfur dioxide, 367, 492
Sulfuric acid, 18, 492
Sumac, 115, 487, 508
Summer wood, 87
Sundew leaves, 113–14
Sunflower Family, 449
Sunflowers, 449
  as day-neutral plants, 205
  domestication of, 244–45
  heliotropism in, 203
Superior ovaries, 127, 436, 459, 462
Supreme Court, U.S. (1893), tomato, 126
Surface area of roots, 65
Survival of the fittest, 268
Swarm cells, 340
Sweet pea flower, 436, 456
Sweet potatoes, 69, 75, 457
Sweet William, 479
Syconium, 133
Symbiosis, 73
  in bacteria, 292, 404
  in lichens, 319, 345, 365
Sympatric species, 270, 271
Synergids, 429, 432–34
Synthetic hormones, 191, 193
Syphilis, 300, 301
Systematics, plant, 7, 285
Systemin, 195

t

2, 4, 5-T, 492
  toxicity to humans, 191
Tachinid flies, A21
Taiga, 500–501
Tamarack trees, 360, 486, 500
Tamarisks, 493
  roots, 64
Tangentially cut wood, 96
Tannin, 424
  in leaves, 115
  wood, 96
Tapetal cells, 429
Tapioca, 457
Taproots, 64, 65
Taro plants, 154, 449
Tarragon, 465
Tasmanian *Eucalyptus* trees, 33
Tasmanian gum, 2, 3
  transpiration by, 149
  water in, 149
Taxa, 279
Taxic movements, 204
Taxol, 424
Taxonomists, 285
Taxonomy, plant, 7
TCA cycle, 177
TCDD, 191
T-DNA (transfer DNA), 250
Tea, 118, 448
Telophase, 46–48
Telophase I (meiosis), 214, 216
Telophase II (meiosis), 214, 216
Tempeh, 349
Temperate deciduous forests, 501–3
Temperature
  circadian rhythms and, 202–3
  in compost, 297
  conversion (°C/°F), A75–A76
  C_3 photosynthesis, 165, 174
  C_4 photosynthesis, 165, 176
  cuttings, propagation by, 255–56
  fungi, growth of, 346
  germination of seeds and, 208
  global rise in, 163–64, 184, 488
  growth and, 193, 207–8
  molecular movement, 15–16
  photosynthesis and, 165
  plant productivity and, 184
  in polymerase chain reaction (PCR), 226
  respiration rate and, 179, 184
  thermotropic responses to, 201
  transpiration rates, 153
Tendrils, 96
  coiling of, 200
  leaves, 109–10
  movement, 110
  twining of, 198
Tendril structure
  collenchyma, 110
  sclerenchyma, 110
Teonanacatl mushrooms, 362
Teosinte, 245, 265
Tequila, from *Agave,* 119
Terminal bud, 83–84, 196
Terminator DNA sequence, 228
Terpenoids, 182
Tertiary period, 388
Tertiary structure polypeptide, 23
Testcross, 237
Test tubes, micropropagation in, 258, 259
Tetanus, 301
Tetrahydrocannabinol (THC), 60, 117
Tetrasporangia, 331
Tetraspores, 331
Tetrasporophyte, 330, 331
Texas paintbrush, 275
Thalloid liverworts, 373–76, 377
Thallus (thalli), 327, 329–31, 365, 373–76
Theory, defined, 9
Thermal cycler, 226
Thermodynamics, laws of, 18–19
Thermoperiodism, 207
Thermophilic bacteria, 308
Thermotropism, 201
Thiamine, 180, 257
Thigmomorphogenesis, 194
Thigmotropism, 200
Thistle, 465
Thorns, 95–96
  leaf, 110
Thylakoid lumen, 171
Thylakoid membranes, 171
  in photosynthesis, 39, 166, 171
  in respiration, 181
Thylakoids, 39, 40
Thymine, 222, 223
Tiger lilies, 96, 467
Tip layering, 256
Tissue culture, 188, 197, 249, 258–60, 404, 424, 469
Tissues, 51–61
  meristematic, 52–53
  primary, 53
  secondary, 53–55
  simple *vs.* complex, 54
Tobacco mosaic virus (TMV), 310
Tobacco plants, 197, 205, 462
  leaves, 117
  root cap cell, 41
  roots of, 200
Tomato plants, 205, 462–63
  legal definition, 126
  seedless, 435
  seeds of, 254
  stem, 60
  VFN, A22
  wild, 208, 262
Tonoplasts (vacuolar membranes), 41
Torus, bordered pits, 56–57
Totipotency, 258
Touch, plant response to, 194, 197, 201, 202
Toxic wastes, bioremediation of, 302, 492, 493
Toxins
  from bacteria, 300, 301
  from cyanobacteria, 306
  from dinoflagellates, 333
  from fungi, 351, 363–64
  from insecticides, 302
Trace, 84
Tracheids, 55, 56, 68, 85, 89–91, 150, 151, 154, 410, 418, 421

Trade-offs, evolutionary principle of, 479
Transcript, 228
Transcription, 228, 231
Transduction, 293
Transduction phase (tropism), 198
Trans fats, 21
Transfer DNA (T-DNA), 250
Transfer RNA (tRNA), 228, 230
Transformation, in bacteria, 293
Transformation techniques, 250–51
Transgenic plants, 249–52
   biological pest control and, 301–2
   pros and cons of, 252–54
Translation, RNA, 227–30
Translocation by phloem
   food transportation in plants, 154–55
Translocations, of chromosomes, 231, 268
Transmission electron micrograph, 48
Transmission electron microscope, 29–30, 46
Transpiration, 149–54
   in leaves, 103
   light intensity and, 165
   reducing in cuttings, 255
   regulation of, 151
   of water, 146
   in water cycle, 480
Transport proteins, 148
Transposable genetic elements, 221–22, 238
Transposition, 221
Transposons, 221
Traumotropism, 201
Travertine, 304
Tree ferns, 404, 409
Tree-of-heaven, 70
Trees
   acid depositions and, 492
   annual rings of, 411
   apical dominance in, 196
   and erosion prevention, 490
   fall color change, 41, 115–16
   fall leaf coloration of, 165
   propagation by grafting, 256–57
   redwoods, 2
   reproductive strategies of, 479
   resistance to fire, 487–88
   rusts and, 360–61
   in secondary succession, 487
Tricarboxylic acid (TCA) cycle, 177
Trichocysts, 334
Trichodesmium, 304
Trichogramma wasps, A20–A21

Trichogyne, 330, 331
Trichomes, 58
Trilliums, 209, 502
Triple bonds (electron), 17
Triploid plants, 216, 232
Trophic efficiency, 477
Tropical figs, 71
Tropical rain forests, 2, 8, 376, 506–8
Tropism, 198
   asymmetric growth phase, 198
   perception phase, 198
   transduction phase, 198
True anaerobic respiration, 179
   in methane production, 179
True bacteria, 296–97
True berry, fruit, 127
True firs, 422
True mosses, 378
Truffles, 350–51
Tryptophan, 189
Tschermak, Eric von, 232
Tsunami, 490
Tube nucleus, of flowering plants, 432, 433
Tuberculosis (TB), 298–99, 301, 313, 367, 397
Tubers, 95, 258
Tubulins, 42
Tularemia, 301
Tulips, 258, 479
Tulip tree, 502
Tumors, 193, 250, 359. *See also* Cancers
Tundra, 499–500
Tung, 458, 471
Turgid cell, 146
Turgor pressure, 146, 149, 151, 152
   in guard cells, 152
   movements involving, 197, 201–4
Turnips, 69, 448, 454
Turpentine, 420, 422
Twine
   use of leaves for, 116
Twining movements, 198
Tyloses, 90
Type specimen, 278
Typhus, 301

U

Ubiquinol, 181
Ulothrix, 317, 318
   life cycle of, 320
Ultraviolet radiation, 490
Ulva (sea lettuce), 325, 343
Underground flowering orchid, 121
Undulant fever, 301
Unit characters, law of, 234

Unsaturated fat, 21
Upper cortex, of lichens, 366
Uracil, 228
Uranium, 14, 266
Urediniospores, 360

V

Vaccination, 309
Vacuolar membranes
   in cell model, 32
   tonoplasts, 41
Vacuoles, 41–42
   in algae, 319
   cell sap, 41
   function, 41
   plants *vs.* animals, 48
   role in water transport, 147
Valence (atomic), 16
Vallecular canals, 395
Valley fever, 365
Vanadium, 157
Van Helmont, J. B. (1577–1644), 7
Vanilla, 449
Vanilla orchid, 71, 469
Van Overbeek, Johannes, 193
Vascular bundles, 87, 93, 102
   composition, 88
Vascular cambium, 43, 51, 53, 55, 68, 69, 84–86, 88
   in grafting, 257
   in pines, 411
   ray initials, 56
Vascular cylinder, in roots, 67
Vascular plants, 371, 385–86. *See also* Seedless vascular plants
Vascular rays
   composition, 90
   function, 90
Vavilov, Nikolai Ivanovich, 245, 449
Vegetables
   harvested, respiration in, 179
   legal definition, 126
   storage of, hormones used in, 193
Vegetable sponge, 464
Vegetarian diet, 477
Vegetative nucleus, of pollen grains, 432, 433
Veins, 105–8
   of leaves, 150
   structure, 107
Velamen roots, 58, 70
Venation, leaves, 103
Veneer, 98
Venter, of mosses, 378, 379
Venus's flytraps, 114, 201
Vernal pools, 504
Vesicles, 38, 47–48
Vessel elements, 55, 85, 89
Vessels, plant, 55

Vetches, 204, 457
Viability, of seeds, 254
Vinegar weed, 461
Violets, 205, 502. *See also* African violets
Virchow, Rudolf, 29
Virginia creeper, 71
Viroids, 313
Viruses, 280, 292, 299, 309–13
   as biological pest controls, 313, A21
   characteristics of, 309–10
   cyanobacteria attacked by, 306
   diseases caused by, 299, 309–10
   distribution of, 309–10
   human relevance of, 311–13
   mutations in, 312
   plant, naming, 310–11
   reproduction of, 311
   size and structure of, 309–10
Vitamin A, 188, 252
Vitamin B, 180
Vitamin C, 303, 337, 362, 455
Vitamins, 188–96
   bacterial synthesis of, 303
   in growth medium, 258
   in pollen grains, 433
   sources of, 337
Volcanic eruption (1783), 90
Volcanoes, 482, 483, 485
Volva, 357
Volvox, green algae, 324
Vries, Hugo de, 232

W

Wakerobins, 209
Walking fern, 112
Wallace, Alfred R., 266
Walnuts, 448. *See also* Black walnut tree
Wandering Jew epidermis, 151–52
Wasps, 261, A20
Waste disposal, leaves, 297
Water, 144
   in cells, 149, 179–80, 480
   conduction, 56
   deficits of, 491
   diffusion, 146
   essential role in plants, 147
   ferns and, 398
   food transportation in plants, 154–56
   germination of seeds, 254
   guttation drops, 153–54
   molecular cohesion, 16
   molecular structure, 15–16
   molecules of, 18
   in mosses, 377, 381
   movement in cells, 144

Water—Cont.
  movement in plants, 149–51
  in mushrooms, 361
  in natural cycles, 479
  pH of, 18
  plant cooling by evaporation, 150
  plant transition to land from, 317
  recirculation, 155
  release in respiration, 177, 182
  respiration rate and, 179
  role in photosynthesis, 149, 152, 161–63, 182
  transpiration, 146, 154
  transport in plants, 147
  use by plants, 149–51
Water conservation movements, 203
Water contamination, 492
Watercress, 341, 454
Water cycle, 480
Water dispersal of seeds, 136
Water hemlock roots, 69
Water hemlocks, 463, 464
Water hyacinths, 64
Water imbibition
  role in germination, 147
Water lilies, 486
Water loss, ABA and, 193
Watermelons, 216, 232, 435
Water molds, 341–43, 347
Water movement in plants, 146, 149
Water nets (*Hydrodictyon*), 323–324
Water pollution, 491, 492
Water potentials, 146
  water transport in plants, 151
Watersilk (*Spirogyra*), 321
Water-splitting, 171
Water table, 480
Water transport
  by capillary action, 150
  by osmosis, 150
  by root pressure, 150
Water-vapor, role in transpiration, 151
Watson, James, 221, 225
Wattle, 456
Wax deposits, 58
Waxes, 22, 421
  from plant leaves, 117
  of *in vitro* grown plants, 259
Weeds
  biological control of, A20–A23
  as day-neutral plants, 205
  hormones used in control of, 191
  seedlings affected by, 254
  seeds of, 208

thermotropic responses in, 201
Weinberg, W., 241
Welwitschia, 419
  leaf fibers, 56
Went, Frits, 189–90, 207
Wetlands
  destruction of, 492–93
  methane bacteria in, 308, 489
  peat mosses in, 371
  succession in, 485
Wheat, 93, 466
  black stem rust of, 360
  dwarf varieties of, 246
  endosperm in, 433
  gene-edited variety, 253
  length of, hormones and, 193
  as long-day plant, 205
  origins of, 448
  polyploidy in, 216, 232, 273
Wheat gluten, 23
Whisk ferns, 87, 386, 402, 406
  life cycle of, 388
White ash, uses, 97
White cedar, 424, 486
White fir, 502
White oak, 51, 97
White pine, 97, 419, 420, 502
White pine blister rust, 360, 420
White spruce, 421
Whittaker, R. H., 280
Whorled, flower, 124
Whorled leaf structure, 83–84, 104
Wiggers, Fredericus Henricus, 277
Wild edible plants and fungi, A27–A29
Wild rice, 469
Wilkins, Maurice, 223
Willow trees
  biomes of, 499, 500
  root section, 68
  secondary compounds of, 183
Wilson, E. O., 495
Wind dispersal
  of bryophyte fragments, 382
  of fern spores, 401
  of seeds, 134–35
Window leaves, 111, 112
Winged bean, 457
Wing petals, 456
Witch hazel, 204
Woese, Carl, 280, 307
Wolfsbane, 451
Wollemi pine, 423
Wood
  annual rings, 87
  bird's nest fungi on, 359
  cellulose content, 96

conifer, 89, 410, 422
decay, 96
density, 96
detailed uses, 96–99
durability, 96
energy stored in, 176
fuel use, 99
growth of, 87
hardwood, 91
heartwood, 91–92
human uses, 96
knots, 96–97
of leguminous trees, 456
lignin content, 96
petrified, 406
products, 98–99
properties, 969
pulp, 98
sapwood, 91
sawing methods, 96
seasoning, 96
secondary products, 99
secondary xylem, 88
shelf (bracket) fungi on, 356, 358
soft *vs.* hard, 410
softwood, 91
spring, 87
sugar converted to, 162
summer, 87
surfaces, 98
tracheid production, 89
tyloses, 90
vascular cambium, 88
veneer, 98
Wood products, 98–99
Woody dicots, 52, 88–92
Woody plants. *See also* Trees
  dormancy of, 209
  fires and, 487–88
  micropropagation of, 260
  mycorrhizal fungi, 478
  perennials, 479
  photoperiodism in, 206
  secondary compounds of, 183
  in tropical rain forests, 507
  twigs, 83

X

Xanthophyceae (yellow-green algae), 325
Xanthophylls, 115, 333
Xerophytes, 475
Xylem, 55–57, 88, 93
  absence in bryophytes, 372
  absence in mosses, 378
  environmental adaptation, 57
  evolution of, 386
  of gnetophytes, 409
  hormone transport, 192, 193

of horsetails and scouring rushes, 394
leaves, 107, 109
of pines, 411
in roots, 68
structure, 56
synopsis, 156
water transport, 146, 150
in whisk ferns, 387
Xylem rays, 89

Y

Yams, 69, 449
Yarrow, 466
Yeasts, 29, 350–51, 354
Yellow-green algae, 325
Yellow pines, 420
Yellowstone National Park, 304, 483
Yellow water weed, 70
Yews, 415, 424
Yosemite National Park, 494
Yucca, 508

Z

Z (oxidation-reduction system), 171
Zinc, 19
Zoospores
  of chytrids, 348
  of golden-brown algae, 326
  of green algae, 320, 324
  of *Oedogonium*, 322
  of *Ulothrix*, 320, 321
  of water molds, 342–43
  of yellow-green algae, 325
Zygosporangium, 349
Zygospore, 349
Zygotes, 213, 217
  of brown algae, 329
  chromosomes in, 213
  of chytrids, 348
  of conifers, 415
  of diatoms, 327
  diploid, 216
  of ferns, 399, 402
  of flowering plants, 433, 434
  of ground pines, 390, 391
  of mosses, 378, 380
  of *Oedogonium*, 322
  of quillworts, 393
  of red algae, 330, 331
  of *Spirogyra* (watersilk), 321
  in sporophyte generation, 217
  of *Ulothrix*, 317, 320
  of water molds, 341–42
  of whisk ferns, 387, 388